Biopesticides

Woodhead Publishing Series in Food Science, Technology and Nutrition

Biopesticides

Volume 2: Advances in Bio-inoculants

Edited by

Amitava Rakshit

Vijay Singh Meena

P.C. Abhilash

B.K. Sarma

H.B. Singh

Leonardo Fraceto

Manoj Parihar

Anand Kumar Singh

Woodhead Publishing is an imprint of Elsevier
The Officers' Mess Business Centre, Royston Road, Duxford, CB22 4QH, United Kingdom
50 Hampshire Street, 5th Floor, Cambridge, MA 02139, United States
The Boulevard, Langford Lane, Kidlington, OX5 1GB, United Kingdom

Library of Congress Cataloging-in-Publication Data
A catalog record for this book is available from the Library of Congress

British Library Cataloguing-in-Publication Data
A catalogue record for this book is available from the British Library

ISBN: 978-0-12-823355-9

For information on all Woodhead Publishing publications visit our website at https://www.elsevier.com/books-and-journals

Publisher: Charlotte Cockle
Acquisitions Editor: Nancy J. Maragioglio
Editorial Project Manager: Allison Hill
Production Project Manager: Surya Narayanan Jayachandran
Cover Designer: Miles Hitchen

Typeset by TNQ Technologies

Contents

Contributors xiii

1. *Bacillus thuringiensis* based biopesticides for integrated crop management

Aurelio Ortiz and Estibaliz Sansinenea

1.1 Introduction 1
1.2 The early beginning of *Bacillus thuringiensis* as biopesticide 1
1.3 The past last twenty years of *B. thuringiensis* as biopesticide 2
1.4 The present and the future of *B. thuringiensis* as biopesticide 3
1.5 Conclusions and perspectives 5
References 5

2. Biopesticides for management of arthropod pests and weeds

Muhammad Razaq and Farhan Mahmood Shah

2.1 Agriculture and pests 7
2.1.1 Synthetic pesticides and challenges 7
2.1.2 From pesticides to biopesticides 8
2.2 Inconsistencies in understanding the term "biopesticides" 8
2.2.1 Microbial biopesticides 9
2.2.2 Nematodes biopesticides 11
2.3 Plant-incorporated protectants (PIPs) 11
2.4 Biochemical pesticides 12
2.4.1 Pheromones 12
2.4.2 Plant essential oils 12
2.5 The biopesticides market and challenges 13
2.5.1 Bioinsecticides 13
2.5.2 Bioherbicides 14
References 14

3. Biopesticide formulations - current challenges and future perspectives

Marian Butu, Steliana Rodino and Alina Butu

3.1 Introduction 19
3.2 A view through history 19
3.3 Regulatory framework 20
3.4 Diversity of biopesticides 21
3.5 Formulation of biopesticides 23
3.6 Biopesticides market 26
3.7 Challenges and future perspectives 27
References 28

4. Application technology of biopesticides

Adriano Arrué Melo and Alexandre Swarowsky

4.1 Introduction 31
4.2 Coverage 31
4.2.1 Crop 32
4.2.2 Biological target 32
4.2.3 Biopesticide 32
4.3 Biopesticides and adjuvants 32
4.4 Mixture of biopesticides and pesticides 33
4.5 Influence of climatic factors on the application of biopesticides 33
4.5.1 Relative humidity and temperature 33
4.5.2 Wind speed and direction 34
4.5.3 Timing of biopesticides application 35
4.6 Final considerations 36
References 36

5. Microbial pesticides: trends, scope and adoption for plant and soil improvement

Pooja Singh and Purabi Mazumdar

5.1 Introduction 37
5.2 Types of microbial pesticides 37
5.2.1 Bacteria 41
5.2.2 Fungi 43
5.2.3 Microsporidia 44
5.2.4 Virus 44
5.2.5 Genetically modified microbes 45
5.2.6 Microbes supporting plants and soil health 47
5.3 Trends and market demand for microbial pesticide 53

5.3.1 Global market reports on use of microbial pesticides 54
5.4 Registration and regulation of microbial pesticide globally 57
5.5 Conclusion and future prospects 58
Acknowledgments 59
References 59

6. Entomopathogenic nematodes: a sustainable option for insect pest management

Ashish Kumar Singh, Manish Kumar, Amit Ahuja, B.K. Vinay, Kiran Kumar Kommu, Sharmishtha Thakur, Amit U. Paschapur, B. Jeevan, K.K. Mishra, Rajendra Prasad Meena and Manoj Parihar

6.1 Introduction 73
6.2 Baiting, isolation, multiplication of EPNs 74
6.3 Identification of EPNs 75
6.4 The liaison between EPNs and mutualistic bacteria and their identification 76
6.5 Lifes cycle, pathogenicity and host range of EPNs 80
6.5.1 Pathogenicity 80
6.5.2 Host range 81
6.6 Mass production, formulation development and application 81
6.6.1 Types of formulations 82
6.7 Application of EPN genomics to enhance the field efficacy 86
6.8 Conclusion and future perspectives 87
References 87

7. Scientific and technological trajectories for sustainable agricultural solutions: the case of biopesticides

Alejandro Barragán-Ocaña, Paz Silva-Borjas and Samuel Olmos-Peña

7.1 Introduction 93
7.2 Method 95
7.3 Discussion and analysis of results 96
7.3.1 Scientific trajectory 96
7.4 Technological trajectory 100
7.5 Conclusions 103
Acknowledgments 103
References 103

8. Biopesticides: a genetics, genomics, and molecular biology perspective

Pawan Basnet, Rajiv Dhital and Amitava Rakshit

8.1 Introduction 107
8.1.1 Advantages of application biopesticides in pest management 107
8.2 Market trends of biopesticides 109
8.3 Factors for increasing trends toward biopesticides 109
8.4 Constraints for the applications of biopesticides 110
8.5 Role of genetic engineering in context of biopesticides 110
8.5.1 *Bacillus thuringiensis* (Bt) 110
8.5.2 Entomopathogenic nematodes (EPNs) 111
8.5.3 Baculoviruses 111
8.5.4 RNAi based biopesticides 111
8.5.5 Plant-incorporated protectants (PIPs) 112
8.5.6 Entomopathogenic fungi 113
8.5.7 Botanical biopesticides 113
8.6 Conclusion 113
References 114

9. *Bacillus thuringiensis*, a remarkable biopesticide: from lab to the field

Igor Henrique Sena da Silva, Marcelo Mueller de Freitas and Ricardo Antônio Polanczyk

9.1 Introduction 117
9.2 Isolation and epizootic potential of *Bacillus thuringiensis* (Bt) 118
9.3 Nomenclature and characterization of *Bacillus thuringiensis* (Bt) Cry pesticidal proteins 119
9.4 Mode of action of *Bacillus thuringiensis* Cry toxins 120
9.5 Development of *Bacillus thuringiensis* formulations 121
9.6 *Bacillus thuringiensis* compatibility with natural enemies and Bt plants 122
9.6.1 Final considerations 124
References 124

10. Biopesticides for management of arthropod pests and weeds

Josef Jampílek and Katarína Kráľová

10.1 Introduction 133

10.2 Bioherbicides 135
10.2.1 Plant extracts and essential oils with herbicidal activity 135
10.2.2 Bioherbicides produced by microorganisms 139
10.3 Biopesticides against harmful arthropodes 140
10.4 Nanoscale biopesticide formulations against arthropod pests and weeds 144
10.4.1 Bioherbicides in nanoformulations 144
10.4.2 Biopesticide nanoformulations against anthropodes 145
10.5 Conclusions 148
Acknowledgement 148
References 149

11. *Salvia leucantha* essential oil encapsulated in chitosan nanoparticles with toxicity and feeding physiology of cotton bollworm *Helicoverpa armigera*

Devakumar Dinesh, Kadarkarai Murugan, Jayapal Subramaniam, Manickam Paulpandi, Balamurugan Chandramohan, Krishnasamy Pavithra, Jaganathan Anitha, Murugan Vasanthakumaran, Leonardo Fernandes Fraceto, Lan Wang, Jiang Shoiu-Hwang and Hans-Uwe Dahms

11.1 Introduction 159
11.2 Materials and methods 160
11.2.1 Plant material 160
11.2.2 Extraction of *S. leucantha* essential oil 160
11.3 Qualitative analysis 161
11.3.1 Phytochemical analysis 161
11.4 Test for flavonoids 161
11.5 Test for alkaloids 161
11.6 Test for tannins 161
11.7 Test for phenolics 161
11.8 Test for terpenoids 161
11.9 Test for saponins 161
11.10 Test for glycosides 161
11.11 GCMS analysis of essential oil of *S. leucantha* 161
11.11.1 GC–MS specification 161
11.11.2 GC–MS analysis 162
11.12 Collection and processing of crab shells 162
11.13 Isolation and extraction of chitosan from crab shell 162
11.13.1 Structure of chitosan 162
11.14 Chitosan nanoparticles preparation with essential oil 162
11.15 Characterization of essential oil loaded chitosan nanomaterials 163
11.16 *H. armigera* and *S. litura* rearing 163
11.17 Rearing of *P. xylostella* 163
11.18 Toxicity against the *H. armigera*, *S. litura* and *P. xylostella* 163
11.19 Impact on longevity and fecundity of *H. armigera*, *S. litura* and *P. xylostella* 164
11.20 Quantitative food utilization efficiency measures 164
11.21 Amylase, protease, proteinase, and lipase assay 164
11.22 Statistical analysis 165
11.23 Results and discussion 165
11.23.1 Phytochemical screening for essential oil of *Salvia leucantha* 165
11.24 GC–MS analysis 165
11.25 Characterization of essential oil loaded chitosan nanoparticles 166
11.25.1 UV-VIS spectral analysis of essential oil loaded chitosan nanoparticles 166
11.26 SEM analysis 167
11.27 Energy-dispersive X-ray spectroscopy analysis 168
11.28 FTIR analysis of essential oil chitosan nanoparticles 168
11.29 Zeta potential measurements 169
11.30 Larvicidal and pupicidal toxicity against *H. armigera*, *S. litura* and *P. xylostella* 170
11.31 Impact of *S. leucantha* essential oil and encapsulated chitosan nanoparticles on insect longevity and fecundity 173
11.32 Food utilization measures 175
11.33 Gut digestive enzymes of *H. armigera*, *S. litura* and *P. xylostella* larvae 177
11.34 Conclusion 178
References 178

12. Microbial bio-pesticide as sustainable solution for management of pests: achievements and prospects

Udayashankar C. Arakere, Shubha Jagannath, Soumya Krishnamurthy, Srinivas Chowdappa and Narasimhamurthy Konappa

12.1 Introduction 183

12.1.1 Biochemical pesticides 184
12.1.2 Microbial pesticides 184
12.1.3 Plant incorporated protectants 184
12.2 Biochemical pesticides 184
12.2.1 Insect pheromones 184
12.3 The few examples of pheromones used in agricultural pest management are as follows 184
12.3.1 Chitosan 184
12.3.2 Plant extract biopesticides 185
12.4 Microbial biopesticides 185
12.5 Bacteria as biopesticides 185
12.6 Members of *Bacilliaceae* as biopesticides (spore formers) 185
12.6.1 *Paenibacillus popilliae* (*Bacillus popillae*) and *B. lentimorbus* 190
12.6.2 *Lysinibacills sphaericus* (*Bacillus sphaericus*) 190
12.6.3 Bacillus subtilis 191
12.6.4 Bacillus firmus 191
12.6.5 *Bacillus thuringiensis* (Bt) 191
12.6.6 Antimicrobial activity of *B. thuringiensis* based biopesticides 192
12.6.7 *Bacillus thuringiens* is used as nano pesticides 192
12.7 Members of *Pseudomonadaceae* and *Enterobacteriaceae* as biopesticides (non-spore formers) 193
12.7.1 Pseudomonadaceae 193
12.8 Enterobacteriaceae 193
12.8.1 Fungi as biopesticides 194
12.8.2 *Trichoderma* spp. as biopesticide 194
12.9 *Coniothyrium minitans* as biopesticide 195
12.10 *Gliocladium catenulatum* as biopesticide 195
12.10.1 *Purpureocillium lilacinum* as biopesticide 195
12.10.2 *Beauveria bassiana* as biopesticide 195
12.10.3 *Lecanicillium* (*Verticillium*) *lecanii* as biopesticide 195
12.10.4 Endophytic fungi as biocontrol agents 196
12.11 Yeast as biocontrol agents 196
12.11.1 Insect viruses as biopesticides 196
12.11.2 Protozoans as biopesticides 197
12.12 Plant incorporated protectants: genetically modified (GM) crops 197
12.13 Advantages of microbial biopesticides 198
12.14 Disadvantages of microbial biopesticides 198
References 198

13. Nano bio pesticide: today and future perspectives

Camelia Ungureanu

Acknowledgments 204
References 205

14. Current development, application and constraints of biopesticides in plant disease management

Shweta Meshram, Sunaina Bisht and Robin Gogoi

14.1 Introduction 207
14.2 History of synthetic pesticides used in plant disease evolution 208
14.3 Current global scenario 208
14.4 Biopesticides 209
14.5 Classification of biopesticides 209
14.6 Microbial biopesticides 211
14.7 Insight into popular fungal and bacterial biopesticides used in plant disease management 212
14.7.1 *Trichoderma* spp 212
14.8 Mass production of *Trichoderma* for commercial purpose 213
14.8.1 Pseudomonas fluorescens 213
14.9 Formulations for *P. fluorescens* 213
14.9.1 Organic carriers 213
14.9.2 Inorganic carriers 213
14.10 Methods 213
14.10.1 Powder formulations 213
14.11 Liquid formulation 213
14.11.1 *Bacillus* sp. 214
14.12 Improvement of formulation efficacy 214
14.13 Molecular approach for improvement of formulation efficacy 214
14.13.1 Protoplast fusion 214
14.13.2 Genetic recombination 215
14.13.3 Mutation 215
14.14 Development of compatible consortia for improvement of formulation efficiency 215
14.14.1 Combining various microbes 215
14.14.2 Combining different mode of action 216
14.14.3 Development of strain mixtures 216
14.15 General mode of actions of microbial pesticides against plant pathogens 216
14.16 Nanobiopesticides 216
14.17 Biopesticides and their association with growth promoter 217

14.18 Inducer of systemic resistance in plant against plant pathogen 218
14.19 Botanical biopesticides usage against plant pathogen 219
14.20 Essential oils 219
14.21 Advantages and limitations of biopesticides 219
14.21.1 Advantages 219
14.21.2 Limitations 220
14.22 Factors affecting biopesticides marketing 220
14.23 Conclusion 221
References 221

15. Insights into the genomes of microbial biopesticides

A.B. Vedamurthy, Sudisha Jogaiah and S.D. Shruthi

15.1 Introduction 225
15.1.1 Entomopathogenic bacteria 225
15.1.2 Entomopathogenic fungi 228
15.1.3 Viral biopesticides 229
15.1.4 Entomopathogenic nematodes 230
15.1.5 Entomopathogenic protozoans 231
15.2 Advantages of genetic manipulation and their commercialization 231
15.3 Conclusions 232
References 232

16. Genetic engineering intervention in crop plants for developing biopesticides

Shambhu Krishan Lal, Sahil Mehta, Sudhir Kumar, Anil Kumar Singh, Madan Kumar, Binay Kumar Singh, Vijai Pal Bhadana and Arunava Pattanayak

16.1 Biopesticides 237
16.2 Engineering of Bt genes for insect resistance 238
16.3 Bt cotton adoption in India 239
16.4 Genetic engineering approaches for combating aphid infestation in crop plants 239
16.5 Applications of RNA interference (RNAi) to control pests 240
16.6 Applications of genome editing to control pests 241
16.7 Future perspectives 242
References 242

17. Medicinal plants associated microflora as an unexplored niche of biopesticide

Ved Prakash Giri, Shipra Pandey, Satyendra Pratap Singh, Bhanu Kumar, S.F.A. Zaidi and Aradhana Mishra

17.1 Introduction 247
17.1.1 Medicinal plant diversity in India 247
17.1.2 Niche of microflora 248
17.2 Plant-microbe association 248
17.2.1 Rhizospheric association of microbes 248
17.2.2 Phyllospheric association of microbes 251
17.2.3 Endophytic microbiome association with medicinal plants 251
17.3 Relative factors between microflora and plants 254
17.4 Conclusion and future perspectives 254
References 255

18. *Trichoderma*: a potential biopesticide for sustainable management of wilt disease of crops

Narasimhamurthy Konappa, Nirmaladevi Dhamodaran, Soumya Satyanand Shanbhag, Manjunatha Amitiganahalli Sampangi, Soumya Krishnamurthy, Udayashankar C. Arakere, Srinivas Chowdappa and Sudisha Jogaiah

18.1 Introduction 261
18.2 *Trichoderma* in the control of wilt disease 263
18.3 Mechanism of biocontrol by *Trichoderma* in the control of wilt pathogens 264
18.3.1 Competition 264
18.3.2 Mycoparasitism 264
18.3.3 Cell wall degrading enzymes 267
18.3.4 Antibiosis by antimicrobial metabolites 267
18.3.5 Induced systemic resistance 268
18.4 Conclusion 270
References 270

19. Biological inoculants and biopesticides in small fruit and vegetable production in California

Surendra K. Dara

19.1 Bioinoculants in strawberry 277
19.2 Bioinoculants in tomato 279
19.3 Biopesticides in strawberry and grapes 279
19.4 Biopesticides in vegetables 280
19.5 Non-entomopathogenic roles of hypocrealean entomopathogenic fungi 280
19.6 Strategies and implications for sustainable food production 281
19.7 Conclusions 281
References 282

20. Development and regulation of microbial pesticides in the post-genomic era

Anirban Bhar, Akansha Jain and Sampa Das

20.1 Introduction 285
20.2 Development of the microbial biopesticide 286
20.2.1 Plant growth regulators play crucial role in development of biopesticides 286
20.2.2 Siderophores causes iron limiting conditions for many pathogenic pests 287
20.2.3 Antibiosis, an important criterion for development of the microbial biopesticides 288
20.3 Microbial pesticides: brief description 288
20.3.1 Bacteria as biopesticides 288
20.3.2 Viruses as biopesticides 289
20.3.3 Fungi as biopesticides 289
20.3.4 Nematodes as biopesticides 289
20.3.5 Protozoan as biopesticides 290
20.4 Genetic improvements of microbial pesticides 290
20.5 Regulation and commercialization of microbial pesticides 292
20.6 Microbial pesticides in the post-genomic era 293
20.7 Future prospects 293
Acknowledgments 294
References 294

21. Microbial biopesticides for sustainable agricultural practices

Indu Kumari, Razak Hussain, Shikha Sharma, Geetika and Mushtaq Ahmed

21.1 Introduction 301
21.2 Microbial biopesticides 302
21.2.1 Bacterial biopesticides 302
21.2.2 Viral biopesticides 304
21.2.3 Fungal biopesticides/ mycopesticides 306
21.2.4 Nematode biopesticides 307
21.2.5 Protozoan biopesticides 308
21.2.6 Algal biopesticides 309
21.3 Microbial products in biopesticides 309
21.4 Current status of biopesticides in India 309
21.4.1 Registration norms and regulation of microbial biopesticides 311
21.4.2 Evolution of microbial biopesticides for the management of insect pest in India 311
21.5 Current advancement in the microbial biopesticides in the field of genomics, transcriptomics and proteomics 312
21.6 Conclusion and future directions 314
References 314

22. Use of microbial consortia for broad spectrum protection of plant pathogens: regulatory hurdles, present status and future prospects

Ratul Moni Ram, Ashim Debnath, Shivangi Negi and H.B. Singh

22.1 Introduction 319
22.2 Biological control 320
22.3 Microbial consortium 321
22.4 Characteristics of microbial consortium 322
22.5 Microbial consortium mediated plant defense mechanism in biological control 322
22.6 Different types of microbial consortium 322
22.6.1 Fungal and fungal 322
22.6.2 Bacterial and bacterial 323
22.6.3 Fungal and bacterial 323
22.6.4 Algae and bacteria 325

22.7 **Need for development of biopesticides containing microbial consortium** 325
22.7.1 Biopesticide 325
22.7.2 Microbial pesticides 325
22.8 **Current status of Indian biopesticide sector** 328
22.9 **Hurdles in commercialization of microbial based products in India** 329
22.9.1 Regulatory framework and challenges for biopesticides in India 330
22.9.2 Future prospects 331
22.10 **Conclusion** 332
References 332

23. Biocides through pyrolytic degradation of biomass: potential, recent advancements and future prospects

Avedananda Ray, Sabuj Ganguly and Ardith Sankar

23.1 **Introduction** 337
23.1.1 Bio-pesticides: a green alternative to synthetic pesticide 338
23.2 **Pyrolysis-an efficient technology** 338
23.3 **Pyrolytic feedstock** 339
23.4 **Products of pyrolysis** 339
23.5 **Acetic acid as potential product** 341
23.5.1 Chemical composition of wood vinegar 341
23.5.2 Eco-toxicology of pyrolytic products 341
23.6 **Acetic acid eco-toxicology** 341
23.7 **Quinone eco-toxicology** 348
23.8 **Catechol eco-toxicology** 348
23.9 **Phenol eco-toxicology** 348
23.10 **Other alcohol** 348
23.10.1 Disadvantage 348
23.11 **Future prospects** 349
References 350

24. *Trichoderma*: agricultural applications and beyond

R.N. Pandey, Pratik Jaisani and H.B. Singh

24.1 **Introduction** 353
24.2 **Achieving UN sustainable development goals (SDGs)** 353
24.3 **Pesticides consumption in the management of pests** 353
24.4 **Benefits of microbes in rhizosphere** 354
24.5 **Soil borne diseases and plant pathogens** 354
24.5.1 *Trichoderma*—a fungus of unique characteristics 354
24.5.2 *Trichoderma* spp. in agricultural application 357
24.5.3 *Trichoderma* spp. in sustainable environment 368
24.5.4 Commercialization of *Trichoderma* spp. 368
24.5.5 Advantages, challenges, constaints in sustainability of *Trichoderma* based disease management technology and future course of action 370
References 371

25. Exploring the potential role of *Trichoderma* as friends of plants foes for bacterial plant pathogens

Narasimhamurthy Konappa, Udayashankar C. Arakere, Soumya Krishnamurthy, Srinivas Chowdappa and Sudisha Jogaiah

25.1 **Introduction** 383
25.2 **Mechanisms** 384
25.2.1 Competition with pathogens for space and nutrients 384
25.2.2 Antibiosis 385
25.2.3 Cell wall degrading enzymes 386
25.2.4 Plant growth promotion 387
25.2.5 Induced systemic resistance (ISR) 389
25.3 **Trichogenic-nanoparticles and its application in crop protection** 391
25.4 **Conclusions** 392
References 392

26. Advance molecular tools to detect plant pathogens

R. Kannan, A. Solaimalai, M. Jayakumar and U. Surendran

26.1 **Introduction** 401
26.2 **Molecular techniques of plant disease detection** 401
26.3 **Spectroscopic and imaging techniques** 402
26.4 **Fluorescence spectroscopy** 403
26.5 **Visible and infrared spectroscopy** 403
26.6 **Fluorescence imaging** 405
26.7 **Hyper spectral imaging** 406
26.8 **Other imaging techniques** 406
26.9 **Profiling of plant volatile organic compounds** 407
26.10 **Electronic nose system** 407

26.11 GC–MS 408
26.12 Fluorescence in-situ hybridization 409
26.13 Hyper spectral techniques 409
26.14 Biosensor platforms based on nonmaterials 409
26.15 Affinity biosensors 409
26.16 Antibody-based biosensors 410
26.17 DNA/RNA-based affinity biosensor 410
26.18 Enzymatic electrochemical biosensors 410
26.19 Bacteriophage based biosensors 410
26.20 Affinity-based biosensors 411
26.21 Genetically-encoded biosensors 411
26.22 Spectroscopic and imaging techniques 411
26.22.1 Fluorescence spectroscopy 411
26.22.2 Visible and infrared spectroscopy 411
26.23 Conclusion 412
References 412

Index 417

Contributors

Mushtaq Ahmed, Centre for Molecular Biology, Central University of Jammu, Jammu, Jammu & Kashmir, India

Amit Ahuja, Division of Nematology, ICAR-IARI, New Delhi, Delhi, India

Jaganathan Anitha, Department of Zoology, School of Life Sciences, Bharathiar University, Coimbatore, Tamil Nadu, India

Udayashankar C. Arakere, Department of Studies in Biotechnology, University of Mysore, Mysore, Karnataka, India

Alejandro Barragán-Ocaña, National Polytechnic Institute, Center for Economic, Administrative and Social Research, Mexico City, Mexico

Pawan Basnet, University of Missouri-Columbia, Division of Plant Science and Technology, Columbia, MO, United States

Vijai Pal Bhadana, School of Genomics and Molecular Breeding, ICAR-Indian Institute of Agricultural Biotechnology, Ranchi, Jharkhand, India

Anirban Bhar, Department of Botany, Ramakrishna Mission Vivekananda Centenary College, Kolkata, West Bengal, India; Division of Plant Biology, Bose Institute, Kolkata, West Bengal, India

Sunaina Bisht, Rani Lakshmi Bai Central Agricultural University, Jhansi, Uttar Pradesh, India

Alina Butu, National Institute of Research and Development for Biological Sciences, Bucharest, Romania

Marian Butu, National Institute of Research and Development for Biological Sciences, Bucharest, Romania

Balamurugan Chandramohan, Department of Zoology, School of Life Sciences, Bharathiar University, Coimbatore, Tamil Nadu, India

Srinivas Chowdappa, Department of Microbiology and Biotechnology, Bangalore University, Bengaluru, Karnataka, India

Hans-Uwe Dahms, Department of Biomedical Science and Environmental Biology, Kaohsiung Medical University, Kaohsiung, Kaohsiung, Taiwan

Surendra K. Dara, University of California Cooperative Extension, San Luis Obispo, CA, United States

Sampa Das, Department of Botany, Ramakrishna Mission Vivekananda Centenary College, Kolkata, West Bengal, India

Ashim Debnath, Department of Genetics and Plant Breeding, A.N.D. University of Agriculture and Technology, Ayodhya, Uttar Pradesh, India

Nirmaladevi Dhamodaran, Department of Microbiology, Ramaiah College of Arts, Science and Commerce, Bengaluru, Karnataka, India

Rajiv Dhital, University of Missouri-Columbia, Food Science Program, Columbia, MO, United States

Devakumar Dinesh, Department of Zoology, School of Life Sciences, Bharathiar University, Coimbatore, Tamil Nadu, India

Leonardo Fernandes Fraceto, Institute of Science and Technology of Sorocaba, São Paulo State University – Unesp, São Paulo, São Paulo, Brazil

Sabuj Ganguly, Department of Entomology and Agricultural Zoology, Institute of Agricultural Sciences, Banaras Hindu University, Varanasi, Uttar Pradesh, India

Geetika, Department of Environmental Sciences, School of Earth and Environmental Sciences, Central University of Himachal Pradesh, Kangra, Himachal Pradesh, India

Ved Prakash Giri, Division of Microbial Technology, CSIR-National Botanical Research Institute, Lucknow, Uttar Pradesh, India; Department of Botany, Lucknow University, Lucknow, Uttar Pradesh, India

Robin Gogoi, Division of Plant Pathology, ICAR-Indian Agricultural Research Institute, New Delhi, Delhi, India

Razak Hussain, Department of Botany, Aligarh Muslim University, Aligarh, Uttar Pradesh, India

Shubha Jagannath, Department of Botany, Molecular Biology division, Jnana Bharathi Campus, Bangalore University, Bengaluru, Karnataka, India

Akansha Jain, Department of Botany, Ramakrishna Mission Vivekananda Centenary College, Kolkata, West Bengal, India

Pratik Jaisani, Department of Plant Pathology, B.A. College of Agriculture, Anand Agricultural University, Anand, Gujarat, India

Josef Jampílek, Department of Analytical Chemistry, Faculty of Natural Sciences, Comenius University, Bratislava, Slovakia; Department of Chemical Biology, Faculty of Science, Palacky University, Olomouc, Czech Republic

M. Jayakumar, Department of Agronomy, Regional Coffee Research Station, Dindigul, Tamil Nadu, India

B. Jeevan, Crop Protection Section, ICAR-Vivekananda Parvatiya Krishi Anusandhan Sansthan, Almora, Uttarakhand, India

Sudisha Jogaiah, Laboratory of Plant Healthcare and Diagnostics, PG Department of Biotechnology and Microbiology, Karnatak University, Dharwad, Karnataka, India

R. Kannan, Department of Plant Pathology, Faculty of Agriculture, Annamalai University, Chidambaram, Tamil Nadu, India

Kiran Kumar Kommu, ICAR-Central Citrus Research Institute, Nagpur, Maharashtra, India

Narasimhamurthy Konappa, Department of Microbiology and Biotechnology, Bangalore University, Bengaluru, Karnataka, India

Soumya Krishnamurthy, Department of Microbiology, Field Marshal K.M. Cariappa College, A Constituent College of Mangalore University, Madikeri, Karnataka, India

Katarína Kráľová, Institute of Chemistry, Faculty of Natural Sciences, Comenius University, Bratislava, Slovakia

Manish Kumar, Division of Nematology, ICAR-IARI, New Delhi, Delhi, India

Sudhir Kumar, School of Genomics and Molecular Breeding, ICAR-Indian Institute of Agricultural Biotechnology, Ranchi, Jharkhand, India

Bhanu Kumar, Pharmacognosy and Ethnopharmacology Division, CSIR-National Botanical Research Institute, Lucknow, Uttar Pradesh, India

Madan Kumar, School of Genomics and Molecular Breeding, ICAR-Indian Institute of Agricultural Biotechnology, Ranchi, Jharkhand, India

Indu Kumari, National Institute of Pathology, New Delhi, Delhi, India

Shambhu Krishan Lal, School of Genetic Engineering, ICAR-Indian Institute of Agricultural Biotechnology, Ranchi, Jharkhand, India; Crop Improvement Group, International Centre for Genetic Engineering and Biotechnology, New Delhi, Delhi, India

Purabi Mazumdar, Centre for Research in Biotechnology for Agriculture (CEBAR), Universiti Malaya, Kuala Lumpur, Malaysia

Rajendra Prasad Meena, Crop Production Division, ICAR-Vivekananda Parvatiya Krishi Anusandhan Sansthan, Almora, Uttarakhand, India

Sahil Mehta, Crop Improvement Group, International Centre for Genetic Engineering and Biotechnology, New Delhi, Delhi, India

Adriano Arrué Melo, Universidade Federal de Santa Maria, Department of Crop Protection, Santa Maria, Rio Grande do Sul, Brazil

Shweta Meshram, Division of Plant Pathology, ICAR-Indian Agricultural Research Institute, New Delhi, Delhi, India

K.K. Mishra, Crop Protection Section, ICAR-Vivekananda Parvatiya Krishi Anusandhan Sansthan, Almora, Uttarakhand, India

Aradhana Mishra, Division of Microbial Technology, CSIR-National Botanical Research Institute, Lucknow, Uttar Pradesh, India; Academy of Scientific and Innovative Research (AcSIR), Ghaziabad, Uttar Pradesh, India

Marcelo Mueller de Freitas, Department of Agricultural Production Sciences, Paulista State University (Unesp), School of Agricultural and Veterinary Sciences, Jaboticabal, Sao Paulo, Brazil

Kadarkarai Murugan, Department of Zoology, School of Life Sciences, Bharathiar University, Coimbatore, Tamil Nadu, India

Shivangi Negi, Department of Seed Technology, A.N.D. University of Agriculture and Technology, Ayodhya, Uttar Pradesh, India

Samuel Olmos-Peña, Autonomous University of the State of Mexico, Mexico State, Mexico

Aurelio Ortiz, Facultad de Ciencias Químicas, Benemérita Universidad Autónoma de Puebla, Puebla, Mexico

R.N. Pandey, Department of Plant Pathology, B.A. College of Agriculture, Anand Agricultural University, Anand, Gujarat, India

Shipra Pandey, Division of Microbial Technology, CSIR-National Botanical Research Institute, Lucknow, Uttar Pradesh, India; Academy of Scientific and Innovative Research (AcSIR), Ghaziabad, Uttar Pradesh, India

Manoj Parihar, Crop Production Division, ICAR-Vivekananda Parvatiya Krishi Anusandhan Sansthan, Almora, Uttarakhand, India

Amit U. Paschapur, Crop Protection Section, ICAR-Vivekananda Parvatiya Krishi Anusandhan Sansthan, Almora, Uttarakhand, India

Arunava Pattanayak, School of Genetic Engineering, ICAR-Indian Institute of Agricultural Biotechnology, Ranchi, Jharkhand, India; School of Genomics and Molecular Breeding, ICAR-Indian Institute of Agricultural Biotechnology, Ranchi, Jharkhand, India

Manickam Paulpandi, Department of Zoology, School of Life Sciences, Bharathiar University, Coimbatore, Tamil Nadu, India

Krishnasamy Pavithra, Department of Zoology, School of Life Sciences, Bharathiar University, Coimbatore, Tamil Nadu, India

Ricardo Antônio Polanczyk, Department of Agricultural Production Sciences, Paulista State University (Unesp), School of Agricultural and Veterinary Sciences, Jaboticabal, Sao Paulo, Brazil

Amitava Rakshit, Banaras Hindu University, Institute of Agricultural Science, Department of Soil Science & Agricultural Chemistry, Varanasi, Uttar Pradesh, India

Ratul Moni Ram, Department of Plant Pathology, A.N.D. University of Agriculture and Technology, Ayodhya, Uttar Pradesh, India

Avedananda Ray, Department of Agricultural & Environmental Sciences, Tennessee State University, Nashville, TN, United States

Muhammad Razaq, Department of Entomology, Faculty of Agricultural Sciences & Technology, Bahauddin Zakariya University, Multan, Punjab, Pakistan

Steliana Rodino, National Institute of Research and Development for Biological Sciences, Bucharest, Romania

Manjunatha Amitiganahalli Sampangi, Department of Microbiology, Ramaiah College of Arts, Science and Commerce, Bengaluru, Karnataka, India

Ardith Sankar, Department of Agronomy, Institute of Agricultural Sciences, Banaras Hindu University, Varanasi, Uttar Pradesh, India

Estibaliz Sansinenea, Facultad de Ciencias Químicas, Benemérita Universidad Autónoma de Puebla, Puebla, Mexico

Igor Henrique Sena da Silva, Department of Agricultural Production Sciences, Paulista State University (Unesp), School of Agricultural and Veterinary Sciences, Jaboticabal, Sao Paulo, Brazil

Farhan Mahmood Shah, Department of Entomology, Faculty of Agricultural Sciences & Technology, Bahauddin Zakariya University, Multan, Punjab, Pakistan

Soumya Satyanand Shanbhag, Department of Microbiology, Ramaiah College of Arts, Science and Commerce, Bengaluru, Karnataka, India

Shikha Sharma, Department of Environmental Sciences, School of Earth and Environmental Sciences, Central University of Himachal Pradesh, Kangra, Himachal Pradesh, India

Jiang Shoiu-Hwang, Institute of Marine Biology, National Taiwan Ocean University, Keelung, Keelung, Taiwan; Center of Excellence for the Oceans, National Taiwan Ocean University, Keelung, Keelung, Taiwan

S.D. Shruthi, Microbiology and Molecular Biology Lab, BioEdge Solutions, Bangalore, Karnataka, India

Paz Silva-Borjas, National Polytechnic Institute, Center for Economic, Administrative and Social Research, Mexico City, Mexico

Ashish Kumar Singh, Crop Protection Section, ICAR-Vivekananda Parvatiya Krishi Anusandhan Sansthan, Almora, Uttarakhand, India

Pooja Singh, School of Science, Monash University Malaysia, Bandar Sunway, Selangor, Malaysia

H.B. Singh, Department of Plant Pathology, Institute of Agricultural Sciences, Banaras Hindu University, Varanasi, Uttar Pradesh, India; Department of Biotechnology, GLA University, Mathura, Uttar Pradesh, India

Satyendra Pratap Singh, Division of Microbial Technology, CSIR-National Botanical Research Institute, Lucknow, Uttar Pradesh, India; Pharmacognosy and Ethnopharmacology Division, CSIR-National Botanical Research Institute, Lucknow, Uttar Pradesh, India

Anil Kumar Singh, School of Genetic Engineering, ICAR-Indian Institute of Agricultural Biotechnology, Ranchi, Jharkhand, India

Binay Kumar Singh, School of Genomics and Molecular Breeding, ICAR-Indian Institute of Agricultural Biotechnology, Ranchi, Jharkhand, India

A. Solaimalai, Department of Agronomy, ARS, Tamil Nadu Agricultural University, Kovilpatti, Tamil Nadu, India

Jayapal Subramaniam, Department of Zoology, School of Life Sciences, Bharathiar University, Coimbatore, Tamil Nadu, India; Division of Vector Biology and Control, Department of Zoology, Faculty of Science, Annamalai University, Chidambaram, Tamil Nadu, India

U. Surendran, Water Management (Agriculture) Division, Centre for Water Resources Development and Management (CWRDM), Kunnamangalam, Kerala, India

Alexandre Swarowsky, Universidade Federal de Santa Maria, Department of Sanitary and Environmental Engineering, Santa Maria, Rio Grande do Sul, Brazil

Sharmishtha Thakur, Chaudhary Sarwan Kumar Himachal Pradesh Krishi Vishvavidyalaya, Palampur, Himachal Pradesh, India

Camelia Ungureanu, University POLITEHNICA of Bucharest, Bucharest, Romania

Murugan Vasanthakumaran, Department of Zoology, Kongunadu Arts and Science College, Coimbatore, Tamil Nadu, India

A.B. Vedamurthy, Department of Biotechnology and Microbiology, Karnatak University, Dharwad, Karnataka, India

B.K. Vinay, Division of Nematology, ICAR-IARI, New Delhi, Delhi, India

Lan Wang, School of Life Science, Shanxi University, Taiyuan, Shanxi, China

S.F.A. Zaidi, Department of Soil Science and Agriculture Chemistry, Acharya Narendra Deva University of Agriculture and Technology, Faizabad, Uttar Pradesh, India

Chapter 1

Bacillus thuringiensis based biopesticides for integrated crop management

Aurelio Ortiz and Estibaliz Sansinenea
Facultad de Ciencias Químicas, Benemérita Universidad Autónoma de Puebla, Puebla, Mexico

1.1 Introduction

The world population is growing very fast; therefore, the agriculture needs an improvement improving the yield and quality of the crops. Integrated Crop Management (ICM) is a pragmatic approach to crop production which includes Integrated Pest Management (IPM) focusing on crop protection. ICM and IPM strategies combine a range of complementary methods to reduce pest populations below economic injury level while minimizing impacts on other components of the agro-ecosystem and environmental conditions of the area (Youssef and Eissa, 2014). According to Environmental Protection Agency (EPA), biopesticides are certain types of pesticides derived from living organisms. The use of biopesticides has gained acceptance worldwide (Bhattacharjee and Dey, 2014).

In this sense, *Bacillus thuringiensis* has been widely used as a biopesticide in agriculture since this bacterium has specific insecticidal activity against target insects and is safety to non-target organisms (Sansinenea, 2012). It is extensively used in bio-pesticides across the globe and these products represent 1% of the global market for agrochemicals including herbicides, fungicides, and insecticides. *B. thuringiensis* products are available in powder form which contains a mixture of toxin crystals and dried spores. They aid crop growth by reducing harmful toxins and providing stability to the soil by fighting the root diseases. *B. thuringiensis* contains nine different types of toxins, among which delta endotoxin is majorly studied for bio-pesticide applications. These are either applied on the plant leaves or mixed with the soil to inhibit the growth of pests. The insecticidal properties are due to the presence of insecticidal proteins, Cry proteins, produced during sporulation stage. The efficacy of *B. thuringiensis* is related with its narrow spectrum of toxicity, since it can be active against lepidopteran, coleopteran or dipteran insects depending of the Cry proteins presents on each subspecie. Initially, the crops were sprayed with *B. thuringiensis* formulations based on spores and proteins, after that and as consequence of several problems with the formulations, and having genetic engineering as an advantage, transgenic crops were the market for an international industry with billion dollars of gain. This chapter reviewed the past, the present and the future of *B. thuringiensis* as biopesticide for integrated crop management.

1.2 The early beginning of *Bacillus thuringiensis* as biopesticide

B. thuringiensis was first discovered in 1901 in Japan by Shigetane Ishiwata who noted that the larvae of silkworms got sick and died after exposed to this bacterium (Sansinenea, 2012). He reported that the intoxication of the larvae seems to be caused by some toxine secreted by the *Bacillus* bacterium, since the death of the insects occurs before the multiplication of the bacterium. However, the complete identification of this bacterium was made by the German Ernst Berliner who named the bacterium as *B. thuringiensis*, since the bacterium was isolated in Thuringia, a German town. In 1915, Berliner reported the presence of parasporal inclusions within bacterium, although not knowing of its insecticide activity (Sansinenea, 2012). In 1920 was applicated for the first time to control insects in Hungary and in Yugoslavia at the beginning of the 1930 to control European corn borer, being in 1938 when the first formulation called Sporeine was commercialized by Laboratoire Libec in France. For the next two decades several studies were conducted with the development of some formulations against lepidopterans, such as Thuricide produced in 1957 by Sandoz. In those years

Biopesticides. https://doi.org/10.1016/B978-0-12-823355-9.00015-8

(1956) some researchers found that the insecticidal activity of this bacterium was mainly due to the presence of crystalline inclusions formed during sporulation (Sansinenea, 2012). This discovery increased interest in the study of this bacterium being registered as a pesticide to the EPA by 1961.

Some subspecies were discovered until 1977, with insecticidal activity focused only in lepidopteran insects which affect diverse important crops such as tobacco and cabbage. By then different countries had developed different formulations with different names. Then, Goldberg and Margalit (Goldberg and Margalit, 1977) discovered a new subspecie, called *B. thuringiensis* subsp. *israelensis* due to the discovery was in the Negev desert of Israel, which was highly toxic to larvae of several mosquito species, making this bacterium of public health importance to control human diseases such as malaria. Some years later, in 1983, a new subspecie was discovered which was toxic to larvae of coleopteran insects, such as potato beetle (Krieg et al., 1983). Therefore, this isolate was developed to control beetle pests on potatoes principally through Mycogen company. After that, several studies were done realizing combinations between several subspecies to expand their potential against some insect's types at the same time. For example, in 1992 Ecogen company developed Foil, a product for use both against lepidopteran and coleopteran pests. With the isolation of new *B. thuringiensis* strains new genes of Cry proteins were discovered making necessary a robust system of nomenclature (Crickmore et al., 1998).

Another important fact is that biopesticide production depends on high quality and efficient processes. The formulations must be safe, easy to use and have a long shelf life. Sprayable formulations have been widely used; however, they have several disadvantages such as cannot be applied uniformly to the plant, cannot be delivered to pests inside plant tissues and are susceptible to rapid degradation by UV light, making them not useful in crops compared with chemical pesticides. To overcome the first two problems and with the help of the molecular tools, transgenic plants expressing *B. thuringiensis* toxins were developed. Major *B. thuringiensis* transgenic crops worldwide include tomato, tobacco, corn, cotton, potatoes and rice (Ely, 1993). Genetic engineering of plants was helpful against the pests that attack parts of the plant that are not susceptible to biopesticide application.

To understand in a better way how transgenic crops were designed and how the next stage of development of *B. thuringiensis* toxins was given, it is necessary to explain the mode of action of Cry (Crystal δ-Endotoxins) or Cyt (cytolytic toxins) toxins, which are pore forming proteins. The Cry proteins comprise more than 200 members; however, despite different toxins have shown be toxic against specific targets the structure of all proteins is the same containing three structural domains connected by single linkers (Bravo et al., 2007). Domain I contains seven-helix amphipathic bundle, with six helices surrounding a central helix and is involved in the toxin insertion into the plasmatic membrane of the epithelial midgut cells. Domain II contains three loop structures in the β-sheets that are responsible of the initial recognition and binding of the protein to binding site on the microvillar membrane. Domain III consists of two antiparallel β-sheets providing structural integrity to the molecule and this domain plays a role in receptor binding and pore formation (de Maagd et al., 2003). The insecticidal activity starts when the insects' larvae ingest endotoxins crystals. In a short way after ingestion the crystals are solubilized by alkaline conditions of the insect midgut and proteolytically converted into a toxic core. The activated toxin binds to receptors of the apical microvillus membranes of epithelial midgut cells. Then, the toxin adopts a conformation to allow its insertion into the cell membrane and form a cation-selective channel. Oligomerization occurs forming a pore induced by an increase in cationic permeability (Bravo et al., 2004). After, toxin aggregation occurs at the membrane surface and once enough number of channels have formed, the cations enter the cell, causing an osmotic imbalance until the cell ruptures. When enough cells have been destroyed, the midgut epithelium loses its integrity resulting in death caused by bacteremia and tissue colonization. The mode of action of Cry proteins has been widely studied and review (Pardo-Lopez et al., 2013).

With the information of the Cry proteins structure and their mode of action in hands, the researchers were able to design new improved proteins opening a new era of *B. thuringiensis* biopesticides.

1.3 The past last twenty years of *B. thuringiensis* as biopesticide

Due to the exit of *B. thuringiensis* biopesticides, they were continuously spread over the infected crops to control pests in the field. This practice generated resistance of insect populations both to a particular toxin or to other toxins to which they have not previously been exposed (Pereira et al., 2008; Gong et al., 2010; Xu et al., 2010). The main objective for several years was to improve the toxicity of the proteins against insects avoiding generating resistance. The economic viability of *B. thuringiensis* biopesticides depends on the potency of their toxins and the environmental stability of the formulations after sprayed.

Therefore, the search for new *B. thuringiensis* strains expressing novel toxins with improved activity was a goal for the researchers and leader companies to reduce their susceptibility to UV light, heat, extreme pH, and proteolytic degradation, and to increase their toxicity (Sanahuja et al., 2011). The first attempt to improve the toxicity of toxins was the isolation of new strains with better and broader insecticidal activity against a particular insect than previously known strains. This topic was described in the above section. To overcome the problem of the susceptibility of the formulation to UV light there were some research work about the improved formulations, being melanin the more suitable due to its natural condition (Sansinenea and Ortiz, 2015; Sansinenea et al., 2015).

As we state before, genetically modified crops that constitutively express toxins was an initial solution to some problems such as the inability of the toxins to target insects feeding on internal tissues of the plants. Therefore, in this section we will focus on other strategies to increase the production or the toxicity of the toxins using recombinant DNA technology.

Product of molecular genetic protein engineering was employed to alter or broaden the activity of a toxin. With this strategy is possible to create new toxins with modified properties such as toxicity, binding affinity or insect specificity. Some aminoacidic residues are the key for the correct functionality of each domain, so replacing some residues by others in each domain the binding properties or affinities or pore formation can be affected with minimal structural alterations (Florez et al., 2012). Multiple techniques can be used to design and develop modified toxins, including molecular modeling, bioinformatics and site-directed mutagenesis. Mutagenesis was used to explore the participation of specific protein regions in the toxicity mechanism and some mutations resulted in the generation of improved Cry toxins (Bravo et al., 2007). This knowledge has served to create chimeric toxins with a wider target spectrum than the parental toxins from which were derived (Walters et al., 2010).

For the construction of new strains conjugal plasmid exchange system was used by conjugation phenomenon (Vilas-Boas et al., 2000), which was utilized by Ecogen company to construct a strain that was the active ingredient of the Condor biopesticide. However, this strategy was only applicable to *cry* genes carried by conjugative plasmids from the same compatibility group. The construction of genetically engineered *B. thuringiensis* strains can be achieved through shuttle vectors carrying *cry* genes and introducing them into the chromosome or resident plasmids by homologous recombination (Sansinenea et al., 2010). Transposition was also used to construct recombinant strains (Sanchis, 2012). In this way using DNA technology many strains have been constructed and many products have been commercialized in the past years (Eski et al., 2017). A second strategy that was used for strain improvement consisted on the expression of *cry* genes in other microbial hosts, such as *Pseudomonas fluorescens*, which was used by Mycogen company to produce two commercial products MVP to control lepidopterans and M-TRAK to control coleopterans (Gaertner et al., 1993).

Ultimately several works have highlighted the importance of nanotechnology to enhance in the efficacy of certain biological substances on pests, decrease in toxicity toward humans and the environment, and reduce losses due to physical degradation through the encapsulation of biopesticide substances in nanoparticulate systems (De Oliveira et al., 2014; Agrawal and Rathore, 2014; Mishra et al., 2017).

1.4 The present and the future of *B. thuringiensis* as biopesticide

The history of Bt as biopesticide explained above puts us in the current context today. New products are registering and marketing day by day. For example, EPA is proposing to register a pesticide product containing *Bacillus thuringiensis*, subsp. *aizawai* strain ABTS 1857 (Bta ABTS, 1857) to prevent and control wax moths in beehives. This product offers beekeepers a new tool against destructive wax moth larvae. The active ingredient in this pesticide product (Bta ABTS, 1857) is part of a large group of bacteria, *Bacillus thuringiensis*, that occur naturally in soil. Bta ABTS 1857 controls wax moth infestations by producing a crystallized protein that is toxic to wax moth larvae. The Greater Wax Moth (*Galleria mellonella*) is a significant pest of honeybees. Adult female moths enter hives at night and deposit eggs in cracks and crevices within the hive. The moth larvae then burrow through and destroy the honeycombs as they feed on the wax, pollen, and larval honeybees. The moth larvae will similarly damage stored honeycomb frames under the appropriate conditions (e.g., temperature, lighting, and ventilation) in short order. To use this product, commercial and hobbyist beekeepers would apply a dilute solution of Bta ABTS 1857 to empty honeycomb frames prior to winter storage. When wax moth larvae attempt to feed on the honeycomb, they would also ingest some Bta ABTS 1857, which will release a protein into the larva's digestive system that attaches to the gut, eventually causing it to rupture. The toxicological data for Bta ABTS 1857 demonstrated a lack of toxicity, pathogenicity, or infectivity to humans. Bta ABTS 1857 has a tolerance exemption for use in or on honey and honeycomb and all other raw agricultural commodities (40 CFR §180.1011). EPA expects minimal to no exposure to honeybees and other nontarget organisms because of the method and timing of application. As noted, beekeepers would make a one-time treatment directly to empty honeycomb frames prior to winter storage. And hives maintain temperatures above 35°C, thus preventing Bta ABTS 1857 spore viability (which declines at 30°C) when hives are returned to the treated frames in the spring (EPA, 2020).

Global production of biopesticides has been estimated to be over 3000 tons per year, which is increasing rapidly by 10% every year (Kumar and Singh, 2015). More than 200 products are being sold in the US market, compared to only 60 comparable products in the EU. More than 225 microbial biopesticides are manufactured in 30 OECD countries (Damalas and Koutroubas, 2018). The NAFTA countries (USA, Canada, and Mexico) use about 45% of the biopesticides sold, while Asia lacks behind with the use of only 5% of biopesticides sold world over. It is expected that biopesticides will equalize with synthetics, in terms of market size, between the late 2040s and the early 2050s, but major uncertainties in the rates of uptake, especially in areas like Africa and Southeast Asia account for a major portion of the flexibility in those projections (Kumar and Singh, 2015).

Based on crop, the market has been segregated into cereals, fruits, vegetables, oils, and other types. In terms of the target pests covered, the market is divided into the following categories: Beetles, Caterpillars, Mosquitos, Black flies Fungus and others (Bacillus Thuringiensis Market). The market is also divided by dry and solid formulations, soil treatment, seed treatment, foliar spray and post-harvest applications and North America, Europe, Asia Pacific, South America, Middle East and Africa regions.

B. thuringiensis holds a significant share in the global bio-pesticide market and is anticipated to expand at a promising pace owing to the increasing need to consume contamination-free food. The arthropods segment constitutes a substantial share in the market on account of the effectiveness of Bt products against arthropods in comparison to microorganisms. Other factors propelling this market over the forecast period include the easy availability of Bt products and their environment-friendliness. Moreover, Bt is required in small quantities to suppress pests, leading to limited exposure. Additionally, it decomposes faster which leaves negligible amounts of residue. All these factors boost the usage of Bt products, propelling this market over the next few years (Bacillus Thuringiensis Market, 2020).

Geographically, North America accounted for a promising share in the market in 2015 which is anticipated to rise over the forecast period owing to the wide acceptance of these products among farmers in this region and efforts taken by government to promote their usage. The market in Europe ranks second in terms of dominance, due to the presence of prominent manufacturers in the region. France, Italy, and Spain have emerged as key countries where Bt is widely accepted whereas Germany is an important testing market for new products. The market in Asia Pacific is anticipated to witness significant growth as a result of the escalating population coupled with increasing health consciousness among consumers here, as it is shown in Fig. 1.1 (Intellectual Research Partners, 2019).

Some of the leading players in the global *Bacillus thuringiensis* market are:

- Basf Se (Germany)
- Bayer Cropscience (Germany)
- Camson Biotechnologies Limited (India)
- Certis Usa (USA)
- Isagro Spa (Italy)
- Koppert B.V (Netherlands)
- Marrone Bio Innovations Inc. (USA)
- Valent Biosciences Corporation (USA)
- Neudorff Gmbh (Germany)
- Novozymes (Denmark)

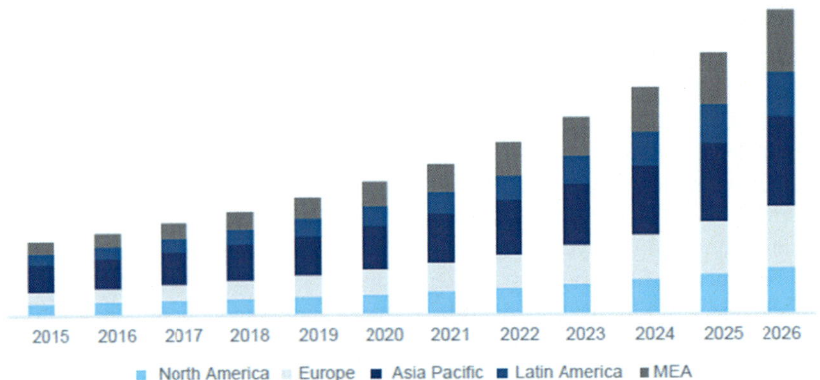

FIG. 1.1 Global *B. thuringiensis* market by region, 2015–26.

These companies have been the leaders on biopesticides market based on *B. thuringiensis* improving their toxicity since they invest in research and development of biopesticides during some decades.

1.5 Conclusions and perspectives

What's next for Bt? Although scientific and technical knowledge about Bt has advanced during the last decades, research in production, formulation and delivery may greatly assist in commercialization of biopesticides. More research is need integrating biological agents into production system, improving capability of developing countries to manufacture and use biopesticides. The regulatory mechanisms are equally important to maintain the quality and availability of the biopesticides at reasonable cost in the developing countries. Thus, various aspects of biopesticides covering the current status, constraints, prospects and regulatory network toward their effective utilization for the benefit of human beings need to be reviewed regularly. The development of the biopesticide market in the future is strongly related to research on biological control agents. Several scientists from diverse research institutes have done some preliminary research in the field, but complete and systematic reports are scarce. Therefore, it is necessary to strengthen the collaboration of enterprises and research institutes on this topic.

A most promising area of investigation is the discovery, identification and validation of molecular targets for development of new insecticides. There are extensive databases for genome sequences of insects and other organisms which afford valuable information for identification of new targets. Recently, a web-based computational pipeline platform was developed for automated large-scale gene mining and insecticide target identification. The platform utilizes bioinformatics, genomics and proteomics for high-throughput gene and protein target identification. All targets selected for consideration can then be analyzed in silico by docking calculations, molecular dynamics simulations and other techniques to characterize appropriate target interactions with chemically or genetically altered Cry toxins. Such an approach should facilitate protein design for the creation of Cry protein and peptide mimics that might be more effective than the natural toxins themselves and less able to induce insect host resistance.

Definitively, the world production and utilization of biopesticides are increasing at a rapid scale. The interest in organic farming and pesticide free agricultural products would certainly warrant increased adoption of biopesticides by the farmers. A good production and quality control of biopesticides from manufacturers, may be essential for better adoption of this technology. It is needed to create a raised concern about environmental safety as a global concern, among the farmers, manufacturers, government agencies, policy makers and the common men to switch-over to biopesticides for pest management requirements. It is also believed that biological pesticides may be less vulnerable to genetic variations in plant populations that cause problems related to pesticide resistance. If deployed appropriately, biopesticides have potential to bring sustainability to global agriculture for food and feed security.

References

Agrawal, S., Rathore, P., 2014. Nanotechnology pros and cons to agriculture: a review. Int. J. Curr. Microbiol. Appl. Sci. 3, 43–55.

Bacillus thuringiensis Market - Global Industry Analysis, Size, Share, Growth, Trends, and Forecast 2016 – 2024, 2020. Transparency Market Research.

Bhattacharjee, R., Dey, U., 2014. Biofertilizer, a way towards organic agriculture: a review. Afr. J. Microbiol. Res. 8 (24), 2232–2342.

Bravo, A., Gomez, I., Conde, J., Muñoz-Garay, C., Sánchez, J., Miranda, R., Zhuang, M., Gill, S.S., Soberón, M., 2004. Oligomerization triggers binding of a *Bacillus thuringiensis* Cry1Ab pore-forming toxin to aminopeptidase N receptor leading to insertion into membrane microdomains. Biochem. Biophys. Acta 1667, 38–46.

Bravo, A., Gill, S.S., Soberón, M., 2007. Mode of action of *Bacillus thuringiensis* Cry and Cyt toxins and their potential for insect control. Toxicon 49, 423–435.

Crickmore, N., Zeigler, D.R., Feitelson, J., Schnepf, E., Van Rie, J., Lereclus, D., Baum, J., Dean, D.H., 1998. Revision of the nomenclature for the *Bacillus thuringiensis* pesticidal crystal proteins. Microbiol. Mol. Biol. Rev. 62, 807–813.

Damalas, C.A., Koutroubas, S.D., 2018. Current status and recent developments in biopesticide use. Agriculture 8, 13.

de Maagd, R.A., Bravo, A., Berry, C., Crickmore, N., Schnepf, H.E., 2003. Structure, diversity, and evolution of protein toxins from spore forming entomopathogenic bacteria. Annu. Rev. Genet. 37, 409–420.

De Oliveira, J.L., Campos, E.V.R., Bakshi, M., Abhilash, P.C., Fraceto, L.F., 2014. Application of nanotechnology for the encapsulation of botanical insecticides for sustainable agriculture: prospects and promises. Biotechnol. Adv. 32, 1550–1561.

Ely, S., 1993. The engineering of plants to express *Bacillus thuringiensis* delta-endotoxins. In: Entwistle, P.F., Cory, J.S., Bailey, M.J., Higgs, S. (Eds.), *Bacillus Thuringiensis*, an Experimental Biopesticide: Theory and Practice. Wiley, Chichester, pp. 105–124.

EPA, 2020. First Beehive Uses of the Currently Registered Active Ingredient *Bacillus thuringiensis*, Subsp. aizawai Strain ABTS 1857. EPA-HQ-OPP-2019-0247.

Eski, A., Demir, I., Sezen, K., Demirbağ, Z., 2017. A new biopesticide from a local *Bacillus thuringiensis* var. *tenebrionis* (Xd3) against alder leaf beetle (Coleoptera: chrysomelidae). World J. Microbiol. Biotechnol. 33, 95.

Florez, A.M., Osorio, C., Alzate, O., 2012. Protein engineering of *Bacillus thuringiensis* δ-endotoxins. In: Sansinenea, E. (Ed.), *Bacillus thuringiensis* Biotechnology. Springer, Netherlands, pp. 93–113.

Gaertner, F.H., Quick, T.C., Thompson, M.A., 1993. CellCap: an encapsulation system for insecticidal biotoxin proteins. In: Kim, L. (Ed.), Advanced Engineered Pesticides. Marcel Dekker Inc, New York.

Goldberg, L.J., Margalit, J.A., 1977. Bacterial spore demonstrating rapid larvicidal activity against *Anopheles sergentii, Uranotaenia unguiculata, Culex univeritattus, Aedes aegypti*, and *Culex pipiens*. Mosq. News 37, 355–358.

Gong, Y.J., Wang, C.L., Yang, Y.H., Wu, S.W., Wu, Y.D., 2010. Characterization of resistance to *Bacillus thuringiensis* toxin Cry1Ac in *Plutella xylostella* from China. J. Invertebr. Pathol. 104, 90–96.

Intellectual Research Partners, 2019. Global *Bacillus thuringiensis* Market: Forecast till 2026. Orian Research Inc.

Krieg, A., Huger, A., Lagenbruch, G., Schnetter, W., 1983. *Bacillus thuringiensis* var. tenebrionis: a new pathotypes effective against larvae of Coleoptera. Z. Angew. Entomol. 96, 500–508.

Kumar, S., Singh, A., 2015. Biopesticides: present status and the future prospects. J. Fertil. Pestic. 6, e129.

Mishra, S., Keswani, C., Abhilash, P.C., Fraceto, L.F., Singh, H.B., 2017. Integrated approach of agri-nanotechnology: challenges and future trends. Front. Plant Sci. 8, 471.

Pardo-Lopez, L., Soberon, M., Bravo, A., 2013. *Bacillus thuringiensis* insecticidal three-domain Cry toxins: mode of action, insect resistance and consequences for crop protection. FEMS Microbiol. Rev. 37, 3–22.

Pereira, E.J.G., Lang, B.A., Storer, N.P., Siegfried, B.D., 2008. Selection for Cry1F resistance in the European corn borer and cross-resistance to other Cry toxins. Entomol. Exp. Appl. 126, 115–121.

Sanahuja, G., Banakar, R., Twyman, R.M., Capell, T., Christou, P., 2011. *Bacillus thuringiensis*: a century of research, development and commercial applications. Plant Biotechnol. J 9, 283–300.

Sanchis, V., 2012. Genetic Improvement of Bt strains and development of novel biopesticides. In: Sansinenea, E. (Ed.), *Bacillus thuringiensis* Biotechnology. Springer, Netherlands, pp. 215–228.

Sansinenea, E., Ortiz, A., 2015. Melanin: a photoprotection for *Bacillus thuringiensis* based biopesticides. Biotechnol. Lett. 37, 483–490.

Sansinenea, E., Vazquez, C., Ortiz, A., 2010. Genetic manipulation in *Bacillus thuringiensis* for strain improvement. Biotechnol. Lett. 32, 1549–1557.

Sansinenea, E., Salazar, F., Ramirez, M., Ortiz, A., 2015. An ultraviolet tolerant wild-type strain of melanin-producing *Bacillus thuringiensis*. Jundishapur J. Microbiol. 8 (7), e20910.

Sansinenea, E., 2012. Discovery and description of *Bacillus thuringiensis*. In: Sansinenea, E. (Ed.), *Bacillus thuringiensis* Biotechnology. Springer, Netherlands, pp. 3–18.

Vilas-Boas, L.A., Vilas-Boas, G.F.L.T., Saridakis, H.O., et al., 2000. Survival and conjugation of *Bacillus thuringiensis* in a soil microcosm. FEMS Microbiol. Ecol. 31, 255–255.

Walters, F.S., de Fontes, C.M., Hart, H., Warren, G.W., Chen, J.S., 2010. Lepidopteran-active variable-region sequence imparts coleopteran activity in eCry3.1Ab, an engineered *Bacillus thuringiensis* hybrid insecticidal protein. Appl. Environ. Microbiol. 76, 3082–3088.

Xu, L., Wang, Z., Zhang, J., He, K., Ferry, N., Gatehouse, A.M.R., 2010. Cross-resistance of Cry1Ab selected Asian corn borer to other Cry toxins. J. Appl. Entomol. 134, 429–438.

Youssef, M.M.A., Eissa, M.F.M., 2014. Biofertilizers and their role in management of plant parasitic nematodes. A review. E3 J. Biotechnol. Pharm. Res. 5, 1–6.

Chapter 2

Biopesticides for management of arthropod pests and weeds

Muhammad Razaq and Farhan Mahmood Shah

Department of Entomology, Faculty of Agricultural Sciences & Technology, Bahauddin Zakariya University, Multan, Punjab, Pakistan

2.1 Agriculture and pests

Human population is expanding very rapidly, currently standing at 7 billion mark, which is expected to reach 9.7 billion by 2050. For continuing population, crop production has to be considerably increased to meet the consumption growth (Ash et al., 2010; Godfray et al., 2010). An outstanding challenge is to accomplish this without destructing environment and human health. In earlier times, food amplification was fairly possible for ancient people due to surplus availability of cultivatable land and other agricultural resources (Abrol and Shankar, 2012). Nowadays, population growth is disproportional to the production supply and agricultural resources are shrinking with increased industrialization, making food amplification a formidable forthcoming challenge for humanity (Godfray et al., 2010). This challenge is most likely to stay improbable if worldwide food losses by certain crop pests are not adequately protected.

Humans are facing crop pest challenges ever since they learnt agriculture and adopted it for their livelihood. A pest is considered any organism or species, which is undesired by humans because of its competing nature with humanity for food, space and shelter and for creating negative consequences for human health and well-being (Sawicka and Egbuna, 2020). Globally, there are about 67,000 divergent species of different weeds, pathogens and invertebrates that are major challenges for agriculture. Of these species, 10,000 are arthropod insects, 30,000 are weeds, about 100,000 are various plant pathogenic microorganisms (fungi, viruses, bacteria etc.) and about 1000 are nematodes (Hall, 1995), altogether causing 40% global food losses (Oerke, 2006) despite protection measures adoption. Within this complex, weeds are often considered to be the most notorious yield reducers, which are economically more harmful than arthropod insects and plant pathogens (Gharde et al., 2018). Potential losses by weeds (34%) are ahead of those by invertebrates (18%), pathogens (16%) or by other invertebrates (Oerke, 2006). Losses by agricultural pests in poor countries (averaging 40%−50%) are much greater compared with 25%−30% losses in developed countries (Thacker, 2002). These agricultural pest losses are projected to grow even bigger in coming future due to trade globalization (Perrings et al., 2005), agricultural intensification (Wilby and Thomas, 2002), and climate change (Gregory et al., 2009). Further discussion in this chapter will detail insect pests and weed challenges to manage them in context of biopesticides.

2.1.1 Synthetic pesticides and challenges

Ancient people lived with pests without controlling them; however increasing encounters urged humans to become intolerant. They suppressed pests via manipulation of biological, cultural and mechanical approaches. Also, research began for developing quicker acting and time saving protection strategies. It was mid-twenties, when humans discovered synthetic pesticides as chemical weapons which they extensively used against pest problems (Metcalf, 1980; Razaq et al., 2019). These products for their effectiveness received broad acceptance among pest managers as substitutes of earlier applied protection strategies. In the long run, too extensive dependence on pesticides proved problematic as it created negative consequences for pest management, human health and environment (Casida and Quistad, 1998). For instance, pests became adapted under natural selection, which made them even more difficult to control. Frequent exposure by pesticides continuously selected pest populations, which increased their resistance status and caused other antagonistic

Biopesticides. https://doi.org/10.1016/B978-0-12-823355-9.00005-5

effects such as emergence of multiple resistance (Davis and Frisvold, 2017; French et al., 1992; Pimentel et al., 1993). Pest resistance is a long standing and ever-expanding problem in crop protection, which is responsible for declining once-promising pesticides, rendering them ineffective subsequently against pests that have become resistant (Sparks and Nauen, 2015). There are, indeed, over 600 cases of insect arthropods developing resistance to pesticidal products, worldwide (Sparks and Nauen, 2015). Not only arthropods but also the weeds and plant pathogens are becoming resistant against herbicides and fungicides (Davis and Frisvold, 2017; Hollomon, 2015; Staub, 1991). Resistance confirming reports in weeds and pathogens are increasing at an alarming rate. As a result, pest management challenges are even growing bigger than before and despite 15–20 fold increase in pesticide usage since 1960s; global losses from agricultural pests continue to remain unsustainably high or even are rising in some instances (Culliney, 2014; Oerke, 2006). These chemical by being broad-spectrum pesticides are potentially dangerous for pollinators (Goulson et al., 2015) and for agents of biological control (Santos et al., 2017). These organisms upon frequent exposure to these chemicals experience undesired physiological and behavioral impairments, which consequence ecosystem services offered by these beneficial organisms (Biondi et al., 2012; Xiao et al., 2016). Most of these chemicals are known for their longer persistence in the biospheres causing residual contamination (Bhanti and Taneja, 2007; Leong et al., 2007) therefore are cursed for being largely responsible for causing carcinogenicity, teratogenicity and hormonal disruption in humans (Alavanja et al., 2004; Papadakis et al., 2015).

2.1.2 From pesticides to biopesticides

Since recent past, realization has grown up on pesticide concerns among consumers all around the world. This concern is reflective as stringent pesticide regulations are being imposed on pesticide Maximum Residue Levels (MRLs) in the agricultural produce that has to be consumed (Amoah et al., 2006). Some countries have even increasingly lowered their criteria of MRLs for agriculture imports (Chandler et al., 2011; Kumar, 2015). Due to such enforcement of safety standards, some of the toxic products are banned from normal use or even are withdrawn from agricultural niche markets (Czaja et al., 2015). Newer and safer chemical development, typically from synthetic origin, is often challenging due to costly nature of the manufacturing process and tighter regulatory standards (Sparks and Nauen, 2015). In urge to explore safer and efficient alternatives, researchers shifted their interest toward "biopesticides" that are inherently less harmful than synthetic counterparts (Khater, 2012; Perrings et al., 2005). Biopesticides are often highly target specific and responsive toward a narrow spectrum of agriculture pests, they are therefore considered less likely to cause serious effects in non-target organisms (Seiber et al., 2014). For example, insect growth regulators show high selective toward caterpillar feeding on a plant and do not adversely target beneficial insects or other organisms residing in that environment. Biopesticides application requires small quantities and their biodegradation and decomposition is quick, meaning lower environmental exposure and subsequent concerns (Hall and Menn, 1999; Seiber et al., 2014). As being inherently safer, their registration is quick (Hassan and Gökçe, 2014), requiring one or two years, compared to conventional pesticides that require almost five to seven years. Also, biopesticides possess versatile modes of action, hence may be extremely important in regaining control of resistant pests (Isman, 2000; Isman and Seffrin, 2014; Koul and Dhaliwal, 2000).

These favorable properties create avenue for biopesticides to be considered in Integrated Pest Management (IPM) programs (Isman, 2019a,b). There are about 241 active ingredients, possessing insecticidal and acaricidal potential, recognized in 1155 products in biopesticide category. IPM programs works in a way that several pest control strategies interact to regulate pest populatoins in order to lower pesticide loads. Research has shown that biopesticides are potentially suitable for IPM programs (Soares et al., 2019), and are efficient both as alone treatment or in integration with other approaches involving synthetic pesticides (Reddy, 2011; Shah et al., 2017), which means that biopesticides integration with synthetic pesticides can reduce overall burden on synthetics (Srinivasan, 2012). Biopesticide products are gaining enormous attention (Srinivasan et al., 2019), newer products are being explored (Bell et al., 2016; Isman, 2019b; Stevenson et al., 2017), and efforts are being concentrated for improving their pest control potential (Pascoli et al., 2019, 2020).

2.2 Inconsistencies in understanding the term "biopesticides"

Biopesticides, a new category of pesticides is derived from natural materials such as microbes, plants, animals and minerals. This new pesticide category from natural materials was created and named by US Environment Protection Agency (USEPA). Biopesticides are pesticidal products, natural in occurrence or derived from natural resources, and possess properties to destroy undesired pests (Seiber et al., 2014). Biopesticides may be clearly distinguished from their synthetic counterparts for their natural origin, unique action modes, high specificity, narrow pest range, and also for their less use quantities. Those natural products possessing any or all of these characteristics are generally recognized as biopesticides, with more a product possessing of these characteristics; greater are its chances of being placed in this category.

Narrowly, biopesticides may be defined as specific separations carrying living microorganisms, or broadly defined as botanicals, semiochemicals and transgene products. The EPA recognizes biopesticides in three main categories (Seiber et al., 2014) depending on the active ingredient: (1) products containing biocontrol agents as active ingredients, including (e.g. *Bacillus thuringiensis*), fungi, viruses or protozoa; (2) Plant-Incorporated-Protectants (PIPs), these are transgenes incorporated into plant materials and responsible for producing pest management compounds, the protein and its genetic material; (3) biochemical pesticides (e.g. based on naturally occurring substances having same structure and function and control pests by non-toxic mechanisms). At global level, there exists an inconsistency in understanding the term biopesticides (Liu et al., 2019a; Moshi and Matoju, 2017); which means that definition of biopesticides given by USEPA is not necessarily followed as such across the entire world, and due to which, International Biocontrol Manufacturer's Association (IBMA) and the International Organization for Biological Control promoted the term biocontrol agents (BCAs) instead of biopesticides (Guillon, 2003; Mishra et al., 2015).

2.2.1 Microbial biopesticides

Microbial biopesticides carry active ingredients from microbial origin including bacteria, viruses, protozoans, fungi, or algae. These microbes have great potential to act like bioinsecticides and bioherbicides in pest management events (Glare and O'Callaghan, 2019; Verma et al., 2020). Active ingredients from these pathogens are highly species-specific, with some being specific herbicides and other being specific bioinsecticides. As being too specific, biopesticides are considered to be safer pesticides for non-target organisms present in the ecosystem. Microbial biopesticides have shown their promising potential in achieving control against many insects, pathogens and weeds. Some biopesticides from soil Actinomycete and some isolates from *Baccillus thuringiensis* (*Bt*) are popular microbial products that are available as avermectins, emamectins, spinosad, and *B. thuringiensis* formulations. Some biopesticides (e.g., bilanafos) are promising weed killers, providing efficient control against many species of perennial weed plants (Liu et al., 2019a).

Entomopathogenic microbes get invasion through the host integument or gut. Once inside, multiplication begins and pathogens start releasing toxins causing host mortality. These mortality causing toxins from various microbes may vary greatly in their structure, specificity and toxicity; however, characterized mainly as being peptides (Mehrotra et al., 2017). In their role as bioherbicides, microbial toxins (toxic substances and lytic enzymes) from attacking pathogen are active weed killers (Radhakrishnan et al., 2018). These substances that are mainly toxins degrade weed seed coat and help the agent to exploit seed endosperm for survival and prevent affected seed germination. In the root core, the invader pathogen deposited toxins or metabolites affect cell division and cellular functioning and cause diseases, chlorosis and necrosis in that host, hence lower seed germination. Also, under microbial attack, weed plants enhance abscisic acid and ethylene activities but lower gibberellin activity, which adversely affects plant photosynthesis, growth, and development (Radhakrishnan et al., 2018).

2.2.1.1 Bacterial bioinsecticides

Bacteria are microbial agents that multiply rapidly, possess easily modifiable genetic makeup and require simple and easy to uptake nutrients for growth. As being cheap, the manufacturing of bacterial biopesticides requires much less cost; hence bacterial formulations are relatively much common than other biopesticides of viral or fungal origins (Liu et al., 2019a; Verma et al., 2020). Many bacterial species possess insect killing properties including those that belong to genera *Pseudomonas*, *Clostridium*, *Proteus*, *Enterobacter*, *Serratia* and *Bacillus* etc., (Verma et al., 2020). Genus *Bacillus* is very much famous and products from *B. thuringiensis* hold wider acceptability among pest managers for the control of several arthropod pests (Glare and O'Callaghan, 2019).

2.2.1.2 Bacterial bioherbicide

There are a number of bacterial genera that possess bioherbicidal potential. They can be used in managing weed pests. The common mechanism of action for these herbicides, as stated above, is the production of certain metabolites that are toxic and affect germination and growth of seeds from infected plants (Radhakrishnan et al., 2018). *Pseudomonas* and *Xanthomonas* are most described genera of bacterial bioherbicides that provide promising control against different species of weed pests (Banowetz et al., 2008). In genus *Pseudomonas*, two strains including *Pseudomonas fluorescens* strain D7 and

P. fluorescens strain BRG100 have shown promising potential against *Bromus tectorum* and *Setaria viridis*, respectively (Kennedy et al., 2001; Quail et al., 2002). Likewise, *Xanthomonas campestris* cv. *poae* JTP482 and *X. campestris* LVA-987 have shown promising potential against *Poa annua* and *Conyza canadensis*, respectively (Imaizumi et al., 1997).

2.2.1.3 Fungal bioinsecticides

Fungi are another group of biocontrol agents that provide control against insect pests on diverse agricultural crops (Kumar et al., 2019). About 700 species of fungi from 90 genera (Hajek and St. Leger, 1994; Moorhouse et al., 1992) possess potential to inhibit, inactivate or kill agricultural pest insects. These insect killing fungi are known as entomopathogenic fungi. These insect killers are easy to apply, need not to be ingested; only their physical contact is sufficient to trigger action in their targeted host. The key paths serving as fungal entries are host cuticle, trachea or open wounds (Holder and Keyhani, 2005), or pathogen may also get entry via ingestion. Fungal spores that fungi contain for their proliferation require humidity, moist soil and cooler environments for their optimum response. Among fungal pathogens, *Beauveria bassiana*, a soil fungus, is a major member of famous bioinsecticides that provide control against a wide array of arthropod pest insects. It kills the host within few days of acquiring infection by the host. Other species, *B. brongniartii* and *Metarhizium anisopliae* are also the well-known fungal biopesticide. These species are applied worldwide for obtaining control against many pests, including the pine caterpillar, *Dendrolimus punctatus*, green vegetable bug, *Nezara viridula* and green and brown mirids, *Creontiades* species (Mehrotra et al., 2017). Other than these species, many other sucking insects are vulnerable to infection by fungal bioinsecticides, mainly *B. bassiana*.

Following entry through host cuticle, *B. bassiana* starts producing an array of fungal hypha and releases toxic secondary metabolites causing mortality of the infected host. The secondary metabolites from the fungus include peptides (non-ribosomally synthesized; e.g., beauveriolides, beauvericin, bassianolides and), polyketides and non-peptide pigments (e.g., oosporein, bassianin and tenellin) and other metabolites (e.g., oxalic acid) that may have role in pathogenesis and virulence (Xu et al., 2009). In USA, *B. bassiana* is commercially available as Mycotrol ES® (Mycotech, Butte) and Naturalis L® (Troy Biosciences) for the control of several sucking insect pests like whitefly, mealybugs, aphids, leafhoppers and many others (Mehrotra et al., 2017).

2.2.1.4 Fungal bioherbicides

Over the years, many fungal pathogens have become popular as bioherbicides that are employed in management of weeds in crops and turfs. Fungal bioherbicides produce some acids which cause cell wall break down and weakening of defense mechanism of the attacked host by inhibiting activities of defense inducing polyphenol oxidases (Ahluwalia, 2007; Cessna et al., 2000). Fungal pathogens in genus *Colletotrichum*, including *Colletotrichum gloeosporioides* f. sp. *malvae* and *C. gloeosporioides* f. sp. *Aeschynomene* are promising suppressors of two weed plants, *Malva pusilla* and *Aeschynomene virginica*, respectively (Ahluwalia, 2007; Mortensen, 1988). Some other products carrying *C. truncatum* and *C. orbiculare* provide control of weeds including *Sesbania exaltata* and *Xanthium spinosum* (Auld et al., 1990). Some species of fungi in genus *Sclerotinia*, for instance, *Sclerotinia minor* are promising bioherbicides, providing control of weeds associated with turfs, including *Plantago major, Taraxacum officinal*, and *Trifolium repens* (Harding and Raizada, 2015).

2.2.1.5 Viral bioinsecticides

Viral bioinsecticides are microbial biopesticides that play a vital role in insect pest management. There are, indeed, over 1000 arthropod insect species belonging to at least 13 insect orders that are known for their viral isolates (Srivastava and Dhaliwal, 2010). There are two major classes of endogenous viruses, viz., inclusion viruses (IV; these generate inclusion bodies in their host cells) and non-inclusion viruses (NIV; these do not generate inclusion bodies). Most insect viruses found in nature belong to Baculoviruses family (Baculoviridae). In them, nuclear polyhedrosis viruses (NPVs) and granuloviruses (GVs) are considered to be the major members of baculoviruses family (Copping and Menn, 2000), valuable in pest management programs, and safer for off-targets (Entwistle, 1983; Krieg et al., 1980) plants, animals, birds, fish and beneficial insects (Hewson et al., 2011). These viral bioinsecticides are diverse, with over 60 baculoviruses acting as bioinsecticides (Beas-Catena et al., 2014; Liu et al., 2019b) against an extensive array of serious insect pests (Cunningham, 1988; Moscardi et al., 2011; Sood et al., 2019; Yang et al., 2012).

Baculoviruses are needed to be ingested by the target host in order to enter their host body and start killing action. Once inside their host, baculoviruses take over replication machinery of the infected host. Then, in the gut cells of their host, viral protein overcoat rapidly disintegrates and viral DNA proceeds and causes infection in the digestive cells. Following this, new multiple infectious viral particles are liberated that are ready for further spread of infection in the same host or among other related organisms. As colonization spreads in the gut cells, infected larvae becomes unable to digest food and ultimately dies within few days (Mehrotra et al., 2017).

Although viral biopesticides are highly target specific, despite that their utility in pest management has been limited due to their narrow pest range, high cost; low efficiency and high sensitivity to UV light (Inceoglu et al., 2001, 2006; Possee et al., 1997). These deficiencies are being overcome in order to enhance insecticidal potential, for instance, they are being genetically modified (Kamita and Hammock, 2010) with recombinant baculoviruses expressing certain types of insect-specific toxins, hormones and enzymes (Wan et al., 2015; Yu et al., 2017). Insect feeding on these genetically modified crops are determined to be vulnerable to toxin proteins, as infected host may also die from protein action in addition to direct viral actions.

2.2.1.6 Viral bioherbicides

Viruses are, for long, known to act as bioherbicides, and some established examples of viruses are described here. For example, tobacco mild green mosaic and Araujo mosaic virus are the potential growth suppressors of the weeds, *Solanum vivarium* and *Araujo moratorium*, respectively (Diaz et al., 2014; Elliott et al., 2009). Some other potential weed growth suppressor viruses include tobacco rattle virus against *Impatiens glandulifera* (Kollmann et al., 2007), and *Pepino mosaic virus* and *Obuda pepper virus* against *Solanum nigrum* (Kazinczi et al., 2006).

2.2.2 Nematodes biopesticides

Nematodes are non-segmented and soft bodied roundworms that naturally occur in soil. Some nematodes species interact with insects, most likely parasitically as obligate or facultative parasites and cause pathogenicity, hence called as entomopathogenic nematodes (EPNs). EPNs from Heterorhabditidae and Steinernematidae family are promising biocontrol agents and widespread in pest control events (Grewal et al., 2005). The pathogenicity of EPNs is mainly by the activities of their bacterial symbionts, for instance, *Steinernematids* are symbiotically associated with bacterium, *Photorhabdus* spp, and *Heterorhabditids* are associated with bacterium, *Xenorhabdus* spp (Kaya and Gaugler, 1993; Poinar Jr, 1990). For tracing their host, EPNs rely on chemical or physical (e.g., vibration) cues from their potential host (Kaya and Gaugler, 1993). EPNs infection starts when infective juveniles (IJs) enter their host body through host cuticle, spiracles, mouth or anal openings (Kaya and Gaugler, 1993). On reaching to the hoemocoel, EPNs release their bacterial symbionts, whose activities are mainly responsible for causing host mortality, which likely occurs within 24–48 h of their release.

EPNs production may be carried out under in-situ or ex-situ environments in solid media or in liquid fermentation (Grewal and Georgis, 1999). Some successfully produced nematodes from fermentation technique include *Steinernema carpocapsae*, *S. riobrave*, *S. glaseri*, *S. scapterisci* and *Heterorhabditis bacteriophora*. Entomopathogenic nematodes are suitable for IPM programs because they are non-toxic to humans, have narrow range of target arthropods, and require standard pesticide spraying equipments for their field application (Shapiro-Ilan et al., 2006).

2.3 Plant-incorporated protectants (PIPs)

Plant-incorporated protectants (PIPs) are another class of biopesticides that are certain pesticidal constituents, produced by the crop plants from genetic material that was inserted into the plant genome. These plants are, basically, genetically modified, also called as transgenic plants, as their nuclear material is stably integrated with genetic material from a naturally occurring microorganism. As a result, transgenic plants possess typical characteristics of added genetic material, which they express to kill their pests. Many PIPs are developed so far. In them, transgenic plants carrying cry gene (producing toxin) from *Bt* bacterium, *B. thuringiensis,* are most widespread examples. Several agricultural crops are genetically-modified and possess *B. thuringiensis* characteristics. This transformation, where a simple plant becomes a transgenic plant, is carried out via manipulation of biological or physical means. For example for transgenic *Bt*, crop plants that are to modify are inserted with corresponding gene from the *Bt* bacterium by the use of recombinant DNA technology with the aid of *Agrobacterium tumefaciens*. This technology was first time used in genetic modification of tomato plants in 1987, and after accomplishment, its application extended to benefit many other economically important agricultural crops including cotton, tobacco etc., (Verma et al., 2020).

Bacteria in genus *Bacillus* produce parasporal, proteinaceous and crystal inclusion bodies (cry toxins) (Copping and Menn, 2000), which possess pesticidal properties against more than 150 species of arthropod insects (Verma et al., 2020).

Several toxins are isolated and described from *Bt* strains including a-exotoxin, b-exotoxin, c-exotoxin, d-endotoxin, louse factor exotoxin, mouse factor exotoxin, water-soluble toxin, Vip3A and enterotoxin (El-Bendary, 2006; Liu et al., 2019a). Each toxin possesses its own pesticidal characteristics, depending upon intensity and receptor specificity (Copping and Menn, 2000).

In general, *Bt* toxins kill insect larvae by their action on the stomach cells of the larvae ingesting toxin. Several members of class Insecta, particularly those from insect orders Lepidoptera and Coleoptera are highly susceptible to *Bt* toxin. When fed, insects along with food also ingest *Bt* toxin from the transgenic plant host, which reaches inside insect gut, where the alkaline nature of gut denatures insoluble *Bt* crystals into soluble crystals. In the presence of gut proteases enzymes, the soluble crystals release active toxins. These active toxins reach to the cell membrane, attach, and start producing spores. This activity induces a condition, known as gut paralysis, which forces insects to cease consumption and starve until insect dies (Dean, 1984).

The interest and adoption of PIPs has dramatically increased in past decades (Broderick et al., 2009), with greater acceptability in developing countries, where chemical pesticides risks have been severe (Amoabeng et al., 2013) and consumer health and safety concern are top priorities. Except USA where PIPs are included as biopesticides, many countries regulatory authorities do not consider PIPs as biopesticides due to consumer concerns for GM crops (Seiber et al., 2014). Frequent intake of produce from GM crops may cause health problems (hormonal and neuronal disruption) in their consumers (Dona and Arvanitoyannis, 2009; Paparini and Romano-Spica, 2004).

2.4 Biochemical pesticides

These are considered as naturally occurring constituents; providing control against pests via non-toxic actions. Unlike chemical pesticides that rapidly kill or inactivate pests, biochemical pesticides comprise such substances that obstruct with normal insect behavior. Female insect sex pheromones and scented plant extracts that can attract insect pests to traps or repel them from the source are popular examples of biochemical pesticides. For their preparation, mainly naturally occurring constituents such as diatomaceous earth, baking soda, various oils extracted from tree or plants (e.g., neem oil, canola oil etc.) are used. These substances are either plant-derivatives or carry microorganisms, thus are non-toxic and rapidly biodegradable pesticides. Except pheromones, which are highly species-specific, other biochemical biopesticides are relatively broad-spectrum pesticides; with activity against an extensive range of arthropod insect pests as compared to the narrow range microbial biopesticides. Some examples of lethal biochemical pesticides that are non-toxic include abrasives (e.g., diatomaceous earth), desiccants (e.g., acetic acid) and suffocants (e.g., soybean oil) (Liu et al., 2019a).

2.4.1 Pheromones

These compounds are chemical signals for communication that are emitted by insects, causing behavioral changes in conspecifics (Chandler et al., 2011). For instance, sex pheromones, which are released by female adults for passing information to their male partners, are being commercially synthesized as protection products for monitoring pests. These synthetic pheromones are efficient mating disruptors and mass trapping agents to be used for lure-kill systems (Izawa et al., 2000; Jung et al., 2007; Reddy et al., 2009; Witzgall et al., 2010). Pheromones are further categorized based on their biological role in an insect life as alarm pheromones, sex pheromones, aggregation pheromones and dispersal pheromones. As being a blend of chemicals, pheromones are difficult for insects to develop resistance (Copping and Menn, 2000) thus are desirable IPM components to lower pesticide risks.

2.4.2 Plant essential oils

Plant essential oils are safer and easy to prepare pest killer products of natural origin (Isman, 2019b). Essential oils majorly constitute plant terpenoids (e.g., mono- and sesquiterpenoids). Terpenoids are defensive compounds, which on exposure are known to interact with different receptors (e.g., octopamine or nicotinic acetylcholine receptors or GABA-gated chloride channels) of the insect nervous system (Price and Berry, 2006; Priestley et al., 2003; Tong et al., 2013). As a result of these interactions, plant essential oils act as pesticides causing toxic and sublethal effects and modifying insect behavior. Plant biopesticides are not new because they are being used in pest management since long, for instance pyrethroids pesticides are plant derivatives of natural pyrethrins, originally extracted from *Chrysanthemum cinerariifolium* and *Chrysanthemum coccineum.* Another example of plant derived pesticides includes rotenone, which is derived from Genus *Derris* of plants. Many plant derived products are active bioinsecticides against many agricultural crop pests. For example essential oil *citronella*, extracted from lemongrass, *Cymbopogon* spp., which is registered in USA since 1948, has shown promising potential as an insect repellent (Atanasova and Leather, 2018). Some plant essential oils biopesticides are comparable to pyrethroids, pyridines and spirotetramat pesticides. Some common examples of such promising essential oils are orange oil, *Chenopodium ambrosioides* extract and oils carrying active ingredient, azadirachtin A (Atanasova and Leather, 2018; Smith et al., 2018). Some biopesticides are applied in the form of extracts (Boursier et al., 2011; Isman, 2008), prepared from various plant parts of the plant possessing pesticidal potential, and are important for developing countries for their properties like easy availability and preparation, less application cost and novel action modes (Amoabeng et al., 2013; Isman, 2020; Shah et al., 2019).

Some essential oils (e.g., *Cinnamomum zeylanicum* L., *Artemisia scoparia* Waldst et Kit. and *Eucalyptus cladocalyx*) or plant extracts (e.g., *Ammi visnaga* (L.) Lam. and *Echinochloa colona*) are allelopathic materials, producing phytotoxic chemicals, which are natural suppressor of weed growth and seed germination. These materials are important biopesticides desired for natural suppression of weed without causing harm to crop plants. These plant extracts constitute several metabolites including lactones, aldehydes, alcohols, phenolics, ketones, flavonoids, quinones and tannins etc., (Soltys et al., 2013). In them some compounds are highly weed-specific, and only responsive toward weed carrying specific receptors (Hosni et al., 2013). On release, metabolites are absorbed by weeds, as a result of which, the weeds become stressed and compromise their germination potential as these metabolites damage cell membrane, disturb DNA activities, and affect plant physiology through disturbing normal release of enzymes or biochemical processes important for weed plant development (Radhakrishnan et al., 2018).

2.5 The biopesticides market and challenges

Biopesticides are increasingly becoming an important and sustainable source of alternative pesticides, offering low risks and greater safety to the environment. About 1400 biopesticides are known, worldwide (Marrone, 2009). Though this amount signifies biopesticides increasing importance, but they are only a small portion (5%–6%) of the global pesticide market (Dunham and Trimmer, 2019; Ndolo et al., 2019). This share is far less but more likely to expand as a result of evolving interest and implementation of rules for adoption of safer pesticides. As a result of safety concerns, agrochemical industry is becoming oriented toward biopesticides, which is why, biopesticide market is growing up. Just recently, India has registered 15 microbial species and 970 microbial biopesticides formulations (Kumar et al., 2019). Research for developing newer biopesticides is becoming a subject of increasing interest, and some new pest control agents of natural origin have been explored, however challenge is to determine their effects, largely unclear, on specific pest problems across a range of cropping systems. These newer pesticidal agents are plant materials (e.g., *Clitoria ternatea* leaf extract, commonly called as butterfly pea) (Mensah et al., 2014) like some alkaloids *oxymatrine* (Rao and Kumari, 2016) and other are products carrying *Bacillus thuringiensis* var. *tenebrionis* strain Xd3 (Btt-Xd3) bacterium products (Eski et al., 2017) and stilbenes isolates from grapevines extract (Pavela et al., 2017). If growth in biopesticide expansion continues, these ecofriendly alternatives are expected to catch up synthetic pesticides, in terms of market size, in about next 20 or 30 years.

2.5.1 Bioinsecticides

Some biopesticides, for instance microbial pesticides and plant essential oils, are highly promising but their take up and adoption rate has been far less in niche markets (Sampson et al., 2005), and up take also differ greatly among regions and localities (Sampson et al., 2005). There are several contributable reasons behind this such as strict legislation, high cost and specificity and narrow pest spectrum (Chandler et al., 2011). In Europe for instance, commercially available biopesticides are far less than those in other agriculture regions, elsewhere (Balog et al., 2017). In USA, biopesticides are given approval as long as they stay safe for environment and humans. In Europe and United Kingdom, there are almost 200 types of defined environmental regulatory laws or acts restricting biopesticides acceptance in the niche markets (Balog et al., 2017). Moreover, microbial biopesticides are often slower acting than are chemical pesticides, hence less attractive for growers. In addition, biopesticides narrow spectrum activity combined with variable efficiencies under different environments represents another big cause of low adoption rate among farmers. Another core reasons is lack of publicly available data addressing phytotoxic and sublethal effects of biopesticides. Also, there is scarcity in literature on biopesticides modes of action on their targets, which is desirable in order to develop resistance countering strategies for sustained use of biopesticides (Leather and Pope, 2019).

2.5.2 Bioherbicides

At present, there are about twenty-four different types of bioherbicides under commercial usage, worldwide. Of the registered products, only handfuls are successful bioherbicides and others are not, therefore bioherbicides fail to gain as wider acceptability as do chemical counterparts. This lag in commercial markets is attributable to several challenges, including, narrow pest range, high specificity, rather less environmental persistence, and incorrect formulation. These challenges may be countered by extending host range, correcting formulations, increasing persistence, and also by including advance techniques (Radhakrishnan et al., 2018). Host range expansion, which is among the most desired aspects, remains a complex issue to resolve because of high-host specificity of most microorganisms found in bioherbicides. This specificity has been complex and difficult to understand due to lack of any clear evidence regarding host-phylogeny and microbe-specificity (Harding and Raizada, 2015). Therefore selection should be careful, otherwise latent colonization by microbes in the plant, may lower bioherbicide viability (Casella et al., 2010). Bioherbicides effectiveness is also negatively affected when different microorganisms compete antagonistically, which may cause loss of microbes and their toxins. Another major limitation is the inappropriate abiotic environment. Environmental conditions have a significant role in determining and regulating initial infection, its spread and development, and infection rate in the target weed host. The commercial development and registration of biopesticides largely depends on possibility of their mass production from the living, pathogenic and genetically stable propagules (e.g., spores, fragments, or pellets) (Boyette and Hoagland, 2015). Many are developed and many are under evaluations (for instance many fungal agents). For their wider acceptability in commercial markets, these products need to be improved via overcoming biological and environmental constraints (Aneja et al., 2017) and by introducing some specific technological and policy based approaches that may make bioherbicides economical and popular among farming communities.

References

Abrol, D.P., Shankar, U., 2012. History, overview and principles of ecologically based pest management. In: Integrated Pest Management: Principles and Practice, pp. 1–26.

Ahluwalia, A.D., 2007. Bioherbicides: an eco-friendly approach to weed management. Curr. Sci. 92, 10–11.

Alavanja, M.C., Hoppin, J.A., Kamel, F., 2004. Health effects of chronic pesticide exposure: cancer and neurotoxicity. Annu. Rev. Publ. Health 25, 155–197.

Amoabeng, B.W., Gurr, G.M., Gitau, C.W., Nicol, H.I., Munyakazi, L., Stevenson, P.C., 2013. Tri-trophic insecticidal effects of African plants against cabbage pests. PloS One 8.

Amoah, P., Drechsel, P., Abaidoo, R., Ntow, W., 2006. Pesticide and pathogen contamination of vegetables in Ghana's urban markets. Arch. Environ. Contam. Toxicol. 50, 1–6.

Aneja, K., Khan, S., Aneja, A., 2017. Bioherbicides: Strategies, Challenges and Prospects, Developments in Fungal Biology and Applied Mycology. Springer, pp. 449–470.

Ash, C., Jasny, B.R., Malakoff, D.A., Sugden, A.M., 2010. Feeding the future. Science 327 (5967), 797.

Atanasova, D., Leather, S., 2018. Plant essential oils: the way forward for aphid control? Ann. Appl. Biol. 173, 175–179.

Auld, B., Say, M., Ridings, H., Andrews, J., 1990. Field applications of *Colletotrichum orbiculare* to control *Xanthium spinosum*. Agric. Ecosyst. Environ. 32, 315–323.

Balog, A., Hartel, T., Loxdale, H.D., Wilson, K., 2017. Differences in the progress of the biopesticide revolution between the EU and other major crop-growing regions. Pest Manag. Sci. 73, 2203–2208.

Banowetz, G.M., Azevedo, M.D., Armstrong, D.J., Halgren, A.B., Mills, D.I., 2008. Germination-Arrest Factor (GAF): biological properties of a novel, naturally-occurring herbicide produced by selected isolates of rhizosphere bacteria. Biol. Contr. 46, 380–390.

Beas-Catena, A., Sánchez-Mirón, A., García-Camacho, F., Contreras-Gómez, A., Molina-Grima, E., 2014. Baculovirus biopesticides: an overview. J. Anim. Plant Sci. 24, 362–373.

Bell, H.A., Cuthbertson, A.G., Audsley, N., 2016. The potential use of allicin as a biopesticide for the control of the house fly, *Musca domestica* L. Int. J. Pest Manag. 62, 111–118.

Bhanti, M., Taneja, A., 2007. Contamination of vegetables of different seasons with organophosphorous pesticides and related health risk assessment in northern India. Chemosphere 69, 63–68.

Biondi, A., Desneux, N., Siscaro, G., Zappalà, L., 2012. Using organic-certified rather than synthetic pesticides may not be safer for biological control agents: selectivity and side effects of 14 pesticides on the predator *Orius laevigatus*. Chemosphere 87, 803–812.

Boursier, C.M., Bosco, D., Coulibaly, A., Negre, M., 2011. Are traditional neem extract preparations as efficient as a commercial formulation of azadirachtin A? Crop Protect. 30, 318–322.

Boyette, C.D., Hoagland, R.E., 2015. Bioherbicidal potential of *Xanthomonas campestris* for controlling *Conyza canadensis*. Biocontrol Sci. Technol. 25, 229–237.

Broderick, N.A., Robinson, C.J., McMahon, M.D., Holt, J., Handelsman, J., Raffa, K.F., 2009. Contributions of gut bacteria to *Bacillus thuringiensis*-induced mortality vary across a range of Lepidoptera. BMC Biol. 7, 11.

Casella, F., Charudattan, R., Vurro, M., 2010. Effectiveness and technological feasibility of bioherbicide candidates for biocontrol of green foxtail (*Setaria viridis*). Biocontrol Sci. Technol. 20, 1027–1045.

Casida, J.E., Quistad, G.B., 1998. Golden age of insecticide research: past, present, or future? Annu. Rev. Entomol. 43, 1–16.

Cessna, S.G., Sears, V.E., Dickman, M.B., Low, P.S., 2000. Oxalic acid, a pathogenicity factor for *Sclerotinia sclerotiorum*, suppresses the oxidative burst of the host plant. Plant Cell 12, 2191–2199.

Chandler, D., Bailey, A.S., Tatchell, G.M., Davidson, G., Greaves, J., Grant, W.P., 2011. The development, regulation and use of biopesticides for integrated pest management. Philos. Trans. R. Soc. Lond. B Biol. Sci. 366, 1987–1998.

Copping, L.G., Menn, J.J., 2000. Biopesticides: a review of their action, applications and efficacy. Pest Manag. Sci. Former. Pestic. Sci. 56, 651–676.

Culliney, T.W., 2014. Crop Losses to Arthropods, Integrated Pest Management. Springer, pp. 201–225.

Cunningham, J.C., 1988. Baculoviruses: their status compared to *Bacillus thuringiensis* as microbial insecticides. Outlook Agric. 17, 10–17.

Czaja, K., Góralczyk, K., Struciński, P., Hernik, A., Korcz, W., Minorczyk, M., Łyczewska, M., Ludwicki, J.K., 2015. Biopesticides—towards increased consumer safety in the European Union. Pest Manag. Sci. 71, 3–6.

Davis, A.S., Frisvold, G.B., 2017. Are herbicides a once in a century method of weed control? Pest Manag. Sci. 73, 2209–2220.

Dean, D.H., 1984. Biochemical genetics of the bacterial insect-control agent *Bacillus thuringiensis*: basic principles and prospects for genetic engineering. Biotechnol. Genet. Eng. Rev. 2, 341–363.

Diaz, R., Manrique, V., Hibbard, K., Fox, A., Roda, A., Gandolfo, D., Mckay, F., Medal, J., Hight, S., Overholt, W., 2014. Successful biological control of tropical soda apple (Solanales: Solanaceae) in Florida: a review of key program components. Fla. Entomol. 97, 179–190.

Dona, A., Arvanitoyannis, I.S., 2009. Health risks of genetically modified foods. Crit. Rev. Food Sci. Nutr. 49, 164–175.

Dunham, W., Trimmer, M., 2019. Biological products around the world. In: Bioproducts Industry Alliance Spring Meeting & International Symposium.

El-Bendary, M.A., 2006. *Bacillus thuringiensis* and *Bacillus sphaericus* biopesticides production. J. Basic Microbiol. 46, 158–170.

Elliott, M., Massey, B., Cui, X., Hiebert, E., Charudattan, R., Waipara, N., Hayes, L., 2009. Supplemental host range of Araujia mosaic virus, a potential biological control agent of moth plant in New Zealand. Australas. Plant Pathol. 38, 603–607.

Entwistle, P., 1983. Viruses for Insect Pest Control.

Eski, A., Demir, İ., Sezen, K., Demirbağ, Z., 2017. A new biopesticide from a local *Bacillus thuringiensis* var. *tenebrionis* (Xd3) against alder leaf beetle (Coleoptera: Chrysomelidae). World J. Microbiol. Biotechnol. 33, 95.

French, N.M., Heim, D.C., Kennedy, G.G., 1992. Insecticide resistance patterns among Colorado potato beetle, *Leptinotarsa decemlineata* (Say) (Coleoptera: Chrysomelidae), populations in North Carolina. Pestic. Sci. 36, 95–100.

Gharde, Y., Singh, P., Dubey, R., Gupta, P., 2018. Assessment of yield and economic losses in agriculture due to weeds in India. Crop Protect. 107, 12–18.

Glare, T.R., O'Callaghan, M., 2019. Microbial biopesticides for control of invertebrates: progress from New Zealand. J. Invertebr. Pathol. 165, 82–88.

Godfray, H.C.J., Beddington, J.R., Crute, I.R., Haddad, L., Lawrence, D., Muir, J.F., Pretty, J., Robinson, S., Thomas, S.M., Toulmin, C., 2010. Food security: the challenge of feeding 9 billion people. Science 327, 812–818.

Goulson, D., Nicholls, E., Botías, C., Rotheray, E.L., 2015. Bee declines driven by combined stress from parasites, pesticides, and lack of flowers. Science 347, 1255957.

Gregory, P.J., Johnson, S.N., Newton, A.C., Ingram, J.S., 2009. Integrating pests and pathogens into the climate change/food security debate. J. Exp. Bot. 60, 2827–2838.

Grewal, P., Ehlers, R., Shapiro-Ilan, D., 2005. Nematodes as Biocontrol Agents. CABI Publishing, Cambridge, MA USA.

Grewal, P., Georgis, R., 1999. Entomopathogenic Nematodes, Biopesticides: Use and Delivery. Springer, pp. 271–299.

Guillon, M., 2003. Regulation of biological control agents in Europe. In: International Symposium on Biopesticides for Developing Countries. CATIE, Turrialba, pp. 143–147.

Hajek, A., St Leger, R., 1994. Interactions between fungal pathogens and insect hosts. Annu. Rev. Entomol. 39, 293–322.

Hall, F.R., Menn, J.J., 1999. Biopesticides: Use and Delivery. Springer.

Hall, R., 1995. Challenges and prospects of integrated pest management. In: Reuveni, R. (Ed.), Novel Approaches to Integrated Pest Management. Lewis Publishers, Boca Raton, Florida, pp. 1–99.

Harding, D.P., Raizada, M.N., 2015. Controlling weeds with fungi, bacteria and viruses: a review. Front. Plant Sci. 6, 659.

Hassan, E., Gökçe, A., 2014. Production and Consumption of Biopesticides, Advances in Plant Biopesticides. Springer, pp. 361–379.

Hewson, I., Brown, J.M., Gitlin, S.A., Doud, D.F., 2011. Nucleopolyhedrovirus detection and distribution in terrestrial, freshwater, and marine habitats of Appledore Island, Gulf of Maine. Microb. Ecol. 62, 48–57.

Holder, D.J., Keyhani, N.O., 2005. Adhesion of the entomopathogenic fungus *Beauveria* (Cordyceps) *bassiana* to substrata. Appl. Environ. Microbiol. 71, 5260–5266.

Hollomon, D.W., 2015. Fungicide resistance: facing the challenge-a review. Plant Protect. Sci. 51, 170–176.

Hosni, K., Hassen, I., Sebei, H., Casabianca, H., 2013. Secondary metabolites from *Chrysanthemum coronarium* (Garland) flowerheads: chemical composition and biological activities. Ind. Crop. Prod. 44, 263–271.

Imaizumi, S., Nishino, T., Miyabe, K., Fujimori, T., Yamada, M., 1997. Biological control of annual bluegrass (*Poa annua* L.) with a Japanese Isolate of *Xanthomonas campestris* pv. *poae* (JT-P482). Biol. Contr. 8, 7–14.

Inceoglu, A.B., Kamita, S.G., Hammock, B.D., 2006. Genetically modified baculoviruses: a historical overview and future outlook. Adv. Virus Res. 68, 323–360.

Inceoglu, A.B., Kamita, S.G., Hinton, A.C., Huang, Q., Severson, T.F., Kang, K.d., Hammock, B.D., 2001. Recombinant baculoviruses for insect control. Pest Manag. Sci. 57, 981–987.

Isman, M., 2019a. Challenges of pest management in the twenty first century: new tools and strategies to combat old and new foes alike. Front. Agron. 1, 2. https://doi.org/10.3389/fagro.

Isman, M.B., 2000. Biopesticides Based on Phytochemicals, Phytochemical Biopesticides. CRC Press, pp. 11–21.

Isman, M.B., 2008. Botanical insecticides: for richer, for poorer. Pest Manag. Sci. 64, 8–11.

Isman, M.B., 2019b. Commercial development of plant essential oils and their constituents as active ingredients in bioinsecticides. Phytochem. Rev. 1–7.

Isman, M.B., 2020. Botanical insecticides in the twenty-first century—fulfilling their promise? Annu. Rev. Entomol. 65, 233–249.

Isman, M.B., Seffrin, R., 2014. Natural insecticides from the Annonaceae: a unique example for developing biopesticides. Adv. Plant Biopestic. 21–33. Springer.

Izawa, H., Fujü, K., Matoba, T., 2000. Control of multiple species of lepidopterous insect pests using a mating disruptor and reduced pesticide applications in Japanese pear orchards. Jpn. J. Appl. Entomol. Zool. 44, 165–171.

Jung, S.-C., Park, C.-W., Park, M.-W., Kim, Y.-G., 2007. Field assessment of two commercial sex pheromone mating disruptors on male orientation of oriental fruit moth, *Grapholita molesta* (Busck). Korean J. Pestic. Sci. 11, 46–51.

Kamita, S.G., Hammock, B.D., 2010. Juvenile hormone esterase: biochemistry and structure. J. Pestic. Sci. 35, 265–274.

Kaya, H.K., Gaugler, R., 1993. Entomopathogenic nematodes. Annu. Rev. Entomol. 38, 181–206.

Kazinczi, G., Lukacs, D., Takacs, A., Horvath, J., Gaborjanyi, R., Nadasy, M., Nadasy, E., 2006. Biological decline of *Solanum nigrum* due to virus infections. J. Plant Dis. Prot. 20, 5–330.

Kennedy, A.C., Johnson, B.N., Stubbs, T.L., 2001. Host range of a deleterious rhizobacterium for biological control of downy brome. Weed Sci. 49, 792–797.

Khater, H.F., 2012. Prospects of botanical biopesticides in insect pest management. Pharmacologia 3, 641–656.

Kollmann, J., Bañuelos, M.J., Nielsen, S.L., 2007. Effects of virus infection on growth of the invasive alien *Impatiens glandulifera*. Preslia 79, 33–44.

Koul, O., Dhaliwal, G., 2000. Phytochemical Biopesticides. CRC Press.

Krieg, A., Franz, J.M., Gröner, A., Huber, J., Miltenburger, H.G., 1980. Safety of entomopathogenic viruses for control of insect pests. Environ. Conserv. 7, 158–160.

Kumar, K.K., Sridhar, J., Murali-Baskaran, R.K., Senthil-Nathan, S., Kaushal, P., Dara, S.K., Arthurs, S., 2019. Microbial biopesticides for insect pest management in India: Current status and future prospects. J. Invertebr. Pathol. 165, 74–81.

Kumar, V., 2015. A review on efficacy of biopesticides to control the agricultural insect's pest. Int. J. Agric. Sci. Res. 4, 168–179.

Leather, S.R., Pope, T.W., 2019. Botanical biopesticides–where to now? Outlooks Pest Manag. 30, 75–77.

Leong, K.H., Tan, L.B., Mustafa, A.M., 2007. Contamination levels of selected organochlorine and organophosphate pesticides in the Selangor River, Malaysia between 2002 and 2003. Chemosphere 66, 1153–1159.

Liu, X., Cao, A., Yan, D., Ouyang, C., Wang, Q., Li, Y., 2019a. Overview of mechanisms and uses of biopesticides. Int. J. Pest Manag. https://doi.org/10.1080/09670874.2019.1664789.

Liu, Z., Wang, X., Dai, Y., Wei, X., Ni, M., Zhang, L., Zhu, Z., 2019b. Expressing double-stranded RNAs of insect hormone-related genes enhances Baculovirus insecticidal activity. Int. J. Mol. Sci. 20, 419.

Marrone, P.G., 2009. Barriers to Adoption of Biological Control Agents and Biological Pesticides. Integrated Pest Management. Cambridge University Press, Cambridge, UK, pp. 163–178.

Mehrotra, S., Kumar, S., Zahid, M., Garg, M., 2017. Biopesticides, Principles and Applications of Environmental Biotechnology for a Sustainable Future. Springer, pp. 273–292.

Mensah, R., Moore, C., Watts, N., Deseo, M.A., Glennie, P., Pitt, A., 2014. Discovery and development of a new semiochemical biopesticide for cotton pest management: assessment of extract effects on the cotton pest *Helicoverpa* spp. Entomol. Exp. Appl. 152, 1–15.

Metcalf, R.L., 1980. Changing role of insecticides in crop protection. Annu. Rev. Entomol. 25, 219–256.

Mishra, J., Tewari, S., Singh, S., Arora, N.K., 2015. Biopesticides: where we stand?. In: Plant Microbes Symbiosis: Applied Facets. Springer, pp. 37–75.

Moorhouse, E., Gillespie, A., Sellers, E., Charnley, A., 1992. Influence of fungicides and insecticides on the entomogenous fungus *Metarhizium anisopliae* a pathogen of the vine weevil, *Otiorhynchus sulcatus*. Biocontrol Sci. Technol. 2, 49–58.

Mortensen, K., 1988. The potential of an endemic fungus, *Colletotrichum gloeosporioides*, for biological control of round-leaved mallow (*Malva pusilla*) and velvetleaf (*Abutilon theophrasti*). Weed Sci. 36, 473–478.

Moscardi, F., de Souza, M.L., de Castro, M.E.B., Moscardi, M.L., Szewczyk, B., 2011. Baculovirus Pesticides: Present State and Future Perspectives, Microbes and Microbial Technology. Springer, pp. 415–445.

Moshi, A.P., Matoju, I., 2017. The status of research on and application of biopesticides in Tanzania. Review. Crop Prot 92, 16–28.

Ndolo, D., Njuguna, E., Adetunji, C.O., Harbor, C., Rowe, A., Den Breeyen, A., Sangeetha, J., Singh, G., Szewczyk, B., Anjorin, T.S., 2019. Research and development of biopesticides: challenges and prospects. Outlooks Pest Manag. 30, 267–276.

Oerke, E.-C., 2006. Crop losses to pests. J. Agric. Sci. 144, 31–43.

Papadakis, E.N., Vryzas, Z., Kotopoulou, A., Kintzikoglou, K., Makris, K.C., Papadopoulou-Mourkidou, E., 2015. A pesticide monitoring survey in rivers and lakes of Northern Greece and its human and ecotoxicological risk assessment. Ecotoxicol. Environ. Saf. 116, 1–9.

Paparini, A., Romano-Spica, V., 2004. Public health issues related with the consumption of food obtained from genetically modified organisms. Biotechnol. Annu. Rev. 10, 85–122.

Pascoli, M., de Albuquerque, F.P., Calzavara, A.K., Tinoco-Nunes, B., Oliveira, W.H.C., Gonçalves, K.C., Polanczyk, R.A., Della Vechia, J.F., de Matos, S.T.S., de Andrade, D.J., 2020. The potential of nanobiopesticide based on zein nanoparticles and neem oil for enhanced control of agricultural pests. J. Pest. Sci. 93, 793–806.

Pascoli, M., Jacques, M.T., Agarrayua, D.A., Avila, D.S., Lima, R., Fraceto, L.F., 2019. Neem oil based nanopesticide as an environmentally-friendly formulation for applications in sustainable agriculture: an ecotoxicological perspective. Sci. Total Environ. 677, 57–67.

Pavela, R., Waffo-Teguo, P., Biais, B., Richard, T., Mérillon, J.-M., 2017. *Vitis vinifera* canes, a source of stilbenoids against *Spodoptera littoralis* larvae. J. Pest. Sci. 90, 961–970.

Perrings, C., Dehnen-Schmutz, K., Touza, J., Williamson, M., 2005. How to manage biological invasions under globalization. Trends Ecol. Evol. 20, 212–215.

Pimentel, D., Acquay, H., Biltonen, M., Rice, P., Silva, M., Nelson, J., Lipner, V., Giordano, S., Horowitz, A., D'amore, M., 1993. Assessment of Environmental and Economic Impacts of Pesticide Use, the Pesticide Question. Springer, pp. 47–84.

Poinar Jr., G.O., 1990. Taxonomy and biology of Steinernematidae and Heterorhabditidae. In: Entomopathogenic Nematodes in Biological Control, 54.

Possee, R.D., Barnett, A.L., Hawtin, R.E., King, L.A., 1997. Engineered baculoviruses for pest control. Pestic. Sci. 51, 462–470.

Price, D.N., Berry, M.S., 2006. Comparison of effects of octopamine and insecticidal essential oils on activity in the nerve cord, foregut, and dorsal unpaired median neurons of cockroaches. J. Insect Pysiol. 52, 309–319.

Priestley, C.M., Williamson, E.M., Wafford, K.A., Sattelle, D.B., 2003. Thymol, a constituent of thyme essential oil, is a positive allosteric modulator of human GABAA receptors and a homo-oligomeric GABA receptor from *Drosophila melanogaster*. Br. J. Pharmacol. 140, 1363–1372.

Quail, J.W., Ismail, N., Pedras, M.S.C., Boyetchko, S.M., 2002. Pseudophomins A and B, a class of cyclic lipodepsipeptides isolated from a *Pseudomonas* species. Acta Crystallogr. Sect. C Cryst. Struct. Commun. 58, o268–o271.

Radhakrishnan, R., Alqarawi, A.A., Abd Allah, E.F., 2018. Bioherbicides: current knowledge on weed control mechanism. Ecotoxicol. Environ. Saf. 158, 131–138.

Rao, P., Kumari, A., 2016. Effect of oxymatrine 0.5% EC on predators and parasites of important pests on certain vegetable crops cultivated in Ranga Reddy District (Telangana). Pestology 40, 15–18.

Razaq, M., Shah, F.M., Ahmad, S., Afzal, M., 2019. Pest Management for Agronomic Crops, Agronomic Crops. Springer, pp. 365–384.

Reddy, G., Cruz, Z., Guerrero, A., 2009. Development of an efficient pheromone-based trapping method for the banana root borer *Cosmopolites sordidus*. J. Chem. Ecol. 35, 111–117.

Reddy, G.V., 2011. Comparative effect of integrated pest management and farmers' standard pest control practice for managing insect pests on cabbage (*Brassica* spp.). Pest Manag. Sci. 67, 980–985.

Sampson, B.J., Tabanca, N., Kirimer, N.E., Demirci, B., Baser, K.H.C., Khan, I.A., Spiers, J.M., Wedge, D.E., 2005. Insecticidal activity of 23 essential oils and their major compounds against adult *Lipaphis pseudobrassicae* (Davis)(Aphididae: Homoptera). Pest Manag. Sci. 61, 1122–1128.

Santos, K.F.A., Zanardi, O.Z., de Morais, M.R., Jacob, C.R.O., de Oliveira, M.B., Yamamoto, P.T., 2017. The impact of six insecticides commonly used in control of agricultural pests on the generalist predator *Hippodamia convergens* (Coleoptera: Coccinellidae). Chemosphere 186, 218–226.

Sawicka, B., Egbuna, C., 2020. Pests of Agricultural Crops and Control Measures, Natural Remedies for Pest, Disease and Weed Control. Elsevier, pp. 1–16.

Seiber, J.N., Coats, J., Duke, S.O., Gross, A.D., 2014. Biopesticides: state of the art and future opportunities. J. Agric. Food Chem. 62, 11613–11619.

Shah, F.M., Razaq, M., Ali, A., Han, P., Chen, J., 2017. Comparative role of neem seed extract, moringa leaf extract and imidacloprid in the management of wheat aphids in relation to yield losses in Pakistan. PloS One 12, e0184639.

Shah, F.M., Razaq, M., Ali, Q., Shad, S.A., Aslam, M., Hardy, I.C., 2019. Field evaluation of synthetic and neem-derived alternative insecticides in developing action thresholds against cauliflower pests. Sci. Rep. 9, 1–13.

Shapiro-Ilan, D.I., Gouge, D.H., Piggott, S.J., Fife, J.P., 2006. Application technology and environmental considerations for use of entomopathogenic nematodes in biological control. Biol. Contr. 38, 124–133.

Smith, G.H., Roberts, J.M., Pope, T.W., 2018. Terpene based biopesticides as potential alternatives to synthetic insecticides for control of aphid pests on protected ornamentals. Crop Protect. 110, 125–130.

Soares, M.A., Campos, M.R., Passos, L.C., Carvalho, G.A., Haro, M.M., Lavoir, A.-V., Biondi, A., Zappalà, L., Desneux, N., 2019. Botanical insecticide and natural enemies: a potential combination for pest management against *Tuta absoluta*. J. Pest. Sci. 92, 1433–1443.

Soltys, D., Krasuska, U., Bogatek, R., Gniazdowska, A., 2013. Allelochemicals as Bioherbicides—Present and Perspectives, Herbicides-Current Research and Case Studies in Use. IntechOpen.

Sood, P., Choudhary, A., Prabhakar, C.S., 2019. Granuloviruses in Insect Pest Management, Microbes for Sustainable Insect Pest Management. Springer, pp. 275–298.

Sparks, T.C., Nauen, R., 2015. IRAC: mode of action classification and insecticide resistance management. Pestic. Biochem. Physiol. 121, 122–128.

Srinivasan, R., 2012. Integrating biopesticides in pest management strategies for tropical vegetable production. J. Biopestic. 5, 36.

Srinivasan, R., Sevgan, S., Ekesi, S., Tamò, M., 2019. Biopesticide based sustainable pest management for safer production of vegetable legumes and brassicas in Asia and Africa. Pest Manag. Sci. 75, 2446–2454.

Srivastava, K., Dhaliwal, G., 2010. A Textbook of Applied Entomology. Kalyani Publishers.

Staub, T., 1991. Fungicide resistance: practical experience with antiresistance strategies and the role of integrated use. Annu. Rev. Phytopathol. 29, 421–442.

Stevenson, P.C., Isman, M.B., Belmain, S.R., 2017. Pesticidal plants in Africa: a global vision of new biological control products from local uses. Ind. Crop. Prod. 110, 2–9.

Thacker, J.R., 2002. An Introduction to Arthropod Pest Control. Cambridge University Press.

Tong, F., Gross, A.D., Dolan, M.C., Coats, J.R., 2013. The phenolic monoterpenoid carvacrol inhibits the binding of nicotine to the housefly nicotinic acetylcholine receptor. Pest Manag. Sci. 69, 775–780.

Verma, D., Banjo, T., Chawan, M., Teli, N., Gavankar, R., 2020. Microbial Control of Pests and Weeds, Natural Remedies for Pest, Disease and Weed Control. Elsevier, pp. 119–126.

Wan, H., Zhang, Y., Zhao, X., Ji, J., You, H., Li, J., 2015. Enhancing the insecticidal activity of recombinant baculovirus by expressing a growth-blocking peptide from the beet armyworm *Spodoptera exigua*. J. Asia Pac. Entomol. 18, 535–539.

Wilby, A., Thomas, M.B., 2002. Natural enemy diversity and pest control: patterns of pest emergence with agricultural intensification. Ecol. Lett. 5, 353–360.

Witzgall, P., Kirsch, P., Cork, A., 2010. Sex pheromones and their impact on pest management. J. Chem. Ecol. 36, 80–100.

Xiao, D., Zhao, J., Guo, X., Chen, H., Qu, M., Zhai, W., Desneux, N., Biondi, A., Zhang, F., Wang, S., 2016. Sublethal effects of imidacloprid on the predatory seven-spot ladybird beetle *Coccinella septempunctata*. Ecotoxicology 25, 1782–1793.

Xu, Y., Orozco, R., Wijeratne, E.K., Espinosa-Artiles, P., Gunatilaka, A.L., Stock, S.P., Molnár, I., 2009. Biosynthesis of the cyclooligomer depsipeptide bassianolide, an insecticidal virulence factor of *Beauveria bassiana*. Fungal Genet. Biol. 46, 353–364.

Yang, M.M., Li, M.L., Zhang, Y., Wang, Y.Z., Qu, L.J., Wang, Q.H., Ding, J.Y., 2012. Baculoviruses and insect pests control in China. Afr. J. Microbiol. Res. 6, 214–218.

Yu, H., Zhou, B., Meng, J., Xu, J., Liu, T.X., Wang, D., 2017. Recombinant *Helicoverpa armigera* nucleopolyhedrovirus with arthropod-specific neurotoxin gene RjAa17f from *Rhopalurus junceus* enhances the virulence against the host larvae. Insect Sci. 24, 397–408.

Chapter 3

Biopesticide formulations - current challenges and future perspectives

Marian Butu, Steliana Rodino and Alina Butu
National Institute of Research and Development for Biological Sciences, Bucharest, Romania

3.1 Introduction

Agriculture is a very important component of a country's economy. In recent years, biopesticides developed from substances of microbiological and botanical origin received great attention as valuable environmentally-friendly replacements for synthetic and chemical pesticides (Glare et al., 2012). The scientific studies and the practice results of the last decades convincingly proved that the massive use of chemicals for the control of pathogens and pests of crops does not give the best results, when taking into account environmental, health and economic issues. It leads to a violation of the biological balance of agrolandscapes, environmental pollution, chemical loaded crops and an increase in the production costs.

To prevent an ecological damage, a gradual transfer of the agricultural sector to less aggressive methods and technologies is desired. Therefore, the development and application of new plant protection products that are non-toxic to humans, animals and environment is of priority importance.

Pesticides are generally defined as chemical products used to prevent, control and destroy the animal, plant or harmful microorganism action, and may take the form of: insecticides, herbicides or fungicides. Biopesticides are a specific type of pesticides that derive from natural substances and materials.

Biopesticides are produced from substances present in nature, such as fungi, bacteria, plant extracts, fatty acids or pheromones. The use of biopesticides is growing rapidly worldwide and is increasingly present in Integrated Pest and Disease Management in order to increase the efficiency and quality of the control programs, together with their low environmental impact. Biopesticides are biological pest control agent applied such as chemical pesticides but achieve pest management in an eco-friendly manner (Dutta, 2015). The aim of these biopesticides is to ensure protection of plants or crops, having at the same time the advantage of avoiding inducing any harm to humans, crops or ecosystems, while being highly pathogenic to most insect pests (Damalas and Koutroubas, 2018; Sala et al., 2020). Biopesticides offer additional benefits, such as complex and novel modes of action for managing resistance and extending the life span of conventional pesticide products.

3.2 A view through history

The progressive decline of agricultural production, caused by the presence of pests, has been solved by the treatment over the years with synthetic pesticides (Kvakkestad et al., 2020). The inevitable consequences of the extended use of pesticides are pollution of groundwater, crops, and reducing the biodiversity. It was reported that 90% of the applied chemical pesticides enter the various environmental resources as a result of run-off, exposing the farmers and consumers to severe health issues (Chaudhary et al., 2017). However, even if are known the dramatic effects as becoming in time a health hazard for humans and environment, pest management is still reliant on chemical and synthetic pesticides to many agricultural practices due to their efficacy, ease and relatively low price (Costa et al., 2019; Kumar et al., 2019).

Biopesticides were used since the 1800s, starting with fungal spores to control insect pests. For the registration of biopesticides, it was first in 1994 when US set up a specialized division within Environmental Protection Agency (EPA),

Biopesticides. https://doi.org/10.1016/B978-0-12-823355-9.00010-9

named Biopesticides and Pollution Prevention division. Later on, in 2000, United States department of Agriculture created regional integrated Pest management centers. As far as European initiatives, in 2009 the EU member states adopted the directive EC 2009/128 on Sustainable Use of Pesticide, aiming to reduce the risks and impacts of pesticide use on human health and the environment. This directive is also promoting the use of Integrated Pest Management (IPM) and of non-chemical alternatives to pesticides.

In the present day, development of biopesticides industry has to be treated as a strategic task (Olson, 2015). Overall, it was estimated that around 50% of biopesticides are frequently used in horticultural trees and crops, 12% in field crops and 30% on grazing and dry lands (Glare et al., 2012). Due to molecular biology, biotechnology and genetics studies, it has become possible to improve some of the critical properties that hold up the usefulness of many biocontrol agents. Therefore, genes from unrelated organisms could be used for biological control purposes.

Control of pest insects with chemical pesticides has generated several problems over the years including insecticide resistance, safety risks for humans and animals, contamination of water as well as decrease in biodiversity. These problems and sustainability of programs based on conventional insecticides have stimulated an expanded interest in integrated pest management. In future, sustainable agriculture will rely on alternative procedures for pest management that are environmentally friendly and reduce the human contact with chemical pesticides (Leng et al., 2011; Moazami, 2008).

3.3 Regulatory framework

The most important organization with responsibilities in biopesticides area is Organization for Economic and Co-operative Development (OECD), through its Expert Group on BioPesticides (formerly known as the BioPesticides Steering Group). The beforementioned group is leading the actions to promote a harmonized approach to biopesticide regulatory framework, to set up guidance documents and in the same time to facilitate communication, dissemination and knowledge share across various stakeholders in the field.

Other key players include the United Nations Food and Agriculture Organization (FAO) and World Health Organization (WHO), the International Organization of Biological Control (IOBC), the European and Mediterranean Plant Protection Organization (EPPO).

However, the world leader in biopesticide regulation is US Environmental Protection Agency (EPA), through Biopesticides and Pollution Prevention division, established back in 1994. This division has successfully set up the registration process for biopesticide products through the development of modified test methodologies with reduced data requirements that have significantly lowered registration costs and timescales. The registration procedure includes three steps:

(1) The EPA checks whether the application is complete enough to be assigned to a division for review in the initial screen for completeness, which takes 21 days.

(2) A preliminary technical screen is done to determine if the data are (i) accurate and complete, (ii) consistent with proposed labeling and any tolerance and tolerance exemption, such that (iii) subject to full review, could result in the granting of the application. If information is not sufficient in the second step, the applicant has 10 business days to provide the required information. Failure to comply with the response period results in rejection of the application.

(3) After receiving the meeting summaries, the Biopesticide sand Pollution Prevention Division (BPPD) has a maximum of 19 months from receipt of a complete application to the registration decision according to the PRIA 3 timelines ("Biopesticide Registration, Pesticide Registration, USEPA," 2020).

At present, there are over 430 biopesticide active ingredients and 1320 products registered and commercially available for use by US farmers.

The cornerstone of legislative force regarding pesticides in US Federal Insecticide, Fungicide, and Rodenticide Act (FIFRA)- 1947. The most recent amendments are set up by Food Quality Protection Act of 1996 (FQPA) and by the Pesticide Registration Improvement Act of 2003 (PRIA), which was reauthorized by the Pesticide Registration Improvement Extension Act of 2012 ("Biopesticide Registration, Pesticide Registration, US EPA," 2020).

In the European Union, public institutions such as KEMI in Sweden, ANSES in France, CRD in the UK, or INIA in Spain evaluate scientific documentation submitted by companies within the agrochemical sector for the approval of the pesticide active substances and authorization of their commercial formulations (Villaverde et al., 2016).

In the EU, biopesticides identified as microbial, biochemical and semiochemical are registered under the same regulatory framework as chemical pesticides. The legislation provides for measures that favor the registration of products defined as "low risk" (most of the biopesticides).

Initially, in EU, the biopesticides use was regulated under the Directive 91/414/EEC (EU 1991), regarding the use of chemical pesticides. The Directive 91/414 was amended by 2001/36/EC (EC 2001) and 2005/25/EC (EC 2005) that added the specific requirements for microorganisms.

In the same year other three legislations: (1) Regulation (EC) No 1107/2009, (2) Directive 2009/128/EC, (3) Directive 2009/127/EC and (4) Regulation (EC) No 1185/2009, were released. The new Regulation (EC) No 1107/2009 applies in all member states from 2011 and replaces Directive 91/414/EEC.

The Directive on Sustainable Use of Chemical Pesticides ("Directive 2009/128/ec of the European Parliament and of the Council of October 21, 2009 establishing a framework for community action to achieve the sustainable use of pesticides," 2009) that aims to promote the use of non-chemical alternatives to pesticides such as microbial based biopesticides and restrict the use of pesticides that may pose risks to health or environment.

Under Regulation (EC) No1107/2009 regarding the placing on the market of Plant protection products, the registrations of products is made by three zones following geographic and climatic criteria. These zones are:

- Zone A (North): Denmark, Estonia, Latvia, Lithuania, Finland and Sweden;
- Zone B (Central): Belgium, Czech Republic, Germany, Ireland, Luxembourg, Hungary, the Netherlands, Austria, Poland, Romania, Slovenia, Slovakia and the UK;
- Zone C (South): Bulgaria, Spain, Greece, France, Italy, Cyprus, Malta and Portugal.

Therefore, as expected, the number of authorized microbial plant protection products has increased during the last decade but, on the other hand, several active substances will not be reapproved because of the new registration criteria. Although biological pest control techniques show an increased growth in the world market, farmers are still lacking satisfactory practical alternatives that may be good alternatives to synthetic ones (Kvakkestad et al., 2020; Villaverde et al., 2014).

Recognizing the harmful effects of chemical pesticides, such as the emergence of resistance, the reappearance of pests, the outbreak of secondary pests, pesticide residues in products, soil, air and water, which deteriorate human health and ecological imbalance, most countries have changed their policies to a minimum. Chemical pesticides and promote the use of biopesticides. The relatively small range of biological means, the lack of network integration in terms of the form, quality and frequency of the interaction determines the immaturity of the policy network, the limited capacities and the lack of trust between the regulatory authorities, the producers of environmentally friendly means and agricultural producers, which constitutes a range of serious problems.

3.4 Diversity of biopesticides

Biopesticide use add flexibility in a classic program of control, with reduced intervals from the last treatment to the harvest, a very good management of the residues of pesticides for the exported products and an excellent eco-toxicological profile for humans, animals and the useful entomofauna.

Biopesticides include a generic range of products based on various different methods of controlling crop pests, such as:

- Microorganisms (viruses, bacteria, fungi)
- Bacterial metabolites (antibiotics)
- Natural pesticides derived from plants
- Insect pheromones
- Entomophagous nematodes

Biopesticides include preparations for the biological control of pests, isolated or produced from substances of natural origin (microorganisms, plants, animals and minerals). The advantage of biopesticides is that affect only the target pest and closely related organisms (Dutta, 2015).

The biopesticide group can be more generally divided into three major categories.

Microbial biopesticides are prepared based on *microorganisms* (bacteria, fungi, viruses and protozoa or alga) and their metabolic products. Biological means are formulations made of natural ingredients, which control harmful organisms through non-toxic and ecologically harmless mechanisms. Biopesticides are microorganisms or their derivatives and include living organisms or their products.

Microbial biopesticides can control many different types of pests, although each active ingredient separately is relatively specific to a certain type of pests. As a quick example, we can specify *Bacillus thuringiensis* one of the most popular ingredients of microbial pesticides. It can control potatoes and cabbage insects by producing a harmful protein against insect pests. Some fungi may control the occurrence of target pathogens or the growth of target weeds, while other fungi kill specific insects. Other biopesticides from this class act by competing against pests. However, microbial biopesticides need to be carefully and constantly monitored to ensure that they do not become harmful to non-target organisms, including microbes, animals or humans.

The second group includes *vegetal extracts* and other natural substrates. Herbal biopesticides or Plant based biopesticides are produced from pesticide substances that are in plants composition. These substances can also be produced from the genetic material that has been added to the plant.

The third category consists *biologically active substances* based on natural compounds that have no toxic effect on pests, but only affect their behavior. While microbial biopesticides use microorganisms (bacteria, fungi, viruses, protozoa) as active ingredients, biochemical biopesticides are substances obtained from plants and animals. These are naturally occurring substances that control pests by non-toxic mechanisms.

Biochemical pesticides include substances that interfere with growth or mating, such as plant growth regulators, or substances that repel or attract parasites, such as pheromones. On the other hand, the conventional pesticides, are synthetic materials that generally kill or inactivate the pest.

Worldwide demand and use of biopesticides is increasing, and therefore research in this area is intensifying, in search for new beneficial substances to be used.

The Environmental Protection Agency (EPA) defines biopesticides as "certain types of pesticides derived from natural sources, such as animals, plants, fungi, bacteria, and certain minerals." Currently, EPA recognizes three major classes of biopesticides: (*i*) *microbial,* (*ii*) *biochemical,* (*iii*) *Plant-Incorporated-Protectants* (*PIPs*). Canola oil, garlic, peppermint or chrysanthemum oil, for example, have pesticide applications and are considered biopesticides. Biopesticides are already recognized as an effective option to control diseases, pests and even weeds for organic crops, especially for vegetable crops, fruit trees, vines, ornamental apartment or garden plants, lawns, but also for most crops field where the objective is to obtain organic crops.

At present, many active substances with a biopesticide profile are registered at European level, used in various products, such as repellents for deer in forestry, repellents for mosquitoes or ticks, but also biopesticide products for commercial agriculture, in particular for pest control.

As several natural sources of biopesticides are identified and synthesized, the number of such registered products will continue to increase. However, the farmers should respect the same rules as for conventional plant protection products: the approved dose per hectare, the optimal application time, the number of treatments recommended and the alternation of the active substances, the interval from the last treatment until the harvest.

In commercial terms there are three types of biopesticides derived from natural materials: microbial pesticides, naturally-occurring substances that control pests and plant incorporated protectants (PIPs) (Dutta, 2015; Kachhawa, 2017; Raja, 2013). Plant Incorporated Protectants are substances with pesticide properties produced by genetic modified plants. For example, a gene from *Bacillus thuringiensis* that codes for pesticide protein is introduced into the plant's genetic material and therefore, the respective plant will be able to manufacture the pesticide substance (Dutta, 2015). Microbial plant protection products (PPPs) are generally used to preserve crops from plant pathogens, harmful pest organisms and weeds. Microbial Pest control Agents (MPCAs) are used worldwide and adverse effects on human health and the environment have not been reported. Moreover, microbial pesticides are considered to pose a low risk to the environment when compared to chemical ones. Many of them can have a high level of selectivity and generate low or no toxic residue. Although, one of the issues of concern is that are living organisms which can vary somewhat in composition and may affect in time non-target organisms (Mudgal et al., 2017).

Other classification identifies the following two types of biopesticides: (i) **microbiological preparations** *containing living microorganisms*; (ii) biochemical (substances extracted naturally) and their synthesized analogs.

The biochemical preparations may be dived into:

- **semi - chemical** - substances of organic origin or their synthesized analogs, including, in particular, pheromones;
- **non - conventional** - materials of natural origin, such as garlic and icing sugar, vinegar, etc., which can also be used to control pests and weeds (including such "folk remedies" that are familiar to us for controlling wireworms, blights and other diseases, etc., like ashes, soapy water).

The category of biochemical preparations includes:

- plant growth regulators (gibberellins, cytokinins, abscisic acid, ethylene, auxins);
- growth regulators of insects that do not kill pests, but interrupt the growth process, preventing the pest from moving to the next stage of development. They are divided into 2 large groups - inhibitors of hormonal regulation of insect

degeneration and inhibitors of chitin synthesis. The most common insect growth regulator is azadirachtin. It mimics the natural hormone molting insect ecdysone. From the action of this substance, insects molt immature and eventually die.
- organic acids (in particular, peroxides, which are used to sterilize, for example, greenhouses or irrigation water tanks against eutrophication);
- pheromones;
- plant extracts. Among the plant extracts we distinguish:
  - insect growth regulators,
  - food inhibitors (eating insects that stop eating),
  - repellents
  - confusants (substances that look like food familiar to the pest and act as bait, for example, lure insects to the other side of the crop),
  - allelopaths (for example, the juglone secreted by black walnut - one of the few bioherbicides),
  - mechanical agents (for example, D-limonin, which "burns" wax coating on the leaves of the weeds, causing necrosis, dehydration and dying),
  - fungicides (penetrate the cell membrane of fungi, neutralizing key enzymes)
- stability enhancers (substances that provoke crop plants to release and accumulate high doses of special proteins and other substances that inhibit the development of bacterial and fungal diseases).

3.5 Formulation of biopesticides

Formulation and fermentation of biopesticides are connected processes in order to determine the well performance of the product. The ideal conditions required for development of efficient biopesticides are considered selection of potent strains, long shelf life, storage, biosafety, quality control and application technology (Keswani et al., 2016). Formulation of biopesticides are various, being composed of microorganisms, spores, an active ingredient, carriers, adjuvants and an inert material which is the substrate for growing microorganisms and deliver the active ingredient to the target. These components should have the ability to maintain the stability of biopesticides during production, processing, storage as well as to promote their activity, the spread of the biopesticides and to protect the final product from unfavorable environmental conditions. When developing a strategy for biopesticides formulation and application it may take into account the ecology of the biopesticides and pest interaction, optimizing the concentration of ingredients, the fermentation parameters, the rate and timing of harvest and post-harvest treatments (Hynes and Boyetchko, 2006; Kachhawa, 2017).

Dry formulations in the form of granules or powders are generally preferred over liquid formulations. This is due to extended shelf-lives and more facile storage and transportation. Furthermore, most granular or powder formulations can also be made into liquid- or water-based suspensions as required for drench, spray, or root-dip applications (Tamreihao et al., 2016).

Biopesticides based on bacteria (*Bacillus* sp., *Pseudomonas fluorescens*, *Enterobacter* sp., *Streptomyces* sp., *Serratia marcescens*, *Burkholderia cepacia*, *Agrobacterium radiobacter*, *Agrobacterium tumefaciens*, *Alcaligenes* sp., *Erwinia amylovora*), fungi (*Trichoderma* sp., *Beauveria bassiana*, *Fusarium oxysporum*, *Verticillium chlamydosporium*, *Verticillium lecanii*, *Streptomyces griseoviridis*, *Streptomyces lydicus*, *Piriformospora indica*, *Pythium oligandrum*, *Candida oleophila*, *Aspergillus niger*), viruses (Zucchini Yellow Mosaik Virus, Baculoviruses, Nuclear polyhedrosis viruses), microalge (*Anabaena laxa*, *Chlorella vulgaris*, *Fischerella ambigua*, *Haematococcus pluviallis*, *Nostoc* sp., *Spirulina platensis*, *Lyngbya* sp.,) plants (*Allium sativum*, *Euphorbia* sp., *Cinnamomum zeylanicum*, *Azadirachta indica*) as well as nematodes (*Steinernema carpocapsae*, *Steinernema feltiae*, *Steinernema kraussei*, *Heterohabditis bacteriophora*, *Heterohabditis downesi*) and their bioactive compounds have been developed and tested as alternatives to synthetic pesticides to control pests, which may harm crops, eventually leading to a decrease in production yield (Costa et al., 2019; Gasic and Tanovic, 2013; Kachhawa, 2017; Keswani et al., 2016; Lengai and Muthomi, 2018; Ruiu, 2018).

Biopesticides can be obtained using various materials and methods. Their biological activity is determined by its biologically active microorganisms or metabolite. According to scientific literature microorganisms, among other living organisms, can be used to produce efficient biopesticides.

In the present day, among the most used bacteria as a component in biopesticides design is *Bacillus thuringiensis* (Kachhawa, 2017). For instance, biopesticides formulations based on *Bacillus thuringiensis* were produced using starch industry wastewater as a substrate. Since its discovery in 1901 to date, over one hundred *B. thuringiensis* based biopesticides have been developed. Most of the Bt based biopesticides are targeted for controlling lepidopteran, dipteran and coleopteran larvae. In addition, the genes that code for the insecticidal crystal proteins have been successfully transferred into different crops plants which have led to significant economic benefits (Agbo et al., 2015). *Bacillus thuringiensis* occurs in the soil and on plants. The most used subspecies of this bacteria, for producing biopesticides are the following: *B.t. aizawai, B.t. islaelensis, B.t. kurstaki, B.t. tenebrionsis* and *B.t. sphaericus*. The scientists selected *Bacillus thuringiensis* due their property of secreting endotoxin protein crystals which are lethal to many pests, being a promising alternative to chemical pesticides. Their remarkable properties have been noticed and started to dominate the microbial pesticide market occupying almost 97% of the world's biopesticides market. Biopesticides are effective to agricultural industries only if they have a potential impact on the target pest, market penetration and lastly, proper performance under variable field conditions (Arthurs and Dara, 2019; Kumar et al., 2019). Commercially available biopesticides based on *B. thuringiensis* take the form of sprays, wettable powders, liquid concentrates and dusts. The bioproduct remains active several days to weeks.

A study of the mechanisms of the phytoprotective action of selected cultures of the *Bacillus* genus showed that their antifungal effect is most pronounced at the stages of spore germination and the formation of growth tubes of phytopathogenic fungi and leads to morphological changes in hyphae (vacuolization, the formation of tumor-like structures) and weakened growth mycelium (Table 3.1).

An important role in the control of pathogens produced by antagonist bacteria is due to antibiotics and lytic enzymes.

Various actinobacteria have been reported for their promising action against phytopathogens, and thus have been used for formulation of biopesticides. The biopesticides developed from actinobacteria take the form of solid, liquid and powder obtained from bacterial spores (Tamreihao et al., 2016). Fresh microbial cultures are not suitable for agronomic use. Optimization of the technological scheme and the formulation process represent a key step in the design of biopesticides, as it will assure the efficacy of the product, aiming to reach a high efficacy and survival rate of the antimicrobial substance. Marketed formulations based on actinobacterial spores as active ingredients are used in the agricultural sector as an alternative to chemical pesticides.

Actinobacteria comprise Streptomyces species that produce macrocyclic lactone derivatives, with insecticidal effect by disruption of insect peripheral nervous system. More than 60% of the sources of antifungal and antibacterial compounds or plant growth-promoting substances that have been used for agricultural purposes originated from this genus (Reddy et al., 2016).

Saccharopolyspora spinosa produces different insecticidal toxins known as spinosins. Natural and semisynthetic derivatives of spinosis were the base for development of commercial biopesticides with great success (Ruiu, 2018).

Antagonistic activity of actinomycetes against fungal pathogens is mainly due to production of antifungal metabolites, volatile compounds and cell wall degrading enzymes such as chitinase, glucanase (Tamreihao et al., 2016). They act by colonizing plant roots before the disease organisms get there, thus depriving them of space and nourishment.

TABLE 3.1 Several examples of microbial biopesticides commercially available.

Bacterial strain	Formulation of biopesticide	Action of biopesticide	Commercial name
Bacillus firmus NCIM 2637	Wettable powder	Biological nematicide	BioNemaGon
Bacillus firmus I582	Liquid	Biological nematicide	Chancellor
Bacillus subtilis (NCIM 2063)	Wettable powder	Biofungicide	Biotilis
Saccharopolyspora spinosa	Suspension concentrate		Tracer 120
Bacillus thuringiensis subsp. kurstaki	Liquid concentrate	Bioinsecticide	BioT plus
Bacillus amyloliquefaciens D747	Water dispersible granules	Biofungicide	Double Nickel 55
Streptomyces griseoviridis K61	Wettable powder	Biofungicide	Mycostops
Streptomyces lydicus WYEC108	Wettable powder	Biofungicide	Actinovates
S. lydicus	Granules	Biofungicide	Actinovate, ActinoGrow
Streptomyces hygroscopicus		Biofungicide	Arzentt
S. hygroscopicus & *S. viridochromeogenes*	Powder	Herbicide	Bialaphoss
Saccharopolyspora spinosa	Powder/suspension concentrate	Insecticide	Tracer
S. atrovirens	Powder		Incide SP
S. atrovirens	Liquid	Biofungicide	Actin

Research on the formulation of biopesticides based on a mixture of endotoxins, spores, microorganism's cells, nutrients and substrate from starch industry wastewater and adjuvants which are responsible for the toxic effect against the insects and pests, was conducted by Kumar et al. (2019). The advantages of choosing this type of formula are represented by the use of wastewater industry as an inexpensive carbon source reducing cost of raw materials for fermentation (Kumar et al., 2019). Moreover, the benefic effect of biopesticides was evaluated by treatment of *Leptinotarsa decemlineata* larvae (Colorado potato beetle) using a commercialized biopesticides based on *Beauveria bassiana* (formulated as Mycotrol) and *Bacillus thuringinesis* (formulated as Novodor), applied on small plots of potatoes over three field seasons, evaluating the pest control efficacy. The scientists noticed that the interaction between these two species of microorganisms produced a 6%–35% reduction in larval populations than the treatment alone (Wraight and Ramos, 2005).

Entomopathogenic fungi are considered to be essentials biological control agents of insect pests (Kachhawa, 2017). Fungal strains used for formulation pf biopesticides include: *Beauveria bassiana, B. brongniartii, Metarhizium anisopliae, Verticillium, Lecanicillium, Hirsutella, Paecilomyce*s, and *Isaria species*.

Thus, fungal biopesticides were applied on plants to evaluate their effect. This type of biopesticides formula is based on the solid-state fermentation using rice husk as substrate with entomopathogenic strains such as *Trichoderma harzianum* or *Beauveria bassiana.* Example of fungal bio pesticides are *Muscodor albus* used in fields, greenhouses, and warehouses *and Aspergillus flavus* targeted for *Aedes fluviatilis* and *Culex quinquefasciatus* (Agbo et al., 2015). Fungal biocontrol agents have been used in recent years in agricultural industry, as being pathogen agents to more than 1000 insect species. The toxic effect induced by fungal species is due to various types of infective propagules, the most common being blastospores and aerial conidia. Entomopathogenic fungi are useful against a wide range of agricultural pests and do not have negative effects on human health. They can infect their hosts, mainly insects, through the external cuticle. The rapid penetration and infection of a susceptible host is enhanced by high humidity, but spores are able to stay viable on the cuticle until reaching favorable conditions and penetrate when humidity rises (Mascarin and Jaronski, 2016; Sala et al., 2020; Sinha et al., 2016).

Microalga biomass is also employed into the production process of biopesticides. The main reason of their use is because are organisms capable of biosynthesize a large number of metabolites with potential toxic action and also can be considered a biological agent for the control of harmful microorganisms and organisms to soils and plants. For instance, cyanobacteria produce biologically active compounds that can be used as a control of pathogenic fungi and soil-borne diseases in plants, exerting, on the other hand, antifungal, antibacterial activities and toxic activities against nematodes (Costa et al., 2019).

Besides bacteria, fungi, microalgae and viruses, plants extract or their metabolites can be efficiently applied in the technology of producing biopesticides.

Botanical pesticides are produced from plat extracts that possess biocidal properties. Botanical pesticides are based on plant metabolites, mainly essential oils which are a mixture of volatile compounds localized in seeds, flowers or leaves (Céspedes et al., 2014; Raja and Masresha, 2015). Monoterpenes are metabolites produced in approximatively 17,500 aromatic species belonging to *Apiaceae, Asteraceae, Myrtaceae, Lamiacease, Lauraceae*, *Lamiaceae* families, showing good biological activity and providing insecticidal, fungicidal, bactericidal effects against pests and pathogens that are important to consider in agricultural yield (Pavela and Benelli, 2016).

For instance, Doty (Sundy Aisha Doty, 2012), through her invention, realized a biopesticide formulation based on a mixture of natural spices that have the ability to kill insect through direct contact. The spices selected were *Piper nigrum* (25.75%), *Cuminum cyminum* (15.45%), *Coriandrum sativum* (12.80%), *Allium sativum* (12.03%), *Cinnamomum verum* (11.05%), *Curcuma sp.* (7.44%), *Trigonella foenum graecum* (3.65%), *Zingiber officinale* (3.36%), *Syzygium aromaticum* (2.58%), *Foeniculum vulgare* (2.37%), *Murraya koenigii* (1.60%), *Capsicum sp.* (1.38%), *Brassica nigra* (0.37%), mixed together with sodium chloride (0.17 %) and potting soil than applied to a plant. All botanical constituents work in a synergism as an effective biopesticide affecting the insects inducing suffocation and paralysis (Sundy Aisha Doty, 2012).

Neem (*Azadirachta indica*) is one of the most effective ingredients for botanical pesticide. All the parts of the neem tree are medicinal. Over 60 different types of biochemical products including, Nimbolide, Margolone, Mahoodin, Margolonone have been purified from neem (Agbo et al., 2015).

Another important commercial class of botanical pesticides include pyrethrum products. Pyrethrum is a natural insecticide extracted from the flowers of *Chrysanthemum cinerariaefolium* and *Chrsanthemum cineum* that is currently used controlling field, household, and storage pests, and parasites in livestock and humans (Fernández-Grandon et al., 2020).

The composition of pyrethrum contains six entomotoxic compounds, namely: cinerin I and II, pyrethrin I and II and jasmolin I and II. Pyrethrin effect takes place by altering the central nervous system causing disruptive nerve function, which causes paralysis in target insect pests, leading to their death (Chen et al., 2018).

Pyrethrins were the cornerstone for development pyrethroids. Pyrethroids represent synthetic insecticides acting in a similar manner to pyrethrins, provoking rapid death of target insects. Pyrethroids formulations was optimized to increase the pesticide stability in the environment, (for example, protect from sunlight) resulting in improved control of pests.

However, the active components in pyrethrum are highly labile in ultraviolet (UV) light, non-persistent, and are less toxic to humans and the environment (Fernández-Grandon et al., 2020).

As previously stated, biopesticides based on plants metabolites are a promising alternative to the conventional ones. For instance, the efficacy of terpene based biopesticides was tested on *Aphis gossypii* (melon and cotton aphid) and *Myzus persicae* (peach-potato aphid) from ornamental crops as foliar spray. This biopesticide formula consisted on orange oil, essential oil from *Chenopodium ambrosioides* and neem oil. Compared to a synthetic insecticide (flonicamid), the biopesticides manifested good results in pest control of *Myzus persicae*, except from neem oil on *A. gossypii* (Smith et al., 2018). Plant extracts of *Calceolaria sp*, used in traditional medicine, were reported to provide defense mechanisms against Gram positive and Gram-negative bacteria or fungi pathogens, as well as against insects such as *Spodoptera frugiperda* and *Drosophila melanogaster* or herbivore predators (Céspedes et al., 2014).

Biopesticides optimally suited for control of insect pests, parasites, weeds and soil borne infecting phytopathogens are soil applied in the form of granules, liquids, microcapsules and powder. Formulations applied on soil demand an active ingredient that is dormant during storage and is promptly adapted to the changes of environmental conditions following application (Hynes and Boyetchko, 2006).

Formulation of biopesticides, in the most cases, is the same as a synthetic pesticide with respect to the equipment needed for different treatments. These formulations can be obtained in both liquid and dry physical state. For instance, liquid formulations can be water-based (suspension concentrate, capsule suspension), oil-based, polymer-based or combination, while dry formulations can be dusts, powders for seed dressing, granules, micro granules or water granules (Gasic and Tanovic, 2013). Another study highlighted the fact that through the use of active ingredients from *Azadirachta indica* (neem plant) exerting insecticidal and immunomodulatory properties may be pesticides substitutes. Azadirachtin was described to be an antifeedant, repellent and repugnant agent inducing sterility in insects (Chaudhary et al., 2017).

3.6 Biopesticides market

The first bioinsecticide in history was Nicotine, followed by Pyrethrin. With 80% of sales, pyrethrin is the world market leader in plant insecticides. The demand for biopesticides is constantly increasing in all countries of the world, and at European level it is a safe alternative to replace many active substances from the rather extensive list of those that will be excluded in the medium and long term.

Key drivers of the growth of biopesticide market include.

- political and societal pressure for greener, safer and more sustainable crop protection technologies;
- population demands for low or no residues on food crops;
- an increasingly tough regulatory framework for chemical plant protection products;
- resistance development to existing conventional chemical pesticides

Public opinion, as well as conventional pesticide users, are increasingly asking themselves pertinent questions about their environmental impact and potential health risks, but also worker safety, bird toxicity, surface and groundwater contamination.

Due to the population awareness of the damaging and adverse effects of pesticides on foods, the tendency is to avoid exploiting them and embrace eco-friendly alternatives (Damalas and Koutroubas, 2018). For instance, worldwide are about 1400 biopesticides products that are being sold. In the United States are currently 356 registered biopesticides based on 57 species or strains of microorganisms used against insects and nematodes while in Europe there are registered 68 active substances and 34 microorganisms. Moreover, it has been reported that ten fungal species are being used against aphids, whiteflies and parasitic plant nematodes from greenhouse and field crops (Arthurs and Dara, 2019; Chandler et al., 2011).

One thing is certain, and farmers and final consumers of agricultural products, for the most part, are aware that the conventional products (pesticides) and the biological formulations used for the plants protection, as an effective pests, weeds and disease control technology, should be complementarily used to ensure a reasonable global production, which in the medium and long term must satisfy the food of humanity in exponential growth.

Regarding the conventional pesticides, large multinational companies, which invest in research hundreds of millions of euros per year, provide all stakeholders, including the officials, farmers and consumers with thorough studies on the ecotoxicological profile of the substances placed on the market and assume major responsibilities regarding their management, providing clear information on usage recommendations and user protection.

The global market for biopesticides is estimated at about 3 billion dollars per year (1% of the pesticide market). They are mainly used in developed countries, where are set in place by coherent regulatory framework and well adapted technologies for use. Generally, biopesticides are used for organic farming, but they are also utilized in tandem with conventional pesticides, for decreasing the current use of chemicals.

In 2012 biopesticides represented 3.5% of the overall pesticides market. Further, biopesticides value in Western Europe pesticides markets was estimated to be 594.2 million dollars in 2008 (Glare et al., 2012) reaching to 1.02 billion dollars in 2015. In 2013 were registered almost 400 biopesticides active ingredients and 1250 actively biopesticide products (Raja, 2013). Worldwide biopesticides producers are Marrone Bio Innovations, Valent BioSciences Corporation, Certis USA, BioWorks, FMC Corporation, Vestaron, Provivi, Boost Biomes (United States), Terramera PlantHealth (Canada), Bayer CropScience, BASF (Germany), Syngenta, Biocontrol (Switzerland), Isagro, Biogard a Division of CBC (Europe) SRL (Italy), Koppert, Imants BV (Netherlands), Seipasa, Inden Biotechnology, Probodelt (Spain), Aphea. Bio, Biobest Group NV, Mebrom NV (Belgium), Center for Process, Agralan Ltd. (United Kingdom), Stockton Group (Israel), Bioorient Biotechnology (Turkey) and UPL (India).

Compared with classic pesticides, many commercially available biopesticides targets only a single major pest which limits their potential market size. Commercially biopesticides Contans and Bioshield™ based on the fungus *Coniothyrium minitans*, which targets fungal pathogen *Sclerotinia spp.* and biopesticides based on *Serratia entomophila* which attacks insect pests respectively, are examples of single-target biopesticides. On the other hand, biopesticides based on *Bacillus thuringiensis* have achieved impressive market penetration due to their extended mechanisms of actions targeting many invertebrate pests (Glare et al., 2012). Biopesticides based on fungus *Verticilliun lecanii* proved to be highly specific which could be noted as a disadvantage when compared with chemical products with broad spectrum activity (Villaverde et al., 2014).

The global biopesticides market has achieved remarkable successes, reaching $ 3.42 billion in 2019 and is forecast to reach $ 14.62 billion in 2025 (Sinha et al., 2016).

If in 1993 the world market of biological control products for harmful organisms amounted to $ 100 million, then in 2020 its volume will grow, according to the Biological Products Industry Alliance (BPIA), to $ 4 billion.

According to AGROW report on biopesticides, released in 2019, the greatest market share of biopesticides for the year 2020 is forecasted to bioinsecticides (47%), closely followed by biofungicides (44%). Only a small share of global biopesticide market will be attributed to bionematicides (3%) and even less to bioherbicides (1%) (Bioinsecticides 2019, 2019). It is expected more than half of the biopesticides that will enter the market will be microbial biopesticides, while Plant based extracts will provide around 30% of the marketed biopesticides (Markets and Markets, 2019).

More than three-quarters of biopesticides, according to BPIA estimates, were used on vegetables, fruits and berries in 2016. In particular, in gardens, according to BPIA, 55% of the total volume of biopesticides is used.

3.7 Challenges and future perspectives

Currently, biopesticides are facing a lot of challenges regarding the improvement of formulations, slower pest control and registration for commercialization, acceptance and proper use in agriculture (Villaverde et al., 2014). However, they offer better alternatives due to their biodegradability in short periods of time and low presence in the natural environment, are easy to handle, provide effective control of target pest and also, the main materials are available and natural (Céspedes et al., 2014; Gasic and Tanovic, 2013; Lengai and Muthomi, 2018). Also, the use of biopesticides is an alternative method for pest control as they have lower toxicity and the pest's resistance development is reduced (Smith et al., 2018). Bioinsecticides, biofungicides and bionematocides are expected to be more expanded on the pesticide market (Kumar, 2012). In order to commercialize a microbial based product, it requires to have a long shelf life (minimum 18 months) and to be stabile when stored at room temperature (Keswani et al., 2016). Microorganisms employed as active substances in pest management are generally recognized as safe for the environment and also on non-target organisms (Ruiu, 2018).

It is generally acknowledged that the use of biopesticides in current agricultural practices comes with a lot of advantages such as:

- *Less toxic* than chemical crop protection products;
- *Rapid degradation* that leads to low exposure and residues, though addressing consumer safety concerns;
- *Specific action* on target pests;
- *Ease of use* since biopesticides do not have a waiting period, people can be displayed in the field immediately after treatment;

- *Highly efficient - for example*, bioinsecticides can act in a complex way - poison insects and discourage their appetite or inhibit the transition to a new stage of development;
- *No danger of acquired resistance* - the same type of exposure to chemical plant protection agents stimulates harmful organisms to develop resistance in the area affected by the drug. In the context of a rapid change in the generations of harmful organisms, this leads to the formation of resistance in a matter of years. Since biological products act on harmful organisms in several directions at once, this accordingly complicates the process of resistance.

In addition to the benefits, biopesticides have their drawbacks. The main ones are:

- the need for frequent crop surveys, since biopesticides have stringent requirements for application dates;
- they are often more expensive to produce than the chemical products;
- biopesticides use need careful analysis of pest identity, behavior and environmental conditions;
- due to the narrow focus of biopesticides, it is often necessary to combine them with chemical plant protection products in order to control the entire spectrum of harmful organisms in the area.

One of the greatest challenges when speaking of biopesticides is that biopesticides are not against all harmful organisms, so there are times when there is no alternative to chemical plant protection agents. Another important challenge is that there is a need to develop biopesticides with a long shelf life.

References

Agbo, B., Nta, A., Ajaba, M., 2015. A review on the use of neem (*Azadirachta indica*) as a biopesticide. J. Biopestic. Environ. 2.

Arthurs, S., Dara, S.K., 2019. Microbial biopesticides for invertebrate pests and their markets in the United States. J. Invertebr. Pathol. https://doi.org/10.1016/j.jip.2018.01.008.

Bioinsecticides 2019, 2019. Biofungicides, Bioinsecticides, Bionematicides & Bioherbicides.

Pesticide Registration Biopesticide Registration, 2020. US PA [WWW Document]. https://www.epa.gov/pesticide-registration/biopesticide-registration (Accessed 15 March 20).

Céspedes, C.L., Salazar, J.R., Ariza-Castolo, A., Yamaguchi, L., Ávila, J.G., Aqueveque, P., Kubo, I., Alarcón, J., 2014. Biopesticides from plants: Calceolaria integrifolia s.l. Environ. Res. 132, 391–406. https://doi.org/10.1016/j.envres.2014.04.003.

Chandler, D., Bailey, A.S., Mark Tatchell, G., Davidson, G., Greaves, J., Grant, W.P., 2011. The development, regulation and use of biopesticides for integrated pest management. Philos. Trans. R. Soc. B Biol. Sci. 366, 1987–1998. https://doi.org/10.1098/rstb.2010.0390.

Chaudhary, S., Kanwar, R.K., Sehgal, A., Cahill, D.M., Barrow, C.J., Sehgal, R., Kanwar, J.R., 2017. Progress on *Azadirachta indica* based biopesticides in replacing synthetic toxic pesticides. Front. Plant Sci. https://doi.org/10.3389/fpls.2017.00610.

Chen, M., Du, Y., Zhu, G., Takamatsu, G., Ihara, M., Matsuda, K., Zhorov, B.S., Dong, K., 2018. Action of six pyrethrins purified from the botanical insecticide pyrethrum on cockroach sodium channels expressed in *Xenopus oocytes*. Pestic. Biochem. Physiol. 151, 82–89. https://doi.org/10.1016/j.pestbp.2018.05.002.

Costa, J.A.V., Freitas, B.C.B., Cruz, C.G., Silveira, J., Morais, M.G., 2019. Potential of microalgae as biopesticides to contribute to sustainable agriculture and environmental development. J. Environ. Sci. Health Part B Pestic. Food Contam. Agric. Wastes. https://doi.org/10.1080/03601234.2019.1571366.

Damalas, C.A., Koutroubas, S.D., 2018. Current status and recent developments in biopesticide use. Agric. For. https://doi.org/10.3390/agriculture8010013.

Directive 2009/128/EC of the European Parliament and of the Council of 21 October 2009 Establishing a Framework for Community Action to Achieve the Sustainable Use of Pesticides, 2009. OJ L 309. https://eur-lex.europa.eu/legal-content/EN/TXT/HTML/?uri=CELEX:02009L0128-20091125&from=EN. [WWW Document]. (Accessed 15 March 20).

Dutta, S., 2015. Biopesticides: an ecofriendly approach for pest control. World J. Pharm. Pharmaceut. Sci. 4, 250–265.

Fernández-Grandon, G.M., Harte, S.J., Ewany, J., Bray, D., Stevenson, P.C., 2020. Additive effect of botanical insecticide and entomopathogenic fungi on pest mortality and the behavioral response of its natural enemy. Plants 9. https://doi.org/10.3390/plants9020173.

Gasic, S., Tanovic, B., 2013. Biopesticide formulations, possibility of application and future trends. Pestic. i Fitomedicina 28, 97–102. https://doi.org/10.2298/pif1302097g.

Glare, T., Caradus, J., Gelernter, W., Jackson, T., Keyhani, N., Köhl, J., Marrone, P., Morin, L., Stewart, A., 2012. Have biopesticides come of age? Trends Biotechnol. https://doi.org/10.1016/j.tibtech.2012.01.003.

Hynes, R.K., Boyetchko, S.M., 2006. Research initiatives in the art and science of biopesticide formulations. Soil Biol. Biochem. 38, 845–849. https://doi.org/10.1016/j.soilbio.2005.07.003.

Kachhawa, D., 2017. Microorganisms as a biopesticides. J. Entomol. Zool. Stud. 468.

Keswani, C., Bisen, K., Singh, V., Sarma, B.K., Singh, H.B., 2016. Formulation technology of biocontrol agents: present status and future prospects. In: Bioformulations: For Sustainable Agriculture. https://doi.org/10.1007/978-81-322-2779-3_2.

Kumar, L.R., Ndao, A., Valéro, J., Tyagi, R.D., 2019. Production of *Bacillus thuringiensis* based biopesticide formulation using starch industry wastewater (SIW) as substrate: a techno-economic evaluation. Bioresour. Technol. 294. https://doi.org/10.1016/j.biortech.2019.122144.

Kumar, S., 2012. Biopesticides: a need for food and environmental safety. J. Biofert. Biopestic. 03. https://doi.org/10.4172/2155-6202.1000e107.

Kvakkestad, V., Sundbye, A., Gwynn, R., Klingen, I., 2020. Authorization of microbial plant protection products in the Scandinavian countries: a comparative analysis. Environ. Sci. Pol. 106, 115–124. https://doi.org/10.1016/j.envsci.2020.01.017.

Leng, P., Zhang, Z., Pan, G., Zhao, M., 2011. Applications and development trends in biopesticides. Afr. J. Biotechnol. https://doi.org/10.5897/AJBX11.009.

Lengai, G.M.W., Muthomi, J.W., 2018. Biopesticides and their role in sustainable agricultural production. J. Biosci. Med. 06, 7–41. https://doi.org/10.4236/jbm.2018.66002.

Markets and Markets, 2019. Biopesticides Market by Type, Source, Mode of Application, Formulation, Crop Application and Region - Global Forecast 2023 [WWW Document]. https://www.marketsandmarkets.com/Market-Reports/biopesticides-267.html (Accessed 15 March 20).

Mascarin, G.M., Jaronski, S.T., 2016. The production and uses of *Beauveria bassiana* as a microbial insecticide. World J. Microbiol. Biotechnol. https://doi.org/10.1007/s11274-016-2131-3.

Moazami, N., 2008. Biopesticide production. In: Biotechnology.

Mudgal, S., De Toni, A., Tostivint, C., Hokkanen, H., Chandler, D., 2017. Scientific support, literature review and data collection and analysis for risk assessment on microbial organisms used as active substance in plant protection products – Lot 1 Environmental Risk characterisation. EFSA Support. Publ. 10. https://doi.org/10.2903/sp.efsa.2013.en-518.

Olson, S., 2015. An analysis of the biopesticide market now and where it is going. Outlooks Pest Manag. 26, 203–206. https://doi.org/10.1564/v26_oct_04.

Pavela, R., Benelli, G., 2016. Essential oils as ecofriendly biopesticides? Challenges and constraints. Trends Plant Sci. https://doi.org/10.1016/j.tplants.2016.10.005.

Raja, N., 2013. Biopesticides and biofertilizers: ecofriendly sources for sustainable agriculture. J. Fertil. Pestic. 4. https://doi.org/10.4172/2155-6202.1000e112.

Raja, N., Masresha, G., 2015. Plant based biopesticides: safer alternative for organic food production. J. Fertil. Pestic. 06. https://doi.org/10.4172/2471-2728.1000e128.

Reddy, K.R.K., Jyothi, G., Sowjanya, C., Kusumanjali, K., Malathi, N., Reddy, K.R.N., 2016. Plant growth-promoting actinomycetes: mass production, delivery systems, and commercialization. In: Plant Growth Promoting Actinobacteria: A New Avenue for Enhancing the Productivity and Soil Fertility of Grain Legumes, pp. 287–298. https://doi.org/10.1007/978-981-10-0707-1_19.

Ruiu, L., 2018. Microbial Biopesticides in Agroecosystems. Agronomy. https://doi.org/10.3390/agronomy8110235.

Sala, A., Artola, A., Sánchez, A., Barrena, R., 2020. Rice husk as a source for fungal biopesticide production by solid-state fermentation using *B. bassiana* and *T. Harzianum*. Bioresour. Technol. 296. https://doi.org/10.1016/j.biortech.2019.122322.

Sinha, K.K., Choudhary, A.K., Kumari, P., 2016. Entomopathogenic fungi. In: Omkar (Ed.), Ecofriendly Pest Management for Food Security. Academic Press, Cambridge, Edmonton, Alberta, pp. 475–505. https://doi.org/10.1016/B978-0-12-803265-7.00015-4.

Smith, G.H., Roberts, J.M., Pope, T.W., 2018. Terpene based biopesticides as potential alternatives to synthetic insecticides for control of aphid pests on protected ornamentals. Crop Protect. 110, 125–130. https://doi.org/10.1016/j.cropro.2018.04.011.

Sundy Aisha Doty, 2012. Bio-pesticide and Methods of Making and Using the Same. US8202557B1.

Tamreihao, K., Ningthoujam, D.S., Nimaichand, S., Singh, E.S., Reena, P., Singh, S.H., Nongthomba, U., 2016. Biocontrol and plant growth promoting activities of a Streptomyces corchorusii strain UCR3-16 and preparation of powder formulation for application as biofertilizer agents for rice plant. Microbiol. Res. 192, 260–270. https://doi.org/10.1016/j.micres.2016.08.005.

Villaverde, J.J., Sandín-España, P., Sevilla-Morán, B., López-Goti, C., Alonso-Prados, J.L., 2016. Biopesticides from natural products: current development, legislative framework, and future trends. BioResources 11, 5618–5640. https://doi.org/10.15376/biores.11.2.Villaverde.

Villaverde, J.J., Sevilla-Morán, B., Sandín-España, P., López-Goti, C., Alonso-Prados, J.L., 2014. Biopesticides in the framework of the European Pesticide Regulation (EC) No. 1107/2009. Pest Manag. Sci. 70, 2–5. https://doi.org/10.1002/ps.3663.

Wraight, S.P., Ramos, M.E., 2005. Synergistic interaction between *Beauveria bassiana-* and *Bacillus thuringiensis* tenebrionis-based biopesticides applied against field populations of Colorado potato beetle larvae. J. Invertebr. Pathol. 90, 139–150. https://doi.org/10.1016/j.jip.2005.09.005.

Chapter 4

Application technology of biopesticides

Adriano Arrué Melo[a] and Alexandre Swarowsky[b]

[a]Universidade Federal de Santa Maria, Department of Crop Protection, Santa Maria, Rio Grande do Sul, Brazil; [b]Universidade Federal de Santa Maria, Department of Sanitary and Environmental Engineering, Santa Maria, Rio Grande do Sul, Brazil

4.1 Introduction

Biopesticides are already a reality in the world agricultural market, occupying a prominent place in pest control. These products have an even greater growth potential in agriculture, and may become the basis for pest control, as the use of biopesticides can result in pest control similar to those achieved with chemical pesticides. Perini et al. (2016) in a study on soybean, evaluating the control of *Helicoverpa armigera*, reports that biological insecticides have an economic return similar to chemical insecticides. One of the main differences is that biopesticides present less risk to applicators, consumers and the environment (Gan-Mor and Matthews, 2003).

However, in order for these products to express their full potential, certain factors need to be taken into account. The application technology used is one of the biggest limitations. To understand how application technology can interfere in biopesticides efficiency, we initially need to understand this science. According to Matuo (1990), pesticide application technology is the use of all scientific knowledge, providing the correct placement of the product on the target, in the necessary quantity, economically, at the appropriate time and with a minimum possible environmental contamination. Therefore, application technology requires the integration of several areas of knowledge.

The biopesticides efficiency can be influenced by several factors related to application technology (Silva and Moscardi, 2002). Errors related to application technology can cause economic loss, due to control inefficiency. In addition, it can cause problems related to the concept that biopesticides have to technicians and farmers. Their recommendation in an inappropriate manner, and consequent control failure, may end up damaging the image that these products have before their users.

Therefore, it is extremely important that everyone involved in the process of using biopesticides is aware of the potential and specificities that these products need. In addition, it is necessary to take into account that there are several ways of applying biopesticides, the main ones being seed treatment, seedling dipping and foliar application (Tijjani et al., 2016), with foliar application in the field being the most used and the focus in that chapter. Thus, the following topics will be addressed: coverage, biopesticides and adjuvants, Mixture of biopesticides and pesticides and influence of climatic factors on the application of biopesticidas.

4.2 Coverage

One of the application technology pilars is to obtain maximum coverage on the target to be reached. When it comes to biopesticides application, this need for coverage is even greater, since most of these products have little translocation in the plant. Therefore, it is essential to have an adequate spray volume, so that the product can express its full control potential.

The coverage is directly influenced by the spray volume, which can be a limiting factor. Therefore, the spray volume is essential for applications to be efficient in pest control (Antuniassi, 2006). The discussion about the ideal spray volume has existed for a long time and in recent years, there has been a search for working with increasingly smaller volumes. However, it is necessary to take into account that this reduction in the spray volume can affect the coverage and consequently the control efficiency of biopesticides.

In addition, when it comes to spray volume it is necessary to understand that it is the spray nozzle that will produce the spectrum of drops, the uniformity of the application and consequently the spray volume determined. In addition, the

Biopesticides. https://doi.org/10.1016/B978-0-12-823355-9.00012-2

TABLE 4.1 Theoretical density of droplets (cm^2).

	Spray volume (L/ha)			
Drop (μm)	**25**	**60**	**100**	**150**
100	480	1150	1910	2870
150	140	340	565	830
200	60	150	240	360
250	30	72	120	180
300	18	42	71	106

droplet spectrum will have a direct influence on the target's coverage, because the larger the droplet size, the smaller the number of droplets produced, as shown in Table 4.1.

The choice of spray volume and nozzle are fundamental for success in agricultural applications, and these choices must take into account the following factors:

4.2.1 Crop

There are a large number of cultivated plants that use biopesticides in which we carry out the foliar application in the field. However, these plants have differences in their architecture and morphology, as well as, they present very different leaf area indexes among themselves and throughout their development cycle. Thus, the choice of spray volume and nozzle should consider these aspects. These variations in the field can vary from very low volumes (less than 30 L/ha) to extremely high volumes (above 400 L/ha) with drops from very fine (<60 μm) to extremely coarse (>500 μm). This spray volume and nozzle adjustment should be carried out considering the culture that is being sought to protect.

4.2.2 Biological target

When applying a biopesticide, it is necessary to know the biological target that is being sought. This fact is due to the differences in the targets behavior (from insects that have high mobility to weeds that are immobile and are normally more exposed to agricultural applications). Therefore, the choice of nozzle and spray volume should take into account the behavior of the target to be controlled.

4.2.3 Biopesticide

We currently have a wide variety of biopesticides on the market, with fungal, bacteria, viruses and nematodes being the most common. Each of these microorganisms has a specific characteristic and specific need when applied in the field. As for example, we know that viruses are small organisms, whereas nematodes (normally because they have a larger diameter) need a larger droplet size, so that it can pass through the nozzle, which consequently leads to the need increased spray volume.

After defining the spray volume and nozzle to be use, the next step is the correct adjustment of the equipment for the application, which is the process of adapting the spray to the operation that are going to be performed. Immediately after adjustment, the equipment must be calibrated, when adjusting the spray volume to be applied and the amount of product to be placed in the spray tank (Ramos, 2002).

4.3 Biopesticides and adjuvants

In agriculture, water is used as a transport vehicle for the application of pesticides. It is used because it is cheaper and has wide availability. However, the water has a high surface tension. Surface tension is a physical effect that occurs at the interface between two chemical phases. It causes the surface layer of a liquid to behave like a membrane. In water this property is caused by the cohesive forces between hydrogen molecules in water.

The resulting force that attracts molecules from the surface of a liquid to its interior becomes an obstacle and these cohesive forces tend to decrease the surface area occupied by the liquid, thus forming spherical drops (Behring et al., 2004). In agricultural spraying, the formation of spherical drops makes wetting difficult. Spherical droplets tend to prevent contact between the droplet and the leaf surface, thus resulting in an increase in the contact angle between the leaf and the product, reducing its penetration rate (Hazen, 2000). One of the ways that we can reduce the surface tension of spray mixes is the addition of adjuvants.

The use of adjuvants with biopesticides has been studied in recent years, as in Costa et al. (2017). They demonstrated that the addition of adjuvants to the product based on *Bacillus thuringiensis* (Best HD®), resulted in the reduction of the surface tension of the spray. Caye et al. (2019) evaluated the surface tension and the control of *Bacillus thuringiensis*, var. *Kurstaki* (Dipel®) for the caterpillar *Chrysodeixis includens* in soybean, demonstrating that the use of adjuvants in addition to reducing surface tension increased the efficiency of biopesticide control.

The application of adjuvants in conjunction with biopesticides can be an alternative to improve the efficiency of these products. However, it is necessary to be careful, mostly because adjuvants are of chemical origin and can interact in a negative way with biopesticides. In this context, the development of adjuvants of biological origin has been sought, as for example, in a work carried out by Reis et al. (2018) that managed to produce a biosurfactant based on *Fusarium fujikuroi* that has the ability to reduce the surface tension at 20 mN/m. Therefore, more work is needed to developed bioadjuvants to help biopesticides application.

4.4 Mixture of biopesticides and pesticides

Product mixing in the spray tank is a reality worldwide. However, according to Gandini et al. (2020), compatibility and interactions between products needs to be better evaluated. When talking about the mixture of chemical pesticides and biopesticides, this assessment is even more important, so that there will be no problems in the field.

Gonçalves (2020) carried out a study in vitro, to understand the compatibility of two bioinsecticides, both based on *Bacillus thuringiensis* (Dipel SC® and Thuricide PM®) and eight insecticides: chlorantraniliprole + lambdacialotrina (Ampligo SC®), indoxacarbe (Avatar CE®), flubendiamide (Belt SC®), lufenuron + profenofós (Curyom CE®), spinetoram (Exalt SC®), metomil (Lannate SL®), tiodicarb (Larvin 800 WG®) and chlorantraniliprole (Premio SC®). The results of this work showed that all insecticides were compatible with Thuricide PM® and 45% of them were compatible with Dipel SC®, while eight were moderately toxic and very toxic and 10 were toxic. However, it should be noted that in vitro compatibility does not confirm the insecticidal activity for *Chrysodeixis includens*.

On the other hand, we have results from studies that show that there may be problems when mixing biopesticides and pesticides. For example, Agostini et al. (2014) evaluated the compatibility of biopesticides based on *B. thuringiensis* (Dipel® and Agree®) with the herbicide glyphosate in different concentrations, resulted in a high degree of incompatibility between the herbicide and the tested products.

4.5 Influence of climatic factors on the application of biopesticides

In the process of applying biopesticides it is essential to understand climatic factors and how they can interfere in the quality of applications. Climatic factors crucially interfere with the product's arrival at the target and its permanence. We can quote air temperature, relative humidity, the displacement of air in the form of wind and rainfall. Furthermore, as these products are applied worldwide, regional realities can interfere in their efficiency.

4.5.1 Relative humidity and temperature

The air moisture content can vary dramatically, both in space and time. This variation is dependent on the atmosphere circulation, the relative location of water vapor sources and sinks, the supply of solar energy, among others. Therefore, the values obtained through the parameters used to obtain air humidity close to the Earth's surface are punctual, for that location, at that moment.

When talking about relative humidity, we must not forget that it is dependent on temperature values. This means that an increase or decrease in relative humidity (RH) does not exactly mean a change in the concentration of water vapor. This variation may be associated only with an increase or decrease in the air temperature. The air RH increases when the temperature decreases and vice versa. Therefore, it is expected that the UR decreases after the sun rises, reaching the lowest values in the hottest hours of the day, and then increases again, just due to the thermal effect. This is the normal and expected behavior for this phenomenon. However, it can be modified under atmospheric situations capable of changing the temperature, the mixing ratio (water vapor mass/dry air mass) or both.

Another important factor is the average air temperature (either monthly or yearly), which increases from the poles to the equator, while its amplitude grows in the opposite direction, for both hemispheres (latitude effect). Higher temperatures tend to be closer to the equator, due to the higher incidence of solar energy in this range. In the areas closer to the poles, the average annual isotherms follow approximately the parallels, while in the tropical zones this trend is only observed over the oceans (Fig. 4.1). This occurs because in the continental area there is great heterogeneity in the land surface, thus making the distribution quite irregular. Another factor of great influence on the temperature is the apparent annual movement of the sun in the north-south direction, since given a certain latitude, the values are higher in summer than in winter (Christidis et al., 2009).

Therefore, the conditions of relative humidity and temperature are variable in different regions of the planet. Thus, the measurements of these variables must be performed at the time and place of the applications, knowing and respecting the limitations imposed by the climate of each region.

In order for us to analyze how air RH and temperature interfere with biopesticide applications, we initially need to understand that these factors are crucial to the success or failure of drop deposition. According to Santos (2006), excessively high temperatures associated with a very rapid evaporation of moisture (plants and soil) cause the formation of upward thermal currents, negatively affecting the deposition of drops in an adequate amount, causing them to be maintained in suspension for a long time, being easily carried by the winds before reaching the target.

Another affected factor is the speed of evaporation of a drop, as it occurs faster or slower due to the percentage of moisture in the air, causing a drop of a certain size, after being released by the spraying process, to lose liquid and weight through the evaporation process, making it lighter and lighter, being able to be dragged or diverted from its predicted trajectory (Santos, 2006).

The logical trend is that the higher the temperature, the lower the relative humidity of the air. Thus, for smaller droplets, the extinction time is shorter and the fall distance is also shorter. However, with the increase of 10°C in the temperature, this relationship triples, that is, the time of fall is reduced and the time of extinction also (Table 4.2).

In this way, air RH and temperature are two fundamental factors when defining the droplet spectrum of a biopesticide application. At times of the day when the RH of the air is low and the temperature is higher, you should work with spray tips that produce larger drops. It is no longer feasible nowadays to think about product application technology for agriculture, without considering the temporal and spatial variation of climatic conditions. There is a need to develop technologies that automate this type of decision, for example, sensors in agricultural sprayers that issue alerts about the high risk of loss of efficiency in biopesticide applications.

4.5.2 Wind speed and direction

The wind is the displacement of atmospheric air. It arises with the movement of some parts of the atmosphere, caused by differences in atmospheric pressure resulting from changes in temperature. These differences play a very important role in

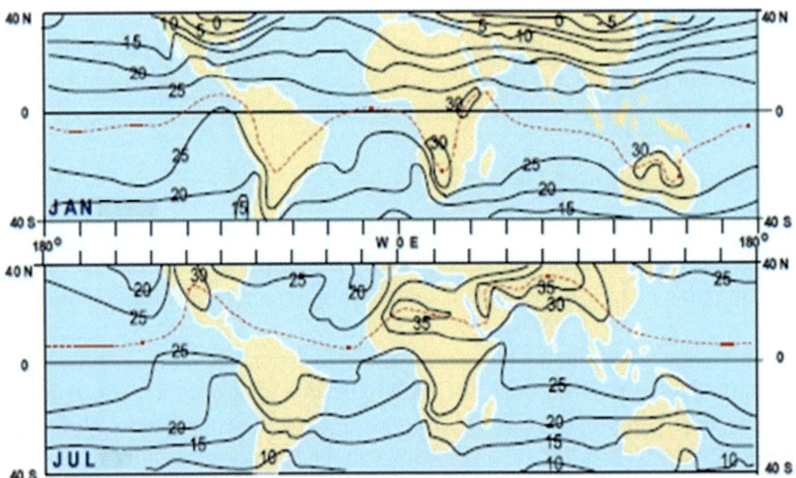

FIG. 4.1 Average surface air temperature distribution in the tropical region in January and July. The *dotted line* indicates the position of the thermal equator (line connecting the points corresponding to the highest average temperatures). *Adapted: Varejão-Silva M.A (2006).*

TABLE 4.2 Lifetime and fall distance of water droplets at different temperatures and humidities.

Environmental conditions	Temperature = 20°C relative humidity = 80%		Temperature = 30°C relative humidity = 50%	
Initial droplet size (μm)	Life time to extinction (s)	Fall distance (m)	Life time to extinction (s)	Fall distance (m)
50	14.0	0.50	4.0	0.15
100	57.0	8.50	16.0	2.4
200	227.0	136.4	65.0	39.0

Adapted: Matthews G.A (2014).

the movement of air masses and winds, as air displacements occur from an area of high pressure (low temperature) to an area of low pressure (high temperature).

The heated air in the low latitude zones near the equator expands, becomes light and rises (ascends), creating an area of low pressure. The colder, denser air in the areas of medium and high latitudes descends, giving rise to an area of high pressure. Since there is a tendency for air masses to equalize these pressures, an atmospheric dynamic is thus established, that is, a general circulation of hot air between the tropics and the poles passing through the areas of medium latitudes (Holton, 2004). The wind must, therefore, be considered as moving air, because it results from the displacement of air masses derived from the effects of differences in atmospheric pressure between two distinct regions, however, being influenced by local effects such as the orography and the roughness of the ground.

Wind is a factor that can directly affect the quality of biopesticide applications. This factor is the cause of the greatest concerns of farmers and technicians, and may even be limiting, causing a stop in the application. When associated with the low relative humidity of the air, it favors the transport of the lighter drops to a different location from the desired target. Wind is a major contributor to spray drift in agricultural applications. The spray drift can be considered as the part of an application of the product that is deflected from the target by the action of the wind.

Spray drift is one of the main problems in chemical pesticides application, as it results in environmental contamination of adjacent areas to the application site. In the application of biopesticides, this is less a problem, as these products have little effect on the environment. However, the spray drift ends up reducing the dose of biopesticides that will effectively reach our target, thus being able to reduce the control efficiency of the product.

According to Bouse et al. (1990), the spray drift increases with the wind, in low humidity and when small droplets make up most of the droplet spectrum. In this way, the wind is a concern when applying biopesticides and the higher the speed, the greater the risk of product loss.

However, applications should not be carried out in the absence of wind. Antuniassi and Cunha (2019) said that when there is no wind, two phenomena that can be harmful to the deposition of droplets can occur, these being thermal inversion and convective currents. Therefore, according to the same authors, agricultural applications must be carried out when we have a minimum wind of 3 km/h.

4.5.3 Timing of biopesticides application

The environmental conditions directly interfere with the biopesticides application, therefore, the choice of the application moment is fundamental to the efficiency of these products. According to Moscardi et al. (2011) the activity of baculovirus-based biopesticides can be affected in the first 24 h, when it is exposed to solar radiation. Botelho et al. (2018) evaluated the efficiency of the Chin-IA (I-A) isolate (ChinSNPV), in two concentrations, in the control of *C. includens*. The results showed that they had an efficient control after the application, but lost, on average, 70% of its efficiency, 4 days after its application. Application time and solar radiation may have influenced these results. According to Hanich et al. (2018) the application technology directly interferes in the biopesticide efficiency based on the baculovirus ChinNPV (Chrysogen®) and the applications carried out at times with milder temperature and relative humidity resulted in better efficiency.

The biopesticide application timing directly influences the efficiency of these products. However, it is important to remember that not all biopesticides have the same response with respect to the application time, and the formulation can also directly interfere in this process. Thus, the timing of applications is an extremely important factor that needs to be taken into account when using biopesticides.

4.6 Final considerations

The application technology can be the difference between success and failure of biopesticides. Therefore, in the recommendation process, the way in which it will be applied is fundamental. Ideal spray volume, appropriate drop size, climatic conditions and the moment of application will be the most important factors.

There are differences between the technology for applying chemical pesticides and biopesticides. Therefore, these aspects need to be considered, so that biopesticides can express their full potential in the field. Thus, in this chapter, the main aspects related to the application technology were addressed, seeking to clarify some that are important for the better use of biopesticides in the field.

References

Agostini, L.T., Otuka, A.K., Silva, E.A., Baggio, M.V., DE Laurentis, V.L., Duarte, R.T., Agostini, T.T., Polanczyk, R.A., 2014. Compatibilidade de produtos à base de *Bacillus thuringiensis* (Berliner, 1911) com glifosato em diferentes dosagens, utilizado em soja (*Glycine max* (L.) Merrill). Ciencia Et Praxis 6, 37–40.

Antuniassi, U.R., 2006. Tecnologia de aplicação de defensivos. Revista Plantio Direto 15, 17–22. Pelotas.

Antuniassi, U.R., Cunha, J.P.A.R., 2019. Boas práticas na tecnologia de aplicação de defensivos agrícolas. In: Tecnologia de Aplicação para culturas anuais. Passo Fundo: Aldeia Norte, p. 373.

Behring, J.L., Lucas, M., Machado, C., Barcellos, I.O., 2004. Adaptação no método do peso da gota para determinação da tensão superficial: um método simplificado para a quantificação da CMC de surfactantes no ensino da química. Quím. Nova 27.

Botelho, A.B.R.Z., Silva, I.F., Ávila, C.J., 2018. Effectiveness reduction of nucleopolyhedrovirus against *Chrysodeixis includens* days after application in soybean plants. Revista de Agricultura Neotropical 5, 94–99. Cassilândia-MS.

Bouse, L.F., Kirk, I.W., Bode, L.E., 1990. Effect of spray mixture on droplet size. Trans. ASAE (Am. Soc. Agric. Eng.) 33, 783–788.

Caye, M., Melo, A.A., Bernardi, O., Hettwer, B.L., Hanich, M.R., Hahn, L., 2019. Efeito da aplicação e do uso de adjuvantes com o inseticida Dipel® no controle de *Chrysodeixis includens* (Lepidoptera: Noctuidae) em soja. In: IX - SINTAG - Simpósio Internacional de Tecnologia de Aplicação, 2019, Campo Grande - MS. Anais do IX - SINTAG - Simpósio Internacional de Tecnologia de Aplicação.

Christidis, N., Stott, P.A., Zwiers, F.W., Shiogama, H., Nozawa, T., 2009. Probalistic estimates of recent changes in temperature: a multi-scale attributions analysis. Clim. Dynam. 34, 1139–1156.

Costa, L.L., Silva, H.J.P.S., Almeida, D.P., Ferreira, M.C., Pontes, N.C., 2017. Droplet spectra and surface tension of spray solutions by biological insecticide and adjuvants. Eng. Agrícola 37, 292–301.

Gandini, E.M.M., Costa, E.S.P., Dos Santos, J.B., Soares, M.A., Barroso, G.M., Corrêa, J.M., Carvalho, A.M., Zanuncio, J.C., 2020. Compatibility of pesticides and/or fertilizers in tank mixtures. J. Clean. Prod. 122152.

Gan-Mor, S., Matthews, G.A., 2003. Recent developments in sprayers for application of biopesticides an overview. Biosyst. Eng. 84, 119–125.

Gonçalves, K.C., 2020. Compatibilidade, efeitos letais e subletais de misturas de bioinseticidas à base de *Bacillus thuringiensis* e inseticidas em *Chrysodeixis includens*. Tese (Doutorado) - Universidade Estadual Paulista Júlio de Mesquita Filho, Faculdade de Ciências Agrárias e Veterinárias, p. 123.

Hanich, M.R., Machado, M.C., Hahn, L.E., Luchese, E.F., Melo, A.A., 2018. Efeito do e Chrysogen no controle de *Chrysodeixis includens* em soja. In: XXVII Congresso Brasileiro/X Congresso Latino Americano de Entomologia, 2018, Gramado. XXVII Congresso Brasileiro/X Congresso. Latino Americano de Entomologia.

Hazen, J.L., 2000. Adjuvants - terminology, classification, and chemistry. Weed Technol 14, 773–784.

Holton, J., 2004. An Introduction to Dynamic Meteorology, Fourth ed. Elsevier Academic Press, San Diego, p. 503p.

Matthews, G.A., 2014. Pesticide Application Methods, Fourth ed. John Wiley & Sons, Ltd, Oxford, p. 405p.

Matuo, T., 1990. Técnicas de aplicação de defensivos agrícolas. FUNEP, Jaboticabal, p. 139.

Moscardi, F., Cunha, F., Moscardi, M.L., 2011. Vírus entomopatogênicos como componentes de programas de manejo integrado de pragas. Ciência & Ambiente (43).

Perini, C.R., Arnemann, J.A., Melo, A.A., Pes, M.P., Valmorbida, I., Beche, M., Guedes, J.V.C., 2016. How to control *Helicoverpa armigera* on soybean in Brazil? What we have learned since its detection. Afr. J. Agric. Res. 11, 1426–1432.

Ramos, H.H., 2002. Regulagem errada. Cultivar Máquinas 14, 28–31. Londrina.

Reis, C.B.L.D., Morandini, L.M.B., Bevilacqua, C.B., Bublitz, F., Ugalde, G., Mazutti, M.A., Jacques, R.J.S., 2018. First report of the production of a potent biosurfactant with α,β-trehalose by *Fusarium fujikuroi* under optimized conditions of submerged fermentation. Braz. J. Microbiol. 49, 185–192.

Santos, J.M.F., 2006. Princípios básicos da aplicação de agrotóxicos. Revista Visão Agrícola. ESALQ. Piracicaba.

Silva, M.T.B., Moscardi, F., 2002. Field efficacy of the nucleopolyhedrovirus of *Anticarsia gemmatalis* (Lepidoptera: Noctuidae); effect of formulations, water pH, volume and time of application, and type of spray nozzle. Neotrop. Entomol. 31, 75–83.

Tijjani, A., Bashir, K.A., Mohammed, I., Muhammad, A., Gambo, A., Habu, M., 2016. Biopesticides for pests control: a review. J. Biopestic. & Agric. 3, 6–13.

Varejão-Silva, M.A., 2006. Meteorologia e Climatologia. INMET, Brasília, p. 463.

Chapter 5

Microbial pesticides: trends, scope and adoption for plant and soil improvement

Pooja Singh[a] **and Purabi Mazumdar**[b]

[a]*School of Science, Monash University Malaysia, Bandar Sunway, Selangor, Malaysia;* [b]*Centre for Research in Biotechnology for Agriculture (CEBAR), Universiti Malaya, Kuala Lumpur, Malaysia*

5.1 Introduction

One of the major challenges' world is facing today is the loss of crop production due to plant diseases and crop pests. Chemical pesticides are considered to be an efficient remedy to tackle the pest problem (Jallow et al., 2017). Therefore, the use of pesticide has become the default selection for many farmers to intensify crop production. The market for pesticides has grown steadily over the year. Global pesticides market is projected to reach USD 90 billion by 2023 from USD 75 billion in 2017 (Global Pesticides Market, 2018). Although chemical pesticides are proven to be very efficient in tackling pest infestation, the usage of pesticide comes along with serious risk to human health and negative effect on the environment (Hajjar, 2012; Verger and Boobis, 2013). Other than the health and environment, intensive use of chemical pesticides was also reported to be associated with the emergence of resistance in target pests (reviewed in Singh et al., 2019), targeting non-specific beneficial natural enemies (predators and parasitoids) and insect and pest (Hassan and Gökçe, 2014). In the last few decades, chemical pesticides have caused irreversible damage to productivity, health and environment, which has initiated to look for safe alternatives. One of the eco-friendly approaches is biopesticides which include microbial pesticides, plant-incorporated protectants (PIPs) and biochemical/herbal pesticides. Among biopesticides, microbial pesticide market constitutes about 90% of total biopesticides (Koul, 2011). Microbial pesticides are naturally occurring or genetically modified microbes including bacteria, fungi, virus, or microsporidia. Microbial pesticides when applied to the crops, they either directly target the pest or stimulate the endogenous plant defense mechanism (Preininger et al., 2018). Microbial pesticides are an important part of the biopesticide industry and have gained high popularity because of rapid and cost-effective production, efficient target specificity, quick decomposition, and ecological safety (Pathak et al., 2017). Although currently, microbial pesticides share a smaller fraction of the global pesticide market, growth is much rapid compared to chemical pesticides. The global market for microbial pesticides was valued at USD 3.48 billion in 2018 and is predicted to reach a value of USD 7.38 billion in the year 2023 with a Compound Annual Growth rate (CAGR) of 17.02% (Market Data Forecast, 2019). The steadily growing market share at the expense of chemical pesticides is mainly because of expanding systematic research to enhanced their effectiveness, mass production, storage in recent years. This chapter is focused on trends and scope of microbial pesticides and their utility in plant protection and improving soil health.

5.2 Types of microbial pesticides

The active ingredient in the microbial pesticide is present in the form of microbes such as bacteria, fungi, viruses or microsporidia which kill pests either by secreting toxins or causing infection (Yadav and Devi, 2017, Table 5.1). Following are the types of microbes used effectively for microbial pesticide production.

Biopesticides. https://doi.org/10.1016/B978-0-12-823355-9.00023-7

TABLE 5.1 Types of microbial pesticides and their target organisms.

Active ingredient	Family	Protein/toxin/enzymes	Type of inoculum	Target organism	References
Bacteria					
Bacillus amyloliquefaciens (MBI 600)	*Bacillaceae*	Lipopolypeptides Surfactin and Iturin A	Whole cell/Spore	Gray mold, *Botrytis cinereal*	EFSA (2016)
Bacillus thuringiensis subsp. *Kurstaki* (ABTS-351)		Cry	Whole cell/Spore	Caterpillers	Czaja et al. (2015), EFSA (2012)
Bacillus thuringiensis subsp. *Aizawai* (ABTS-1857)		Cry	Whole cell/Spore	Caterpillers	EFSA (2013a)
Bacillus thuringiensis subsp. *Galleriae* (SDS-502)		Cry8Da	Whole cell/Spore	Beetles	Bauer and Londoño (2011)
Bacillus thuringiensis subsp. *Tenebrionis* (SA-10)		Cry IIIAa	Whole cell/Spore	Coleoptera foliar feeding beetle larvae, Colorado potato beetle	EFSA (2013b)
Bacillus thuringiensis subsp. *israelensis* (BMP 144)		Cry4Aa, Cry4Ba, Cry11Aa, Cyt1Aa,	Whole cell/Spore	Mosquito, black fly, fungus gnats	Roh et al. (2007), Ben-Dov (2014)
Bacillus firmus (I-1582)		Protease (Serine protease 1), secondary metabolites	Whole cell/Spore	Plant parasitic nematodes	Peleg et al. (2002), Susič et al. (2020)
Bacillus sphaericus (ABTS 1743)		BinA, BinB and Mtx	Whole cell/Spore	Black fly and mosquito larvae	Charles et al. (1996), Koul (2011), Arthurs and Dara (2019)
Burkholderia sp. (A396)	*Burkholderiaceae*	Toxoflavin, fervenulin rhizobitoxin	Whole cell	Beet armyworm, *Spodoptera exigua*	Cordova-Kreylos et al. (2013)
Chromobacterium subtsugae (PRAA4-1T)	*Neisseriaceae*	Heat stable endotoxin	Whole cell	Insect and mites	Martin et al. (2007), Arthurs and Dara (2019)
Pasteuria nishizawae (Pn1)	*Pasteuriaceae*	–	Endospore	Plant parasitic cyst nematodes	EFSA (2018)
Serratia entomophila (A1MO2)	*Yersiniaceae*	Sep protein (toxin complex)	Whole cell, endospore	*Costelytra zealandica* (Coleoptera)	Hurst et al. (2007)
Serratia marcescens		Chitinase, siderophore	Whole cell, endospore	*Rhizoctonia solani* AG3	Khaldi et al. (2015)
Xenorhabdus nematophila (A24)	*Enterobacteriaceae*	Txp40	Whole cell	*Galleria mellonella* and *Helicoverpa armigera*	Brown et al. (2006)
Xenorhabdus nematophila (HB310)		PirAB	Whole cell	*Galleria mellonella*	Yang et al. (2017)

Paenibacillus popilliae	*Paenibacillaceae*	Chitinase	Whole cell and spore	*Japanese beetle (Popillia japonica)*	Grady et al. (2016)
Brevibacillus laterosporus (UNISS 18)		Chitinase (chiA, and chiD), collagenase-like protease (prtC), bacillolysin (Bl18), and insecticidal toxin (mtx) genes	Whole cell and spore	Diptera *(Musca domestica and Aedes aegypti), Coleoptera, Lepidoptera, Diptera and against nematodes and mollusks.*	Ruiu et al. (2007), Ruiu et al. (2013), Marche et al. (2018)
Streptomyces sp.	*Streptomycetaceae*	Macrocyclic lactone, nanchanmycin, tartrolon, staurosporine, Quinomycin A, 2-Hydroxy-3,5,6 trimethylolactan-4-one, flavensomycin, antimycin A, piericidins, macrotetralides, and prasinons, neomycin, cypemycin, grisemycin, bottromycins and chloramphenicol	Whole cell	*Helicoverpa armigera, Drosophila melanogaster.*	Vijayabharathi et al. (2014), Ruiu et al. (2013), Xiong et al. (2004), Obeidat et al. (2017)
Saccharopolyspora spinosa	*Pseudonocardiaceae*	Spinosads	Fermentation product	Lepidoptera, Hymenoptera (Leafminer)	Dara (2017)
Photorhabdus luminescens	*Morganellaceae*	Toxin complexes (Tc), makes caterpillars floppy (Mcf), Photorhabdus virulence cassettes (PVCs) and Photorhabdus insect-related protein (PirAB)	Whole cell	*Spodoptera exigua, Aedes aegypti* and *Aedes albopictus*	Waterfield et al. (2005), Daborn et al. (2002), Yang et al. (2006), Ffrench-Constant et al. (2007), Chaston et al. (2011)
Photorhabdus temperata					
Pseudomonas entomophila	*Pseudomonadaceae*	Insecticidal toxin (PSEEN2485, PSEEN2697, and PSEEN2788, serine protease (PSEEN3027, PSEEN3028, PSEEN4433) and an alkaline protease (PSEEN1550).2, Monalysin (ß-Pore-Forming Toxin)	Whole cell	Nematodes (*Caenorhabditis elegans*) and pathogenic bacteria such as *Mycobacterium marinum* and the fungus *Candida albicans* and Drosophila	Vodovar et al. (2006), Opota et al. (2011), Dieppois et al. (2015)
Pseudomonas fluorescens		phenazines, siderophores, pyoluteorin (antibiotic), and 2,4 diacetylphloroglucinol (antifungal), FIT toxin	–	*Manduca sexta* and *Drosophila melanogaster.*	Gleeson et al. (2010) Rangel et al. (2016)
Fungus					
Beauveria bassiana	*Cordycipitaceae*	Beauvericin, bassianin, bassianolide, beauverolides, beauveriolides, tenellin, oosporein Bassiacridin	Conidia	Caterpillars, mosquito, thrips, aphids, whiteflies, plant bugs, mites and other arthropods	Strasser et al. (2000), Vey et al. (2001)
		Oxalic acid			Roberts (1981)
		Bassiacridin			Quesada-Moraga and Vey (2004)
Beauveria brongniartii		2-piperridinone and 2-coumaranone	Conidia	Pine Caterpillar (*Dendrolimus tabulaeformis*)	Fan et al. (2013)
Lecanicillium lecanii		Ctininase	Conidia	Aphids, leafminers, mealybugs, scale insects, thrips, whiteflies	Askary et al. (1997), Ruiu et al. (2018)
Verticillium lecanii (Vertalec and strain 198499)		Endochitinase (*Vlchit1*)	Mycelia/conidia	Sucking pests such as aphids, Jassids, whitefly, leaf hoppers and mealy bugs	Askary et al. (1998), Ruiu et al. (2018), Zhu et al. (2008)

Continued

TABLE 5.1 Types of microbial pesticides and their target organisms.—cont'd

Active ingredient	Family	Protein/toxin/enzymes	Type of inoculum	Target organism	References
Hirsutella thompsonii (FI JAB-04)	*Ophiocordycipitaceae*	Hirsutella A	Mycelia/conidia	Spider mites	Liu et al. (1996)
Purpureocillium lilacinum (6029)		Nematocidal toxin	Mycelia/conidia	Root-knot nematodes (*Meloidogyne incognita*).	Sharma et al. (2016)
Isaria fumosorosea	*Clavicipitaceae*	Beauvericin, pyridine-2,6-dicarboxylic acid, farinosones A and B	Mycelia/conidia	Whitefly, aphids, thrips, leafminers, plant bugs, mites	Zimmermann (2007, 2008), Arthurs and Dara (2019)
Metarhizium anisopliae (1080)		Proteases subtilisin (Pr1), trypsin (Pr2), cysteine protease (Pr4) and metaloprotease	Conidia	Thrips, whiteflies, mites, weevils and ticks	St Leger et al. (1996), Schrank and Vainstein (2010)
Metarhizium acridum		Trypsins, subtilisin proteases and endochitinase (Hog 1 kinase)	Mycelia/conidia	Locusts and grasshoppers	Aw and Hue (2017)
Myrothecium verrucaria (KACC 40321)	*Stachybotryaceae*	Verrucain A and roridin A	Dead fungal hyphae	Nematodes	Nguyen et al. (2018)
Paecilomyces lilacinus (251)	*Ophiocordycipitaceae*	Paecilotoxins	Mycelia/blasto-spores/conidia	Nematodes	Khan et al. (2003)
Trichoderma harzianum	*Hypocreaceae*	Chitinases, glucanases, and proteases	Mycelia/conidia	Root-knot nematodes, plant parasitic fungus	Elad (2000), Sharon et al. (2001)
Verticillium lecanii (Vertalec and strain 198499)	*Cordycipitaceae*	Endochitinase (*Vlchit1*)	Mycelia/conidia	Sucking pests such as aphids, Jassids, whitefly, leaf hoppers and mealy bugs	Askary et al. (1998), Ruiu et al. (2018), Zhu et al. (2008)
Virus					
Cydia pomonella granulosis virus (*CpGVs*)	*Baculoviridae*	–	Occlusion body	Codling moth	Luque et al. (2001), Koul (2011)
Plodia interpunctella granulosis virus (*PiGVs*)		–	Occlusion body	Indian meal moth	Burden et al. (2002)
Helicoverpa zea nucleopolyhedrovirus		–	Occlusion body	Corn earworm, cotton ballworm	Washburn et al. (2001), Ruiu et al. (2018)
Spodoptera litura nucleopolyhedrovirus		–	Occlusion body	Tobacco cutworm or cotton leafworm,	Nathan and Kalaivani (2005)
Lymantria dispar multiple nucleopolyhedrovirus (*LdMNPV*)		–	Occlusion body	Gypsy moth	Slavicek and Popham (2005)
Autographa californica nucleopolyhedrovirus		–	Occlusion body	Cabbage looper	Pijlman (2001)
Orgyia pseudotsugata nucleopolyhedrovirus		–	Occlusion body	Douglas fir tussock moth	Thorne (2008)

5.2.1 Bacteria

Entomopathogenic bacteria are mainly crystalliferous-spore forming (e.g. *Bacillus thuringiensis*), obligate pathogens (e.g. *Bacillus popilliae*), potential pathogen (e.g. *Serratia marcesens*) and facultative pathogen (e.g. *Pseudomonas aeruginosa*) (reviewed in Koul, 2011). Although more than 100 entomopathogenic bacteria have been identified as insect pathogen, only a few of them are used in the commercial pest management system (reviewed in Chattopadhyay et al., 2017). Major bacterial entomopathogens species used for commercial pest management system belong to the phyla: Firmicutes (*Bacillaceae, Paenibacillaceae* and *Pasteuriaceae*), Actinobacteria (*Pseudonocardiaceae* and *Actinomycetaceae*) and Proteobacteria (*Enterobacteriaceae, Pseudomonadaceae, Yersiniaceae, Burkholderiacea* and *Neisseriaceae*) (Table 5.1). Below we have discussed about the some most popular bacterial entomopathogens:

5.2.1.1 Bacillaceae

Bacillus thuringiensis (Bt) and *Bacillus sphaericus* (Glare et al., 2017) are the most commonly used bacteria in *Bacillaceae* family for microbial pesticide production. Among them, Bt is the most extensively used microbial pesticide worldwide (Peralta and Palma, 2017, Table 5.1). Bt is Gram negative spore-forming, motile, entomopathogenic bacteria and produces insecticidal proteins as parasporal crystals during the sporulation phase (reviewed in Ruiu, 2018; Raymond and Federici, 2017). The crystals are cry (crystal) and cyt (cytolytic) proteins also known as δ-endotoxins (Bravo et al., 2007). These crystal toxins are largely regulated by the insecticidal activity of Bt. Some of the beneficial properties of these toxins are specificity to the target organism, non-toxic nature to vertebrates, plants and human, and biodegradable (Argôlo-Filho and Loguercio, 2014). Bt generally enters the host organisms through ingestion, the Bt toxin gets dissolve in the high pH in the insect gut and becomes active. The toxins then attack the gut cells of the insect, punching holes in the lining (Glare et al., 2017) which alters the epithelial membrane permeability by disrupting the intestinal barrier and inducing septicemia and finally leading to death (Bravo et al., 2007; Soberón et al., 2016, Fig. 5.1). Bt was found to be effective against a wide range of organism as shown in Table 5.1.

Unlike, Bt, *B. sphaericus* is a Gram-positive bacterium, which produces round spores in a bulged club-like terminal or subterminal sporangium (Park et al., 2010). *B. sphaericus* produces intracellular protein toxin and a parasporal crystalline toxin BinA and BinB and mosquitocidal Mtx at the time of sporulation (reviewed in Koul, 2011). Mode of action *B. sphaericus* is similar like Bt, and act as an active parasite against black flies and mosquitoes (Filha et al., 2014).

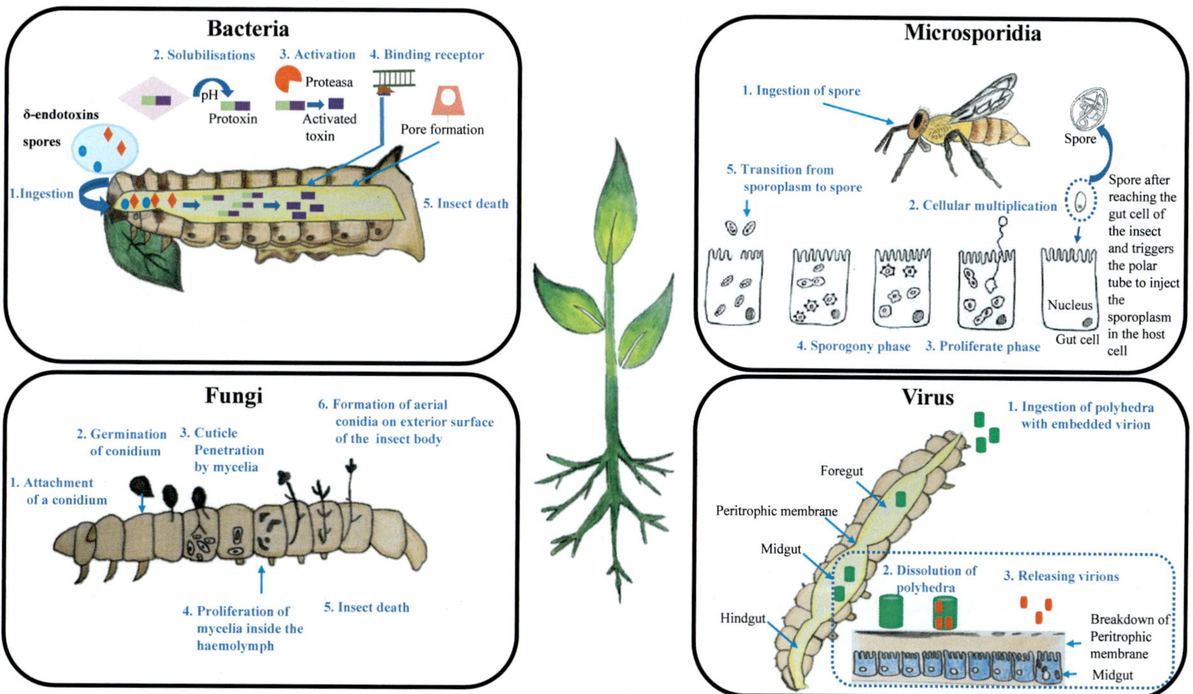

FIG. 5.1 Schematic representation of mode of action of entomopathogenic microbes (a) Bacteria (b) Fungus (c) Microsporidia (d) Virus. *Modified from Fernández-Chapa et al. (2019), Senthil-Nathan (2015), Araújo and Hughes (2016).*

5.2.1.2 Paenibacillaceae

Paenibacillus species have been known as a pathogen of lepidopteran insects (Sharma et al., 2013). Among the *Paenibacillaceae*, the most popularly used species is *Paenibacillus popilliae. P. popilliae* was found to infect Japanese beetle and was registered as the first microbial control agent in US (Klein, 1988). *P. popilliae* is a gram-positive, rod-shaped non-motile bacterium. It produces chitinase enzyme which hydrolyses chitin, a structural polysaccharide of insect exoskeleton and gut lining, leading to insect death (Grady et al., 2016). Another species in *Paenibacillaceae*, is *Brevibacillus laterosporus* which also possess a wider spectrum of pesticidal activity. This species has a unique morphology characterized with typical swollen sporangium structure surrounded by a firmly attached canoe-shaped parasporal body (Marche et al., 2018). The bacteria possess biocontrol potential against insects, nematodes and mollusks, phytopathogenic bacteria and fungi (Ruiu, 2013). Broad-spectrum of activity of these bacteria were found to be associated with a wide variety of molecules, including surface layer proteins of cell envelope and antibiotics (Gramicidin S and D laterosporamine, loloatins, bogorols and the lipopeptide tauramamide), it produces (Ruiu, 2013; Marche et al., 2018).

5.2.1.3 Streptomycetaceae

Microbes belonging to family *Streptomycetaceae* is recognized as the source of several bioactive metabolites that are useful in agriculture (reviewed in Anandan et al., 2016). Among members of Streptomycetacea, *Streptomyces* are the most widely studied genus and several of them were reported to possess insecticidal, pesticidal, larvicidal, acaricidal and nematocidal activity (reviewed in Vijayabharathi et al., 2014). *Streptomyces* are Gram-positive, non-motile, filamentous bacteria that produce branched vegetative hyphae and the long chain of spore (often >50 spores) from specialized aerial hyphae (Chater, 1984). Streptomyces species produce various insecticidal compounds such as macrocyclic lactone, nanchanmycin, tartrolon, staurosporine, Quinomycin A, 2-Hydroxy-3,5,6 trimethylolactan 4-one, flavensomycin, antimycin A, piericidins, macrotetralides, and prasinons which act on the peripheral nervous system inducing paralysis in the target insect (reviewed in Vijayabharathi et al., 2014; Ruiu, 2013).

5.2.1.4 Pseudonocardiaceae

Microbes belonging to the family *Pseudonocardiaceae* are known to possess insecticidal activity (reviewed in Dhakal et al., 2017). Among them, *Saccharopolyspora spinosa* is the most popular one. *Saccharopolyspora spinosa* is a gram-positive, non-motile, filamentous bacteria. *S. spinosa* produces metabolites named "Spinosad" which is a source of broad-spectrum insecticides. Spinosads derived from *Saccharopolyspora spinosa* are commercially available active ingredients, generally used against various vegetable pests (Dara, 2017). Contact and ingestion of spinosad are two major ways of exposure to the pesticide (reviewed in Dhakal et al., 2017). Spinosad activates nicotinic acetylcholine receptor and affects γ-aminobutyric acid (GABA) receptor in the nervous system leading to the death of the target insect (Sporleder and Lacey, 2013). Spinosad target insects like caterpillars, leafminers, foliage-feeding beetles and thrips (Sporleder and Lacey, 2013).

5.2.1.5 Morganellaceae *and* enterobacteriaceae

Most popular microbes belonging to the family *Morganellaceae* includes entomopathogens such as *Photorhabdus luminescens, P. temperata* and family *Enterobacteriaceae* includes *Xenorhabdus nematophila* (reviewed in Ruiu, 2018). The uniqueness of these species is that all the species of this genus live in symbiosis with soil entomopathogenic nematodes (*Steinernema* sp.) (Chaston et al., 2011). These bacteria are motile gram-negative and non-spore forming (Easom and Clarke, 2008; Givaudan and Lanois, 2000). When nematode invades the host, *P. luminescens* are released from the gut of the nematode directly into the hemocoel (blood circulatory system) of the insect which kills the insects with their toxins (reviewed in Ruiu, 2018). Once the bacteria released in the hemocoel of the insect, it rapidly replicates and produces a large array of virulence factors, including toxins, proteases lipases, hemolysins, and various antimicrobial substances which kill the insects while dead insect serves as a source of food for the bacteria and for the nematode (Lang et al., 2011). When the insect cadaver is depleted, the reassociation of the bacteria and nematodes occurs and, finally, leave the insect cadaver and invade new host (Brown et al., 2006). Various toxins have been identified and characterized for *Photorhabdus* and *Xenorhabdus*. Toxin complexes (Tcs) (Ffrench-Constant and Waterfield, 2006), makes caterpillars floppy (Mcf) (Daborn et al., 2002), Photorhabdus virulence cassettes (PVC) (Yang et al., 2006) and Photorhabdus insect-related protein (PirAB) (Ffrench-Constanta et al., 2007) are the main toxin protein identified for *Photorhabdus*, whereas, for *Xenorhabdus*, *Xenorhabdus* protein toxins (Xpt) (Morgan et al., 2001), *Xenorhabdus nematophila* GroEL (XnGroEL) (Shi et al., 2012), toxin protein 40 (Txp40) (Brown et al., 2004) and Xenorhabdus alpha-xenorhabdolysin A,B (XaxAB) toxins (Vigneux et al., 2007).

5.2.1.6 *Yersiniaceae*

Microbes belonging to the family *Yersiniaceae* includes several entomopathogens such as *Serratia* are recognized as a versatile biocontrol agent against phytopathogenic bacteria and fungus. Serratia are gram-negative, rod-shaped, motile bacteria, which produces notable secondary metabolites including the β-lactam antibiotic carbapenem or the antifungal compound oocydin (Soenens and Imperial, 2019). Among *Serratia* species, *S. marcescens* is most widely studied bacterium that has the ability to induce systemic resistance to phytopathogen (Table 5.1).

5.2.1.7 *Pseudomonadaceae*

Microbes belonging to family *Pseudomonadaceae* is also recognized as an important source of the insecticidal toxin. Bacteria from the family *Pseudomonadaceae* is gram-negative, motile and rods shaped (Jurat-Fuentes and Jackson, 2012). *Pseudomonas entomophila* (Dieppois et al., 2015) and *Pseudomonas fluorescens* (Gull and Hafeez, 2012) are the most common bacteria of this family which is widely used for microbial pesticide production. *Pseudomonas entomophila*, is pathogenic to nematodes (*Caenorhabditis elegans*) and pathogenic bacteria such as *Mycobacterium marinum* and the fungus *Candida albicans* (reviewed in Dieppois et al., 2015). Upon ingestion, *P. entomophila* activates a systemic immune response in target insects. *P. entomophila* produces an array of secondary metabolites and several of them are associated with their entomopathogenicity such as insecticidal toxins, putative hemolysins, proteases, hydrogen cyanide (Vodovar et al., 2006). Similar to *P. entomophila, P. fluorescens* strains are known to produce a wide range of secondary metabolites including phenazines, siderophores, pyoluteorin (antibiotic), and 2,4 diacetylphloroglucinol (antifungal) (Gleeson et al., 2010). Among these, antifungal metabolite diacetylphloroglucinol is of particular interest because its immense potential to inhibit several phytopathogenic fungi (Gong et al., 2016, Table 5.1).

5.2.2 Fungi

Fungal biopesticides (Table 5.1) are possibly one of the highly adopted biological control methods because the mode of action is mainly depended on the contact and not ingestion (Kumar et al., 2019). In addition, it can be mass-produced easily and have a broad host range including insects, other fungi, bacteria, nematodes and weeds (Becher et al., 2018). Entomopathogenic fungi are obligate or facultative, commensal or symbionts of the host insect (reviewed in Koul, 2011). Major fungal entomopathogens species mainly belong to the phyla Ascomycota (*Cordycipitaceae, Clavicipitaceae, Ophiocordycipitaceae, Ophiocordycipitaceae, Stachybotryacea*e and *Hypocreaceae*) (Vega and Kaya, 2012). Entomopathogenic fungi infect their hosts often by their surface layer (Fig. 5.1). Mechanical force and enzymatic processes along with metabolic acids mediate the initial interaction. The attachment of infectious spore to the surface of hosts and germination is supported by free amino acids and peptides on the cuticle of the host insect (Khachatourians, 1996). Upon establishment on the cuticle layer of insects, the fungus produces germ tubes or appressoria and release different enzymes including protease, chitinase and lipases to support the penetration of the host cuticle (reviewed in Zimmermann, 2007). The whole process takes around 24–48 h (de la Cruz Quiroz et al., 2019). After penetration within 5–7 days, mycelium develops and colonize the joints and integuments of host and starts producing spores (de la Cruz Quiroz et al., 2019; Kumar et al., 2019, Fig. 5.1). Following the vegetative growth, fungus produces various metabolites and toxins to the host which accelerates the fast growth of the fungus and ultimately leads to the death of the insect (Nisa et al., 2015; Kumar et al., 2014). Meanwhile, fungus keeps producing new spores outside the host body to ease the infection outside the host during the process (reviewed in Ruiu, 2018). Below we have discussed some of the most popular fungal entomopathogens:

5.2.2.1 *Cordycipitaceae*

Members of the *Cordycipitaceae* are entomopathogens, parasites of slime molds or fungi as well as endophytic colonist of grasses and fruit crops (reviewed in McKinnon et al., 2017; Vega, 2008). *Beauveria bassiana* and *B. brongniartii* belong to family *Cordycipitaceae* are the most popular one (Sung et al., 2007). Spores of *B. bassiana* are mainly used in microbial pest formulation to control insect pests like caterpillars (Liu et al., 2013) and mosquito (Lynch et al., 2012, Table 5.1). *B. bassiana* causes insect disease known as white muscardine (Miranpuri and Khachatourians, 1995). Unlike bacterial microbial pesticides, it does not need to be ingested by the host; exposure with a host is enough. Once the host insect is infected, the fungus spread rapidly inside of the insect hemolymph utilizing the nutrients present in the host's body and producing toxins which ultimately kills the insect (reviewed in Blumberg et al., 2016). The organism produces a number of cyclodepsipeptides such as beauvericin and bassianolide which constitutes the active ingredient of toxin complex, which possess antimicrobial, cytotoxic, and apoptotic activity (Klaric and Pepeljnjak, 2005). After the death of the host insect, the *B. bassiana* cover up the carcass with a layer of white mold that produces more infective spores to start new cycles of infection. Some of the listed hosts of *B. bassiana* are Heteroptera, Homoptera, Coleoptera (Rashki et al., 2009). *B. brongniartii* also possess similar toxins and mode of actions like *B. bassiana* and used in controlling caterpillars (Fan et al., 2013).

Lecamicillium lecanii also known as *Verticillium lecanii* Viegas is another well-known entomopathogen of *Cordycipitaceae* family used for controlling a wide host range of insects such as whiteflies, mealy bugs, aphids, mites, jassids (Wang et al., 2007). Mode of infection is common as in other entomopathogenic fungi i.e. fungal conidial spore attachment on the insect surface and penetration of the integuments by hyphae and rapid spread in insect hemocoel followed by secretion of toxin and death of the insects (Sugimoto et al., 2003; Kim et al., 2008).

5.2.2.2 Clavicipitaceae

Metarhizium anisopliae and *Metarhizium acridum* are the entomopathogens of the *Clavicipitaceae* family which were well explored in biopesticide formulation. Mode of action is common as with other entomopathogenic fungi where infectious spores (conidium/blastospore) enters through the cuticle, spiracle or the mouth or the anal opening and passes to insect hemocoel, release toxin causing insect death. *Metarhizium* produces a group of enzymes responsible for cuticle degradation known as subtilisin-like serine protease, metalloproteases, trypsin, chymotrypsin, dipeptidyl peptidase, aminopeptidase and chitinase (reviewed in Zimmermann, 2007). The species has been effectively used in controlling whiteflies, mites, weevils and thrips (Arthurs and Dara, 2019). *Isaria fumosorosea* (formerly known as *Paecilomyces lilacinus*) another member of Clavicipitaceae also commonly used for microbial pesticide preparation. *I. fumosorosea* are well known for their activity against aphids, thrips, leafminers, plant bugs (Zimmermann, 2008). Products of this pathogen are available as blastospore/conidial formulations (Avery et al., 2013). It also possesses similar mode of action like other entomopathogen fungi (Fiedler and Sosnowska, 2007).

5.2.2.3 Ophiocordycipitaceae

Purpureocillium lilacinum belongs to the family *Ophiocordycipitaceae*, is a nematode pathogenic fungus and has been utilized to parasitize juveniles and females' nematodes (Atkins et al., 2003). Mode of action described for the fungus is that it forms an appressorium and release chitinase enzyme which degrades chitosan and penetrate the eggshell of the host nematode *Meloidogyne* spp. (Palma-Guerrero et al., 2010; Swarnakumari and Kalaiarasan, 2017). The products of this fungus are bionematicides which are effective against a wide range of nematodes including root-knot, burrowing, cyst, root lesion, reniform and false root (reviewed in Arthurs and Dara, 2019). The products are commercialized as fermentation solids or spore and are widely used for cereals and vegetable crops (Baidoo et al., 2017).

5.2.3 Microsporidia

Microsporidia are a diverse group of spore-forming unicellular obligate parasites to vertebrates and invertebrate (Han et al., 2020). Microsporidia are broadly distributed in nature and comprise of more than 200 genera and 1400 species (Cali et al., 2017). They were previously considered as protozoans, but are now grouped under fungi (Hibbett et al., 2007). Approximately 1000 species in microsporidia have been known to attack invertebrates with many insect species such as Heliothine moth and grasshoppers (Solter and Becnel, 2000). Microsporidia are reported to be an obligate parasite and can only reproduce in living cells. Among the microsporidia, member of *Nosematidae* are highly used in microbial pesticide formulations (Table 5.1).

5.2.3.1 Nosematidae

The most successful microsporidia entomopathogens are *Nosema* spp. and *Vairimorpha necatrix* belonging to the family *Nosematidae* are effective against Lepidoptera, beetles and locusts (reviewed in Bjørnson and Oi, 2014). *Nosema locustae* was the first microbial agent to be developed as a biopesticide for locust and grasshopper control (reviewed in Lange and Sokolova, 2017). Microsporidia infection initiates when spore (3−5 μm) are ingested by host insect, which replicates inside gut of insects and releases the sporoplasm which invade the target cells, causing massive infection leading to damage of tissues and organs (Burges and Jones, 1998, Fig. 5.1).

5.2.4 Virus

Entomopathogenic viruses which infect and kill insects are used as microbial control agents (Table 5.1). Seven families of viruses, including *Baculoviridae, Reoviridae, Iridoviridae, Poxviridae, Parvoviridae, Picornaviridae* and *Rhabdoviridae* are known to cause diseases in insects and among them *Baculoviridae* and *Reoviridae* are highly used for microbial pesticides formulation because of their high virulence (Kalawate, 2014).

5.2.4.1 Baculoviridae

Baculoviruses are one of the major groups of viruses comprising 76 species, which infects insect order of arthropods, lepidoptera, hymenoptera, diptera, and decapoda (Harrison et al., 2018). Based on the phenotype of the occlusion bodies Baculoviruses are divided into two major groups: the nucleopolyhedroviruses (NPVs), in which these bodies are polyhedron-shaped occlusion body carrying over 100 virions and the granuloviruses (GVs), in which these bodies are granular-shaped occlusion body (granuloviruses, GVs) carrying only one virion (Jehle et al., 2006). Mode of infection for baculovirus is both of occlusion body are specific to the larval stage of the host insect. Occlusion body on the plants is ingested by the insect and solubilized in the midgut, directly interacting with the membrane of the microvillar epithelial cells through envelopes proteins (Fig. 5.1). Once the NPVs reach inside the nuclease it forms budded virus to ensure the infection of nearby tissues and progressively liquefies the dead insect body (Szewczyk et al., 2006). Finally, NPVs are spread by excrements of infected insects (Vasconcelos, 1996). It is also reported that viruses alter the behavior of the host insect for their own benefit by modifying the gene expression mechanism (Katsuma et al., 2012). One example of such behavior modification was seen in baculovirus-induced *Wipfelkrankheit* disease that causes caterpillars to migrate to upper foliage of the plant where they die. This climbing behavior of the baculovirus-infected caterpillars was controlled by two genes *ptp* and *egt* which induced enhanced locomotory activity (ELA) in the insect (Katsuma et al., 2012). For commercialization, baculovirus produced in vivo from viral occlusion bodies has been used for several years (Szewczyk et al., 2006).

5.2.4.2 Reoviridae

Reoviridae is a family of segmented double-stranded RNA virus which infects vertebrates, invertebrates, plants, protists and fungi (King et al., 2011). Among them, Cypovirus genus infects a variety of insect species including lepidoptera, diptera and hymenoptera (reviewed in Rincón-Castro and Ibarra, 2011). Similar to baculoviruses, Cypovirus (CPVs) also produce virions encapsulated in polyhedron-shaped occlusion body during infection (Mori and Metcalf, 2010). The most studied and economically important CPV is *Bombyx mori* cytoplasmic polyhedrosis virus (BmCPV) as it causes mortality in silkworm (Jiang and Xia, 2014). CPV infection is initiated when larvae consume the occlusions. The highly alkaline condition of the larval gut facilitates the release of the virions which later penetrates the midgut cells (reviewed in Harrison and Hoover, 2012). Transcription of the double-stranded RNA genomic segments takes place within the capsid and transcripts are extruded from the capsid into the cytoplasm (Yu et al., 2011). Progeny genomic segments and capsids are assembled into virions in the cytoplasm of infected cells, likely by mechanisms common with other reoviruses (Lemay, 2018). Infections by CPVs often do not lead to cell lysis, its effect ranges from the death of the infected host to larval developmental retardation, reduction in adult fecundity, and the transmission of the disease to next generation (reviewed in Harrison and Hoover, 2012).

5.2.5 Genetically modified microbes

Several of the natural entomopathogenic microorganisms are not suitable for commercial microbial pesticide formulation. This is mainly because of inefficient colonization (Zhang et al., 2020), competition with natural rhizoflora (Downing and Thomson, 2000), repressor gene associated with pathogenicity (Barahona et al., 2011), the slow killing of the host (Gongora, 2004). In such cases, genetic engineering offers an efficient option to improve performance by increasing the rate of reproduction, speed of transmission and increasing the quantity of toxin production in those natural biocontrol agents (Spadaro and Gullino, 2005; reviewed in Weller and Thomashow, 2015; Pathak et al., 2017).

Efficient colonization is an important attribute of bacterial biopesticide. Introduction of a four-gene operon which encodes antibiotic pyrronitrin from *Pseudomonas protegens Pf-5* to *P. synxantha*, an efficient root colonizer showed elevated suppression of Rhizoctonia root rot of wheat (Zhang et al., 2020). Other than that, the transfer of toxin genes from entomopathogenic bacteria to plant growth-promoting bacteria were showed to improved plant protection with the additional benefit of providing nutrition to the host plants. Ectopic expression of *cry1Ac* gene of *Bacillus thuringiensis* to *Bacillus polymyxa*, nitrogen-fixing bacteria provided protection against yellow stem borer of rice while enhancing the plant growth (Sudha et al., 1999). Also, manipulation of the way toxin produced in bacteria also showed enhancement in the self-life of toxin. The expression of Bt toxin (codon optimized cryIAc) in *Pseudomonas fluorescens* leads to formation encapsulated toxin unlike original crystal protein in original Bt which has short field life (Peng et al., 2003). The encapsulated toxin retains its effectiveness two to three times higher than other Bt formulations.

In addition, genetic engineering was also used to exploits endophyte's lifestyle to overcome the competition between the introduced microorganism and natural microorganism of the rhizosphere. A study showed transfer of *Serratia marcescens* a chitinase gene (*chiA*) gene into endophyte *Pseudomonas fluorescens* elevated the inhibition against *Rhizoctonia solani* infection in bean seedling (Downing and Thomson, 2000). Chitinases enzymes degrade chitin which is an important constituent of the fungal cell wall (Langner and Gohre, 2016).

In addition to that genome shuffling also proved to be an efficient tool in enhancing the antagonistic ability of entomopathogenic bacteria. Genome shuffling is a process of recursive protoplast fusion of a phenotypically selected population which can lead to improved strains for a given phenotype (reviewed in Gong et al., 2009). Two rounds of genome shuffling of *Streptomyces melanosporofaciens* EF76, resulted increased antagonistic activity against potato pathogen *Streptomyces scabies* and *Phytophthora infestans* (Clermont et al., 2011). Similarly, two rounds of genome shuffling of *Bacillus subtilis* strain BS14 enhanced antagonistic activity against *Fusarium oxysporum* f. sp. *melonis* (Chen and Chen, 2009).

Another approach was the mutation of genes which represses pathogenicity associated traits. Motility is an important trait for rhizosphere colonization by pseudomonads and motility is a polygenic trait which is repressed by at least three independent pathways including the Gac posttranscriptional system, the Wsp chemotaxis-like pathway, and the SadB pathway (Navazo et al., 2009). KinB is a signal transduction protein that participates in swimming motility repression through the Gac pathway (Narayanasamy, 2013). A triple mutation in the genes *sadB*, *wspR* and *kinB* of *Pseudomonas fluorescens* F113 resulted in hypermotility, improved root colonization and enhanced biocontrol activity against *Phytophthora cactorum* (on strawberry) and *Fusarium oxysporum* f. sp. *Radicis-lycopersici* (on tomato) (Barahona et al., 2011). Another mutation study on retS gene, a sensor kinase, negatively regulator of antibiotic production, showed an increase in the of 2,4-diace-tylphloroglucinol (antibiotic) production, which contributed to the enhanced antifungal activity (Jing et al., 2020). Also, chemical mutagenesis was observed to enhance biocontrol activity in entomopathogenic bacteria. Nitrosoguanidine mutagenesis in *Pseudomonas aurantiaca* strain B-162 resulted in overproduction of phenazine antibiotics (Feklistova and Maksimova, 2008). An advantage of chemical mutagenesis is that it does not fall under the same regulations as of genetically modified organism and can bypass the regulatory restrictions in several countries (O'Brien, 2017). However, it comes with the limitation of lack of consistency, as mutations of the target genes may also result in the mutation of other genes resulting in undesirable consequences (O'Brien, 2017). The recent emerging technique CRISPR/Cas genome editing is a better alternative than chemical mutagenesis. CRISPR/Cas mediated mutations are site-directed and accurate (Barrangou and van Pijkeren, 2016) and can bypass the regulatory restrictions as no genomic material is integrated (O'Brien, 2017). Successful example for genome editing of bacteria is *Bacillus licheniformis* using CRISPR-Cas9 nickase. As genetic manipulation of bacteria is difficult, this genome editing tool offers an efficient method for single-gene deletion, multiple-gene disruption, large DNA fragment deletion and single-gene integration in *Bacillus licheniformis via* Cas9 nickase (Li et al., 2018).

Like entomopathogenic bacteria, traits of natural entomopathogenic fungus were also modified using genetic engineering methods such as overexpression or ectopic expression of pathogenesis-related gene or expression of insect protein gene in the pathogen. Overexpression or ectopic expression of pathogenicity-related genes in entomopathogenic fungus showed to improve virulence. Constitutive overexpression subtilisin-like protease (Pr1A) gene in *Metarhizium anisopliae* increased the virulence to insect (*Manduca sexta*) compared to the parent wild-type strain (St. Leger et al., 1996). Similarly, expression of Pr*1A* gene of *M. anisopliae* to *Beauveria bassiana* showed enhancement in the virulence capacity (Gongora, 2004). In addition to the native gene, chimeric gene such as hybrid *chitinase* expression in *B. bassiana* (Chitinase from *B. bassiana* fused with chitin-binding domain from the silkworm *Bombyx mori*) improved virulence of recombinant strain by 23% (Fang et al., 2005). Ectopic expression of a phytotoxic gene from *Fusarium* spp. *NEP1* into *Colletotrichum coccodes* increased the virulence of fungus to nine-folds against weed velvetleaf (Amsellem et al., 2002).

Other than that, manipulating the way entomopathogenic fungi exploit their hosts for nutrition is also showed promise to improve virulence. Trehalose in the insect hemolymph is the major nutrient source for entomopathogenic fungi (Thompson and Borchardt, 2003). Overexpression of *the acid trehalase ATM1* an endogenous hydrolase of trehalose in *M. acridum* showed improvement in virulence mediated by accelerated growth of the fungus in host hemolymph (Peng et al., 2015). In addition, genetic engineering is also used to expand the host range in addition to the virulence. Expression of an esterase gene (*Mest1*) from the *Metarhizium robertsii* to *M. acridum* enabled the latter strain to expands its host range to caterpillars (Wang et al., 2011). In addition to the pathogenesis-related gene, expression of insect gene also found to improve the virulence. Expression of an inhibitory regulator of a key immune-related signal pathway showed improvement in virulence of *B. bassiana's* against two insect species; *Galleria mellonella* (wax moth) and adult *Myzus persicae* (green peach aphid) (Yang et al., 2014).

Similarly, like entomopathogenic bacteria chemical mutagenesis was also observed to enhance biocontrol activity in entomopathogenic fungus. UV mediated mutagenesis of *Trichoderma harzianum* yielded tolerance to fusaric acid which enhanced biocontrol activity against *Fusarium oxysporum* f. Sp *lycopersici* (Marzano et al., 2013). Other than UV mediated mutagenesis, methods recent establishment of CRISPR-Cas9 mediated mutagenesis system in entomopathogenic fungi such as *Metarhizium robertsii* (Zhao et al., 2014) and *Beauveria bassiana* (Chen et al., 2017) showed promise for developing precise mutagenesis which can be exploited for rapid improvement in biocontrol activity of entomopathogenic fungi.

Several entomopathogenic viruses take a longer time to kill the host. Application of genetic-engineering technologies in such cases had shown promising improvement in the speed of killing action. The main strategies used for the improvement of the speed of killing are; (i) expression of insect-specific toxin gene, (ii) interference with host physiology. Examples of improvement in the speed of killing action expressing insect-specific toxin gene are baculoviruses expressing HD-73 δ endotoxin derived from the scorpion *Buthus eupeus*, *Cry1AB* derived from the *Bacillus thuringiensis* and *AaIT* toxin gene derived from the North African scorpion *Androctonus australis* (Bonning and Hammock, 1996). Other strategies which are to interfere with host physiology was the introduction of lepidopteran hormones (diuretic and antidiuretic) to disrupt the normal physiology of the larvae (Maeda, 1989; Coast et al., 2002), inactivation of viral ecdysteroid UDP-glycosyltransferase (egt) gene (O'Reilly, 1995) to increase in the baculovirus speed of kill (about 20%–30%). Other than that expression of the enzyme genes also showed enhancement in the speed of killing action. Examples are the expression of *Helicoverpa armigera cathepsin B-like proteinase* in baculovirus (*Autographa californica* multiple nucleocapsid nucleopolyhedrovirus; AcMNPV) (Shao et al., 2008) and cathepsin L-like cysteine protease from the flesh fly (*Sarcophaga peregrina*) to baculovirus (*Helicoverpa armigera* nucleopolyhedrovirus) (Sun et al., 2009), *ScathL* gene from *Sarcophaga peregrina*, *Keratinase* gene from the fungus *Aspergillus fumigatus* (Gramkow et al., 2010) and *Cydia pomonella* granulovirus matrix metalloprotease in baculovirus (AcMNPV) (Ishimwe et al., 2015).

Major advantages of using microbial pesticides are that microbial pesticides are inherently less toxic and mainly affect the target pest in contrast to broad-spectrum chemical pesticides that mostly affect several non-target organisms including human (Hassan and Gökçe, 2014; Preininger et al., 2018). Microbial pesticides are effective in low dosage and often decompose rapidly, resulting in lower contamination problems to the environment (Pathak et al., 2017). Also, the use of microbial pesticides is not limited to any particular habitat (such as lake stream, border of watersheds, near public residences, etc.) and some potential microorganism establish themselves in one season and provide control activity in the subsequent season, in case of pest generation (reviewed in Kabaluk et al., 2010; Koul, 2011; Mnif and Ghribi, 2015). Although, microbial pesticides provides enormous advantages over chemical pesticides there are still several limitations which need to be addressed to support widescale commercial adoption of microbial pesticides; the limitations are owing to specificity of action, the active ingredient of microbial pesticide sometimes control over only a portion of the pest in the field, while other types of pest can continue damage. Also, stresses like heat, UV light and desiccation reduces the efficacy of microbial pesticides. Short shelf life and storage is another important issue. Other than that, the microbial pesticide has a narrow market scope as the terms and condition for the development and registration is not flexible (reviewed in Kabaluk et al., 2010; Koul, 2011; Mnif and Ghribi, 2015).

5.2.6 Microbes supporting plants and soil health

Microorganisms in the rhizosphere are versatile and serves the dual purpose of suppressing soil-borne pathogen along with improving soil health (Glick, 2012). Hence, plant growth-promoting rhizobacteria (PGPR) and fungi (PGPF) have become a popular ingredient of modern microbial pesticide, as several of them, possesses entomopathogenic and plant protection properties, in addition to plant growth promoting properties. In this respect, some of the important activities performed by PGPR and PGPF are plant protection against abiotic and biotic stress (Vacheron et al., 2015), nutrient acquisition and hormonal stimulation (Besset-Manzoni et al., 2018), expansion of beneficial rhizoflora growth (reviewed in Berg, 2009), improvement of soil texture and bioremediation of the polluted soil by sequestering toxic and heavy metals (Rajkumar et al., 2010) and the degradation of xenobiotic compounds such as chemical pesticides (Braud et al., 2009). Soil fertility also depends on another component i.e. root exudates released as plant's metabolites in the process called exudation. The process of exudation is important for the circulation of carbon and nitrogen that can be taken up by the microorganisms to enhance soil fertility (Jones et al., 2009). The exudates are a mixture of various primary (carbohydrates, amino acids, organic acids, vitamins, nucleosides, mucilage) and secondary (flavonoids, auxins, glucosinolates) metabolites (reviewed in Vives-Peris et al., 2019). The root metabolites work as signals to attract microbial populations, especially those which can metabolize plant exudates compounds and propagates microbial habitat (Shukla et al., 2011; Drogue et al., 2013). The microbial community which gets attracted to the plant roots are referred as rhizo-microbiome (Chaparro et al., 2013). It is important to note that rhizo-microbiome composition changes with the root exudates composition (Bouffaud et al., 2012; Bulgarelli et al., 2013). The root exudate composition changes along with the root system depending on stages of the plant development and plant genotypes, soil type and pH (Bouffaud et al., 2012; Bulgarelli et al., 2013). Some of the important types of rhizosphere microorganisms performing plant growth-promoting activity and improving soil quality are explained as under:

5.2.6.1 Plant growth-promoting rhizobacteria

PGPR contain heterogeneous groups of non-pathogenic, root-colonizing bacteria that support plant growth and several of them possess biocontrol activities. A summary of biocontrol activities against soil-borne pathogen and plant promoting activities of popular plant growth-promoting bacteria (PGPR) is summarized in Table 5.2. PGPR are found in the narrow region of soil surrounding the root. This region of the rhizosphere is highly competitive for diverse groups of microbes to obtain nutrients and proliferative growth helping in plant growth-promoting activities and suppressing plant pathogens (Verma et al., 2019). PGPR performs a symbiotic interaction with the plant roots which can be categorized into two types, first is mutualistic interaction which mostly corresponds to intimate or obligate interaction where a structure is formed and is dedicated to the interaction (e.g. nodule formation during symbiotic interaction between *Fabaceae* and rhizobia, Masson-Boivin et al., 2009). Second is cooperation or associative symbioses which is a more

TABLE 5.2 Biocontrol and plant growth promoting activities of PGPR as active ingredient of microbial pesticide.

Active ingredient of microbial pesticide	Test crop	Specific effect	References
Biocontrol activities			
Bacillus amyloliquefaciens (SS-12.6 and SS-38.4) and *Bacillus pumilus* (SS-10.7)	Sugar beet	Production of surfactin, fengycin A, iturin A to overcome *P. syringae* pv. aptata (leaf spot)	Nikolić et al. (2019)
Bacillus velezensis (S3-1)	Tomato	Production of surfactin, fengycin A, iturin A to overcome *Botrytis cinerea*	Jin et al. (2017)
Bacillus amyloliquefaciens PGPBacCA1	Common bean	Production of surfactin, iturin and fengycin to overcome *Sclerotium rolfsii, Sclerotinia sclerotiorum, Rhizoctonia solani, Fusarium solani, Penicillium* spp.	Torres et al. (2017)
Bacillus thuringiensis (UM96)	*Medicago truncatula*	Production of chitinase to inhibit *Botrytis cinerea* (gray mold)	Martínez-Absalón et al. (2014)
Bacillus cereus (AR156)	Arabidopsis	Induction of ISR-SA, jasmonic acid and ethylene to overcome *Pseudomonas syringae* pv. tomato DC3000	Niu et al. (2011)
Bacillus mojavensis (RRC101)	Maize	Secretion of VOCs to inhibit *Fusarium verticillioides*	Rath et al. (2018)
Chryseobacterium wanjuense (KJ9C8)	Pepper	Production of protease and HCN to inhibit *Phytophthora capsici* (blight)	Kim et al. (2012)
Pseudomonas putida, P. fluorescens, P. aeruginosa	Apple	Production of Phenazine, Pyrrolnitrin, DAPG to overcome *Dematophora necatrix, Fusarium oxysporum, Phytophthora cactorum and Pythium ultimum*	Sharma et al. (2017a)
Pseudomonas brassicacearum (J12)	Tomato	Production of DAPG, HCN, siderophore, protease inhibit *Ralstonia solanacearum* (wilt)	Zhou et al. (2012)
Pseudomonas spp., *Paenibacillus* spp. (Pb28), *Enterobacter sp.* (En38), *Serratia sp.* (Se40)	Potato	Production of siderophore, HCN, protease production to inhibit *Ralstonia solanacearum* (wilt)	Kheirandish and Harighi (2015)
Pseudomonas fluorescens (Psd)	Tomato	Production of PCA, pyrrolnitrin to inhibit *Fusarium oxysporum*	Upadhyay and Srivastava (2011)

TABLE 5.2 Biocontrol and plant growth promoting activities of PGPR as active ingredient of microbial pesticide.—cont'd

Active ingredient of microbial pesticide	Test crop	Specific effect	References
Pseudomonas fluorescens (MGR12)	Cereals	Secretion of VOCs to overcome *Fusarium proliferatum* (head blight)	Cordero et al. (2014)
Pseudomonas stutzeri (E25), *Stenotrophomonas maltophilia* (CR71)	Tomato	Secretion of VOCs to inhibit *Botrytis cinerea* (gray mold)	Rojas-Solís et al. (2018)
Pantoea agglomerans (E325)	Pome fruits	Production of Pantocin A, herbicolins, microcins, phenazines to overcome *Erwinia amylovora* (fire blight)	Braun-Kiewnick et al. (2012)
Plant growth promoting traits			
Bacillus subtilis (BHUPSB13)	Chickpea	Enhanced shoot and root length	Yadav et al. (2010)
Pseudomonas aeruginosa (BHUPSB02)			
Pseudomonas putida strain (BHUPSB04)			
Bacillus sphaericus (UPMB10)	Rice	Seedling emergence, seedling vigor and enhanced root growth	Mia et al. (2012)
Rhizobium spp. (SB16, UPMR1006 and UPMR1102)			
Azospirillum lipoferum	Maize	Seed germination, biomass and crop yield	Noumavo et al. (2013)
Pseudomonas fluorescens			
Pseudomonas putida			
Azospirillum brasilence (Sp7, Sp7-S and Sp245)	Tomato and lettuce	Enhanced germination rate of tomato, longer and heavier roots, superior vigor.	Mangmang et al. (2014)
Burkholderia phytofirmans (PsJN)			
Pseudomonas moraviensis	Wheat	Improved seeds establishment by increasing number of seeds/spike and spike length, combination of tryptophan enhanced PGPR efficiency	Ul Hassan and Bano (2015)
Bacillus cereus			
Pseudomonas spp.	Tomato	Enhanced time of germination and growth of the seedling	Widnyana and Javandira (2016)
Bacillus spp.			
Pseudomonas fluoresecens (PF2s5)	Maize	Enhanced seed germination and plant height	Boominathan (2020)
Bacillus megaterium (BM29)			
Bacillus cereus and Pseudomonas rhodesiae combined bioformulation	Tomato, cauli-flower, chili and brinjal	Improved shoot height, number of leaves, early fruiting and total biomass content	Kalita et al. (2015)
Pseudomonas spp. (PS_2)	Chrysanthemum	Improved number of flowers, flower yield, number of suckers, stalk length and flower size	Kumari et al. (2016)
Bacillus spp. (BS_3)			
Bacillus cereus KI-2	Strawberry	Enhanced yield and number of fruits	Kurokura et al. (2017)
Bacillus polymyxa (and *Saccharomyces cereivisea*)	Olive	Increased leaf density, leaf surface area, number of flowers per inflorescences, fruit yield and oil content	El Taweel et al. (2011)
Pseudomonas fluorescens N21.4	Blackberry	Increased flower bud, fruit quality together with an increased productivity	García-Seco et al. (2013)
Burkholderia caribensis	Mango	Enhanced floral buds, number of flowers and fruits	de los Santos-Villalobos et al. (2013)
Rhizobium spp.			

Continued

TABLE 5.2 Biocontrol and plant growth promoting activities of PGPR as active ingredient of microbial pesticide.—cont'd

Active ingredient of microbial pesticide	Test crop	Specific effect	References
Bacillus pummilus (S4)	Runner bean	Increased photosynthesis, transpiration, water use efficiency, leaves chlorophyll content and grain yield.	Stefan et al. (2013)
Bacillus mycoides (S7)			
Bacillus cereus (K46)	Pepper	Enhanced photosynthetic mechanism in pepper plants to increase chlorophyll fluorescence and gas exchange parameters.	Samaniego-Gámez et al. (2016)
Bacillus spp. (M9, *B. subtilis* and *B. amyloliquefaciens* combination)			
Bacillus amyloliquefaciens LL2012	Soybean	Phytohormone (IAA, GA3, SA) production and improved nodulation	Masciarelli et al. (2014)
Bradyrhizobium japonicum			
Klebsiella sp. Br1	Maize	Enhanced vegetative growth, nitrogen fixation and nitrogen remobilization	Kuan et al. (2016)
Klebsiella pneumoniae Fr1			
Bacillus pumilus S1r1			
Acinetobacter sp. S3r2			
Azospirillum spp. (N4)	Sorghum	Enhanced plant growth, nitrogen fixation and IAA production	Ashraf et al. (2011)
Pseudomonas spp. (K1)			
Pseudomonas fluorescens (Ms-01)	Wheat	Improved proline accumulation and POD and APX antioxidant enzyme activity	Kadmiri et al. (2018)
Azosprillum brasilense (DSM1690)			
Enterobacter hormaechei (KSB-8) (combined with *Aspergillus terreus* KSF-1)	Okra	Enhanced root and shoot growth and potassium mobilization efficiency	Prajapati et al. (2013)
Pantoea ananatis (KM977993)	Rice	Improved potassium and phosphorus solubilization, enhanced plant growth	Bakhshandeh et al. (2017)
Rahnella aquatilis (KM977991)			
Enterobacter sp. (KM977992)			
Enterobacter ludwigii (FB Endo 135)	Rice	Enhanced IAA producing capability and plant height	Susilowati et al. (2018)
Pseudomonas fragi, Bacillus cereus, and *Rhizobium*		Improve plant height	
Bacillus aerius, Pseudomonas fragi, and *Bacillus cereus*		Improved dry weight of the rice grains	
Bacillus amyloliquenfaciens		Improved dry weight of the root	

DAPG, 2,4-diacetylphloroglucinol; *HCN*, hydrocyanic acid; *ISR*, induced systemic resistance; *PCA*, phenazine1-carboxylic acid; *SA*, salicylic acid; *VOCs*, volatile organic compounds.

specific interaction of PGPR and plant root (Drogue et al., 2013). The mechanism involves colonization of PGPR to the root surface and sometimes the inner tissues and stimulate plant growth. In general, PGPR exhibit various direct and indirect mechanism which influence the plant growth. Direct mechanism includes facilitating resource acquisition such as essential minerals and nutrients from the surrounding environment or by providing synthesized compounds; and indirect mechanism includes suppressing the harmful effects of phytopathogens by synthesizing antibiotics, lytic enzymes and chelation of available iron at the root interface (Verma et al., 2019, Table 5.2). PGPR can be categorized into two main types namely extracellular PGPR (ePGPR-symbiotic) and intracellular PGPR (iPGPR–fee living) (Martínez-Viveros et al., 2010). ePGPR are found in the rhizosphere or in the spaces between the cells of the root cortex belonging to the genera *Azotobacter, Azospirillum, Agrobacterium, Bacillus, Caulobacter, Chromobacterium, Erwinia, Serratia, Flavobacterium, Arthrobacter, Micrococcus, Pseudomonas* and *Burkholderia* (Gray and Smith, 2005;

Verma et al., 2019). On the other hand, iPGPR inhabits inside the specialized nodular structure of the root cells belonging to the genera *Allorhizobium, Bradyrhizobium, Mesorhizobium, Rhizobium* and *Frankia* (Wang and Martinez-Romero, 2000; reviewed in Bhattacharyya and Jha, 2012; Verma et al., 2019).

5.2.6.2 Plant growth-promoting fungi

The rhizo-microbiome community constitute of non-pathogenic naturally occurring saprophytes which help to increase plant growth and soil fertility and induce frontline defense response against pathogen infection, commonly known as plant growth-promoting fungi (PGPF) (Hossain et al., 2017). It is interesting to note that not all the fungi that promote plant growth are considered as PGPF, such as symbiotic mycorrhizal fungi improve plant growth but it is not believed as PGPF (Hossain et al., 2017). Mycorrhizal fungi have been reported to act as obligate biotrophs establishing a close association with the roots of most of the plants, while on the other hand, PGPF is non-symbiotic saprophytes which live freely in the rhizosphere on the root surface or inside the roots (Mehrotra, 2005; Corradi and Bonfante, 2012). Therefore, the term PGPF is considered as not an absolute term, rather it is an operational term (Bent, 2006). PGPF communities present in the rhizosphere are either free-living or in a symbiotic association, exist as unicellular yeast or present inside plant as an endophyte. PGPF which are present in the subaerial and subsoil (free-living) region is influenced by the presence of root exudates (Kour et al., 2019). Root exudate helps in the colonization of the PGPF with the plant roots to establish the association. PGPF enhances plant growth, suppress plant diseases and systemic resistance (Verma et al., 2019). Reports suggest that PGPF mimics the PGPR in the way they interact with the host plants (Hossain et al., 2017); however, there are certain activities PGPF performs better than PGPR such as tolerance of acidic conditions and mobilization of phosphates (Kumar et al., 2018; Wahid and Mehana, 2000). Popular PGPF belongs to the genera *Aspergillus, Piriformospora, Fusarium, Penicillium, Phoma, Rhizoctonia, and Trichoderma* (Hossain et al. 2007, 2014; Jaber and Enkerli, 2017; Kour et al., 2019). A summary of biocontrol activities of popular PGPF is summarized in Table 5.3. Apart from plant growth promotion, PGPFs also maintain the soil structure through filamentous branching growth and exopolymers (Kour et al., 2019).

TABLE 5.3 Biocontrol and Plant growth promoting activities of PGPF as active ingredient of microbial pesticide.

Active ingredient of microbial pesticide	Test crop	Specific effect	References
Biocontrol activities			
Aspergillus terreus JF27	Tomato	ISR to overcome *Pseudomonas syringae* pathovar (pv.) tomato DC3000 (speck)	Yoo et al. (2018)
Chaetomium globosum	*Ginkgo biloba*	Production of gliotoxin to inhibit *Fusarium sulphureum, Alternaria alternate, Cercospora sorghi, Fusarium oxysporum f. sp. vasinfectum, Botrytis cinerea, Fusarium graminearum*	Li et al. (2011)
Trichoderma harzianum (Ths97)	Olive	Mycoparasitism to control *Fusarium solani* (root rot)	Amira et al. (2017)
Trichoderma harzianum	Soybean	ISR against *Fusarium oxysporum*	Zhang et al. (2017)
Trichoderma asperellum (CWD CHF 78)	Tomato	Secretion of chitinase, proteases and siderophore to reduce *Fusarium wilt* and promote plant growth and nutrient uptake	Li et al. (2018)
Penicillium simplicissimum (CEF-818), *Leptosphaeria* spp. (CEF-714), *Talaromyces flavus* (CEF-642), *Acremonium* spp. (CEF-193)	Cotton	CEF-818 and CET-714 induced systemic resistance to inhibit *Verticillium dahlia* (Verticillium wilt) and improve seed cotton yield in cotton fields	Yuan et al. (2017)
Trichoderma M10	*Vitis vinifera*	ISR and production of secondary metabolite to overcome *Uncinula necator* (powdery mildew)	Pascale et al. 2017

Continued

TABLE 5.3 Biocontrol and Plant growth promoting activities of PGPF as active ingredient of microbial pesticide.—cont'd

Active ingredient of microbial pesticide	Test crop	Specific effect	References
Trichoderma harzianum T-aloe	Soybean	Secretion of chitinase, 1,3-β-glucanase and hyphal parasitism against *Sclerotinia sclerotiorum* (stem rot)	Zhang et al. (2016)
Trichoderma asperellum (CCTCC-RW0014)	Cucumber	Secretion of cell wall degrading enzymes (*chitinase, protease, glucanase*) against *Fusarium oxysporum* f. sp. cucumerinum	Saravanakumar et al. (2016)
Streptomyces spp. (CB-75)	Banana	Production of Type I polyketide synthase, nonribosomal peptide synthetase to inhibit *Colletotrichum musae*	Chen et al. (2018)
Streptomyces spp. (UPMRS4)	Rice	Secretion of chitinase, glucanase and PR1 to overcome *Pyricularia oryzae* (blast)	Awla et al. (2017)
Streptomyces plicatus (B4-7)	Bell pepper	Secretion of borrelidin antibiotic to inhibit *Phytophthora capsici* (damping off, root rot, leaf blight)	Chen et al. (2016)
Streptomyces sp.	Rice	Production of chitinase, phosphatase, and siderophore to overcome *Xanthomonas oryzae* pv. oryzae (Xoo), (leaf blight)	Hastuti et al. (2012)
	Sugar beet	Production of protease, chitinase, α-amylase to inhibit *Rhizoctonia solani* AG-2, *Fusarium solani, Phytophthora drechsleri* (root rot)	Karimi et al. (2012)
Plant growth promoting activities			
Aspergillus spp. (PPA1)	Cucumber	Increased seed germination and seedling vigor	Islam et al. (2014a)
Fusarium spp. (PPF1)	Indian spinach	Enhanced germination percentage and increased vigor index	Islam et al. (2014b)
Talaromyces wortmannii (FS2)	*Brassica campestris* L. var. perviridis	Secrete β-caryophyllene to improve plant growth and inhibition of *Colletotrichum higginsianum*	Yamagiwa et al. (2011)
Trichoderma harzianum	Maize	Reduction in *F. verticillioides* and fumonisin incidence and increasing the seed germination, vigor index, field emergence, yield, seed weight	Chandra Nayaka et al. (2010)
	Rice	Enhanced number of seedlings, length and dry matter of shoot and root and yield	Rahman et al. (2015)
Trichoderma viride (BBA 70239)	*Arabidopsis thaliana*, tomato	Produced volatile organic compounds to enhance plant biomass, large plant size, lateral roots	Lee et al. (2016)
Piriformospora indica	*Aloe vera*	Increase in plant biomass, shoot and root length and number, photosynthetic pigment (Chl a, Chl b and total Chl) and gel content	Sharma et al., 2017(b)
	Centella asiatica	Enhanced root and shoot biomass, and medicinally active compounds (asiaticoside)	Satheesan et al. (2012)
	Chlorophytum spp.	Enhanced plant growth and saponin content	Gosal et al. (2010)

TABLE 5.3 Biocontrol and Plant growth promoting activities of PGPF as active ingredient of microbial pesticide.—cont'd

Active ingredient of microbial pesticide	Test crop	Specific effect	References
Alternaria spp. (A7, A38)	Tobacco	Increased plant growth and total leaf chlorophyll content	Zhou et al. 2014
Phomopsis spp. (H25)			
Cladosporium spp. B50			
Epichloë typhina	Orchard grass	Enhanced chlorophyll *b* contents, abundance of LHCI and LHCII proteins and photosynthesis efficiency	Rozpądek et al. (2015)
Alternaria alternata	*Arabidopsis thaliana*	VOCs enhanced photosynthesis, accumulation of cytokinins, sugars and flowering	Sánchez-López et al. (2016)
Trichoderma harzianum T-3 (combined with mustard oil cake)	Pea	Inhibited *Rhizoctonia solani* and increased seed yield	Akhter et al. (2015)
Fusarium oxysporum (V5W2, Eny 7.11°, and Emb 2.4°)	Banana	Inhibited plant nematodes *Pratylenchus goodeyi* and *Helicotylenchus multicinctus*, reduction in root necrosis and improved yield (Emb 2.4° by 35% and V5W2 by 36%)	Waweru et al. (2014)
Aspergillus niger	Tomato	Increased accumulation of salicylic acid, total phenolic and chlorophyll contents of plant, as well as lycopene, ascorbic acid (Vitamin C), and Brix index of fruit	Anwer and Khan (2013)
Westerdykella aurantiaca (FNBR-3)	Rice and pea	Enhanced total chlorophyll, carotenoids and protein contents, plant growth and soil fertility	Srivastava et al. (2012)
Trichoderma longibrachiatum (FNBR-6)			
Trichoderma atroviride D16	*Salvia miltiorrhiza*	Improved hairy root and production of tanshinone I (T-I) and tanshinone IIA (T-IIA) content	Ming et al. (2013)
Curvularia geniculata	Pigeon pea	Enhanced growth, IAA production, phosphorus solubilization	Priyadharsini and Muthukumar (2017)
Haematonectria ipomoeae (CML 3249)	Corn and cowpea	Enhanced plant growth and phosphate solubilization	Gudiño Gomezjurado et al. (2015)
Pochonia chlamydosporia var. catenulate (CML 3250)			
Porostereum spadiceum (AGH786)	Soybean	Production of phytohormone GA_3, isoflavones, and enhanced tolerance to NaCl	Hamayun et al. (2017)

AA, Indole-3-acetic acid; *DAPG*, 2,4-diacetylphloroglucinol; *GA_3*, Gibberellic acid; *IHCN*, hydrocyanic acid; *ISR*, induced systemic resistance; *PCA*, phenazine1-carboxylic acid; *SA*, salicylic acid; *VOCs*, volatile organic compounds.

5.3 Trends and market demand for microbial pesticide

The global market of biopesticide has grown significantly during recent years and is predicted to reach nearly USD 7.38 billion in the year 2023 with a CAGR of 17.02% (Market Data Forecast, 2019). The CAGR value was calculated using the online source (https://cagrcalculator.net/) (Fig. 5.2). Subsequently, the market for microbial pesticides has evolved with the growing demand for the high yield and efficient crop protection inputs. The development of microbial pesticide utilizing specific forms of microorganisms as an active ingredient is the fastest-growing trends creating a niche demand within the global market. The products are mainly available as liquid concentrates, wettable powders and ready-to-use dust and granules (Mishra et al., 2015). Over the years Bt-based products have dominated the global market with 50% of the market

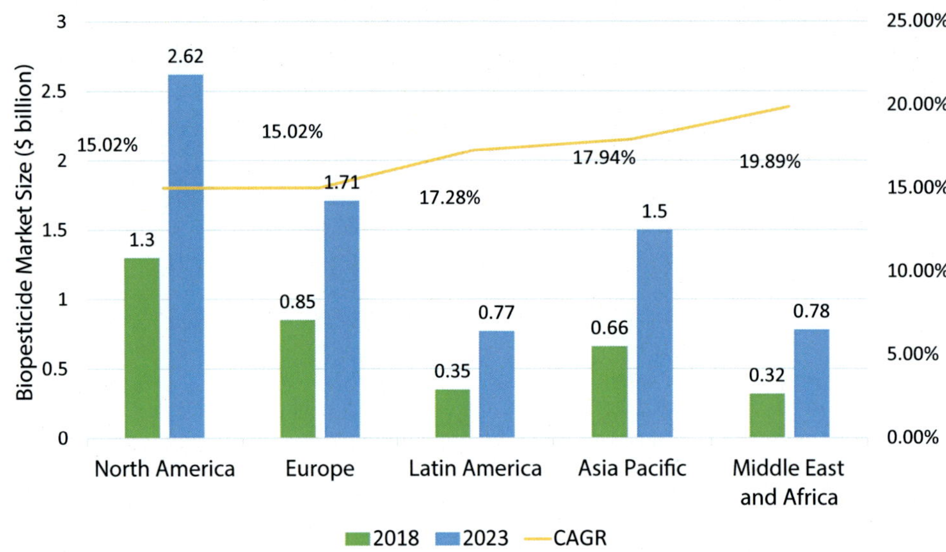

FIG. 5.2 The biopesticide market 2018 and 2023. *Modified from Market Data Forecast (2019).*

share and around 200 products (CABI, 2010), followed by fungi, viruses and nematodes (GBMAIF, 2017). Nearly 3000 tons/year of microbial pesticides is being produced globally and the number is constantly increasing (Saritha and Tollamadugu, 2019). According to data on biopesticide agents from Agriculture and Agri-Food Canada (Kabaluk and Gazdik, 2005) and the USEPA, the United States tops the list in biopesticide (including microbial-based products) use with the sale of nearly 200 products while nearly 60 products in European Union (Hubbard et al., 2014), UK has been reported to sell only 5 products, 10 products in Germany and 15 each in France and Netherlands (Chandler et al., 2008). Additionally, the demand for moving toward safe crop, protection strategy has attracted OECD (Organization for Economic Co-operation and Development, 2003) member countries and they have registered over 250 biological plant protection products based on microorganism and over 1000 different biopesticides are produced around the world comprising living organisms, plant growth regulators and plant extracts (Egbuna et al., 2020). Insight into the product and application of microbial pesticide globally is discussed below:

5.3.1 Global market reports on use of microbial pesticides

5.3.1.1 North America

North America (The USA, Canada and Mexico) leads the global biopesticide including microbial pesticides market with 46% of the share (Mishra et al., 2015), it has registered a CAGR of 15.02% over the period of 2018–23. The market is valued at USD 1.3 million in the year 2018 and is expected to reach USD 2.62 billion in the year 2023 (North America Biopesticide Market, 2019). The country has the world's largest organic food market with a sale of USD 45.8 billion in 2018. The increasing demands of organic food in the market of North America has gained attention to promote microbial pesticides. The first registered microbial pesticide was prepared using *Bacillus popilliae* in 1948 to control the Japanese beetle (Schneider, 2006). In the USA, the development of microbial pesticide is a challenging process and has to pass strict regulations before its marketing (reviewed in Mishra et al., 2015). In Canada, the end-user sale has valued USD 1.4 billion in 2010 (CPL Business Consultant, 2010). Mexico and Canada are behind the USA in terms of usage of microbial pesticide as there is less prevalence of pests. According to EPA data in the USA, 102 microbial pesticides are being used (USEPA, 2011). Until 2017, 41 microbial pesticides were registered in North America, including 21 bacterial, 12 fungal and 8 virus-based products (reviewed in Arthurs and Dara, 2019). Some of the major key player companies in North America are Valent Biosciences, Certis, Becker Microbial Products, Bayer, Andermatt/Sylvar, Marrone Bio Innovations.

5.3.1.2 Europe

Europe includes Germany, France, Spain, United Kingdom, Italy, Netherland, Russia and rest of Europe. Europe ranks second with 19% of the biopesticide including microbial pesticide market share. The market is expected to grow from USD 0.85 million in 2018 to USD 1.71 million in 2023 with a CAGR value of 15.02% (Europe Biopesticide Market, 2019). The European market is estimated to be one of the fastest-growing markets with France and Germany as the major contributors. The regulatory pressure and ban on the use of glyphosate, neonicotinoids and paraquat, farmers are forced to shift to biopesticides and hence the market is expected to enhance furthermore. The first Bt product, Thuricide was approved in 1964, and first registration for an entomopathogenic fungus *Lecanicillium longisporum* was approved in 1981 and given to Tate and Lyle in the UK (Quinlan, 1990). Europe registered 68 biopesticides (EUPD, 2010) in which 39 were microbial pesticides (reviewed in Mishra et al., 2015). Europe also observed a decline in Bt based products in the year 2000 because of the arrival of other bacterial (*Bacillus subtilis*), fungal (*Trichorderma* spp.) and viral (Zucchini Yellow Mosaic Virus, weak strain Virus) biocontrol agents (Business Wire, 2010). Major market players are AGRAUXINE SA, BASF SE, Bayer CropSciences Ltd., Koppert Biological Systems and Isagro.

5.3.1.3 Asia Pacific

Asia-Pacific countries, China, Japan, South Korea, India, Australia, Indonesia, Thailand, Malaysia, Philippines and Vietnam are together called APAC. APAC biopesticide including microbial pesticide market is expected to register a CAGR of 17.94% during the forecast period of 2018–23. The market is valued at USD 0.66 billion in the year 2018 and expected to reach USD 1.5 billion by 2023 (Asia Pacific Biopesticide Market, 2019). APAC is in third position with 12% of the biopesticide market share. APAC countries with largest land size and highly biodiverse regions have opportunities for the development of a number of biopesticides. The economy of most countries is heavily depended upon agriculture (Mishra et al., 2015) and farmers practice both traditional and smart farming. Most of the regions grow cereals crops including rice, wheat, maize, barley, sorghum, millet, corn; cash crops such as cotton, tobacco, sugarcane and vegetables using biopesticides.

Some of the top contributors in the market include China, where biopesticides were first produced as unformulated dried culture in 1960 (Xu et al., 1987). China has registered 270 biopesticides, among which microbial pesticide registered are 11 Bt-based biopesticides, 22 fungal and 35 viral (14 of which are from *Heliothis armigera* NPV) (ICAMA, 2008).

In India rise in microbial pesticide was noted after chemical insecticide failed to control Cotton bollworm, Tobacco cutworm, and other pests of cotton (Armes et al., 1992; Kranthi et al., 2002). India has more than 100,000 ha land occupied by organic farming (Sekhar et al., 2016) supported by nearly 100 biopesticides products registered with the Central Insecticides Board and Registration Committee (CIBRC) (Mishra et al., 2020). Out of which 66 are fungal (*Trichoderma spp., Beauveria bassiana, Verticillium anisopliae, Paecilomyces lilacinus* and *Hirsutella thompsonii*), 29 are bacterial (*Bacillus* spp. and *Pseudomonas fluorescens*), 4 are viral (Nucleopolyhedrovirus) (reviewed in Mishra et al., 2020).

South Korea started using microbial pesticide in the year 1970 and included use of bacteria, fungi, virus and nematodes to control pests in agriculture, forestry and golf courses. But the first commercial biopesticide was Bt-based (subspecies aizawai) named as Solbichae, and was registered in the year 2003 to control diamondback moth and beetle armyworm in Chinese cabbage (Jeong et al., 2010). Approximately 24 microbial pesticides have been registered by the year 2009 (Jeong et al., 2010).

Japan is one of the pioneers to start the use of Bt-based product in early 1901, but it was observed that the earlier application created a negative impact on the sericulture (Aizawa and Shigetane, 2001). Later, it was reported that preventing farmers from spraying of microbial pesticide directly in the mulberry fields can stop the threats to silkworm (Aizawa and Shigetane, 2001). Japan has one of the best-established systems for registration and commercialization of biopesticides. The first registered bacterial biopesticide product was Bt-based launched in 1981 (Kunimi, 2007), and subsequently, other subspecies (kurstaki, japonensis and aizawai) were introduced (Ohba et al., 1992). Fungal biopesticide was derived from *Beauveria bassiana* against pine caterpillar in 1933 (Hidaka, 1933) and for viral biopesticide Nucleopolyhedrovirus was introduced to control cabbage armyworm in 1962 (Kunimi, 1998). In total, 11 microbial pesticides were registered in Japan (Asia and Pacific Plant Protection Commission and Food and Agriculture Organization of the United Nations Regional Office for Asia and the Pacific, 2007–2008).

Australia started using microbial pesticide in late 1960 with Granulovirus of codling moth and Nucleopolyhedrovirus of *Helicoperva zea* observing limited success in the field (Mishra et al., 2015). However, in the later years, Granulovirus was reported to be effective against potato tuber moth (Reeda and Springetta, 1971). Bt-based (subspecies kurstaki) products were registered after increased incidents of the cotton pest were reported in the year 1987–88 (Powles and Rogers, 1989). *Metarhizium*-based product was first to be registered as a fungal biopesticide (BioGreen) and were found to be very effective against a range of pests including canegrubs, termites and locusts (reviewed in Milner, 2000). In total, 11 microbial pesticides are registered in Australia in which 7 are bacterial (*Bacillus* spp., *Streptomyces Lydicus* and *Pseudomonas fluorescens*), 3 fungal (*Trichoderma* spp.) and 1 viral (Nucleopolyhedrovirus of *Helicoperva* spp.) (Biopesticides in Australia, 2018).

Other small countries in the Asia Pacific region such as Malaysia has a negative perception for biopesticide due to its slow response as compared to chemical pesticides. Therefore, farmers were reluctant to use it in the fields. But the increased use of chemicals created a huge impact which urged farmers to use alternatives. By the year 2016, 19 microbial pesticides were registered (Krishnen et al., 2016). In Vietnam, organic agriculture was not promoted until 2017, when the state provided VND 156.3 billion (approximately USD 6.7 million) to practice smart farming (Vietnam Biopesticide Market by Product Group & Application Type (2019—2024), 2019). Microbial pesticides registered by the year 2003, included 6 bacterial (*Bacillus* spp.), 2 fungal (*Beauveria bassiana* and *Metarhizium anisopilae*) and 3 viral (Nucleopolyherovirus) (Jäkel, 2003). In Philippines, the agricultural sector drives the economy using a big proportion of chemicals, as small-scale farmers considered it as effective, cheap and easy for application. In spite of lukewarm success, there are microbial pesticides which are available includes bacterial (Bt-based), fungal (*Beauveria bassiana, Aspergillus versicolor,* and *Metarhizium anisopliae*) (Brown et al., 2016). In Indonesia also the use of biopesticide was started in the beginning of 1960 after the introduction of Bt-based products. By the year 2005, 13 bacterial (Bt-based products) and 6 fungal (3 *Trichoderma koningii*, 2 *Beauveria bassiana* and 1 *Gliocladium* spp.) microbial pesticides were registered. Thailand market introduced and commercialized first Bt-based biopesticide in the year 1965 but due to its slow and highly selective action, it was not accepted by farmers (Rushtapakomchai, 2003). Thereafter in 1969, Bt kurstaki was introduced to control the lepidopteron of cruciferous crops (Prasetphol et al., 1969). The active research started in 1980, and by the year 2003, bacterial Bt-based (including subspecies kurstaki, aizawai, and tenebrionis), fungal (*Trichoderma harzianum, Beauveria bassiana*) and viral (Nucleopolyhedrovirus) microbial pesticides were registered (Jäkel, 2003).

Some of the major key players in the Asia Pacific market are BASF SE, Valent BioSciences Corporation, Bayer CropScience, Emery Oleochemicals LLC, ISHIHARA SANGYO KAISHA LTD.

5.3.1.4 Latin America

In Latin America, biopesticide including microbial pesticide market is expected to reach a CAGR of 17.28% during the forecast period of 2018—23. The market is valued at USD 0.35 million in the year 2018 and expected to reach USD 0.77 billion by 2023 (Latin America Biopesticide Market, 2019). The biopesticide market is dominated by Bt-based products which cover about 40% of the market (reviewed in Mishra et al., 2015). It was first introduced in the year 1950 against *Colias lesbian* in alfalfa in Argentina (Botto, 1996). In Brazil, Bt based products came in the year 1990 and were reported to be only bacterial biopesticide in the market. Later by 2000, fungal product *Beauveria bassiana* against *Triatoma infestans* and *Musca domestica* and the virus-based products were registered (Alves et al., 2008). The registration process involves high cost and length of time and therefore many products go unregistered in the market (Bettiol, 2011). However, by the year 2014, countries like Argentina, Brazil and Cuba have taken initiatives in setting the pest management system by biological means in Latin America (reviewed in Mishra et al., 2015). Some of the leading players covered include Bayer CropScience, BASF, Marrone Bio Innovations, De Sangosse and Valent Biosciences.

5.3.1.5 Middle East and Africa

The Middle East and Africa biopesticide including microbial pesticide market are expected to reach a CAGR of 19.79% during the forecast period of 2018—23. The market is valued at USD 0.32 billion in the year 2018 and expected to reach USD 0.78 billion by 2023 (Middle-East and Africa Biopesticide Market, 2019). The rapid growth in the market is driven by the high prevalence of crop diseases and demand for organic crops, which has attracted foreign investors into this segment. Being an underdeveloped market, this region has provided an immense scope of exploration in the field of microbial pesticide. And therefore, substantial growth in the biopesticide including microbial pesticide market with increasing trade in horticulture producers into the European Union (EU) and The North American Free Trade Agreement (NAFTA) regions have been observed (Koul et al., 2012). One of the earliest uses of microbial pesticide to control pests was established in 1960 (Kunjeku et al., 1998) and is still one of the top priorities in the African research field. It was recorded that by the year 2010, 77% of the overall microbial pesticide market of Africa and Middle East was dominated by Bt-based products (reviewed in Koul et al., 2012), which were marketed by international companies (Grzywacz et al., 2009). Application of fungal based products of *Metarhizium anisopliae* was also reported to be very effective in pest management (Hunter et al., 2001). Some of the key market players in this region are Perry America, Andermatt Biocontrol Ag, Marrone Bio Innovations Inc, Agbitech Pty Limited, Valent Biosciences Corp, Biocare.

5.4 Registration and regulation of microbial pesticide globally

The law and policies regulating the use and production of microbial pesticides vary from country to country and there is no standardized regulatory model that can simplify the regulation and registration process worldwide (Arora et al., 2016). Although there is variability in regulation, the aim of all the regulations is similar that is to ensure the protection of the community health and the environment from the possible risk associated with microbial pesticide production and use (Ravensberg, 2011). Even though few global international organization such as the European and Mediterranean Plant Protection Organization (EPPO), International Organization for Biological Control (IOBC), and Organization for Economic and Co-operative Development (OECD), Biopesticide and Pollution Prevention Division (BPPD) have put their efforts to standardize the microbial pesticide regulation, but in comparison to chemical pesticides, which has a stable market and established nonoverlapping laws, microbial pesticides are still dawdling (WHO, 2018; Arora et al., 2016). Despite the complicated regulatory systems governing the use and production of microbial pesticides, the demand for microbial pesticides are steadily increasing in the pesticide market (Sinha and Biswas, 2008). Several countries easing the regulation issues to promote a more friendly business environment for commercialization of the microbial pesticides, whereas in a few countries, still, there are no clear guidelines available for regulations, which is creating problems to introduce new products in the market (Ochieng, 2015).

In North America, particularly in the USA, Environment Protection Act (EPA) has the complex regulatory system for registration and regulation process of microbial pesticides and it comes under the term biopesticides along with other plant protection products (USEPA, 2011). The registration is the responsibility of the Office of Chemical Safety and Pollution Prevention (OCSPP) and Office of Pesticide Programs (OPP). OPP has three divisions including Antimicrobial division, Registration division and Biopesticide and Pollution Prevention Division (BPPD) which are involved in the microbial pesticide registration process (Matthews, 2014). Biopesticide Industry Alliance (BIP) along with EPA helps to quality check on the active ingredient, efficacy, toxicity and safety to health and environment and promotes industry standards, and also ensures transparent and appropriate registration and regulatory requirements (http://www.biopesticideindustryalliance.org/). Canada closely follows USEPA protocols for registration of microbial pesticides. Pest Management Regulatory Agency (PMRA) regulates microbial pesticides involving registration, evaluation and has developed joint guidelines for the registration of microbial pest control agents and products (PMRA, 2001) with EPA and Organization for Economic and Co-operative Development (OECD) (AGBR, 2015).

On the other hand, the European Union follows very stringent regulation for microbial pesticide registration. In early 1990, microorganisms, botanicals and pheromones were all regulated under the Directive 91/414/EEC (EU, 1991), which was originally developed for chemical pesticides (reviewed in Czaja et al., 2015). The Directive 91/414 was amended in the year 2009 by 2001/36/EC (EC, 2001) and 2005/25/EC (EC, 2005) to add specific requirements and uniform principles for microbial plant protection products. The new Regulation (EC, 2009) No 1107/2009 was applied in all the member states from 2011 and made the registration of the microbial pesticide stricter when compared to rest of the world as the application undergoes toxicology, environmental and efficacy evaluation (Arora et al., 2016). The applicants have to submit the dossier for the registration under the microbial plant protection product to a zonal rapporteur member state where it gets further evaluated (EC, 2013) under the Plant Protection Product Regulations (PPPR), European Directive (91/414/EEC).

Asia presents a large distribution of microbial pesticide market across the region. In China, the Chinese Ministry of Agriculture along with the Institute for the Control of Agrochemicals (ICAMA) is responsible for microbial pesticide registration (ICAMA, 2008; Fang, 2014). The governing body only allows registered and authorized companies to file the application (no research institutes, universities and other groups). The microbial pesticide also undergoes quality control check for acute and chronic toxicology and ecological safety during the registration process by ICAMA. Chinese central and local government encourages farmers to use microbial pesticides for crop pest management and ensure safe agricultural practices. India also follows very strict regulations under the Insecticide Act 1968, which states that any microbial organism manufactured or sold for pest and disease control should be registered with the Central Insecticide Board (CIB). CIB and Registration Committee (RC) ensures the quality standards with reference to content, the virulence of the organism, moisture content, shelf life, secondary non-pathogenic microbial load, safety to human and environment (Rabindra, 2005; Kabaluk et al., 2010). The use of microbial pesticide in crop protection has been promoted under National Agricultural Technology Project (NATP) which governs Integrated Pest Management (IPM) (1998–2005) and the National Farmer Policy (2007). Unlike China, in India the National Agricultural Research System comprising of various Indian Council of Agricultural Research institutes and state Universities paly lead role in promoting microbial pesticides.

In Japan, Ministry of Agriculture, Forestry and Fisheries (1997) released guidelines for the registration of microbial pesticides (Kunimi, 2007). Since that time the Japanese Agricultural Standard Law (2000) is responsible for registration and regulation of microbial pesticides which is considered along with other chemical and soil improvement components. In South Korea, the microbial pesticide regulations were modified in 2009 replacing the one from 2005 (Wang and Zengzhi, 2010). In the regulation, the microbial pesticide was defined as agricultural pest control agents which include bacteria, fungi, viruses and protozoa as an active ingredient. The Agromaterials Management Division (AMD) along with Rural Development Administration (RDA, 2009) is responsible for the registration of a microbial pesticide. This is further evaluated for quality check for efficacy and safety to human and environment and by Pesticide Safety Evaluation (PSED) and National Academy of Agricultural Science (NAAS). In Malaysia, the term microbial pesticide is considered under the term biopesticide which covers various kinds of pesticides with different biological functions depending on different regions. The registration guidelines for microbial pesticides and botanicals products (such as natural plant extract) was released by Ministry of Agriculture in the year 2012. The guidelines issued were in line with Food and Agriculture Organization (FAO) Guidance for Harmonizing Pesticide Regulatory Management in Southeast Asia (2012), and the Association of Southeast Asian Nations (ASEAN) Guidelines on the Regulation, Use and Trade of Biological Control Agents (2014) (Biopesticide registration data requirement issued by Malaysia MOA, 2016). In Australia, the National Registration Scheme (NRS) governs the microbial pesticide registration and commercialization process along with Australian Pesticide and Veterinary Medicines Authority 1992 (http://apvma.gov.au/). The guidelines issued by APVMA (2005) (https://apvma.gov.au/node/33081) for the registration is strict and has been focused on toxicology testing including characterization of the active ingredient of the microbial pesticide, evaluation on the potential hazard, toxin production, pathogenicity, infectivity, host range and effect on flora and fauna (Kabaluk et al., 2010).

In Latin America, Brazil approved the reformation (2009) to register biocontrol agents for organic agriculture. This also included microbial pesticides and was governed by the Brazilian Ministry of Agriculture, Livestock and Food Supply (MAPA) which works along with National Health Surveillance Agency (Cotes, 2010; Arora et al., 2016). MAPA has been strict in registration and regulation and performs a quality check on the stable minimum concentration of viable active ingredient with longer shelf life, UV protection, simplified application procedure and efficacy in the fields (Kabaluk et al., 2010). Argentina has the Coordination of Agrochemical and Biological Products which comes under the Servicio Nacional de Sanidad y Calidad Agroalimentaria (SENASA) to perform task related to microbial pesticide registration and commercialization. Other important regulatory bodies involved in the regulatory process are National Agriculture Department and Environmental Policy Secretary (NADEPS), the National Agriculture Department (NAD) and Environmental Policy Secretary (EPS) (Cotes, 2010; Arora et al., 2016). Specifications for the registration of biological (including microbiological) products are provided in the Resolution 350/1999 (Chapter 12, Agents for Microbial Control), available from SACPyA (at www.infoleg.gov.ar/infoleg/internet/anexos/55000-59999/59812/texact.htm).

In the Middle East and Africa region, some of the countries are proactive and use various guidelines for registration and regulation of microbial pesticide. Countries such as Iran follows Plant Protection Act 1996 under the jurisdiction of Plant Protection Organization (PPO, 2017) for microbial pesticide registration, quality control, product efficacy and safety issues while regulation is governed by Iran Food and Drug Administration (Morteza et al., 2017). While in Africa, the demand for the microbial pesticides is driven by the need to support high-value export of horticultural crops with Integrated Pest Management standards. Earlier microbial pesticides were mainly available at the horticulture hotspots Kenya and South Africa. The success of microbial pesticides has attracted other countries in Africa to follow the guidelines for the microbial pesticide production and usage. Kenya has the regulatory authority Pest Control Product Board (PCPB) which is supported by foreign stakeholders UK Department for International Development (UK DFID). The governing body along with The Kenya Plant Health Inspectorate Service Act, Cap 512/2013 and Plant Protection Act work together for registration of microbial pesticides as well as export and import of it (Kimani, 2014). On the other hand, in South Africa, the registrations of microbial pesticides are regulated by the Department of Agriculture, Forestry and Fisheries under the Act 36 of 1947 (www.daff.gov.za) (DAFF, 2010). Similarly, Nigeria follows the protocol from the National Agency for Food and Drug Administration and Control (NAFDAC) for manufacturing, sale and distribution of microbial pesticide (reviewed in Arora et al., 2016). While Tanzania has Tropical Pesticide Research Institute (TPRI), Ministry of Agriculture which governs the registration and regulation for microbial pesticide (Stadlinger et al., 2013). Rest of the countries in this region are still improving the registration and regulation protocols of microbial pesticides (Simiyu et al., 2013).

5.5 Conclusion and future prospects

Microbial pesticides comprising entomopathogenic bacteria, fungus, virus and microsporidia are emerging as an efficient alternative to chemical pesticides. The biodegradable nature, efficiency, target specificity, low toxicity, cost-effective production has increased the demands of the microbial pesticide steadily in the global market which predicted to reach USD 7.8 billion by the year 2023. Although a tremendous surge in the knowledge and technology has been observed in recent years on microbial pesticides with respect to the mode of action, formulation development and commercialization; there are still several constraints ahead to compete with the chemical pesticides market in terms of developing stable formulations, supply and storage, standardizing appropriate delivery methods, regulation on usage, advertising, sale transportation and the establishment of market-friendly legislation and registration procedures.

Other than that, as microbial pesticides are pest-specific which makes the potential market for these products limited. Research must be focused toward the development of microbes which can withstand the changing environmental conditions. Product formulation should encompass more on microbial ingredients which can target the few important pests and diseases of major crops in the region in order to sustain the microbial pesticide industry. Enhancing product availability for the farmers and technology transfer and extension services are important steps for increasing the awareness for microbial pesticides among all stakeholders. In addition, strong networking should be established to introduce specific policies and provide technical support to all stakeholders involved in research, production, regulation, export, quarantine, marketing and use of microbial pesticide. Most of the countries have revised their policies to reduce the use of chemical pesticides and promote microbial pesticides. Standardization of regulation and registration requirements globally would further support the hassle-free trade between the countries.

Acknowledgments

Authors gratefully acknowledge, Ms. Dharane A/P Kethiravan, CEBAR, Universiti Malaya, Malaysia for her assistance with artwork.

References

AGBR, 2015. Agrow Global Biopesticide Regulations. Commodity Analysis. www.agra-net.com (Assessed on 15 May 2020).

Aizawa, K., Shigetane, I., 2001. Discovery of sottokin (*Bacillus thuringiensis*) in 1901 and subsequent investigations in Japan. In: Proceedings of a Centennial Symposium Commemorating Ishiwata's Discovery of *Bacillus thuringiensis*. Bioresource Technology, vol. 99, pp. 959–964.

Akhter, W., Bhuiyan, M.K.A., Sultana, F., Hossain, M.M., 2015. Integrated effect of microbial antagonist, organic amendment and fungicide in controlling seedling mortality (*Rhizoctonia solani*) and improving yield in pea (*Pisum sativum* L.). Comptes Rendus Biol. 338 (1), 21–28.

Alves, S.B., Lopes, R.B., Vieira, S., Tamai, M.A., 2008. Fungos entomopatogénicos usados no controle de pragasna America Latina. In: Alves, S.B., Biaggioni, L.R. (Eds.), Controle microbiano de pragasna America Latina. FEALQ, Piracicaba, pp. 69–110.

Amira, M.B., Lopez, D., Mohamed, A.T., Khouaja, A., Chaar, H., Fumanal, B., Gousset-Dupont, A., Bonhomme, L., Label, P., Goupil, P., Ribeiro, S., 2017. Beneficial effect of *Trichoderma harzianum* strain Ths97 in biocontrolling *Fusarium solani* causal agent of root rot disease in olive trees. Biol. Contr. 110, 70–78.

Amsellem, Z., Cohen, B.A., Gressel, J., 2002. Engineering hypervirulence in a mycoherbicidal fungus for efficient weed control. Nat. Biotechnol. 20 (10), 1035–1039.

Anandan, R., Dharumadurai, D., Manogaran, G.P., 2016. An introduction to actinobacteria. In: Actinobacteria-Basics and Biotechnological Applications. Intechopen.

Anwer, M.A., Khan, M.R., 2013. *Aspergillus niger* as tomato fruit (*Lycopersicum esculentum* Mill.) quality enhancer and plant health promoter. J. Postharvest Technol. 1 (1), 36–51.

Araújo, J.P., Hughes, D.P., 2016. Diversity of entomopathogenic fungi: which groups conquered the insect body?. In: Advances in Genetics, vol. 94. Academic Press, pp. 1–39.

Argôlo-Filho, R.C., Loguercio, L.L., 2014. *Bacillus thuringiensis* is an environmental pathogen and host-specificity has developed as an adaptation to human-generated ecological niches. Insects 5 (1), 62–91.

Armes, N.J., Jadhav, D.R., Bond, G.S., King, A.B., 1992. Insecticide resistance in *Helicoverpa armigera* in South India. Pestic. Sci. 34 (4), 355–364.

Arora, N.K., Verma, M., Prakash, J., Mishra, J., 2016. Regulation of biopesticides: global concerns and policies. In: Bioformulations: for Sustainable Agriculture. Springer, New Delhi, pp. 283–299.

Arthurs, S., Dara, S.K., 2019. Microbial biopesticides for invertebrate pests and their markets in the United States. J. Invertebr. Pathol. 165, 13–21.

Ashraf, M.A., Rasool, M., Mirza, M.S., 2011. Nitrogen fixation and indole acetic acid production potential of bacteria isolated from rhizosphere of sugarcane (*Saccharum officinarum* L.). Adv. Biol. Regul. 5 (6), 348–355.

n.d. Asia and Pacific Plant Protection Commission and Food and Agriculture Organization of the United Nations Regional Office for Asia and the Pacific. 2007–2008. http://www.fao.org/3/AG123E00.htm#Contents. (Assessed on 16 May 2020).

Asia-pacific BioPesticide Market, August 2019. https://www.marketdataforecast.com/market-reports/asia-pacific-bio-pesticide-market (Assessed on 16 May 2020).

Askary, H., Benhamou, N., Brodeur, J., 1997. Ultrastructural and cytochemical investigations of the antagonistic effect of *Verticillium lecanii* on cucumber powdery mildew. Phytopathology 87 (3), 359–368.

Askary, H., Carriere, Y., Belanger, R.R., Brodeur, J., 1998. Pathogenicity of the fungus *Verticillium lecanii* to aphids and powdery mildew. Biocontrol Sci. Technol. 8 (1), 23–32.

Atkins, S.D., Hidalgo-Diaz, L., Kalisz, H., Mauchline, T.H., Hirsch, P.R., Kerry, B.R., 2003. Development of a new management strategy for the control of root-knot nematodes (*Meloidogyne* spp) in organic vegetable production. Pest Manag. Sci. Form. Pestic. Sci. 59 (2), 183–189.

Avery, P.B., Pick, D.A., Aristizábal, L.F., Kerrigan, J., Powell, C.A., Rogers, M.E., Arthurs, S.P., 2013. Compatibility of *Isaria fumosorosea* (Hypocreales: Cordycipitaceae) blastospores with agricultural chemicals used for management of the Asian citrus psyllid, *Diaphorina citri* (Hemiptera: Liviidae). Insects 4 (4), 694–711.

Aw, K.M.S., Hue, S.M., 2017. Mode of infection of *Metarhizium* spp. fungus and their potential as biological control agents. J. Fungi 3 (2), 30.

Awla, H.K., Kadir, J., Othman, R., Rashid, T.S., Hamid, S., Wong, M.Y., 2017. Plant growth-promoting abilities and biocontrol efficacy of *Streptomyces* sp. UPMRS4 against *Pyricularia oryzae*. Biol. Contr. 112, 55–63.

Baidoo, R., Mengistu, T., McSorley, R., Stamps, R.H., Brito, J., Crow, W.T., 2017. Management of root-knot nematode (*Meloidogyne incognita*) on *Pittosporum tobira* under greenhouse, field, and on-farm conditions in Florida. J. Nematol. 49 (2), 133.

Bakhshandeh, E., Pirdashti, H., Lendeh, K.S., 2017. Phosphate and potassium-solubilizing bacteria effect on the growth of rice. Ecol. Eng. 103, 164–169.

Barahona, E., Navazo, A., Martínez-Granero, F., Zea-Bonilla, T., Pérez-Jiménez, R.M., Martín, M., Rivilla, R., 2011. *Pseudomonas fluorescens* F113 mutant with enhanced competitive colonization ability and improved biocontrol activity against fungal root pathogens. Appl. Environ. Microbiol. 77 (15), 5412–5419.

Barrangou, R., van Pijkeren, J.P., 2016. Exploiting CRISPR–Cas immune systems for genome editing in bacteria. Curr. Opin. Biotechnol. 37, 61–68.

Bauer, L.S., Londoño, D.K., 2011. Effects of *Bacillus thuringiensis* SDS-502 on adult emerald ash borer. Annapolis, MD. Gen. Tech. Rep. NRS-P-75. In: McManus, K.A., Gottschalk, K.W. (Eds.), 2010. Proceedings. 21st US Department of Agriculture Interagency Research Forum on Invasive Species 2010; 2010 January 12–15. US Department of Agriculture, Forest Service, Northern Research Station: 74–75, Newtown Square, PA, pp. 74–75.

Becher, P.G., Jensen, R.E., Natsopoulou, M.E., Verschut, V., Henrik, H., 2018. Infection of *Drosophila suzukii* with the obligate insect-pathogenic fungus *Entomophthora muscae*. J. Pest. Sci. 91 (2), 781–787.

Ben-Dov, E., 2014. *Bacillus thuringiensis* subsp. israelensis and its dipteran-specific toxins. Toxins 6 (4), 1222–1243.

Bent, E., 2006. Induced systemic resistance mediated by plant growth-promoting rhizobacteria (PGPR) and fungi (PGPF). In: Tuzun, S., Bent, E. (Eds.), Multigenic and Induced Systemic Resistance in Plants. Springer, New York, pp. 225–258.

Berg, G., 2009. Plant–microbe interactions promoting plant growth and health: perspectives for controlled use of microorganisms in agriculture. Appl. Microbiol. Biotechnol. 84 (1), 11–18.

Besset-Manzoni, Y., Rieusset, L., Joly, P., Comte, G., Prigent-Combaret, C., 2018. Exploiting rhizosphere microbial cooperation for developing sustainable agriculture strategies. Environ. Sci. Pollut. Control Ser. 25 (30), 29953–29970.

Bettiol, W., 2011. Biopesticide use and research in Brazil. Outlooks Pest Manag. 22, 280–283.

Bhattacharyya, P.N., Jha, D.K., 2012. Plant growth-promoting rhizobacteria (PGPR): emergence in agriculture. World J. Microbiol. Biotechnol. 28 (4), 1327–1350.

Biopesticide Registration Data Requirement Issued by Malaysia MOA, 2016. https://agrochemical.chemlinked.com/news/news/biopesticide-registration-data-requirement-issued-malaysia-moa. (Accessed 19 May 2020).

Biopesticides in Australia, 2018. In Integrated Crop Protection. https://www.soilwealth.com.au/imagesDB/news/ICP-SW_Biopesticidesfactsheetv6.pdf (Assessed on 18 May 2020).

Bjørnson, S., Oi, D., 2014. Microsporidia biological control agents and pathogens of beneficial insects. In: Microsporidia: Pathogens of Opportunity, vol. 1, pp. 635–670.

Blumberg, B.J., Short, S.M., Dimopoulos, G., 2016. Employing the mosquito microflora for disease control. In: Genetic Control of Malaria and Dengue. Academic Press, pp. 335–362.

Bonning, B.C., Hammock, B.D., 1996. Development of recombinant baculoviruses for insect control. Annu. Rev. Entomol. 41 (1), 191–210.

Boominathan, U., 2020. Screening PGPR for improving seed germination, seedling growth and yield of maize (*Zee Mays* L.). In: Studies in Indian Place Names, vol. 40, pp. 425–430 (20).

Botto, E.N., 1996. Control biológico de plagas en La Argentina: informe de la situación actual. El control biológico en América Latina, Buenos Aires, pp. 1–8.

Bouffaud, M.L., Kyselková, M., Gouesnard, B., Grundmann, G., Muller, D., Moënne-Loccoz, Y.V.A.N., 2012. Is diversification history of maize influencing selection of soil bacteria by roots? Mol. Ecol. 21 (1), 195–206.

Braud, A., Jézéquel, K., Bazot, S., Lebeau, T., 2009. Enhanced phytoextraction of an agricultural Cr-and Pb-contaminated soil by bioaugmentation with siderophore-producing bacteria. Chemosphere 74 (2), 280–286.

Braun-Kiewnick, A., Lehmann, A., Rezzonico, F., Wend, C., Smits, T.H., Duffy, B., 2012. Development of species-, strain-and antibiotic biosynthesis-specific quantitative PCR assays for *Pantoea agglomerans* as tools for biocontrol monitoring. J. Microbiol. Methods 90 (3), 315–320.

Bravo, A., Gill, S.S., Soberon, M., 2007. Mode of action of *Bacillus thuringiensis* Cry and Cyt toxins and their potential for insect control. Toxicon 49 (4), 423–435.

Brown, S.E., Cao, A.T., Hines, E.R., Akhurst, R.J., East, P.D., 2004. A novel secreted protein toxin from the insect pathogenic bacterium *Xenorhabdus nematophila*. J. Biol. Chem. 279 (15), 14595–14601.

Brown, M.B., Brown, C.M.B., Nepomuceno, R.A., 2016. Regulatory requirements and registration of biopesticides in the Philippines. In: Singh, H., Sarma, B., Keswani, C. (Eds.), Agriculturally Important Microorganisms. Springer, Singapore, pp. 183–195.

Brown, S.E., Cao, A.T., Dobson, P., Hines, E.R., Akhurst, R.J., East, P.D., 2006. Txp40, a ubiquitous insecticidal toxin protein from *Xenorhabdus* and *Photorhabdus* bacteria. Appl. Environ. Microbiol. 72 (2), 1653–1662.

Bulgarelli, D., Schlaeppi, K., Spaepen, S., Van Themaat, E.V.L., Schulze-Lefert, P., 2013. Structure and functions of the bacterial microbiota of plants. Annu. Rev. Plant Biol. 64, 807–838.

Burden, J.P., Griffiths, C.M., Cory, J.S., Smith, P., Sait, S.M., 2002. Vertical transmission of sublethal granulovirus infection in the Indian meal moth, *Plodia interpunctella*. Mol. Ecol. 11 (3), 547–555.

Burges, H.D., Jones, K.A., 1998. Formulation of bacteria, viruses and protozoa to control insects. In: Formulation of Microbial Biopesticides. Springer, Dordrecht, pp. 33–127.

Business Wire, 2010. Research and Markets: The 2010 Biopesticides Market in Europe & Company Index – Opportunities Exist Which Could Raise the Total Market to $200 Million by 2020. New York.

CABI, 2010. The 2010 Worldwide Biopesticides: Market Summary. CPL Business Consultants, London, p. 40.

Cali, A., Becnel, J.J., Takvorian, P.M., 2017. Microsporidia. In: Archibald, J.M., Simpson, A.G.B., Slamovits, C.H., Margulis, L., Melkonian, M., Chapman, D.J., et al. (Eds.), Handbook of the Protists Ebook. Springer International Publishing, Berlin.

Chandler, D., Davidson, G., Grant, W.P., Greaves, J., Tatchell, G.M., 2008. Microbial biopesticides for integrated crop management: an assessment of environmental and regulatory sustainability. Trends Food Sci. Technol. 19 (5), 275–283.

Chandra Nayaka, S., Niranjana, S.R., Uday Shankar, A.C., Niranjan Raj, S., Reddy, M.S., Prakash, H.S., Mortensen, C.N., 2010. Seed biopriming with novel strain of *Trichoderma harzianum* for the control of toxigenic *Fusarium verticillioides* and fumonisins in maize. Arch. Phytopathol. Plant Protect. 43 (3), 264–282.

Chaparro, J.M., Badri, D.V., Bakker, M.G., Sugiyama, A., Manter, D.K., Vivanco, J.M., 2013. Root exudation of phytochemicals in *Arabidopsis* follows specific patterns that are developmentally programmed and correlate with soil microbial functions. PloS One 8 (2).

Charles, J.F., Nielson-LeRoux, C., Delecluse, A., 1996. *Bacillus sphaericus* toxins: molecular biology and mode of action. Annu. Rev. Entomol. 41 (1), 451–472.

Chaston, J.M., Suen, G., Tucker, S.L., Andersen, A.W., Bhasin, A., Bode, E., Bode, H.B., Brachmann, A.O., Cowles, C.E., Cowles, K.N., Darby, C., 2011. The entomopathogenic bacterial endosymbionts *Xenorhabdus* and *Photorhabdus*: convergent lifestyles from divergent genomes. PloS One 6 (11).

Chater, K.F., 1984. Morphological and Physiological Differentiation in Streptomyces. Microbial development, pp. 89–115.

Chattopadhyay, P., Banerjee, G., Mukherjee, S., 2017. Recent trends of modern bacterial insecticides for pest control practice in integrated crop management system. 3 Biotech 7 (1), 60.

Chen, L., Chen, W., 2009. Genome shuffling enhanced antagonistic activity against *Fusarium oxysporum* f. sp. melonis and tolerance to chemical fungicides in *Bacillus subtilis* BS14. J. Food Agric. Environ. 7 (2), 856–860.

Chen, Y.Y., Chen, P.C., Tsay, T.T., 2016. The biocontrol efficacy and antibiotic activity of *Streptomyces plicatus* on the oomycete *Phytophthora capsici*. Biol. Contr. 98, 34–42.

Chen, J., Lai, Y., Wang, L., Zhai, S., Zou, G., Zhou, Z., Cui, C., Wang, S., 2017. CRISPR/Cas9-mediated efficient genome editing via blastospore-based transformation in entomopathogenic fungus *Beauveria bassiana*. Sci. Rep. 7, 45763.

Chen, Y., Zhou, D., Qi, D., Gao, Z., Xie, J., Luo, Y., 2018. Growth promotion and disease suppression ability of a *Streptomyces* sp. CB-75 from banana rhizosphere soil. Front. Microbiol. 8, 2704.

Clermont, N., Lerat, S., Beaulieu, C., 2011. Genome shuffling enhances biocontrol abilities of *Streptomyces* strains against two potato pathogens. J. Appl. Microbiol. 111 (3), 671–682.

Coast, G.M., Orchard, I., Phillips, J.E., Schooley, D.A., 2002. Insect diuretic and antidiuretic hormones. Adv. Insect Physiol. 29, 279–409.

Cordero, P., Príncipe, A., Jofré, E., Mori, G., Fischer, S., 2014. Inhibition of the phytopathogenic fungus *Fusarium proliferatum* by volatile compounds produced by *Pseudomonas*. Arch. Microbiol. 196 (11), 803–809.

Cordova-Kreylos, A.L., Fernandez, L.E., Koivunen, M., Yang, A., Flor-Weiler, L., Marrone, P.G., 2013. Isolation and characterization of *Burkholderia rinojensis* sp. nov., a non-*Burkholderia cepacia* complex soil bacterium with insecticidal and miticidal activities. Appl. Environ. Microbiol. 79 (24), 7669–7678.

Corradi, N., Bonfante, P., 2012. The arbuscular mycorrhizal symbiosis: origin and evolution of a beneficial plant infection. PLoS Pathog. 8 (4).

Cotes, A.M., October 2010. Registry and regulation of biocontrol agents on food commodities in South America, 905. In: International Symposium on Biological Control of Postharvest Diseases: Challenges and Opportunities, pp. 301–306.

CPL Business Consultants, 2010. The 2010 Worldwide Biopesticides Market Summary, vol. 1. CPL Business Consultants, Wallingford.

Czaja, K., Góralczyk, K., Struciński, P., Hernik, A., Korcz, W., Minorczyk, M., Łyczewska, M., Ludwicki, J.K., 2015. Biopesticides–towards increased consumer safety in the European Union. Pest Manag. Sci. 71 (1), 3–6.

Daborn, P.J., Waterfield, N., Silva, C.P., Au, C.P.Y., Sharma, S., 2002. A single *Photorhabdus* gene, makes caterpillars floppy (mcf), allows *Escherichia coli* to persist within and kill insects. Proc. Natl. Acad. Sci. U. S. A. 99 (16), 10742–10747.

DAFF, 2010. Department of Agriculture, Forestry and Fisheries. Act No. 36 of 1947, Guidelines on the Data Required for Registration of Biological/biopesticides Remedies in South Africa, Republic of South Africa.

Dara, S.K., 2017. Microbial control of arthropod pests in small fruits and vegetables in temperate climate. In: Microbial Control of Insect and Mite Pests. Academic Press, pp. 209–221.

de la Cruz Quiroz, R., Maldonado, J.J.C., Alanis, M.D.J.R., Torres, J.A., Saldívar, R.P., 2019. Fungi-based biopesticides: shelf-life preservation technologies used in commercial products. J. Pest. Sci. 92 (3), 1003–1015.

de los Santos-Villalobos, S., De Folter, S., Délano-Frier, J.P., Gómez-Lim, M.A., Guzmán-Ortiz, D.A., Pena-Cabriales, J.J., 2013. Growth promotion and flowering induction in mango (*Mangifera indica* L. cv "Ataulfo") trees by Burkholderia and Rhizobium Inoculation: morphometric, biochemical, and molecular events. J. Plant Growth Regul. 32 (3), 615–627.

Dhakal, D., Pokhrel, A.R., Jha, A.K., Thuan, N.H., Sohng, J.K., 2017. *Saccharopolyspora* species: laboratory maintenance and enhanced production of secondary metabolites. Curr. Protoc. Microbiol. 44 (1), 10H-1.

Dieppois, G., Opota, O., Lalucat, J., Lemaitre, B., 2015. *Pseudomonas entomophila*: a versatile bacterium with entomopathogenic properties. In: Pseudomonas. Springer, Dordrecht, pp. 25–49.

Downing, K.J., Thomson, J.A., 2000. Introduction of the *Serratia marcescens chiA* gene into an endophytic *Pseudomonas fluorescens* for the biocontrol of phytopathogenic fungi. Can. J. Microbiol. 46 (4), 363–369.

Drogue, B., Combes-Meynet, E., Moënne-Loccoz, Y., Wisniewski-Dyé, F., Prigent-Combaret, C., 2013. Control of the cooperation between plant growth-promoting rhizobacteria and crops by rhizosphere signals. Mol. Microbial. Ecol. Rhizosphere 1, 279–293.

Easom, C.A., Clarke, D.J., 2008. Motility is required for the competitive fitness of entomopathogenic *Photorhabdus luminescens* during insect infection. BMC Microbiol. 8 (1), 168.

EC, 2001. European Commission Directive 2001/36/EC amending Council Directive91/414/EEC concerning the placing of plant protection products on the market. Off. J. Eur. Union 164, 1–38.

EC, 2005. European Commission. Council directive 2005/25/EC of 14 march 2005 amending annex VI to directive 91/414/EEC as regards plant protection products containing micro-organisms. Off. J. Eur. Union 90, 1–34.

EC, 2009. European Commission. Regulation (EC) No 1107/2009 of the European parliament and of the council of 21 October 2009 concerning the placing of plant protection products on the market and repealing council directive 79/117/EEC and 91/414/EEC. Off. J. Eur. Union 309, 1–50.

EC, 2013. European Commission. No. 283/2013 setting out the data requirements for active substances, in accordance with regulation (EC) No. 1107/2009 of the European parliament and of the council concerning the placing of plant protection products on the market. Off. J. Eur. Union 93, 1–84.

European Food Safety Authority (EFSA), 2016. Peer review of the pesticide risk assessment of the active substance *Bacillus amyloliquefaciens* strain MBI 600. EFSA J. 14 (1), 4359.

European Food Safety Authority (EFSA), Arena, M., Auteri, D., Barmaz, S., Bellisai, G., Brancato, A., Brocca, D., Bura, L., Byers, H., Chiusolo, A., Court Marques, D., 2018. Peer review of the pesticide risk assessment of the active substance *Pasteuria nishizawae* Pn1. EFSA J. 16 (2), 05159.

Egbuna, C., Sawicka, B., Tijjani, H., Kryeziu, T.L., Ifemeje, J.C., Skiba, D., Lukong, C.B., 2020. Biopesticides, safety issues and market trends. In: Natural Remedies for Pest, Disease and Weed Control. Academic Press, pp. 43–53.

El Khaldi, R., Remadi, M.D., Hamada, W., Somai, L., Cherif, M., 2015. The potential of serratia marcescens: an indigenous strain isolated from date palm compost as biocontrol agent of *Rhizoctonia solani* on potato. J. Plant Pathol. Microbiol. 3, 006.

El Taweel, A.A., Omar, M.N.A., Shaheen, S.A., 2011. Effect of inoculation by some plant growth promoting rhizobacteria (PGPR) on production of'manzanillo'olive trees, 1018. In: International Symposium on Organic Matter Management and Compost Use in Horticulture, pp. 245–254.

Elad, Y., 2000. Biological control of foliar pathogens by means of *Trichoderma harzianum* and potential modes of action. Crop Protect. 19 (8–10), 709–714.

EU., 1991. European Union. Council directive 91/414/EEC of 15 July 1991 concerning the placing of plant protection products on the market. Off. J. Eur. Union 230, 1–32.

EUPD, 2010. European Union Pesticides Database. http://ec.europa.eu/food/plant/protection/evaluation/database. Assessed on 19 May 2020).

Europe BioPesticide Market, 2019. https://www.marketdataforecast.com/market-reports/europe-bio-pesticide-market (Assessed on 15 May 2020).

European Food Safety Authority, 2012. Conclusion on the peer review of the pesticide risk assessment of the active substance *Bacillus thuringiensis* subsp. kurstaki (strains ABTS 351, PB 54, SA 11, SA 12, EG 2348). EFSA J. 10 (2), 2540.

European Food Safety Authority, 2013a. Conclusion on the peer review of the pesticide risk assessment of the active substance *Bacillus thuringiensis* subsp. aizawai (strains ABTS 1857, GC-91). EFSA J. 11 (1), 3063.

European Food Safety Authority, 2013b. Conclusion on the peer review of the pesticide risk assessment of the active substance *Bacillus thuringiensis* ssp. tenebrionis strain NB-176. EFSA J. 11 (1), 3024.

Fan, J., Xie, Y., Xue, J., Liu, R., 2013. The effect of *Beauveria brongniartii* and its secondary me-tabolites on the detoxification enzymes of the pine caterpillar, *Dendrolimus tabulaeformis*. J. Insect Sci. 13 (1), 44.

Fang, L., 2014. Development and Management of Biopesticides in china. https://pesticide.chemlinked.com/news/newsagrochemical-news/developmentand-management biopesticides-china. (Accessed 20 May 2020).

Fang, W., Leng, B., Xiao, Y., Jin, K., Ma, J., Fan, Y., Feng, J., Yang, X., Zhang, Y., Pei, Y., 2005. Cloning of *Beauveria bassiana* chitinase gene *Bbchit1* and its application to improve fungal strain virulence. Appl. Environ. Microbiol. 71 (1), 363–370.

Feklistova, I.N., Maksimova, N.P., 2008. Obtaining *Pseudomonas aurantiaca* strains capable of overproduction of phenazine antibiotics. Microbiology 77 (2), 176–180.

Fernández-Chapa, D., Ramírez-Villalobos, J., Galán-Wong, L., 2019. Toxic potential of *Bacillus thuringiensis*: an overview. In: Protecting Rice Grains in the Post-Genomic Era. IntechOpen.

Ffrench-Constant, R., Waterfield, N., 2006. An ABC guide to the bacterial toxin, complexes. Adv. Appl. Microbiol. 58, 169–183.

Ffrench-Constant, R.H., Dowling, A., Waterfield, N.R., 2007. Insecticidal toxins from *Photorhabdus* bacteria and their potential use in agriculture. Toxicon 49 (4), 436–451.

Fiedler, Ż., Sosnowska, D., 2007. Nematophagous fungus *Paecilomyces lilacinus* (Thom) Samson is also a biological agent for control of greenhouse insects and mite pests. Bio Control 52 (4), 547–558.

Filha, M.H.N.L.S., Berry, C., Regis, L., 2014. *Lysinibacillus sphaericus*: toxins and mode of action, applications for mosquito control and resistance management. In: Advances in Insect Physiology, vol. 47. Academic Press, pp. 89–176.

García-Seco, D., Bonilla, A., Algar, E., García-Villaraco, A., Mañero, J.G., Ramos-Solano, B., 2013. Enhanced blackberry production using *Pseudomonas fluorescens* as elicitor. Agron. Sustain. Dev. 33 (2), 385–392.

GBMAIF, September 22, 2017. Global Biopesticides Market Analysis and Industry Forecasts 2017–2022—A $7.62 Billion Opportunity. News provided by Research and Markets, Dublin, 07.

Givaudan, A., Lanois, A., 2000. *flhDC*, the flagellar master operon of *Xenorhabdus nematophilus*: requirement for motility, lipolysis, extracellular hemolysis, and full virulence in insects. J. Bacteriol. 182 (1), 107–115.

Glare, T.R., Jurat-Fuentes, J.L., O'callaghan, M., 2017. Basic and applied research: entomopathogenic bacteria. In: Microbial Control of Insect and Mite Pests. Academic Press, pp. 47–67.

Gleeson, O., O'Gara, F., Morrissey, J.P., 2010. The Pseudomonas fluorescens secondary metabolite 2, 4 diacetylphloroglucinol impairs mitochondrial function in *Saccharomyces cerevisiae*. Antonie van Leeuwenhoek 97 (3), 261–273.

Glick, B.R., 2012. Plant growth-promoting bacteria: mechanisms and applications. Scientifica 2012, 1–15.

Global Pesticides Market by Type (Synthetic Pesticides & Bio Pesticides), By Application (Cereal, Fruits, Plantation Crops, Vegetables & Others), By Formulation (Dry & Liquid), By Region, Competition Forecast & Opportunities, 2013 – 2023, June 2018. https://www.techsciresearch.com/report/global-pesticides-market/1311.html. (Accessed 24 April 2020).

Gong, L., Tan, H., Chen, F., Li, T., Zhu, J., Jian, Q., Yuan, D., Xu, L., Hu, W., Jiang, Y., Duan, X., 2016. Novel synthesized 2, 4-DAPG analogues: antifungal activity, mechanism and toxicology. Sci. Rep. 6, 32266.

Gong, J., Zheng, H., Wu, Z., Chen, T., Zhao, X., 2009. Genome shuffling: progress and applications for phenotype improvement. Biotechnol. Adv. 27 (6), 996–1005.

Gongora, B., 2004. Transformacion de *Beauveria bassiana* cepa Bb9112 con los genes de la proteina verde fluorescente y la proteasa pr1A de *Metarhizium anisopliae*. Sociedad Botanica de Entomologia, Bogota (Colombia).

Gosal, S.K., Karlupia, A., Gosal, S.S., Chhibba, I.M., Varma, A., 2010. Biotization with Piriformospora Indica and Pseudomonas Fluorescens Improves Survival Rate, Nutrient Acquisition, Field Performance and Saponin Content of Micropropagated *Chlorophytum* Sp.

Grady, E.N., MacDonald, J., Liu, L., Richman, A., Yuan, Z.C., 2016. Current knowledge and perspectives of *Paenibacillus*: a review. Microb. Cell Factories 15 (1), 203.

Gramkow, A.W., Perecmanis, S., Sousa, R.L.B., Noronha, E.F., Felix, C.R., Nagata, T., Ribeiro, B.M., 2010. Insecticidal activity of two proteases against *Spodoptera frugiperda* larvae infected with recombinant baculoviruses. Virol. J. 7 (1), 143.

Gray, E.J., Smith, D.L., 2005. Intracellular and extracellular PGPR: commonalities and distinctions in the plant–bacterium signaling processes. Soil Biol. Biochem. 37 (3), 395–412.

Grzywacz, D., Cherry, A., Gwynn, R., 2009. Biological pesticides for Africa: why has so little of the research undertaken to date resulted in new products to help Africa's poor? Outlooks Pest Manag. 20 (2), 77.

Gudiño Gomezjurado, M.E., de Abreu, L.M., Marra, L.M., Pfenning, L.H., de, S., Moreira, F.M., 2015. Phosphate solubilization by several genera of saprophytic fungi and its influence on corn and cowpea growth. J. Plant Nutr. 38 (5), 675–686.

Gull, M., Hafeez, F.Y., 2012. Characterization of siderophore producing bacterial strain *Pseudomonas fluorescens* Mst 8.2 as plant growth promoting and biocontrol agent in wheat. Afr. J. Microbiol. Res. 6 (33), 6308–6318.

Hajjar, M.J., 2012. The persisted organic pesticides pollutant (POPs) in the Middle East Arab countries. Int. J. Agron. Plant Prod. 3, 11–18.

Hamayun, M., Hussain, A., Khan, S.A., Kim, H.Y., Khan, A.L., Waqas, M., Irshad, M., Iqbal, A., Rehman, G., Jan, S., Lee, I.J., 2017. Gibberellins producing endophytic fungus *Porostereum spadiceum* AGH786 rescues growth of salt affected soybean. Front. Microbiol. 8, 686.

Han, B., Takvorian, P.M., Weiss, L.M., 2020. Invasion of host cells by microsporidia. Front. Microbiol. 11.

Harrison, R., Hoover, K., 2012. Baculoviruses and other occluded insect viruses. Insect Pathol. 73–131.

Harrison, R.L., Herniou, E.A., Jehle, J.A., Theilmann, D.A., Burand, J.P., Becnel, J.J., Krell, P.J., van Oers, M.M., Mowery, J.D., Bauchan, G.R., 2018. ICTV virus taxonomy profile: Baculoviridae. J. Gen. Virol. 99 (9), 1185–1186.

Hassan, E., Gökçe, A., 2014. Production and consumption of biopesticides. In: Advances in Plant Biopesticides. Springer, New Delhi, pp. 361–379.

Hastuti, R.D., Lestari, Y., Suwanto, A., Saraswati, R., 2012. Endophytic *Streptomyces* spp. as biocontrol agents of rice bacterial leaf blight pathogen (*Xanthomonas oryzae* pv. oryzae). HAYATI J. Biosci. 19 (4), 155–162.

Hibbett, D.S., Binder, M., Bischoff, J.F., Blackwell, M., Cannon, P.F., Eriksson, O.E., Huhndorf, S., James, T., Kirk, P.M., Lücking, R., Lumbsch, H.T., 2007. A higher-level phylogenetic classification of the Fungi. Mycol. Res. 111 (5), 509–547.

Hidaka, Y., 1933. Utilization of natural enemies for control of the pine caterpillar. J. Jpn. For. Soc. 15, 1221–1231.

Hossain, M.M., Sultana, F., Kubota, M., Koyama, H., Hyakumachi, M., 2007. The plant growth-promoting fungus *Penicillium simplicissimum* GP17-2 induces resistance in *Arabidopsis thaliana* by activation of multiple defense signals. Plant Cell Physiol. 48 (12), 1724–1736.

Hossain, M.M., Sultana, F., Miyazawa, M., Hyakumachi, M., 2014. The plant growth-promoting fungus *Penicillium* spp. GP15-1 enhances growth and confers protection against damping-off and anthracnose in the cucumber. J. Oleo Sci. 63 (4), 391–400.

Hossain, M.M., Sultana, F., Hyakumachi, M., 2017. Role of ethylene signalling in growth and systemic resistance induction by the plant growth-promoting fungus *Penicillium viridicatum* in *Arabidopsis*. J. Phytopathol. 165 (7–8), 432–441.

Hubbard, M., Hynes, R.K., Erlandson, M., Bailey, K.L., 2014. The biochemistry behind biopesticide efficacy. Sustain. Chem. Process. 2 (1), 18.

Hunter, D.M., Milner, R.J., Spurgin, P.A., 2001. Aerial treatment of the Australian plague locust, *Chortoicetes terminifera* (Orthoptera: Acrididae) with *Metarhizium anisopliae* (Deuteromycotina: Hyphomycetes). Bull. Entomol. Res. 91 (2), 93–99.

Hurst, M.R., Jones, S.M., Tan, B., Jackson, T.A., 2007. Induced expression of the *Serratia entomophila* Sep proteins shows activity towards the larvae of the New Zealand grass grub *Costelytra zealandica*. FEMS (Fed. Eur. Microbiol. Soc.) Microbiol. Lett. 275 (1), 160–167.

ICAMA, 2008. Pesticide manual, the institute for the control of agrochemicals. Ministry of Agriculture, China (in Chinese). Taken from secondary reference. In: Arora, N.K. (Ed.), Plant Microbes Symbiosis: Applied Facets, vol. 147. Springer, New Delhi, 2015.

Ishimwe, E., Hodgson, J.J., Passarelli, A.L., 2015. Expression of the *Cydia pomonella* granulovirus matrix metalloprotease enhances *Autographa californica* multiple nucleopolyhedrovirus virulence and can partially substitute for viral cathepsin. Virology 48, 166–178.

Islam, S., Akanda, A.M., Sultana, F., Hossain, M.M., 2014a. Chilli rhizosphere fungus *Aspergillus* spp. PPA1 promotes vegetative growth of cucumber (*Cucumis sativus*) plants upon root colonisation. Arch. Phytopathol. Plant Protect. 47 (10), 1231–1238.

Islam, S., Akanda, A.M., Prova, A., Sultana, F., Hossain, M.M., 2014b. Growth promotion effect of *Fusarium* spp. PPF1 from bermudagrass (*Cynodon dactylon*) rhizosphere on Indian spinach (*Basella alba*) seedlings are linked to root colonisation. Arch. Phytopathol. Plant Protect. 47 (19), 2319–2331.

Jaber, L.R., Enkerli, J., 2017. Fungal entomopathogens as endophytes: can they promote plant growth? Biocontrol Sci. Technol. 27 (1), 28–41.

Jäkel, T., 2003. Biopesticides and pest management systems: recent developments and future needs in developing countries of Southeast Asia. In: International Symposium on Biopesticides for Developing Countries. Bib. Orton IICA/CATIE, pp. 187–193.

Jallow, M.F., Awadh, D.G., Albaho, M.S., Devi, V.Y., Thomas, B.M., 2017. Pesticide risk behaviors and factors influencing pesticide use among farmers in Kuwait. Sci. Total Environ. 574, 490–498.

n.d. Japanese Agriculture Standard (JAS) Law (2000) Notification No. 1005 of the Ministry of Agriculture, Forestry and Fisheries. www.maff.go.jp/e/jas/specific/pdf/1401qa-organic-plants.pdf.

Jehle, J.A., Blissard, G.W., Bonning, B.C., Cory, J.S., Herniou, E.A., Rohrmann, G.F., Theilmann, D.A., Thiem, S.M., Vlak, J.M., 2006. On the classification and nomenclature of baculoviruses: a proposal for revision. Arch. Virol. 151 (7), 1257–1266.

Jeong, J.K., Sang, G.L., Siwoo, L., Hyeong, J.J., 2010. South Korea. In: Kabaluk, J.T., Antonet, M.S., Mark, S.G., Stephanie, G.W. (Eds.), The Use and Regulation of Microbial Pesticides in Representative Jurisdictions Worldwide. IOBC Global.

Jiang, L., Xia, Q., 2014. The progress and future of enhancing antiviral capacity by transgenic technology in the silkworm Bombyx mori. Insect Biochem. Mol. Biol. 48, 1–7.

Jin, Q., Jiang, Q., Zhao, L., Su, C., Li, S., Si, F., Li, S., Zhou, C., Mu, Y., Xiao, M., 2017. Complete genome sequence of *Bacillus velezensis* S3-1, a potential biological pesticide with plant pathogen inhibiting and plant promoting capabilities. J. Biotechnol. 259, 199–203.

Jing, X., Cui, Q., Li, X., Yin, J., Ravichandran, V., Pan, D., Fu, J., Tu, Q., Wang, H., Bian, X., Zhang, Y., 2020. Engineering *Pseudomonas protegens* Pf-5 to improve its antifungal activity and nitrogen fixation. Microbial Biotechnol. 13 (1), 118–133.

Jones, D.L., Kielland, K., Sinclair, F.L., Dahlgren, R.A., Newsham, K.K., Farrar, J.F., Murphy, D.V., 2009. Soil organic nitrogen mineralization across a global latitudinal gradient. Global Biogeochem. Cycles. 23 (1).

Jurat-Fuentes, J.L., Jackson, T.A., 2012. Bacterial entomopathogens. In: Insect Pathology. Academic Press, pp. 265–349.

Kabaluk, T., Gazdik, K., 2005. Directory of microbial pesticides for agricultural crops in OECD countries. Agriculture and Agri-Food Canada.

Kabaluk, J.T., Svircev, A.M., Goettel, M.S., Woo, S.G. (Eds.), 2010. The Use and Regulation of Microbial Pesticides in Representative Jurisdictions Worldwide. International Organization for Biological Control of Noxious Animals and Plants (IOBC), p. 99.

Kadmiri, I.M., Chaouqui, L., Azaroual, S.E., Sijilmassi, B., Yaakoubi, K., Wahby, I., 2018. Phosphate-solubilizing and auxin-producing rhizobacteria promote plant growth under saline conditions. Arab. J. Sci. Eng. 43 (7), 3403–3415.

Kalawate, A.S., 2014. Microbial viral insecticides. In: Basic and Applied Aspects of Biopesticides. Springer, New Delhi, pp. 47–68.

Kalita, M., Bharadwaz, M., Dey, T., Gogoi, K., Dowarah, P., Unni, B.G., Ozah, D., Saikia, I., 2015. Developing novel bacterial based bioformulation having PGPR properties for enhanced production of agricultural crops. Indian J. Exp. Biol. 53 (1), 56–60.

Karimi, E., Sadeghi, A., Abbaszadeh Dahaji, P., Dalvand, Y., Omidvari, M., Kakuei Nezhad, M., 2012. Biocontrol activity of salt tolerant *Streptomyce*s isolates against phytopathogens causing root rot of sugar beet. Biocontrol Sci. Technol. 22 (3), 333–349.

Katsuma, S., Koyano, Y., Kang, W., Kokusho, R., Kamita, S.G., Shimada, T., 2012. The baculovirus uses a captured host phosphatase to induce enhanced locomotory activity in host caterpillars. PLoS Pathog. 8 (4).

Khachatourians, G.G., 1996. Biochemistry and molecular biology of entomopathogenic fungi. In: Human and Animal Relationships. Springer, Berlin, Heidelberg, pp. 331–363.

Khan, A., Williams, K., Nevalainen, H., 2003. Testing the nematophagous biological control strain *Paecilomyces lilacinus* 251 for paecilotoxin production. FEMS (Fed. Eur. Microbiol. Soc.) Microbiol. Lett. 227 (1), 107–111.

Kheirandish, Z., Harighi, B., 2015. Evaluation of bacterial antagonists of *Ralstonia solanacearum*, causal agent of bacterial wilt of potato. Biol. Contr. 86, 14–19.

Kim, J.J., Goettel, M.S., Gillespie, D.R., 2008. Evaluation of *Lecanicillium longisporum*, Vertalec® for simultaneous suppression of cotton aphid, *Aphis gossypii*, and cucumber powdery mildew, *Sphaerotheca fuliginea*, on potted cucumbers. Biol. Contr. 45 (3), 404–409.

Kim, H.S., Sang, M.K., Jung, H.W., Jeun, Y.C., Myung, I.S., Kim, K.D., 2012. Identification and characterization of *Chryseobacterium wanjuense* strain KJ9C8 as a biocontrol agent of *Phytophthora* blight of pepper. Crop Protect. 32, 129–137.

Kimani, V., May 2014. Bio-Pesticides development, use and regulation in Kenya. In: Regional Experts Workshop on Development, Regulation and Use of Bio-Pesticides in East Africa, Nairobi, Kenya, pp. 22–23.

King, A.M., Lefkowitz, E., Adams, M.J., Carstens, E.B. (Eds.), 2011. Virus Taxonomy: Ninth Report of the International Committee on Taxonomy of Viruses, vol. 9. Elsevier.

Klarić, M.S., Pepeljnjak, S., 2005. Beauvericin: chemical and biological aspects and occurrence. Arh. Hig. Rad. Toksikol. 56 (4), 343–350.

Klein, M.G., 1988. Pest management of soil-inhabiting insects with microorganisms. Agric. Ecosyst. Environ. 24 (1–3), 337–349.

Koul, O., 2011. Microbial biopesticides: opportunities and challenges. CAB Rev. 6, 1–26.

Koul, O., Dhaliwal, G.S., Khokhar, S., Singh, R., 2012. Biopesticides in Environment and Food Security: Issues and Strategies. Scientific Publishers.

Kour, D., Rana, K.L., Yadav, N., Yadav, A.N., Singh, J., Rastegari, A.A., Saxena, A.K., 2019. Agriculturally and industrially important fungi: current developments and potential biotechnological applications. In: Recent Advancement in White Biotechnology through Fungi. Springer, Cham, pp. 1–64.

Kranthi, K.R., Jadhav, D.R., Kranthi, S., Wanjari, R.R., Ali, S.S., Russell, D.A., 2002. Insecticide resistance in five major insect pests of cotton in India. Crop Protect. 21 (6), 449–460.

Krishnen, G., Noor, M.R.M., Jack, A., Haron, S., 2016. Research, development and commercialisation of agriculturally important microorganisms in Malaysia. In: Agriculturally Important Microorganisms. Springer, Singapore, pp. 149–166.

Kuan, K.B., Othman, R., Rahim, K.A., Shamsuddin, Z.H., 2016. Plant growth-promoting rhizobacteria inoculation to enhance vegetative growth, nitrogen fixation and nitrogen remobilisation of maize under greenhouse conditions. PloS One 11 (3).

Kumar, S., Aharwal, R.P., Shukla, H., Rajak, R.C., Sandhu, S.S., 2014. Endophytic fungi: as a source of antimicrobials bioactive compounds. World J. Pharm. Pharmaceut. Sci. 3 (2), 1179–1197.

Kumar, C.S., Jacob, T.K., Devasahayam, S., Thomas, S., Geethu, C., 2018. Multifarious plant growth promotion by an entomopathogenic fungus *Lecanicillium psalliotae*. Microbiol. Res. 207, 153–160.

Kumar, D., Singh, M., Singh, H.K., Singh, K., 2019. Fungal biopesticides and their Uses for control of insect pest and diseases. Biofertil. & Biopestic. Sustain. Agric. 43.

Kumari, A., Goyal, R.K., Choudhary, M., Sindhu, S.S., 2016. Effects of some plant growth promoting rhizobacteria (PGPR) strains on growth and flowering of chrysanthemum. J. Crop Weed 12 (1), 7–15.

Kunimi, Y., 1998. Japan. In: Hunter-Fujita, H.R., Entwistle, P.F., Evans, H.F., Crook, N.F. (Eds.), Insect Viruses and Pest Management. Wiley, Chichester, pp. 269–279.

Kunimi, Y., 2007. Current status and prospects on microbial control in Japan. J. Invertebr. Pathol. 95 (3), 181–186.

Kunjeku, E., Jones, K.A., Moawad, G.M., 1998. Africa, the near and middle East. In: Hunter-Fujita, F.R., Entwhistle, P.E., Evans, H.F., Crook, N.E. (Eds.), Insect, Viruses and Pest Management. Wiley, Chichester, pp. 280–302.

Kurokura, T., Hiraide, S., Shimamura, Y., Yamane, K., 2017. PGPR improves yield of strawberry species under less-fertilized conditions. Environ. Control Biol. 55 (3), 121–128.

Lang, A.E., Schmidt, G., Sheets, J.J., Aktories, K., 2011. Targeting of the actin cytoskeleton by insecticidal toxins from *Photorhabdus luminescens*. N. Schmied. Arch. Pharmacol. 383 (3), 227–235.

Lange, C.E., Sokolova, Y.Y., 2017. The development of the microsporidium Paranosema (Nosema) locustae for grasshopper control: John Henry's innovation with worldwide lasting impacts. Protistology 11 (3).

Langner, T., Göhre, V., 2016. Fungal chitinases: function, regulation, and potential roles in plant/pathogen interactions. Curr. Genet. 62 (2), 243–254.

Latin America BioPesticide Market, 2019. https://www.marketdataforecast.com/market-reports/latin-america-bio-pesticide-market (Assessed on 24 May 2020).

Lee, S., Yap, M., Behringer, G., Hung, R., Bennett, J.W., 2016. Volatile organic compounds emitted by *Trichoderma* species mediate plant growth. Fungal Biol. & Biotechnol. 3 (1), 7.

Lemay, G., 2018. Synthesis and translation of viral mRNA in reovirus-infected cells: progress and remaining questions. Viruses 10 (12), 671.

Li, H.Q., Li, X.J., Wang, Y.L., Zhang, Q., Zhang, A.L., Gao, J.M., Zhang, X.C., 2011. Antifungal metabolites from *Chaetomium globosum*, an endophytic fungus in *Ginkgo biloba*. Biochem. Systemat. Ecol. 4 (39), 876–879.

Li, K., Cai, D., Wang, Z., He, Z., Chen, S., 2018. Development of an efficient genome editing tool in *Bacillus licheniformis* using CRISPR-Cas9 nickase. Appl. Environ. Microbiol. 84 (6), 02608–02617.

Li, Y.T., Hwang, S.G., Huang, Y.M., Huang, C.H., 2018. Effects of *Trichoderma asperellum* on nutrient uptake and Fusarium wilt of tomato. Crop Protect. 110, 275–282.

Liu, J.C., Boucias, D.G., Pendland, J.C., Liu, W.Z., Maruniak, J., 1996. The mode of action of hirsutellin A on eukaryotic cells. J. Invertebr. Pathol. 67 (3), 224–228.

Liu, Y.J., Liu, J., Ying, S.H., Liu, S.S., Feng, M.G., 2013. A fungal insecticide engineered for fast per os killing of caterpillars has high field efficacy and safety in full-season control of cabbage insect pests. Appl. Environ. Microbiol. 79 (20), 6452–6458.

Luque, T., Finch, R., Crook, N., O'Reilly, D.R., Winstanley, D., 2001. The complete sequence of the *Cydia pomonella* granulovirus genome. J. Gen. Virol. 82 (10), 2531–2547.

Lynch, P.A., Grimm, U., Thomas, M.B., Read, A.F., 2012. Prospective malaria control using entomopathogenic fungi: comparative evaluation of impact on transmission and selection for resistance. Malar. J. 11 (1), 383.

Maeda, S., 1989. Increased insecticidal effect by a recombinant baculovirus carrying a synthetic diuretic hormone gene. Biochem. & Biophys. Res. Commun. 165 (3), 1177–1183.

Mangmang, J.S., Deaker, R., Rogers, G., 2014. Effects of plant growth promoting rhizobacteria on seed germination characteristics of tomato and lettuce. J. Trop. Crop Sci. 1 (2), 35–40.

Marche, M.G., Camiolo, S., Porceddu, A., Ruiu, L., 2018. Survey of *Brevibacillus laterosporus* insecticidal protein genes and virulence factors. J. Invertebr. Pathol. 155, 38–43.

Market Data Forecast for Microbial Pesticides Market Segmented By Product, Type, Application, and Region - Global Growth, Trends, and Forecast to 2024, August 2019. https://www.marketdataforecast.com/market-reports/microbial-pesticides-market (Assessed on 24 May 2020).

Martin, P.A., Hirose, E., Aldrich, J.R., 2007. Toxicity of *Chromobacterium subtsugae* to southern green stink bug (Heteroptera: Pentatomidae) and corn rootworm (Coleoptera: Chrysomelidae). J. Econ. Entomol. 100 (3), 680–684.

Martínez-Absalón, S., Rojas-Solís, D., Hernández-León, R., Prieto-Barajas, C., Orozco-Mosqueda, M.D.C., Peña-Cabriales, J.J., Sakuda, S., Valencia-Cantero, E., Santoyo, G., 2014. Potential use and mode of action of the new strain *Bacillus thuringiensis* UM96 for the biological control of the grey mould phytopathogen *Botrytis cinerea*. Biocontrol Sci. Technol. 24 (12), 1349–1362.

Martínez-Viveros, O., Jorquera, M.A., Crowley, D.E., Gajardo, G.M.L.M., Mora, M.L., 2010. Mechanisms and practical considerations involved in plant growth promotion by rhizobacteria. J. Soil Sci. Plant Nutr. 10 (3), 293–319.

Marzano, M., Gallo, A., Altomare, C., 2013. Improvement of biocontrol efficacy of *Trichoderma harzianum* vs. *Fusarium oxysporum* f. sp. lycopersici through UV-induced tolerance to fusaric acid. Biol. Contr. 67 (3), 397–408.

Masciarelli, O., Llanes, A., Luna, V., 2014. A new PGPR co-inoculated with *Bradyrhizobium japonicum* enhances soybean nodulation. Microbiol. Res. 169 (7–8), 609–615.

Masson-Boivin, C., Giraud, E., Perret, X., Batut, J., 2009. Establishing nitrogen-fixing symbiosis with legumes: how many rhizobium recipes? Trends Microbiol. 17 (10), 458–466.

Matthews, K.A., 2014. Regulation of biopesticides by the environmental protection agency. In: Coats, J.R. (Ed.), Biopesticides: State of the Art and Future Opportunities, Vol 18, Symposium Series, 1172. Sidley Austin, Washington, DC, pp. 267–279.

McKinnon, A.C., Saari, S., Moran-Diez, M.E., Meyling, N.V., Raad, M., Glare, T.R., 2017. *Beauveria bassiana* as an endophyte: a critical review on associated methodology and biocontrol potential. Bio Control 62 (1), 1–17.

Mehrotra, V.S. (Ed.), 2005. Mycorrhiza: Role and Applications. Allied Publishers.

Mia, M.B., Shamsuddin, Z.H., Mahmood, M., 2012. Effects of rhizobia and plant growth promoting bacteria inoculation on germination and seedling vigor of lowland rice. Afr. J. Biotechnol. 11 (16), 3758–3765.

Middle-East And Africa BioPesticide Market, August 2019. https://www.marketdataforecast.com/market-reports/middle-east-and-africa-bio-pesticide-market. (Accessed 16 May 2020).

Ministry of Agriculture, Forestry and Fisheries, 1997. Plant Protection Division, Agricultural Production Bureau, Ministry of Agriculture, Forestry and Fisheries. The notification No. 9-Seisan-5090, issued on August 29, 1997. Guidelines for preparation of data necessary for safety evaluation of microbial pesticides, Japan

Milner, R.J., 2000. Current status of *Metarhizium* as a mycoinsecticide in Australia. Biocontrol News Inf. 21 (2), 47N–50N.

Ming, Q., Su, C., Zheng, C., Jia, M., Zhang, Q., Zhang, H., Rahman, K., Han, T., Qin, L., 2013. Elicitors from the endophytic fungus *Trichoderma atroviride* promote *Salvia miltiorrhiza* hairy root growth and tanshinone biosynthesis. J. Exp. Bot. 64 (18), 5687–5694.

Miranpuri, G.S., Khachatourians, G.G., 1995. Entomopathogenicity of *Beauveria bassiana* toward flea beetles, *Phyllotreta cruciferae* Goeze (Col., Chrysomelidae). J. Appl. Entomol. 119 (1-5), 167–170.

Mishra, J., Tewari, S., Singh, S., Arora, N.K., 2015. Biopesticides: where we stand?. In: Plant Microbes Symbiosis: Applied Facets. Springer, New Delhi, pp. 37–75.

Mishra, J., Dutta, V., Arora, N.K., 2020. Biopesticides in India: technology and sustainability linkages. 3 Biotech 10, 1–12.

Mnif, I., Ghribi, D., 2015. Potential of bacterial derived biopesticides in pest management. Crop Protect 77, 52–64.

Morgan, J.A.W., Sergeant, M., Ellis, D., Ousley, M., Jarrett, P., 2001. Sequence analysis of insecticidal genes from *Xenorhabdus nematophilus* PMFI296. Appl. Environ. Microbiol. 67 (5), 2062–2069.

Mori, H., Metcalf, P., 2010. Cypoviruses. In: Asgari, S., Johnson, K.N. (Eds.), Insect Virology. Caister Academic Press, United Kingdom.

Morteza, Z., Mousavi, S.B., Baghestani, M.A., Aitio, A., 2017. An assessment of agricultural pesticide use in Iran, 2012–2014. J. Environ. Health Sci. & Eng. 15 (1), 10.

Narayanasamy, P., 2013. Mechanisms of action of bacterial biological control agents. In: Hokkanen, H.M.T (Ed.), Biological Management of Diseases of Crops. Springer, Dordrecht, pp. 295–429.

Nathan, S.S., Kalaivani, K., 2005. Efficacy of nucleopolyhedrovirus and azadirachtin on *Spodoptera litura* Fabricius (Lepidoptera: Noctuidae). Biol. Contr. 34 (1), 93–98.

Navazo, A., Barahona, E., Redondo-Nieto, M., Martínez-Granero, F., Rivilla, R., Martín, M., 2009. Three independent signalling pathways repress motility in *Pseudomonas fluorescens* F113. Microbial Biotechnol. 2 (4), 489–498.

Nguyen, L.T.T., Jang, J.Y., Kim, T.Y., Yu, N.H., Park, A.R., Lee, S., Bae, C.H., Yeo, J.H., Hur, J.S., Park, H.W., Kim, J.C., 2018. Nematicidal activity of verrucarin A and roridin A isolated from *Myrothecium verrucaria* against *Meloidogyne incognita*. Pestic. Biochem. Physiol. 148, 133–143.

Nikolić, I., Berić, T., Dimkić, I., Popović, T., Lozo, J., Fira, D., Stanković, S., 2019. Biological control of *Pseudomonas syringae* pv. aptata on sugar beet with *Bacillus pumilus* SS-10.7 and *Bacillus amyloliquefaciens* (SS-12.6 and SS-38.4) strains. J. Appl. Microbiol. 126 (1), 165–176.

Nisa, H., Kamili, A.N., Nawchoo, I.A., Shafi, S., Shameem, N., Bandh, S.A., 2015. Fungal endophytes as prolific source of phytochemicals and other bioactive natural products: a review. Microb. Pathog. 82, 50–59.

Niu, D.D., Liu, H.X., Jiang, C.H., Wang, Y.P., Wang, Q.Y., Jin, H.L., Guo, J.H., 2011. The plant growth–promoting rhizobacterium *Bacillus cereus* AR156 induces systemic resistance in *Arabidopsis thaliana* by simultaneously activating salicylate-and jasmonate/ethylene-dependent signaling pathways. Mol. Plant Microbe Interact. 24 (5), 533–542.

North America Biopesticides Market - Industry Growth, Trends and Forecasts (2020–2025), 2019. https://www.mordorintelligence.com/industry-reports/north-america-biopesticides-market-industry. (Accessed 16 May 2020).

Noumavo, P.A., Kochoni, E., Didagbé, Y.O., Adjanohoun, A., Allagbé, M., Sikirou, R., Gachomo, E.W., Kotchoni, S.O., Baba-Moussa, L., 2013. Effect of different plant growth promoting rhizobacteria on maize seed germination and seedling development. Am. J. Plant Sci. 4 (5), 1013.

O'Reilly, D.R., 1995. Baculovirus-encoded ecdysteroid UDP-glucosyltransferases. Insect Biochem. Mol. Biol. 25 (5), 541–550.

Obeidat, M., Abu-Romman, S., Odat, N., Haddad, M., Al-Abbadi, A., Hawari, A., 2017. Antimicrobial and insecticidal activities of n-butanol extracts from some *Streptomyces* isolates. Microbiology 12 (4), 218–228.

Ochieng, J.R.A., 2015. Towards a Regulatory Framework for Increased and Sustainable Use of Bio-Fertilizers in Kenya. Doctoral dissertation. University of Nairobi.

Ohba, M., lwahana, H., Asano, S., Suzuki, N., Sato, R., Hori, H., 1992. A unique isolate of *Bacillus thuringiensis* serovar japonensis with a high larvicidal activity specific for scarabaeid beetles. Lett. Appl. Microbiol. 14, 54–57.

Opota, O., Vallet-Gély, I., Vincentelli, R., Kellenberger, C., Iacovache, I., Gonzalez, M.R., Roussel, A., Van Der Goot, F.G., Lemaitre, B., 2011. Monalysin, a novel ß-pore-forming toxin from the *Drosophila* pathogen *Pseudomonas entomophila*, contributes to host intestinal damage and lethality. PLoS Pathog. 7 (9).

Organization for Economic Cooperation and Development, 2003. Guidance for Registration Requirements for Microbial Pesticides. Environment Directorate. ENV/JM/MONO(2003)5. www.oecd.org/dataoecd/4/23/28888446.pdf. (Accessed 16 May 2020).

O'Brien, P.A., 2017. Biological control of plant diseases. Australas. Plant Pathol. 46 (4), 293–304.

Palma-Guerrero, J., Gómez-Vidal, S., Tikhonov, V.E., Salinas, J., Jansson, H.B., Lopez-Llorca, L.V., 2010. Comparative analysis of extracellular proteins from *Pochonia chlamydosporia* grown with chitosan or chitin as main carbon and nitrogen sources. Enzym. Microb. Technol. 46 (7), 568–574.

Park, H.W., Bideshi, D.K., Federici, B.A., 2010. Properties and applied use of the mosquitocidal bacterium, *Bacillus sphaericus*. J. Asia Pac. Entomol. 13 (3), 159–168.

Pascale, A., Vinale, F., Manganiello, G., Nigro, M., Lanzuise, S., Ruocco, M., Marra, R., Lombardi, N., Woo, S.L., Lorito, M., 2017. Trichoderma and its secondary metabolites improve yield and quality of grapes. Crop Prot. 92, 176–181.

Pathak, D.V., Yadav, R., Kumar, M., 2017. Microbial pesticides: development, prospects and Popularization in India. In: Plant-Microbe Interactions in Agro-Ecological Perspectives. Springer, Singapore, pp. 455–471.

Peleg, I., Feldman, K., Minrav Industries Ltd, 2002. Bacillus Firmus CNCM I-1582 or Bacillus Cereus CNCM I-1562 for Controlling Nematodes. U.S. Patent 6: 406690.

Peng, R., Xiong, A., Li, X., Fuan, H., Yao, Q., 2003. A δ-endotoxin encoded in *Pseudomonas fluorescens* displays a high degree of insecticidal activity. Appl. Microbiol. Biotechnol. 63 (3), 300–306.

Peng, G., Jin, K., Liu, Y., Xia, Y., 2015. Enhancing the utilization of host trehalose by fungal trehalase improves the virulence of fungal insecticide. Appl. Microbiol. Biotechnol. 99 (20), 8611–8618.

Peralta, C., Palma, L., 2017. Is the insect world overcoming the efficacy of *Bacillus thuringiensis*? Toxins 9 (1), 39.

Pijlman, G.P., van den Born, E., Martens, D.E., Vlak, J.M., 2001. *Autographa californica* baculoviruses with large genomic deletions are rapidly generated in infected insect cells. Virology 283 (1), 132–138.

PMRA, 2001. DIR2001-02: Guidelines for the Registration of Microbial Pest Control Agents and Products. www.hc-sc.gc.ca. (Accessed 16 May 2020).

Powles, R.J., Rogers, P.L., 1989. Bacillus toxin for insect control – a review. Aust. J. Biotechnol. 3, 223–228.

PPO (Plant Protection Organization), 2017. Guideline for Registration of Microbial Pesticides. http://old.ppo.ir/English/Pages/EnPageContent.aspx?id=53&portal=1. (Accessed 16 May 2020).

Prajapati, K., Sharma, M.C., Modi, H.A., 2013. Growth promoting effect of potassium solubilizing microorganisms on *Abelmoscus esculantus*. Int. J. Agric. Sci. 3 (1), 181–188.

Prasetphol, S., Areekul, P., Buranarerk, A., Kritpitayaavuth, M., 1969. Life history of orange dog butterfly and its microbial control. Tech. Bull. 10.

Preininger, C., Sauer, U., Bejarano, A., Berninger, T., 2018. Concepts and applications of foliar spray for microbial inoculants. Appl. Microbiol. Biotechnol. 102 (17), 7265–7282.

Priyadharsini, P., Muthukumar, T., 2017. The root endophytic fungus *Curvularia geniculata* from *Parthenium hysterophorus* roots improves plant growth through phosphate solubilization and phytohormone production. Fungal Ecol. 27, 69–77.

Quesada-Moraga, E., Vey, A., 2004. Bassiacridin, a protein toxic for locusts secreted by the entomopathogenic fungus *Beauveria bassiana*. Mycol. Res. 108, 441–452.

Quinlan, R.J., 1990. Registration requirements and safety considerations for microbial pest control agents in the European economic community. In: Laird, M., Lacey, L.A., Davidson, E.W. (Eds.), Safety of Microbial Insecticides. CRC Press, Boca Raton, pp. 11–18.

Rabindra, R.J., 2005. Current status of production and use of microbial pesticides in India and the way forward. In: Rabindra, R.J., Hussaini, S.S., Ramanujam, B. (Eds.), Microbial Biopesticde Formulations and Application. Technical Document No.55. Project Directorate of Biological Control.

Rahman, M., Ali, J., Masood, M., 2015. Seed priming and *Trichoderma* application: a method for improving seedling establishment and yield of dry direct seeded boro (winter) rice in Bangladesh. Univ. J. Agric. Res. 3 (2), 59–67.

Rajkumar, M., Ae, N., Prasad, M.N.V., Freitas, H., 2010. Potential of siderophore-producing bacteria for improving heavy metal phytoextraction. Trends Biotechnol. 28 (3), 142–149.

Rangel, L.I., Henkels, M.D., Shaffer, B.T., Walker, F.L., Davis, E.W., Stockwell, V.O., Bruck, D., Taylor, B.J., Loper, J.E., 2016. Characterization of toxin complex gene clusters and insect toxicity of bacteria representing four subgroups of *Pseudomonas fluorescens*. PloS One 11 (8), 0161120.

Rashki, M., Kharazi-Pakdel, A., Allahyari, H., Van Alphen, J.J.M., 2009. Interactions among the entomopathogenic fungus, *Beauveria bassiana* (Ascomycota: Hypocreales), the parasitoid, *Aphidius matricariae* (hymenoptera: braconidae), and its host, *Myzus persicae* (Homoptera: Aphididae). Biol. Contr. 50 (3), 324–328.

Rath, M., Mitchell, T.R., Gold, S.E., 2018. Volatiles produced by *Bacillus mojavensis* RRC101 act as plant growth modulators and are strongly culture-dependent. Microbiol. Res. 208, 76–84.

Ravensberg, W.J., 2011. A Roadmap to the Successful Development and Commercialization of Microbial Pest Control Products for Control of Arthropods, vol. 10. Springer Science & Business Media.

Raymond, B., Federici, B.A., 2017. In defense of *Bacillus thuringiensis*, the safest and most successful microbial insecticide available to humanity—a response to EFSA. FEMS (Fed. Eur. Microbiol. Soc.) Microbiol. Ecol. 93 (7).

RDA., 2009. Rural Development Administration. The Acts for Management of Agricultural Chemicals. www.rda.go.kr. (Accessed 16 May 2020).

Reed, E.M., Springett, B.P., 1971. Large-scale field testing of a granulosis virus for the control of the potato moth (*Phthorimaea operculella* (Zell.)(Lep., Gelechiidae)). Bull. Entomol. Res. 61 (2), 223–233.

Rincón-Castro, M.C.D.R., Ibarra, J.E., 2011. Entomopathogenic viruses. In: Ninfa, R. (Ed.), Book: Biological Control of Insect Pests Edition: First Chapter: Entomopathogenic Viruses. Publisher: Studium Press LLC, USA.

Roberts, D.W., 1981. Toxins of entomopathogenic fungi. In: Burges, H.D. (Ed.), Microbial Control of Pests and Plant Diseases 1970–1980. Academic Press, London, pp. 441–464.

Roh, J.Y., Choi, J.Y., Li, M.S., Jin, B.R., Je, Y.H., 2007. *Bacillus thuringiensis* as a specific, safe, and effective tool for insect pest control. J. Microbiol. Biotechnol. 17 (4), 547.

Rojas-Solís, D., Zetter-Salmón, E., Contreras-Pérez, M., del Carmen Rocha-Granados, M., Macías-Rodríguez, L., Santoyo, G., 2018. *Pseudomonas stutzeri* E25 and *Stenotrophomonas maltophilia* CR71 endophytes produce antifungal volatile organic compounds and exhibit additive plant growth-promoting effects. Biocatal. & Agric. Biotechnol. 13, 46–52.

Rozpądek, P., Wężowicz, K., Nosek, M., Ważny, R., Tokarz, K., Lembicz, M., Miszalski, Z., Turnau, K., 2015. The fungal endophyte *Epichloë typhina* improves photosynthesis efficiency of its host orchard grass (*Dactylis glomerata*). Planta 242 (4), 1025–1035.

Ruiu, L., 2013. *Brevibacillus laterosporus*, a pathogen of invertebrates and a broad-spectrum antimicrobial species. Insects 4 (3), 476–492.

Ruiu, L., 2018. Microbial biopesticides in agroecosystems. Agronomy 8 (11), 235.

Ruiu, L., Floris, I., Satta, A., Ellar, D.J., 2007. Toxicity of a *Brevibacillus laterosporus* strain lacking parasporal crystals against *Musca domestica* and *Aedes aegypti*. Biol. Contr. 43 (1), 136–143.

Rushtapakornchai, W., 2003. Use and production of biopesticides in Thailand. In: Roettger, U., Reinhold, M. (Eds.), International Symposium on Biopesticides for Developing Countries. CATIE, Turrialba, pp. 126–130.

Samaniego-Gámez, B.Y., Garruña, R., Tun-Suárez, J.M., Kantun-Can, J., Reyes-Ramírez, A., Cervantes-Díaz, L., 2016. *Bacillus* spp. inoculation improves photosystem II efficiency and enhances photosynthesis in pepper plants. Chil. J. Agric. Res. 76 (4), 409–416.

Sánchez-López, Á.M., Bahaji, A., De Diego, N., Baslam, M., Li, J., Muñoz, F.J., Almagro, G., García-Gómez, P., Ameztoy, K., Ricarte-Bermejo, A., Novák, O., 2016. *Arabidopsis* responds to *Alternaria alternata* volatiles by triggering plastid phosphoglucose isomerase-independent mechanisms. Plant Physiol. 172 (3), 1989–2001.

Saravanakumar, K., Yu, C., Dou, K., Wang, M., Li, Y., Chen, J., 2016. Synergistic effect of *Trichoderma*-derived antifungal metabolites and cell wall degrading enzymes on enhanced biocontrol of *Fusarium oxysporum* f. sp. cucumerinum. Biol. Contr. 94, 37–46.

Saritha, M., Tollamadugu, N.P., 2019. The status of research and application of biofertilizers and biopesticides: global scenario. In: Recent Developments in Applied Microbiology and Biochemistry. Academic Press, pp. 195–207.

Satheesan, J., Narayanan, A.K., Sakunthala, M., 2012. Induction of root colonization by *Piriformospora indica* leads to enhanced asiaticoside production in *Centella asiatica*. Mycorrhiza 22 (3), 195–202.

Schneider, W., 2006. September. US EPA Regulation of biopesticides, microbial and biochemical pesticide regulation. In: REBECA Workshop on Current Risk Assessment and Regulation Practice. Salzau, Germany.

Schrank, A., Vainstein, M.H., 2010. Metarhizium anisopliae enzymes and toxins. Toxicon 56 (7), 1267–1274.

Sekhar, M., Riotte, J., Ruiz, L., Jouquet, P., Braun, J.J., 2016. Influences of climate and agriculture on water and biogeochemical cycles: Kabini critical zone observatory. In: Proceedings of the Indian National Science Academy, vol. 82, pp. 833–846.

Senthil-Nathan, S., 2015. A review of biopesticides and their mode of action against insect pests. In: Environmental Sustainability. Springer, New Delhi, pp. 49–63.

Shao, H.L., Dong, D.J., Hu, J.D., Wang, J.X., Zhao, X.F., 2008. Construction of the recombinant baculovirus AcMNPV with cathepsin B-like proteinase and its insecticidal activity against *Helicoverpa armigera*. Pestic. Biochem. Physiol. 91 (3), 141–146.

Sharma, A., Thakur, D.R., Kanwar, S., Chandla, V.K., 2013. Diversity of entomopathogenic bacteria associated with the white grub, *Brahmina coriacea*. J. Pest. Sci. 86 (2), 261–273.

Sharma, A., Sharma, S., Mittal, A., Naik, S.N., 2016. Evidence for the involvement of nematocidal toxins of *Purpureocillium lilacinum* 6029 cultured on Karanja deoiled cake liquid medium. World J. Microbiol. Biotechnol. 32 (5), 82.

Sharma, P., Verma, P.P., Kaur, M., 2017a. Identification of secondary metabolites produced by fluorescent Pseudomonads for controlling fungal pathogens of apple. Indian Phytopatholog. Soc. 70 (4), 452–456.

Sharma, P., Kharkwal, A.C., Abdin, M.Z., Varma, A., 2017b. *Piriformospora indica*-mediated salinity tolerance in Aloe vera plantlets. Symbiosis 72, 103–115.

Sharon, E., Bar-Eyal, M., Chet, I., Herrera-Estrella, A., Kleifeld, O., Spiegel, Y., 2001. Biological control of the root-knot nematode *Meloidogyne javanica* by *Trichoderma harzianum*. Phytopathology 91 (7), 687–693.

Shi, H., Zeng, H., Yang, X., Zhao, J., Chen, M., Qiu, D., 2012. An insecticidal protein from *Xenorhabdus ehlersii* triggers prophenoloxidase activation and hemocyte decrease in *Galleria mellonella*. Curr. Microbiol. 64 (6), 604–610.

Shukla, K.P., Sharma, S., Singh, N.K., Singh, V., Tiwari, K., Singh, S., 2011. Nature and role of root exudates: efficacy in bioremediation. Afr. J. Biotechnol. 10 (48), 9717–9724.

Simiyu, N.S.W., Tarus, D., Watiti, J., Nang'ayo, F., 2013. Effective Regulation of Bio-Fertilizers and Biopesticides: A Potential Avenue to Increase Agricultural Productivity, Policy Series, COMPRO II. IITA.

Singh, P., Mazumdar, P., Harikrishna, J.A., Babu, S., 2019. Sheath blight of rice: a review and identification of priorities for future research. Planta 1–21.

Sinha, B., Biswas, I., 2008. Potential of Bio-Pesticides in Indian Agriculture Vis-À-Vis Rural Development. India Science and Technology, SSRN. https://ssrn.com/abstract=1472371.

Slavicek, J.M., Popham, H.J., 2005. The *Lymantria dispar* nucleopolyhedrovirus enhancins are components of occlusion-derived virus. J. Virol. 79 (16), 10578–10588.

Soberón, M., Monnerat, R., Bravo, A., 2016. In: Gopalakrishnakone, P., Stiles, B., Alape-Girón, A., Dubreuil, J.D., Mandal, M. (Eds.), Mode of Action of Cry Toxins from *Bacillus thuringiensis* and Resistance Mechanisms. Microbial Toxins. Springer Netherlands, Dordrecht, the Netherlands, pp. 1–13.

Soenens, A., Imperial, J., 2019. Biocontrol capabilities of the genus *Serratia*. Phytochem. Rev. 1–11.

Solter, L.F., Becnel, J.J., 2000. Entomopathogenic microsporidia. In: Field Manual of Techniques in Invertebrate Pathology. Springer, Dordrecht, pp. 199–221.

Spadaro, D., Gullino, M.L., 2005. Improving the efficacy of biocontrol agents against soilborne pathogens. Crop Protect. 24 (7), 601–613.

Sporleder, M., Lacey, L.A., 2013. Biopesticides. In: Alyokhin, A., Vincent, C., Giordanengo, P. (Eds.), Insect Pests of Potato. Elsevier, New York, pp. 463–497.

Srivastava, P.K., Shenoy, B.D., Gupta, M., Vaish, A., Mannan, S., Singh, N., Tewari, S.K., Tripathi, R.D., 2012. Stimulatory effects of arsenic-tolerant soil fungi on plant growth promotion and soil properties. Microb. Environ. ME11316.

St Leger, R.J., Joshi, L., Bidochka, M.J., Roberts, D.W., 1996. Construction of an improved mycoinsecticide overexpressing a toxic protease. Proc. Natl. Acad. Sci. U. S. A. 93 (13), 6349–6354.

Stadlinger, N., Mmochi, A.J., Kumblad, L., 2013. Weak governmental institutions impair the management of pesticide import and sales in Zanzibar. Ambio 42 (1), 72–82.

Stefan, M.A.R.I.U.S., Munteanu, N., Stoleru, V., Mihasan, M.A.R.I.U.S., 2013. Effects of inoculation with plant growth promoting rhizobacteria on photosynthesis, antioxidant status and yield of runner bean. Rom. Biotechnol. Lett. 18 (2), 8132–8143.

Strasser, H., Vey, A., Butt, T.M., 2000. Are there any risks in using entomopathogenic fungi for pest control, with particular reference to the bioactive metabolites of *Metarhizium, Tolypocladium* and *Beauveria* species? Biocontrol Sci. Technol. 10 (6), 717–735.

Sudha, S.N., Jayakumar, R., Sekar, V., 1999. Introduction and expression of the cry1Ac gene of *Bacillus thuringiensis* in a cereal-associated bacterium, *Bacillus polymyxa*. Curr. Microbiol. 38 (3), 163–167.

Sugimoto, M., Koike, M., Nagao, H., Okumura, K., Tani, M., Kuramochi, K., 2003. Genetic diversity of the entomopathogen *Verticillium lecanii* on the basis of vegetative compatibility. Phytoparasitica 31 (5), 450–457.

Sun, X., Wu, D., Sun, X., Jin, L., Ma, Y., Bonning, B.C., Peng, H., Hu, Z., 2009. Impact of *Helicoverpa armigera* nucleopolyhedroviruses expressing a cathepsin L-like protease on target and nontarget insect species on cotton. Biol. Contr. 49 (1), 77–83.

Sung, G.H., Hywel-Jones, N.L., Sung, J.M., Luangsa-Ard, J.J., Shrestha, B., Spatafora, J.W., 2007. Phylogenetic classification of Cordyceps and the clavicipitaceous fungi. Stud. Mycol. 57, 5–59.

Susič, N., Janežič, S., Rupnik, M., Stare, B.G., 2020. Whole genome sequencing and comparative genomics of two nematicidal *Bacillus* strains reveals a wide range of possible virulence factors. G3: Genes, Genomes, Genetics 10 (3), 881–890.

Susilowati, D.N., Riyanti, E.I., Setyowati, M., Mulya, K., 2018. Indole-3-acetic acid producing bacteria and its application on the growth of rice. In: AIP Conference Proceedings, vol. 2002. AIP Publishing LLC, p. 020016. No. 1.

Swarnakumari, N., Kalaiarasan, P., 2017. Mechanism of nematode infection by fungal antagonists, *Purpureocillium lilacinum* (Thom) samson and *Pochonia chlamydosporia* (Goddard) Zare & Gams 2001. Pest Manag. Hortic. Ecosyst. 23 (2), 165–169.

Szewczyk, B., Hoyos-Carvajal, L., Paluszek, M., Skrzecz, I., De Souza, M.L., 2006. Baculoviruses—re-emerging biopesticides. Biotechnol. Adv. 24 (2), 143–160.

Thompson, S.N., Borchardt, D.B., 2003. Glucogenic blood sugar formation in an insect Manduca sexta L.: asymmetric synthesis of trehalose from 13C enriched pyruvate. Comp. Biochem. Physiol. B Biochem. Mol. Biol. 135 (3), 461–471.

Thorne, C.M., Levin, D.B., Otvos, I.S., Conder, N., 2008. Virus loads in Douglas-fir tussock moth larvae infected with the *Orgyia pseudotsugata* nucleopolyhedrovirus. Can. Entomol. 140 (2), 158–167.

Torres, M.J., Brandan, C.P., Sabaté, D.C., Petroselli, G., Erra-Balsells, R., Audisio, M.C., 2017. Biological activity of the lipopeptide-producing *Bacillus amyloliquefaciens* PGPBacCA1 on common bean *Phaseolus vulgaris* L. pathogens. Biol. Contr. 105, 93–99.

Ul Hassan, T., Bano, A., 2015. The stimulatory effects of L-tryptophan and plant growth promoting rhizobacteria (PGPR) on soil health and physiology of wheat. J. Soil Sci. Plant Nutr. 15 (1), 190–201.

Upadhyay, A., Srivastava, S., 2011. Phenazine-1-carboxylic acid is a more important contributor to biocontrol Fusarium oxysporum than pyrrolnitrin in *Pseudomonas fluorescens* strain Psd. Microbiol. Res. 166 (4), 323–335.

USEPA, 2011. Pesticide News Story: EPA Releases Report Containing Latest Estimates of Pesticide Use in the United States, USA.

Vacheron, J., Renoud, S., Muller, D., Babalola, O.O., Prigent-Combaret, C., 2015. Alleviation of abiotic and biotic stresses in plants by *Azospirillum*. In: Handbook for *Azospirillum*. Springer, Cham, pp. 333–365.

Vasconcelos, S.D., 1996. Alternative routes for the horizontal transmission of a nucleopolyhedrovirus. J. Invertebr. Pathol. 68 (3), 269–274.

Vega, F.E., 2008. Insect pathology and fungal endophytes. J. Invertebr. Pathol. 98 (3), 277–279.

Vega, F.E., Kaya, H.K., 2012. Insect pathology. In: Vega, F.E., Kaya, H.K. (Eds.), Insect Pathology, second ed. Academic Press, San Diego, pp. 171–220.

Verger, P.J.P., Boobis, A.R., 2013. Reevaluate pesticides for food security and safety. Science 341, 717–718.

Verma, P.P., Shelake, R.M., Das, S., Sharma, P., Kim, J.Y., 2019. Plant growth-promoting rhizobacteria (PGPR) and fungi (PGPF): potential biological control agents of diseases and pests. In: Microbial Interventions in Agriculture and Environment. Springer, Singapore, pp. 281–311.

Vey, A., Hoagland, R., Butt, T.M., 2001. Toxic metabolites of fungal biocontrol agents. In: Butt, T.M., Jackson, C.W., Magan, N. (Eds.), Fungi as Biocontrol Agents: Progress, Problems and Potential. CAB International, Wallingford, pp. 311–346.

Vietnam Biopesticide Market by Product Group & Application Type (2019-2024), March 2019. ResearchAndMarkets.com (Assessed on 18 May 2020).

Vigneux, F., Zumbihl, R., Jubelin, G., Ribeiro, C., Poncet, J., Baghdiguian, S., Givaudan, A., Brehélin, M., 2007. The *xaxAB* genes encoding a new apoptotic toxin from the insect pathogen *Xenorhabdus nematophila* are present in plant and human pathogens. J. Biol. Chem. 282 (13), 9571–9580.

Vijayabharathi, R., Kumari, B.R., Gopalakrishnan, S., 2014. Microbial agents against *Helicoverpa armigera*: where are we and where do we need to go? Afr. J. Biotechnol. 13 (18).

Vives-Peris, V., de Ollas, C., Gómez-Cadenas, A., Pérez-Clemente, R.M., 2019. Root exudates: from plant to rhizosphere and beyond. Plant Cell Rep. 1–15.

Vodovar, N., Vallenet, D., Cruveiller, S., Rouy, Z., Barbe, V., Acosta, C., Cattolico, L., Jubin, C., Lajus, A., Segurens, B., Vacherie, B., 2006. Complete genome sequence of the entomopathogenic and metabolically versatile soil bacterium *Pseudomonas entomophila*. Nat. Biotechnol. 24 (6), 673–679.

Wahid, O.A.A., Mehana, T.A., 2000. Impact of phosphate-solubilizing fungi on the yield and phosphorus-uptake by wheat and faba bean plants. Microbiol. Res. 155 (3), 221–227.

Wang, E.T., Martinez-Romero, E., 2000. *Sesbania herbacea–Rhizobium huautlense* nodulation in flooded soils and comparative characterization of S. herbacea-nodulating Rhizobia in different environments. Microb. Ecol. 40 (1), 25–32.

Wang, L., Huang, J., You, M., Guan, X., Liu, B., 2007. Toxicity and feeding deterrence of crude toxin extracts of Lecanicillium (Verticillium) lecanii (Hyphomycetes) against sweet potato whitefly, *Bemisia tabaci* (Homoptera: Aleyrodidae). Pest Manag. Sci. Form. Pestic. Sci. 63 (4), 381–387.

Wang, B., Zengzhi, L., 2010. Use and regulation of biopesticides in China. In: Kabaluk, J.T., Svircev, A.M., Goettel, M.S., Woo, S.G. (Eds.), Use and Regulation of Microbial Pesticides in Representative Jurisdictions Worldwide Edition: 1st Chapter: Use and Regulation of Biopesticides in ChinaPublisher: IOBC Global.

Wang, S., Fang, W., Wang, C., Leger, R.J.S., 2011. Insertion of an esterase gene into a specific locust pathogen (*Metarhizium acridum*) enables it to infect caterpillars. PLoS Pathog. 7 (6).

Washburn, J.O., Wong, J.F., Volkman, L.E., 2001. Comparative pathogenesis of *Helicoverpa zea* S nucleopolyhedrovirus in noctuid larvae. J. Gen. Virol. 82 (7), 1777–1784.

Waterfield, N., Hares, M., Yang, G., Dowling, A., Ffrench-Constant, R., 2005. Potentiation and cellular phenotypes of the insecticidal toxin complexes of Photorhabdus bacteria. Cell Microbiol. 7 (3), 373–382.

Waweru, B., Turoop, L., Kahangi, E., Coyne, D., Dubois, T., 2014. Non-pathogenic *Fusarium oxysporum* endophytes provide field control of nematodes, improving yield of banana (*Musa* sp.). Biol. Contr. 74, 82–88.

Weller, D.M., Thomashow, L.S., 2015. January. Phytosanitation and the development of transgenic biocontrol agents. In: Biosafety and the Environmental Uses of Micro-organisms: Conference Proceedings, pp. 35–45.

Widnyana, I.K., Javandira, C., 2016. Activities *Pseudomonas* spp. and *Bacillus* sp. to stimulate germination and seedling growth of tomato plants. Agric. & Agric. Sci. Procedia 9, 419–423.

World Health Organization, 2018. International code of conduct on pesticide management: guidelines for the registration of microbial, botanical and semiochemical pest control agents for plant protection and public health uses. Food & Agric. Org. VII, 76. https://apps.who.int/iris/handle/10665/259601.

Xiong, L., Li, J., Kong, F., 2004. *Streptomyces* sp. 173, an insecticidal micro-organism from marine. Lett. Appl. Microbiol. 38 (1), 32–37.

Xu, Q.F., Song, Y.L., Du, C.X., Zun, S.L., Wang, W.X., Xu, B.S., 1987. An investigation of culturing the fungus pathogen, *Beauveria bassiana* in maize whorl against corn borer, *Ostrinia furnacalis*. J. Jilin Agric. Sci. 4, 25–27.

Yadav, I.C., Devi, N.L., 2017. Pesticides classification and its impact on human and environment. Environ. Sci. Eng. 6, 140–158.

Yadav, J., Verma, J.P., Tiwari, K.N., 2010. Effect of plant growth promoting rhizobacteria on seed germination and plant growth chickpea (*Cicer arietinum* L.) under in vitro conditions (No. 2). In: Biological Forum, vol. 2, pp. 15–18.

Yamagiwa, Y., Inagaki, Y., Ichinose, Y., Toyoda, K., Hyakumachi, M., Shiraishi, T., 2011. *Talaromyces wortmannii* FS2 emits β-caryphyllene, which promotes plant growth and induces resistance. J. Gen. Plant Pathol. 77 (6), 336–341.

Yang, G., Dowling, A.J., Gerike, U., Waterfield, N.R., 2006. Photorhabdus virulence cassettes confer injectable insecticidal activity against the wax moth. J. Bacteriol. 188 (6), 2254–2261.

Yang, L., Keyhani, N.O., Tang, G., Tian, C., Lu, R., Wang, X., Pei, Y., Fan, Y., 2014. Expression of a toll signalling regulator serpin in a mycoinsecticide for increased virulence. Appl. Environ. Microbiol. 80 (15), 4531–4539.

Yang, Q., Zhang, J., Li, T., Liu, S., Song, P., Nangong, Z., Wang, Q., 2017. PirAB protein from *Xenorhabdus nematophila* HB310 exhibits a binary toxin with insecticidal activity and cytotoxicity in *Galleria mellonella*. J. Invertebr. Pathol. 148, 43–50.

Yoo, S.J., Shin, D.J., Won, H.Y., Song, J., Sang, M.K., 2018. *Aspergillus terreus* JF27 promotes the growth of tomato plants and induces resistance against *Pseudomonas syringae* pv. tomato. Mycobiology 46 (2), 147–153.

Yu, X., Ge, P., Jiang, J., Atanasov, I., Zhou, Z.H., 2011. Atomic model of CPV reveals the mechanism used by this single-shelled virus to economically carry out functions conserved in multishelled reoviruses. Structure 19 (5), 652–661.

Yuan, Y., Feng, H., Wang, L., Li, Z., Shi, Y., Zhao, L., Feng, Z., Zhu, H., 2017. Potential of endophytic fungi isolated from cotton roots for biological control against verticillium wilt disease. PloS One 12 (1), 0170557.

Zhang, F., Ge, H., Zhang, F., Guo, N., Wang, Y., Chen, L., Ji, X., Li, C., 2016. Biocontrol potential of *Trichoderma harzianum* isolate T-aloe against *Sclerotinia sclerotiorum* in soybean. Plant Physiol. Biochem. 100, 64–74.

Zhang, F., Chen, C., Zhang, F., Gao, L., Liu, J., Chen, L., Fan, X., Liu, C., Zhang, K., He, Y., Chen, C., 2017. Trichoderma harzianum containing 1-aminocyclopropane-1-carboxylate deaminase and chitinase improved growth and diminished adverse effect caused by *Fusarium oxysporum* in soybean. J. Plant Physiol. 210, 84–94.

Zhang, J., Mavrodi, D.V., Yang, M., Thomashow, L.S., Mavrodi, O.V., Kelton, J., Weller, D.M., 2020. *Pseudomonas synxantha* 2-79 transformed with pyrrolnitrin biosynthesis genes has improved biocontrol activity against soilborne pathogens of wheat and canola. Phytopathology. PHYTO-09.

Zhao, H., Xu, C., Lu, H.L., Chen, X., Leger, R.J.S., Fang, W., 2014. Host-to-pathogen gene transfer facilitated infection of insects by a pathogenic fungus. PLoS Pathog. 10 (4).

Zhou, T., Chen, D., Li, C., Sun, Q., Li, L., Liu, F., Shen, Q., Shen, B., 2012. Isolation and characterization of *Pseudomonas brassicacearum* J12 as an antagonist against *Ralstonia solanacearum* and identification of its antimicrobial components. Microbiol. Res. 167 (7), 388–394.

Zhou, Z., Zhang, C., Zhou, W., Li, W., Chu, L., Yan, J., Li, H., 2014. Diversity and plant growth-promoting ability of endophytic fungi from the five flower plant species collected from Yunnan, Southwest China. J. Plant Interact. 9 (1), 585–591.

Zhu, Y., Pan, J., Qiu, J., Guan, X., 2008. Isolation and characterization of a chitinase gene from entomopathogenic fungus *Verticillium lecanii*. Braz. J. Microbiol. 39 (2), 314–320.

Zimmermann, G., 2007. Review on safety of the entomopathogenic fungus *Metarhizium anisopliae*. Biocontrol Sci. Technol. 17 (9), 879–920.

Zimmermann, G., 2008. The entomopathogenic fungi *Isaria farinosa* (formerly *Paecilomyces farinosus*) and the Isaria fumosorosea species complex (formerly *Paecilomyces fumosoroseus*): biology, ecology and use in biological control. Biocontrol Sci. Technol. 18 (9), 865–901.

Chapter 6

Entomopathogenic nematodes: a sustainable option for insect pest management

Ashish Kumar Singh[a], Manish Kumar[b], Amit Ahuja[b], B.K. Vinay[b], Kiran Kumar Kommu[c], Sharmishtha Thakur[d], Amit U. Paschapur[a], B. Jeevan[a], K.K. Mishra[a], Rajendra Prasad Meena[e] and Manoj Parihar[e]

[a]*Crop Protection Section, ICAR-Vivekananda Parvatiya Krishi Anusandhan Sansthan, Almora, Uttarakhand, India;* [b]*Division of Nematology, ICAR-IARI, New Delhi, Delhi, India;* [c]*ICAR-Central Citrus Research Institute, Nagpur, Maharashtra, India;* [d]*Chaudhary Sarwan Kumar Himachal Pradesh Krishi Vishvavidyalaya, Palampur, Himachal Pradesh, India;* [e]*Crop Production Division, ICAR-Vivekananda Parvatiya Krishi Anusandhan Sansthan, Almora, Uttarakhand, India*

6.1 Introduction

Nematodes are the most diversified group of organisms on the planet. They are an inhabitant of a wide range of environments including the deserts, the mountains, the Polar Regions, marines, various soils, lakes, rivers and as parasites inhabiting other animals and plants (Ettema, 1998; De Ley, 2006; Borgonie et al., 2011). Despite of being regarded as a threat to agriculture some nematodes are beneficial for agriculture in some ways. Entomopathogenic nematodes parasitize pest insects of agriculture crops and kill them, thus prove to be suitable as biocontrol agents.

Entomopathogenic nematodes (EPNs) are the nematodes belonging to the families *Steinemernatidae* and *Heterorhabditidae* (Ley and Blaxter, 2002). EPNs possess tremendous potential to kill the wide range of insect pest due to the associated mutualistic bacteria of *Enterobacteraceae* like *Xenorhabdus* and *Photorhabdus* with genera *Steinernema* and *Heterorhabditis* respectively (Boemare, 2002). The infective juveniles (IJs) of EPNs reside in the soil looking for susceptible insect host to infect. Once the host insects come in contact with the IJs, they enter the host body through the natural openings (anus, mouth, spiracles) or directly through the cuticle. In the internal environment (haemocoel) of insect host the IJsregurgitates bacterial symbionts which rapidly proliferate logarithmically producing several exoenzymes, metabolites toxins and virulence factors to kill insect host within 24—48 h (Ciche and Ensign, 2003). EPN completes its lifecycle in host and recruits symbiont by feeding on it before dispersing to the new host. In present days the EPNs has gained a wide interest as excellent bioagent for insect pest management due to its attributes such as, motility, chemoreceptor to find host, wide host range, being environment friendly with negligible off-target effects, easily applicable through spray machines, compatible with pesticides, high virulence which kills host quickly and easy mass production under in-vitro and in-vivo condition. Despites of several traits which favor the EPNs to parasitize insect, there are few traits which require improvements, such as tolerance to temperature, desiccation, UV light tolerance enhanced host finding ability and resistance to pesticides. Improvement in these traits is possible through the identification of specific locally adapted populations, selective breeding, selection and using reverse genetic tools such as RNAi to validate the role of certain critical genes (Glazer, 2014). Researchers across the world are working to improve the efficacy of EPNs against insect pests in the field conditions through the artificial selection; isolation, identification and collection of new native species and genetic improvement in isolates (Gaugler, 1997; Hiltpold, 2015).

Here we provide information about EPNs isolation and identification, their symbionts and their multiplication along with a guide formulation development and recent biotechnological advances made to enhance the effectiveness of EPNs biology and insect management.

Biopesticides. https://doi.org/10.1016/B978-0-12-823355-9.00007-9

6.2 Baiting, isolation, multiplication of EPNs

IJs of EPNs is the free living stage in the soil hence the best method to find the EPNs is the collection of random soil samples from different areas. IJs of EPNs present in the soil and can be collected easily by sampling of the crop fields, forest lands, riverside, pastures and orchards from diverse climatic regions using different tools such as augers, tubes, shovels and trowels. Generally, two sampling strategies are used, stratified sampling and random sampling. In the stratified sampling method, samples are taken at fixed time interval across the period for intensive study and the random sampling method issued for larger geographical regions (Stuart and Gaugler, 1994; Stock et al., 2006). Soil samples are collected from a depth of up to 25 cm and from the distance of 4–5 m^2 area. The collected sub-samples (3–5/sample) are pooled in a plastic bag. Collected samples needed to be labeled with the information of the locality/site by its GPS coordinates, sampling date, habitat, existing vegetation, climatic conditions, and altitude and insect pests prevalent in those areas etc. It is highly recommended to place collected soil samples in Styrofoam boxes maintained at a temperature of 8–10°C to preserve moisture during long distance transports. A fraction of the collected soil samples are subjected to analysis of soil moisture, texture, structure, organic matter contents and EC etc., Chances of finding EPN affected insect cadavers naturally is very low therefore; insect baiting is a convenient approach to isolate EPNs from soil. For isolation of EPNs insect baiting technique is one of the simplest, low cost methods to attract insect pathogenic nematodes as well fungus (Bedding and Akhurst, 1975). In this method up to 100 g soil samples are placed in clean containers, 4–5 larvae of Galleria are added as baits and covered containers are placed upside down in incubator at 25°C in dark and observed for mortality after 2–3 days. The dead cadavers that attain yellow to brown coloration are infected with Steinernematids while the ones which attain red color are infected by *Heterorhabditids* (Fig. 6.1 A,B). Dead cadavers are then rinsed in sterile water are placed in white trap for the recovery of IJs and other stages (White, 1927). The white trap consists of a large dish (diameter-100 mm) containing a small petridish (diameter-50 mm) lined with filter paper. The dead cadavers are placed over small petridish and filled with small amount of sterile water and subsequently placed in incubator at 25°C. The fresh IJs and other stages can be harvested after 10–15 days from the suspensions (Fig. 6.1 C, D). Isolated EPNs can be successfully reared through in vitro and in vivo approach under laboratory conditions. In vivo approaches require living susceptible hosts of insect orders Coleopteran, Lepidoptera and Diptera for culture maintenance, growth and reproduction of EPNs. If suitable insect host are not available then the in vitro method can be used for rearing EPNs. In vitro method son the use of different Medias (NBTA, Liver kidney agar and Lipid agar media) as a source of nutrients to nematodes and symbionts (McMullen and Stock, 2014).

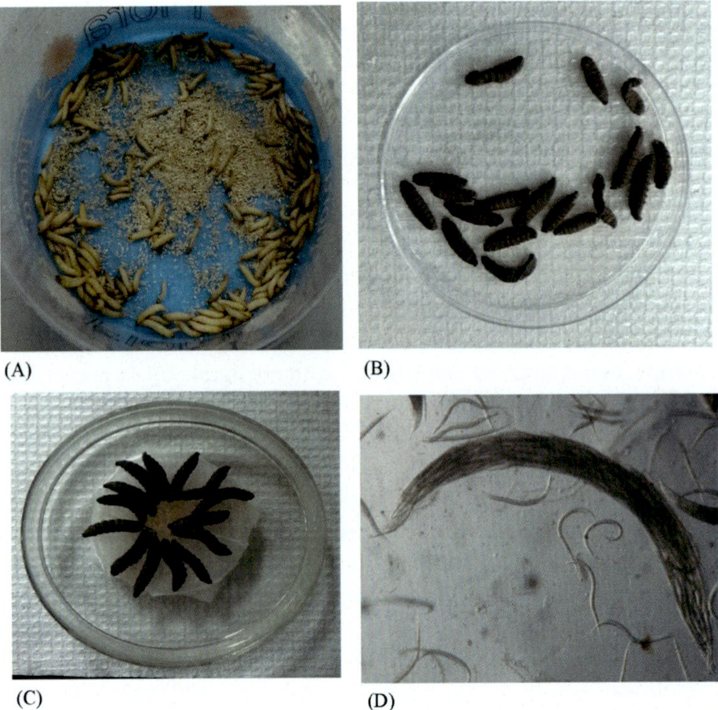

FIG. 6.1 (A): Healthy larvae of Galleria, (B): EPN infected larvae, (C): White trap and (D): IJs emerging from body of adult.

6.3 Identification of EPNs

Generally, EPNs belongs to three genera *Steinernema* (Travassos, 1927), *Heterorhabditis* (Poinar, 1976) and *Neosteinernema* (Nguyen and Smart, 1994). Identification of EPNs can be done easily using morphological parameters, morphometry and molecular information. Morphological and morphometry based identification depends on taxonomic keys of IJs, male, female and hermaphrodites for most of the species. These details can be observed under high magnification using a compound microscope as well under Scanning electron microscopes (SEMs). Specimen for identification of *Steinernema* requires to be reared in vivo on susceptible hosts like Galleria and first generation adult male, female, IJs should be dissected out from cadaver to meet the description criterias for an accurate identification. Diagnosis of *Steinernema* can be done using the following parameters: (1) obligate nature of insect parasitism, (2) presence of bacterial pouch in the anterior region of intestine, (3) presence of male and female population, (4) females are usually large, cuticle or smooth or annulated, absence of lateral field, presence of distinct excretory pores, head either rounded or truncate with 6 lips each having a labial papillae, presence of 4 cephalic papillae, amphids, collapsed stoma, rhabditoid esophagus with pronounced oesphagointestinal valve, swollen metacarpus, narrow isthmus. Didelphic amphidelphic, reflexed reproductive systems, median vulva and are oviparous or ovoviviparous. Tail could be longer or shorter with or without phasmids. (5) Males steinernematids are smaller than females, with 6 labial papillae, 4 large cephalic papillae, preoral disc, monarchic, reflexed, paired spicule, long gubernaculums, bursa absent, round, digitate or mucronated tail terminus and 10–14 pairs of genital papillae are present. (6) IJs can be identified by presence of collapsed stoma, slender body, annulated cuticle with or without sheath, lateral field with 4–9 incisures and 3–8 ridges, reduced esophagus and intestine, distinct excretory pores conoid or filiform tail position of phasmid at mid tail (Nguyen and Smart, 1994, 1996).

Similarly, *Neosteinernema* can be diagnosed by observing morphological features of female such as presence of prominent phasmids at posterior half of the tail, longer tail, ovoviviparity. Male are smaller than females, posterior part with 13–14 pair of ventral genital papillae, prominent phasmids, digitate tail tips, foot-shaped spicule and long gubernaculums almost up to the length of spicules. IJs with swollen heads, large phasmid, filliform, elongated tail (Nguyen and Smart, 1994, 1996).

Heterorhabditids are the important genera of EPNs and can be identified using following morphological parameters; (1) obligate insect parasitism, presence of symbiotic bacteria in anterior region of intestine and hermaphrodite/amphimictic both types of population found to be present. (2) Female hermaphrodites possess truncate to round head with 6 conical lips having terminal papillae at each lips, small amphidial openings, small and wide stoma, presence of refractile dots like cheilorhabdions, esophagus without metacarpus, slender isthmus, swollen basal bulb, excretory pores are located at posterior of esophagus, median vulva, amphidelphic reflexed. Tail is pointed longer than the body width and presence of post anal swelling. (2) Amphimictic femalesare smaller than hermaphrodite females with amphidelphic reproductive system, nonfunctional vulva for egg laying. (3) Male are monarchic, reflexed, paired spicule, ventrally curved, gubrnaculum are half of the spicule length and peloderan bursa with 9 pairs of genital papillae. (4) Bodies of IJs of Heterorhabditis are covered with sheath with tessellate pattern in anterior region; presence of dorsal cuticular tooth in the head region is the most prominent morphological feature of IJs of *Heterorhabditids*. Mouth anus is atrophied, reduced stoma, oeophagus and intestine, excretory pore is located posterior to nerve ring and tail are pointed. EPN species are usually identified based on the morphological characters but their identity should be verified by combining morphometric (Table 6.1) and molecular data and further by comparing with the original descriptions (Nguyen and Smart, 1994, 1996).

Characterization of EPNs based on morphological and morphometric traits requires higher skills and proficiency in identification, presence of low morphological variations; the overlapping characters hamper the accurate identifications. To overcome these challenges, molecular characterization using species specific primer sequences of rDNA unit (28S) offers an easiest approach to identify EPNs (Thanwisai et al., 2012; De Brida et al., 2017; Cimen et al., 2016; De Brida et al., 2017, 2017). Molecular diagnostic requires genomic DNA of EPNs which can be extracted using Holterman lysis buffer which contains components like 800 μg proteinase K/mL, β-mercaptoethanol 1% (v/v), 0.2 M NaCl and 0.2 M Tris HCl of pH 8 (Holterman et al., 2006). Following this protocol DNA from 3 to 5 IJs can be extracted easily. The extracted DNA is subjected to PCR using universal primers of D2/D3 segments of 28S rDNA (Table 6.2). In the next step the amplified fragments of 28S rDNA region need to be sequenced using advanced sequencing platforms. The obtained sequences need to be aligned and edited using Bioedit program and further used for homology based genetic similarity search in GenBank database (http://www.ncbi.nlm.nih.gov).

TABLE 6.1 Morphometric parameters for the EPNs identification.

S. No.	IJs of *Steinernema* and *Neosteinernema* and *Heterorhabditids* species	Identification of first-generation males of *Steinernema*, *Neosteinernema* and *Heterorhabditids* species
1	L (length)	Spicule
2	W (greatest width)	Gubernaculum length
3	EP (distance from anterior end to excretory pore)	W (greatest body width)
4	NR (distance from anterior end to nerve ring)	D% (distance from anterior end to excretory pore/esophagus length × 100)
5	ES (esophagus length)	SW (spicule length/anal body width)[a]
6	T (tail length)	GS (gubernaculum length/spicule length)
7	a (L/W)	MUC (mucron)[a]
8	b (L/ES)	N (number of specimens measured)
9	c (L/T)	
10	D% (EP/ES)	
11	E% (EP/T × 100.)	

[a]*Morphometric characters not applicable in* Heterorhabditids.

TABLE 6.2 Universal primers for EPN identification.

S.No.	Primer	Sequence	PCR conditions	References
1	28S rDNA	D2A F (5′-CAAGTACCGTGAGGGAAAGTTG3′) D3B R (5′TCGGAAGGAACCAGCTACT A-3′)	94°C-7 min; 35 cycles-94°C -60 s 55°C—60 s, 72°C - 60 s 72°C-10 min	Al-Banna, L. et al. (2004)
2	ITS	TW81 F (5′-GTTTCCGTAGGTGAACCTGC-3′) AB28 R (5′-ATATGCTTAAGTTCAGCGGGT-3′)	1 cycle-94°C-1 min 35 cycles-94°C -1 min 55°C-1 min 72°C-2 min. 1 cycle-72°C-10 min.	Joyce et al. (1994)

6.4 The liaison between EPNs and mutualistic bacteria and their identification

Heterorhabditis and *Steinernema* genera of nematodes have evolved to establish mutualistic relationship with the bacteria *Photorhabdus* and *Xenorhabdus* respectively (Forst and Clarke, 2002). Tables 6.3 and 6.4 reveals the diversity of EPNs and their associated symbionts. Both nematode and bacterial partners perform interdependent activities to kill their insect's hosts and nourish themselves (Ciche et al., 2001). These bacteria reside inside the anterior part of the intestine of its nematode partners. The *Photorhabdus* resides by forming a film while *Xenorhabdus* resides inside a specialized structure, called vesicle at the anterior part of the intestine of infective juvenile stages. The nematodes provide shelter and protection to its bacterial partner and transport it to the midgut of the insects. While in return the bacteria kills the insect hosts, degrade the tissues of their midgut, protect the dead cadaver from other microbe inhabitants of the soil and protect the nematode partner from host immune responses (Clarke, 2008).

The infective juvenile stages of entomopathogenic nematodes maintain amutualistic relationship with their entomopathogenic bacterial partner. IJs of entomopathogenic nematode can be isolated from the soil by using insect baiting techniques. In this technique, larvae of the wax moths are allowed to infect by adding them in soil samples. After infestation, these insect larvae die and support the growth and development of entomopathogenic nematodes for two-three generations. Now at this stage the dead cadaver remains filled with the young generation of infective juveniles and bacterial spores.

TABLE 6.3 Associated species of *Heterorhabditis* nematode and *Photorhabdus* bacteria and their distribution.

S.No.	Nematode species	Bacterial species	Distribution	References
1	*Heterorhabditis indica*	*Photorhabdus luminescens* subsp. *akhurstii*	India	Thomas and Poinar (1979)
2	*Heterorhabditis bacteriophora*	*Photorhabdus luminescens* subsp. *laumondii*	Australia	Fischer-Le Sauxet al. (1999a,b)
3	*Heterorhabditis bacteriophora*	*Photorhabdus luminescens* subsp. *luminescens*	Australia, USA	Boemare et al. (1993)
4	*Heterorhabditis* sp.	*Photorhabdus luminescens* Subsp. *kayaii.*	Turkey, China, France, Italy, Korea	Hazir et al. (2004)
5	*Heterorhabditis* sp.	*Photorhabdus temperata* subsp. *thracensis*	Turkey, France	Hazir et al. (2004)
6	*Heterorhabditis* sp.	*Photorhabdus asymbiotica* subsp. *australis*	Australia	Akhurst et al. (2004)
7	*Heterorhabditis* sp.	*Photorhabdus temperata* subsp. *cinerea*	Hungary	Toth and Lakatos (2008)
8	*Heterorhabditis megidis*	*Photorhabdus temperata* subsp. *temperate*	Russia, Belgium, Ireland	Tailliez et al. (2010)
9	*Heterorhabditis heliothi-dis USA*	*Photorhabdus temperata* subsp. *khanii*	USA, Cuba	Tailliez et al. (2010)
10	*Heterorhabditis megidis*	*Photorhabdus temperata* subsp. *khanii*	USA	Tailliez et al. (2010)
11	*Heterorhabditis zealandica*	*Photorhabdus temperata* subsp. *tasmaniensis*	New Zealand, Australia	Tailliez et al. (2010)
11	*Heterorhabditis marelatus*	*Photorhabdus temperata* subsp. *tasmaniensis*	USA	Tailliez et al. (2010)
12	*Heterorhabditis sp.*	*Photorhabdus luminescens* subsp. *hainanensis*	China	Tailliez et al. (2010)
13	*Heterorhabditis sp. St Martin*	*Photorhabdus luminescens* subsp. *car-ibbeanen sis*	St Martin, Guadeloupe	Tailliez et al. (2010)
14	*Heterorhabditis bacteriophora*	*Photorhabdus temperata* subsp. *stackebrandti*	USA	An and Grewal (2010)
15	*Heterorhabditis georgiana*	*Photorhabdus luminescens*	USA	An and Grewal (2011)
16	*Heterorhabditis* sp.	*Photorhabdus luminescens* subsp. *noenieputensis*	South Africa	Ferreira et al. (2013a,b)
17	*Heterorhabditis zealandica*	*Photorhabdus heterorhabditis* strain SF41	South Africa	Ferreira et al. (2014)
18	*Heterorhabditis safricana*	*Photorhabdus luminescens* subsp. *laumondii*	South Africa	Geldenhuys et al. (2016)
19	*Heterorhabditis bacteriophora*	*Photorhabdus luminescens* strain DSPV002N	Argentina	Palma et al. (2016)
20	*Heterorhabditis sonorensis*	*Photorhabdus luminescens* subsp. *sonorensis*	USA	Duong et al. (2019)
21	*Heterorhabditis atacamensis*	*Photorhabdus khanii subsp. guanajuatensis*	Mexico	Machado et al. (2019)

TABLE 6.4 Associated species of *Steinernema* nematode and *Xenorhabdus* bacteria and their distribution.

S. No.	Nematode's species	Bacteria's species	Distribution	References
1	*Steinernema feltiae*	*Xenorhabdus bovienii*	Australia	Akhurst (1982)
2	*Steinernema glaseri*	*Xenorhabdus poinarii*	USA, Portugal	Akhurst and Bedding (1986)
3	*Steinernema kushidai*	*Xenorhabdus japonica*	Japan	Yamanaka et al. (1992)
4	*Steinernema intermedium*	*Xenorhabdus bovienii*	USA	Boemare et al. (1993)
5	*Steinernema abbasi*	*Xenorhabdus indica*	Sultanate of Oman	Elawad et al. (1997)
6	*Steinernema cubanum*	*Xenorhabdus poinarii*	Cuba	Fischer-Le Saux et al. (1999a,b)
7	*Steinernema thermophilum*	*Xenorhabdus indica*	India	Ganguly and Singh (2000)
8	*Steinernema kraussei*	*Xenorhabdusbovienii*	USA	Burnell and stock (2000)
9	*Steinernema scapterisci*	*Xenorhabdus innexi*	Uruguay	Lengyel et al. (2005)
10	*Steinernema rarum*	*Xenorhabdus szentirmaii*	Argentina	Lengyel et al. (2005)
11	*Steinernema bicornutum*	*Xenorhabdus budapestensis*	Serbia	Lengyel et al. (2005)
12	*Steinernema serratum*	*Xenorhabdus ehlersii*	China	Lengyel et al. (2005)
13	*Steinernema sichuanense*	*Xenorhabdus bovienii*	China	Mracek et al. (2006)
14	*Steinernema weiseri*	*Xenorhabdus bovienii*	Czech Republic	Tailliez et al. (2006)
15	*Steinernema apuliae*	*Xenorhabdus species*	Italia	Tailliez et al. (2006)
16	*Steinernema arenarium*	*Xenorhabdus kozodoii*	Russia	Tailliez et al. (2006)
17	*Steinernema karii*	*Xenorhabdus hominickii*	Kenya	Tailliez et al. (2006)
18	*Steinernema riobrave*	*Xenorhabdus cabanillasii*	USA	Tailliez et al. (2006)
19	*Steinernema hermaphroditum*	*Xenorhabdus griffiniae*	Indonesia	Tailliez et al. (2006)
20	*Steinernema puertoricense*	*Xenorhabdus romanii*	Puerto Rico	Tailliez et al. (2006)
21	*Steinernema* sp.	*Xenorhabdus miraniensis*	Australia	Tailliez et al. (2006)
22	*Steinernema siamkayai*	*Xenorhabdus stockiae*	Thailand	Tailliez et al. (2006)
23	*Steinernema* sp.	*Xenorhabdus mauleonii*	St. Vincent	Tailliez et al. (2006)
24	*Steinernema scarabaei*	*Xenorhabdus koppenhoeferi*	USA	Tailliez et al. (2006)
25	*Steinernema australe*	*Xenorhabdus magdalenensis*	Chile	Tailliez et al. (2006)
26	*Steinernema aciari*	*Xenorhabdus ishibashii*	China and Japan	Kuwata et al. (2013)
27	*Steinernema rarum*	*Xenorhabdus szentirmaii*	France	Gualtieri et al. (2014)
28	*Steinernema* sp.	*Xenorhabdus khoisanae*	South Africa	Naidoo et al. (2015)
29	*Steinernema biddulphi*	*Xenorhabdus* sp.	South Africa	Cimen et al. (2016)
30	*Steinernema beitlechemi*	*Xenorhabdus* sp.	South Africa	Cimen et al. (2016)
31	*Steinernema yirgalemense*	*Xenorhabdus indica*	USA	Ferreira et al. (2016)
32	*Steinernema diaprepesi*	*Xenorhabdus doucetiae*	Martinique	Ogier et al. (2016)
33	*Steinernema surkhetense*	*Xenorhabdus stockiae*	India	Bhat et al. (2017)
34	*Steinernema sangi*	*Xenorhabdus vietnamensis*	Vietnam	Lalramnghaki and Vanlalhlimpuia (2017)

TABLE 6.4 Associated species of *Steinernema* nematode and *Xenorhabdus* bacteria and their distribution.—cont'd

S. No.	Nematode's species	Bacteria's species	Distribution	References
35	*Steinernema monticolum*	*Xenorhabdus hominickii*	Korea	Park et al. (2017)
37	*Steinernema poinar*	*Xenorhabdus bovienii*	Poland	Sajnaga et al. (2018)
38	*Steinernema pakistanense*	*Xenorhabdus indica*	India	Bhat et al. (2019)
39	*Steinernema longicaudum*	*Xenorhabdus ehlersii*	Philippines	Ubaub et al. (2018)

Before isolation of bacterial spores from the dead cadaver, it must be sterilized by washing with 70%–80 % ethyl alcohol. After sterilization, the bacteria can be isolated by dissecting the dead insect cadaver. From dissected cadaver, 1–2 μL of hemolymph can be streaked onto an NBTA agar medium for the multiplication of the bacteria. The Petri plates should be incubated at 25–28̊C temperature to support the bacterialcolony formation. Both *Photorhabdus* and *Xenorhabdus* bacteria are gram-negative and rod-shaped, belonging to the family *Enterobacteriaceae*. Initial microscopic observations are required to reveal their shape and gram reaction. *Photorhabdus* and Xenorhabdus bacteria can be identified by observing the colony-forming patterns and colors on different agar plates. Both of these bacteria make specific colonies, which helps in easy differentiation. Biological assays are also utilized to identify bacterial species. In this assays, the whole bacteria is injected into the hemocoel of wax moth larva. These bacteria kill the insect's larvae by causing septicemia within 24–48 h. The color of the dead cadaver is indicative of the presence of specific bacteria.

There is certain biochemical and physiological tests activities are performed to reveal and confirm the identity of entomopathogenic bacteria. These tests include gelatin liquefaction, catalase test, urease test, motility test, and lactose fermentation test, etc. For the exact identity of the species, these assays are followed by the molecular identification of the bacteria. Species identification for both entomopathogenic bacteria i.e. *Photorhabdus* and *Xenorhabdus* is done based on PCR amplification of *recA* sequences from total genomic DNA (Table 6.5). The amplified segment of DNA is extracted and purified from the gel and then subjected to Sangers sequencing. The identity of the species is further confirmed by using BLASTn search of sequenced recA across the NCBI redundant database. The retrieved othologous sequences are then subjected multiple sequence alignment using MEGA software and the similarity score can be calculated by using

TABLE 6.5 characterization parameters of symbiotic bacteria.

		Primary microscopic observation	
S. No.	Bacteria	Nutrient bromothymol blue-triphenyltetrazolium chloride agar (NBTA)	Mac conkey agar medium
1	*Photorhabdus*	Dark green colony with convex surface and umbonated	Reddish colonies
2	*Xenorhabdus*	Light or dark red or blue colony with convex surface and umbonated	Bright pink colonies
Identification based on biological assay:			
1	*Photorhabdus*	Pale or brown colored dead cadaver	
2	*Xenorhabdus*	Red or pink colored dead cadaver	
Molecular identification			
1	Recombinase A gene (*recA*)	F 5′-GCTATTGATGAAAATAAACA-3′ R 5′-GCTATTGATGAAAATAAACA-3′	Tailliez et al. (2010)
2	*16S RNA*	F: 5′-AGTTTGATCATGGCTCAGATTG-3′ R: 5′-TACCTTGTTACGACTTCACCCCAG-3′	Sandström et al. (2001)
3	Gyrase B gene (*gyrB*)	F 5′-TACACGAAGAAGAAGGTGTTTCAG-3′ R 5′-TACTCATCCATTGCTTCATCATCT-3′ (reverse)	Tailliez et al. (2010)

Clustal W platform. Phylogenetic trees and bootstrap analysis performed by these software exhibit the identity of the bacterial species. Till date, three species have been identified and reported in genus *Photorhabdus*, which includes *P. temperata, P. luminescens, and P. asymbiotica*. The species *P. asymbiotica* is a human parasitic but also reported from the *Heterorhabditis* host. In case of genus *Xenorhabdus*, more than 22 species have been reported from the different species of *Steinernema* (Tailliez et al., 2010).

6.5 Lifes cycle, pathogenicity and host range of EPNs

Entomopathogenic nematodes (EPN) belonging to the genera *Heterorhabditis*, *Neosteinernema* and *Steinernema* (Nematoda: Rhabditidae) occur naturally in soil and locate their host in response to carbon dioxide, vibration and other chemical cues, and they react to chemical stimuli or sense the physical structure of insect's integument (Kaya and Gaugler, 1993). EPNs complete most of their life cycle in insects with an exception of IJs, this stage is developmentally arrested third larval stage which is free-living and able to target the insect host. Both *Heterorhabditis* and *Steinernema* are mutually associated with bacteria of the genera *Photorhabdus* and *Xenorhabdus*, respectively (Ferreira and Malan, 2014). In the IJs *Steinernematidae*, *Xenorhabdus* bacterial cells are housed in the anterior intestinal caecum, whereas in the IJs of *Heterorhabditid*, *Photorhabdus* cells are found throughout in the intestinal tract. The IJs enter the host through natural openings (oral cavity, anus, and spiracles) or infective juveniles of *Heterorhabditis* also penetrate through intersegmental membranes (Kaya and Gaugler, 1993) and through the midgut wall or tracheae and reach into the haemocoel (Bedding and Molyneux, 1982). Inside the insect's haemocoel, the IJs release their symbiotic bacteria through the anus by defecation for *Steinernematids* (Martens et al., 2003) or through the mouth for *Heterorhabditids* (Ciche and Ensign, 2003). The nematodes after entering the insect's body release the associated bacteria, which rapidly multiply and kill the host by inducing septicemia within 24−72 h. Development and reproduction of EPNs within the cadaver can take 1−3 weeks (Stock, 1995). *Steinernematids* and *Heterorhabditids* from these families have similar life cycles, and the only difference between the life cycles of *Heterorhabditis* and *Steinernema* occurs in the first generation. *Steinernema* species are amphimictic whereas *Heterorhabditis* species are hermaphroditic. The nematodes initiate its development and then feed on the bacterial cells and the host tissues until the food resources in the host cadaver are depleted. The nematodes provide shelter to the bacteria, which, in return, kill the insect host and provide nutrients to the nematode. Together, the nematodes and bacteria feed on the liquefying host and reproduce for several generations inside the cadaver maturing through the growth stages of J2, J4 and finally into adults. Finally, they emerge as a new generation of IJs in the soil in search of new hosts (Hazir et al., 2003).

6.5.1 Pathogenicity

The invasion and evasion of host defenses are the important steps for pathogenicity. The ability of the nematode to penetrate into the insect haemocoel is enabled by the release of proteolytic enzymes. Nematode-insect relationship is the ability of the nematodes to evade insect defenses enabling failure of recognition and/or by destruction of insect's antibacterial factors. Toxins and extracellular enzymes are important virulence factors released by these nematodes (Simoes and Rosa, 2010). Symbiotic bacteria provide a protected niche for themselves and host nematodes by producing antibiotics (Akhurst, 1982; Hu and Webster, 2000), nematicides (Hu et al., 1999) and even compounds which deter scavenging ants (Zhou et al., 2002). *Xenorhbdus* and *Photortrabdus* species produce endotoxins (Dunphy and Wetster, 1988) and exotoxins (Ehlers, 1991) which leads to pathogenicity. The *X. nematophilus* endotoxins are toxic to the haemocytes of *G. mellonella* (Dunphy and Wetster, 1988). Xenorhabdin 2, an antibiotic, produced by *Xenorhabdus* spp. had larvicidal activity against *Heliothis punctigera* (McInerney et al., 1991). It is suggested that xenorhabdins might be involved in the mechanism of pathogenic action against the host insect through their insecticidal properties.

Initially the bacteria *Photorhabdus* colonise the anterior of the midgut then subsequently spread along its length and begins to destroy the midgut epithelium via secretion of toxins. Prior to insect death, bacteria are mainly found in the haemoceol which after death spread to every available tissue in the cadaver and begin bioconversion to generate a nutrient source for nematodes to feed upon (Fig. 6.2). *Xenorhabdus* bacteria initially grow within the hemolymph and then migrate to the anterior region of the midgut where they associated with the connective tissues surrounding the muscle fibers and trachea. No fluorescent bacteria were seen in contact with or within haemocytes at any stage during infection (Hinchliffe et al., 2010). During bioconversion, the variety of antibiotics are secreted by *Photorhabdus* and *Xenorhabdus* spp. which protect the cadaver from contaminating organisms. *Xenorhabdus* has been reported to produce a whole range of antimicrobials, such as benzylideneacetone, nematophin, xenocoumacins 1 and 2, xenortides A and B, xenematide, and the bicornutins A, B and C (Bode, 2009). *P. luminescens* strains have been reported to produce the anti-microbials, such as 2-isopropyl-5-(3-phenyl-oxiranyl)-benzene-1, 3-diol, 3,5,-dihydroxy-4 isopropyl-stilbene and carbapenem (Hu et al., 2006).

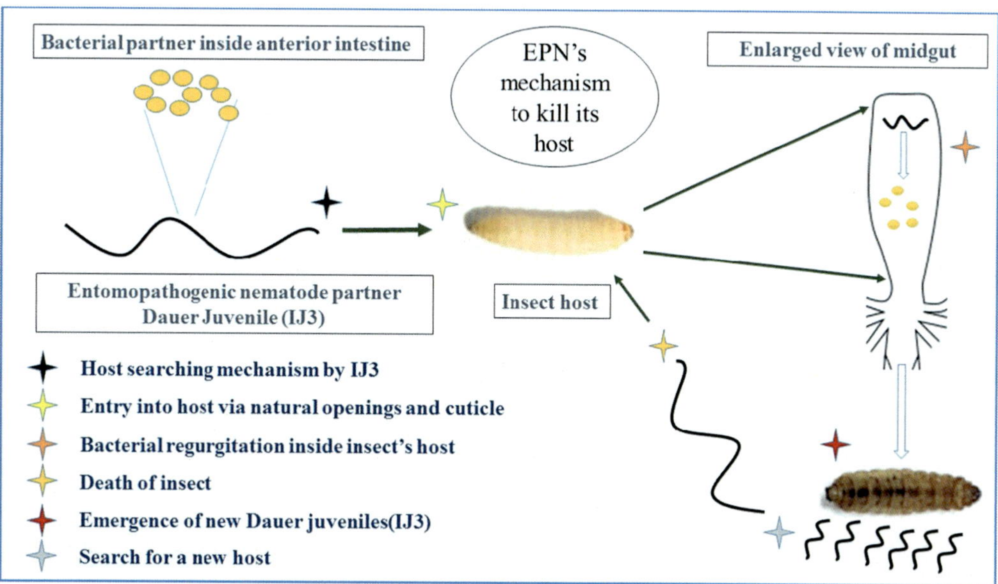

FIG. 6.2 Mode of action of EPNs.

The insect cadaver becomes red if the insects are killed by *Heterorhabditids* and brown or tan if killed by *Steinernematids*. The color of the insect host body is indicative of the pigments produced by the mutualistic bacteria growing in the host insects (Kaya and Gaugler, 1993). After about a week, thousands of infective juveniles emerge and leave in search of new hosts, carrying with them symbiotic bacteria, received from the internal host environment. In nature, the infective juveniles invariably contain only phase-one bacteria.

6.5.2 Host range

Nematode-bacterium complex kills insects so rapidly that the nematodes do not form the intimate, highly adapted, host-parasite relationship which is a characteristic of other insect-nematode associations, e.g., mermithids. This rapid mortality permits the nematodes to exploit a range of hosts that spans nearly all insect orders, a spectrum of activity well beyond that of any other microbial control agent. In laboratory conditions, *S. carpocapsae* alone infected more than 250 species of insects from over 75 families in 11 orders (Poinar, 1975). The nematodes attack a far wider spectrum of insects in the laboratory where host contact is assured, while the environmental conditions are optimal, and no ecological or behavioral barriers to infection exist (Kaya and Gaugler, 1993). Some nematode species search for hosts at or near the soil surface (*S. carpocapsae* and *S. scapterisci*), referred to as "ambusher", which remains nearly sedentary while waiting for the mobile surface-dwelling hosts (Campbell and Gaugler, 1993), whereas others are adapted to search deeper in the soil profile (*H. bacteriophora* and *S. glaseri*) referred to as "cruiser" which is highly mobile, responds strongly to long-range host chemical cues, and is therefore best adapted to find sedentary hosts (Grewal et al., 1994). More than 100 species of *Steinernema* and 16 species of *Heterorhabditis* have been recorded (Shapiro-Ilan et al., 2017).

Seventeen insect species belonging to three orders, *viz.*, Lepidoptera, Coleoptera, Hemiptera and one slug were recorded as hosts for *H. indica* and *S. glaseri*. *H. indica* and *S. glaseri* killed all the 18 insect species tested under laboratory condition. Both these species infected and completed their life cycle on the larvae of *Spodoptera litura*, *Helicoverpa armigera*, *Plutella xylostella*, *Leucinodes orbonalis*, *Earias vittella*, *Orthaga exvinascea*, *Eublemma versicolor*, *Papilio polytes*, *Exelastis atomosa*, *Oryctes rhinoceros*, *Hymenia recurvalis*, *Anomala communis*, *Agrotis ipsilon*, *Cosmopolites sordidus*, *Aulacophora faveicollis*, *Ferrisia virgate* and *Deroceras reticulatum* (Sharmila et al., 2018) (Fig. 6.3).

6.6 Mass production, formulation development and application

The development of commercialized EPN formulation is an important factor for their successful use against target insect pests (Grewal, 2002). The infective stages of nematodes are formulated into products to increase their shelf life, storage, transport and application (Grewal, 2002; Strauch et al., 2000; Hussein and Abdel-Aty, 2012; Askary and Ahmad, 2017).

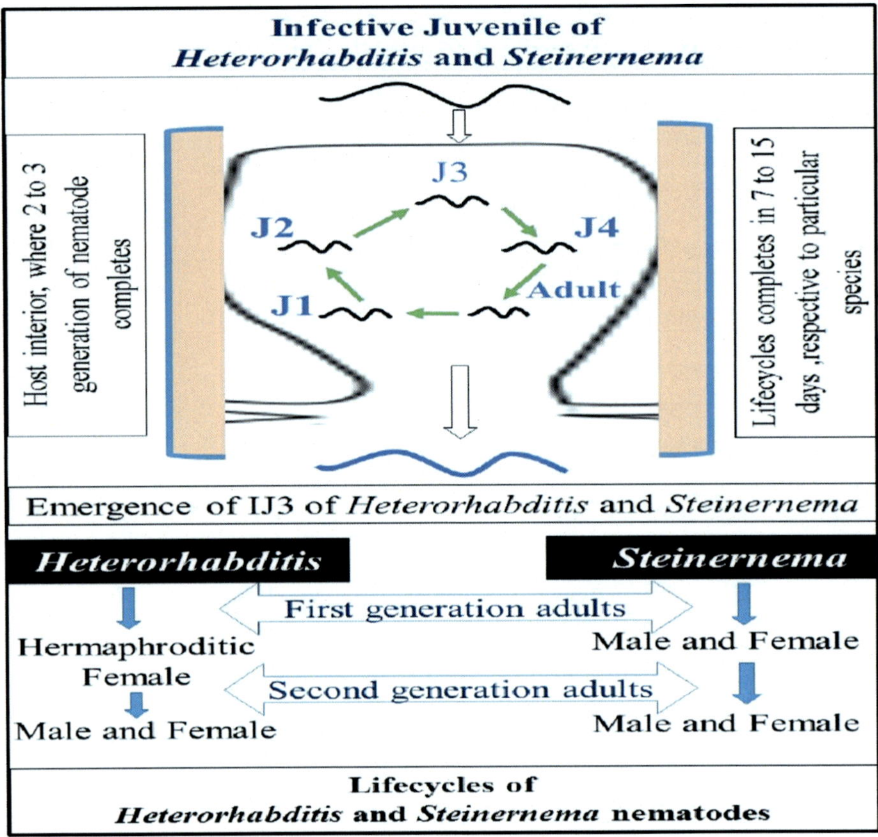

FIG. 6.3 Life cycle of *Steinernema* and *Heterorhabditis*.

However, several factors such as carriers, additives, temperature, moisture content, oxygen, UV-ray protectors and IJ concentrations play a crucial role in successful EPN formulations (Strauch et al., 2000; Grewal, 2002; Feng and Han, 2005; Grewal and Peters, 2005; Shapiro-Ilan et al., 2012; Peters, 2016; Cruz-Martínez et al., 2017; Guo et al., 2017). Tables 6.6 and 6.7 showed a list of commercial EPN formulations developed in the world against target insect pests. The commercially developed EPN formulations include aqueous suspension, sponge, alginate capsules, dispersible granules and clays. However, the formulation type should be optimized according to the nematode strain and culture methods (Hussein and Abdel-Aty, 2012).

6.6.1 Types of formulations

6.6.1.1 Aqueous suspensions

These are the most common type of EPN formulations mainly used for storage, transportation and application (Chen and Glazer, 2005). Hazir et al. (2003) reported the survival times of 6–12 months for *steinernema* spp. and 3–6 months for *Heterorhabditis* spp. at a storage temperatures of 4–15°C. However, sedimentation, aeration, temperature and susceptibility to microbial contamination, refrigeration and specific requirement in application equipment limited the multiplication of these type formulations by agro companies for mass production (Grewal and Peters, 2005; Toepfer et al., 2010; Brusselman et al., 2012; Lacey and Georgis, 2012; Beck et al., 2014; Shapiro-Ilan et al., 2015).

6.6.1.2 Gels

Initially several workers have developed various gel formulations using activated carbon powder, polyacrylamide, hydrogenated vegetable oil paste containing mono and diglycerides. However, these formulations were unsuccessful due to low stability, less survival time, desiccation tolerance and viability of IJs (Yukawa and Pitt, 1985; Georgis, 1990; Chang and Gehert, 1991, 1992; Bedding and Butler, 1994; Bedding et al., 2000). In gel formulations, the mobility of IJs become reduced and remains infective for up to 9 months (Ganguly et al., 2008). Later introduction of alginate and flowable gel

TABLE 6.6 Commercial EPN products available in different countries.

Species	Trade name	Formulation type	Target insect species	Producer/country
Steinernema carpocapsae	Bouncer	WP	Root grubs, cutworms, root weevils, various lepidopterans and termites	Multiplex/India
	Nemastar	Clay powder	*Capnodis tenebrionis*, mole crickets and black cutworms	e-nema GmbH, Germany
	OPTIMEN C	WP	Codling moth	Agrifutur Srl/Italy
	Nemabact	Water suspension	Western flower thrips, wireworm, soil-inhabiting insects	OOO Biometodika/Russia
	Exhibitline Sc	WP	*Agrostis* spp., *Tipula* spp.	Syngenta
	Capsanem	WP	Larvae of Noctuidae, Pyralidae, Tipulidae	Koppert Italia Srl/Italy
	NEMAPACK SC	WP	Codling moth, Palm weevil, Palm moth	Bioplanet s.c.a./Italy
	Nemastar	WP	Palm weevil, Palm moth, Codling moth, *Spodoptera* spp., *Agrotis* spp.,	Biogard Divisone CBC (Europe) Srl/Italy
	Ecomask		Caterpillars	BioLogic/USA
	Guardian		Caterpillars	HydroGardens/USA
S. feltiae	Nemapom	Clay powder	Codling moth	e-nema GmbH/Germany
	Nemapom	WP	Codling moth	Biogard Divisone CBC (Europe) Srl/Italy
	Nemaplus	Clay powder	Glasshouse sciarids	e-nema GmbH/Germany
	OPTIMEN F	WP	*Cydia pomonella*	Agrifutur Srl/Italy
	Entonem F	Water suspension	Western flower thrips, wireworm and other soil-inhabiting insects	OOO Biometodika/Russia
	Entonem	WP	Western flower thrips	Koppert Italia Srl/Italy
	SCIARID	WP	Western flower thrips	Koppert
	Nemasys F	WP	Western flower thrips	Syngenta
	Exhibitline Sf	WP	Western flower thrips	Syngenta
	Nemax F	WP	Codling moth, Twig borer, Palm weevil	Serbios Srl/Italy
	Xedanema	WP	Codling moth, Twig borer, Palm weevil	Xeda Italia/Italy
	NEMAPACK SF	WP	Cutworms, flies	Bioplanet s.c.a./Italy
	Scanmask		Fungus Gnats	BioLogic/USA
S. kraussei	Nemasys L	WP	Vine weevil	Syngenta
S. braziliense	Bio Steinernema	sponge	Billbugs	Bio Controle/Brazil
S. pakistanense	PakNema-1	Sponge	American, spotted, pink bollworm	NNRC/Pakistan
S. maqbooli	PakNema-2	Sponge	Root knot nematodes	NNRC/Pakistan

Continued

TABLE 6.6 Commercial EPN products available in different countries.—cont'd

Species	Trade name	Formulation type	Target insect species	Producer/country
S. bifurcatum	PakNema-3	Sponge	Pupal stages in soil and termite	NNRC/Pakistan
S. balochiense	PakNema-4	sponge	Fruit fly pupa	NNRC/Pakistan
Heterorhabditis indica	Soldier	WP	White grubs, termites	Multiplex/India
	Nema power	WP	White grubs, termites	KN Biosciences/ Hyderabad, India
	BCS-Grub terminator	WP	White grubs, termites	Benzor Crop Science/India
	Grub Cure	WP	White grubs, termites	SRI Biotech
	Armour	WP	White grubs, termites	Ponalab/India
	Calterm	WP	White grubs, termites	Camson Biotech/ Bengaluru, India
H. bacteriophora	Larvanem	Inert carrier, infection larvae mixed with inert carrier or on clay	Weevils	Biocont Laboratory spol. s r. o.
	Nematop	Clay powder	Weevils	e-nema GmbH/ Germany
	Nemagreen	Clay powder	Chafer grubs	e-nema GmbH/ Germany
	OPTINEM H	WP	Vine weevil and other soil beetles	Agrifutur Srl/Italy
	LARVANEM	WP	Larvae of Vine weevil, grubs of Chafer beetle	Koppert Italia Srl/ Italy
	B-GREEN	WP	Grubs of Garden chafer or foliage beetle	Biobest
	Nemax H	WP	Vine weevil larvae and Western corn rootworm	Serbios Srl/Italy
	Terranem	WP	Garden chafer or foliage beetle, *Hoplia* spp., *Aphodius* spp.	Koppert Italia Srl/ Italy
	NEMAPACK HB	WP	Vine weevil larvae and other soil beetles	Bioplanet s.c.a./Italy
	Nematop	WP	Vine weevil larvae	Biogard Divisone CBC (Europe) Srl/ Italy
	Cryptonem	WP	Codling moth, weevils	River Bioscience/ South Africa
	Heteromask		Weevils, grubs	BioLogic/USA
	Lawn patrol		Weevils, grubs	HydroGardens/USA
H. megidis	Heterorhabditis system	WP	Vine weevil larvae	Biobes
	Nemasys HM	WP	Vine weevil larvae	Syngenta
	Exhibilinehm	WP	Vine weevil larvae	Syngenta
S. carpocapsae/ H. bacteriophora	NOBUG	Sponge	Borers	The National Research Center/ Egypt

TABLE 6.7 Genomic feature of EPNs.

S. No.	EPN species	Genome size	GC%	N50	Number of predicted gene models	References
1	*Heterorhabditis bacteriophora*	77.0	32.2	312,328	21,250	Bai et al. (2013)
2	*Steinernema carpocapsae*	85.6	45.5	299,566	28,313	Dillman et al. (2015)
3	*Steinernema scapterisci*	79.4	48	90,783	31,378	Dillman et al. (2015)
4	*Steinernema feltiae*	82.4	47	47,472	33,459	Dillman et al. (2015)
5	*Steinernema glaseri*	92.9	47.6	37,444	34,143	Dillman et al. (2015)
6	*Steinernema monticolum*	89.3	42	11,556	36,007	Dillman et al. (2015)

formulations, and alginate capsules resulted in major breakthrough toward the utilization of these formulations in different market segments by agrochemical companies (Chen and Glazer, 2005; Hussein and Abdel-Aty, 2012; Goud et al., 2010). Grewal (2002) reported the storage of *S. carpocapsae* for 3–4 months at 25°C and *S. feltiae* for 2–4 weeks in alginate gel formulations. Shapiro-Ilan et al. (2015) reported direct mixing of gel formulation in tank at lower concentration (<2%) results in enhancing the efficacy of nematodes subsequently to protect the nematodes from desiccation and UV rays after soil application.

6.6.1.3 Sponges

The sponge based formulations are generally prepared by applying an aqueous nematode suspension at the rate of 500–1000 IJs per cm^2 surface area resulting a quantity of 5–25 × 10^6 IJs per sponge placed in a plastic bag for storage for a period of 1–3 months at 5–10°C (Georgis, 1990; Grewal, 2002). Ramakuwela et al. (2015) reported that the survival of IJs were highest (84%–88%) and most stable at 15°C and the mortality of *Galleria mellonella* larvae was >90% by IJs after 84 days. The nematodes are removed from the sponge by soaking in water before field application. However, these formulations are not suitable for large scale applications since it needs refrigeration for storage, laborious and large quantity of sponges are needed (Cruz-Martínez et al., 2017).

6.6.1.4 Clay and powder

Initially Bedding (1988) developed a formulation with a layer of nematodes between two layers of clay termed as sandwich type, but due to poor storage stability, clogging of spray nozzles, and a low nematode-to-clay ratio products with this formulated were withdrawn from market (Grewal, 2002). Further, several wettable powder formulations have been developed that improve the shelf life of *Steinernematids* and *Heterorhabditids* at room temperature. The production of *H. indica* as a WP formulation with a shelf life of 12 months marked a remarkable success for the control of white grubs and other soil borne pests in arecanut, banana, brinjal, cardamom, groundnut and sugarcane crops (NBAIR, 2017).

6.6.1.5 Application

Entomopathogenic nematodes can be applied with equipments such as sprayers, mist blowers, and electrostatic sprayers or as aerial sprays (Georgis, 1990; Wright et al., 2005; Shapiro-Ilan et al., 2006). However the usage of equipment depends on the cropping system, area of application, agitation, volume and nozzle type etc. (Grewal, 2002; Fife et al., 2003, 2005; Wright et al., 2005; Shapiro-Ilan et al., 2006; Lara et al., 2008). Several biotic and abiotic factors play an important role in success of nematode application. Among biotic factors, the appropriate nematode species against target insect pest is crucial. The nematode species should have high virulence, host finding ability, recycling potential and better environmental tolerance (Shapiro-Ilan et al., 2002). The relationship between nematodes and other entomopathogens can vary depending upon the nematode species, relative timing and rate of application. Among abiotic factors, the soil type and moisture content influence the survival, activity and movement of nematodes (Kagimu et al., 2017). In general, the survival and activity of nematodes is higher in sandy-loam soils than clay soils (Kaya, 1990). Soil temperature can also affect the efficacy of nematodes. However, soil temperature between 12 and 28°C is favorable for nematode application (Kaya, 2002). Further, UV radiations are detrimental to nematodes when applied to the soil. Therefore it is necessary to apply nematodes during evening or morning hours. Generally, EPNs are applied to soil at a minimum rate of 2.5 × 10^9 IJs/ha or higher depending on the target pest (Georgis and Hague, 1991; Georgis et al., 1995; Shapiro-Ilan et al., 2002).

6.7 Application of EPN genomics to enhance the field efficacy

EPNs are globally being used as successful bio agent to control insect pest and most of the EPN research deals with the applied aspects for management of insect pests. Recent developments in genomic perspectives have provided new opportunities to improve the potential of EPNs as successful bio control agents. To establish the EPN genomics as an important tool we need genomic, transcriptomic, ESTs and proteomics data of EPNs and their symbiotic bacteria to decipher the function of critical genes involved in various biological process of EPNs and for further manipulation to enhance the effectiveness. In order to fulfill the need of genomic data the various project have been developed and are underway to sequence and elucidate the complex biological processes of EPNs and their symbionts. There are many numbers of putative genes involved in parasitism and survival has been identified in the genome sequencing effort of EPNs and their symbionts.

EPNs depend upon associated symbiont to become parasite of insect pest and the information about genomic background enables to trace the genetic basis of various fundamental processes of mutualism to parasitism. Phylogenetically, EPNs are a transitional taxon among the sub-class Rhabditina and exhibits shared traits with *C. elegans* like microbivorous ancestors. EPNs have evolved parasitism and shares most recent ancestry with hookworm and lungworm like mammalian parasites. Phylogenetic position of the EPNs can be exploit to explore the evolutionary changes in genetics happened between microbivorous to obligate parasitism in mammals in the evolutionary history (Bai et al., 2013).

The announcement of first high quality draft genome of *H. bacteriophora* was made by *H. bacteriophora* genome consortium in June 2005 (Ciche et al., 2006). The flow cytometry technique based estimated size of the genome was nearly 111.4 Mb (Bennett et al., 2003). The cDNA sequencing project of *H. bacteriophora* revealed 1246 ESTs in IJs and further these ESTs annotated to 1072 functional groups based on KEGG pathways, Genbank. Among these ESTs, 417 having significant similarity to *C. elegans* protein including genetic information group, metabolism, environmental stress pathways. Many genes of survival and infectivity were found which includes proteases, dauer formation (*akt-1, pdk-1 & daf-7*), stress and aging related genes such as heat shock genes (*hsp-4 & hsp-6*), superoxide dismutase (*sod-4*) and *eat* genes, GPCRs and serine/threonine kinases for signaling. Presence of RNAi related genes such as *sid-1, rnh-2 & rnc*, and Tc3A shows the opportunities of RNAi tools for functional genomics (Sandhu et al., 2006). The recently published high quality draft genome of *H. bacteriophora* predicted 21,250 protein coding genes in genome size of 77 Mbp. Majority of the predicted genes (11,207) had no homologues to *C. elegans* and may be considered as novel. It has 19 insulin/IGF-1 signaling pathway genes having prominent role in dauer formation, longevity, innate immunity and stress resistance in *C. elegans*. A functional genomics study of these genes may be helpful to enhance longevity, stress tolerance of IJs for pest control. Presence of RNAi related genes homologue such as *drsh-1, ego-1, rsd-3, smg-2, sid 1* and *sid-3* in *H. bacteriophora* genome suggests that the RNAi mechanism successfully can be harness as a toll for functional genomics to study the function of various genes. Protein domain analysis revealed 82 GPCRs and 24 NHR (nuclear hormone receptor) gene families were found in the *H. bacteriophora* genome. A total of 3794 microsatellite predicted in draft genome which could be useful as genetic marker to discriminate closely related populations in phylogeography studies (Bai et al., 2013). Presence of insulin/IGF-1 signaling pathways in *H. bacteriophora* suggests that their function may be the same as in *C. elegans* such as dauer formation, longevity, stress tolerance and innate immunity. The 19 genes involved in insulin/IGF-1 signaling pathways can be manipulated to improve the specific trait for successful application in field condition. Large number of predicted GPCRs (82) are another most important gene families of interest to enhance/alter the heritable trait of olfaction, sensory response, host specificity and host seeking beahviour of IJs to affect the field efficacy positively (Robertson and Thomas, 2006). The protease secreted by EPNs plays an important role in penetration into the host and suppression of immune systems however, *H. bacteriophora* known to have less than 30 protease and protease inhibitors which could be improved by introducing additional copy of such genes through the genetic engineering approach (Leger et al., 1996; Bai et al., 2013).

Genome of 5 *Steinernema* species (*S. carpocapsae, S. feltiae, S. glaseri, S. monticolum, S. scapterisci*) have been sequenced and analyzed to have various gene associated to parasitism that can be useful to improve important traits (Dillman et al., 2015). All 5 genome have been found to be similar in size but differs in G = C content that can affect the genetic alteration using recombinant DNA tools (Lu et al., 2016). In comparison to Genome of *H. bactereophora* Genome of Steinernematids revealed a large number of genes coding for signal peptides, proteases and protease inhibitors which plays a significant role in host invasion, breaking host immunity and mortality. FAR (Fatty acids and retinol binding) proteins coding gene family have been predicted to be expanding in steinernematids genome. These proteins have been known to sequester the host retinoid and host immune suppression.

The genome of entomopathogenic bacteria, *Photorhabdus luminescens* published in 2003 and revealed to be 5,688,987 base pairs (bp) long, 42.8% GC with 4839 predicted protein-coding genes encoding for toxins, adhesions, hemolysins, lipases and proteases and antibiotics. The identified putative protein coding genes can possibly be linked to play a wide range of roles in enhancing immunity, colonization in host, invasion and bioconversion of host cadavers. Comparative genomics of *P. luminescens* with other bacteria revealed the HGT (horizontal transfer events) to acquire virulence during evolution. The newly identified toxin proteins in the genome can be helpful in devising new strategies to control insect pests (Duchaud et al., 2003).

Our understanding of developmental genetics, molecular biology and functional genomics of EPNs is still low despite of the successful demonstration of In-vitro cultivation of EPNs, transformation through microinjection, induced mutagenesis, RNAi by soaking and cryopreservation (Lunau et al., 1993; Koltai et al., 1994; Hashmi, 1995; Hashmi et al., 1997; Nugent et al., 1996). However, EPNs can be an excellent model to gain insight into the fundamental process of mutualism between EPNs and their symbionts, basis of parasitism and a useful comparator to free living, *Caenorhabditis elegans* and Many PPNs (Ciche et al., 2007).

The availability of more and more genomic resources of EPNs can widen our understanding at molecular level for various biological process like symbiotic transmission of bacteria to insect host, insect parasitism, dauer formation and recovery, longevity of IJs, heterogenic mode of reproduction (eggs develops into male, female or hermaphrodites) stress tolerance under dynamic soil environment. The genomic resource availability coupled with genetic and molecular tools offers an avenue to improve the potential of EPN strains to kill broad range of insect pests. The availability of new genomic resources and deep understanding of regulatory pathways and genetic engineering like tools may help to improve the trait of interest such as host finding, longer shelf life (longevity) and stress tolerance. The first successful attempt was made by the introduction of *hsp70A* gene into *H. bacteriophora* from *C. elegans* to enhance the thermal tolerance however the tolerance didn't persist under field condition (Hashmi, 1995). The available genome sequences of EPNs provide an archive of putative genes involved in trait of interest that can be useful to alter/improve through the transformation as well RNAi like tools for enhanced effectiveness of EPNs as bioagent under field condition.

6.8 Conclusion and future perspectives

EPNs are nematode with excellent applied value for the successful biological control of insect pest and can be a viable alternative for integrated pest management program. In order to determine the full potential as an alternative to chemical pesticides, certain improvement such as longevity of IJs, retention of bacteria, tolerance to extreme environmental condition and immunity against haemocoelic encapsulation in EPN strains are necessary. To overcome these barriers, identification of new EPNs adapted to the various climatic conditions and identifying the genes involved in specific traits are highly important in improvement of strain through conventional genetics (crossing) or genetic engineering approach. They also provide valuable opportunity as a model system to study the biological processes which has not yet been studied in *C. elegans*, including parasitism, symbiosis and heterogenic sex determination etc. Information obtained from the EPN genome sequencing project, along with comparative genomics with other nematode genomes, will surely provide us a better and deeper understanding of what kind of genes make a worm a parasite and suitable host for symbiotic bacteria to develop mutualism. Finally, these magnificent animals have potential to contribute much more to the basic science than a biocontrol agent of insect pest.

References

Akhurst, R.J., 1982. Antibiotic activity of *Xenorhabdus* spp, bacteria symbiotically associated with insect pathogenic nematodes of the families Heterorhabditidae and Steinernematidae. J. Gen. Microbiol. 128, 3061–3065.

Akhurst, R.J., Bedding, R.A., 1986. Natural occurrence of insect pathogenic nematodes (Steinernematidae and Heterorhabditidae) in soil in Australia. Aust. J. Entomol. 25 (3), 241–244.

Akhurst, R.J., Boemare, N.E., Janssen, P.H., Peel, M.M., Alfredson, D.A., Beard, C.E., 2004. Taxonomy of Australian clinical isolates of the genus *Photorhabdus* and proposal of *Photorhabdus asymbiotica* subsp. asymbiotica subsp. nov. and *P. asymbiotica* subsp. Australis subsp. nov. Int. J. Syst. Evol. Microbiol. 54, 1301–1310.

Al-Banna, L., et al., 2004. Discrimination of six *Pratylenchus* species using PCR and species-specific primers. J. Nematol. 36, 142–146.

An, R., Grewal, P.S., 2010. Molecular mechanisms of persistence of mutualistic bacteria *Photorhabdus* in the entomopathogenic nematode host. PloS One 5, e13154.

An, R., Grewal, P.S., 2011. *purL* gene expression affects biofilm formation and symbiotic persistence of *Photorhabdus temperata* in the nematode *Heterorhabditis bacteriophora*. Microbiology 157, 2595–2603.

Askary, T.H., Ahmad, M.J., 2017. Entomopathogenic nematodes: mass production, formulation and application. In: Abd-Elgawad, M.M.M., Askary, T.H., Coupland, J. (Eds.), Biocontrol Agents: Entomopathogenic and Slug Parasitic Nematodes. CAB International, Wallingford, UK, pp. 261–287.

Bai, X., Adams, B.J., Ciche, T.A., Clifton, S., Gaugler, R., Kim, K.S., Spieth, J., Sternberg, P.W., Wilson, R.K., Grewal, P.S., 2013. A lover and a fighter: the genome sequence of an entomopathogenic nematode *Heterorhabditis bacteriophora*. PloS One 8 (7), e69618.

Beck, B., Brusselman, E., Nuyttens, D., Moens, M., Temmerman, F., Pollet, S., Van Weyenberg, S., Spanoghe, P., 2014. Improving the biocontrol potential of entomopathogenic nematodes against *Mamestra brassicae*: effect of spray application technique, adjuvants and an attractant. Pest Manag. Sci. 70, 103–112.

Bedding, R.A., 1988. Storage of Insecticidal Nematodes. World Patent No. WO 88/08668.

Bedding, R.A., Akhurst, R.J., 1975. A simple technique for the detection of insect parasitic rhabditid nematodes in soil. Nematologica 21, 109–110.

Bedding, R.A., Butler, K.L., 1994. Method for the Storage of Entomopathogenic Nematodes. WIPO Patent No. WO 94/05150.

Bedding, R., Molyneux, A., 1982. Penetration of insect cuticle by infective juveniles of *Heterorhabditis* spp. (Heterorhabditidae: Nematoda). Nematologica 28, 354–359.

Bedding, R.A., Clark, S.D., Lacey, M.J., Butler, K.L., 2000. Method and Apparatus for the Storage of Entomopathogenic Nematodes. WIPO Patent No. WO 00/18887.

Bennett, M.D., Leitch, I.J., Price, H.J., Johnston, J.S., 2003. Comparisons with Caenorhabditis (~100 Mb) and Drosophila (~175 Mb) Using Flow Cytometry Show Genome Size in Arabidopsis to be ~157 Mb and thus ~25 % Larger than the Arabidopsis Genome Initiative Estimate of ~125 Mb. Ann. Bot. 91, 547–557. Abstract Article.

Bhat, A.H., Istkhar Chaubey, A.K., Půža, V., San-Blas, E., 2017. First report and comparative study of *Steinernema surkhetense* (Rhabditida: Steinernematidae) and its symbiont bacteria from subcontinental India. J. Nematol. 49, 92–102. https://doi.org/10.21307/jofnem-2017-049.

Bhat, A., Chaubey, A., Půža, V., 2019. The first report of *Xenorhabdus indica* from *Steinernema pakistanense*: co-phylogenetic study suggests co-speciation between *X. indica* and its Steinernematid nematodes. J. Helminthol. 93 (1), 81–90. https://doi.org/10.1017/S0022149X17001171.

Bode, H.B., 2009. Entomopathogenic bacteria as a source of secondary metabolites. Curr. Opin. Chem. Biol. 13, 1–7.

Boemare, N., 2002. Interactions between the partners of the entomopathogenic bacterium nematode complexes, Steinernema-Xenorhabdus and Heterorhabditis-Photorhabdus. Nematology 4 (5), 601–603.

Boemare, N.E., Akhurst, R.J., Mourant, R.G., 1993. DNA relatedness between *Xenorhabdus* spp. (Enterobacteriaceae), symbiotic bacteria of entomopathogenic nematodes, and a proposal to transfer *Xenorhabdus luminescens* to a new genus, *Photorhabdus* gen. nov. Int. J. Syst. Evol. Microbiol. 43 (2), 249–255.

Borgonie, G., García-Moyano, A., Litthauer, D., Bert, W., Bester, A., van Heerden, E., Möller, C., Erasmus, M., Onstott, T.C., 2011. Nematode from the terrestrial deep subsurface of South Africa. Nature 474 (7349), 79.

Brusselman, E., Beck, B., Pollet, S., Temmerman, F., Spanoghe, P., Moensa, M., Nuyttens, D., 2012. Effect of the spray application technique on the deposition of entomopathogenic nematodes in vegetables. Pest Manag. Sci. 68, 444–453.

Burnell, A.M., Stock, S.P., 2000. *Heterorhabditis, Steinernema* and Their Bacterial Symbionts—Lethal Pathogens of insects Nematology, vol. 2, pp. 1–12.

Campbell, J.F., Gaugler, R., 1993. Nictation behaviour and its ecological implications in the host search strategies of entomopathogenic nematodes (*Heterorhabditidae* and *Steinernematidae*). J. Behav. 126, 3–14.

Chang, F.N., Gehret, M.J., 1991. Insecticide Delivery System and Attractant. WIPO Patent No. WO 91/01736.

Chang, F.N., Gehret, M.J., 1992. Stabilizer Insect Nematode Compositions. WIPO Patent No. WO 92/10170.

Chen, S., Glazer, I., 2005. A novel method for long-term storage of the entomopathogenic nematode steinernema feltiae at room temperature. Biol. Contr. 32, 104–110.

Ciche, T.A., Ensign, J.C., 2003. For the insect pathogen *Photorhabdus luminescens*, which end of a nematode is out? Appl. Environ. Microbiol. 69, 1890–1897.

Ciche, T.A., Bintrim, S.B., Horswill, A.R., Ensign, J.C., 2001. A phosphopantetheinyl transferase homolog is essential for *Photorhabdus luminescens* to support growth and reproduction of the entomopathogenic nematode *Heterorhabditis bacteriophora*. J. Bacteriol. 183, 3117–3126.

Ciche, T.A., Darby, C., Ehlers, R.-U., Forst, S., Goodrich-Blair, H., 2006. Dangerous liaisons: The symbiosis of entomopathogenic nematodes and bacteria. Biol. Control. 38, 22–46.

Ciche, TA, Sternberg, PW, 2007. Postembryonic RNAi in Heterorhabditis bacteriophora: a nematode insect parasite and host for insect pathogenic symbionts. BMC Dev Biol 7, 101.

Cimen, H., Půža, V., Nermu, J., Hatting, J., Ramakuwela, T., Faktorova, L., Hazir, S., 2016. *Steinernema beitlechemi* n. sp., a new entomopathogenic nematode (Nematoda: Steinernematidae) from South Africa. Nematology 18 (4), 439–453.

Clarke, D.J., 2008. *Photorhabdus*: a model for the analysis of pathogenicity and mutualism. Cell Microbiol. 10, 2159–2167.

Cruz-Martinez, H., Ruiz-Vega, J., Matadamas-Ortiz, P.T., Cortes-Martinez, C.I., Rosas-Diaz, J., 2017. Formulation of entomopathogenic nematodes for crop pest control: a review. Plant Protect. Sci. 53, 15–24.

De Brida, A.L., Rosa, J.M.O., De Oliveira, C.M.G., e Castro, B.M.D.C., Serrão, J.E., Zanuncio, J.C., Leite, L.G., Wilcken, S.R.S., 2017. Entomopathogenic nematodes in agricultural areas in Brazil. Sci. Rep. 7, 45254.

De Ley, P., 2006. A quick tour of nematode diversity and the backbone of nematode phylogeny. In: WormBook.

Dillman, A.R., et al., 2015. Comparative genomics of *Steinernema* reveals deeply conserved gene regulatory networks. Genome Biol. 16, 200.

Duchaud, E., Rusniok, C., Frangeul, L., Buchrieser, C., Givaudan, A., Taourit, S., Bocs, S., Boursaux-Eude, C., Chandler, M., Charles, J.F., et al., 2003. The genome sequence of the entomopathogenic bacterium Photorhabdus luminescens. Nat. Biotechnol. 21, 1307–1313.

Dunphy, G.B., Wetster, J.M., 1988. Lipopolysaccharides of *Xenorhabdus nematophilus* (Emterobacteriaceae) and their haemocyte toxicity in non-immune *Galleria mellonella* (Insecta: Lepidoptera) larvae. J. Gen. Microbiol. 134, 1017–1028.

Duong, D.A., Espinosa-Artiles, P., Orozco, R.A., Molnár, I., Stock, S.P., 2019. Draft genome assembly of the entomopathogenic bacterium *Photorhabdus luminescens* subsp. *Sonorensis caborca*. Microbiol. Resour. Announc. 8 (36) e00692-19.

Ehlers, R.U., 1991. Interactions in the entomopathogenic nematode-bacteria complex *Steinernema/Heterorhabditis/Xenorhbhs*. In: Proceedings of the 3rd European Meeting for Microbiological Control of Insect Pests, Wageningen, pp. 36–44.

Elawad, S., Ahmad, W., Reid, A.P., 1997. *Steinernema abbasi* sp. n. (Nematoda: Steinernematidae) from the sultanate of Oman. Fundam. Appl. Nematol. 20, 435–442.

Ettema, C.H., 1998. Soil nematode diversity: species coexistence and ecosystem function. J. Nematol. 30, 159–169.

Feng, S.P., Han, R.C., 2005. Advances in storage and formulation of entomopathogenic nematodes *Steinernema* and *Heterorhabditis*. Plant Prot. 31, 20–25.

Ferreira, T., Malan, A.P., 2014. *Xenorhabdus* and *Photorhabdus*, bacterial symbionts of the entomopathogenic nematodes *Steinernema* and *Heterorhabditis* and their in vitro liquid mass culture: a review. Afr. Entomol. 22, 1–14.

Ferreira, T., Van Reenen, C.A., Endo, A., Spröer, C., Malan, A.P., Dicks, L.M., 2013a. Description of *Xenorhabdus khoisanae* sp. nov., the symbiont of the entomopathogenic nematode *Steinernema khoisanae*. Int. J. Syst. Evol. Microbiol. 63, 3220–3224.

Ferreira, T., Van Reenen, C., Pagès, S., Tailliez, P., Malan, A.P., Dicks, L.M., 2013b. *Photorhabdus luminescens* subsp. *noenieputensis* subsp. nov., a symbiotic bacterium associated with a novel *Heterorhabditis* species related to *Heterorhabditis indica*. Int. J. Syst. Evol. Microbiol. 63, 1853–1858.

Ferreira, T., Van Reenen, C.A., Endo, A., Tailliez, P., Pages, S., Spröer, C., Malan, A.P., Dicks, L.M., 2014. Photorhabdusheterorhabditis sp. nov., a symbiont of the entomopathogenic nematode *Heterorhabditis zealandica*. Int. J. Syst. Evol. Microbiol. 64 (5), 1540–1545.

Ferreira, T., Addison, M.F., Malan, A.P., 2016. Development and population dynamics of *Steinernema yirgalemense* (Rhabditida: Steinernematidae) and growth characteristics of its associated *Xenorhabdus indica* symbiont in liquid culture. J. Helminthol. 90, 364–371.

Fife, J.P., Derksen, R.C., Ozkan, H.E., Grewal, P.S., 2003. Effects of pressure differentials on the viability and infectivity of entomopathogenic nematodes. Biol. Contr. 27, 65–72.

Fife, J.P., Ozkan, H.E., Derksen, R.C., Grewal, P.S., Krause, C.R., 2005. Viability of a biological pest control agent through hydraulic nozzles. Transac. ASAE 48, 45–54.

Fischer-Le Saux, M., Arteaga-Hernández, E., Mrácek, Z., Boemare, N.E., 1999a. The bacterial symbiont *Xenorhabdus poinarii* (Enterobacteriaceae) is harbored by two phylogenetic related host nematodes: the entomopathogenic species *Steinernema cubanum* and *Steinernema glaseri* (Nematoda: Steinernematidae). FEMS (Fed. Eur. Microbiol. Soc.) Microbiol. Ecol. 29, 149–157.

Fischer-Le Saux, M., Viallard, V., Brunel, B., Normand, P., Boemare, N.E., 1999b. Polyphasic classification of the genus *Photorhabdus* and proposal of new taxa: *P. luminescens* subsp. *luminescens* subsp. nov., *P. luminescens* subsp. *akhurstii* subsp. nov., *P. luminescens* subsp. *laumondii* subsp. nov., *P. temperata* sp. nov., *P. temperata* subsp. *temperata* subsp. nov. and *P. asymbiotica* sp. nov. Int. J. Syst. Evol. Microbiol. 49 (4), 1645–1656.

Forst, S., Clarke, D., 2002. Bacteria-nematode symbiosis. Entomopathogen. Nematol. 31, 57–77.

Ganguly, S., Singh, L.K., 2000. *Steinernema thermophilum* sp. n. *(Rhabditida: Steinernematidae)* from India. Int. J. Nematol. 10, 183–191.

Ganguly, S., Kumar, A., Parmar, B.S., 2008. Nemagel – a formulation of the entomopathogenic nematode *Steinernema thermophilum* mitigating the shelf-life constraint of the tropics. Nematol. Mediterr. 36, 125–130.

Gaugler, R., 1997. Ecology in the service of biological control: the case of entomopathogenic nematodes. Oecologia 109, 483–489.

Geldenhuys, J., Malan, A.P., Dicks, L.M.T., 2016. First report of the isolation of the symbiotic bacterium *Photorhabdus luminescens* subsp. *laumondii* associated with *Heterorhabditis safricana* from South Africa. Curr. Microbiol. 73 (6), 790–795.

Georgis, R., 1990. Formulation and application technology. In: Gaugler, R., Kaya, H.K. (Eds.), Entomopathogenic Nematodes in Biological Control. CRC Press Inc, Boca Raton, pp. 173–191.

Georgis, R., Hague, N.G.M., 1991. Nematodes as biological insecticides. Pestic. Outlook 3, 29–32.

Georgis, R., Dunlop, D.B., Grewal, P.S., 1995. Formulation of entomopathogenic nematodes. In: Hall, F.R., Barry, J.W. (Eds.), Biorational Pest Control Agents: Formulation and Delivery. American Chemical Society, Washington, DC, pp. 197–205.

Glazer, I., 2014. Genetic improvement and breeding of EPN: the race for the "super nematode". J. Nematol. 46, 168.

Goud, S., Hugar, P.S., Prabhuraj, A., 2010. Effect of temperature, population density and shelf life of EPN *Heterorhabditis indica* (RCR) in sodium alginate gel formulation. J. Biopestic. 3, 627–632.

Grewal, P.S., 2002. Formulation and application technology. In: Gaugler, R. (Ed.), Entomopathogenic Nematology, Oxfordshire. CABI, pp. 265–287.

Grewal, P.S., Peters, A., 2005. Formulation and quality. In: Grewal, P.S., Ehlers, R.-U., Shapiro-Ilan, D.I. (Eds.), Nematodes as Bio-Control Agents. CAB International, Wallingford, UK, pp. 79–90.

Grewal, P.S., Lewis, E.E., Gaugler, R., Campbell, J.F., 1994. Host finding behavior as a predictor of foraging strategy in entomopathogenic nematodes. Parasitology 108, 207–215.

Gualtieri, M., Ogier, J.-C., Pagès, S., Givaudan, A., Gaudriault, S., 2014. Draft genome sequence and annotation of the entomopathogenic bacterium *Xenorhabdus szentirmaii* strain DSM16338. Genome Announc. 2 (2) e00190-14.

Guo, W., Yan, X., Han, R., 2017. Adapted formulations for entomopathogenic nematodes, *Steinernema* and *Heterorhabditis* spp. Nematology 19 (5), 587–596.

Hashmi, S., 1995. Genetic transformation of an entomopathogenic nematode by microinjection. J. Invertebr. Pathol. 66, 293–296.

Hashmi, S., Hatab, M.A.A., Gaugler, R.R., 1997. GFP: Green fluorescent protein a versatile gene marker for entomopathogenic nematodes. Fundam. Appl. Nematol. 20, 323–327.

Hazir, S., Kaya, H.K., Stock, S.P., Keskin, N., 2003. Entomopathogenic nematodes (Steinernematidae and Heterorhabditidae) for biological control of soil pests. Turk. J. Biol. 27, 181–202.

Hazir, S., Kaya, H.K., Stock, S.P., Keskin, N., 2004. Entomopathogenic nematodes (Steinernematidae and Heterorhabditidae) for biological control of soil pests. Turk. J. Biol. 27 (4), 181–202.

Hiltpold, I., 2015. Prospects in the application technology and formulation of entomopathogenic nematodes for biological control of insect pests. In: Campos-Herrera, R. (Ed.), Nematode Pathogenesis of Insects and Other Pests: Ecology and Applied Technologies for Sustainable Plant and Crop Protection. Springer, pp. 187–206.

Hinchliffe, S.J., Hares, M.C., Dowling, A.J., ffrench-Constant, R.H., 2010. Insecticidal toxins from the *Photorhabdus* and *Xenorhabdus* bacteria. Open Toxinol. J. 3, 83–100.

Holterman, M., et al., 2006. Phylum-wide analysis of SSU rDNA reveals deep phylogenetic relationships among nematode and accelerated evolution toward crown clades. Mol. Biol. Evol. 23, 1792–1800.

Hu, K., Webster, J.M., 2000. Antibiotic production in relation to bacterial growth and nematode development in *Photorhabdus Heterorhabditis* infected *Galleria mellonella* larvae. FEMS (Fed. Eur. Microbiol. Soc.) Microbiol. Lett. 189, 219–223.

Hu, K.J., Li, J.X., Webster, J.M., 1999. Nematicidal metabolites produced by *Photorhabdus luminescens* (Enterobacteriaceae), bacterial symbiont of entomopathogenic nematodes. Nematology 1, 457–469.

Hu, K., Li, J., Li, B., Webster, J.M., Chen, G., 2006. A novel antimicrobial epoxide isolated from larval *Galleria mellonella* infected by the nematode symbiont, *Photorhabdus luminescens* (Enterobacteria-ceae). Bio-org. & Med. Chem. 14, 4677–4681.

Hussein, M.A., Abdel-Aty, M.A., 2012. Formulation of two native entomopathogenic nematodes at room temperature. J. Biopestic. 5, 23–27.

Joyce, S.A., Burnell, A.M., Powers, T.O., 1994. Characterization of *Heterorhabditis* isolates by PCR amplification of segments of mtDNA and rDNA genes. J. Nematol. 26 (3), 260.

Kagimu, N., Ferreira, T., Malan, A.P., 2017. The attributes of survival in the formulation of entomopathogenic nematodes utilised as insect bio-control agents. Afr. Entomol. 25 (2), 275–292.

Kaya, H.K., 1990. Soil ecology. In: Gaugler, R., Kaya, H.K. (Eds.), Entomopathogenic Nematodes in Biological Control. CRC Press, Boca Raton, FL, pp. 93–116.

Kaya, H.K., 2002. Natural enemies and other antagonists. In: Gaugler, R. (Ed.), Entomopathogenic Nematology. CABI Publishing, Wallingford, UK, pp. 189–204.

Kaya, H.K., Gaugler, R., 1993. Entomopathogenic nematodes. Annu. Rev. Entomol. 38, 181–206.

Koltai, H., Glazer, I., Segal, D., 1994. Phenotypic and genetic characterization of two new mutants of Heterorhabditis bacteriophora. J. Nematol. 26, 32–39.

Kuwata, R., Qiu, L.H., Wang, W., Harada, Y., Yoshida, M., Kondo, E., Yoshiga, T., 2013. *Xenorhabdus ishibashii* sp. nov., isolated from the entomopathogenic nematode *Steinernema aciari*. Int. J. Syst. Evol. Microbiol. 63 (5), 1690–1695.

Lacey, L.A., Georgis, R., 2012. Entomopathogenic nematodes for control of insect pests above and below ground with comments on commercial production. J. Nematol. 44, 218–225.

Lalramnghaki, H.C., Vanlalhlimpuia, V., 2017. Lalramliana characterization of a new isolate of entomopathogenic nematode, *Steinernema sangi* (Rhabditida, Steinernematidae), and its symbiotic bacteria *Xenorhabdus vietnamensis* (γ-Proteobacteria) from Mizoram, North-Eastern India. J. Parasit. Dis. 41, 1123–1131. https://doi.org/10.1007/s12639-017-0945-z.

Lara, J.C., Dolinski, C., Fernandes de Sousa, E., Figueiredo Daher, E., 2008. Effect of mini-sprinkler irrigation system on *Heterorhabditis baujardi* LPP7 (Nematoda: Heterorhabditidae) infective juvenile. Sci. Agric. 65, 433–437.

Leger, R.J.S., et al., 1996. Construction of an improved mycoinsecticide over-expressing a toxic protease. In: Proceedings of. National Academy of Science of the United States of America, vol. 93, pp. 6349–6354.

Lengyel, K., Lang, E., Fodor, A., Szállás, E., Schumann, P., Stackebrandt, E., 2005. Description of four novel species of Xenorhabdus, family Enterobacteriaceae: *Xenorhabdus budapestensis* sp. nov., *Xenorhabdus ehlersii* sp. nov., *Xenorhabdus innexi* sp. nov., and *Xenorhabdus szentirmaii* sp. nov. Syst. Appl. Microbiol. 28 (2), 115–122.

Ley, P.D., Blaxter, M., 2002. Systematic position and phylogeny. In: Lee, D. (Ed.), The Biology of Nematodes. Taylor & Francis, pp. 1–30.

Lunau, S., Stoessel, S., Schmidt-Peisker, A.J., Ehlers, R.U., 1993. Establishment of monoxenic inocula for scaling up in vitro cultures of the entomopathogenic nematodes Steinernema spp and Heterorhabditis spp. Nematologica 39, 385–399.

Lu, D., Baiocchi, T., Dillman, A.R., 2016. Genomics of entomopathogenic nematodes and implications for pest control. Trends Parasitol. 32, 588–598.

Machado, R.A., Bruno, P., Arce, C.C., Liechti, N., Köhler, A., Bernal, J., Bruggmann, R., Turlings, T.C., 2019. *Photorhabdus khanii* subsp. *guanajuatensis* subsp. nov., isolated from *Heterorhabditis atacamensis*, and *Photorhabdus luminescens* subsp. mexicana subsp. nov., isolated from *Heterorhabditis mexicana* entomopathogenic nematodes. Int. J. Syst. Evol. Microbiol. 69 (3), 652–661.

Martens, E.C., Heungens, K., Goodrich-Blair, H., 2003. Early colonizationevents in the mutualistic association between *Steinernema carpocapsae* nematodes and *Xenorhabdus nematophila* bacteria. J. Bacteriol. 185, 3147–3154.

Mclnerney, B.V., Gregson, R.P., Lacey, M.J., Akhurst, R.J., Lyons, C.R., Rhodes, S.H., Smith, R.J., Engelhardt, L.M., White, A.H., 1991. Biologically active metabolites tiom *Xenorhabdus* spp., part 1. Dithiolopyrrolone derivatives with antibiotic activity. J. Nat. Proahrcts 54, 774–784.

McMullen, J.G., Stock, S.P., 2014. In vivo and in vitro rearing of entomopathogenic nematodes (*Steinernematidae* and *Heterorhabditidae*). JoVE 91, e52096.

Mráček, Z., Nguyen, K.B., Tailliez, P., Boemare, N., Chen, S., 2006. *Steinernema sichuanense* n. sp. (Rhabditida Steinernematidae) a novel species of Sichuan east Tibetan Mts China. J. Invertebr. Pathol. 93, 157–169.

Naidoo, S., Featherston, J., Gray, V.M., 2015. Draft whole-genome sequence and annotation of the entomopathogenic bacterium *Xenorhabdus khoisanae* strain MCB. Genome Announc. 3 (4) e00872-15.

NBAIR, 2017. ICAR- NBAIR Success story. Greentech with Entomopathogenic Nematodes (EPN) for Securing Crop Care and Soil Health. National Bureau of Agricultural Insect Resources, Bengaluru, India. http://www.nbair.res.in/Success-storeis/.

Nguyen, K.B., Smart, G.C., 1994. *Neosteinernema longicurvicauda* n. gen., n. sp. (Rhabdifida: Steinernematidae), a parasite of the termite *Reticulitermes flavipes* (Koller). J. Nematol. 26, 162–174.

Nguyen, K.B., Smart, G.C., 1996. Identification of entomopathogenic nematodes in the Steinernematidae and Heterorhabditidae (Nemata: Rhabditida). J. Nematol. 28, 286–300.

Nugent, M.J., O'Leary, S.A., Burnell, A.M., 1996. Optimised procedures for the cryopreservation of different species of Heterorhabditis. Fundam. Appl. Nematol 19, 1–6.

Ogier, J.C., Duvic, B., Lanois, A., Givaudan, A., Gaudriault, S., 2016. A new member of the growing family of contact-dependent growth inhibition systems in *Xenorhabdus doucetiae*. PloS One 11 (12), e0167443.

Palma, L., Del Valle, E.E., Frizzo, L., Berry, C., Caballero, P., 2016. Draft genome sequence of *Photorhabdus luminescens* strain DSPV002N isolated from Santa Fe, Argentina. Genome Announ. 4 (4). https://doi.org/10.1128/genomeA.00744-16. e00744-16.

Park, Y., Kang, S., Sadekuzzaman, M., Kim, H., Jung, J.K., Kim, Y., 2017. Identification and bacterial characteristics of *Xenorhabdus hominickii* ANU101 from an entomopathogenic nematode, *Steinernema monticolum*. J. Invertebr. Pathol. 144, 74–87.

Peters, A., 2016. Formulation of nematodes. In: Glare, T.R., Moran-Diez, M.E. (Eds.), Microbial-based Biopesticides: Methods and Protocols. Springer, New York, USA, pp. 121–135.

Poinar Jr., G.O., 1975. Entomogenous nematodes: a manual and host list of insect-nematode associations. Brill Archive 254.

Poinar Jr., G.O., 1976. Description and biology of a new insect parasitic rhabditoid, *Heterorhabditis bacteriophora* n. gen. n. sp. (Rhabditida; Heterorhabditidae n. ram.). Nematologica 21, 463–470.

Ramakuwela, T., Hatting, J., Laing, M.D., Hazir, S., Thiebaut, N., 2015. Effect of storage temperature and duration on survival and infectivity of *Steinernema innovationi* (Rhabditida: Steinernematidae). J. Nematol. 47 (4), 332–336.

Robertson, H.M., Thomas, J.H., 2006. Theputative chemoreceptor families of *C. elegans*. In: The *C. elegans* Research Community. WormBook.

Sajnaga, E., et al., 2018. Arch. Microbiol. 200, 1307–1316.

Sandhu, S.K., Jagdale, G.B., Hogenhout, S.A., Grewal, P.S., 2006. Comparative analysis of the expressed genome of the infective juvenile entomopathogenic nematode, *Heterorhabditis bacteriophora*. Mol. Biochem. Parasitol. 145, 239–244.

Sandstrom, J.P., Russel, J.A., White, J.P., Moran, N.A., 2001. Independent origins and horizontal transfer of bacterial symbionts of aphids. Mol. Ecol. 10, 217–228.

Shapiro-Ilan, D.I., Gouge, D.H., Koppenhofer, A.M., 2002. Factors affecting commercial success: case studies in cotton, turf and citrus. In: Gaugler, R. (Ed.), Entomopathogenic Nematology. CABI Publishing, Wallingford, UK, pp. 333–356.

Shapiro-Ilan, D.I., Gouge, D.H., Piggott, S.J., Fife, J.P., 2006. Application technology and environmental considerations for use of entomopathogenic nematodes in biological control. Biol. Contr. 38, 124–133.

Shapiro-Ilan, D.I., Han, R., Dolinksi, C., 2012. Entomopathogenic nematode production and application technology. J. Nematol. 44, 206–217.

Shapiro-Ilan, D.I., Cottrell, T.E., Mizell, R.F., Horton, D.L., Zaid, A., 2015. Field suppression of the peach tree borer, *synanthedon exitiosa*, using *steinernema carpocapsae*: effects of irrigation, a sprayable gel and application method. Biol. Contr. 82, 7–12.

Shapiro-Ilan, D., Hazir, S., Glazer, I., 2017. Basic and applied research: entomopathogenic nematodes. In: Lacey, L.A. (Ed.), Microbial Control of Insect and Mite Pests. Elsevier, London, pp. 91–105.

Sharmila, R., Subramanian, S., Poornima, K., 2018. Host range of entomopathogenic nematodes. J. Entomol. Zool. Stud. 6 (3), 1310–1312.

Simoes, N., Rosa, J.S., 2010. Pathogenicity and Host Specificity of Entomopathogenic Nematodes, pp. 403–412.

Stock, S.P., 1995. Natural populations of entomopathogenic nematodes in the Pampean region of Argentina. Nematropica 25, 143–148.

Stock, S.P., Gress, J.C., 2006. Diversity and phylogenetic relationships of entomopathogenic nematodes (Steinernematidae and Heterorhabditidae) from the Sky Islands of Southern Arizona. J. Inverte br. Pathol. 92, 66–72.

Strauch, O., Niemann, I., Neumann, A., Schmidt, A.J., Peters, A., Ehlers, R.U., 2000. Storage and formulation of the entomopathogenic nematodes *Heterorhabditis indica* and *H. bacteriophora*. BioControl 45, 483–500.

Stuart, R.J., Gaugler, R., 1994. Patchiness in populations of entomopathogenic nematodes. J. Invertebr. Pathol. 64, 39–45.

Tailliez, P., Pagès, S., Ginibre, N., Boemare, N., 2006. New insight into diversity in the genus *Xenorhabdus,* including the description of ten novel species. Int. J. Syst. Evol. Microbiol. 56, 2805–2818.

Tailliez, P., Laroui, C., Ginibre, N., Paule, A., Pages, S., Boemare, N., 2010. Phylogeny of *Photorhabdus* and *Xenorhabdus* based on universally conserved protein-coding sequences and implications for the taxonomy of these two genera. Proposal of new taxa: *X. vietnamensis* sp. nov., *P. luminescens* subsp. *caribbeanensis* subsp. nov., *P. luminescens* subsp. *hainanensis* subsp. nov., *P. temperata* subsp. *khanii* subsp. nov., *P. temperata* subsp. *tasmaniensis* subsp. nov., and the reclassification of *P. luminescens* subsp. *thracensis* as *P. temperata* subsp. *thracensis* comb. nov. Int. J. Syst. Evol. Microbiol. 60 (8), 1921–1937.

Thanwisai, A., Tandhavanant, S., Saiprom, N., Waterfield, N.R., Long, P.K., Bode, H.B., Peacock, S.J., Chantratita, N., 2012. Diversity of *Xenorhabdus* and *Photorhabdus* spp. and their symbiotic entomopathogenic nematodes from Thailand. PloS One 7 (9), e43835.

Thomas, G.M., Poinar, G.O., 1979. *Xenorhabdus* gen nov., a genus of entomopathogenic nematophilic bacteria of the family Enterobacteriaceae. J. Syst. & Evol. Microbiol. 29, 352–360.

Toepfer, S., Hatala-Zseller, I., Ehlers, R.U., Peters, A., Kuhlmann, U., 2010. The effect of application techniques on field-scale efficacy: can the use of entomopathogenic nematodes reduce damage by western corn rootworm larvae? Agric. For. Entomol. 12, 389–402.

Tóth, T., Lakatos, T., 2008. A new subspecies of *Photorhabdus temperata*, isolated from *Heterorhabditis nematodes*: *Photorhabdus temperata* subsp. cinerea subsp. nov. Int. J. Syst. Evol. Microbiol. (in press).

Travassos, L., 1927. Sobre o genera *Oxysomatium*. Boletim Biologico 5, 20–21.

Ubaub, et al., 2018. First report of Steinernema longicaudum and its bacterial symbionts, Xenorhabdus species, in pummelo orchards of Davao region, Philippines. The Philipine scientist 53, 112–154.

White, G.F., 1927. A method for obtaining infective nematode larvae from cultures. Science 66, 302–303.

Wright, D.J., Peters, A., Schroer, S., Fife, J.P., 2005. Application technology. In: Grewal, P.S., Ehlers, R.U., ShapiroIlan, D.I. (Eds.), Nematodes as Biocontrol Agents. CABI, New York, NY, pp. 91–106.

Yamanaka, S., Hagiwara, A., Nishimura, Y., Tanabe, H., Ishibashi, N., 1992. Biochemical and physiological characteristics of *Xenorhabdus* species, symbiotically associated with entomopathogenic nematodes including *Steinernema kushidai* and their pathogenicity against *Spodoptera litura* (Lepidoptera: Noctuidae). Arch. Microbiol. 158 (6), 387–393.

Yukawa, T., Pitt, J.M., 1985. Nematode storage and transport. WIPO Patent No. WO 85/03412.

Zhou, X., Kaya, H.K., Heungens, K., Goodrich-Blair, H., 2002. Response of ants to a deterrent factor(s) produced by the symbiotic bacteria of entomopathogenic nematodes. Appl. Environ. Microbiol. 68, 6202–6209.

Chapter 7

Scientific and technological trajectories for sustainable agricultural solutions: the case of biopesticides

Alejandro Barragán-Ocaña[a], Paz Silva-Borjas[a] and Samuel Olmos-Peña[b]
[a]*National Polytechnic Institute, Center for Economic, Administrative and Social Research, Mexico City, Mexico;* [b]*Autonomous University of the State of Mexico, Mexico State, Mexico*

7.1 Introduction

The adoption of sustainable agriculture is a matter of considerable interest in many economies. On the other hand, the development of sustainable agriculture involves an adequate use of material resources, environmental awareness, a focus on product quality, job creation, and the promotion of social inclusion. All of these features can take advantage of biophysical, economic, and social indicators measuring progress in the field of sustainability, as well as global indicators that have been used to assess sustainable agriculture and other related disciplines (Rezaei-Moghaddam and Karami, 2008; Lancker and Nijkamp, 2000; Rigby et al., 2001). In this context, the different elements of knowledge management can be combined; for example, in developing countries, ancestral knowledge recuperates sustainable practices to be used by rural agricultural producers while the challenges to implement these practices correctly are assessed and addressed (Lwoga et al., 2010). This approach can be related to other decision-making systems to enable optimal technical support on best practices and maximize the benefits of sustainable agriculture initiatives (Kurlavičius, 2009).

The benefits expected from the implementation of sustainable agriculture include the use of environmentally friendly agricultural practices; minimizing soil erosion and desiccation; increasing humidity, organic content, and nutrients in soils; mitigating costs for farmers; decreasing pollution due to chemical fertilizers, and reducing carbon dioxide emissions (Kassam and Brammer, 2013). In addition, certifications for production represent an alternative to support economic and social development and protect the environment, as is the case with sustainably produced coffee in Latin America (Kilian et al., 2004). However, not only developing economies face challenges; in the United States, for instance, the agricultural sector is currently characterized by increasing competition and the need to enable sustainable development, make rational use of natural resources, and increase research and development (R&D) in the area (Hanson et al., 2008). Hence the importance of using biotechnological tools as an alternative. For instance, bioremediation (the practical use of microorganisms or plants) has been shown to decrease soil contamination that affects the enzymatic activity of the soils and the physiology of non-target microorganisms, as well as water contamination caused by the use of conventional pesticides (Anjum et al., 2012; Ahemad and Khan, 2011). Nevertheless, it is also necessary to study and select many still unknown microorganisms that may have important potential to provide environmental, agricultural, and health solutions (Ahmad et al., 2011).

The field of biotechnology provides many alternatives that can contribute to the development of sustainable agriculture. For example, the activity of microorganisms in the soil can favor the availability and assimilation of micronutrients and macronutrients needed by plants (Meena et al., 2016). These technologies include the use of biofertilizers, defined as micro bialinoculants that can improve a crop's nutrition and productivity, for example, *Azotobacter* and *Azospirillum* (Lesueur et al., 2016). In the broadest sense, bioinoculants are classified into two groups, bacteria (intra-, inter-, and extra-cellular) and fungi. The latter include those associated with the root (ectomycorrhizas and arbuscular mycorrhizas), which have an important effect in root strength and plant growth and improve agricultural crop

Biopesticides. https://doi.org/10.1016/B978-0-12-823355-9.00011-0

performance. In addition, these bioinoculants can produce different metabolites, such as phytohormones, antibiotics, and antimicrobials, that influence root development or provide other benefits such as protection against pathogens. The use of arbuscular mycorrhizal fungi, found in agricultural, horticultural, and forestry plants, pose a compelling case; they improve the plant's tolerance to environmental stress without disturbing the soil and contribute to the uptake of nutrients, such as phosphorus, via hyphae (Owen et al., 2015; Kapoor et al., 2002; Ortaş and Varma, 2007; Priyadharsini and Muthukumar, 2015).

Thus, arbuscular mycorrhizal fungi and growth-promoting rhizobacteria are closely associated with three fundamental requirements for crops: nutrition, growth, and health (Bona et al., 2016). Other technological developments have used formulations based on different strains of *Bacillus* with the purpose of forming endospores useful to promote biocontrol and crop growth (Wu et al., 2015). Despite their diversity, the use of microbial bioinoculants still has important challenges concerning their *in situ* effectiveness, formulations, and application before it can be compared with chemical alternatives, especially in developing economies (Mishra and Arora, 2016). Another biotechnological alternative are biopesticides. The main disadvantages of biopesticides in comparison with chemical pesticides have to do with effectiveness, cost, and market share, although due to the negative impact of chemical pesticides on the environment, biotechnological alternatives represent an opportunity for sustainable agriculture (Mishra et al., 2015).

Agricultural and food systems need to change their focus and intensify sustainability by taking actions aimed at increasing or maintaining production while the construction of agroecological territories moves forward. These actions include preserving biodiversity and natural habitats, planting trees, conducting integral pest control management, improving irrigation, and avoiding water pollution and soil overexploitation, among others (Pretty, 2018; Wezel et al., 2016). Given the new global approach to legislation and the resistance to chemical pesticides developed by pests, crop-destroying plagues need to be controlled integrally, including the use of biopesticides. In this regard, research is focused on the development of genetic innovations under an ecological approach, and the regulatory challenges include market barriers, technical issues (effectiveness), and economic considerations (Chandler et al., 2011).

Biopesticides have been shown to have different positive features in terms of effectiveness, mitigation of environmental pollution, agricultural modernization, and gradual substitution of chemical products, although in some cases, the literature reports toxicity problems that need to be addressed (Leng et al., 2011). This group of technologies includes the use of living microorganisms (bacteria and fungi) and bioactive compounds derived from them; entomopathogenic viruses; botanical products; beneficial insects; semiochemicals (for example, insect pheromones); entomophagous nematodes; genetically-engineered plants that resist the action of insects or various microorganisms or herbicides; plant-incorporated protectants (PIPs), as well as other efforts aimed at the production of bioherbicides and bionematicides. However, these technologies have not been capable of replacing chemical-based pesticides, mainly due to the lack of technical capability and performance when using biotechnological alternatives, in addition to their prohibitive costs, making them feasible to specialized markets only, although recent improvements aimed at broadening their spectrum of activity, presentation, application, and the duration of their effects present a favorable outlook for their increased use in the long term (Glare et al., 2012; Copping and Menn, 2000; Seiber et al., 2014; Senthil-Nathan, 2015).

There are many examples of biopesticides. Baculoviruses (Baculoviridae) are used to control insect plagues by infecting the undesired organisms while humans and wildlife remain safe. Due to their specificity, they are also safe for non-target arthropods, vertebrates, plants, and the environment, but three important problems have limited their use: 1. Large-scale in vitro production is complicated; 2. They are slow-acting; and 3. Their specificity is limited. Future advances will be focused on the improvement of diagnostic techniques, formulations, and production or, where appropriate, genetic modification (Szewczyk et al., 2006; Moscardi et al., 2011). Tick control by entomogenous fungi has shown promising results (Kaaya and Hassan, 2000). Individual and combined experiments using botanical biopesticides and *Bacillus thuringiensis* (Berliner) have also been carried out, reducing larvae growth and feeding in *Cnaphalocrocis medinali*, a moth that attacks the foliage of rice crops, presenting good results in vitro (Nathan et al., 2005). Another promising application is the use of *Pseudomonas* for biological control in agriculture, an ecologically-friendly option due to its capacity to produce elements such as antibiotics, fungal enzymes, and siderophores (Pandya and Saraf, 2015).

Biopesticide formulations using *Bacillus thuringiensis* (Bt) have a very important market share among this group of products; they can be safely used for health, environmental, and insect control purposes (Lepidoptera, Diptera, and Coleoptera). The main issues with existing formulations are derived from environmental conditions at the time of their use, viability (potency), and costs (formulation and production), but their enterotoxicity has been improved by microencapsulation or microgranule formation. In addition, extensive research experience in the field has resulted in new strains and the development of recombinant organisms presenting increased toxicity and action spectrum and the enhancement of insecticidal crystalline proteins (ICPs), and Bt transgenic plants provide a wide range of technologies associated with this bacterium (Brar et al., 2006; Kaur, 2000). In the case of *Bacillus thuringiensis* (Bt), the use of recombinant DNA technology in the production of insecticide proteins has faced regulatory limitations, but specific products have been developed despite these obstacles (Schnepf, 2012).

The entomopathogens *Bacillus thuringiensis* (Bt) and *Bacillus sphaericus* (Bs) have been shown to be safe, but due to their high costs, their massive application in developing economies is unviable (El-Bendary, 2006). Other pest control applications of promising potential and epidemiological importance will likely be produced in the future, among them, the use of pathogenic fungi to kill populations of malaria-transmitting mosquitoes; these technologies have had a limited response in agriculture and medicine, and R&D, regulatory aspects, and the development of *ad hoc* markets need to be strengthened (Thomas and Read, 2007). Plant essential oils have become an alternative in arthropod control due to their effectiveness, mechanisms of action, low toxicity against non-target vertebrates, as well as their potential use in the synthesis of nanopesticides; however, these applications still face critical technical challenges concerning the procurement process, the development of formulations, and the construction of normative frameworks (Pavela and Benelli, 2016).

In addition to those associated with the development of effective formulations, the following issues associated with the use of biopesticides stand out: 1.- Acute toxicity has been evaluated in the short term, but long-term sublethality is yet to be evaluated for physiological and behavioral effects on non-target species that may result in demographic growth changes and disruptions in the environmental services provided by these species, whose consideration is paramount as part of an integrated pest management (IPM) program or when planting organic crops (Biondi et al., 2013); 2.- Although the use of safe active substances is promoted in Europe as a measure of food safety (for instance, lignocellulosic materials are an option in biopesticide manufacturing), they are treated in the same way as traditional pesticides, and their approval and registration tend to be complicated, requiring a number of adjustments (Villaverde et al., 2014; Czaja et al., 2015; Villaverde et al., 2016); 3.- The design of more efficient technologies is required to facilitate the *in situ* use of biopesticides (Gan-Mor and Matthews, 2003); and 4.- Positive experiences (Canada) show that the main challenges are funding for the innovation and commercialization of new products, purity control, biosecurity, cost and price reduction, compatibility with conventional techniques, scale-up and industrial development (Bailey, 2010).

Other problems associated with the development of agricultural biotechnology requiring further study and evaluation suggest that physical contact with microbial biofertilizers could be associated with the appearance of allergic reactions that increase farmers' sensitivity to these products (mediated by IgE) (Doekes et al., 2004). Other studies have reported lethal and sublethal effects against native pollinators (bees) due to biopesticide exposure, compromising the development of non-target species (Barbosa et al., 2015). Consequently, more studies are essential in order to evaluate the possible adverse effects of biopesticides in greater depth and develop *ad hoc* solutions to address and mitigate problems associated with their use. Additionally, any technological development entering the market must integrate studies guaranteeing their biosecurity as a sine *qua non* condition, which undoubtedly requires the definition of an *ex professo* regulatory framework for each specific context and reality in every country or region.

The use of biopesticides for insect control began in the nineteenth century, and their use would be marked by constant growth, although their market share is still limited (Olson, 2015). However, thanks to the potential contributions of biopesticides to sustainable agriculture, their participation in the global market is now expected to increase, although this will require substantial research and development and the creation of *ad hoc* policy to promote their incorporation into agricultural activities (Ansari et al., 2012). As an essential part of agricultural biotechnology, biopesticides encompass a wide range of technologies that can contribute to the development and modernization of low-environmental impact agriculture as part of integrated pest management systems, although issues arise in many fronts, such as politics, regulation, technology, and the social, economic, and market spheres; consequently, future efforts must take a sustainability-based approach from their inception (see Fig. 7.1).

7.2 Method

The purpose of the present study was to define scientific and technological trajectories by analyzing publications and patent documents using the terms "biopesticides" and "sustainability." Our goal was to carry out a first approach to the current status of R&D on biopesticides and their sustainability so that the economic, social, and environmental progress in the field becomes apparent. Therefore, we carried out two different procedures: (1) identification of scientific trajectory, based on a bibliometric and network analysis of publications including both terms, and (2) identification of technological trajectory, based on the analysis of patent documents using both terms.

In the first case, we searched the Scopus database (Scopus, 2020) to locate target publications, which were then subjected to network analysis (keywords). 263 publications were obtained using the search terms biopesticide* and sustainab* in the title, abstract, or keywords were found for the period from 2011 to 2020, which provided an overview of

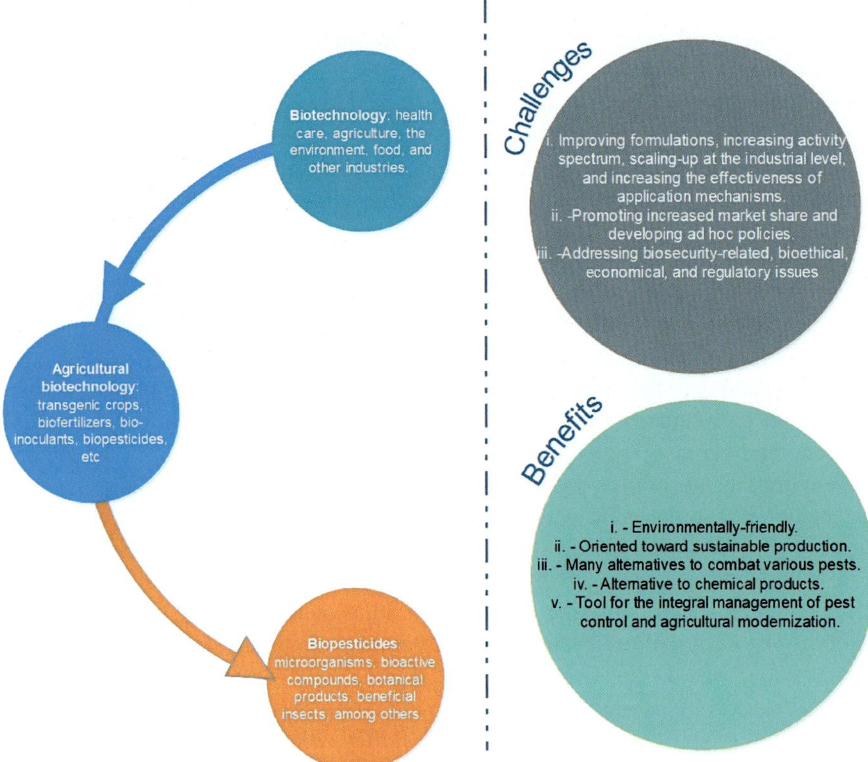

FIG. 7.1 Biopesticides: benefits and challenges.

the most important institutions, countries, regions, publication sources, and knowledge areas carrying out basic research on biopesticides and sustainability. Additionally, we were able to identify the main thematic clusters and nodes associated with these terms (frequency and proximity).

The technological trajectory was approached by conducting an advanced search for patent documents in the Patenscope database, operated by the World Intellectual Property Organization (WIPO) (WIPO, 2020b), using the terms biopesticide and sustainability, including lexemes. This search was limited to English-language documents in all intellectual property offices, but patent families were excluded. A total of 1156 patent documents were found for the period from 1992 to 2020 (the items/groups were based on the 50 results available in Patentscope's analysis bar). This information was analyzed using the following criteria: 1. Number of documents per year; 2. Country or territory; 3. Applicants (applicants were ranked based on reported name, e.g., company or subdivision, but not both); and 4. Main technological sectors for the development of this field of knowledge. This analysis revealed the scientific and technological trajectory about the relationship between biopesticides and sustainability and allowed for a first approach to the current progress and future challenges in this field.

7.3 Discussion and analysis of results

7.3.1 Scientific trajectory

As can be appreciated in Fig. 7.2, the number of research documents related to biopesticides and sustainability that were published during the study period was below 50 in any given year. Although the amount is small, a marginal increase can be observed from 2016 to 2019, with 35 documents published in 2016 and 2018, reaching 40 documents in 2017, an amount which is exceeded in 2019. However, the total number of publications was only 263 for the entire period. Although biopesticide research is relevant and in constant evolution, basic research with an eye on sustainability must be promoted.

Many of these studies are published in highly prestigious journals. Based on number of publications, Journal of Citation Reports (JCRs) impact factor (Journal Citation Reports -JCR-, 2020), and ranking on the ESI Total Citations Rank in 2018, the following journals stood out: 1. Crop Protection (impact factor: 2.172 and Q1 for Agricultural Sciences, position

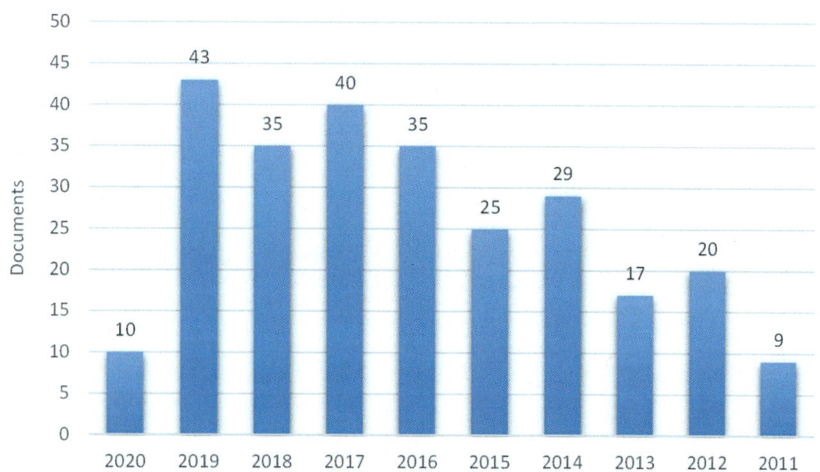

FIG. 7.2 Production of academic documents. *Elaborated by the authors based on Scopus (2020).*

46 of 352); 2. Pest management Science (impact factor: 3.255 and Q1 for Agricultural Sciences, position 38 out of 352); 3. Industrial Crops and Products (impact factor: 4.191 and Q1 for Agricultural Sciences, position 13 of 352); 4. Journal of Applied Entomology (impact factor: 1.827 and Q2 for Plant and Animal Science, position 211 of 804); and 5. Journal of Invertebrate Pathology (impact factor: 2.101 and Q1 for Plant and Animal Science, position 99 out of 804) (see Fig. 7.3).

Four Brazilian universities and one public organization from this country are among the top ten institutions producing publications involving the terms biopesticides and sustainability: Paulista State University, University of Sorocaba, Campinas State University, the Federal University of Parana, and the Brazilian Agricultural Research Corporation, which reflects the interest of academia and the government. Considering the limited amount of publications within the study period, the total number of documents produced by these organizations is high. The rest of the ranking includes organizations from India (Banaras Hindu University); Kenya (International Center of Insect Physiology and Ecology Nairobi); the United States (USDA Agricultural Research Service, Washington DC., the University of California, Davis), and Canada (University of Guelph) (see Table 7.1).

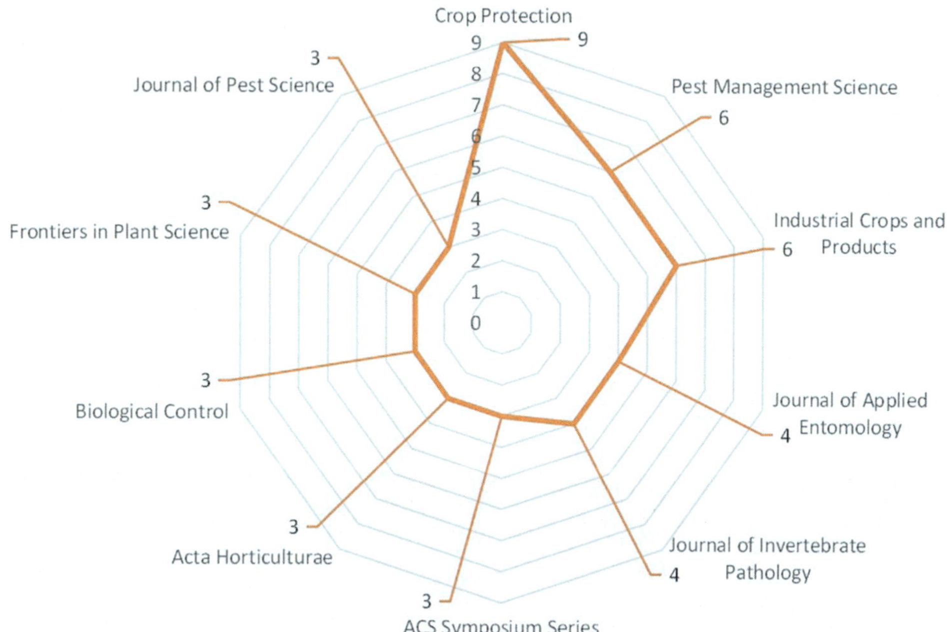

FIG. 7.3 Documents by publication source. *Elaborated by the authors based on Scopus (2020).*

TABLE 7.1 Documents by affiliation.

No.	Affiliation	Documents
1	Paulista State University (Universidade Estadual Paulista)	11
2	Banaras Hindu University	9
3	University of Sorocaba (Universidade de Sorocaba)	5
4	Campinas State University (Universidade Estadual de Campinas)	5
5	International Center for Insect Physiology and Ecology Nairobi	5
6	USDA Agricultural Research Service, Washington DC	4
7	Brazilian Agricultural Research Corporation (Empresa Brasileira de Pesquisa Agro-pecuária - Embrapa)	4
8	University of California, Davis	4
9	Federal University of Parana (Universidade Federal do Parana)	4
10	University of Guelph	4

Source: Elaborated by the authors based on Scopus (2020).

However, when analyzing by country or territory (Top Ten), the first position is held by India, whose results are more than two times those from economies such as the United States, Brazil, and China, among others. Despite that, India and China are the only Asian countries in this ranking. Brazil is the only country from Latin America in this ranking; the rest of the productivity is concentrated in the United States, Canada, Australia, and European countries, all of them developed economies. Therefore, emerging economies and developing countries must increase their investment in basic research for their organizations to generate scientific knowledge that they can apply in the development of biopesticide technology with a focus on sustainability and local problems, without losing sight of global issues (see Fig. 7.4).

Most of the publications belong to different technical areas (82.2%) primarily agricultural and biological science; the rest fall in the environmental science category (13.5%) and only a small percentage in disciplines such as social science, multidisciplinary science, arts and humanities, economics, econometrics, and finance (4.3%). This indicates that these topics are mainly addressed by technical publications and that the participation of environmental science and the rest of the areas is considerably less significant. Therefore, social and economic sciences must engage these topics to enable an integral analysis (see Fig. 7.5).

Most of this academic production focuses on three main sources of publication (91.6%): academic journals (55.9%), book chapters (18.6%), and review articles (17.1%). The lowest proportion (8.4%) consists of conference articles, books,

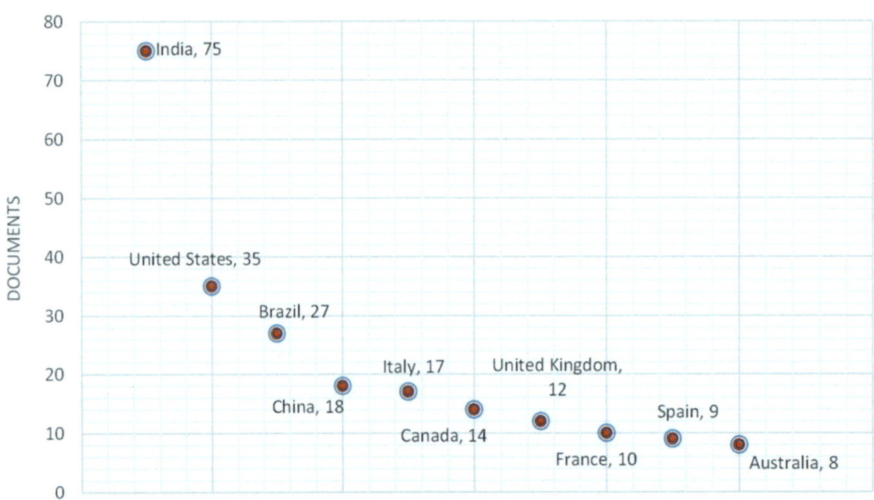

FIG. 7.4 Documents by country or territory. *Elaborated by the authors based on Scopus (2020).*

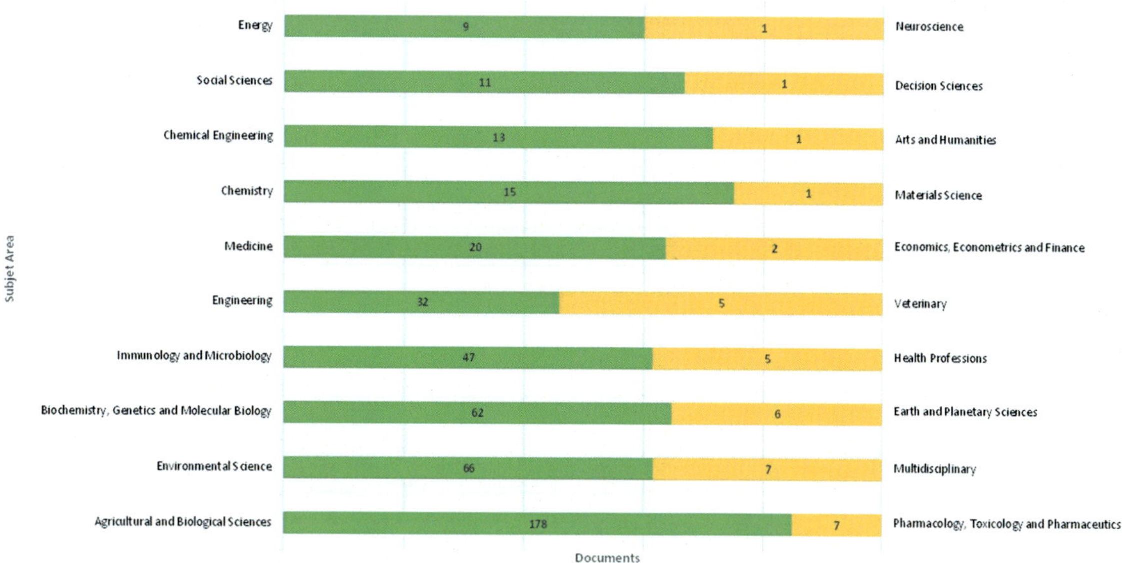

FIG. 7.5 Documents by area of knowledge. *Elaborated by the authors based on Scopus (2020).*

notes, and short surveys. Therefore, most of this information was published as a paper, review article, or book chapter. These results underscore the need for specialized books and indexed conference proceedings where the scientific community can disseminate and discuss cutting-edge research (see Fig. 7.6).

The document search allowed for a co-occurrence analysis to identify keywords associated with the terms "biopesticides" and "sustainability." Based on this analysis, five clusters with multiple associated nodes of varying size (frequency) and proximity (distance) were detected. The red cluster stands out among these clusters; its central node refers to biopesticides and is associated with concepts such as sustainability, antimicrobial activity, integrated pest management, sustainability, and insecticides, among others. The green cluster is also significant; it identifies three significant nodes centered around agriculture, sustainable agriculture, and biopesticides. Terms such as fungi, biocontrol, bacteria, fertilizers, biofertilizers, pesticides, disease control, and microorganisms, among others, can also be appreciated, although their weight is lower.

The central nodes in the blue cluster refer to animals, hexapods, biological control of pests, Lepidoptera, procedures, physiology, and effects of drugs and insecticides, and less markedly to other concepts associated with pesticide development. For its part, the yellow cluster includes terms associated with microbiology, soil microbiology, metabolism, biological control agents, and microorganisms, as well as plant diseases, nitrogen fixation, the rhizosphere, biofertilizers, and plant growth, among others. The purple cluster shares several terms with the other four clusters; therefore, the lack of a clear definition of this cluster prevents the identification of its dominant elements (see Fig. 7.7).

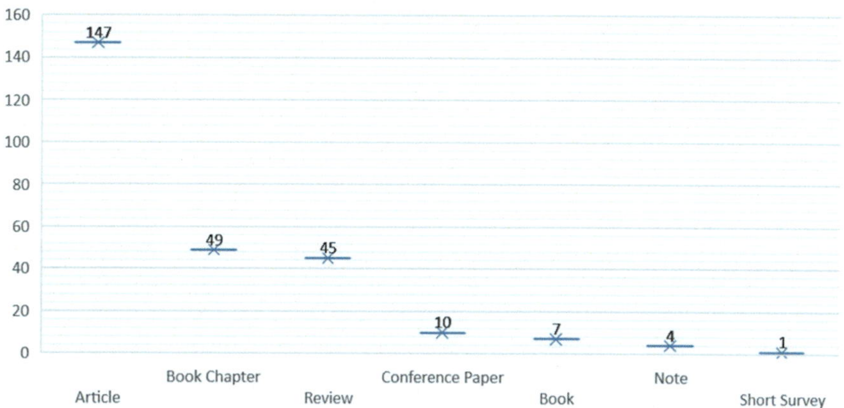

FIG. 7.6 Types of published documents. *Elaborated by the authors based on Scopus (2020).*

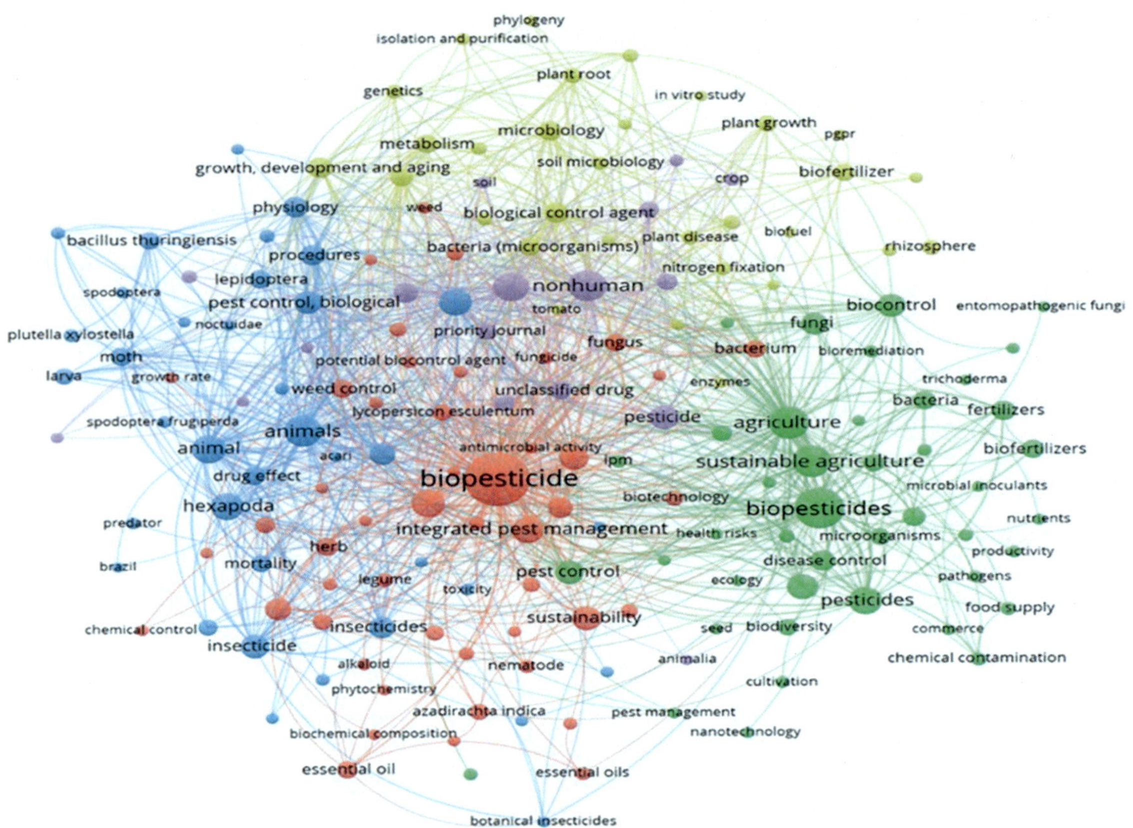

FIG. 7.7 Co-occurrence analysis (all keywords-author and index) total count. *Elaborated by the authors based on Scopus (2020) and VOSviewer (2020). Vosviewer (Software). Version 1.6.14.*

7.4 Technological trajectory

As shown in graphic 6, the majority of patent documents were submitted to patent registration offices in United States, Australia, Europe, Canada, and Israel in addition to the Patent Cooperation Treaty (PCT), most documents are concentrated in the Top Ten (1992–2020). China and New Zealand have filed a significantly lower number of documents than the leading countries, followed by emerging economic actors such as South Africa and the Philippines. However, Latin American and Asian countries such as India, Singapore, Brazil, Argentina, and Mexico are absent from this ranking in the English search, and therefore, they are far from prepared to participate in the global economy and its markets (see Fig. 7.8).

American companies Dow Agrosciences, DuPont, FMC Corporation, and The Board of Regents of the University of Nebraska, hold the first four places, totaling 505 patent documents. From this same country are Cool Planet Energy Systems (sixth place), Pioneer Hi-Bred International (seventh place), and Agrinos (eighth place). German companies Fraunhofer-Gesellschaft, Monsanto Technology, and Basf submitted 31, 20, and 14 documents, respectively; therefore, as shown by the analysis of documents by country or territory, developed economies are clearly leading the field. As a consequence, despite the difficulties, the entry of Latin American, Asian, and African companies into the market is desirable to promote the scientific and technological development of these regions in the area of biopesticides and sustainability (see Fig. 7.9).

The ten categories in which most of these documents fall, according to the International Patent Classification (IPC), are shown in Table 7.2. The most important category by number of documents is A01N, which focuses on biocides, such as pesticides and herbicides, as well as plant growth regulators (pesticide-fertilizer combination). This category is followed by C12N, related to the composition and conservation of microorganisms and enzymes, as well as aspects related to genetic engineering, mutation, and culture media, and finally by category C07D, which focuses on heterocyclic compounds. The remaining categories, with less than 200 documents, refer to different areas, most remarkably pest attractants and repellents, the development of new plants and their obtaining processes, and plant reproduction (tissue culture techniques), among others (see Table 7.2).

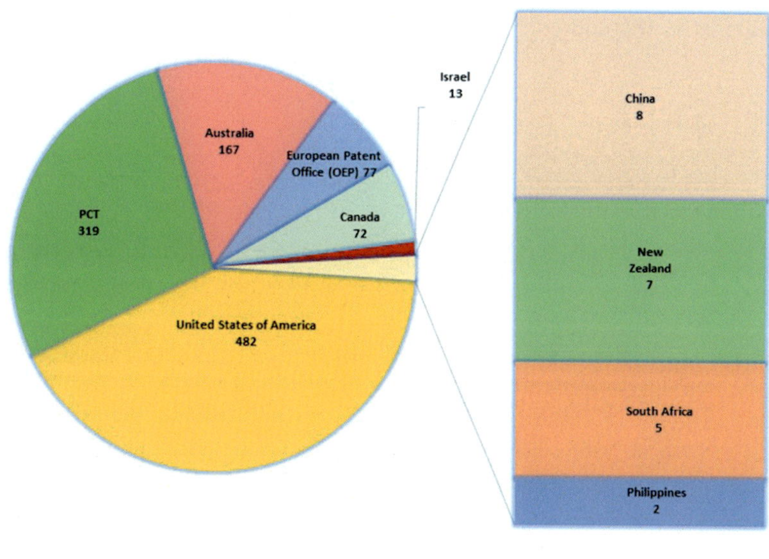

FIG. 7.8 Patent documents by country or territory. *Elaborated by the authors based on WIPO (2020b).*

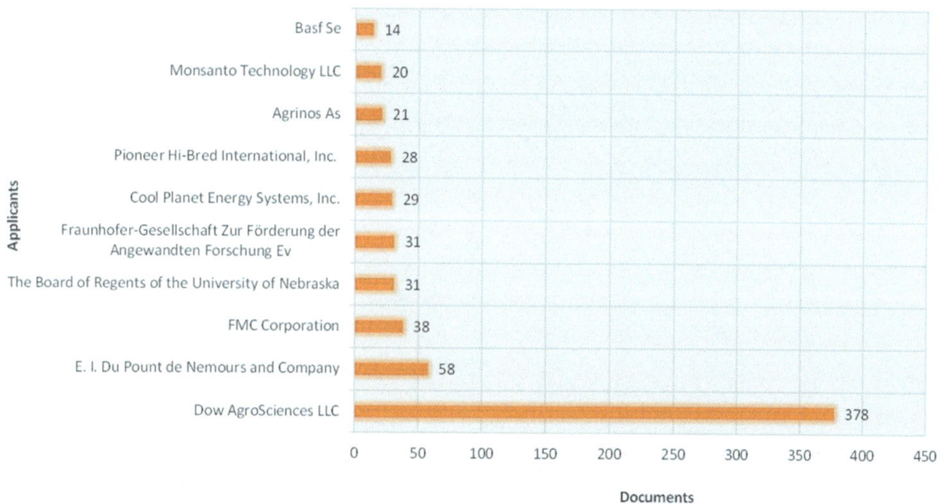

FIG. 7.9 Patent documents by applicant. *Elaborated by the authors based on WIPO (2020b).*

During the period from 2011 to 2020, the number of patent documents increased steadily from 31 documents to a total of 190 from 2011 to 2016; however, after this point, subsequent values show a decrease until the year 2019, but never below 100 documents. In the case 2020, a total of 32 documents had been submitted at the time of the search. However, although 1156 patent documents related to the terms "biopesticides" and "sustainability" were found for the study period (1992–2020), this number is still relatively small, and more efforts are needed to develop technology integrating the economic and social aspects, as well as environmental perspectives. In the context of globalization and climate change, this condition is not only desirable but also necessary to mitigate their adverse effects, especially in the less developed economies (see Fig. 7.10).

TABLE 7.2 Patent documents by IPC code.

IPC code			
No.	**Code**	**Classification**	**Number**
1	A01N	"Preservation of bodies of humans or animals or plants or parts thereof …; biocides, e.g. as disinfectants, pesticides, herbicides …; pest repellants or attractants; plant growth regulators" …	823
2	C12N	"Microorganisms or enzymes; compositions thereof; propagating, preserving, or maintaining microorganisms; mutation or genetic engineering; culture media …".	377
3	C07D	"Heterocyclic compounds (macromolecular compounds)".	236
4	A61K	"Preparations for medical, dental, or toilet purposes …; chemical aspects of, or use of materials for deodorization of air, for disinfection or sterilization, or for bandages, dressings, absorbent pads or surgical articles; soap compositions".	167
5	A01P	"Biocidal, pest repellant, pest attractant or plant growth regulatory activity of chemical compounds or preparations".	142
6	C07K	"Peptides …; cyclic dipeptides not having in their molecule …; ergot alkaloids of the cyclic peptide …; single cell proteins, enzymes; genetic engineering processes for obtaining peptides …".	133
7	A01H	"New plants or processes for obtaining them; plant reproduction by tissue culture techniques".	128
8	C07C	"Acyclic or Carbocyclic Compounds …; production of organic compounds by electrolysis or electrophoresis …".	98
9	C05F	"Organic fertilisers not covered by subclasses … Fertilisers from waste or refuse".	90
10	C12R	"Indexing scheme associated with subclasses …, relating to microorganisms".	77

Elaborated by the authors based on WIPO (WIPO, 2020a,b).

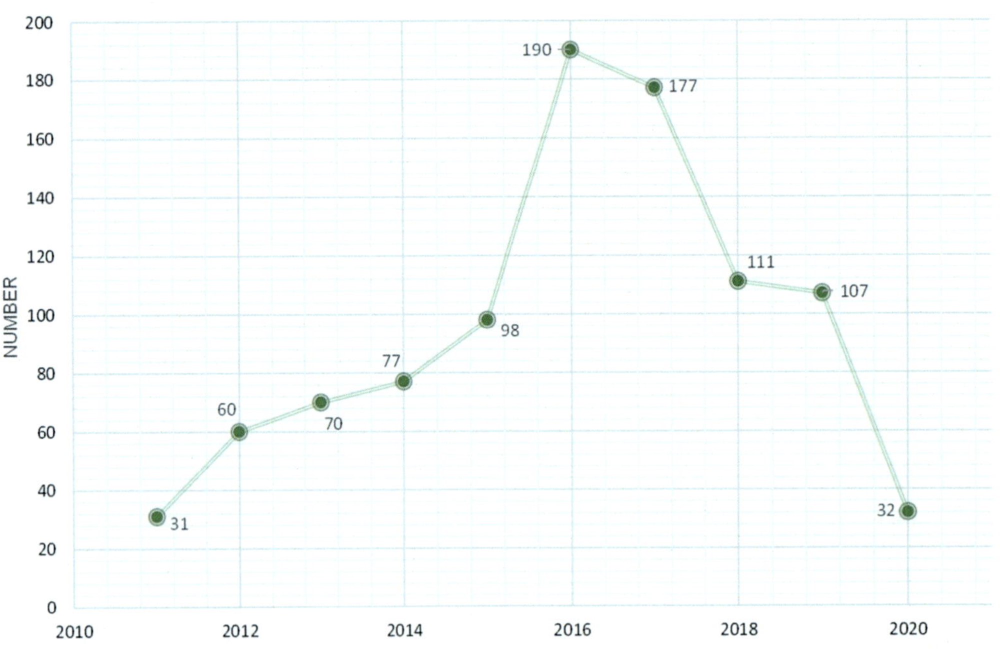

FIG. 7.10 Patent documents by publication date. *Elaborated by the authors based on WIPO, (2020a,b).*

7.5 Conclusions

Biotechnology is an emerging scientific and technological discipline relying on the intensive use of knowledge. Consequently, its progress and potential can be appreciated in different areas of knowledge related to health, agriculture, food, the environment, and energy, in addition to different areas of industrial interest. In the specific case of agricultural biotechnology, the progress is focused on the development of transgenic crops (a field characterized by opposite opinions), biofertilizers, bioinoculants, and biopesticides, among other technologies. Biopesticides already provide a wide range of applications for combating agricultural pests and have an excellent potential for the development of further alternatives. Therefore, increasing the use of biopesticides can not only to mitigate the environmental impact of agriculture and reduce the use of chemical pesticides, but also contribute to the creation of integral pest control management systems, the modernization of agriculture, and the attainment of sustainable development goals.

Some of the technical challenges of this group of technologies have to do with the improvement of formulations, increasing the spectrum of activity, using engineering design to enable applications based on these alternatives, industrial scale-up, and biosecurity testing for each technological development. In addition, policy must be geared toward the promotion of biopesticides and other non-conventional alternatives to address agricultural problems that, at the same time, contribute to the attainment of sustainable development targets. Finally, the approach to the economic, regulatory, bioethical, and market-related dimensions of the field must be addressed in greater depth in order to increase the market share of biopesticides. Marketable products should be effectual, reliable, cost-effective, and easy to use, and they must meet the technical and legal requirements established for this category of technological innovations by local and global markets.

In this regard, the scientific trajectory described in the present study shows that the number of documents published during the study period is limited, and therefore, basic research related to biopesticides and sustainability must be promoted. However, although the number of documents produced during the study period is low, many of these studies were published in academic journals with a significant impact factor, classified in the upper two quartiles. The present study also revealed that several Brazilian organizations stand out at the institutional level, as well as organizations from India, the United States, and Canada. However, at the country level, the leading countries are India, the United States, Brazil, China, European countries, Canada, and Australia, which evinces the meager participation of developing economies. The co-occurrence analysis showed five thematic clusters and nodes referring to areas traditionally associated with the development of biopesticides. Additionally, this analysis pointed out emerging areas of interest that will most certainly have an impact on the future development of sustainable technologies.

Concerning the technological trajectory analysis, the number of patent documents was found to be noticeably higher than the number of academic publications (953 documents from 2011 to 2020); nevertheless, patent documents related to the terms of biopesticides and sustainability must still be further advanced, especially in developing countries. Most patent documents are submitted in the United States, the Patent Cooperation Treaty (PCT), Australia, the European Patent Office, and Canada, which evinces the importance of patenting in these countries. American and German organizations are the leading applicants of patents in this area of knowledge. Finally, the present study shows the technological sectors in which knowledge on biopesticides and sustainability is concentrated according to the IPC categories.

Acknowledgments

We wish to acknowledge the support provided by the National Polytechnic Institute (Instituto Politécnico Nacional) and the Secretariat for Research and Postgraduate Studies (Secretaría de Investigación y Posgrado), grant numbers 20195587 and 20200773.

References

Ahemad, M., Khan, M.S., 2011. Pesticide interactions with soil microflora: importance in bioremediation. In: Ahmad, I., Ahmad, F., Pichtel, J. (Eds.), Microbes and Microbial Technology. Agricultural and Environmental Applications. Springer, Science+Business Media, pp. 393–413.

Ahmad, I., Sajjad, M., Khan, A., Aqil, F., Singh, M., 2011. Microbial applications in agriculture and the environment: a broad perspective. In: Ahmad, I., Ahmad, F., Pichtel, J. (Eds.), Microbes and Microbial Technology. Agricultural and Environmental Applications. Springer, Science+Business Media, pp. 1–27.

Anjum, R., Rahman, M., Masood, F., Malik, A., 2012. Bioremediation of pesticides from soil and wastewater. In: Malik, A., Grohmann, E. (Eds.), Environmental Protection Strategies for Sustainable Development. In: Lozano, R. (Ed.), Strategies for Sustainability Series. Springer, Science+-Business Media, pp. 295–328.

Ansari, M.S., Ahmad, N., Hasan, F., 2012. Potential of biopesticides in sustainable agriculture. In: Malik, A., Grohmann, E. (Eds.), Environmental Protection Strategies for Sustainable Development. In: Lozano, R. (Ed.), Strategies for Sustainability Series. Springer, Science+Business Media, pp. 529–595.

Bailey, K.L., 2010. Canadian innovations in microbial biopesticides. J. Indian Dent. Assoc. 32 (2), 113–121.

Barbosa, W.F., Tomé, H.V.V., Bernardes, R.C., Siqueira, M.A.L., Smagghe, G., Guedes, R.N.C., 2015. Biopesticide-induced behavioral and morphological alterations in the stingless bee Melipona quadrifasciata. Environ. Toxicol. Chem. 34 (9), 2149–2158.

Biondi, A., Zappalà, L., Stark, J.D., Desneux, N., 2013. Do biopesticides affect the demographic traits of a parasitoid wasp and its biocontrol services through sublethal effects? PloS One 8 (9), 1–11.

Bona, E., Lingua, G., Todeschini, V., 2016. Effect of bioinoculants on the quality of crops. In: Arora, N.K., Mehnaz, S., Balestrini, R. (Eds.), Bioformulations: for Sustainable Agriculture. Springer, India, pp. 93–124.

Brar, S.K., Verma, M., Tyagi, R.D., Valéro, J.R., 2006. Recent advances in downstream processing and formulations of *Bacillus thuringiensis* based biopesticides. Process Biochem. 41 (2), 323–342.

Chandler, D., Bailey, A.S., Tatchell, G.M., Davidson, G., Greaves, J., Grant, W.P., 2011. The development, regulation and use of biopesticides for integrated pest management. Philos. Trans. R. Soc. Lond. B Biol. Sci. 366 (1573), 1987–1998.

Copping, L.G., Menn, J.J., 2000. Biopesticides: a review of their action, applications and efficacy. Pest Manag. Sci. 56 (8), 651–676.

Czaja, K., Góralczyk, K., Struciński, P., Hernik, A., Korcz, W., Minorczyk, M., et al., 2015. Biopesticides – towards increased consumer safety in the European Union. Pest Manag. Sci. 71 (1), 3–6.

Doekes, G., Larsen, P., Sigsgaard, T., Baelum, J., 2004. IgE sensitization to bacterial and fungal biopesticides in a cohort of Danish greenhouse workers: the BIOGART study. Am. J. Ind. Med. 46 (4), 404–407.

El-Bendary, M.A., 2006. *Bacillus thuringiensis* and Bacillus sphaericus biopesticides production. J. Basic Microbiol. 46 (2), 158–170.

Gan-Mor, S., Matthews, G.A., 2003. Recent developments in sprayers for application of biopesticides-an overview. Biosyst. Eng. 84 (2), 119–125.

Glare, T., Caradus, J., Gelernter, W., Jackson, T., Keyhani, N., Köhl, J., et al., 2012. Have biopesticides come of age? Trends Biotechnol. 30 (5), 250–258.

Hanson, J.D., Hendrickson, J., Archer, D., 2008. Challenges for maintaining sustainable agricultural systems in the United States. Renew. Agric. Food Syst. 23 (4), 325–334.

Villaverde, J.J., Sevilla-Morán, B., Sandín-España, P., López-Goti, C., Alonso-Prados, J.L., 2014. Biopesticides in the framework of the European pesticide regulation (EC) No. 1107/2009. Pest Manag. Sci. 70 (1), 2–5.

Journal Citation Reports -JCR-, 2020. Incites Journal Citation Reports. Available online. https://jcr.clarivate.com. (Accessed 10 April 2020).

Kaaya, G.P., Hassan, S., 2000. Entomogenous fungi as promising biopesticides for tick control. Exp. Appl. Acarol. 24, 913–926.

Kapoor, R., Giri, B., Mukerji, K.G., 2002. Glomus macrocarpum: a potential bioinoculant to improve essential oil quality and concentration in Dill (*Anethum graveolens* L.) and Carum (*Trachyspermum ammi* (Linn.) Sprague). World J. Microbiol. Biotechnol. 18, 459–463.

Kassam, A., Brammer, H., 2013. Combining sustainable agricultural production with economic and environmental benefits. Geogr. J. 179 (1), 11–18.

Kaur, S., 2000. Molecular approaches towards development of novel *Bacillus thuringiensis* biopesticides. World J. Microbiol. Biotechnol. 16, 781–793.

Kilian, B., Pratt, L., Jones, C., Villalobos, A., 2004. Can the private sector be competitive and contribute to development through sustainable agricultural business? A case study of coffee in Latin America. Int. Food Agribus. Man. 7 (3), 21–45.

Kurlavičius, A., 2009. Sustainable agricultural development: knowledge-based decision support. Technol. Econ. Dev. Econ. 15 (2), 294–309.

Lancker, E., Nijkamp, P., 2000. A policy scenario analysis of sustainable agricultural development options: a case study for Nepal. Impact Assess. Proj. Apprais. 18 (2), 111–124.

Leng, P., Zhang, Z., Pan, G., Zhao, M., 2011. Applications and development trends in biopesticides. Afr. J. Biotechnol. 10 (86), 19864–19873.

Lesueur, D., Deaker, R., Herrmann, L., Bräu, L., Jansa, J., 2016. The production and potential of biofertilizers to improve crop yields. In: Arora, N.K., Mehnaz, S., Balestrini, R. (Eds.), Bioformulations: For Sustainable Agriculture. Springer, India, pp. 71–92.

Lwoga, E.T., Ngulube, P., Stilwell, C., 2010. Managing indigenous knowledge for sustainable agricultural development in developing countries: knowledge management approaches in the social context. Int. Inf. Libr. Rev. 42 (3), 174–185.

Meena, V.S., Bahadur, I., Maurya, B.R., Kumar, A., Meena, R.K., Kumari, S., et al., 2016. Potassium-solubilizing microorganism in evergreen agriculture: an overview. In: Meena, V.S., Maurya, B.R., Verma, J.P., Meena, R.S. (Eds.), Potassium Solubilizing Microorganisms for Sustainable Agriculture. Springer, India, pp. 1–20.

Mishra, J., Arora, N.K., 2016. Bioformulations for plant growth promotion and combating phytopathogens: a sustainable approach. In: Arora, N.K., Mehnaz, S., Balestrini, R. (Eds.), Bioformulations: for Sustainable Agriculture. Springer, India, pp. 3–33.

Mishra, J., Tewari, S., Singh, S., Arora, N.K., 2015. Biopesticides: where we stand? In: Arora, N.K. (Ed.), Plant Microbes Symbiosis: Applied Facts. Springer, New Delhi, pp. 37–75.

Moscardi, F., Lobo de Souza, M., Batista de Castro, M.E., Lara Moscardi, M., Szewczyk, B., 2011. Baculovirus pesticides: present state and future perspectives. In: Ahmad, I., Ahmad, F., Pichtel, J. (Eds.), Microbes and Microbial Technology. Agricultural and Environmental Applications. Springer, Science+Business Media, pp. 415–445.

Nathan, S.S., Chung, P.G., Murugan, K., 2005. Effect of biopesticides applied separately or together on nutritional indices of the rice leaffolder *Cnaphalocrocis medinalis*. Phytoparasitica 33 (2), 187–195.

Olson, S., 2015. An analysis of the biopesticide market now and where it is going. Outlooks Pest Manag. 26 (5), 203–206.

Ortaş, I., Varma, A., 2007. Field trials of bioinoculants. In: Varma, A., Oelmüller, R. (Eds.), Advanced Techniques in Soil Microbiology. In: Varma, A. (Ed.), Soil Biology Series, vol. 11. Springer-Verlag, Berlin Heidelberg, pp. 397–413.

Owen, D., Williams, A.P., Griffith, G.W., Withers, P.J.A., 2015. Use of commercial bio-inoculants to increase agricultural production through improved phosphorus acquisition. Appl. Soil Ecol. 86, 41–54.

Pandya, U., Saraf, M., 2015. Antifungal compounds from pseudomonads and the study of their molecular features for disease suppression against soil borne pathogens. In: Arora, N.K. (Ed.), Plant Microbes Symbiosis: Applied Facets. Springer, New Delhi, pp. 179–192.

Pavela, R., Benelli, G., 2016. Essential oils as ecofriendly biopesticides? Challenges and constraints. Trends Plant Sci. 21 (12), 1000–1007.

Pretty, J., 2018. Intensification for redesigned and sustainable agricultural systems. Science 362 (6417), 1–7.

Priyadharsini, P., Muthukumar, T., 2015. Insight into the role of arbuscular mycorrhizal fungi in sustainable agriculture. In: Thangavel, P., Sridevi, G. (Eds.), Environmental Sustainability Role of Green Technologies. Springer, New Delhi, pp. 3–37.

Rezaei-Moghaddam, K., Karami, E., 2008. A multiple criteria evaluation of sustainable agricultural development models using AHP. Environ. Dev. Sustain. 10, 407–426.

Rigby, D., Woodhouse, P., Young, T., Burton, M., 2001. Constructing a farm level indicator of sustainable agricultural practice. Ecol. Econ. 39 (3), 463–478.

Schnepf, H.E., 2012. *Bacillus thuringiensis* recombinant insecticidal protein production. In: Sansinenea, E. (Ed.), *Bacillus thuringiensis* Biotechnology. Springer, Science+Business Media, pp. 259–281.

Scopus, 2020. Abstract and Citation Database. Available online: https://www.scopus.com. (Accessed 6 February 2020).

Seiber, J.N., Coats, J., Duke, S.O., Gross, A.D., 2014. Biopesticides: state of the art and future opportunities. J. Agric. Food Chem. 62 (48), 11613–11619.

Senthil-Nathan, S., 2015. A review of biopesticides and their mode of action against insect pests. In: Thangavel, P., Sridevi, G. (Eds.), Environmental Sustainability Role of Green Technologies. Springer, New Delhi, pp. 49–63.

Szewczyk, B., Hoyos-Carvajal, L., Paluszek, M., Skrzecz, I., Lobo de Souza, M., 2006. Baculoviruses—re-emerging biopesticides. Biotechnol. Adv. 24 (2), 143–160.

Thomas, M.B., Read, A.F., 2007. Can fungal biopesticides control malaria? Nat. Rev. Microbiol. 5, 377–383.

Villaverde, J.J., Sandín-España, P., Sevilla-Morán, B., López-Goti, C., Alonso-Prados, J.L., 2016. Biopesticides from natural products: current development, legislative framework, and future trends. Bioresources 11 (2), 5618–5640.

Wezel, A., Brives, H., Casagrande, M., Clément, C., Dufour, A., Vandenbroucke, P., 2016. Agroecology territories: places for sustainable agricultural and food systems and biodiversity conservation. Agroecol. Sust. Food 40 (2), 132–144.

WIPO, 2020a. International Patent Classification (IPC). Available online: https://www.wipo.int/classifications/ipc/en/. (Accessed 10 April 2020).

WIPO, 2020b. Patentscope. In: Search International and National Patent Collections. Available online: https://patentscope.wipo.int/search/en/search.jsf. (Accessed 23 May 2020).

Wu, L., Wu, H.J., Qiao, J., Gao, X., Borriss, R., 2015. Novel routes for improving biocontrol activity of *Bacillus* based bioinoculants. Front. Microbiol. 6 (1395), 1395 1–13.

Chapter 8

Biopesticides: a genetics, genomics, and molecular biology perspective

Pawan Basnet[b], **Rajiv Dhital**[c] and **Amitava Rakshit**[a]

[a]*Banaras Hindu University, Institute of Agricultural Science, Department of Soil Science & Agricultural Chemistry, Varanasi, Uttar Pradesh, India;* [b]*University of Missouri-Columbia, Division of Plant Science and Technology, Columbia, MO, United States;* [c]*University of Missouri-Columbia, Food Science Program, Columbia, MO, United States*

8.1 Introduction

To overcome the increasing food demand due to global population growth, the crop production will have to increase in a substantial amount. However, the crop production should be done in a manner so that it would not cause any harm to the humans and environment. To address this issue, an innovative method should be developed which would meet the global food requirements as well maintain the balance of environment (Clements and Ditommaso, 2011). Agricultural yield and quality are adversely affected by pests such as bacteria, fungi, weeds and insects (Kumar, 2012). These organisms are responsible for about 40% reduction in crop yield (Oerke et al., 2012). Since 1960s there has been wide use of synthetic pesticides for controlling pests (Nicholson, 2007). In combination with improved plant varieties, mechanization, irrigation methods, synthetic pesticides helped in increasing the agricultural productivity (Paarlberg, 2013). But, the haphazard use of synthetic pesticides have gained attention due to several factors such as: the excessive use of chemical pesticides can cause adverse effect on human health and environment (Assessment, 2005); the uncontrolled use of pesticides can result in the development of pesticide resistance (Abudulai et al., 2001); pesticide residues in plant, soil and water (Al-Zaidi et al., 2011). Therefore, there is an urgent need to develop alternatives to replace the use of synthetic pesticides in pest management. To address these issues, biopesticides can be one of the promising alternatives.

Biopesticides include pesticides derived from microorganisms (microbial pesticides), naturally occurring substances (biochemical pesticides), and pesticidal substances containing added genetic material (plant-incorporated protectants) (Kachhawa, 2017). The development of biopesticides employs microorganisms, biochemical substances, genes, and their interactions with the targeted crop species. Biopesticides are the mass-produced agents which are derived from animals (nematodes), plants and microorganisms (bacteria, algae or fungi) or their products (phytochemicals, microbial products) or byproducts used for the control of plant pests (Chandler et al., 2011; Mazid, 2011). Biopesticides are comparatively less toxic than chemical pesticides and are often target specific. Due to this fact, these preparations pose less threat to the environment and human health. Moreover, they have little or no residual effects and can be used in organic farming (Kumar, 2012).

8.1.1 Advantages of application biopesticides in pest management

- Less toxic than conventional pesticides.
- Effective in low concentration.
- Decompose quickly resulting lower exposure to the environment.
- Easier for applications in pest management.
- Have a broad-spectrum for controlling the pests.
- Can reduce the use of chemical pesticides when used in Integrated Pest Management (IPM) programs.
- Less prone to resistance by pests.

Broadly, biopesticides fall into three major categories based on the type of active ingredient used, namely, (a) microbial pesticides (b) plant-incorporated protectants (PIPs) and (c) biochemical pesticides (Kumar, 2012).

Biopesticides. https://doi.org/10.1016/B978-0-12-823355-9.00019-5

8.1.1.1 Microbial biopesticides

Microbial biopesticides, also known as biological control agents, are derived from naturally occurring bacteria, fungi, protozoa and viruses. These are one of the major groups of biopesticides used for the control of insects, plant pathogens and weeds (Chandler et al., 2011; Arthurs and Dara, 2019). Globally, the production of microbial biopesticides is one of the fastest growing segment of biocontrol industry (Dunham and DunhamTrimmer, 2015). Microbial pesticides are available commercially in the form of dust, emulsifiable concentrate, suspensions, granules and powder etc. (Arthurs and Dara, 2019). One of the most widely used microbial biopesticide is bacterium *Bacillus thuringiensis* (Bt), which produces a protein crystal the Bt d-endotoxin capable of causing lysis of gut cells when consumed by susceptible insects (Jurat-Fuentes and Crickmore, 2017). Currently, about 90% of the microbial biopesticides are derived from *Bacillus thuringiensis.* Some of the other examples of microbial biopesticides include products based on baculoviruses and fungi, the biopesticides derived from *Cydia pomonella* granulovirus (CpGV) against codling moth on apples, the Nuclear Polyhedrosis Virus (NPV) of the soya bean caterpillar *Anticarsia gemmatalis* (Chandler, 2011). Table 8.1 summarizes the most commonly used microorganisms (bacteria, fungi and viruses) for the commercial production of biopesticides, their active ingredients and their target pests.

8.1.1.2 Plant-incorporated protectants (PIPs)

Plant-incorporated protectants include a broad range of pesticidal substances, or the genetic materials produced by plants incorporated into the other plant for imparting desirable functions. The most common example of PIPs includes the genetically modified organisms (GMOs) which contain foreign genes responsible for desirable traits in crops (US EPA, 2020). The PIP-producing crops are genetically modified by biological or physical means of transformation and these crops are called as "genetically modified" or "genetically engineered" or "transgenic" (Mehrotra et al., 2017). One of the popular examples of development of transgenic plant which produce PIPs is plants incorporated with *cry* genes from *Bacillus thuringiensis* (Bt). The *cry* gene express Crystals of proteinaceous insecticidal δ-endotoxins (Crystal proteins or cry proteins) during sporulation. The toxin is highly lethal to insects of the Lepidoptera (moths and butterflies), Coleoptera, and

TABLE 8.1 Types of microbial biopesticides that have been approved by EPA.

Types	Active ingredients	Target pests
Bacterial	*Agrobacterium radiobacter*	Crown gall caused by *Agrobacterium tumefaciens*
	Bacillus firmus	Plant parasitic nematodes
	Bacillus thuringiensis subsp. *aizawai*	Caterpillars
	Bacillus thuringiensis subsp. *kurstaki*	Caterpillars
	Bacillus thuringiensis subsp. *galleriae*	Beetles
	Bacillus thuringiensis subsp. *tenebrionis*	Colorado potato and elm leaf larvae
	Bacillus thuringiensis subsp. *israelensis*	Mosquitoes, blackflies, fungus gnats
	Bacillus sphaericus	Mosquito larvae
	Paenibacillus popilliae	Japanese beetle
	Pasteuria spp.	Plant parasitic nematodes
	Pseudomonas fluorescens	Mussels
	Trichoderma harzianum	*Rhizoctonia, Pythium, Fusarium* and soil-borne pathogens
Fungal	*Beauveria bassiana*	Aphids, whiteflies, plant bugs and mites
	Metarhizium brunneum (anisopliae)	Thrips, whiteflies, mites, weevils and ticks
	Isaria fumosorosea	Whiteflies, aphids, thrips, plant bugs, mites and some soil pests
	Pandora delphacis	Brown plant hopper and green hopper.
	Purpureocillium lilacinum	Plant parasitic nematodes
	Paranosema locustae	Grasshoppers and Mormon crickets
	Verticillium lecanii	Aphids, whiteflies and scales
Viral	**Baculovirus**	Codling moth and oriental fruit moth
	(a). Granuloviruses (GV)	Indianmeal moth
	Cydia pormonella	Corn earworm, tobacco budworm
	Plodia interpunctella GV	Beet armyworm
	(b). Nuclear Polyhedrosis Viruses (NPV)	Gypsy moth
	Helicoverpa spp.	
	Spodoptera exigua	
	Lymantria dispar	

Hymenoptera and nematodes (Dean, 1984). In 1995, Bt cotton was the first commercialized transgenic plant (Shelton et al., 2000). In different crops plants the insecticidal crystal proteins coding genes have been transferred, some of the examples of these crops are corn, soybeans, potatoes and plums. The production of PIPs in transgenic plants helps to develop resistance against viruses, bacteria and insects.

8.1.1.3 Biochemical biopesticides

Biochemical pesticides are natural products that are distinguished with the conventional synthetic

chemical pesticides by their non-toxic mode of action to the target plants and are usually species specific (O'Brien et al., 2009). These chemicals are known to interfere the growth and mating cycle of the pest insects, bacterial, fungi and weeds (Mehrotra et al., 2017). Use of plant extracts have long been used to control insects. Currently, there are number of plant extracts used as a biopesticides to control the pests (Fradin and Day, 2002). One of the well-known plant-based biopesticide is neem oil which is seed extract of *Azadirachta indica* (Schmutterer, 1990). The active ingredient of most neem-based biopesticides is a compound named azadiractin. This compound has shown to have a biological activity on more than 400 insect species such as ants, aphids, roaches, fleas, flies and ticks (Schmutterer and Singh, 1995). Similarly, biochemical biopesticides pyrethrins produced by *Chrysanthemum cinerariaefolium*, which have insecticidal properties and low toxicity to humans (Alvarenga et al., 2012). Other plant based biopesticides such as Limonene and Linalool are used against fleas, aphids, mites, ants, flies and crickets (Sarwar, 2015). Rotenone, an isoflavonoid obtained from tropical legumes *Derris elliptica*, *Amorpha fructicosa* or *Tephrosia vogeli* exhibits insecticidal activity aphids, thrips, suckers, and other insects on fruits and vegetables (Isman, 2006; Martin et al., 2013).

8.2 Market trends of biopesticides

In global context, the use of biopesticides is increasing by 10% every year (Markets and Markets, 2020). The biopesticide market was valued at 3.06 billion USD in 2018 and is expected to grow at compound annual growth rate (CAGR)of approximately 15.99 % to reach USD 6.42 billion 2023 (Markets and Markets, 2020). Among the continents, north America represents the highest share of 41.6% of global market for biopesticides (Mordor Intelligence, 2020) while the fastest growth of market of biopesticides is in Latin America. Among three different types of biopesticides, microbial pesticides account for the highest i.e. 58% of the total pesticides market (Dunham and Trimmer, 2019). The use of biopesticides in the United States has been more than a century, but their application agriculture has been increasing currently. As of July 2018, there are 366 biopesticide active ingredients (biochemical and microbial) registered by US Environmental Protection Agency (EPA) which comprises bio-insecticides, acaricides, nematicides, fungicides, bactericides, herbicides, behavior-modifying biochemicals and plant health promoters (US EPA, 2020).

8.3 Factors for increasing trends toward biopesticides

The growing demand of organic foods among consumers and their awareness toward the safety regarding the harmful effects of chemical pesticide residue in foods is one of the major factors contributing the growth of biopesticide market. Similarly, the demand of natural pesticides and competitiveness among the industries is also one of the key factors. Similarly, the stringent regulatory actions implemented by several countries on the use of chemical pesticides have helped for the growth of biopesticides industries (Global Market Insights, 2020). In the United States, EPA has eased in registration of biopesticides requiring less data while it requires a lot of data related to the safe use of conventional chemical pesticides (US EPA, 2020). This has provided an incentive to the companies to shift their research in the development of biopesticides rather than conventional chemical pesticides.

Sustainable agriculture relies on the appropriate use of integrated crop management measures which are mediated predominantly by using plant genetic resistances, cultural approach, and crop protection management practices. One of the key components of crop management is the use of chemical pesticides, however, it possesses risks to the human and the environment. Chemical pesticides have undesirable targeted and non-targeted effects on the living organisms. With increasing issues and restrictions in the use of chemicals, there is an immense opportunity in the utilization of biopesticides with integrated crop management approaches for sustainable agricultural crop production.

8.4 Constraints for the applications of biopesticides

Although, biopesticides have a wide range of beneficial properties which makes them favorable in pest management, however, there are some disadvantages which should be taken into consideration while application of biopesticides. These include:

- Slower killing rate than conventional pesticides
- Short shelf life
- Less stability which causes rapid degradation of biopesticides.
- Frequency of application higher compared to conventional pesticides.
- High cost and limited availability to the farmers.

8.5 Role of genetic engineering in context of biopesticides

The constraints posed by the biopesticides for their effective implementation in agricultural practices can be addressed by the use genetic engineering in the development and production of biopesticides. Advancement in the genetics, genomics and molecular biology are great resources for reconstructing biopesticides and their extensive uses. Recent technological advancements in genetics and genomics have led to a better understanding of the evolution of natural microbial enemies, host-pathogen interactions, gene-gene interactions, proteomics, and microbial secretions (Marrone, 2019; Anderson et al., 2019). Genetic engineering and synthetic biology have emerged as a powerful tool to overcome barriers of biopesticides developments. Some of the advancements in the biopesticide development due to advanced genetics, genomics, and molecular biology are explained in the following section for each group of microorganisms, biochemical pesticides, and plant-incorporated protectants.

8.5.1 *Bacillus thuringiensis* (Bt)

Bacillus thuringiensis is a common bacterium found in a wide range of environments from desert to tundra region (Niederhuber, 2015). It produces a crystal protein toxic on several lepidopteran insects targeted specifically to the host. Cry protein when reaches the gut of the insects releases a toxic protein after finding the receptor successfully in the insect gut (Niederhuber, 2015). This leads to a series of events ultimately forming holes in cell membrane and the destructs gut region. As a secondary effect, it also leads to the growth of the other bacterium in the degraded region of the insect gut. This mechanism has been successfully decoded with the advancements in genetics and molecular biology and has been a great success in managing insect pests within vast majority of crops.

Bt crops have been successfully adopted by growers from several countries for a variety of crops. In the United States, Bt cotton against insects such as boll weevil, mirid bug, cotton leaf perforators, cabbage loopers etc. has been managed successfully. Similarly, Bt eggplant in Bangladesh and the Philippines and Bt corn in Brazil are some of the successful adoptions of Bt crops for insect-pests management (Anderson et al., 2019). Historically genetically engineered crops have been developed expressing proteins from Bt however, technological recent advancements in genetic engineering have led to new traits through RNAi and expression of similar proteins from other non-Bt sources (ISAAA, 2019). Additionally, with advancements in genetics and molecular biology, the mechanism of genes identifying specific functions such as genes responsible for insect toxicity and cry protein-producing genes has been much easier. Furthermore, the genetic manipulation of these effective genes by attaching it to the effective promoter and overexpression of the genes has been successful strategies. Besides, identification of other genes that have similar toxic properties and expression of several effective genes against different pests has resulted in the multi-purpose pesticidal effects (Carlton, 1985; Xu et al., 2019).

Bt derived genetically engineered crops have been a great success and has also been well adopted by growers in different countries. However, there are several concerns related to the use of Bt derived crops which might impact crop production. Some of the challenges which limit the effectiveness of Bt derived crops are the evolution of resistance in the targeted species which is brought by the over-reliance on the Bt crops (Anderson et al., 2019). Monotonous usage of Bt derived crops without appropriate integrated practices might lead to the development of resistance through selection pressure. For example, Cry1Ab corn resistance had been developed in African Stalk Borer, Cry1F corn resistance had been reported in the fall armyworm, Cry1Ac cotton resistance had been reported in pink bollworm, etc (Van Rensberg, 2007; Huang et al., 2014; Dhurua and Gujar, 2011). Another challenge with Bt crops is the increased impact of secondary pests which might gradually adapt to the crop and become a primary damaging pest (Anderson et al., 2019). For example, the adoption of Bt cotton increased mirid bugs which were used to be managed by chemicals before Bt crops were adopted (Lu et al., 2010). The development of multi-purpose pesticidal effects through expressing several resistant genes and reduction in over-reliance on Bt crops along with strict adoption of an integrated approach can be useful in mitigating such challenges.

8.5.2 Entomopathogenic nematodes (EPNs)

Entomopathogenic nematodes such as *Heterorhabditis* and *Steinernema* has been successfully used as a biocontrol agent in crops (Torres-Barragan et al., 2011; Dillman et al., 2012; Dillman and Sternberg, 2012; Kaya and Gaugler, 1993; Lu et al., 2016). Mechanism of the pathogenesis with EPNs can be mediated by their association with symbiotic bacteria and host mortality within 72 h of infection (Dillman et al., 2012). However, the crucial aspects of locating the host cells and gaining entry are the most important step in the pathogenesis which is a major role of EPNs (Dillman et al., 2012; Lu et al., 2016).

For the successful deployment of the EPNs as a biocontrol agent in the crops, several important criteria have to be successfully met such as the ability to locate the hosts, the ability to infect the hosts, and the ability to survive in the soil (Burnell and Dowds, 1996; Lu et al., 2016). Also, the complexity of the associated mechanisms with other microorganisms such as bacteria has to be well studied. With regards to EPNs, the recent developments in the genomics have led to the genome sequencing of the agriculturally important EPNs *Heterorhabditis bacteriophora* and *Steinernema carpocapsae* which has been a great breakthrough in the studies related to the pathogenesis of EPNs (Bai et al., 2013; Pena et al., 2015).

Host cell detection is a complex trait as it involves several heritable genes as well as the possible interactions of these genes with the environment. Genetic data will be an important tool to determine the role of the genes responsible for the host cell detection which will improve the efficacy of the biocontrol agents. With current technologies, researchers can also be able to selectively breed for the nematodes with desirables traits and incorporate favorable alleles for the improvement of the host cell detection trait (Gaugler et al., 1991; Lu et al., 2016). Similarly, the ability of nematode survival can also be improved through the knowledge of the genome of these nematodes. Gene detection and their gene regulatory networks can be utilized to improve the ability of the survival of these nematodes (Bai et al., 2013; Pena et al., 2015). The selective breeding approach has been applied in several EPNs such as *H. bacteriophora* for improving the heat tolerance trait in these nematodes (Ehlers et al., 2005; Bal et al., 2014). Although genetics and genomics would be a great tool to detect genes and improve the trait through selective breeding, it would be important and equally challenging to determine the impacts of in the other traits while improving one trait. For instance, heat tolerance trait improvement can compromise other traits such as host-seeking, penetration, and virulence (Anbesse et al., 2013). Hence, the interaction of the important trait regulatory genes and their multiple screening in the field setting is an important aspect for successful application of EPNs.

8.5.3 Baculoviruses

Baculoviruses or NPV are potential biocontrol agents that can be infectious to several insect and pests. Baculoviruses are rod-shaped viruses (Baculo rod) and appear to occur sorbed with proteinaceous crystals known as polyhedra on plants and soils, which when feed upon by insects results in feeding cessation and ultimately leads to insect death (Mishra, 1998; Szewczyk et al., 2011; Popham et al., 2016). A major beneficial aspect of the mechanism of Baculoviruses is its inability to infect arthropod and humans (Kost et al., 2005; Popham et al., 2016). Molecular techniques and genetic engineering have established baculoviruses based expression vectors as an important epitome of viruses as biopesticides (Kost et al., 2005; Szewczyk et al., 2011). From the pragmatic standpoint, baculovirus based biopesticides are applied in the form of spraying the concentrated virus that can suppress the pest along with the establishment of virus for subsequent generations (Szewczyk et al., 2011; Popham et al., 2016).

Virus based biopesticide was applied in the trade name of Elcar™ for insects of the genus *Helicoverpa* and *Heliothis* in crops such as soybean, sorghum, maize, and tomato bean since 1975. (Ignoffo and Couch, 1981). Genomes of these viruses have led to a profound understanding of the genetic variability of these viruses resulting due to several genetic mechanisms such as point mutation, insertions, and deletions (Krell, 1996). Knowledge of transposable elements has further improved the understanding of mutations in these viruses which has led its utilization in genetic engineering and pragmatic use as biopesticides (Szewczyk et al., 2011).

Genetic modifications of baculovirus resulting from the expression of exogenous toxins have also led to the development of recombinant constructs that can code for the toxin of the particular insect (Szewczyk et al., 2011). For instance, a gene that codes for a toxin from a scorpion species, *Androctonus australis*, has led to the development of a construct that can code for the toxin (Inceoglu et al., 2007). Virus based biopesticide can be a safer alternative to chemical protection of the crops, however, profound knowledge of the cloned genes jumping from one to another organism and its subsequent impacts to the biodiversity needs to be well studied (Inceoglu et al., 2001, 2007; Jin et al., 2019; Liu et al., 2019).

8.5.4 RNAi based biopesticides

RNA interference-based biopesticides have shown to be promising tools for the management of insect pests. RNAi is a naturally occurring and conserved innate phenomenon for the regulation and functioning of genes along with the defense mechanism against pathogens (Sledz and Williams, 2005; Muhammad et al., 2019). This mechanism is resultant of the formation of interfering molecules (small interfering RNAs; siRNAs or microRNAs) by dicer enzyme causing sequence-specific silencing of target genes (Fletcher et al., 2020; Mamta and Rajan, 2017). It has proven to be an efficient, environmentally friendly, and flexible pest management technique.

Several functional genomics studies use this technique for silencing the target genes through feeding bacteria expressing dsRNA to the target region and also for examining gene function in several insects (Tenllado et al., 2004; Baulcombe, 2004; Mao et al., 2007; FSANZ, 2013; Christiaens et al., 2018). Through different research studies, researchers have reported impairment of the insects through feeding on the leaf material expressing dsRNA. An orally applied RNAi method against western corn rootworm and expression of RNAi associated with cytochrome P450 gene against cotton bollworm successfully reported the impairment of the insects (Baum et al., 2007; Mao et al., 2007).

Insect pests of several crops such as corn, potato, apple have been successfully managed through this technique. However, targeting off-target regions by RNAi based approach has been a common drawback of this approach (Christiaens et al., 2018; Fletcher et al., 2020). Nevertheless, with reduction in the sequencing costs, genome sequencing of beneficial species decoding their underlying genetic information has been easier. This will provide more information about the pattern of off-target impacts and has the potential to resolve the issues. Besides, whole-genome sequencing will provide more efficiency and broaden the understanding of RNAi machinery which can improve the status of the use of the RNAi approach as the potential effective biopesticides (Fletcher et al., 2020).

8.5.5 Plant-incorporated protectants (PIPs)

GM crops especially herbicide and insect-resistant crops are popular among the growers in the United States and Canada which facilitates effective weeds and insect pest's management strategies (Bonny, 2016; Nandula, 2019). However, the use of such GM crops often comes in conjunction with the use of the herbicides that can selectively kill weeds other than crops. This will increase demand for the chemical herbicide designed as per the objective and also have adverse effects on the environment along with the development of herbicide-tolerant weeds. Most of the common herbicide includes glyphosate and glufosinate where glyphosate has been by far the most used herbicide in GM crops (Funke et al., 2006; Bonny, 2016).

Glyphosate, an active ingredient of Roundup inhibits an enzyme 5-enolpyruvylshikimate-3- phosphate synthase affecting the shikimate pathway (Funke et al., 2006). With a huge investment in research for about three decades, herbicide-resistant crops have been commercialized made possible by the advancement in the genetics and molecular biology techniques. Since 1996, herbicide-resistant crops have transformed the management of weeds in crops such as soybean, corn, and cotton (Green and Owen, 2011). Currently, the innovation from this breakthrough has been adopted by 94 % in soybeans, 92% in corn, and 94% in cotton in the United States (US FDA, 2020).

Since 1980, researchers sought on identifying glyphosate-tolerant enzymes and genes responsible for such traits (Funke et al., 2006; Pollegioni et al., 2011). Glyphosate tolerant microbes were subsequently identified in Agrobacterium sp. CP4 strain and few other microbes from glyphosate rich environment (Barry et al., 1992). A breakthrough was made by identifying 5-enolpyruvylshikimate-3- phosphate synthase which was used for genetically modifying crops (Barry et al., 1992, 1997). Currently, roundup ready crops contain Agrobacterium CP4 derived gene which encodes the glyphosate-tolerant enzyme. When these genes are expressed in the glyphosate-resistant crops, it facilitates effective weed control through post-emergence herbicide uses (Funke et al., 2006; Pollegioni et al., 2011; Nandula, 2019).

The adoption of GM crops itself has been a disputable issue and requires strict regulations. Besides, numerous issues have been encountered after the popular adoption of glyphosate-resistant crops in North America (Green and Owen, 2011; Bonny, 2016). First and foremost, it has threatened the sustainability of GM crops and glyphosate itself. Glyphosate has been the victim of its success through monotonous usage. It has also led to the fact that an agricultural management system should not rely solely on a single management aspect. Another important evolving issue is the development of herbicide-resistant weed species (Bonny, 2016). 262 species of herbicide-resistant weeds have been reported from 92 crops from 70 different countries where a significant portion of which emerged after the adoption of GM crops (International Herbicide-Resistant Weed Database). This has led to the evolution of more competitive and prolific weeds species emphasizing the further need for research on newly evolved species. Also, intensive research on the herbicide with different modes of action that can be utilized as an alternative is inevitable (Funke et al., 2006; Pollegioni et al., 2011). Rapidly advancing molecular science has given hope to researchers as it is becoming more convenient and cheaper for the genome sequencing of an organism. Genome sequences of the crops, weeds, and pathogens will be an important knowledge pool to explore the molecular mechanism of insect pests and weed resistance (Bonny, 2016; Nandula, 2019).

8.5.6 Entomopathogenic fungi

Entomopathogenic fungi have immense potential in managing pests in agriculture. Fungi usually have an advantage over other biopesticides because of their broad host range where a single isolate has been able to manage several species of insects and pests (Lee et al., 2018). Fungi when attached to the target insects germinates and uses the insect as the source of energy causes degradation of insect's exogenous layers which ultimately leads to desiccation and death (Rustiguel et al., 2017; Wang and Wang, 2017).

Several entomopathogenic fungi such as *B. bassiana* and *Metarhizium* has been robustly studied for their gene functions and their mode of actions (Lee et al., 2018; Butt et al., 2016). With genetics and molecular studies, genes for several functions such as adhesion, cuticle degrading, stress, adaptation, nutrients assimilation, and vast groups of transcription factors have been well studied (Butt et al., 2016). Whole-genome sequencing and the profound understanding of the function and regulation of these genes were possible by broadened knowledge of genomics and recent advancement in molecular biology (Lee et al., 2018). Furthermore, comparative studies of the genome of the different isolates have been able to determine the difference in the pathogenesis and their differential genes related to differential pathogenesis of each isolate. For instance, the genome of 10 isolates of the fungi *B. bassiana* has been released and many studies have been conducted on comparative genome analysis of *B. bassiana* which has decoded the understanding of the different isolates of the fungi (Lee et al., 2018). With further technological advancements in molecular biology in the current era, more studies are focused on whole-genome resequencing, gene annotation, comparative genetics, biological characterization, gene expression, sequence variations, de novo genome assembly, etc. These research studies are promising in elevating our understanding of gene function which will help to design highly effective entomopathogenic fungi for the management of insect pests.

8.5.7 Botanical biopesticides

Research studies have proven that the secondary metabolites produced by plants to be effective against insect pests and have the potential to be used as biopesticides. For instance, *C. cinerariaefolium* derived compound Pyrethrin was found to be an effective biopesticide that promoted the production of synthetic pyrethrins (Villaverde et al., 2016; Casida, 2011). Plant-derived essential oils have been another crucial botanical pesticide that is mediated by compounds such as terpenoids or phenylpropanoids rich secondary metabolites (Pavela and Benelli, 2016). Neem *Azadirachta indica* derived insecticidal compound has been the most widely used insecticide and has been popular in insect pest's management (Chaudhary et al., 2017). Neem based biopesticide is an excellent antifeedant, repellant, repugnant, and sterility inducing agents. Due to the limitation of genetics and genomics resources, there had not been a detailed molecular study of genes of Neem (Chaudhary et al., 2017; Kuravadi et al., 2015). However, the rapid development in molecular biology has made it possible to study these genes at molecular levels. Next-generation sequencing techniques thorough genome sequencing and transcriptomics has been able to identify several putative genes for the broader functionality as biopesticides in Neem (Kuravadi et al., 2015).

The molecular approach has also been able to identify the genes responsible in the secondary metabolites pathways which include farnesyl diphosphate synthase, squalene synthase, geranyl diphosphate synthase, mevalonate kinase, etc (Chaudhary et al., 2017; Kuravadi et al., 2015). Rapidly advancing molecular biology techniques and gene knockouts tools such as the CRISPR technique will be a great tool to decode the functionality of the genes responsible for the secondary metabolites. This will help to understand the biochemical mechanism of secondary metabolites and will trigger for synthetic production of those compounds which can have a broader and pragmatic application as biopesticides. Similarly, more research on the molecular analysis of the genes and their pathway is required to annotate the gene functions which will enhance the use of botanical biopesticides as an important strategy of insect pest management.

8.6 Conclusion

Biopesticide is comparatively safer strategy than the use of chemicals or chemical derived pesticides posing low risks to humans and the environment. The major advantage of the use of biopesticide for crop pest's management is environment safety along with their host specificity. Besides, the associated costs for the development and the registration of the biopesticide are comparatively lesser than that of the chemical pesticides. Besides, due to the continuous evolution of the resistance to chemicals and legislations being more restrictive, there is a need for intensive research on the biopesticide development both from the private and the public sectors.

Nevertheless, there are several issues with the development and use of biopesticides in agriculture. Some of the major issues include development of the resistances in the target organism, evolution of the pests, shifts in the pest type, changes in the doses, evaluation of new species, etc. Further research using genetic engineering tools along with synthetic biology will be important to determine the basis of the gene actions and the evolution of the resistance of the target organism. CRISPR/Cas9 is an important technique that has the potential for the development of genetically modified biological control organisms. The use of genome sequencing data along with proteomics data will help to identify genes, gene regulatory actions and has potential to explore the genetic basis of resistance. Further research on the chemical modes of actions of biopesticide can resolve issues of host complexity and functionality of the biopesticides in different targeted and non-targeted organisms. Research on the precision application of the biopesticide by utilizing big data acquired through different sensors and the drone-based system might warrant the effective utilization of biopesticides. Reliance on a few strategies for pest management cannot warrant the successful crop production while a holistic approach of compatible measures is important for successful pest management. A successful biopesticide development can only be obtained through interdisciplinary research involving genetics, genomics, molecular biology, biochemistry, agronomy, plant pathology, physiology, ecology, and social sciences.

References

Abudulai, M., Shepard, B.M., Mitchell, P.L., 2001. Parasitism and predation on eggs of *Leptoglossus phyllopus* (L.) (Hemiptera: Coreidae) in cowpea: impact of endosulfan sprays. J. Agric. Urban Entomol. 18 (2), 105–115.

Alvarenga, E.S.D., Carneiro, V.M., Resende, G.C., Picanço, M.C., Farias, E.D.S., Lopes, M.C., 2012. Synthesis and insecticidal activity of an oxabicyclolactone and novel pyrethroids. Molecules 17 (12), 13989–14001.

Al-Zaidi, A.A., Elhag, E.A., Al-Otaibi, S.H., Baig, M.B., 2011. Negative effects of pesticides on the environment and the farmers awareness in Saudi Arabia: a case study. J. Anim. & Plant Sci. 21 (3), 605–611.

Anbesse, S., Sumaya, N.H., Dorfler, A.V., Strauch, O., Ehlers, R., 2013. Stabilization of heat tolerance traits in *Heterorhabditis bacteriophora* through selective breeding and creation of inbred lines in liquid culture. Biocontrol 58, 85–93.

Anderson, J., Ellsworth, P.C., Faria, J.C., Head, G.P., Owen, M.D., Pilcher, C.D., Meissle, M., 2019. Genetically engineered crops: importance of diversified integrated pest management for agricultural sustainability. Front. Bioeng. Biotechnol. 7 (24).

Arthurs, S., Dara, S.K., 2019. Microbial biopesticides for invertebrate pests and their markets in the United States. J. Invertebr. Pathol. 165, 13–21.

Assessment, M.E., 2005. Ecosystems and Human Well-Being, vol. 5. Island press, Washington, DC, p. 563.

Bai, X., Adams, B.J., Ciche, T.A., Clifton, S., Gaugler, R., Kim, K., Spieth, J., Sternberg, P.W., Wilson, R.K., Grewal, P.S., 2013. A lover and a fighter: the genome sequence of an entomopathogenic nematode *Heterorhabditis bacteriophora*. PloS One 8 (7), e69618.

Bal, H.K., Michel, A.P., Grewal, P.S., 2014. Genetic selection of the ambush foraging entomopathogenic nematode, *Steinernema carpocapsae* for enhanced dispersal and its associated trade-offs. Evol. Ecol. 28, 923–939.

Barry, G.F., Kishmore, G.M., Padgette, S.R., 1992. A Herbicidal Composition Comprising Glyphosate in the Form of a Mixture of the Potassium and Ammonium Salts, 92/04449. International Patent.

Barry, G.F., Kishore, G.M., Padgette, S.R., Stallings, W.C., 1997. Glyphosate-tolerant 5-Enolpyruvylshikimate-3-Phosphate Synthases, 5633435. US Patent.

Baulcombe, D., 2004. RNA silencing in plants. Nature 431, 356–363. https://doi.org/10.1038/nature02874.

Baum, J.A., Bogaert, T., Clinton, W., Heck, G.R., Feldmann, P., Ilagan, O., Johnson, S., Plaentinck, G., Munyikwa, T., Pleau, M., Vaughn, T., Roberts, J., 2007. Control of coleopteran insect pests through RNA interference. Nat. Biotechnol. 25, 1322–1326. https://doi.org/10.1038/nbt1359.

Bonny, S., 2016. Genetically modified herbicide-tolerant crops, weeds, and herbicides: overview and impact. Environ. Manag. 57, 31–48.

Burnell, A.M., Dowds, B.C.A., 1996. The genetic improvement of entomopathogenic nematodes and their symbiont bacteria: phenotypic targets, genetic limitations and an assessment of possible hazards. Biocontrol Sci. Technol. 6, 435–447.

Butt, T., Coates, C., Dubovskiy, I., Ratcliffe, N., 2016. Entomopathogenic fungi: new insights into host–pathogen interactions. Adv. Genet. 94, 307–364.

Carlton, B.C., 1985. Genetic engineering of microbial pesticides. Windsor Locks, CT. Gen. Tech. Rep. NE-100. In: Grimble, D.G., Lewis, F.B. (Eds.), Proceedings, Symposium: Microbial Control of Spruce Budworms and Gypsy Moths; 1984 April 10-12. U.S. Department of Agriculture, Forest Service, Northeastern Forest Experiment Station, Broomall, PA, pp. 133–136.

Casida, J.E., 2011. Curious about pesticide action. J. Agric. Food Chem. 59 (7), 2762–2769. https://doi.org/10.1021/jf102111s.

Chandler, D., Bailey, A.S., Tatchell, G.M., Davidson, G., Greaves, J., Grant, W.P., 2011. The development, regulation and use of biopesticides for integrated pest management. Phil. Trans. Biol. Sci. 366 (1573), 1987–1998.

Chaudhary, S., Kanwar, R.K., Sehgal, A., Cahill, D.M., Barrow, C.J., Sehgal, R., Kanwar, J.R., 2017. Progress on *Azadirachta indica* based biopesticides in replacing synthetic toxic pesticides. Front. Plant Sci. 8, 610.

Christiaens, O., Dzhambazova, T., Kostov, K., Arpaia, S., Joga, M.R., Urru, I., Sweet, J., Smagghe, G., 2018. Literature Review of Baseline Information on RNAi to Support the Environmental Risk Assessment of RNAi-Based GM Plants. EFSA Supporting Publications. https://doi.org/10.2903/sp.efsa.2018.EN-1424.

Clements, D.R., Ditommaso, A., 2011. Climate change and weed adaptation: can evolution of invasive plants lead to greater range expansion than forecasted? Weed Res. 51 (3), 227–240.

Dean, D.H., 1984. Biochemical genetics of the bacterial insect-control agent *Bacillus thuringiensis*: basic principles and prospects for genetic engineering. Biotechnol. Genet. Eng. Rev. 2 (1), 341–363.

Dhurua, S., Gujar, G.T., 2011. Field-evolved resistance to Bt toxin Cry1Ac in the pink bollworm, *Pectinophora gossypiella* (Saunders) (Lepidoptera: *Gelechiidae*), from India. Pest Manag. Sci. 67, 898–903. https://doi.org/10.1002/ps.2127.

Dillman, A.R., Sternberg, P.W., 2012. Entomopathogenic nematodes. Curr. Biol. 22, R430–R431.

Dillman, A.R., Chaston, J.M., Adams, B.J., Ciche, T.A., Goodrich-Blair, H., Stock, S.P., Sternerg, P.W., 2012. An entomopathogenic nematode by any other name. PLoS Pathog. 8, e1002527.

Dunham, W.C., DunhamTrimmer, L.L.C., October 2015. Evolution and future of biocontrol. In: 10th Annual Biocontrol Industry Meeting (ABIM), Basel, Switzerland, October 20th.

Dunham, V., Trimmer, M., 2019. Biological products around the world. Bioproducts Industry Alliance Spring Meeting & International Symposium (www.bpia.org) 20.

Ehlers, R.U., Oestergaard, J., Hollmer, S., Wingen, M., Strauch, O., 2005. Genetic selection for heat tolerance and low temperature activity of the entomopathogenic nematode-bacterium complex *Heterorhabditis bacteriophora-Photorhabdus luminescens*. Biocontrol 50, 699–716.

Fletcher, S.J., Reeves, P.T., Hoang, B.T., Mitter, N., 2020. A perspective on RNAi-based biopesticides. Front. Plant Sci. 11, 51. https://doi.org/10.3389/fpls.2020.00051.

Food Standards Australia New Zealand, 2013. Response to Heinemann et al. on the Regulation of GM Crops and Foods Developed using Gene Silencing.

Fradin, M.S., Day, J.F., 2002. Comparative efficacy of insect repellents against mosquito bites. N. Engl. J. Med. 347 (1), 13–18.

Funke, T., Han, H., Healy-Fried, M.L., Fischer, M., Schonbrunn, E., 2006. Molecular basis for the herbicide resistance of Roundup Ready crops. Proc. Natl. Acad. Sci. U. S. A. 103 (35), 13010–13015.

Gaugler, R., Campbell, J.F., Gupta, P., 1991. Characterization and basis of enhanced host-finding in a genetically improved strain of *Steinernema carpocapsae*. J. Invertebr. Pathol. 57, 234–241.

Global Market Insights, 2020. Biopesticides Market Size by Product (Bioherbicides, Bioinsecticides, Biofungicides), by Application (Seed Treatment, Foliar, Soil Spray), by Crop (Grains & Oil Seeds, Fruits & Vegetables), by Form (Dry, Liquid), by Source (Microbial, Biochemical), Industry Analysis Report, Country Outlook, Growth Potential, Price Trends, Competitive Market Share & Forecast, pp. 2020–2026. https://www.gminsights.com/industry-analysis/biopesticides-market.

Green, J.M., Owen, M.D.K., 2011. Herbicide-resistant crops: utilities and limitations for herbicide-resistant weed management. J. Agric. Food Chem. 59 (11), 5819–5829.

Huang, F., Qureshi, J.A., Meagher Jr., R.L., Reisig, D.D., Head, G.P., Andow, D.A., Ni, X., Kerns, D., Buntin, D., Niu, Y., Yang, F., Dangal, V., 2014. Cry1F resistance in fall armyworm *Spodoptera frugiperda*: single gene versus pyramided Bt maize. PloS One 9, e112958.

Ignoffo, C.M., Couch, T.L., 1981. The nucleopolyhedrosis virus of *Heliothis* species as a microbial pesticide. In: Burges, H.D. (Ed.), Microbial Control of Pests and Plant Diseases. Academic Press, London, pp. 329–362.

Inceoglu, A.B., Kamita, S.G., Hammock, B.D., 2007. Genetically modified baculoviruses: a historical overview and future outlook. Adv. Virus Res. 68, 323–360.

Inceoglu, A.B., Kamita, S.G., Hinton, A.C., Huang, Q., Severson, T.F., Kang, K., Hammock, B.D., 2001. Recombinant baculoviruses for insect control. Pest Manag. Sci. 57 (10), 981–987.

International Herbicide-Resistant Weed Database, n.d. (Accessed on March 20, 2020). http://www.weedscience.org/Home.aspx.

ISAAA, 2019. ISAAA's GM Approval Database. ISAAA. https://www.isaaa.org/gmapprovaldatabase/default.asp.

Isman, M.B., 2006. Botanical insecticides, deterrents, and repellents in modern agriculture and an increasingly regulated world. Annu. Rev. Entomol. 51, 45–66.

Jin, S., Clark, B., Kuznesofa, S., Lin, X., Frewer, L.J., September 2019. Synthetic biology applied in the agri-food sector: public perceptions, attitudes and implications for future studies. Trends Food Sci. Technol. 91, 454–466.

Jurat-Fuentes, J.L., Crickmore, N., 2017. Specificity determinants for cry insecticidal proteins: insights from their mode of action. J. Invertebr. Pathol. 142, 5–10.

Kachhawa, 2017. Microorganisms as a biopesticides. J. Entomol. Zool. Stud. 5 (3), 468–473.

Kaya, H.K., Gaugler, R., 1993. Entomopathogenic nematodes. Annu. Rev. Entomol. 38, 181–206.

Kost, T.A., Condreay, J.P., Jarvis, D.L., 2005. Baculovirus as versatile vectors for protein expression in insect and mammalian cells. Nat. Biotechnol. 23, 567–575.

Krell, P.J., 1996. Passage effect of virus infection in insect cells. Cytotechnology 20, 125–137.

Kumar, S., 2012. Biopesticides: a need for food and environmental safety. J. Biofert. Biopestic. 3 (4), 1–3.

Kuravadi, N.A., Yenagi, V., Rangiah, K., Mahesh, H.B., Rajamani, A., Shirke, M.D., Russiachand, H., Loganathan, R.M., Lingu, C.S., Siddappa, S., Ramamurthy, A., Sathyanarayana, B.N., Gowda, M., 2015. Comprehensive analyses of genomes, transcriptomes and metabolites of neem tree. Peer J. https://doi.org/10.7717/peerj.1066.

Lee, S. J, Lee, M.R., Kim, S., Kim, J.C., Park, S.E., Li, D., Kim, J.S., 2018. Genomic analysis of the insect-killing fungus Beauveria bassiana JEF-007 as a biopesticide. Sci. Rep. 8 (1), 1–12.

Lee, S.J., Lee, M.R., Kim, S., Kim, J.C., Park, S.E., Li, D., Shin, T.Y., Nai, Y.S., Kim, J.S., 2018. Genomic analysis of the insect- killing fungus *Beauveria bassiana* JEF-007 as a Biopesticide. Nat. Sci. Rep. 8, 12388.

Liu, X., Cao, A., Yan, D., Ouyang, C., Wang, Q., Li, Y., 2019. Overview of mechanisms and uses of biopesticides. Int. J. Pest Manag. https://doi.org/10.1080/09670874.2019.1664789.

Lu, D., Baiocchi, T., Dillman, A.R., 2016. Genomics of entomopathogenic nematodes and implications for pest control. Trends Parasitol. 32 (8), 588–598.

Lu, Y., Wu, K., Jiang, Y., Xia, B., Li, P., Feng, H., et al., 2010. Mirid bug outbreaks in multiple crops correlated with wide-scale adoption of Bt cotton in China. Science 328, 1151–1154.

Mamta, B., Rajan, M.V., 2017. RNAi technology: a new platform for crop pest control. Physiol. Mol. Biol. Plants 23 (3), 487–501.

Mao, Y.B., Cai, W.J., Wang, J.W., Hong, G.J., Tao, X.Y., Wang, L.J., et al., 2007. Silencing a cotton bollworm P450 monooxygenase gene by plant-mediated RNAi impairs larval tolerance of gossypol. Nat. Biotechnol. 25, 1307–1313. https://doi.org/10.1038/nbt1352.

Markets and Markets, 2020. Biopesticides Market - Global Forecast to 2022. By Type (Bioinsecticides, Biofungicides, Bioherbicides, and Bionematicides), Origin (Beneficial Insects, Microbials, Plant-Incorporated Protectants, and Biochemicals), Mode of Application, Formulation, Crop Type and Region. https://www.marketsandmarkets.com/Market-Reports/biopesticides-267.html.

Marrone, 2019. Pesticidal natural products—status and future potential. Pest Manag. Sci. 75 (9), 2325–2340.

Martin, L., Liparoti, S., Della Porta, G., Adami, R., Marqués, J.L., Urieta, J.S., Mainar, A.M., Reverchon, E., 2013. Rotenone coprecipitation with biodegradable polymers by supercritical assisted atomization. J. Supercrit. Fluids 81, 48–54.

Mazid, S., Kalita, J.C., Rajkhowa, R.C., 2011. A review on the use of biopesticides in insect pest management. Int. J. Adv. Sci. & Technol. 1 (7), 169–178.

Mehrotra, S., Kumar, S., Zahid, M., Garg, M., 2017. Biopesticides. In: Principles and Applications of Environmental Biotechnology for a Sustainable Future. Springer, Singapore, pp. 273–292.

Mishra, S., 1998. Baculoviruses as biopesticides. Curr. Sci. 75 (10).

Mordor Intelligence, 2020. Biopesticides Market-Growth, Trends, and Forecast (2020–2025). https://www.mordorintelligence.com/industry-reports/global-biopesticides-market-industry.

Muhammad, T., Zhang, F., Zhang, Y., Liang, Y., 2019. RNA Interference: a natural immune system of plants to counteract biotic stressors. Cells 8 (1), 38.

Nandula, V.K., 2019. Herbicide resistance traits in maize and soybean: current status and future outlook. Plants 8, 337.

Nicholson, G.M., 2007. Fighting the global pest problem: preface to the special Toxicon issue on insecticidal toxins and their potential for insect pest control. Toxicon 49 (4), 413–422.

Niederhuber, M., 2015. Insecticidal plants: the tech and safety of GM Bt crops. In: Genetically Modified Organism and Our Food. http://sitn.hms.harvard.edu/flash/2015/insecticidal-plants/. (Accessed 20 May 2020).

O'Brien, K.P., Franjevic, S., Jones, J., 2009. Green chemistry and sustainable agriculture: the role of biopesticides, advancing green chemistry. Ecology 90, 2223–2232.

Oerke, E.C., Dehne, H.W., Schönbeck, F., Weber, A., 2012. Crop Production and Crop Protection: Estimated Losses in Major Food and Cash Crops. Elsevier.

Paarlberg, R., 2013. Food Politics: What Everyone Needs to Know®. Oxford University Press.

Pavela, R., Benelli, G., 2016. Essential oils as ecofriendly biopesticides? Challenges and constraints. Trends Plant Sci. 21 (12).

Pena, J.M., Carrillo, M.A., Hallem, E.A., 2015. Variation in the susceptibility of *Drosophila* to different entomopathogenic nematodes. Infect. Immun. 83, 1130–1138.

Pollegioni, L., Schonbrunn, E., Siehl, D., 2011. Molecular basis of glyphosate resistance: different approaches through protein engineering. Federation Eur. Biochem. Soc. J. 278 (16), 2753–2766.

Popham, H.J.R., Nusawardani, T., Bonning, B.C., 2016. Introduction to the use of baculoviruses as biological insecticides. In: Murhammer, D. (Ed.), Baculovirus and Insect Cell Expression Protocols, Methods in Molecular Biology, vol. 1350. Humana Press, New York, NY.

Rustiguel, C.B., Fernández-Bravo, M., Guimarães, L.H.S., Quesada-Moraga, E., 2017. Different strategies to kill the host presented by *Metarhizium anisopliae* and *Beauveria bassiana*. Can. J. Microbiol. 64 (3), 191–200.

Sarwar, M., 2015. Information on activities regarding biochemical pesticides: an ecological friendly plant protection against insects. Int. J. Eng. & Adv. Res. Technol. 1 (2), 27–31.

Schmutterer, H., 1990. Properties and potential of natural pesticides from the neem tree, *Azadirachta indica*. Annu. Rev. Entomol. 35 (1), 271–297.

Schmutterer, H., Singh, R.P., 1995. List of Insect Pests Susceptible to Neem Products. The Neem Tree: Azadirachta indica A. *Juss* and Other *Meliaceae* Plants. VCH, New York, pp. 326–365.

Shelton, A.M., Tang, J.D., Roush, R.T., Metz, T.D., Earle, E.D., 2000. Field tests on managing resistance to Bt-engineered plants. Nat. Biotechnol. 18 (3), 339–342.

Sledz, C.A., Williams, B.R.G., 2005. RNA interference in biology and disease. Blood 106 (3), 787–794.

Szewczyk, B., De Souza, M.L., De Castro, M.E.B., Moscardi, M.L., Moscardi, F., 2011. Baculovirus biopesticides. In: Baculovirus Biopesticides, Pesticides - Formulations, Effects, Fate, Margarita Stoytcheva. IntechOpen.

Tenllado, F., Llave, C., Diaz-Ruiz, J.R., 2004. RNA interference as a new biotechnological tool for the control of virus diseases in plants. Virus Res. 102, 85–96. https://doi.org/10.1016/j.virusres.2004.01.019.

Torres-Barragan, A., Suazo, A., Buhler, W.G., Cardoza, Y.J., 2011. Studies on the entomopathogenicity and bacterial associates of the nematode *Oscheius carolinensis*. Biol. Contr. 59, 123–129. https://doi.org/10.1016/j.biocontrol.2011.05.020.

U.S. Environmental Protection Agency (EPA), 2020. What Are Biopesticides? https://www.epa.gov/ingredients-used-pesticide-products/what-are-biopesticides.

United States Food and Drug Administration (US FDA), 2020. GMO Crops, Animal Food, and Beyond.

Van Rensburg, J.B.J., 2007. First report of field resistance by stem borer, *Busseola fusca* (Fuller) to Bt-transgenic maize. S. Afr. J. Plant Soil 24, 147–151.

Villaverde, J.J., Sandín-España, P., Sevilla-Morán, B., López-Goti, C., Alonso-Prados, J.L., 2016. Biopesticides from natural products: current development, legislative framework. Future Trends BioResour. 11 (2), 5618–5640.

Wang, C., Wang, S., 2017. Insect pathogenic fungi: genomics, molecular interactions, and genetic improvements. Annu. Rev. Entomol. 62, 73–90.

Xu, Y., Wang, X., Chi, G., Tan, B., Wang, J., 2019. Effects of *Bacillus thuringiensis* genetic engineering on induced volatile organic compounds emission in maize and the attractiveness to a parasitic wasp. Front. Bioeng. & Biotechnol. 7, 160.

Chapter 9

Bacillus thuringiensis, a remarkable biopesticide: from lab to the field

Igor Henrique Sena da Silva, Marcelo Mueller de Freitas and Ricardo Antônio Polanczyk
Department of Agricultural Production Sciences, Paulista State University (Unesp), School of Agricultural and Veterinary Sciences, Jaboticabal, Sao Paulo, Brazil

9.1 Introduction

Pest control in crops of economic interest was based on biological principles that sought to predict the interaction between the host plant and the pest in the ecosystem (Ehler, 1998; El-Shafie, 2018). After World War II, with the development of synthetic insecticides (dichlorodiphenyltrichloroethane [DDT]), there was a significant global change in the philosophy of pest control, guided by the use of synthetic pesticides that initially had success in controlling several pests (El-Shafie, 2018; Pimentel and Burgess, 2014).

The intensive and inadequate use of these insecticides significantly reduced the action of natural enemies. It led to the reemergence of primary pests, secondary pest outbreaks, development of resistance to pesticides, and side effects on humans and other animals (Kogan, 1988). Since the 1960s, the use of sustainable alternative techniques has been discussed, and the concept of Integrated Pest Management (IPM) was consolidated with the application of several tactics alternative to chemical control and the provision of new incentives to research new pest management strategies (El-Shafie, 2018; Pimentel and Burgess, 2014).

The late 1990s were characterized by major advances in molecular biology through genetic engineering techniques for the development of plants with insect resistance characteristics, such as Bt crops expressing *Bacillus thuringiensis* (Bt) Cry toxins and the RNA interference technique (RNAi); which suppresses a target gene and interrupts a vital step in protein synthesis (Dias et al., 2020; Vogel et al., 2019). In recent decades, the Bt crops has been widely used due to many benefits, such as effective control of target insects, decreased use of chemical insecticides, and high specificity, significantly contributing to the increased world production of major crops (Tabashnik and Carrière, 2017).

Despite the benefits of Bt crops for pest management, this tactic alone is not sustainable since the development of specimens resistant to these technologies is inevitable and increasingly rapid (Anderson et al., 2019; Tabashnik and Carrière, 2019). The high cost of discovering new insecticide molecules (about $250 million) (Glare et al., 2012) and the growing reports of an arthropod pest population resistant to conventional insecticides are also considered (Sparks et al., 2020).

In this scenario, compared to chemical insecticides, Bt-based biopesticides stand out for their low cost of development (Glare et al., 2012), slower resistance development than Bt crops (De Bortoli and Jurat-Fuentes, 2019), and selectivity and specificity (Lacey et al., 2015; Kesho, 2020; Steinhaus, 1957). These biopesticides contain mixtures of spores and crystals, which are historically the main microbial products commercialized in the world, used particularly to control lepidopteran pests (Glare and O'Callagham, 2000; Lambert and Peferoen, 1992; Marrone, 2019; Salama, 1984; Sanahuja et al., 2011).

This chapter describes the main aspects that have been addressed by researchers in over 100 years of Bt research, from Cry pesticidal protein characterization responsible for much of the insecticidal activity of this bacterium to its compatibility with other control methods. It concludes with the main challenges Bt biopesticides face to remain and increase their importance in IPM.

Biopesticides. https://doi.org/10.1016/B978-0-12-823355-9.00021-3

9.2 Isolation and epizootic potential of *Bacillus thuringiensis* (Bt)

The isolation of *Bacillus thuringiensis* (Bt) from various substrates is an important step in the process of selecting promising isolates for pest control. Since it is an essential source of new toxins, being possible to select more virulent isolates, consequently increasing the efficiency and use of this microorganism within the context of IPM.

Bacillus thuringiensis is present in soil samples from annual or perennial crop areas, desert areas, aquatic environments, surface and interior parts of plants, plant debris, dead insects and small mammals, cobwebs, and stored grains (Aboussaid et al., 2011; Bernhard et al., 1997; Delgado-Silva et al., 2020; Forsyth and Logan, 2000; Khaleghi et al., 2019; Konecka et al., 2007; Martin and Travers, 1989; Panwar et al., 2018; Patel et al., 2013; Reyaz and Arulselvi, 2016; Thammasittirong and Attathom, 2008).

In quantitative terms, the studies by Martin and Travers (1989) and Bernhard et al. (1997) are noteworthy. In the first study, 1,115 soil samples from 30 countries were analyzed, and 8,916 Bt isolates were obtained from 785 samples, no Bt isolates were found in the other 300 samples. In the second study, 5,303 Bt isolates were obtained, of which 45% was from stored grains and 25% from soil samples. In this case, 2363 samples (from soils, dead insects, stored grains, and plants) from 80 countries were analyzed.

The methods used in Bt isolation are easy to perform (Saleh et al., 1969; Swiecicka et al., 2020; Travers et al., 1987; World Health Organization, 1988). The process involves (1) substrate selection, (2) sample dilution, (3) thermal shock to eliminate nonsporulating bacteria, (4) culture medium plating, (5) visualization of bacterial colonies, (6) liquid antibiotic culture (Penicillin 100 μg/L), and (7) Bt separation from *Bacillus cereus* under phase-contrast microscopy by the presence of crystal. Santana et al. (2008) recommended drying the soil sample for 5 h before isolation to obtain more Bt isolates.

The variation between methods is mainly due to the choice of growing medium, which depends mostly on the available resources. The number of Bt cells obtained from soil samples varies between 10^2 and 10^4 colony forming units (cfu) per gram of soil, while with plant samples, this number varies between 0 and 100 cfu/cm^2 (Damgaard, 2000).

Successful Bt isolation in soil samples may be related to the presence of insects in sample collection places, as it occurs in most tropical countries, facilitating the dispersion and multiplication of this pathogen (Hossain et al., 1997; Raymond, 2017). But the variation in Bt isolation success in various locations can have several reasons, including climatic conditions, agricultural activity, soil types, and isolation methods. These factors make comparisons between results difficult, but constitute a critical theoretical basis in studies on the ecological role of this bacterium in soil, and can predict the toxic potential of isolates in certain regions under specific environmental and soil conditions (Polanczyk, 2004).

Bacillus thuringiensis is not considered an entomopathogen with high epizootic capacity and does not always sporulate in insects before or after death. However, this entomopathogen causes natural epizootic diseases in insects (Burges and Hurst, 1977; Meadows et al., 1992; Porcar and Caballero, 2002; Talalaev, 1956; Vankova and Purrini, 1979). Natural epizootic were reported, especially in Lepidoptera, and occurs under certain specific conditions in the field, insect farms, and grain storage environments (Damgaard, 2000; Hansen and Salamitou, 2000; Raymond, 2017).

This entomopathogen can multiply in favorable microhabitats such as target insects, although it can also grow and sporulate in nutrient-rich soils. However, due to their low epizootic occurrence, it is unlikely that the primary source of toxins and spores are colonized insects. The Bt symbiotic capability in plants could explain the production of specific and efficient Cry pesticidal proteins against phytophagous insects (Aronson and Shai, 2001; Maksimov et al., 2018; Monnerat et al., 2009; Raymond, 2017).

The limitations of Bt persistence in the soil led to the hypothesis that its insecticidal activity is accidental (Martin and Travers, 1989), although Bt has a large arsenal of virulence factors (Malovichko et al., 2019; Raymond et al., 2010; Raymond, 2017). The fact that this bacterium is usually present in the environment regardless of the presence of insects supports this theory. Chak et al. (1994) obtained 93.5% of the total isolates in mountainous sites, where insects are rare compared to lower altitude areas. More recently, Paulino-Lima et al. (2012) isolated Bt from soil samples from the Atacama Desert in Chile, where insects are absent or extremely rare.

Meadows (1993) suggested four possible explanations for Bt presence in the soil: (1) Bt rarely develops in soil, but is deposited in this substrate by insects, leaves, and thus the soil would act as a reservoir of spores that could later be transported over long distances by the wind; (2) Bt may be pathogenic to soil insects of reduced economic importance, which have been scarcely studied; (3) Bt may develop in soil when there are enough nutrients, obtained mainly from decomposing organic waste; and (4) Bt affinity with *B. cereus*, with which it can exchange genetic material, allowing its permanence in the environment.

It should be noted that most studies evaluate the persistence of Cry pesticidal protein expressed in Bt crops. Longer Cry pesticidal protein persistence in the soil can accelerate the insect resistance to Bt crops (Clark et al., 2005; Icoz and Stotzky, 2008). Hung et al. (2016) evaluated the persistence of the biopesticides, Vi-Bt (Bt *kurstaki*) and Cry1Ac toxin in the soil and reported that the persistence of both treatments was influenced by sunlight and temperature and that the purified protein showed lower persistence due to its adsorption to the soil, while the persistence of the Bt biopesticide depends on spore germination in the soil and on the inert formulations that act as a protector against adverse conditions. Cry pesticidal protein bind to humic acids, organic supplements or soil particles that protect them from degradation by microorganisms without, however, losing their insecticide activity.

9.3 Nomenclature and characterization of *Bacillus thuringiensis* (Bt) Cry pesticidal proteins

Bacillus thuringiensis produces a wide range of substances that can be used for various purposes, including the control of fungal diseases in plants (Akram et al., 2013), bioremediation (Wu et al., 2013), and against cancer cells (Poornima et al., 2010). Also, this bacterium is well known for its ability to produce toxins with insecticidal activity, particularly those called Cry pesticidal protein, which has been the focus of researchers worldwide (Mendoza-Almanza et al., 2020; Schnepf et al., 1998). Due to this broad interest, after isolation, it is necessary to characterize the isolates with a direct mode of action and spectrum of action studies. In Brazilian public and private research institutions, the collections are estimated to contain about 10,000 Bt strains stored.

It should be highlighted that although the term *Bacillus thuringiensis* is used for only one species, it belongs to a group known as *Bacillus cereus* (Raymond, 2017). *B. thuringiensis* and *B. cereus* show common phenotypic and biochemical characteristics, however, by definition and agricultural interest, Bt can be differentiated by the presence of crystals (Luthy and Wolfersberger, 2000) visible in phase-contrast microscopy; although this criterion has a nominal taxonomic value (Lysenko, 1983; Raymond, 2017).

The similarity between Bt and *B. cereus* is due to the transfer of plasmids that encode the delta-endotoxins in Bt, therefore, changing Bt to *B. cereus*. Conversely, Bt may lose its ability to produce crystals; thus, "becoming *B. cereus*" (Hansen and Salamitou, 2000; Schnepf et al., 1998). Bagcõoglu et al. (2019) achieved 99.5% success in separating and identifying *Bacillus* spp. using unique spectroscopy (FTIR) associated with the neural network.

Serotyping was a widely used method to differentiate Bt groups; however, the determination of serotype/subspecies does not always predict insecticidal activity. De Barjac and Bonnefoi (1962) and Norris (1964) initiated the Bt serotyping, resulting in six and nine serotypes, respectively. Afterward, Lecadet et al. (1999) identified 82 serotypes. However, they discovered that there were some cross-reaction problems with several *B. cereus* isolates, self-agglutinating strains, and some Bt strains which for not producing crystals, were considered as *B. cereus*. Thus, the biochemical characterization of Bt is laborious, mainly because it shows response variations, and this type of description is often not associated with serotyping results.

Advances in molecular biology have allowed the development of DNA-based methods that can differentiate the inter and intra-species of Bt. These methods can differentiate strains and isolates and can also be used to determine the presence or absence of Cry genes that encode Cry toxins.

Described in the 1980s by Mullis and Faloona (1987), the Polymerase Chain Reaction (PCR) technique allows the in vitro obtainment of several copies of a given DNA segment. Due to the growing interest in this microorganism for agricultural pest control, Hofte and Whiteley (1989) proposed the first classification of Bt Cry pesticidal proteins based on the combination of amino acid sequences and insecticide activity. Consequently, 38 toxins were grouped into 14 different classes. The four main classes contained toxins with activity against Lepidoptera (I), Lepidoptera and Diptera (II), Coleoptera (III), and Diptera (IV). However, this scheme was problematic with its bid to relate toxins with similar amino acid sequences with different insecticide activities.

During the 1990s, a new classification was proposed by Crickmore et al. (1998), based only on the relationships between amino acid sequences. This change improved the relation between toxins and eliminated the need for bioassays on many insects. In the second classification, there were 250 Cry pesticidal proteins grouped into 40 groups. This number increased to 311 in 2005, 592 in 2011, 776 in 2016, and by June 2020, there were 810 Cry toxins (http://www.btnomenclature.info/). Recently, a new nomenclature was proposed and is available at https://www.bpprc.org/ (Crickmore et al., 2020).

Although characterization helps predict the insecticidal activity of Bt isolates (Bravo et al., 1998), another possibility is to test their virulence, generally estimated by CL_{50} (Shapiro-Ilan et al., 2005), without comparing the toxins with those previously described. It thereby led to the discovery of new Cry pesticidal proteins with insecticidal activity that could increase the Bt host spectrum and be used in resistance management.

9.4 Mode of action of *Bacillus thuringiensis* Cry toxins

Coleoptera, Diptera, Hymenoptera, Hemiptera, Isoptera, Lepidoptera and Orthoptera of agricultural importance are remarkable susceptible to Bt (Aboussaid et al., 2010; Bergamasco et al., 2013; Blanco et al., 2010; Buentello-Wong et al., 2015; Cao et al., 2020; De Oliveira Dorta et al., 2018; Kumar et al., 2016; Liu et al., 2017; Lone et al., 2017; Machado et al., 2020; Mushtaq et al., 2017; Oppert et al., 1994; Radosavljevic and Naimov, 2016; Schünemann et al., 2014; Wang et al., 2018; Zhang et al., 2013).

Bacillus thuringiensis biological activity has also been reported for some species of pest mites and nematodes such as *Ascaris suum* (Ascaridida: Ascarididae); *Blatta orientalis* (Blattodea Blattidae); *Leishmania major* (Trypanosomatida: Trypanosomatida); and Schistosoma *japonicum* (Strigeiformes: Schistosomatidae) (Amanchi and Hussain, 2008; Berlitz et al., 2013; El-Sadawy et al., 2008; Gutiérrez and Gonçalves, 2006; Porcar et al., 2006; Radwan, 2007; Silveira et al., 2011; Urban Jr. et al., 2013; Yu et al., 2015).

Despite the broad action spectrum of this bacterium, several aspects related to the mode of action of Cry pesticidal proteins remain unclear. The "classic model" presented in the 1990s states that for an insect to experience toxicity, the Bt crystal must be ingested, solubilized by the intestinal pH to release pro-toxins, cleaved to by proteases present in the intestinal lumen (proteolytic activation), bound to receptors, inserted in the membrane, and form a pore (Carroll and Ellar, 1993; Knowles, 1994). However, the first model used to describe the detailed mode of action of Cry pesticidal protein in insects at the molecular level, known as the sequential binding model, was proposed by Dr. Alejandra Bravo group in 2004 (Bravo et al., 2004). The second model, known as the protein kinase A (PKA) activation model, was proposed by Zhang et al. (2006).

Both models corroborate the steps used in the classical model up to pesticidal protein activation, as reviewed by Vachon et al. (2012). However, after activation, Dr. Bravo's model proposes a complex multi-step process involving the interaction of the pesticidal protein with at least two different receptors, one CAD-type, and the others anchored by GPI, aminopeptidase-N (APN), and alkaline phosphatase (ALP), triggering toxin oligomerization, oligomer insertion into the membrane, pore formation in intestinal cells, osmotic lysis, and insect death (Gómez et al., 2002; Bravo et al., 2007).

In the PKA activation model, after specifically joining the CAD, the Cry1Ab monomeric pesticidal protein starts an Mg^{2+} dependent signaling cascade. This cascade stimulates G protein and adenylate cyclase protein synthesis. It stimulates cyclic AMP (cAMP) accumulation in the cell, thus activating the PKA, which once activated, destabilizes cytoskeleton cells and the ionic channels of the membrane, leading to cell apoptosis or programmed cell death (Zhang et al., 2005, 2006). Additionally, another model was proposed by Jurat-Fuentes and Adang (2006) to explain the mode of action of the Cry1Ac pesticidal protein in *Chloridea virescens* (Lepidoptera: Noctuidae). This model suggests that cytotoxicity is due to the combined effects of osmotic lysis and cellular signaling and, therefore, Bravo and Zhang model elements were incorporated into the mode of action.

The main techniques currently used for studies on the Bt mode of action consist of molecular tools, such as RNA interference (RNAi) and CRISPR (Gómez et al., 2018; Guo et al., 2018). Both tools "knockout" the possible receptors located in the intestinal epithelium of the insect and proves their participation in Cry pesticidal protein toxicity. Other techniques used for receptor identification and binding with Cry pesticidal protein include ELISA binding assays, SPR resonance, ligand blotting, western blotting, pull-down (immunoprecipitation), and LC-MS sequencing (Arenas et al., 2010; Da Silva et al., 2018; Flores-Escobar et al., 2013; Penã-Cardenã et al., 2018; Pigott and Ellar, 2007; Zhou et al., 2016). In addition, cell lines of different susceptible and resistant insects have been successfully used with all these techniques (Soberón et al., 2017).

A fundamental part of further understanding the mechanism of action of Cry pesticidal proteins is to identify the receptors involved in the interaction with the toxin and their participation in toxicity. Since 2010, these advances have been significant, especially in Lepidoptera. Thus, different proteins have been described as Cry pesticidal protein receptors, such as CAD, APN, ALP, a 270-kDa glycoconjugate, P252 (250 kDa protein), an α-amylase and, recently, several ABC-type carrier proteins, such as ABCC2, ABCC3, and ABCA2 (Gómez et al., 2018; Heckel, 2010; Pardo-López et al., 2013; Park et al., 2014; Pigott and Ellar, 2007; Tay et al., 2015). Besides, it has been suggested that other molecules may be involved in this interaction, such as glycolipids and other proteins present in "lipid rafts" and regions of membrane microdomains, such as flotillin, prohibitin, V-ATPase and actin (Bayyareddy et al., 2009; Griffitts et al., 2003; Ochoa-Campuzano et al., 2013).

Currently, several researchers have focused on a better understanding of the Bt mode of action. The studies developed, besides broadening the knowledge of this mechanism, have the objective of (1) studying toxin structure and their different domains; (2) understanding the resistance mechanisms of target insects to Bt pesticidal proteins inserted in transgenic plants; (3) understanding how the specificity of various Bt pesticidal protein occurs, focusing on gene pyramiding against possible target insects; and (4) producing new mutant pesticidal protein with increased activity for target insects and pesticidal proteins with combined domains to broaden the spectrum of action of the toxins (chimera toxins).

To this day, considering the large number of Cry pesticidal proteins described, many of the Bt modes of action studies are focused on Cry1A family toxins in Lepidoptera. Therefore, one of the great challenges to further understand the mechanism of action is the lack of more comprehensive studies that explores a greater number of toxins, in insects of different orders, at different instars. Different insect instars were reported to interact with various receptors (Arenas et al., 2010; Da Silva et al., 2018; Flores-Escobar et al., 2013).

Understanding the mode of action of the pesticidal proteins produced by Bt is fundamental for the development of more powerful toxins with outstanding durability, capable of delay insect ressitance. However, in the last decade, the two models proposed to explain the mechanism of action of Bt pesticidal proteins attracted considerable attention from researchers and generated abundant literature. Moreover, many aspects related to the two models still require data corroboration, and the "classical model" is currently in use to guide Bt mode of action studies.

9.5 Development of *Bacillus thuringiensis* formulations

Bacillus thuringiensis (Bt) was initially called *Bacillus sotto* by the bacteriologist Ishiwata and was isolated in *Bombyx mori* (Lepidoptera: Bombycidae) (by Ishiwata (1901). Afterward, the German microbiologist Ernst Berliner isolated the same bacteria in *Ephestia kuehniella* (Lepidoptera: Pyralidae) (Berliner, 1911) and later named it *Bacillus thuringiensis,* in homage to Thuringia, a German city where the larvae were collected (Berliner, 1915).

Hannay (1953) suggested the association between the pathogenicity of this bacterium and the presence of crystals formed during sporulation. Hannay and Frizt-James (1955) described the crystal, and in 1968, Angus showed that Hannay's hypothesis was valid (Glare and O'Callagham, 2000), which aroused interest in studies on the bacterium mode of action.

Although the importance of crystal for the pathogenicity of the bacterium was only discovered in the 1950s, the first attempts to use Bt in pest control were in Europe (Hungary and Yugoslavia) to control *Ostrinia nubilalis* (Lepidoptera: Noctuidae). However, the commercial production of this pathogen only began in 1938, in France, with the launch of Sporeine (Sanahuja et al., 2011), but was interrupted due to World War II.

In 1951, the efficiency of Sporeine to control *E. kuehniella* was demonstrated. In the USA, the interest in using Bt in agriculture, mainly to control Lepidoptera, increased after 1950 resulting, in the production of Thuricide in liquid, powder, and wettable powder formulations. Two years after this product was launched in 1959, Biotrol was established in the market. The former USSR produced Bt biopesticides based on Bt *thuringiensis*, Bt *dendrolimus,* and Bt *galleriae* to control agricultural and forest pests since 1950 (Beegle and Yamamoto, 1992).

Furthermore, before 1970, most Bt biopesticide formulations were based on Bt *thuringiensis,* with inconsistent results and most of them containing β-exotoxin, pathogenic to vertebrates. With the discovery of Bt *kurstaki* (Dulmage, 1970), a new Bt formulation, much more potent than the previous ones, increased the Bt potential to control insects and had rapid growth and sporulation in a relatively inexpensive environment; a key success factor on the market (Beegle and Yamamoto, 1992). Subsequently, improvements in these formulations were necessary, such as the standardization of products in international power units (IUs); use of wettable powder (WP) formulation, long shelf-life, and the addition of photo protectors in the formulations to increase product persistence in the field (Dulmage et al., 1971; Sanahuja et al., 2011).

With the launch of pyrethroid insecticides in the 1970s, the market for Bt biopesticides was significantly reduced (Beegle and Yamamoto, 1992). Therefore, it should be noted that the consistent concern of the consumers about pesticide residues in food and their effect on the environment (Bogdal et al., 2013; Carson, 1962; Nicolopoulou-Stamati et al., 2016) is a reminder of the opportunity to adopt and increase the use of microbial products in agriculture; owing to their insignificant risks to the environment and beneficial organisms (Glare et al., 2012; Lacey et al., 2015; Raymond and Federici, 2018).

In 1970, Dipel (biopesticide based on Bt *kurstaki*) (Beegle and Yamamoto, 1992) was marketed. This product has a broad spectrum of action against lepidopteran pests, especially noctuids (Glare and O'Callagham, 2000). In 1976, an effective subspecies was discovered against Diptera, called Bt *israelensis*, and in 1983, another one called Bt *tenebrionis*, lethal to Coleoptera, was found. These discoveries increased the interest in the use of these products in pest management, especially in forest pest control in the USA and Canada (Polanczyk et al., 2008). Bt biopesticides represent the largest share of these products in the world (Rao and Jurat-Fuentes, 2020), with significant use in agricultural pest control (Arthurs and Dara, 2019; Fernández-Chapa et al., 2019; Glare and O'Callagham, 2017; Hatting et al., 2018; Huang et al., 2007; Karimi et al., 2018; Kumar et al., 2018).

Since 2000 the use of Bt biopesticides, especially Dipel, to control *Helicoverpa armigera* and *Chrysodeixis includens* noctuid outbreaks in the Brazilian 2013/2014 agricultural crops is highlighted. At that time, Bt soy, which expresses the Cry1Ac pesticidal protein, was not being cultivated in Brazil yet, and Bt biopesticides were used in about nine million hectares of large crops such as soy and cotton (Polanczyk et al., 2017).

Despite the great diversity of Cry pesticidal proteins, only two Bt (Bt *kurstaki* and Bt *aizawai*) subspecies are used as active ingredients for Bt biopesticides in agricultural pest control (Bravo et al., 2011; Sanchis and Bourguet, 2008). This aspect can be observed in Brazil, where, despite the large increase in the number of Bt biopesticides marketed; from nine in 2010 to 26 in 2020 (Agrofit, 2020; Polanczyk et al., 2017), most biopesticides are based on Bt *kurstaki* HD-1 and some on Bt *aizawai* HD-68, aiming at controlling Lepidopteran pest. In 2015, the biopesticide Sympatic, a mixture of Bt *kurstaki* (15.6%) and Bt *aizawai* (10.4%), was launched in the US market. This product contains six pesticidal proteins Cry1Aa, Cry1Ab, Cry1Ac, Cry2A, Cry1C, and Cry1D, being recommended to control 90 lepidopteron pests (EPA, 2015).

Although Bt is easily cultured in a laboratory, the large-scale production of Bt is complex mainly due to the nutrients needed for crystal production (Duarte Neto et al., 2020). Valicente et al. (2018) stated that Bt requires carbon, nitrogen, and salts ($FeSO_4$, $ZnSO_4$, $MnSO_4$, $MgSO_4$) to grow. However, each Bt isolate can have a specific need regarding the amount of carbon and nitrogen, and the different salts used in the culture media for its growth. Typically, the main factors evaluated during Bt growth are cell mass, colony-forming units, viable spores, and mortality of the target insect; these results vary with different bt strains.

A determining factor in Bt-based biopesticide production is the oxygenation of the medium during the fermentation process (Jallouli et al. 2020). Valicente and Zanasi (2005) proposed the fermentation process in a solid medium using 50 g of rice as a substrate for Bt growth. The rice was enriched with carbon, nitrogen, and mineral salts sources and then autoclaved at 120°C for 20 min in polypropylene bags; this process is known as sterilization. The appropriate Bt strain previously grown in a liquid medium in a sterile environment was then was inoculated in the material. This mixture was kept at 30°C for four days for complete sporulation. This type of biopesticide production is inexpensive and easy to use. After four days, the material is frozen until it is needed for use. Moreover, the structure and production must follow proper laboratory procedures, including sterilization of all materials used.

With live bacteria (spores and crystals) being the active ingredient in Bt biopesticides, these products are sensitive to adverse environmental conditions such as temperature, humidity, and ultraviolet light. Ultraviolet radiation acts directly on nucleic acids, changing or even inactivating them, which prevents the growth and multiplication of entomopathogens. In the case of Bt, the amino acids (cysteine, tyrosine, and tryptophan) in the crystals are affected, inactivating the presticidal protein (Batista Filho et al., 1998). This aspect is critical in hot regions with low relative humidity, because, in addition to the lower field persistence, the pests in these conditions accelerate the cycle, resulting in more generations and increased population, which makes its control with microbial products difficult due to the absence of the shock effect. Besides, most field persistence studies were conducted in the northern hemisphere with milder temperatures than in the southern hemisphere.

Bacillus thuringiensis persistence and efficiency provides essential information on the potential of affecting non-target organisms and resistance management. Ultraviolet (UV) radiation is one of the factors that significantly reduces the persistence of Bt biopesticides and, consequently, their efficiency in the field (Lacey et al., 2015). The first studies on this subject were conducted in the 1960s and 1970s, under laboratory conditions (Griego and Spence, 1978; Pinnock et al., 1971, 1974, 1975; Raun et al., 1966; Yamvrias, 1962).

Raun (1963) and Raun and Jackson (1966) reported that Bt encapsulation, i.e., the use of five different techniques to encapsulate lyophilized bacteria, can lead to an increase in Bt persistence. They discovered good control efficiency for *O. nubilalis* under field conditions using granulated and liquid formulations. Encapsulation is a technology that allows the use of an active ingredient (control agent) with protection materials against sunlight, high temperatures, and low relative humidity. Examples of these materials are biopolymers, chemical polymers, and microorganisms (Vemmer and Patel, 2013).

Subsequently, Bt Cry genes were introduced in *Escherichia coli, Bacillus subtilis, Bacillus megaterium,* and *Pseudomonas fluorescens*. Recombinant fermentation processes were used to produce formulations containing crystals encapsulated by dead cells. This type of formulation increases the field persistence of these biopesticides due to protection against ultraviolet radiation (Gaertner et al., 1993; Hernandez-Rodrigues et al., 2013; Schnepf et al., 1998).

Mixed Bt formulations with insecticides were launched in the market. Branscome et al. (2015a,b) highlighted synergism between insecticides in the diamide group with Bt *kurstaki* and Bt *aizawai* to control different pests. They stated that the synergistic combination of biological products and synthetic insecticides is a less explored concept that can benefit from the use of microbial products in IPM. When combined, these products provide distinct modes of action with overlapping action spectrum to ensure mortality, delaying the development of resistance in insect populations.

9.6 *Bacillus thuringiensis* compatibility with natural enemies and Bt plants

The publication by Glare and O'Callagham (2000) presents the complete compilation to date on the spectrum of action of Bt on pests and natural enemies (predator and parasitoid insects). Although this publication, as well as others (Glare et al., 2012; Lacey et al., 2015), highlights the selectivity and specificity of Bt biopesticides as outstanding characteristics. These aspects are less explored by suppliers and extensionists, being almost nonexistent for farmers, due to the complexity of ecological interactions between Bt and other natural enemies and also, insufficient research in this area.

Furthermore, several articles report the importance of natural biological control of agricultural pests. For example, Firake and Behere (2020) reported more than 26 natural enemy species, 10 of which are natural biological control of *Spodoptera frugiperda* (Lepidoptera: Noctuidae) in northeast India. Sharanabasappa et al. (2019) also reported the occurrence of these natural enemies in *S. frugiperda* in southern India. Koffi et al. (2020) reported the occurrence of seven parasitoids species and three predators in *S. frugiperda* in Ghana (Africa). They emphasized that the parasitoid *Chelonus bifoveolatus* (Hymenoptera: Braconidae) and the predator *Pheidole megacephala* (Hymenoptera: Formicidae) were the most abundant species. Molina-Ochoa (2003) reported 150 natural enemy species of *S. frugiperda* in the Americas, with the Ichneumonidae and Braconidae being the most abundant.

Despite the high selectivity of Bt biopesticides for natural enemies, the complex Bt mode of action (Heckel, 2020; Vachon et al., 2012) does not prove that Bt is harmless to parasitoids and predators, and it is always important to compare Bt selectivity with chemical insecticide selectivity. Cunha et al. (2012), for example, pointed out that Bt cotton Cry1Ac presticidal protein can affect the physiology of the predator *Podisus nigrispinus* (Hemiptera: Pentatomidae).

Interactions between Bt, parasitoids, and pests may be classified as harmful to the parasitoid or pest. Harmful interactions with the parasitoid may result from direct Bt infection, early host death, reduced host population, or physiological and nutritional changes in the host. Harmful host interactions are related to increased susceptibility of the host to infections, transmission of pathogens by the parasitoid, difficult discrimination of parasitic and nonparasitic hosts, delay in the biological cycle of the host, and suppression of the host (Magalhães et al., 1998).

Parasitism can stress the host and increase its susceptibility to Bt (Brooks, 1993). However, some reports show varied responses. For example, *Plutella xylostella* (Lepidoptera: Plutellidae) larvae parasitized by *Diadegma* sp. are less susceptible to Bt than non-parasitized larvae (Monnerat, 1995). Nealis and van Frankenhuyzen (1990) observed the same fact in *Choristoneura fumiferana* (Lepidoptera: Tortricidae) larvae parasitized by *Apanteles fumiferanae* (Hymenoptera: Braconidae).

Both cases above report decreased food consumption by parasitized larvae, reducing the amount of Bt ingested. On the other hand, the decreased susceptibility of parasitic hosts to bacteria can be attributed to differences in the mechanism of host use by the parasitoid (Brooks, 1993).

Bacterial infections can change the behavior of host insects, and some parasitoids can notice these changes. When infected by Bt, *P. xylostella* larvae reduce their movements and become less attractive to oviposition by the parasitoid *Diadegma* sp. (Monnerat, 1995). The same occurs with the ectoparasitoid *Bracon brevicornis* (Hymenoptera: Braconidae), which avoids oviposition in *Sesamia cretica* (Lepidoptera: Noctuidae) larvae infected by Bt (Temerak, 1980).

In other cases, Bt increases the biological cycle of the hosts, prolonging larval stages, thus favoring the action of natural enemies. This prolonged larval phase was verified by Moreau and Bauce (2003) with the biopesticide Foray 48B (*Bt* kurstaki), used to control *C. fumiferana*.

Direct parasitoid infection by Bt was the target of some studies that initially concluded that this bacterium does not affect parasitic insects and predators, except indirectly, due to the premature death of the hosts (Krieg and Langenbruch, 1981; Niwa et al., 1987). However, Temerak (1980) reported that the development of *B. brevicornis* in *S. cretica* larvae contaminated with Bt is decreased due to reduced oviposition, pupae, and emerging adults, to decrease adult longevity. Salama et al. (1982) observed decreased emergence and reproductive potential of the parasitoid *Microplitis demolitor* (Hymenoptera: Braconidae) when developed in *Spodoptera littoralis* (Lepidoptera: Noctuidae) larvae fed with Bt.

Brooks (1993) emphasized that the toxins produced by Bt do not directly seem to affect the parasitoids that emerge before the host dies. As this bacterium acts in the insect midgut, the probability that a parasitoid developing in a contaminated host has direct contact with the toxins of this bacterium is low, unless it consumes the entire host, as is the case of *Diadegma* sp. Monnerat (1995) demonstrated in vitro, through immunocytochemical analysis, that Cry1Aa, Cry1Ab and Cry1Ac pesticidal proteins do not bind to the receptors present in the microvilliated apical process in columnar cells of the midgut.

Since the 2000s, due to the extensive global use of Bt crops in agricultural pest control (James, 2013), most studies on the effects of Bt on parasitoids have been conducted with purified Cry proteins or Bt crops (Baur and Boethel, 2003; Bernal, 2010; Chen et al., 2008; Hagenbucher et al., 2014; Sisterson and Tabashnik, 2005) and some with Bt biopesticides (Biondi et al., 2013; Nascimento et al., 2018; Patel and Pramanik, 2012; Pinto et al., 2019).

It is not possible to extrapolate Cry pesticidal proteins results to biopesticides because, although Cry pesticidal proteins are the main virulent factor of this bacterium, the potential toxic effect of spores and their synergism with Cry pesticidal proteins has always aroused the interest of insect pathologists (Crickmore, 2006). Somerville and Pockett (1975) described the toxicity of Bt spores to *Pieres brassicae* (Lepidoptera: Pieridae) larvae and suggested that this may be an essential virulence factor for Bt biopesticides. In the following year, Burges et al. (1976) reported that Bt spores play a vital role, along with crystals, in the death of the greater wax moth (*Galleria mellonella*). Similar results were later reported by Mohd-Salleh (1980) for *O. nubilalis*.

Research on the toxicity of Bt spores was left in the background due to the belief that transgenic plants would be the solution for insect pests. Studies on the pathology of this bacterium were directed to understand the mechanisms of action of Cry pesticidal proteins, mainly aimed at developing resistance management strategies for Bt crops (Tabashnik et al., 2014).

Jakka et al. (2014) stressed the importance of the spore as a Bt virulent factor and its potential to improve the insect resistance management of Bt crops. Conte et al. (2019) pointed out Bt biopesticides should not be used in Bt crop refuge areas, due to the risk of cross-resistance between the Bt soybean pesticidal protein and the Bt biopesticide pesticidal proteins. However, it should be noted that besides the spore effect, these biopesticides have more than one toxin, for example, Dipel has the Cry1Aa, Cry1Ab, Cry1Ac and Cry2Aa pesticidal proteins (Fu et al., 2008), which work as different active ingredients in the product. Also, recent studies demonstrated the absence of cross-resistance between Bt crop and Bt biopesticide pesticidal proteins (Ferral-Pina et al., 2015; Horikoshi et al., 2019; Souza et al., 2019).

9.6.1 Final considerations

Dr. Heckel's publication: "How do toxins from *Bacillus thuringiensis* kill insects? An evolutionary perspective" (https://doi.org/10.1002/arch.21673) using a question mark demonstrates that even about 100 years after the first attempts to use Bt in agriculture, many advances are still needed to increase and preserve the use of this pathogen and its toxins in pest management.

The use of Bt biopesticides in agricultural pest control should be conducted within an integrated management program. Although Bt stands out for its high control efficiency (around 80% mortality), it does not have a shock effect, being more specific and virulent for the early stages of pests. Therefore, samplings to identify the pest and to know its age composition are essential for the success of Bt biopesticides application (Glare et al., 2012; Lacey et al., 2015).

Most Bt biopesticides have the *Bt* kurstaki strain as an active ingredient, which contains the Cry1Aa, Cry1Ab, Cry1Ac and Cry2A pesticidal proteins. Perhaps the easy cultivation of this strain is a justification since fermentation process adjustments can be expensive and time-consuming, delaying the launch of a biopesticide in the market. Another aspect of consideration is the quantification of each Cry pesticidal protein in the product label. This aspect would be interesting for the choice of product according to the pest, also helping in resistance management with the option of biopesticides having different toxins.

The spectrum of action of Cry pesticidal proteins is well known, for example, Cry1Ac has activity on several noctuid pests (van Frankenhuyzen, 2009), but the Bt toxin bank (http://www.btnomenclature.info/) has 39 Cry1Ac (Cry1Ac1 to Cry1Ac39). The same happens for Cry1Aa (Cry1Aa1 to Cry1Aa25) and Cry1Ab (Cry1Ab1 to Cry1Ab36) toxins. These variations should be considered for detailed knowledge of the potential of Cry pesticidal proteins, since small changes in the amino acid sequence of Cry toxins may change insecticide specificity (Bravo et al., 2012; Jurat-Fuentes and Crickmore, 2017).

Bacillus thuringiensis based biopesticides selectivity and specificity can be reduced or lost when these products are used in tank mixtures with other pesticides. The use of tank mixes is increasingly common (Gazziero, 2015; Schreiner et al., 2016) for optimizing agricultural operations and increasing the action spectrum of the products. Most of the studies available so far evaluate the *in vitro* compatibility of pesticides and biological products, but this compatibility does not always imply increased efficiency, since the interaction between the active ingredients is extremely complex and depends on the inert ingredients used in the formulation. Therefore, studies that relate compatibility with the insecticidal activity of the mixture are necessary.

References

Aboussaid, H., El-Aouame, L., El-Messoussi, S., Oufdou, K., 2010. Biological activity of *Bacillus thuringiensis* (Berliner) strains on larvae and adults of *Ceratitis capitata* (Wiedemann) (Diptera: Tephritidae). J. Environ. Protect. 1, 337–345. https://doi.org/10.4236/jep.2010.14040.

Aboussaid, H., Vidal-Quist, J.C., Oufdou, K., El Messoussi, S., Castañera, P., González-Cabrera, J., 2011. Occurrence, characterization and insecticidal activity of *Bacillus thuringiensis* strains isolated from argan fields in Morocco. Environ. Technol. 32, 1383–1391. https://doi.org/10.1080/09593330.2010.536789.

Agrofit, 2020, 1. https://agrofit.agricultura.gov.br/agrofit_cons/principal_agrofit_conse. (Accessed 20 May 2020).

Akram, W., Mahboob, A., Javed, A.A., 2013. *Bacillus thuringiensis* strain 199 can induce systemic resistance in tomato against fusarium wilt. Eur. J. Microbiol. Immunol. 34, 275–280. https://doi.org/10.1556/EuJMI.3.2013.4.7.

Amanchi, N.R., Hussain, M.M., 2008. Cytotoxic effects of Delfin insecticide (*Bacillus thuringiensis*) on cell behavior, phagocytosis, contractile vacuole activity and macronucleus in a protozoan ciliate *Paramecium caudatum*. Afr. J. Biotechnol. 7, 2637–2643.

Anderson, J.A., Ellsworth, P.C., Faria, J.C., Head, G.P., Owen, M.D.K., Pilcher, C.D., Shelton, A.M., Meissle, M., 2019. Genetically engineered crops: importance of diversified integrated pest management for agricultural sustainability. Front. Bioeng. Biotech. 7, 1–14.

Arenas, I., Bravo, A., Soberón, M., Gómez, I., 2010. Role of alkaline phosphatase from *Manduca sexta* in the mechanism of action of *Bacillus thuringiensis* Cry1Ab toxin. J. Biol. Chem. 285, 12497–12503.

Aronson, A.I., Shai, Y., 2001. Why *Bacillus thuringiensis* insecticidal toxins are so effective: unique features of their mode of action. FEMS Microbiol. Lett. 195, 1–8.

Arthurs, S., Dara, S.K., 2019. Microbial biopesticides for invertebrate pests and their markets in the United States. J. Invertebr. Pathol. 165, 13–21.

Bagcõoglu, M., Fricker, M., Johler, S., Ehling-Schulz, M., 2019. Detection and identification of *Bacillus cereus, Bacillus cytotoxicus, Bacillus thuringiensis, Bacillus mycoides* and *Bacillus weihenstephanensis* via machine learning based FTIR spectroscopy. Front. Microbiol. 10, 902. https://doi.org/10.3389/fmicb.2019.00902.

Batista Filho, A., Alves, S.B., Alves, L.F.A., Pereira, R.M., Augusto, N.T., 1998. Formulação de entomopatógenos. In: Alves, S.B. (Ed.), Controle microbiano de insetos. FEALQ, Piracicaba, pp. 917–966.

Baur, M.E., Boethel, D.J., 2003. Effect of Bt-cotton expressing Cry1A(c) on the survival and fecundity of two hymenopteran parasitoids (Braconidae, Encyrtidae) in the laboratory. Biol. Contr. 26, 325–332.

Bayyareddy, K., Andacht, T.M., Abdullah, M.A., Adang, M.J., 2009. Proteomic identification of *Bacillus thuringiensis* subsp. *israelensis* toxin Cry4Ba binding proteins in midgut membranes from *Aedes* (*Stegomyia*) *aegypti* Linnaeus (Diptera, Culicidae) larvae. Insect Biochem. Molec. 39, 279–286.

Beegle, C.B., Yamamoto, T., 1992. Invitation paper (C.P. Alexander Fund): history of *Bacillus thuringiensis* berliner research and development. Can. Entomol. 24, 587–616. https://doi.org/10.4039/Ent124587-4.

Bergamasco, V.B., Mendes, D.R.P., Fernandes, O.A., Desidério, J.A., Lemos, M.V.F., 2013. *Bacillus thuringiensis* Cry1Ia10 and Vip3Aa protein interactions and their toxicity in *Spodoptera* spp. (Lepidoptera). J. Invertebr. Pathol. 112, 152–158. https://doi.org/10.1016/j.jip.2012.11.011.

Berliner, E., 1911. Uber die Schlaffsucht der Mehlmottenraupe. Z. ges. Getreidew. 3, 63–70.

Berliner, E., 1915. Uuber die Schlaffsucht der Mehlmottenraupe (*Ephestia kuehniella* Zell.) und ihren Erreger *Bacillus thuringiensis*. N. Sp. Z. Angew. Entomol. 2, 29–56.

Berlitz, D.L., Saul, D.A., Machado, V., Santin, R.C., Guimarães, A.M., Matsumura, A.T.S., Ribeiro, B.M., Fiuza, L.M., 2013. *Bacillus thuringiensis*: molecular characterization, ultrastructural and nematoxicity to *Meloidogyne* sp. J. Biopestic. 6, 120–128.

Bernal, J.S., 2010. Genetically modified crops and biological control with egg parasitoids. In: Cônsoli, F.L., Parra, J.R.P., Zucchi, R.A. (Eds.), Egg Parasitoids in Agroecosystems with Emphasis on *Trichogramma*. Springer, New York, pp. 443–466.

Bernhard, K., Jarret, P., Meadows, M., 1997. Natural isolates of *Bacillus thuringiensis*: worldwide distribution, characterization, and activity against insects pests. J. Invertebr. Pathol. 70, 59–68.

Biondi, A., Zappala, L., Stark, J.D., Desneux, N., 2013. Do biopesticides affect the demographic traits of a parasitoid wasp and its biocontrol services through sublethal effects? PloS One 8, e76548. https://doi.org/10.1371/journal.pone.0076548.

Blanco, C.A., Gould, F., Groot, A.T., Abel, C.A., Hernandez, G., Perera, O.P., Teran-Vargas, A.P., 2010. Offspring from sequential matings between *Bacillus thuringiensis*- resistant and *Bacillus thuringiensis*-susceptible *Heliothis virescens* moths (Lepidoptera: Noctuidae). J. Econ. Entomol. 103, 861–868. https://doi.org/10.1603/EC09232.

Bogdal, C., Scheringer, M., Abad, E., Abalos, M., van Bavel, B., Hagberg, J., Fiedler, H., 2013. Worldwide distribution of persistent organic pollutants in air, including results of air monitoring by passive air sampling in five continents. Trac. Trends Anal. Chem. 46, 150–161. https://doi.org/10.1016/j.trac.2012.05.011.

Branscome, D., Storey, R., Eldridge, R., Brazil, E., Devisette, B., 2015a. Synergistic *Bacillus thuringiensis* Subsp. *aizawai, Bacillus thuringiensis* Subsp. *kurstaki* and Chlorantraniliprole Mixtures for Diamondback Moth, Beet Armyworm, Sugarcane Borer, Soybean Looper, Corn Earworm, Cabbage Looper, and Southwestern Corn Borer Control. Company: Valent BioSciences LLC. VAL06131P01730US. Deposit: 7 dez. 2017. Concession: 14 abr. 2018.

Branscome, D., Storey, R., Eldridge, R., Brazil, E., 2015b. Synergistic *Bacillus thuringiensis* Subsp. *kurstaki* and Chlorantraniliprole Mixtures for Diamondback Moth, Beet Armyworm, Sugarcane Borer, Soybean Looper and Corn Earworm Control. Company: Valent BioSciences LLC. VAL06131P01710US. Deposit: 14 fev. 2018. Concession: 21 jun. 2018.

Bravo, A., Sarabia, S., Lopez, L., Ontiveros, H., Abarca, C., Ortiz, A., Ortiz, M., Lina, L., Villalobos, F.J., Pena, G., Nunez-Valdez, M.E., Soberon, M., Quintero, R., 1998. Characterization of *cry* genes in a mexican *Bacillus thuringiensis* strain collection. Appl. Environ. Microbiol. 64, 4965–4972.

Bravo, A., Gómez, I., Conde, J., Munoz-Garay, C., Sanchez, J., Miranda, R., Zhuang, M., Gill, S.S., Soberón, M., 2004. Oligomerization triggers binding of a *Bacillus thuringiensis* Cry1Ab pore-forming toxin to aminopeptidase N receptor leading to insertion into membrane microdomains. Biochim. Biophys. Acta 1667, 38–46.

Bravo, A., Gill, S.S., Soberón, M., 2007. Mode of action of *Bacillus thuringiensis* Cry and Cyt toxins and their potential for insect control. Toxicon 49, 423–435.

Bravo, A., Likitvivatanavong, S., Gill, S.S., Soberón, M., 2011. *Bacillus thuringiensis*: a story of a successful bioinsecticide. Insect Biochem. Mol. 41, 423–431. https://doi.org/10.1016/j.ibmb.2011.02.006.

Bravo, A., Gómez, I., Porta, H., García-Gómez, B.I., Rodriguez-Almazan, C., Pardo, L., Soberón, M., 2012. Evolution of *Bacillus thuringiensis* Cry toxins insecticidal activity. Microb. Biotechnol. 6, 17–26. https://doi.org/10.1111/j.1751-7915.2012.00342.x.

Bravo, A., Gómez, I., Porta, H., García-Gómez, B.I., Rodriguez-Almazan, C., Pardo, L., Soberón, M., 2013. Evolution of *Bacillus thuringiensis* Cry toxins insecticidal activity. Microb. Biotechnol. 6, 17–26.

Brooks, W.M., 1993. Host-parasitoid-pathogen interactions. In: Beckage, N., Thompson, S., Federici, B. (Eds.), Parasites and Pathogens of Insects. Academic, San Diego, pp. 231–272.

Buentello-Wong, S., Galán-Wong, L., Arévalo-Niño, K., Almaguer-Cantú, V., Rojas-Verde, G., 2015. Characterization of Cry proteins in native strains of *Bacillus thuringiensis* and activity against *Anastrepha ludens*. Southwest. Entomol. 40, 15–24. https://doi.org/10.3958/059.040.0102.

Burges, H.D., Hurst, J.A., 1977. Ecology of *Bacillus thuringiensis* in storage moths. J. Invertebr. Pathol. 30, 131–139.

Burges, H.D., Thomson, E.M., Latchford, R.A., 1976. Importance of spores and g-endotoxin protein crystals of *Bacillus thuringiensis* in *Galleria mellonella*. J. Invertebr. Pathol. 27, 87–94.

Cao, B., Shu, C., Geng, L., Song, F., Zhang, J., 2020. Cry78Ba1, one novel crystal protein from *Bacillus thuringiensis* with high insecticidal activity against rice planthopper. J. Agric. Food Chem. https://doi.org/10.1021/acs.jafc.9b07429.

Carroll, J., Ellar, D.J., 1993. An analysis of *Bacillus thuringiensis* δ-endotoxin action on insect midgut membrane permeability using a light-scattering assay. Eur. J. Biochem. 214, 771–778.

Carson, R., 1962. Silent Spring. Houghton and Mifflin, New York.

Chak, K.F., Chao, D.C., Tseng, M.Y., Kao, S.S., Tuan, S.J., Feng, T.Y., 1994. Determination and distribution of cry-type genes of *Bacillus thuringiensis* isolates from Taiwan. Appl. Environ. Microbiol. 60, 2415–2420.

Chen, M., Zhao, J.-Z., Collins, H.L., Earle, E.D., Cao, J., Shelton, A.M., 2008. A critical assessment of the effects of bt transgenic plants on parasitoids. PloS One 3, e2284. https://doi.org/10.1371/journal.pone.0002284.

Clark, B.W., Phillips, T.A., Coats, J.R., 2005. Environmental fate and effects of *Bacillus thuringiensis* (Bt) proteins from transgenic crops: a review. J. Agric. Food Chem. 53, 4643–4653.

Conte, O., de Oliveira, F.T., Harger, N., Corrêa-Ferreira, B.S., Roggia, S., Prando, A.M., Possamai, E.J., Reis, E.A., Marx, E.F., 2019. Resultados do manejo integrado de pragas da soja na safra 2018/19 no Paraná. Embrapa, Londrina.

Crickmore, N., 2006. Beyond the spore – past and future developments of *Bacillus thuringiensis* as a biopesticide. J. Appl. Microbiol. 101, 616–619.

Crickmore, N., Zeigler, D.R., Feitelson, J., Schnepf, E., Van Rie, J., Lereclus, D., Baum, J., Dean, D.H., 1998. Revision of the nomenclature for the *Bacillus thuringiensis* pesticidal crystal proteins. Microbiol. Mol. Biol. Rev. 62, 807–813.

Crickmore, N., Baum, J., Bravo, A., Lereclus, D., Narva, K., Sampson, K., Schnepf, E., Sun, M., Zeigler, D.R., 2020. *Bacillus thuringiensis* Toxin Nomenclature. Available at: http://www.btnomenclature.info/. (Accessed 16 June 2020).

Cunha, F.M., Caetano, F.H., Wanderley-Teixeira, V., Torres, J.B., Teixeira, A.A.C., Alves, L.C., 2012. Ultra-structure and histochemistry of digestive cells of *Podisus nigrispinus* (Hemiptera: Pentatomidae) fed with prey reared on Bt-cotton. Micron 43, 245–250. https://doi.org/10.1016/j.micron.2011.08.006.

Da Silva, I.H.S., Gómez, I., Sanchez, J., Martinez De Castro, D.L., Valicente, F.H., Soberón, M., Polanczyk, R.A., 2018. Identification of midgut membrane proteins from different instars of *Helicoverpa armigera* (Lepidoptera: Noctuidae) that bind to Cry1Ac toxin. PloS One 13, 1–16.

Damgaard, P.H., 2000. Natural occurrence and dispersal of *Bacillus thuringiensis* in the environment. In: Charles, J.F., Delécluse, A., Nielsen-Le Roux, C. (Eds.), Entomopathogenic Bacteria: From Laboratory to Field Application. Kluwer Academic Publishers, Netherlands, pp. 23–40.

De Barjac, H., Bonnefoi, A., 1962. Essai de classification biochimique et sérologique de 24 souches de *Bacillus* du type *B. thuringiensis*. Entomophaga 7, 5–31. https://doi.org/10.1007/BF02375988.

De Bortoli, C.P., Jurat-Fuentes, J.L., 2019. Mechanisms of resistance to commercially relevant entomopathogenic bacteria. Curr. Opin. Insect Sci. https://doi.org/10.1016/j.cois.2019.03.007.

De Oliveira Dorta, S., Balbinotte, J., Monnerat, R., Lopes, J.R.S., da Cunha, T., Zanardi, O.Z., de Miranda, M.P., Machadom, M.A., Freitas-Astúa, J., 2018. Selection of *Bacillus thuringiensis* strains in citrus and their pathogenicity to *Diaphorina citri* (Hemiptera: Liviidae) nymphs. Insect Sci. https://doi.org/10.1111/1744-7917.12654.

Delgado-Silva, Y.B., Tarazona, D., Serna, F., Juscamayta, E., Chávez-Galarza, J.C., Farfán-Vignolo, E.R., Delgado, G., Flores, A., Solano, G., Gutierrez, D.L., 2020. Draft genome sequence of *Bacillus thuringiensis* strain UNMSM10RA, isolated from potato crop soil in Peru. Microbiol. Res. Announc. 9. https://doi.org/10.1128/mra.01189-19.

Dias, N.P., Cagliari, D., Dos Santos, E.A., Smagghe, G., Jurat-Fuentes, J.L., Mishra, S., Nava, D.E., Zotti, M.J., 2020. Insecticidal gene silencing by RNAi in the neotropical region. Neotrop. Entomol. 49, 1–11.

Duarte Neto, J.M., Wanderley, M.C., da Silva, T.A., Marques, D.A., da Silva, J.R., Gurgel, J.F., Oliveira, J.P., Porto, A.L., 2020. *Bacillus thuringiensis* endotoxin production: a systematic review of the past 10 years. World J. Microbiol. Biotechnol. 36 (128). https://doi.org/10.1007/s11274-020-02904-4.

Dulmage, H.T., 1970. Insecticidal activity of HD-1, a new isolate of *Bacillus thuringiensis* var. *alesti*. J. Invertebr. Pathol. 15, 232–239.

Dulmage, H.T., Boening, C.S., Rehnborg, G.D., Hansen, O.G., 1971. A proposed standardized bioassay for formulations of *Bacillus thuringiensis* based on the international unit. J. Invertebr. Pathol. 18, 240–245.

Ehler, L., 1998. Conservation biological control: past, present, and future. In: Barbosa, P. (Ed.), Conservation Biological Control. Elsevier, San Diego, pp. 1–8.

El-Sadawy, H.A., El-Hag, H.A., Georgy, J.M., El-Hossary, S., Kassem, H.A., 2008. *In vitro* activity of *Bacillus thuringiensis* (H14) 43 kDa crystal protein against *Leishmania major*. Am.-Eurasian J. Agric. Environ. Sci. 3, 583–589.

El-Shafie, H., 2018. Integrated insect pest management. In: Haouas, D. (Ed.), Pests Control and Acarology. IntechOpen, London, pp. 1–18.

EPA, 2015. Environmental Protection Agency. Office of Pesticide Programs. EPA Reg. Number: 73049-502.

Fernández-Chapa, D., Ramírez-Villalobos, J., Galán-Wong, L., 2019. Toxic potential of *Bacillus thuringiensis*: an overview. In: Jia, Y. (Ed.), Protecting Rice Grains in the Post-Genomic Era. Intechopen. https://doi.org/10.5772/intechopen.85756.

Ferral-Pina, J., Vilas Boas, G.F.L.T., Sosa-Gómez, D.R., 2015. Mortality of a HD-73 resistant strain of velvetbean caterpillar, *Anticarsia gemmatalis* (Lepidoptera: Erebidae) fed on Bt soybean. In: Resistance 2015, Harpenden, Hertfordshire, p. 72.

Firake, D.M., Behere, G.T., 2020. Natural mortality of invasive fall armyworm, *Spodoptera frugiperda* (J. E. Smith) (Lepidoptera: Noctuidae) in maize agroecosystems of northeast India. Biol. Contr. 148, 104303. https://doi.org/10.1016/j.biocontrol.2020.104303.

Flores-Escobar, B., Rodriguez-Magadan, H., Bravo, A., Soberón, M., Gómez, I., 2013. Differential role of *Manduca sexta* aminopeptidase-N and alkaline phosphatase in the mode of action of Cry1Aa, Cry1Ab, and Cry1Ac toxins from *Bacillus thuringiensis*. Appl. Environ. Microbiol. 79, 4543–4550.

Forsyth, G., Logan, N.A., 2000. Isolation of *Bacillus thuringiensis* from northern Victoria Land, Antarctica. Lett. Appl. Microbiol. 30, 263–266. https://doi.org/10.1046/j.1472-765x.2000.00706.x.

Fu, Z., Sun, Y., Xia, L., Ding, X., Mo, X., Li, X., Huang, K., Zhang, Y., 2008. Assessment of protoxin composition of *Bacillus thuringiensis* strains by use of polyacrylamide gel block and mass spectrometry. Appl. Microbiol. Biotechnol. 79, 875–878.

Gaertner, F.H., Quick, T.C., Thompson, M.A., 1993. CellCap: an encapsulation system for insecticidal biotoxin proteins. In: Kim, L. (Ed.), Advanced Engineered Pesticides. Marcel Dekker, New York, pp. 73–83.

Gazziero, D.L.P., 2015. Misturas de agrotóxicos em tanque nas propriedades agrícolas do Brasil. Planta Daninha 33, 83–92.

Glare, T.R., O'Callaghan, M., 2000. *Bacillus thuringiensis*: Biology, Ecology and Safety. John Wiley and Sons, Chichester.

Glare, T.R., O'Callaghan, M., 2017. Microbial biopesticides for control of invertebrates: progress from New Zealand. J. Invertebr. Pathol. https://doi.org/10.1016/j.jip.2017.11.014.

Glare, T., Caradus, J., Gelernter, W., Jackson, T., Keyhani, N., Kohl, J., Marrone, P., Morin, L., Stewart, A., 2012. Have biopesticides come of age? Trends Biotechnol. 30, 250–258.

Gómez, I., Sánchez, J., Miranda, R., Bravo, A., Soberón, M., 2002. Cadherin-like receptor binding facilitates proteolytic cleavage of helix α-1 in domain I and oligomer pre-pore formation of *Bacillus thuringiensis* Cry1Ab toxin. FEBS Lett. 513, 242–246.

Gómez, I., Rodríguez-Chamorro, D.E., Flores-Ramirez, G., Grande, R., Zuniga, F., Portugal, F.J., Sanchéz, J., Pacheco, S., Bravo, A., Soberón, M., 2018. *Spodoptera frugiperda* (J. E. Smith) aminopeptidase N1 is a functional receptor of *Bacillus thuringiensis* Cry1Ca toxin. Appl. Environ. Microbiol. 84, 1–11.

Griego, V.M., Spence, K.D., 1978. Inactivation of *Bacillus thuringiensis* spores by ultraviolet and visible light. Appl. Environ. Microbiol. 35, 906–910.

Griffitts, J.S., Huffman, D.I., Whitacre, J.L., Barrows, B.D., Marroquin, L.D., Muller, R., Brown, J.R., Hennet, T., Esko, J.D., Aroian, R.V., 2003. Resistance to a bacterial toxin is mediated by removal of a conserved glycosylation pathway required for toxin and host interactions. J. Biol. Chem. 278, 45594–45602.

Guo, Z., Sun, D., Kang, S., Zhou, J., Gong, L., Qin, Q., Guo, L., Zhu, L., Bai, Y., Luo, L., Zhang, Y., 2018. CRISPR/Cas9-mediated knockout of both the PxABCC2 and PxABCC3 genes confers high-level resistance to *Bacillus thuringiensis* Cry1Ac toxin in the diamondback moth, *Plutella xylostella* (L.). Insect Biochem. Mol. Biol. 107, 31–38.

Gutiérrez, M.E.M., Gonçálves, E.F., 2006. Selección de cepas de *Bacillus thuringiensis* con efecto nematicida. Manejo Integr. Plagas Agroecol. 78, 63–66.

Hagenbucher, S., Wäckers, F.L., Romeis, J., 2014. Indirect multi-trophic interactions mediated by induced plant resistance: impact of caterpillar feeding on aphid parasitoids. Biol. Lett. 10. https://doi.org/10.1098/rsbl.2013.0795.

Hannay, C.L., Fitz-James, P., 1955. The protein crystals of *Bacillus thuringiensis* Berliner. Can. J. Microbiol. 1, 694–710. https://doi.org/10.1139/m55-083.

Hansen, B.M., Salamitou, S., 2000. Virulence of *Bacillus thuringiensis*. In: Charles, J.F., Delécluse, A., Nielsen-Le Roux, C. (Eds.), Entomopathogenic Bacteria: From Laboratory to Field Application. Kluwer Academic Publishers, Netherlands, pp. 41–64.

Hatting, J.L., Moore, S.D., Malan, A.P., 2018. Microbial control of phytophagous invertebrate pests in South Africa: current status and future prospects. J. Invertebr. Pathol. https://doi.org/10.1016/j.jip.2018.02.004.

Heckel, D.G., 2010. Learning the ABCs of Bt: ABC transporters and insect resistance to *Bacillus thuringiensis* provide clues to a crucial step in toxin mode of action. Pestic. Biochem. Physiol. 104, 103–110.

Heckel, D.G., 2020. How do toxins from *Bacillus thuringiensis* kill insects? An evolutionary perspective. Arch. Insect Biochem. Physiol. 104, e21673. https://doi.org/10.1002/arch.21673.

Hernández-Rodríguez, C.S., Ruiz de Escudero, I., Asensio, A.C., Ferré, J., Caballero, P., 2013. Encapsulation of the *Bacillus thuringiensis* secretable toxins Vip3Aa and Cry1Ia in *Pseudomonas fluorescens*. Biol. Contr. 66, 159–165. https://doi.org/10.1016/j.biocontrol.2013.05.002.

Hofte, H, Whitely H, R, 1989. Insecticidal crystal protein of *Bacillus thuringiensis*. Microbiological Reviews 53 (2), 242–255.

Horikoshi, R.J., Bernardi, O., Amaral, F.S.A.E., Miraldo, L.L., Durigan, M., Bernardi, D., Silva, S.S., Omoto, C., 2019. Lack of relevant cross-resistance to Bt insecticide XenTari in strains of *Spodoptera frugiperda* (J. E. Smith) resistant to Bt maize. J. Invertebr. Pathol. 161, 1–6.

Hossain, M.A., Ahmed, S., Hoque, S., 1997. Abundance and distribution of *Bacillus thuringiensis* in the agricultural soil of Bangladesh. J. Invertebr. Pathol. 70, 221–225.

Huang, D.-F., Zhang, J., Song, F.-P., Lang, Z.-H., 2007. Microbial control and biotechnology research on *Bacillus thuringiensis* in China. J. Invertebr. Pathol. 95, 175–180. https://doi.org/10.1016/j.jip.2007.02.016.

Hung, T.P., Truong, L.V., Binh, N.D., Frutos, R., Quiquampoix, H., Staunton, S., 2016. Fate of insecticidal *Bacillus thuringiensis* Cry protein in soil: differences between purified toxin and biopesticide formulation. Pest Manag. Sci. 72, 2247–2253. https://doi.org/10.1002/ps.4262.

Icoz, I., Stotzky, G., 2008. Fate and effects of insect-resistant Bt crops in soil ecosystems. Soil Biol. Biochem. 40, 559–586.

Jakka, S.R.K., Knight, V.R., Jurat-Fuentes, J.L., 2014. *Spodoptera frugiperda* (J.E. Smith) with field-evolved resistance to Bt maize are susceptible to Bt pesticides. J. Invertebr. Pathol. 122, 52−54.

Jallouli, A., Driss, F., Fillaudeau, L., Rouis, S., 2020. Review on biopesticide production by *Bacillus thuringiensis* subsp. *kurstaki* since 1990: Focus on bioprocess parameters. Process Biochem. 98, 224−232.

James, C., 2013. Global Status of Commercialized Biotech/GM Crops. ISAAA, Ithaca.

Jurat-Fuentes, J.L., Adang, M.J., 2006. Cry toxin mode of action in susceptible and resistant *Heliothis virescens* larvae. J. Invertebr. Pathol. 92, 166−171.

Jurat-Fuentes, J.L., Crickmore, N., 2017. Specificity determinants for Cry insecticidal proteins: insights from their mode of action. J. Invertebr. Pathol. 42, 5−10.

Karimi, J., Dara, S.K., Arthurs, S., 2018. Microbial insecticides in Iran: history, current status, challenges and perspective. J. Invertebr. Pathol. https://doi.org/10.1016/j.jip.2018.02.016.

Kesho, A., 2020. Microbial Bio-Pesticides and Their Use in Integrated Pest Management. Chem. Biomol. Eng. 5 (26). https://doi.org/10.11648/j.cbe.20200501.15.

Khaleghi, M., Khorrami, S., Ravan, H., 2019. Identification of *Bacillus thuringiensis* bacterial strain isolated from the mine soil as a robust agent in the biosynthesis of silver nanoparticles with strong antibacterial and anti-biofilm activities. Biocatal. Agric. Biotechnol. 101047. https://doi.org/10.1016/j.bcab.2019.101047.

Knowles, B.H., 1994. Mechanism of action of *Bacillus thuringiensis* insecticidal δ-endotoxins. In: Evans, P.D. (Ed.), Advances in Insect Physiology. Academic Press, London, pp. 275−308.

Koffi, D., Kyerematen, R., Eziah, V.Y., Agboka, K., Adom, M., Goergen, G., Meagher Jr., R.L., 2020. Natural enemies of the fall armyworm, *Spodoptera frugiperda* (J.E. Smith) (Lepidoptera: Noctuidae) in Ghana. Fla. Entomol. 103, 85−90.

Kogan, M., 1988. Integrated pest management theory and practice. Entomol. Exp. Appl. 49, 59−70. https://doi.org/10.1111/j.1570-7458.1988.tb02477.x.

Konecka, E., Kaznowski, A., Ziemnicka, J., Ziemnicki, K., 2007. Molecular and phenotypic characterisation of *Bacillus thuringiensis* isolated during epizootics in *Cydia pomonella* L. J. Invertebr. Pathol. 94, 56−63. https://doi.org/10.1016/j.jip.2006.08.008.

Krieg, A., Langenbruch, G.A., 1981. Susceptibility of arthropod species to *Bacillus thuringiensis*. In: Burges, H.D. (Ed.), Microbial Control of Pests and Plant Diseases 1970-1980. Academic, New York, pp. 837−898.

Kumar, D.S., Tarakeswari, M., Lakshminarayana, M., Sujatha, M., 2016. Toxicity of *Bacillus thuringiensis* crystal proteins against eri silkworm, *Samia cynthia ricini* (Lepidoptera: Saturniidae). J. Invertebr. Pathol. 138, 116−119. https://doi.org/10.1016/j.jip.2016.06.012.

Kumar, K., Sridhar, J., Kanagaraj Murali-Baskaran, R., Senthil-Nathan, S., Kaushal, P., Dara, S.K., Arthurs, S., 2018. Microbial biopesticides for insect pest management in India: current status and future prospects. J. Invertebr. Pathol. https://doi.org/10.1016/j.jip.2018.10.008.

Lacey, L.A., Grzywacz, D., Shapiro-Ilan, D.I., Frutos, R., Brownbridge, M., Goettel, M.S., 2015. Insect pathogens as biological control agents: back to the future. J. Invertebr. Pathol. 132, 1−41. https://doi.org/10.1016/j.jip.2015.07.009.

Lambert, B., Peferoen, M., 1992. Insecticidal promise of *Bacillus thuringiensis*. Bioscience 42, 112−122. https://doi.org/10.2307/1311652.

Lecadet, M.M., Frachon, E., Dumanoir, V.C., Ripouteau, H., Hamon, S., Laurent, P., Thiery, I., 1999. Updating the H-antigen classification of *Bacillus thuringiensis*. J. Appl. Microbiol. 86, 660−672.

Liu, Y., Wang, Y., Shu, C., Lin, K., Song, F., Bravo, A., Soberon, M., Zhang, J., 2017. Cry64Ba and Cry64Ca, Two ETX/MTX2-Type *Bacillus thuringiensis* Insecticidal proteins active against hemipteran pests. Appl. Environ. Microbiol. 84. https://doi.org/10.1128/aem.01996-17.

Lone, S.A., Malik, A., Padaria, J.C., 2017. Selection and characterization of *Bacillus thuringiensis* strains from northwestern Himalayas toxic against *Helicoverpa armigera*. Microbiology (Read.) 6, e00484. https://doi.org/10.1002/mbo3.484.

Luthy, P., Wolfersberger, M.G., 2000. Pathogenisis of *Bacillus thuringiensis* toxins. In: Charles, J.F., Delécluse, A., Nielsen-Le Roux, C. (Eds.), Entomopathogenic Bacteria: From Laboratory to Field Application. Kluwer Academic Publishers, Netherlands, pp. 167−180.

Lysenko, O., 1983. *Bacillus thuringiensis*: evolution of a taxonomic conception. J. Invertebr. Pathol. 41, 295−298.

Machado, D.H.B., Livramento, K.G. do, Máximo, W.P.F., Negri, B.F., Paiva, L.V., Valicente, F.H., 2020. Molecular characterization of *Bacillus thuringiensis* strains to control *Spodoptera eridania* (Cramer) (Lepidoptera: Noctuidae) population. Rev. Bras. Entomol. 64, e201947. https://doi.org/10.1590/1806-9665-rbent-2019-47.

Magalhães, B., Monnerat, R.G., Alves, S.B., 1998. Interações entre entomopatógenos, parasitoides e predadores. In: Alves, S.B. (Ed.), Controle Microbiano de Insetos, Fealq, Piracicaba, pp. 445−498.

Maksimov, I.V., Maksimova, T.I., Sarvarova, E.R., Blagova, D.K., Popov, V.O., 2018. Endophytic bacteria as effective agents of new-generation biopesticides (Review). Appl. Biochem. Microbiol. 54, 28−140. https://doi.org/10.1134/s0003683818020072.

Malovichko, Y.V., Nizhnikov, A.,A., Antonets, K.S., 2019. Repertoire of the *Bacillus thuringiensis* virulence factors unrelated to major classes of protein toxins and its role in specificity of host-pathogen interactions. Toxins 11, 347.

Marrone, P.G., 2019. Pesticidal natural products − status and future potential. Pest Manag. Sci. 75, 2325−2340. https://doi.org/10.1002/ps.5433.

Martin, P.A.W., Travers, R.S., 1989. Worldwide abundance and distribution of *Bacillus thuringiensis* isolates. Appl. Environ. Microbiol. 55, 2437−2442.

Meadows, M.P., 1993. *Bacillus thuringiensis* in the environment: ecology and risk assessment. In: Entwistle, P.F., Cory, J.S., Bailey, M.J., Higgs, S. (Eds.), *Bacillus thuringiensis*, an Environmental Biopesticide: Theory and Practice. John Wiley, Chichester, pp. 193−220.

Meadows, M.P., Ellis, D.J., Butt, J., Jarret, P., Burges, H.D., 1992. Distribution, frequency, and diversity of *Bacillus thuringiensis* in an animal feed mill. Appl. Environ. Microbiol. 58, 1344−1350.

Mendoza-Almanza, G., Esparza-Ibarra, E.L., Ayala-Luján, J.L., Mercado-Reyes, M., Godina-González, S., Hernández-Barrales, M., Olmos-Soto, J., 2020. The cytocidal spectrum of *Bacillus thuringiensis* toxins: from insects to human cancer cells. Toxins 12, 301. https://doi.org/10.3390/toxins12050301.

Mohd-Salleh, M.B., 1980. Effect of crystals, spores, and exotoxins of six varieties of Bacillus thuringiensis on selected corn insects. Retrosp. Thesis & Diss. 6744. http://lib.dr.iastate.edu/rtd/6744.

Molina-OChoa, J, Carpenter, J.E., Heinrich, E.A., FostJer, J.E., 2003. Parasitoids and parasites of *Spodoptera frugiperda* (Lepidoptera: Noctuidae) in the Americas and Caribbean Basin: an inventory. Fla. Entomol. 86 (3), 254–289.

Monnerat, R.G., 1995. Interrelations entre la teigne des crucifires *Plutella xylostella* (L.) (Lep: Yponomeutidae), son parasitoide Diadegma sp. (Hym: Ichneumonidae) et la bacttrie entomopathogtne *Bacillus thuringiensis* Berliner. These de Doctorat en Sciences Agronomiques. ENSA, Montpellier, p. 162.

Monnerat, R.G., Soares, C.M., Capdeville, G., Jones, G., Martins, E.S., Praça, L., Cordeiro, B.A., Braz, S.V., Santos, R.C., Berry, C., 2009. Translocation and insecticidal activity of *Bacillus thuringiensis* living inside of plants. Microb. Biotechnol. 2, 512–520.

Moreau, C., Bauce, E., 2003. Lethal and sublethal effects of single and double applications of *Bacillus thuringiensis* variety *kurstaki* on spruce budworm (Lepidoptera: Tortricidae) larvae. J. Econ. Entomol. 96, 280–286.

Mullis, K., Faloona, F., 1987. Specific synthesis of DNA in vitro via a polymerase catalysed chain reaction. Methods Enzymol. 55, 335–350.

Mushtaq, R., Behle, R., Liu, R., Niu, L., Song, P., Shakoori, A.R., Jurat-Fuentes, J.L., 2017. Activity of *Bacillus thuringiensis* Cry1Ie2, Cry2Ac7, Vip3Aa11 and Cry7Ab3 proteins against *Anticarsia gemmatalis, Chrysodeixis includens* and *Ceratoma trifurcata*. J. Invertebr. Pathol. 150, 70–72. https://doi.org/10.1016/j.jip.2017.09.009.

Nascimento, P.T., Fadini, M.A.M., Valicente, F.H., Ribeiro, P.E.A., 2018. Does *Bacillus thuringiensis* have adverse effects on the host egg location by parasitoid wasps? Rev. Bras. Entomol. 62, 260–266.

Nealis, V., van Frankenhuyzen, K., 1990. Interactions between *Bacillus thuringiensis* berliner and *Apanteles fumiferanae* vier. (Hymenoptera: Braconidae), a parasitoid of the spruce budworm, *Choristoneura fumiferana* (Clem.) (Lepidoptera: Tortricidae). Can. Entomol. 122, 585–594.

Nicolopoulou-Stamati, P., Maipas, S., Kotampasi, C., Stamatis, P., Hens, L., 2016. Chemical pesticides and human health: the urgent need for a new concept in agriculture. Publ. Health 18. https://doi.org/10.3389/fpubh.2016.00148.

Niwa, C.G., Stelzer, M.J., Beckwith, R.C., 1987. Effects of *Bacillus thuringiensis* on parasites of western spruce budworm (Lepidoptera: Tortricidae). J. Econ. Entomol. 80, 750–753.

Norris, J.R., 1964. The classification of *Bacillus thuringiensis*. J. Appl. Bacteriol. 27, 439–447. https://doi.org/10.1111/j.1365-2672.1964.tb05053.x.

Ochoa-Campuzano, C., Martínez-Ramírez, A.C., Contreras, E., Rausell, C., Dolores Real, M., 2013. Prohibitin, an essential protein for Colorado potato beetle larval viability, is relevant to *Bacillus thuringiensis* Cry3Aa toxicity. Pestic. Biochem. Physiol. 107, 299–238.

Oppert, B., Kramer, K.J., Johnson, D.E., Macintosh, S.C., Mcgaughey, W.H., 1994. Altered protoxin activation by midgut enzymes from a *Bacillus* thuringiensis resistant strain of *Plodia interpunctella*. Biochem. Biophys. Res. Commun. 198, 940–947. https://doi.org/10.1006/bbrc.1994.1134.

Panwar, B.S., Kaur, J., Kumar, P., Kaur, S., 2018. A novel cry52Ca1 gene from an Indian *Bacillus thuringiensis* isolate is toxic to *Helicoverpa armigera* (cotton boll worm). J. Invertebr. Pathol. 159, 137–140. https://doi.org/10.1016/j.jip.2018.11.002.

Pardo-López, L., Soberón, M., Bravo, A., 2013. *Bacillus thuringiensis* insecticidal toxins: mode of action, insect resistance and consequences for crop protection. FEMS Microbiol. Rev. 37, 3–22.

Park, Y., González-Martínez, R.M., Navarro-Cerrillo, G., Chakroun, M., Kim, Y., Ziarsolo, P., Blanca, J., Cañizares, J., Ferré, J., Herrero, S., 2014. ABCC transporters mediate insect resistance to multiple Bt toxins revealed by bulk segregant analysis. BMC Biol. 12, 1–15.

Patel, L.C., Pramanik, A., 2012. Impact of bio-pesticides and insecticides on *Trichogramma chilonis* Ishii. J. Insect Sci. 25, 281–285.

Patel, K.D., Purani, S., Ingle, S.S., 2013. Distribution and diversity analysis of *Bacillus thuringiensis cry* genes in different soil types and geographical regions of India. J. Invertebr. Pathol. 112, 116–121. https://doi.org/10.1016/j.jip.2012.10.008.

Paulino-Lima, I.G., Azua-Bustos, A., Vicuña, R., González-Silva, C., Salas, L., Teixeira, L., Rosado, A., Leitao, A.A.C., Lage, C., 2012. Isolation of UVC-tolerant bacteria from the hyperarid Atacama Desert, Chile. Microb. Ecol. 65, 325–335. https://doi.org/10.1007/s00248-012-0121-z.

Peña-Cardeña, A., Grande, R., Sánchez, J., Tabashnik, B.E., Bravo, A., Soberón, M., Gómez, I., 2018. The C-terminal protoxin region of *Bacillus thuringiensis* Cry1Ab toxin has a functional role in binding to GPI-anchored receptors in the insect midgut. J. Biol. Chem. 293, 20263–20272.

Pigott, C.R., Ellar, D.J., 2007. Role of receptors in *Bacillus thuringiensis* crystal toxin activity. Microbiol. Mol. Biol. Rev. 71, 255–281.

Pimentel, D., Burgess, M., 2014. Pesticides applied worldwide to combat pests. In: Peshin, R., Pimentel, D. (Eds.), Integrated Pest Management. Springer, Dordrecht, pp. 1–12.

Pinnock, D.E., Brand, R.J., Milstead, J.E., 1971. The field persistence of *Bacillus thuringiensis* spores. J. Invertebr. Pathol. 18, 405–411.

Pinnock, D.E., Brand, R.J., Jackson, K.L., 1974. The field persistence of *Bacillus thuringiensis* spores on *Cercis occidentalis* leaves. J. Invertebr. Pathol. 23, 341–346.

Pinnock, D.E., Brand, R.J., Milstead, J.E., 1975. Effect of three species on the coverage and field persistence of *Bacillus thuringiensis* spores. J. Invertebr. Pathol. 25, 209–214.

Pinto, C.P.G., De Azevedo, E.B., Zéro dos Santos, A.L., Camila, P.V.C., Fernandes, F.O., Guilherme, D.R., Polanczyk, R.A., 2019. Immune response and susceptibility to *Cotesia flavipes* parasitizing *Diatraea saccharalis* larvae exposed to and surviving LC25 dosage of Bacillus thuringiensis. J. Invertebr. Pathol. https://doi.org/10.1016/j.jip.2019.107209.

Polanczyk, R.A., 2004. Estudos de *Bacillus thuringiensis* Berliner visando ao controle de Spodoptera frugiperda (J.E.Smith). Tese (Doutorado em Entomologia Agricola). Escola Superior de Agricultura Luiz de Queiroz – USP. Piracicaba, 158 f.

Polanczyk, R.A., Valicente, F.H., Barreto, M.R., 2008. Utilização de *Bacillus thuringiensis* no controle de pragas agrícolas na América Latina. In: Alves, S.B., Lopes, R.B. (Eds.), Controle Microbiano de Pragas na América Latina: avanços e desafios. FEALQ, Piracicaba, pp. 111–136.

Polanczyk, R.A., van Frankenuyzen, K., Pauli, G., 2017. The American *Bacillus thuringiensis* based biopesticides market. In: Fiuza, L.M., Polanczyk, R.A., Crickmore, N. (Eds.), *Bacillus Thuringiensis* and *Lysinibacillus Sphaericus*. Characterization and Use in the Field of Biocontrol. Springer International Publishing, Berlim, pp. 173–184.

Poornima, K., Selvanayagam, P., Shenbagarathai, R., 2010. Identification of native *Bacillus thuringiensis* strain from South India having specific cytocidal activity against cancer cells. J. Appl. Microbiol. 109, 348–354. https://doi.org/10.1111/j.1365-2672.2010.04697.x.

Porcar, M., Caballero, P., 2002. Molecular and insecticidal characterization of a *Bacillus thuringiensis* strain isolated during a natural epizootic. J. Appl. Microbiol. 89, 309–316.

Porcar, M., Navarro, L., Jiménez-Peydró, R., 2006. Pathogenicity of intrathoracically administrated *Bacillus thuringiensis* spores in *Blatta orientalis*. J. Invertebr. Pathol. 93, 63–66. https://doi.org/10.1016/j.jip.2006.05.001.

Radosavljevic, J., Naimov, S., 2016. Toxicity of *Bacillus thuringiensis* (L.) Cry proteins against summer fruit tortrix (*Adoxophyes orana* - Fischer von Rösslerstamm). J. Invertebr. Pathol. 138, 63–65. https://doi.org/10.1016/j.jip.2016.06.004.

Radwan, M.A., 2007. Efficacy of *Bacillus thuringiensis* integrated with other non-chemical materials to control *Meloidogyne incognita* in tomato. Nematol. Mediterr. 35, 69–73.

Rao, T., Jurat-Fuentes, J., 2020. Advances in the use of entomopathogenic bacteria/microbial control agents (MCAs) as biopesticides in suppressing crop insect pests. In: Birch, N., Glarem, T. (Eds.), Biopesticides for Sustainable Agriculture. Burleigh Dodds Science Publishing, Cambridge, pp. 1–37.

Raun, E.S., 1963. Corn borer control with *Bacillus thuringiensis* Berliner. Iowa State J. Sci. 38, 141–150.

Raun, E.S., Jackson, R.D., 1966. Encapsulation as a technique for formulating microbial and chemical insecticides. J. Econ. Entomol. 59, 612–622.

Raun, E.S., Sutter, G.R., Revelo, M.A., 1966. Ecological factors affecting the pathogenicity of *Bacillus thuringiensis* var. *thuringiensis* to the European corn borer and fall armyworm. J. Invertebr. Pathol. 8, 365–375.

Raymond, B., 2017. The biology, ecology and taxonomy of *Bacillus thuringiensis* and related bacteria. In: Fiuza, L.M., Polanczyk, R.A., Crickmore, N. (Eds.), *Bacillus thuringiensis* and *Lysinibacillus sphaericus*. Characterization and Use in the Field of Biocontrol. Springer International Publishing, Berlim, pp. 19–40.

Raymond, B., Federici, B., 2018. An appeal for a more evidence-based approach to biopesticide safety in the EU. FEMS Microbiol. Ecol. 94. https://doi.org/10.1093/femsec/fix169.

Raymond, B., Johnston, P.R., Nielsen-LeRoux, C., Lereclus, D., Crickmore, N., 2010. *Bacillus thuringiensis*: an impotent pathogen? Trends Microbiol. 18, 189–194. https://doi.org/10.1016/j.tim.2010.02.006.

Reyaz, A.L., Arulselvi, P.I., 2016. Cloning, characterization and expression of a novel haplotype *cry2A*-type gene from *Bacillus thuringiensis* strain SWK1, native to Himalayan valley Kashmir. J. Invertebr. Pathol. 136, 1–6. https://doi.org/10.1016/j.jip.2016.02.005.

Salama, H.S., 1984. *Bacillus thuringiensis* Berliner and its role as a biological control agent in Egypt. Z. Angew. Entomol. 98, 206–220. https://doi.org/10.1111/j.1439-0418.1984.tb02702.x.

Salama, H.S., Zaki, F.N., Sharaby, A.F., 1982. Effect of *Bacillus thuringiensis* Berl. on parasites and predators of the cotton leafworm *Spodoptera littoralis* (Boisd.). J. Appl. Entomol. 94, 498–504.

Saleh, S.M., Harris, R.F., Allen, O.N., 1969. Method for determining *Bacillus thuringiensis* var. *thuringiensis* Berliner in soil. Can. J. Microbiol. 15, 1101–1104.

Sanahuja, G., Banakar, R., Twyman, R.M., Capell, T., Christou, P., 2011. *Bacillus thuringiensis*: a century of research, development and commercial applications. Plant Biotechnol. J. 9, 283–300. https://doi.org/10.1111/j.1467-7652.2011.00595.x.

Sanchis, V., Bourguet, D., 2008. *Bacillus thuringiensis*: applications in agriculture and insect resistance management. A review. Agron. Sustain. Dev. 28, 11–20. https://doi.org/10.1051/agro:2007054.

Santana, M.A., Moccia-V, C.C., Gillis, A.E., 2008. *Bacillus thuringiensis* improved isolation methodology from soil samples. J. Microbiol. Methods 75, 357–358. https://doi.org/10.1016/j.mimet.2008.06.008.

Schnepf, E., Crikmore, N., Van Rie, J., Lereclus, D., Baum, J., Feitelson, J., Zeigler, D.R., Dean, D.H., 1998. *Bacillus thuringiensis* and its pesticidal crystal proteins. Microbiol. Mol. Biol. Rev. 62, 775–806.

Schreiner, V.C., Szöcs, E., Bhowmik, A.K., Vijver, M.G., Ralf, B., Schäfer, R.B., 2016. Pesticide mixtures in streams of several European countries and the USA. Sci. Total Environ. 573, 680–689.

Schünemann, R., Knaak, N., Fiuza, L.M., 2014. Mode of action and specificity of *Bacillus thuringiensis* toxins in the control of caterpillars and stink bugs in soybean culture. ISRN Microbiol. https://doi.org/10.1155/2014/135675.

Shapiro-Ilan, D.I., Fuxa, J.R., Lacey, L.A., Onstad, D.W., Kaya, H.K., 2005. Definitions of pathogenicity and virulence in invertebrate pathology. J. Invertebr. Pathol. 88, 1–7. https://doi.org/10.1016/j.jip.2004.10.003.

Sharanabasappa, Kalleshwaraswamy, C.M., Poorani, J., Maruthi, M.S., Pavithra, H.B., Divaviam, J., 2019. Natural enemies of *Spodoptera* frugiperda (J. E. Smith) (Lepidoptera: Noctuidae), a recent invasive pest on maize in South India. Fla. Entomol. 102, 619–623.

Silveira, L.F.V., Polanczyk, R.A., Pratissoli, D., Franco, C.R., 2011. Seleção de isolados de *Bacillus thuringiensis* para *Tetranychus urticae* Koch. Arq. Inst. Biol. 78, 273–278.

Sisterson, M.S., Tabashnik, B.E., 2005. Simulated effects of transgenic Bt crops on specialist parasitoids of target pests. Environ. Entomol. 34, 733–742. https://doi.org/10.1603/0046-225X-34.4.733.

Soberón, M., Portugal, L., Garcia-Gómez, B.I., Sánchez, J., Onofre, J., Gómez, I., Pacheco, S., Bravo, A., 2017. Cell lines as models for the study of Cry toxins from *Bacillus thuringiensis*. Insect Biochem. Mol. Biol. 93, 66–78.

Somerville, H.J., Pockett, H.Y., 1975. An insect toxin from spores of *Bacillus thuringiensis*. J. Gen. Microbiol. 87, 359–369.

Souza, C.S.F., Silveira, L.C.P., Pitta, R.M., Waquil, J.M., Pereira, E.J.G., Mendes, S.M., 2019. Response of field populations and Cry-resistant strains of fall armyworm to Bt maize hybrids and Bt-based bioinsecticides. Crop Protect. https://doi.org/10.1016/j.cropro.2019.01.001.

Sparks, T.C., et al., 2020. Insecticides, biologics and nematicides: updates to IRAC's mode of action classification - a tool for resistance management. Pestic. Biochem. Physiol. https://doi.org/10.1016/j.pestbp.2020.104587.

Steinhaus, E.A., 1957. Concerning the harmlessness of insect pathogens and the standardization of microbial control products. J. Econ. Entomol. 50, 7I5–720.

Swiecicka, I., Fiedoruk, K., Bednarz, G., 2020. The occurrence and properties of *Bacillus thuringiensis* isolated from free-living animals. Lett. Appl. Microbiol. 34, 194–198.

Tabashnik, B.E., Carrière, Y., 2017. Surge in insect resistance to transgenic crops and prospects for sustainability. Nat. Biotechnol. 35, 926–935.

Tabashnik, B.E., Carrière, Y., 2019. Global patterns of resistance to Bt crops highlighting pink bollworm in the United States, China, and India. J. Econ. Entomol. 112, 2513–2523.

Tabashnik, B.E., Mota-Sanchez, D., Whalon, M.E., Hollingworth, R.M., Carrière, Y., 2014. Defining terms for proactive management of resistance to Bt crops and pesticides. J. Econ. Entomol. 107, 496–507.

Talalaev, E.V., 1956. Septicemia of the caterpillars of the Siberian silkworm. Mikrobiologiya 25, 99–102.

Tay, W.T., Mahon, R.J., Heckel, D.G., Walsh, T.K., Downes, S., James, W.J., Lee, S.-F., Reineke, A., Williams, A.K., Gordon, K.H.J., 2015. Insect resistance to *Bacillus thuringiensis* toxin Cry2Ab is conferred by mutations in an ABC transporter subfamily A protein. PLOS Gen. 1–23.

Temerak, S.A., 1980. Detrimental effects of rearing a braconid parasitoid on the pink borer larvae inoculated by different concentrations of the bacterium, *Bacillus thuringiensis* Berliner. Z. Angew. Entomol. 89, 315–319.

Thammasittirong, A., Attathom, T., 2008. PCR-based method for the detection of *cry* genes in local isolates of *Bacillus thuringiensis* from Thailand. J. Invertebr. Pathol. 98, 121–126. https://doi.org/10.1016/j.jip.2008.03.001.

Travers, R.S., Martin, P.A.W., Reichefelder, C.F., 1987. Selective process for efficient isolation soil *Bacillus* sp. Appl. Environ. Microbiol. 53, 1263–1266.

Urban Jr., J.F., Hu, Y., Miller, M.M., Scheib, U., Yiu, Y.Y., Aroian, R.V., 2013. *Bacillus thuringiensis*-derived Cry5B has potent anthelmintic activity against *Ascaris suum*. PloS One. https://doi.org/10.1371/journal.pntd.0002263 e2263.

Vachon, V., Laprade, R., Schwartz, J., 2012. Current models of the mode of action of *Bacillus thuringiensis* insecticidal crystal proteins: a critical review. J. Invertebr. Pathol. 111, 1–12.

Valicente, F.H., Zanasi, R.F., 2005. Uso de meios alternativos para produção de bioinseticida à base de *Bacillus thuringiensis*. Circular Técnica (Embrapa Milho e Sorgo) 1, 1–4.

Valicente, F.H., Lana, U.G.P., Pereira, A.C.P., Martins, J.L.A., Tavares, A.N.G., 2018. Riscos à produção de biopesticida à base de *Bacillus thuringiensis*. Circular Técnica (Embrapa Milho e Sorgo) 239, 1–20.

van Frankenhuyzen, K., 2009. Insecticidal activity of *Bacillus thuringiensis* crystal proteins. J. Invertebr. Pathol. 101, 1–16. https://doi.org/10.1016/j.jip.2009.02.009.

Vankova, J., Purrini, K., 1979. Natural epizootics caused by bacilli of the species *Bacillus thuringiensis* and *Bacillus cereus*. Z. Angew. Entomol. 88, 216–221.

Vemmer, M., Patel, A.V., 2013. Review of encapsulation methods suitable for microbial biological control agents. Biol. Contr. 67, 380–389.

Vogel, E., Santos, D., Mingels, L., Verdonckt, T.W., Broeck, J.V., 2019. RNA interference in insects: protecting beneficials and controlling pests. Front. Physiol. 9, 1912. https://doi.org/10.3389/fphys.2018.01912.

Wang, Y., Liu, Y., Zhang, J., Crickmore, N., Song, F., Gao, J., Shu, C., 2018. Cry78Aa, a novel *Bacillus thuringiensis* insecticidal protein with activity against *Laodelphax striatellus* and *Nilaparvata lugens*. J. Invertebr. Pathol. 158, 1–5. https://doi.org/10.1016/j.jip.2018.07.007.

World Health Organization, 1988. Informal Consultation on the Development of *Bacillus Sphaericus* as a Microbial Larvicide. World Bank/WHO, Geneva/UNDP, p. 24. Special Programme for Research and Training in Tropical Deseases (TDR).

Wu, S., Peng, Y., Huang, Z., Huang, Z., Xu, L., Ivan, G., Guan, X., Zhang, L., Zou, S., 2013. Isolation and characterization of a novel native *Bacillus thuringiensis* strain BRC-HZM2 capable of degrading chlorpyrifos. J. Basic Microbiol. 53, 1–9. https://doi.org/10.1002/jobm.201300501.

Yamvrias, C., 1962. Contribution à l'étude du mode d'action de *Bacillus thuringiensis* Berliner vis-à-vis de la teigne de la farine *Anagasta* (*Ephestia*) *kuehniella* Zeller (Lépidoptère). Entomophaga 7, 101–159.

Yu, Z., Xiong, J., Zhou, Q., Luo, H., Hu, S., Xia, L., Ming, S., Lin, L., Yu, Z., 2015. The diverse nematicidal properties and biocontrol efficacy of *Bacillus thuringiensis* Cry6A against the root-knot nematode *Meloidogyne hapla*. J. Invertebr. Pathol. 125, 73–80. https://doi.org/10.1016/j.jip.2014.12.011.

Zhang, X., Candas, M., Griko, N.B., Rose-Young, L., Bulla Jr., L.A., 2005. Cytotoxicity of *Bacillus thuringiensis* Cry1Ab toxin depends on specific binding of the toxin to the cadherin receptor BT-R1 expressed in insect cells. Cell Death Differ. 12, 1407–1416.

Zhang, X., Candas, M., Griko, N.B., Taussig, R., Bulla Jr., L.A., 2006. A mechanism of cell death involving an adenylyl cyclase/PKA signaling pathway is induced by the Cry1Ab toxin of *Bacillus thuringiensis*. PNSA USA 103, 9897–9902.

Zhang, Y., Ma, Y., Wan, P.J., Mu, L.L., Li, G.Q., 2013. *Bacillus thuringiensis* insecticidal crystal proteins affect lifespan and reproductive performance of *Helicoverpa armigera* and *Spodoptera exigua* adults. J. Econ. Entomol. 106, 614–621. https://doi.org/10.1603/EC12413.

Zhou, Z., Wang, Z., Liu, Y., Liang, G., Shu, C., Song, F., 2016. Identification of ABCC2 as a binding protein of Cry1Ac on brush border membrane vesicles from *Helicoverpa armigera* by an improved pull-down assay. Microbiol. Open 5, 659–669.

Chapter 10

Biopesticides for management of arthropod pests and weeds

Josef Jampílek[a, b] **and Katarína Kráľová**[c]
[a]*Department of Analytical Chemistry, Faculty of Natural Sciences, Comenius University, Bratislava, Slovakia;* [b]*Department of Chemical Biology, Faculty of Science, Palacky University, Olomouc, Czech Republic;* [c]*Institute of Chemistry, Faculty of Natural Sciences, Comenius University, Bratislava, Slovakia*

10.1 Introduction

According to Environmental Protection Agency the biopesticides could be classifies into three major classes: (i) biochemical pesticides, (ii) microbial pesticides, (iii) plant-incorporated-protectants (PIPs). Biochemical pesticides are naturally occurring substances that control pests by non-toxic mechanisms. The active ingredients in microbial pesticides are microorganisms (e.g., a bacterium, fungus, virus or protozoan) and these biopesticides can control many different kinds of pests, whereby each active ingredient exhibit relatively specific activity for its target pest(s). Compounds produced by plant from genetic material that has been added to the plant (e.g., *Bacillus thuringiensis* pesticidal protein) are called plant-incorporated-protectants (PIPs) (EPA, 2020). In contrast to parasitoids, predators or entomopathogenic nematodes, i.e. biocontrol agents that actively seek out the pest, as biopesticides are considered passive biocontrol agents (Dimetry, 2014). Synthetic herbicides frequently used in excess amounts greatly contributed to increasing environmental pollution and rapidly increasing evolution of plant resistance, whereby also the number of chemical herbicides, which were banned due to adverse impact on human health gradually rise. Resistance was recorded in at least 954 species of pests, including 546 arthropods, 218 weeds, and 190 plant pathogens Tabashnik et al. (2014). Therefore new generation of herbicides based on natural compounds and synthetic herbicides with new target sites utilizing the structures of natural phytotoxin is urgently needed (Dayan and Duke, 2014). However, using only one natural herbicide for suppressing weed occurrence usually does not achieve the herbicidal effectiveness of chemical and therefore it is favorable to apply combination of natural herbicides (Saini and Singh, 2019). Ongoing climate change will also impact spread of weeds in certain regions. At elevated CO_2 levels C_3 weeds in C_4 or C_3 crops could become a problem, particularly in tropical regions and temperature increases will expand the geographical range, particularly for C_4 weeds resulting in yields losses. Moreover, due to increasing expansion of invasive weed species at higher temperatures the cost for their control will increase (Korres et al., 2016). According to Perotti et al. (2020) the global climate change may act on the selection of herbicide resistant weeds, particularly those evolving into non-target-site resistance mechanisms.

On the other hand, arthropods destroy an estimated 18%–26% of annual global crop production, at a value exceeding $ 470 billion, whereby the greater proportion of yield losses (13%–16%) occurs in the field, before harvest (Culliney, 2014). More than 10,000 species of insects were reported to damage food crops (Sallam, 2013), causing in some cases yield reduction up to 60%–70% (Singhand and Gandhi, 2012). Temperature increases expected at global climate change conditions will result in further yield lost in wheat, rice, and maize caused by insects, even by 10%–25% per Celsius degree of warming, the temperate zone being mostly affected (Deutsch et al., 2018). Increasing temperatures will pronouncedly affect species population dynamics, due to warmer winter temperatures mortality of insect species will be reduced and will result in to poleward range expansions (Stange and Ayres, 2010) and harmful insects causing immense damages in cereals, grain legumes, vegetables, and fruit crops (e.g., cereal stem borers, the pod borers, aphids, and white flies) may move to temperate regions, whereby the effectiveness of pest-resistant cultivars, natural enemies, and pesticides will be reduced as well (Sharma et al., 2018). Although recently mostly synthetic pesticides greatly contributing to

Biopesticides. https://doi.org/10.1016/B978-0-12-823355-9.00009-2

environmental pollution and having adverse impact on non-target organisms are widely applied (many times also in excess) to control noxious arthropods, concern over utilization of biopesticides gradually increases. It was shown that such environment friendly approach can be more sustainable and provide greater economic benefits, while the return per dollar invested in ecologically-based biological and cultural pest controls is considerably higher compared to control using synthetic pesticides (from $ 30−300 vs. $ 4) (Culliney, 2014).

As biopesticides used to control noxious weeds and harmful arthropods mainly plant extracts (Hassan and Gökçe, 2014; Tembo et al., 2018; Dougoud et al., 2019) and essential oils, eventually their bioactive constituents showing desirable activity (Mkindi et al., 2020; Pavela and Benelli, 2016; Cloyd et al., 2009; Ayvaz et al., 2010; Said-Al Ahl et al., 2017) and effective compounds produced by microorganisms (Hajek et al., 2009; Lacey, 2017; Chattopadhyay et al., 2017; Kachhawa, 2017; Ruiu, 2018; Mnif and Ghribi, 2015) are used. However, similarly to synthetic pesticides, higher effectiveness of biopesticides can be obtained using nanoformulations with encapsulated active ingredients, which increase their stability, show controlled release, whereby lower amount of biopesticide is needed for the same effect than at application of bulk form of respective biopesticide (Jampílek and Králová, 2017a, 2018a, 2019a, Jampílek et al., 2019, 2020; Kumar et al., 2019a; Ibrahim, 2019; Camara et al., 2019; Shang et al., 2019; Kitherian, 2017).

Mosquitoes and similar stinging/sucking insects are widespread from the tropics to the temperate climate zone. In addition to the unpleasant harassment of humans and animals and the sucking of their blood, they are above all important vectors of diseases. Mosquito-borne diseases are diseases caused by viruses, bacteria or parasites, the causative agents of which survive in the mosquito, which releases them inadvertently into the body of its victim while sucking blood. According to statistics, approximately 700 million people get mosquito-borne diseases every year, resulting in more than a million deaths. Mosquito-borne diseases include dengue fever, West Nile virus, chikungunya, yellow fever, Japanese encephalitis, St. Louis encephalitis, Eastern and Western equine encephalitis, La Crosse encephalitis, Venezuelan equine encephalitis, Ross River fever, Barmah forest fever, Zika fever, Rift Valley fever and Keystone virus. Also included are tularemia, malaria, sleeping sickness (tse-tse fly), filariasis, dirofilariasis, etc. (Caraballo and King, 2014; Mosquitoborne Diseases, 2020; AMCA, 2020). In addition to repellents, aromatic plants full of essential oils can be used to fight mosquitoes. These plants, belonging mainly to the families *Lamiaceae*, *Poaceae* and *Pinaceae*, are common all over the world, they can be kept directly in the room, these plants can be spread or the essential oils obtained from them can be used directly in the form of aromatic oils (Mosquito Squad, 2020; Noutcha et al., 2016; Lee, 2018; Dahmana and Mediannikov, 2020; Gillij et al., 2008). As above-mentioned, the efficiency of these essential oils is prolonged and increased, for example, by their encapsulation in nanosystems (Jampílek and Králová, 2017a, 2018a, 2019a; Jampílek et al., 2019, 2020; Kumar et al., 2019a; Ibrahim, 2019; Camara et al., 2019; Shang et al., 2019; Kitherian, 2017). In addition to these terpenoid compounds, nanoparticles of metals, metal oxides, and metalloids also have a significant effect on mosquitoes, wherein this type of nanoparticles acts primarily on mosquito larvae. Recently, there has been frequent preparation of nanoparticles, so-called green synthesis, most often by reduction of metals using various extracts from leaves, flowers, bark, fruits and various "waste" biomaterials (Jampílek and Králová, 2018a, 2019a; Jampílek et al., 2020; Benelli, 2016, 2018a; Benelli et al., 2017; Hajra et al., 2016; Marslin et al., 2018; Soni and Prakash, 2014; Priya and Santhi, 2014; Amutha et al., 2019; Al-Dhabi and Arasu, 2018).

Nanotechnology has begun to evolve with the new millennium, and since the second decade of the 21st century we can say that it has been applied in a wide range of human activities (Wennersten et al., 2008; Drexler and Pamlin, 2013; Plachá and Jampílek, 2019). Nanotechnology takes advantage of the specific properties of materials that begin to emerge, when the size of a particular material is in the range of 1 to 1000 nm. Apart from industry and various materials, nanotechnology can be found in biomedical fields and, of course, in the agriculture and the food industry (Jampílek and Králová, 2017b, 2018b, 2019b; Jampílek et al., 2019). The application of nanotechnologies is an excellent alternative to improving the profile of existing pesticide agents. Thus, nanomaterials are an alternative approach to the treatment and alleviation resistant pathogens (Jampílek and Králová, 2017b, 2019b, 2020a, 2020b, 2020c; Kumar et al., 2019a; Ibrahim, 2019; Camara et al., 2019; Shang et al., 2019; Kitherian, 2017). In addition, the use of various nanosystems/nanoformulations can increase the bioavailability of the active agents and modify the route of administration. As mentioned, specific nanoformulations also provide a controlled release or targeted biodistribution system. Based on these facts, smaller amounts of active substances can be used, i.e. dose-dependent toxicity and thus various side/unfavorable effects can be reduced. Increased efficacy of individual insecticides can also be provided by fixed combinations of various agents, combinations with metal- or metalloid-based nanoparticles, or by biologically active matrices that are used as stabilizers. In addition, formulations also protect bioactive agents from degradation (Jampílek and Králová, 2017b; Jampílek et al., 2019; Camara et al., 2019; Shah et al., 2016; Kumar et al., 2019a; Elabasy et al., 2019; Kitherian, 2017; Meyer et al., 2015).

This chapter gives a comprehensive overview of recent findings related to the activity and mechanisms of action of environmentally friendly biopesticides including plant extracts and essential oils or their constituents, biopesticides produced by microorganisms, peptides obtained in spider venoms, and Cry proteins produced by *Bacillus thuringiensis* in their bulk as well as nanoscale form against noxious weeds and harmful arthropods causing losses of economically important crops. Special attention is also paid to the insecticidal activity of green synthesized metal nanoparticles against mosquitoes, which are important vectors of many diseases causing annually more than a million deaths in tropical regions.

10.2 Bioherbicides

Weeds can cause immense yield losses even reaching in average ca. 20%–37% of the world's agricultural output because they can compete with crop plants for natural resources such as light, moisture, nutrients and space and therefore weed control is indispensable during crop cultivation (Sindhu and Sehrawat, 2017; Jabran et al., 2015). In weed management various strategies could be used including crop rotation, mechanical weeding, harvest weed seed control, seed predation or application of herbicides in order to suppress yield losses (Bajwa, 2014; Bajwa et al., 2015; Cordeau et al., 2016; Dahiya et al., 2019a). The pre-emergence herbicides inhibit the seed germination stage of weeds, while post-emergent herbicides can be applied over crops throughout the growing season (Beckie et al., 2019). However, overuse of chemical pesticides in last 70 years markedly contributed to soil degradation, environmental pollution, adverse impact on non-target organisms and growing number of herbicide resistant weed biotypes, whereby some synthetic herbicides were also found to seriously damage health of humans and their use was banned. Therefore increased attention, mainly in developing countries, was devoted to bioherbicides, i.e. compounds of natural origin showing herbicidal activity (Auld and Morin, 1995; Aktar et al., 2009; Glab et al., 2017; Schütte et al., 2017; Farina et al., 2019; Saini and Singh, 2019; Shang et al., 2019; Maluin and Hussein, 2020; Rajmohan et al., 2020).

Some plants using allelopathic mechanisms can inhibit or destroy the other plants/weeds via the release of secondary metabolites (allelochemicals) into the environment, whether through the volatilization from living parts of the plant, the leaching from aboveground parts of the plant, the decomposition of plant material or the root exudation, which is considered as the most important pathway for the release of allelochemicals (Scavo et al., 2018, 2019a,b; Li et al., 2019; Macias et al., 2019). It was reported that powerful allelopathic effects on plants exhibit more than 2000 plant species (39 families), although from ca 400,000 compounds in plants, which show allelopathic activities, only 3% were found to act as bioherbicides (Einhellig and Leather, 1988; Saini and Singh, 2019). Allelochemicals can be used as natural herbicides, eventually the weed management can utilize the allelopathic interactions (Li et al., 2019). Recent research related to plant-derived allelochemicals and signaling chemicals and their roles in agricultural pest management was overviewed by Kong et al. (2019). Plant extracts or their metabolites showing herbicidal activity following absorption by weed seeds can damage the cell membrane, DNA, adversely affect mitosis and some biochemical processes resulting in delayed or suppressed seed germination and at post-emergence application can reduce weed growth causing impaired nutrient uptake and photosynthetic pigment synthesis, generation of reactive oxygen species (ROSs), etc. (Radhakrishnan et al., 2018). The isolation of plant allelopathic substances showing phytotoxic effects can be then utilized for the discovery of new natural herbicides (Kaab et al., 2020; Kong et al., 2019).

However, also many bacteria and fungi originating from the plant rhizospheres were found to inhibit the growth of weeds and could be used as bioherbicides (Dahiya et al., 2019b; Rakian et al., 2018a,b). Microbes can secrete lytic enzymes and toxic substances able to degrade the weed seed coat and utilizing the endosperm for survival, which results in the inhibition of seed germination and some secondary metabolites of harmful microbes cause disease, necrosis and chlorosis of weeds as well (Radhakrishnan et al., 2018). Bailey (2014) summarized the findings related to the bioherbicide approach to weed control using plant pathogens.

10.2.1 Plant extracts and essential oils with herbicidal activity

Essential oils (EOs) are volatile secondary metabolites of many higher plants, which were recognized as an important natural source of green pesticides. Their constituents are generally small amphiphilic compounds capable to cross the cell wall and directly interact with the plant plasma membrane (Lins et al., 2019), induce generation of reactive oxygen species (ROS) resulting in oxidative stress accompanied with membrane damage and electrolyte leakage (Ahuja et al., 2015) or cause alterations in normal structure of nuclei and chromosome (Aragao et al., 2015). A review paper focused on the phytotoxic effects of EOs and their components, which could be used in weed management, was presented by Amiri et al. (2013). Saad et al. (2019) investigated herbicidal activity of 12 natural compounds belonging to monoterpenes, phenylpropenes, and sesquiterpenes against *Echinochloa crus-galli* under laboratory and glasshouse conditions and found that cinnamaldehyde, eugenol, and thymol caused pronounced inhibition of seed germination and shoot growth of weed; *p*-cymene and *trans*-cinnamaldehyde were found to be the most effective inhibitors of root growth with EC_{50} values 0.22 and 0.34 mM, respectively, while thymol, *trans*-cinnamaldehyde, eugenol, farnesol, and nerolidol were classified as potent pre-emergent herbicides showing serious visible injury symptoms and a complete weed control already at doses 1.0% and 2.0%.

The crude extract of *Medinilla magnifica* Lindl. subjected to acid hydrolysis was found to inhibit the growth of *E. crus-galli*, *Cyperus iria*, and *Ludwigia hyssopifolia* weeds, *E. crus-galli* being the most sensitive weed and its adverse impact on chlorophyll (Chl) production was reflected in reduced biomass of the tested weeds (Tinio et al., 2019).

At 21 days after application of the leaf methanolic extracts at 30% concentration of Merkus pine (*Pinus merkusii* Jungh. et de Vriese) containing isopropyl palmitate (33.45%), isopropyl linoleate (11.89%) and 3,7,11,15-tetramethyl-2-hexadecen-1-ol (7.32%) to weed spiny amaranth (*Amaranthus spinosus* L.), the seedlings growth of weed was inhibited by 64.2%. More than 50% inhibition of *A. spinosus* was also observed at application of 30% leaf methanolic extracts of *Terminalia catappa* L. (containing 25.84% of lupeol, 15.43% of 22,23-dihydro-stigmasterol, and 9.81% of α-amyrin) and *Tectona grandis* L.f. containing 13.04% of d:b-friedo-b′:a′-neogammacer-5-en-3-ol (3β), 13.02% of 22,23-dihydro-stigmasterol, and 8.32% of n-hexadecanoic acid (Erida et al., 2020).

The phytochemical screening of secondary metabolites from *Crambe abyssinica* plants confirmed the presence or absence of total saponins, triterpenoids, flavonoids, coumarins, tannins, phenols and alkaloids, whereby it was found that different compounds were extracted using hexane, ethyl acetate and methanol. For example, flavonoids were found only in the ethyl acetate extract and saponin only in the methanol extract. The hexane and ethyl acetate extracts showed inhibitory effect on the *Bidens pilosa* L. weed but did not adversely affect *Glycin max* plants suggesting that herbicidal compounds of *C. abyssinica* could be used to control the weed hairy beggartick (Spiassi et al., 2019).

Treatment of *Amaranthus retroflexus* L., *Portulaca oleracea* L., *Stellaria media* (L.) Vill., and *Anagallis arvensis* L. weeds with ethanolic extracts prepared from lyophilized leaves of cultivated cardoon (*Cynara carduncculus* L. var. *altilis* DC.) containing as major compounds caffeoylquinic acids, followed by flavones (apigenin and luteolin derivatives) and by the sesquiterpene lactone cynaropicrin completely inhibited seed germination of tested weeds and high inhibitory efficiency was observed also at application of dried leaves extracts (Scavo et al., 2020). Methanolic extract of aerial parts of *C. cardunculus* plant containing as main bioactive compounds syringic acid, *p*-coumaric acid, myricitrin, quercetin and naringenin significantly inhibited germination and seedling growth of *Trifolium incamatum*, *Silybum marianum* and *Phalaris minor* weeds and caused necrosis or chlorosis, myricitrin, naringenin and quercetin being the most active herbicidal compounds. Herbicidal formulation containing the *C. cardunculus* crude methanolic extract exhibited the same herbicidal effect as the standard industrial bioherbicide containing pelargonic acid (Kaab et al., 2020). Comparison of the allelopathic activity of leaf aqueous extracts of the three botanical varieties of *Cynara cardunculus* L. (globe artichoke, cultivated and wild cardoon) on the growth of *A. retroflexus* L. and *P. oleracea* L. weeds showed that it decreased in following order: wild cardoon > cultivated cardoon > globe artichoke, *P. oleracea* being the most sensitive species. In *C. cardunculus* leaf aqueous extract 5 sesquiterpene lactones, 2 caffeoylquinic acids, 6 flavones and 1 lignan were estimated, although apigenin and luteolin 7-*O*-glucoronide were detected only in wild cardoon, 11,13-dihydro-deacylcynaropicrin and 11,13-dihydroxi-8-deoxygrosheimin were characteristic for cultivated cardoon, while apigenin 7-*O*-glucoside was a feature of globe artichoke (Scavo et al., 2019a). As the most relevant allelochemicals of *C. cardunculus* L. sesquiterpene lactones (STLs) were estimated (cynaropicrin being the most abundant), whereby their amount depended on the climatic conditions and was higher in the April harvest. The allelopathic activity, expressed as inhibition of wheat coleoptile elongation, was found to depend on the STL profile, whereby from six genotypes studied the highest STLs levels were observed on wild cardoon (Scavo et al., 2019b).

Extracts of *Reichardia tingitana*, an annual plant growing in different habitats of the Egyptian deserts, containing as major compounds quercetin, naringenin, ellagic, gallic, chlorogenic, and caffeic acids were reported considerably inhibit the germination and growth of *Amaranthus lividius* and *Chenopodium murale* weeds and could be integrated into weed control strategies as environment friendly bioherbicide (Abd-ElGawad et al., 2020). The volatile oils from the Egyptian ecospecies *Lactuca serriola* containing 34 compounds, mainly oxygenated sesquiterpenes and diterpenes, with the major compounds isoshyobunone (64.22%), isocembrol (17.35%), and alloaromadendrene oxide-1 (7.32%) showing pronounced antioxidant activity, were found to show considerable allelopathic activity on germination and seedling growth of the noxious weed, *B. pilosa* (Abd-ElGawad et al., 2019a). EO of Egyptian ecospecies *Cleome droserifolia* (Forssk.), in which sesquiterpene was the major class of constituents with dominant compounds *cis*-nerolidol, α-cadinol, δ-cadinene, and γ-muurolene inhibited germination of *Cuscuta trifolii*, *Melilotus indicus*, and *Chenopodium murale* showing IC_{50} values of 183.5, 159.0, and 157.5 μL/L, respectively (Abd El-Gawad et al., 2018).

Aqueous leaf extract of *Ludwigia hyssopifolia* (G. Don) Exell, a weed abundant in lowland rice fields, containing phenols, tannins, flavonoids, terpenoids, saponins, coumarins and syringic acid as major phenolic compound, pronouncedly inhibited the shoot growth and biomass accumulation of *Amaranthus spinosus* L., *Dactyloctenium aegyptium* L., and *Cyperus iria* L. weeds with reduced adverse impact on rice suggesting the potential of this plant to be used as a bioherbicide (Mangao et al., 2020).

On the other hand, ethanolic leaf extracts from five *Amaranthus* species (*A. spinosum*, *A. viridis*, *A. deflexus*, *A. hybridus* and *A. retroflexus*) applied at doses 0.25−4.0 g/L were found to inhibit the germination and the germination speed index of the lettuce seeds in a dose-dependent manner. Moreover, extracts of *A. spinosum* and *A. viridis* showed mitodepressive effect, while treatment with extracts of *A. deflexus*, *A. hybridus* and *A. retroflexus* resulted in a significant increase in the frequency of mitotic cells with chromosomal alterations, most frequently c-metaphases and sticky chromosomes and at cell exposure to the extracts from *A. spinosum*, *A. deflexus* and *A. retroflexus* considerable increase in condensed nuclei was observed (Carvalho et al., 2019).

From nonvolatile compounds obtained from the flowers of *Acacia dealbata*, one of the most invasive Australian acacias in southern Europe, methyl cinnamate and methyl anisate were identified as potential phytotoxic compounds. Methyl cinnamate was found to reduce guaiacol peroxidase activity in monocotyledonous weed *Lolium rigidum* Gaudin and dicotyledonous crop *Lactuca sativa* by 57% and 85%, respectively and caused also a decrease of α-amylase activity of *L. rigidum* by 6%. In these two tested species having small seeds also pronounced inhibition of early stem and radicle growth was observed in contrast to *Triticum aestivum* showing larger seed size, which was not adversely affected by this phytotoxin suggesting that application of methyl cinnamate could be recommended for wheat crops infested by *L. rigidum* (Lorenzo et al., 2020).

Leaf extracts of *Moringa oleifera*, *Mangifera indica*, *Albizia procera* and *Delonix regia* plants showed phytotoxic effects on *Lepidium sativum* seedlings reflected in delayed germination and inhibition of length of plant organs and plant dry weight (root growth inhibition of $\geq$85% as compared to control and seedling persistence index $<$30% of control), which were due to phenolic allelochemicals present in the extracts (Perveen et al., 2019).

EO of *Ocimum basilicum*, *Salvia officinalis*, *Thymus vulgaris*, *Melissa officinalis* and *Solidago virgaurea* emulsified with Tween 20 and dissolved in distilled water inhibited seed germination and shoot and root length of velvetleaf (*Abutilon theophrasti* Medik.), *O. basilicum*, *T. vulgaris* and *M. officinalis* EOs being more effective than *S. officinalis* and *S. virgaurea* (Saric-Krsmanovic et al., 2019). EO of *Zingiber officinale* containing as key compounds α-zingiberene (24.9 $\pm$ 0.8%), β-sesquiphelladrene (11.7 $\pm$ 0.3%), arcurcumene (10.7 $\pm$ 0.2%), and β-bisabolene (10.5 $\pm$ 0.3%) pronouncedly inhibited the seed germination of *Portulaca oleracea*, *Lolium multiflorum*, and *Cortaderia selloana* using a dose of 1 μL/mL and the hypocotyl and radicle growth of these weeds was inhibited as well. On the other hand, *Curcuma longa* EO consisting of 38.7 $\pm$ 0.8% ar-turmerone, 18.6 $\pm$ 0.6% β-turmerone, and 14.2 $\pm$ 0.9% α-turmerone considerably inhibited the seed germination of *C. selloana* and hypocotyl and radicle growth of tested weeds (*P. oleracea*, *L. multiflorum*, *C. selloana*, *E. crus-galli* and *Nicotiana glauca*), while did not affect either the seed germination or seedling growth of the food crops tomato, cucumber, and rice suggesting that this EO can be used as post-emergent bioherbicide against the tested weeds without causing phytotoxic impact on crops (Ibanez and Blazquez, 2019). EO of *Xanthium strumarium* L. leaves (family: *Asteraceae*) containing as major constituents oxygenated (61.78%) and non-oxygenated (10.62%) sesquiterpenes (25.19%), 1,5-dimethyltetralin (14.27%), eudesmol (10.60%), l-borneol (6.59%), ledene alcohol (6.46%), (−)-caryophyllene oxide (5.36%), isolongifolene, 7,8-dehydro-8a-hydroxy (5.06%), L-bornyl acetate (3.77%), and aristolene epoxide (3.58%) being its main constituents, exhibited pronounced allelopathic effect on the germination and growth of *B. pilosa* weed species (Abd El-Gawad et al., 2019b).

Ocimum gratissimum extract containing 43% of phenylpropanoid eugenol inhibited germination of *Euphorbia heterophylla* L. weed and application of the extract at doses $>$2.5% to adult plants resulted in 100% death of weed plants due to modified cell membrane permeability. Moreover, the extract can damage the mitochondria by mitochondrial oxidative phosphorylation decoupling, reducing the energy production of the germination process and initial growth of the seedlings (Martendal et al., 2018).

EO of *Origanum vulgare* ssp. *hirtum* (Link) Ietswaart containing monoterpenes and sesquiterpenes, with a strong abundance of two monoterpenic phenols, namely carvacrol and thymol, and the monoterpene *o*-cymene impaired the ability of *Arabidopsis* seedlings to incorporate inorganic nitrogen into amino acids and affected the glutamine metabolism resulting in strong accumulation of ammonia in leaf cells, decreased the efficiency of photosystem (PS) II and induced oxidative stress, which was reflected in great reduction of plant growth, leaf necrosis and eventually plant death. Due to its multitarget activity it could be used as effective bioherbicide (Araniti et al., 2018).

Algandaby and El-Darier (2018) tested the inhibitory effects of aqueous extracts of 4 medicinal plants on germination percentage, plumule and radicle lengths and dry weight of *Medicago polymorpha* L. seedlings and found that their allelopathic potential decreased as follows *Artemisia santolina* $>$ *Artemisia monosperma* Del. $>$ *Thymus capitatus* L., whereby the effect of *Pituranthus tortuosus* L. was insignificant and in a growth experiment using crude powder of these plants mixed with clay loam soil the species exhibited pronounced inhibitory allelopathic effect on leaf area index, photosynthetic pigments, total photosynthetic pigment and Chl*a* contents of *M. polymorpha* seedlings.

Eucalyptus leaves that were incorporated into the soil continuously released several phenolic and volatile compounds (VOCs) during a 30-day period of decomposition and it was found that phenolic compounds inhibited germination of lettuce seeds, while VOCs caused growth reduction of *Lactuca sativa* plants (Puig et al., 2018a). EO of *Eucalyptus citriodora* containing as major compounds citronellal (64.7%) and citronellol (10.9%) applied at doses 0.01% and 0.02% strongly reduced germination of *Sinapis arvensis*, *Sonchus oleraceus*, *Xanthium strumarium* and *Avena fatua* weeds, exhibited pronounced allelopathic effect on tested weed seedlings and treatment with 3% EO caused death of *S. arvensis*, *S. oleraceus* and *A. fatua* and caused severe injuries on *X. strumarium* due to serious reduction in total Chl and cell membrane disruption (Benchaa et al., 2018). Aqueous leaf extract of *Eucalyptus camaldulensis* inhibited germination and seedling growth, decreased net photosynthetic rate, caused membrane damage and loss of its integrity resulting in electrolyte leakage and induced oxidative stress in tested weeds *Sinapis arvensis* L., *Eruca vesicaria* and *Scorpiurus muricatus* L. and crops (*Triticum turgidum* subsp. *durum* [Desf.] Husn., *Vicia faba* subsp. *paucijuga* [Alef.] Murat, *Phaseolus vulgare* var. *vulgaris* L.), the phytotoxic effect being higher in weed species (Grichi et al., 2018). *Eucalyptus globulus* Labill. aqueous extract containing 8 phenolic compounds (chlorogenic, two coumaric derivatives, ellagic, hyperoside, rutin, quercitrin, and kaempferol 3-*O*-glucoside) and 5 low weight organic acids (citric, malic, shikimic, succinic and fumaric acids) showed pre-emergence inhibitory effects on *Lactuca sativa* and *Agrostis stolonifera*, which were comparable to that of metolachlor herbicide. However, due to phytotoxic effects on lettuce adult plants, its use as post-emergence herbicide is not desirable (Puig et al., 2018b). As potential bioherbicides to control weeds crabgrass (*Digitaria horizontalis*) and burrgrass (*Cenchrus echinatus*) EOs from *Eucalyptus citriodora* and *Cymbopogon nardus* were recommended also by Ootani et al. (2017).

Allelopathic effects of sunflower and wheat root exudates on *Sinapis arvensis* and *Sinapis alba* weeds was reflected in the reduction of total Chl and carotenoid contents, increased, proline amounts and decreased superoxide dismutase activity of treated weeds suggesting that such plant root exudates could be used as bioherbicides or foliar fertilizers (Unal and Bayram, 2019). Allelopathic effect of the extract of Algerian medicinal plant *Tetraclinis articulata* on germination and growth of *Lactuca sativa* L. was reported by Zohra et al. (2019). Allelopathic properties of *Retama raetam* L. against the weed *Phalaris minor* Retz. growing in *Triticum aestivum* L. fields was investigated by El-Darier et al. (2018). Favaretto et al. (2018) summarized the findings related to allelopathic potential of *Poaceae* species present in Brazil and highlighted that among 47 investigated Brazilian species only *Bothriochloa barbinodis*, *Bothriochloa laguroides*, *Paspalum notatum*, and *Paspalum urvillei* are native to Brazil, whereby phenolic acids represented 67% of identified allelopathic compounds. Considerable allelopathic potential of *Eremanthus erythropappus* (DC.) MacLeish essential oil containing sesquiterpene hydrocarbons (46.6%), oxygenated sesquiterpenes (29.3%) and monoterpene hydrocarbons (18.8%) with the major compounds β-caryophyllene (29.3%), caryophyllene oxide (22.1%) and β-pinene (12.8%) against *Brassica rapa* L. and *B. pilosa* L., but showing minimal effect on germination of crop plants, lettuce and tomato, was reported by Pinto et al. (2018).

Exposure to juglone (5-hydroxy-1,4-naphthoquinone) was found to suppress the germination process and $O_2^{\bullet-}$ generation in *Papaver rhoeas* L., and *Agrostemma githago* L. weeds much more than in *Triticum aestivum* L. and *Avena sa*tiva L. crops (Sytykiewicz et al., 2019).

Bioherbicidal activity of *Sinapis alba* seed meal extracts against greenhouse-grown *Amaranthus powellii* and *Setaria viridis* weeds was reported by Morra et al. (2018) and the most phytotoxic compound from the extract, ionic thiocyanate (SCN^-), when used as pre-emergence herbicide, exhibited effective control of both weeds comparable with that of the extract. On the other hand, at application as post-emergence herbicide the extract efficiency was pronouncedly lower than that of SCN^-.

Exposure of *Digitaria sanguinalis* weed to volatile organic compounds produced and emitted by the legume shrubs *Ulex europaeus* L. and *Cytisus scoparius* (L.) Link. resulted in damaged seedlings, being unable to recover germination capacity after removing the phytotoxin or producing unviable seedlings. Among 20 identified compounds from *U. europaeus* flowering biomass theaspirane and eugenol, while among 28 compounds of *C. scoparius* oxygenated monoterpenes such as terpinen-4-ol, verbenol, α-terpineol, and verbenone were estimated (Pardo-Muras et al., 2018).

Aqueous and ethanolic extracts of three *Baccharis* species, *B. dentata*, *B. uncinella* and *B. anomala*, containing catechin as the most abundant phenolic compound, strongly inhibited germination and seedling growth of *B. pilosa* weed. Germination of *B. pilosa* was reduced by 80% using aqueous extract containing 10% of *B. uncinella*, while treatment with ethanolic extracts containing 2.5%–10% of *B. uncinella* caused 100% mortality of weed seedlings suggesting feasibility of this plant species to be used as alternative bioherbicide to control the weeds (Dias et al., 2017).

Ibanez and Blazquez (2018) compared in vitro and in vivo phytotoxic activity of winter savory, peppermint, and anise EOs against *Portulaca oleracea*, *Lolium multiflorum*, and *Echinochloa crus-galli* weeds and food crops (maize, rice, and tomato) and found that due to the higher in vivo phytotoxicity of winter savory EO components it can provide an eco-friendly and less noxious alternative to weed control.

10.2.2 Bioherbicides produced by microorganisms

Moura et al. (2020) evaluated methanolic extracts of two endophytic fungi as photosynthesis and weed growth inhibitors of *Senna occidentalis* and *Ipomoea grandifolia* weeds and found that *Trichoderma spirale* reduced electron transport efficiency and shoot length, while the site of inhibition caused by *Diaporthe phaseolorum* was situated at the acceptor side of PSII and this fungus was able to reduce the shoot length on germination assay by 50%.

Radicinin, a fungal dihydropyranopyran-4,5-dione isolated from the culture filtrates of the fungus *Cochliobolus australiensis*, a foliar pathogen of buffelgrass (*Cenchrus ciliaris*), an invasive weed in North America, showing target-specific activity against the host plant and no toxicity on zebrafish embryos, could be considered as suitable candidate for natural bioherbicide formulation to manage buffelgrass (Masi et al., 2019a). Masi et al. (2019b) prepared and tested the phytotoxicity of some hemisynthetic derivatives of radicinin and found that the presence of an α,β unsaturated carbonyl group at C-4, as well as the presence of a free secondary hydroxyl group at C-3 and the stereochemistry of the same carbon are essential features for activity.

Secondary metabolite of *Pseudomonas aeruginosa* rhizospheric strain C1501, rhamnolipid biosurfactant, was found to enhance the dry mycelia weight yield of *Lasiodiplodia pseudotheobromae* (strain C1136) producing a phytotoxic metabolite showing detrimental impact on weeds. Combination of mutant strain of C1136 with rhamnolipid (0.003% v/v) enhanced biodegradability and resulted in high cellulase and xylanase activities as well as pronounced necrosis on the tested weeds and it was stated that the use of rhamnolipid as an adjuvant can improve penetrability of bioherbicide active ingredient for controlling weeds (Adetunji et al., 2017). An eco-friendly bioactive 2-(hydroxymethyl) phenol, from beneficial rhizobacteria *Pseudomonas aeruginosa* (C1501) considerably reduced the dry-weight of *Amaranthus hybridus* seedlings and exhibited only a low adverse effect on soil when compared to glyphosate application, whereby its application resulted in substantially higher enzymatic activities and the soil carbon content compared to untreated soil (Adetunji et al., 2019). The cultural filtrate of the bacterial isolate of *P. aeruginosa* from Wadi El Natroun region caused 100% reduction of seed germination, and shoot and root length of *Convolvulus arvensis* and *Portulaca oleracea* and with application of 40 mg/mL of the crude ethyl acetate extract the total biomass fresh weights of *P. oleracea* and *C. arvensis* seedlings were reduced by 71.27% and 39.37%, respectively. The IC_{50} values related to growth inhibition (1.3 and 1.64 mg/mL for *P. oleracea* and *C. arvensis*, respectively) suggested high potential of *P. aeruginosa* to be used as natural bacterial herbicide in broadleaf weed control (Tawfik et al., 2019).

Broth, supernatant culture and crude extract of rhizospheric bacteria, H6, isolated from the rhizosphere of *Momordica charantia* and identified as *P. aeruginosa* considerably inhibited *Pennisetum purpureum*, and *Amaranthus spinosum* weeds, whereby some of its secondary metabolites, quinoline derivatives, were greatly phytotoxic to target weeds (Lawrance et al., 2019). The effectiveness of rhizobacteria, which can form colonies on weed rooting and produce secondary metabolites when used as bioherbicides, depends on their survival at long-term storage, that could be enhanced using effective carriers of rhizobacteria such as powder and chaff charcoal powder, which were able to maintain the viability of rhizobacteria *Bacillus lentus* A05 and *P. aeruginosa* for 5 month and their stability as bioherbicide as well (Rakian et al., 2018b).

Dirhamnolipid (Rha-Rha-C10-C10), the phytotoxin produced by the pathogenic fungus *Colletotrichum gloeosporioides* BWH-1 showed broad herbicidal activity against dicot weeds (*Ageratum conyzoides*, *Celosia argentea*, *B. pilosa*, *Mikania micrantha*, Capsella bursa-pastoris and *A. retroflexus*) and monocot weeds *Alopecurus aequalis* and *Echinochloa crusgalli* with IC_{50} values ranging from 28.91 to 217.71 mg/L, whereby no toxicity on *Oryza sativa* was observed (Xu et al., 2019). Mortality as well as dry weight and plant height reduction of *Aeschynomene virginica*, *A.indica* and *Sesbania exaltata* weed seedlings ranging from 98% to 100% was observed 15 days after inoculation with fungus *Colletotrichum gloeosporioides* f.sp. *aeschynomene* in an invert emulsion, or in Silwet L-77 surfactant, whereby at application of tested fungal spores in Silwet formulation the serious disease development occurred more rapidly than with the use of invert emulsion (Boyette et al., 2019). Field tests showed that invert emulsion formulation of fungus *Colletotrichum coccodes* (NRRL strain 15547) was able to reduce dry weight of treated Eastern black nightshade (*Solanum ptycanthum*) weed plants, ensuring control exceeding 90% (Boyette et al., 2018).

Comparison of the impact of the fungus *Sclerotium rolfsii*, strain SC64 and chemical herbicide glyphosate that were applied after plowing in habitats seriously invaded by *Solidago canadensis* 180 days after treatment showed that bioherbicide treatment produced not only better weed control (89.61% vs. 70.06%) but also caused considerable increases in the total number of local weed species, considerably improving the weed community structure and strongly enhancing the biodiversity in habitats invaded by *S. canadensis* (Zhang et al., 2019).

Seed bacterization with bacterial isolate containing *Bacillus endophyticus* resulted in pronounced increases in root and shoot dry weights of *T. aestivum* plants under pot house conditions 25 and 50 days after sowing compared to control, while inoculation of this bacterial isolate considerably decreased the above mentioned characteristics in *Avena fatua* (wild oat), a damaging grass weed responsible for 17%–62% losses in yield of winter wheat (Dahiya et al., 2019b). The rhizospheric bacterial isolate identified as *Bacillus flexus* showed a growth retardation effect on the fifth and 10th day after seed germination in *Lathyrus aphaca* L. weed and in pot experiment it caused up to 92% reduction in root and shoot dry weight of weed suggesting its potential to be used as bioherbicide also under field conditions (Phour and Sindhu, 2019).

Phytotoxic effects of extracellular biopolymers produced by *Phoma* sp. using submerged fermentation, which were concentrated by hollow fiber membranes and by adsorption were assessed through absorption assays in detached leaves of *Cucumis sativus* and evaluated on the seventh day after application by Luft et al., 2021. The herbicidal activity was proportional to the concentration of biopolymers and it was stated that for development of liquid formulation of bioherbicide membrane processes could be used, while solid formulations could be obtained using silica and activated carbon as adsorbents, which were able to recover >93% of extracellular biopolymers from cell-free fermented broth. Concentration of metabolites from *Phoma* sp. using microfiltration membrane for increasing bioherbicidal activity against *B. pilosa* and *A. retroflexus* was reported also by Todero et al. (2019).

Trichoderma koningiopsis isolated from plants with symptoms of fungal disease caused up to 60% of foliar damage to *Euphorbia heterophylla* (popular name - Mexican fire plant) weed, without adversely affecting *Zea mays* crop and bioherbicidal potential showed also isolates of *Fusarium oxysporum* and *Fusarium proliferatum* (Reichert et al., 2019). Dedjell and Cliquet (2019) published media and culturing protocol for the production of high numbers of UV and drying resistant submerged micro-sclerotia-like structures named aggregates by potential mycoherbicide *Plectosporium alismatis* showing herbicidal activity against weed species *Alisma plantago-aquatica*. Culture filtrate of *Fusarium fujikuroi* fungus, which was isolated from weed plants with infections symptoms completely inhibited *Cucumis sativus* and *Sorghum bicolor* seed germination and its phytotoxic effects on cucumber were reflected in strong reduction of plant height and root length as well as in leaf necrosis and chlorosis (Daniel et al., 2018).

Fungal species *Gibbago trianthemae* Simmons was recommended as a potential mycoherbicidal candidate for the management of *Trianthema portulacastrum* L. (Horse purslane), a noxious weed of crop fields, by Gandipilli and Ratnakumar (2017).

Shi et al. (2020) in their review focused on the structure, herbicidal activity, and modes of action of secondary metabolites from actinomycetes, some of which have been successfully developed as commercial herbicides and emphasized that the secondary metabolites from actinomycetes can be utilized to develop both directly used bioherbicides and synthetic herbicides with new target sites.

10.3 Biopesticides against harmful arthropodes

Crop losses from damage caused by arthropod pests can exceed 15% annually. Plant defense against insect pest can use resistance or tolerance strategies. Expression of traits that restrict the adverse impact of herbivore damage on productivity and yield results in plant tolerance to herbivory. On the other hand, plants resistant to herbivory express traits that deter pests from settling, attaching to surfaces, feeding and reproducing, or that reduce palatability (Mitchell et al., 2016). However, usually to the suppression of detrimental arthropod pests in order to secure desirable crop yields the use of pesticides is inevitable. Although the efficacy of bioinsecticides is lower than that of synthetic insecticides, their use is favorable due to lower environmental pollution and minimized impact on non-target, beneficial arthropods, which is requirement for the sustainable pest management (Bale et al., 2008; Bostanian et al., 2012; Biondi et al., 2012; Orton, 2020). As bioinsecticides predominantly microbial pesticides, plant extracts or plant essential oils or their bioactive constituents, peptides obtained in spider venoms and Cry proteins produced by *Bacillus thuringiensis* are used (Hajek et al., 2009; Windley et al., 2012; King and Hardy, 2013; Siegwart et al., 2015; Chattopadhyay et al., 2017; Jouzani et al., 2017; Kaushik et al., 2020).

George et al. (2014) summarized recent findings related to the potential of plant-derived products to control arthropods of veterinary and medical significance. Benelli and Pavela (2018) summarized the findings related to the toxic and repellent potential of EOs and their selected constituents against bloodsucker insects such as biting midges, black flies, horse and deer flies, horn fly, stable fly, sandflies and tsetse flies. Pavela et al. (2019a) analyzed larvicidal activity of plant extracts from more than 400 species from which 29 showed excellent activity (i.e., $LC_{50} < 10$ ppm) against major vectors belonging to the genera *Anopheles*, *Aedes* and *Culex*. In plant extracts secondary metabolites belonging to alkaloids, alkamides, sesquiterpenes, triterpenes, sterols, flavonoids, coumarins, anthraquinones, xanthones, acetogenonins and aliphatics were detected and the adverse effects on mosquito larvae ranged from neurotoxic effects to inhibition of detoxificant enzymes and larval development and/or midugut damages. Efficacy of homemade botanical insecticides based on traditional knowledge, including *Allium sativum*, *Azadirachta indica*, *Capsicum* spp., *Chromolaena odorata*, *Gliricidia sepium*, *Melia azedarach*, *Moringa oleifera*, *Nicotiana tabacum*, *Ocimum gratissimum*, *Tephrosia vogelii*, *Tithonia diversifolia*, *Vernonia amygdalina*, which could contribute to reducing losses in food production was discussed by Dougoud et al. (2019). Carvacrol and thymol sprayed on the egg masses of the castor bean tick (*Ixodes ricinus*) at doses 0.25%−5% caused considerable hatching decrease and at treatment of larvae with these compounds at doses 1%, 2% and 5% even 100% mortality was observed after 24 h, both compounds showing higher efficacy than permethrin. In contrast, treatments with linalool were not effective. Moreover, while carvacrol and thymol exhibited >90% repellency on *I. ricinus* at doses 0.25%−5%, linalool showed 50.24% repellency only at a dose of 5%. Consequently, carvacrol and thymol could be applied as ingredients in acaricidal formulations to control the populations of *I. ricinus* (Tabari et al., 2017). Ethanol solutions of 5 fractions obtained from *Ocimum basilicum* EO applied to fresh leaves of the host plant caused considerable deterrence of second instar gypsy moth larvae, *Lymantria dispar* L., from feeding and at application of the most effective formulations used at a dose of 0.5% an antifeedant index > 80% after 5 days was observed (Popovic et al., 2013). Industrial EO from *Hyssopus officinalis* showed excellent activity against *Spodoptera littoralis*, while *Santolina chamaecyparissus* EO was found to show powerful antifeedant activity against the aphid *Rhopalosiphum padi*, was toxic to the tick *Hyalomma lusitanicum* and exhibited only moderate impact on *Leptinotarsa decemlineata* and *S. littoralis*. On the other hand, industrial EO from *Lavandula* x *intermedia* var. Super showed highest activity against the insect *S. littoralis* and was also toxic to *H. lusitanicum* (de Elguea-Culebras et al., 2018). Chitosan (CS) NPs functionalized with β-cyclodextrin containing carvacrol and linalool exhibited activity against the species *Helicoverpa armigera* and *Tetranychus urticae* (Acari: *Tetranychidae*), whereas repellent activity and reduction in oviposition were observed for the mites as well (Campos et al., 2018).

Methanol extracts of *Maerua edulis* leaves containing *E*- and *Z*-isomers of cinnamoyl-4-aminobutylguanidine as well as *E*- and *Z*-isomers of 4-hydroxycinnamoyl-4-aminobutylguanidine, stachydrine and 3-hydroxystachydrine were toxic to *Callosobruchus maculatus* and inhibited oviposition even at 0.1% w/v (Stevenson et al., 2018).

The Mexican sunflower (*Tithonia diversifolia*) leaf methanolic extract and its ethyl acetate fraction were tested for acute and chronic toxicity and for oviposition inhibitory effects in two-spotted spider mite *T. urticae* (*Tetranychidae*). The major compounds in the extracts were sesquiterpene lactones, tagitinin C and tagitinin A, showing toxicity against mosquitoes, aphids, and beetles. In chronic toxicity assays, on day 5 from application, the LD_{50}/LD_{90} values of the *T. diversifolia* methanolic extract were 41.3 and 98.7 μg/mL, respectively. The ethyl acetate extract of *T. diversifolia* was found to be potent oviposition inhibitor with ED_{50}/E_{90} values of 44.3 and 121.5 μg/mL, respectively (Pavela et al., 2018). Feeding of neonates of *Copitarsia decolora* Guenee with extracts from the *Trichilia americana* (Besse and Mocino) T.D. Penn. (rind) reduced average larval weight, prolonged duration of the larval stage, caused malformation of pupae and adults, affected adult fertility and fecundity and its application at doses at 1% and 5% resulted in 98% and 100% larval mortality (Brito et al., 2019).

Using the adult immersion test, the plant extracts of *Calotropis procera* and *Taraxacum officinale* considerably decreased the index of egg laying ($P < 0.01$) and increased the percent inhibition of oviposition of adult female of *Rhipicephalus microplus* at a dose 40 mg/mL and at exposure to the same concentration of tested plant extracts the larval mortality reached 96.0% ± 0.57% and 96.7% ± 0.88%, respectively. The LD_{50}/LD_{90} estimated for *C. procera* and *T. officinale* were 3.21/21.15 mg/mL, and 4.04/18.92 mg/mL, respectively (Khan et al., 2019).

Neem seed extract suppressed wheat aphid population less effectively than imidacloprid, while ensured yield protection comparable to that of imidacloprid. Moreover, treatment with neem seed extract resulted in higher population densities of coccinellids and syrphids compared to that of imidacloprid and its insecticidal efficiency depended on degree of synchronization among the application timing, the activity of aphids, crop variety and environmental conditions (Shah et al., 2017). The median effective concentration (EC_{50}-48 h) of neem-based oil formulation for non-target species arthropod *Daphnia magna* was estimated as 0.17 mL/L and in the chronic test lasting 21 days impact on the reproduction and size of *D. magna* was observed suggesting adverse effect of this preparation on aquatic organisms (Maranho et al., 2014). The aqueous extract of weed, *Clerodendrum viscosum* Ventenat (*Verbenaceae*) considerably reduced the tea red spider mite (*Oligonychus coffeae* Nietner) population and mosquito bug, *Helopeltis theivora* Waterhouse on *Camellia sinensis* (L) plants by 68%−95% and 73%−86%, respectively, showing comparable effectiveness to synthetic and neem pesticides (Roy et al., 2010).

The gene expression profiles in rutin-fed nymphs of the grasshopper *Oedaleus asiaticus* Bey-were strongly affected (308 genes were significantly upregulated, while 287 genes were downregulated) and treated insects showed reduced body size, lower survival rate, and reduced growth (Huang et al., 2020).

Microbial pesticides, including bacteria, fungi and viruses or their bioactive compounds designed to control invertebrate pests, can be used as alternatives for synthetic pesticides. In US, 57 species and/or strains of microbes or their derivatives are labeled for use against pestiferous insects, mites and nematodes (Arthurs and Dara, 2019). Approximately a half of registered microbial biopesticides in Brazil are mycoinsecticides and/or mycoacaricides consisting of hypocrealean fungi, mostly based on *Metarhizium anisopliae* and *Beauveria bassian* fungi used predominantly to control spittlebugs in sugarcane fields and whiteflies in row crops, respectively (Mascarin et al., 2019). Investigation lasting 9 years showed that spraying the fungus *Paranosema locustae* (Microsporidia), a potential biopesticide for control of grasshoppers, exhibited beneficial impact on natural enemies (17%−250% increase) and biodiversity in an arthropod community (40%−126% increase in the number of individual species) in Chinese rangeland (Shi et al., 2019). Currently most widely used microbial biopesticides in Iran are products based on *B. thuringiensis* subsp. *kurstaki* applied for lepidopteran pest control and *B. thuringiensis* subsp. *isralensis* against dipteran pests, mycoinsecticides based on *Lecanicillium lecanii* and *Beauveria bassiana* are applied against various arthropod pests, and a mycofungicide based on *Trichoderma harzianum* is used to control soil borne diseases (Karimi et al., 2019). Microbial biopesticides for insect pest management in India comprise approximate to 5% of the Indian pesticide market and at least 15 microbial species and 970 microbial formulations are registered (Kumar et al., 2019b).

Evolution of pest resistance recorded in 546 arthropods represents a global problem. Field-evolved resistance is a genetically based decrease in susceptibility to a pesticide in a population caused by exposure to the pesticide in the field. Field-evolved resistance to five *B. thuringiensis* toxins in transgenic corn and cotton based on monitoring of nine major pest species was analyzed by Tabashnik et al. (2014). Findings related to mutations in insects, ticks and mites resulting in the resistance to insecticides, acaricides and biopesticides were discussed in a review paper by Feyereisen et al. (2015). Genetically engineered (GE) crops producing insecticidal proteins from *B. thuringiensis* (mainly Cry proteins) can effectively control a number of key lepidopteran and coleopteran pests, mainly in maize, cotton, and soybean, however before commercial release of any new GE plant it is necessary to recognize whether the non-target species, including arthropod predators and parasitoids contributing to biological control, will not be negatively affected (Romeis et al., 2019). Cry proteins, which are expressed in rice lines for lepidopteran pest control when move through food webs may reduce fitness of non-target arthropods and impact them at the physiological and biochemical levels resulting in reduced fitness of non-target beneficial predators (Zhou et al., 2014).

A new class of biopesticides is represented by recombinant fusion proteins containing arthropod toxins. The recombinant fusion protein Hv1a/GNA containing the spider venom toxin ω-ACTX-1-Hv1a linked to snowdrop lectin (*Galanthus nivalis* agglutinin; GNA) delivered in artificial diet reduced survival of the peach-potato aphid *Myzus persicae*, whereby after 8 days exposure of *M. persicae* to fusion protein at 1 mg/mL the survival was <10%. Moreover, transgenic *Arabidopsis* expressing Hv1a/GNA was able to induce up to 40% mortality of *M. persicae* after 7 days exposure in detached leaf bioassays suggesting that plant delivered fusion proteins to aphids. Higher susceptibility of grain aphids (*Sitobion avenae*) to the Hv1a/GNA fusion protein in artificial diet bioassays compared with *M. persicae* was connected with slower hydrolyzation of the fusion protein (Nakasu et al., 2014). Optimizing expression of the recombinant fusion protein containing an arthropod toxin, ω-hexatoxin-Hv1a (from funnel web spider *Hadronyche versuta*) linked to snowdrop lectin (GNA) in *Pichia pastoris* using sequence modifications and a simple method for the generation of multi-copy strains without altering its functional properties was reported by Pyati et al. (2014).

Spider venoms include peptides showing insecticidal activities that can be used as bioinsecticides in agricultural applications (Saez and Herzig, 2019). In a field trial in blueberries, spider venom peptide, GS-omega/kappa-Hxtx-Hv1a gave control of spotted-wing *Drosophila* comparable to phosmet, and significantly reduced infestation in fruit. In laboratory experiment the exposure of fruit fly to this spider venom peptide combined with three adjuvants reduced survival of eggs oviposited into blueberries (Fanning et al., 2018).

Crude venom of two wolf spiders species, *Pardosa sumatrana* Thorell and *Pardosa birmanica* Simon, and 35-kDa protein fraction of both spiders were found to cause considerably higher mortality of *Rhopalosiphum padi* compared to the control aphids (Tahir et al., 2018a). Treatment of *Rhopalosiphum erysimi* (Hemiptera: Aphididae) with the protein fractions (similar to 29 kDa) from crude venom of two the jumping spiders, *Plexippus paykulli* (Audouin) and *Theyne imperialis* (Rossi) resulted in mortality rate of 79.5% and 90%, respectively, suggesting bioinsecticide potential of both tested protein fractions (Tahir et al., 2018b).

The simplification of agricultural landscapes due to intensification of agricultural production resulted in a decrease in the abundance and diversity of farmland plant, bird and insect communities and was accompanied also with adverse impact on beneficial insects in contrast to pest populations, which generally took advantage from reducing natural enemies, and greater concentrations of crop host plants (Grab et al., 2018). Average level of pest control was found to be 46% lower in homogeneous landscapes dominated by cultivated land, as compared with more complex landscapes (Rusch et al., 2016).

Recently increased interest is devoted to the insect-specific viruses, bacteria, fungi and nematodes as components of integrated pest management strategies for the control of arthropod pests of crops. Insect pathogenic viruses, e.g. baculoviruses are used particularly for the control of lepidopteran pests. Entomopathogenic bacteria used to control insect pests include several *B. thuringiensis* sub-species, *Lysinibacillus sphaericus*, *Paenibacillus* spp. and *Serratia entomophila*. Some of entomopathogenic fungi are able not only to control arthropod pest, but also simultaneously suppress plant pathogens and plant parasitic nematodes as well as promote plant growth; entomopathogenic nematodes from the genera *Steinernema* and *Heterorhabditis* are potent microbial control agents (Lacey et al., 2015). Kong et al. (2018) systematically reviewed studies focused on the interaction between baculoviruses infecting arthropods via the midgut and their insect hosts. Recent advances in the biological control of stored-grain insects with entomopathogenic fungi were summarized by Batta and Kavallieratos (2018) who also discussed their perspective to be used as bionsecticides and alternatives to synthetic insecticides. Jaber and Ownley (2018) analyzed the recent findings related to the endophytic colonization of different host plants by fungal entomopathogens, summarized the adverse impact of such colonization on insect pests and plant pathogens, discussed the possible mechanisms of protection mediated by endophytic fungi as well as their possible use as dual microbial control agents against both insect and pathogen pests.

Endophytes, i.e. bacteria or fungi living within intercellular spaces, tissue cavities, or vascular bundles of plants (for at least part of their life) are not harmful to the host but often are beneficial for the host. Endophytic survival of entomopathogenic fungus *Beauveria bassiana* inside leaf tissues of 7-week-old potted grapevine *Vitis vinifera* (L.) plants was estimated for at least 3 weeks after inoculation, irrespective of the inoculum used and resulted in reduced infestation rate and growth of the vine mealybug *Planococcus ficus*. In the vineyard this fungus was detected as an endophyte in mature grapevine plants up to 5 weeks after treatment, whereby the infestation with grape leafhopper, *Empoasca vitis,* pronouncedly decreased (Rondot and Reineke, 2018).

Six *Metarhizium anisopliae* isolates, which were tested against eggs and second instar larvae of *Spodoptera frugiperda* showed high cumulated mortality of eggs and neonates ranging from 92% to 97.5% and two of them are already commercialized for spider mites and ticks control, respectively (Akutse et al., 2019). Spraying of the soil surface with conidial suspensions of highly virulent *M. anisopliae* isolates after the release of longhorned ticks resulted in 60%−90% mortality of *Haemaphysalis longicornis* after 30 days suggesting that such treatment with fungal biopesticide could be successfully applied to control the tick population (Lee et al., 2019).

Three strains of *Metarhizium* spp., the commercial strain Ma-43, acaropathogenic strain Ma-7 and a most effective native strain, which were tested against the poultry red mite *Dermanyssus gallinae*, caused efficacy ranging from 85% to 92% under optimal conditions, while under poultry-house conditions, the efficacy reached only 30%−40% (Tomer et al., 2018). The mode of infection of *Metarhizium* spp. fungi on both terrestrial and aquatic insect larvae and their potential as biological control agents was discussed by San Aw and Hue (2017). Efficacy of two entomopathogenic fungi, *Metarhizium brunneum*, strain F52 alone and combined with *Paranosema locustae* against the migratory grasshopper, *Melanoplus sanguinipes* was reported by Dakhel et al. (2019).

A bioinsecticide spinosad and biofungicide *Trichoderma harzianum* were reported to be less disruptive to soil invertebrate fauna in a citrus agroecosystem than synthetic conventional pesticides (Majeed et al., 2018). Although treatment of *T. urticae* at doses of spinosad relevant to field conditions was able to kill certain part of population, the survivors showed considerable aptitude for population recovery (Medo et al., 2017).

Warsaba et al. (2019) in a review paper summarized recent findings related to dicistroviruses, small RNA viruses containing a monopartite positive-sense RNA genome, belonging to the family *Dicistroviridae*, mainly infecting arthropods and causing diseases that impact agriculture including the shrimp and honey bee industries, and discussed their potential use as biopesticides.

Herbivore-induced plant volatiles (HIPVs) play a crucial role in herbivore location by carnivorous arthropods such as parasitoids and the spatial matrix of volatiles ("volatile mosaic") within which parasitoids locate their hosts is affected by the concentration, chemical composition and breakdown of the emitted HIPV blends and by environmental factors (e.g., wind, turbulence) and vegetation that influence transport and mixing of odor plumes (Aartsma et al., 2017).

Constitutively produced and arthropod-induced plant proteins and defense allelochemicals synthesized by resistance gene products ensure plant resistance to arthropods, whereby rice and sorghum arthropod-resistant cultivars and, to a lesser extent, raspberry and wheat cultivars are also involved in integrated pest management (IPM) programs in Asia, Australia, Europe, and North America (Smith and Clement, 2012). On the other hand, pesticide resistance in arthropods could be connected either with enhanced production of metabolic enzymes, which bind to and/or detoxify the pesticide or with mutation of the target protein reducing its sensitivity to the pesticide (Bass and Field, 2011).

Reynolds et al. (2016) analyzed potential of Si to stimulate direct and indirect effects on plant defense against arthropod pests in agriculture and found that besides the resistance to herbivores mediated by Si via physical and biochemical/

molecular mechanisms, Si plays also a role in attracting predators or parasitoids to plants under herbivore attack. It could be assumed that soluble Si may enhance the production of herbivore induced plant volatiles by affecting protein expression (or modify proteins structurally) resulting in higher HIPVs production or modified HIPV profile of plants. The fact that Si-treated plants attacked by arthropod are characterized with increased attractiveness to natural enemies can ensure improved biological control in the field.

Wastewater containing yellowish liquid referred to in Brazil as manipueira obtained as waste product at processing of cassava roots into flour or starch containing great amounts of cyanide originating from the hydrolysis of cyanogenic glycosides was found to show pesticidal activity against nematodes, fungi and arthropods (Pinto Zevallos et al., 2018).

10.4 Nanoscale biopesticide formulations against arthropod pests and weeds

10.4.1 Bioherbicides in nanoformulations

By encapsulation of herbicidal compounds in nanoformulation improved stability and protection of active ingredient from rapid degradation as well as its controlled release could be achieved. However, in studied herbicidal nanoformulations predominantly synthetic herbicides were encapsulated also in order to reduce the amount required for desired effect and suppress adverse impact of the chemical compound on the environment and non-target species (Grillo et al., 2014; Nuruzzaman et al., 2016; Jampílek and Králová, 2017a; Kumaraswamy et al., 2018; Shang et al., 2019; Maluin and Hussein, 2020; Preisler et al., 2020). Therefore, in many formulations biopolymers of natural origin, e.g., alginate (Silva et al., 2010), CS (de Oliveira and Andrade, 2018), alginate/CS (Silva et al., 2011; Maruyama et al., 2016), CS/tripolyphosphate (Grillo et al., 2014; Maruyama et al., 2016); AgNPs−CS (Namasivayam and Aruna, 2014) or pectin (Kumar et al., 2017) were used for encapsulating synthetic herbicides. Mejias et al. (2019) described preparation of fully organic nanotubes composed of human bile acid (lithocholic acid) enabling up to 78% encapsulation of herbicidal compounds and allowing complete water solubilization. On the other hand, in microemulsions (MEs) and nanoemulsions (NEs) also emulsifiers, biosurfactants, or cosurfactants of natural origin could be used, such as *Quillaja* saponin, phospholipids, lecithin, polysaccharides, pectin, whey protein, lactoferrin or lactoferrin/alginate (Ozturk et al., 2014; Bai et al., 2016; McClements and Gumus, 2016; Pinheiro et al., 2016; Artiga-Artigas et al., 2018; Kumari et al., 2018; Zhao et al., 2018). However, it could be noted, that nanoformulations with encapsulated bioherbicidal ingredients were studied rarely, although by application of bio-based NEs problems connected with ecotoxic impact could be pronouncedly reduced (Mishra et al., 2018). Assessment of nano-encapsulated bioherbicides based on biopolymers and essential oil was performed by Taban et al. (2020).

CS NPs loaded with herbicidal necrosis induced protein extracted from fungal strain *F. oxysporum* 07 isolated from soil showed phytotoxic effects against weed species *Peperomia wightiana* (Chandra mohan et al., 2019), while with herbicidal metabolites of *F. oxysporum* soil isolate loaded with CS NPs were phytotoxic against *Ninidam theejan* weed species in vitro, retaining activity also at 70°C (Namasivayam et al., 2015). EO of *Satureja hortensis* formulated into oil-in-water (O/W) NE was found to exhibit effective herbicidal activity against *Amaranthus retroflexus* and *Chenopodium album* causing death of weed species at a dose of 4000 μL/L (Hazrati et al., 2017). Despite of lower effectiveness of bioherbicides compared to conventional synthetic herbicides, considering their above mentioned benefits, it could be expected that in the future dissemination of the investigation of herbicidal metabolites of fungi as well as constituents of plant EOs and their nanoscale formulations for practical application will increase (Mishra et al., 2018).

Whereas selective herbicides kill weeds without harming the crop, the nonselective ones, including toxic metals in bulk or nanoform as well when applied at higher concentrations kill all vegetation. Toxic effects of metal-based NPs on plants were reported by many researchers (e.g., Masarovičová and Králová, 2013; Masarovičová et al., 2014; Jampílek and Králová, 2019c; Králová et al., 2019). The green synthesized metal NPs fabricated using metal extracts or extracts of microorganisms as reducing and capping agents show usually improved biological activity compared to those prepared via chemical way. However, due to the fear that the applied metal nanoherbicide could not only kill weeds but also damage the cultivated plant in the field conditions, their eventual application against harmful weeds would have to be carefully considered. On the other hand, using the suitable plant/microorganism extracts for green synthesis of metal NPs can contribute to selective weed inhibition as was shown by El-Darier et al. (2020) who reported that AgNPs phytosynthesized using *Haplophyllum tuberculatum* crude aqueous extract completely inhibited germination of weed species *Phalaris minor* Retz and treatment with a dose of 20% resulted in decreases of photosynthetic pigments and photosynthetic efficiency as well as alterations in protein profile. *P. minor* showing seeds of smaller sizes was found to be more sensitive to AgNPs than *T. aestivum* suggesting potential of such AgNPs to be used for selective inhibition of weeds.

10.4.2 Biopesticide nanoformulations against anthropodes

Advantages of green MEs and NEs for managing parasites, vectors and pests reflected in higher efficacy and reduced non-target toxicity as well as enhanced stability and controlled release of encapsulated active ingredients were highlighted in a review paper of Pavoni et al. (2019). Moreover, the researchers emphasized, that such nanosystems are easy-to-use and could be prepared in industrial large-scale production. Applications of nanoformulations in insect's pest control were summarized by Sabry and Ragaei (2018). Controlled release systems consisting of polymer micro/nanocapsules, micro/solid lipid NPs, NEs/MEs, liposomes/niosomes, nanostructured hydrogels and cyclodextrins encapsulating repelents such as EOs can be used to prolong repellent action time duration and decrease permeation and systemic toxicity, whereby these environment-friendly nanoformulations are characterized with reduced repellent action time due to rapid evaporation after skin application (Tavares et al., 2018).

10.4.2.1 Nanoemulsions of botanical insecticides

Using high-energy ultrasonication process NE with *Mentha piperita* (wild-type) EO as an active ingredient (a.i.) and particle sizes <10 nm was prepared showing contact toxicity against the cotton aphid with LC_{50} value of 3879.5 ± 16.2 μL a.i./L (Heydari et al., 2020). NEs of *Ocimum basilicum* (L.), *Achillea fragrantissima* (Forssk.) and *Achillea santolina* (L.) EOs with particle sizes ranging from 78.5 to 104.6 nm and tested as fumigants against the mold mite, *Tyrophagus putrescentiae* (Schrank) (Sarcoptiformes: Acaridae), showed acaricidal toxicity reflected in LC_{50} values of 2.2, 4.7, and 9.6 μL/L air, respectively (Al-Assiuty et al., 2019). NE of *Pimpinella anisum* L. (*Apiaceae*) EO containing 81.2% of (*E*)-anethole with mean droplet size 198.9 nm, and zeta potential of -25.4 ± 4.47 mV exhibited toxic impact on *Tribolium castaneum* Herbst (Coleoptera: Tenebrionida) with LC_{50} of 9.3% (v/v) as well as on its progeny. Moreover, morphological and histological damages caused by feeding and exposure to the NE were observed (Hashem et al., 2018). Highly stable MEs of from *P. anisum*, *Trachyspennum cutuni* and *Crithnuan maritimum* EOs exhibited toxicity against third instar larvae of *Culex quinquefasciatus*, with LC_{50} values in the range 1.45–4.01 mL/L, caused high larval mortality and low percentage of hatched adults following short-term exposure to sublethal concentrations, whereby low or no mortality was observed on non-target invertebrates *Daphnia magna* and *Eisenia fetida* (Pavela et al., 2019b). Pascual-Villalobos et al. (2017) investigated the use of NEs of plant EOs as aphid repellents and in a laboratory choice bioassay with *Rhopalosiphum padi* found that at a dose of 0.15 μL/cm^2 the *P. anisum*, *M. piperita* and *Cymbopogon flexuosus* EOs, were found to be repellent for apterous females. Volatile toxicity of EO components, *trans*-anethole and caryophyllene, to the insects expressed by LD_{50} was 0.11 μL/cm^2. On the other hand, the repellency index values after 24 h for farnesol, geraniol, *cis*-jasmone, citral, linalool, estragole, pulegone and caryophyllene ranged from 68.8 to 100. In an experiment, in which NEs with oil droplets <100 nm were sprayed on leaves using a computer-controlled spraying apparatus it was found that carvone enhanced the mobility and *cis*-jasmone and showed repellent activity against *R. padi* already at a dose of 0.02 μL/cm^2 of the treated leaf. O/W NE containing *Rosmarinus officinalis* L. EO with droplet sizes <200 nm showed larvicidal activity against *Aedes aegypti* (Diptera: Culicidae) and at a dose of 250 ppm (related to EO) $80 \pm 10\%$ and $90 \pm 10\%$ mortality levels were observed after 24 and 48 h, respectively (Duarte et al., 2015).

Lippia alba volatile oil (citral chemotype) with droplet size 117.0 ± 1.0 nm, having as major compounds geranial (30.02%) and neral (25.26%), showed insecticidal activity against third instar larvae of *Ae. aegypti* and *Cx. quinquefasciatus* with LC_{50}/LC_{90} values of 38.22/59.42 and 31.02/47.19 ppm, respectively (Ferreira et al., 2019). NEs of *Geranium maculatum* L. EO prepared using ultrasonication exhibited approx. twofold greater insecticidal activity against *Culex pipiens* compared to bulk EO (LC_{50}: 48.27 ppm vs. 80.97 ppm) and similar results were obtained at co-application of NE with β-cypermethrin against *Plodia interpunctella* (Jesser et al., 2020).

It was found that 10% *Cinnamomun zeylanicum* EO and its NE (5%) were able to kill 100% of adult *Musca domestica* after 90 min of exposure and at in vivo application of 5% cinnamon oil on Holstein cows naturally infested by *Haemotobia irritans* considerable reduction of flies was observed on cows sprayed with this NE suggesting repelent effect of tested cinnamon EO (Boito et al., 2018). Whereas mortality in larva and adults of *Alphitobius diaperinus* after treatment with 5 and 10% of cinnamon EO was observed, mortality in both phases of *A. diaperinus* life cycle was estimated at exposure to 1% cinnamon EO NE and 5% cinnamon EO nanocapsules, whereby application of NE minimalized the toxic impact on survival and reproduction of springtails (Volpato et al., 2016).

Treatments with NEs of *Baccharis reticularia* DC EO showing a size 90.0 nm and its major constituent, D-limonene, exhibited larvicidal activity against *Ae. aegypti* with LC_{50} values 118.94 and 81.19 g/mL, respectively after 48 h of treatment (Botas et al., 2017).

After a 12/24 h exposure period of second instar larva of *Ae. aegypti* L. to NEs of *Vitex negundo* L. leaf EO with droplet sizes <200 nm the larvicidal activity was observed at doses $81.00 \pm 0.88/94.33 \pm 1.20$ ppm, while at treatment of third instar larvae with respective NEs it was observed at 79.00 ± 3.70/93.00 ± 1.25 ppm suggesting the suitability of these NEs for the control of dengue fever disease transmitting mosquito vector (Balasubramani et al., 2017).

Botanical insecticides derived from the white pyrethrum daisy, *Tanacetum cinerariifolium,* formulated into NE with globular droplets of 36–37 nm in diameter and larger droplets with diameters > 150 nm dispersed in the aqueous phase exhibited insecticidal effect against the cotton aphid *Aphis gossypii* Glover (Hemiptera: Aphididae) in eggplant exceeding that of the commercial pyrethrin formulation (Kalaitzaki et al., 2015). NEs of *Ageratum conyzoides, Achillea fragrantissima* and *Tagetes minuta* EOs tested as fumigants were highly toxic, the LC_{50} values 96 h after treatment ranged from 16.1 to 40.5 μL/L air against eggs and 4.5–43 μL/L air against adults of *Callosobruchus maculatus*, respectively (Nenaah et al., 2015).

Oil of *Pterodon emarginatus* Vogel, a Brazilian species known as sucupira, formulated into NE showed larvicidal activity against *A. aegypti* and was not toxic to mammals; it was supposed that its mechanism of action might be involved in reversible inhibition of acetylcholinesterase (Oliveira et al., 2016). Comparison of the insecticidal activity of the pulegone (a constituent of the EOs of some plants) applied in form of coarse or NE against stored product insects, rice weevil (*Sitophilus oryzae* L.) and red flour beetle (*T. castaneum* Herbst) showed that NE was able to cause >90% mortality rates for tested insect species for five weeks, while the high effectiveness of the coarse emulsions lasted only one week (Golden et al., 2018). Quercetin nanosuspension prepared by a solvent displacement method followed by solvent evaporation caused mortality of *A. aegypti* larvae between 44% and 100% at applied doses 100 and 500 ppm, respectively, and inhibited the larvae survival to emerge from water, while the mortality at application of bulk-quercetin induced ca 50% mortality regardless the concentration used at the same time-period of 48 h. Moreover, the nanosuspension was not toxic to a non-target organism (Pessoa et al., 2018).

CS nanocomposite fabricated with biocompatible polymer CS and insecticidal metabolites derived from insect-infecting fungus *Nomuraea rileyi* showed larvicidal activity against all larval stages of *Spodoptera litura* reflected in high mortality, drastic reduction of midgut and hemolymph macromolecules biochemical composition and this nanocomposite could be considered as promising candidate for pest control against economically important insect pests (Namasivayam et al., 2018).

In dusting toxicity bioassay a dose of 1.5 mg/500 mL of polyethylene glycol (PEG) nanocapsules loaded with *Matricaria chamomilla* L. oil caused 100% mortality of *Periplaneta americana* (L.) (Dictyoptera: *Blattidae*) after 24 h (El-Khodary et al., 2020). Citrus peel EOs included in PEG NPs showed good insecticidal activity against invasive tomato pest *Tuta absolut*a through ingestion larvae by these NPs, whereby visible toxic effects on the plants were pronouncedly reduced. However, contact exposure of eggs and larvae to EO emulsions resulted in a higher mortality compared to nanoencapsulated EO (Campolo et al., 2017).

NEs formulated with *Eucalyptus* EO and Tween 80 at a ratio of 1:2 and 1:2.5, respectively, showing droplet size of 4.04 and 2.27 nm and zeta potential of 6.20 and 7.69 mV exhibited insecticidal activity of against *S. oryzae* (LC_{50} values: 0.56 and 0.45 $\mu L/cm^2$), and *T. castaneum* of (LC_{50} values: 1.11 and 0.89 $\mu L/cm^2$). On the other hand, the LC_{50} of EO reached 0.795 $\mu L/cm^2$ against *S. oryzae* and 4.178 $\mu L/cm^2$ against *T. castaneum* (Adak et al., 2020).

10.4.2.2 Green-synthesized metal nanoparticles

Green-synthesized metal NPs prepared using plant extracts as capping and reducing agents were found to be effective insecticides and can provide a cost-effective and eco-safe alternative to conventional insecticides (Athanassiou et al., 2018; Marslin et al., 2018; Singh et al., 2018; Gour and Jain, 2019; Salem and Fouda, 2021). The use of botanical insecticides could have great importance mainly in developing countries in tropical regions where the source plants are readily available and conventional products are both expensive and dangerous to users (Isman, 2020). Phytosynthesized metal NPs can act as ovicides, larvicides, pupicides, adulticides, and oviposition deterrents against different mosquito species and at doses corresponding to few ppm they show great toxicity against *Anopheles stephensi* (malaria vector), *Ae. aegypti* (dengue vector), and the filariasis mosquito *Culex quiquefasciatus* (Benelli, 2016). A systematic overview of the impact of green synthesized NPs on both malaria parasites (*Plasmodium* spp.) and relevant vectors was presented by Barabadi et al. (2019).

Spherical AgNPs (35–55 nm) biofabricated using stearic acid from *Catharanthus roseus* leaf extract showed antifeedant and larvicidal activity against insect pest *Earias vittella* (LC_{50} of 45.46 and 25.12 ppm, respectively) and acute toxicity against *C. quinquefasciatus* and *A. aegypti* ($LC_{50} < 40$ ppm) (Pavunraj et al., 2017). The LC_{50} values of AgNPs phytosynthesized using *Habenaria plantaginea* leaf extract related to mosquito larvicidal activity on *An. stephensi*, *Ae. aegypti*, *An. subpictus*, *Ae. albopictus*, *Cx. quinquefasciatus*, and *Cx. tritaeniorhynchus* were estimated as 12.23, 13.38, 14.37, 15.39, 14.78 and 16.89 μg/mL, respectively, whereby the LC_{50} values of the *H. plantaginea* extract were considerably higher, namely 102.51, 111.99, 123.96, 136.56, 123.47 and 149.42 μg/mL. Moreover, low toxicity of *H. plantaginea* aqueous extract and AgNPs against the non-target species *Anisops bouvieri*, *Diplonychus indicus*, *Poecilia reticulata*, and *Gambusia affinis* was reflected in LC_{50} values ranging from 831.82 to 36,212.67 μg/mL (Aarthi et al., 2018). Larvicidal activities of Ag nanocomposite eco-friendly formulated using *Achyranthes aspera* leaf extract and 4 mM $AgNO_3$ showing spherical shape and polydispersed structure with diameters ranging from 1 to 25 nm expressed by LC_{50} values were 6.262, 1.412 and 1.302 μg/mL after 24, 48 and 72 h and no oxicity on non-target organisms such as *G. affinis*, *Daphnia magna* and *Moina macrocopa* was observed (Sharma et al., 2020). The larvicidal activity of green synthesized AgNPs using *Carmona retusa* (Vahl) Masam leaf extract against *An. stephensi*, *Ae. aegypti* and *Cx. quinquefasciatus* expressed by LC_{50} values was 116.681, 83.553 and 198.766 ppm, respectively (Rajkumar et al., 2018). AgNPs prepared using *Acacia caesia* leaf extract exhibited high acute larvicidal activity against larvae of *An. subpictus*, *Ae. albopictus* and *Cx. tritaeniorhynchus* with LC_{50} values ranging from 10.33 to 12.35 μg/mL, while in adulticidal assays the LC_{50} values varied from 18.66 to 22.63 μg/mL and for the complete egg hatchability inhibition of these three vectors doses 60, 75, and 90 μg/mL, respectively, were sufficient, whereby the susceptibility to AgNPs decreased as follows *An. subpictus* > *Ae. albopictus* > *Cx. tritaeniorhynchus* (Benelli et al., 2018).

AgNPs biofabricated using jujube leaf aqueous extract, which were applied to whitefly infested Al-Mustakbal eggplant hybrid grown in a greenhouse at concentrations 1000, 2000 and 3000 ppm decreased population density of whitefly (*Bemisia tabaci*) nymphs; at treatment with 3000 ppm the insecticidal activity percentage 7 days after treatment reached 100%, while 21 days after treatment it was 80% (Al Shammari et al., 2018).

Biocomposites of CS-coated AgNPs phytosynthesized using *Carmona retusa* (Vahl) Masam aqueous leaf extract showed excellent larvicidal activity against *An. stephensi*, *Ae. aegypti*, and *Cx. quinquefasciatus* (Rajkumar et al., 2019). AgNPs phytosynthesized using leaf extract of *Annona reticulata* of 17.33 nm showing spherical shape were found to exhibit high repellency against *S. oryzae* and feeding deterrence as well as powerful larvicidal activity against fourth instar larvae of mosquitoes (Malathi et al., 2019).

AgNPs generated via the surface functionalization by the root extract of *Cyprus rotundas* showed excellent larvicidal activity against *Ae. albopictus*, *An. stephensi* and *Cx. quinquefasciatus* already at doses 0.001−1.00 mg/L, they were able to enter the cuticle membrane of larvae and subsequently devasted their complete intestinal system and caused DNA damage (Sultana et al., 2020). The acaricidal activity of AgNPs prepared using *Saponaria officinalis* root extract against *T. urticae* Koch exceeded that of the root extract alone (LC_{50}/LC_{90}: 1.2/2.8 vs. 7.8/11.9 g/L), adults being less sensitive to both treatments (LC_{50}: 6.1 vs. 19.9 g/L) and AgNPs showed also high ovicidal toxicity (LC_{50}: 3.1 vs. 13.8 g/L); a pronounced inhibition of oviposition in females of *T. urticae* by spray residues from the treatments were observed as well (Pavela et al., 2017).

AgNPs biofabricated using *Garcinia mangostana* bark showed larvicidal activity against fourth instar larvae of *Ae. aegypti* with LC_{50} of 5.93 mg/L and it was supposed that AgNPs passed through the insect cuticle and into individual cells where they interfered with molting and other physiological processes (Karthiga et al., 2018).

The mortality rate of mosquito larvae due to treatment with biosynthesized AgNPs using seaweed *Sargassum polycystum* extract showing cubical shape and sizes ranging from 20 to 88 nm was the highest for *Ae. aegypti* (ca 80% and 90% after 48 h and 72 h, respectively) followed by *Cx. quinquefasciatus* (80% after 72 h), while the larvae of *An. stephensi* and *Cx. tritaeniorhynchus* were less affected (Vinoth et al., 2019).

Chitin rich *Periplaneta americana* (American cockroach) wings' extract was used for green synthesis of AgNPs (<50 nm), which exhibited insecticidal against *Aphis gossypii* under laboratory conditions and the mortality rate percentage in aphids treated with 100 μg AgNPs/mL 48 h after treatments was ca 40%, while in the control sample it was only ca 10% (Khatami et al., 2019).

AgNPs biosynthesized using entomopathogenic isolates of *Beauveria bassiana* fungus were reported to be effective against mustard aphid *Lipaphis erysimi* Kalt (Kamil et al., 2017).

An overview related to the impact of AuNPs on parasites and insect vectors, highlighting the antiparasitic role of Au NPs at combating diseases such as malaria, leishmaniosis, toxoplasmosis, trypanosomiasis, cryptosporidiosis, and microsporidian parasites was presented by Benelli (2018b). CuNPs biofabricated using the whole cell biomass of *Fusarium proliferatum* as a catalyst exhibited larvicidal activity against the larvae of *An. stephensi*, *Ae. aegypti* and *Cx. quinquefasciatus* with LC_{50} values of 81.34, 39.25, and 21.84 μg/mL, respectively (Kalaimurugan et al., 2019).

Green synthesized ZnO NPs using *Ulva lactuca* seaweed extract showed not only good antimicrobial activity but at application of a dose of 50 μg/mL they were able to kill 100% of *Ae. aegypti* fourth instar larvae (Ishwarya et al., 2018a). ZnO NPs phytosynthesized using hot water extract prepared from *Sargassum wightii* showed larvicidal activity against *Aedes aegypti* third instar larvae with LC_{50}/LC_{90} of 9.22/86.96 mg/mL exceeding that of the seaweed extract alone (Ishwarya et al., 2018b). *Pongamia pinnata* leaf extract coated ZnO NPs with average particle size 21.3 nm and zeta potential of −12.45 mV reduced the fecundity (eggs laid) and hatchability of *Callosobruchus maculatus* in a dose-dependent manner and decreased the mid-gut α-amylase, cysteine protease, β-glucosidase, glutathione *S*-transferase and lipase activity in the insects (Malaikozhundan and Vinodhini, 2018). Similar impact on *C. maculatus* exhibited also *Bacillus thuringiensis* coated ZnO NPs with mean particle size of 20 nm and zeta potential of −12.7 mV (Malaikozhundan et al., 2017).

In larvicidal and pupicidal experiments on *Cx. quinquefasciatus* performed by Murugan et al. (2018) the LC_{50} of Fe^0NPs prepared using a *Ficus natalensis* aqueous extract ranged from 20.9 (first instar larvae) to 43.7 ppm (pupae), while those of pure plant extract were by one order higher, i.e. 234.6 ppm for first instar larvae and 504.1 ppm for pupae and the predation efficiency of the guppy fish, *P. reticulata*, after a single treatment with sub-lethal doses of green synthesized Fe^0NPs was magnified. PdNPs (16−73 nm) green synthesized using aqueous extract of *Lagenaria siceraria* peel were reported to exhibit pronounced insecticidal activity against *S. oryzae* (Kalpana and Rajeswari, 2018).

TiNPs biosynthesized using an aqueous solution of banana peel extract as a bioreductant with mean diameter of 88.45 nm showed inhibitory effect against several pathogenic bacteria and caused high mortality of three larval stages of house fly (Hameed et al., 2019). TiO_2 NPs photosynthesized using *Vitex segundo* leaf aqueous extract were reported to show superb larvicidal activity against the fourth instar larvae of *An. subpictus* Grassy and filariasis vector *Cx. quinquefasciatus* as well as anti-lice activity against the head louse, *Pediculus humanus capitis* De Geer (Phthiraptera: Pediculidae) (Gandhi et al., 2016). The pediculocidal activity against the adult head louse, *P. humanus capitis*, acaricidal activity against larvae of cattle tick *Hyalomma anatolicum* (a.) *anatolicum* Koch (Acari: *Ixodidae*) and larvicidal activity against fourth instar larvae of *An. subpictus* Grassi was observed with TiO_2 NPs (70 nm) biofabricated using leaf aqueous extract of *Solanum trilobatum* (Rajakumar et al., 2014).

Thymus eriocalyx and *Thymus kotschyanus* EOs nanoencapsulated by a mesoporous alumosilicate MCM-41 against *T. urticae* showed higher stability and extended persistence up to 20 and 18 d for *T. eriocalyx* and *T. kotschyanus* and increased mortality of *T. urticae* from 80 to 203 mites at treatment with nanoencapsulated *T. eriocalyx* oil and from 58 to 186 mites at exposure to *T. kotschyanus* nanoformulation (Ebadollahi et al., 2017).

10.5 Conclusions

The changing climate and soil degradation due to the widespread cultivation of monocultures result in the reduced performance of important crops, which become an easy target for many pests. Chemical pesticides are used as standard for pest control and crop prevention. Unfortunately, they are often not sufficiently selective to target harmful pests and represent a serious risk for non-target organisms resulting in great ecological damage. Moreover, their entry in the food chain could injure the health of humans and animals. In addition, resistance to these chemical compounds develops very often. One of the ways to prevent the risks arising from the overuse or the incorrect application of synthetic pesticides is the use of nanotechnologies. By incorporation of pesticides into nanoformulations, their effects are modified and their dose-dependent toxicity is reduced. However, at the use of nanopesticides, their often unknown or insufficiently researched toxicity associated with very small particle sizes should be considered. The stability and the half-life of such nanoformulations after their application are critical not only for the target species, but also for their persistence in soil, water, etc. The use of nanoformulated highly effective pesticidal compounds should be approved only after thorough determination of all their physicochemical characteristics and impact on the ecosystem and human health. Another modern approach currently developing and extensively used particularly in the countries of the third world is the use of biopesticides, i.e. natural active ingredients of various origins acting specifically on target pests and being less toxic to non-target organisms. The use of biopesticides against weeds and harmful arthropods, whether in bulk or nanoscale form, can pronouncedly contribute to the improvement of the environment devastated by anthropogenic activity. Therefore, the future research should be focused on searching new biologically active compounds from natural sources such as plants or microorganisms, showing excellent selective herbicidal or insecticidal activity against target pests and acting via a unique mode of action. Encapsulating biopesticides in nanoscale formulations, similarly to synthetic pesticides, will pronouncedly reduce their dose-dependent toxicity enabling pest control without permanent contamination of the environment.

Acknowledgement

This study was supported by the Slovak Research and Development Agency (project APVV-17-0318).

References

Aarthi, C., Govindarajan, M., Rajaraman, P., Alharbi, N.S., Kadaikunnan, S., Khaled, J.M., Mothana, R.A., Siddiqui, N.A., Benelli, G., 2018. Eco-friendly and cost-effective Ag nanocrystals fabricated using the leaf extract of *Habenaria plantaginea*: toxicity on six mosquito vectors and four non-target species. Environ. Sci. Pollut. Res. 25 (11), 10317−10327.

Aartsma, Y., Bianchi, F.J.J.A., van der Werf, W., Poelman, E.H., Dicke, M., 2017. Herbivore-induced plant volatiles and tritrophic interactions across spatial scales. New Phytol. 216 (4), 1054−1063.

Abd-ElGawad, A.M., El-Amier, Y.A., Bonanomi, G., 2018. Essential oil composition, antioxidant and allelopathic activities of *Cleome droserifolia* (Forssk.). Delile. Chem. Biodivers. 15 (12), e1800392.

Abd-ElGawad, A.M., Elshamy, A.I., El Gendy, A., Al-Rowaily, S.L., Assaeeda, A.M., 2019a. Preponderance of oxygenated sesquiterpenes and diterpenes in the volatile oil constituents of *Lactuca serriola* L. revealed antioxidant and allelopathic activity. Chem. Biodivers. 16 (8), e1900278.

Abd-ElGawad, A., Elshamy, A., El Gendy, A.E., Gaara, A., Assaeed, A., 2019b. Volatiles profiling, allelopathic activity, and antioxidant potentiality of *Xanthium strumarium* leaves essential oil from Egypt: evidence from chemometrics analysis. Molecules 24 (3), 584.

Abd-ElGawad, A.M., El-Amier, Y.A., Assaeed, A.M., Al-Rowaily, S.L., 2020. Interspecific variations in the habitats of *Reichardia tingitana* (L.) Roth leading to changes in its bioactive constituents and allelopathic activity. Saudi J. Biol. Sci. 27 (1), 489−499.

Adak, T., Barik, N., Patil, N.B., Govindharaj, G.P.P., Gadratagi, B.G., Annamalai, M., Mukherjee, A.K., Rath, P.C., 2020. Nanoemulsion of eucalyptus oil: an alternative to synthetic pesticides against two major storage insects (*Sitophilus oryzae* (L.) and *Tribolium castaneum* (Herbst) of rice. Ind. Crop. Prod. 143, 111849.

Adetunji, C., Oloke, J., Kumar, A., Swaranjit, S., Akpor, B., 2017. Synergetic effect of rhamnolipid from *Pseudomonas aeruginosa* C1501 and phytotoxic metabolite from *Lasiodiplodia pseudotheobromae* C1136 on *Amaranthus hybridus* L. and *Echinochloa crus-galli* weeds. Environ. Sci. Pollut. Res. 24 (15), 13700−13709.

Adetunji, C.O., Oloke, J.K., Bello, O.M., Pradeep, M., Jolly, R.S., 2019. Isolation, structural elucidation and bioherbicidal activity of an eco-friendly bioactive 2-(hydroxymethyl) phenol, from *Pseudomonas aeruginosa* (C1501) and its ecotoxicological evaluation on soil. Environ. Technol. Innov. 13, 304−317.

Ahuja, N., Singh, H.P., Batish, D.R., Kohli, R.K., 2015. Eugenol-inhibited root growth in *Avena fatua* involves ROS-mediated oxidative damage. Pestic. Biochem. Physiol. 118, 64−70.

Aktar, M.W., Sengupta, D., Chowdhury, A., 2009. Impact of pesticides use in agriculture: their benefits and hazards. Interdiscipl. Toxicol. 2 (1), 1−12.

Akutse, K.S., Kimemia, J.W., Ekesi, S., Khamis, F.M., Ombura, O.L., Subramanian, S., 2019. Ovicidal effects of entomopathogenic fungal isolates on the invasive fall armyworm *Spodoptera frugiperda* (Lepidoptera: *Noctuidae*). J. Appl. Entomol. 143 (6), 626−634.

Al-Assiuty, B.A., Nenaah, G.E., Ageba, M.E., 2019. Chemical profile, characterization and acaricidal activity of essential oils of three plant species and their nanoemulsions against *Tyrophagus putrescentiae*, a stored-food mite. Exp. Appl. Acarol. 79 (3−4), 359−376.

Al-Dhabi, N.A., Arasu, M.V., 2018. Environmentally-friendly green approach for the production of zinc oxide nanoparticles and their anti-fungal, ovicidal, and larvicidal properties. Nanomaterials 8 (7), 500.

Algandaby, M.M., El-Darier, S.M., 2018. Management of the noxious weed; *Medicago polymorpha* L. via allelopathy of some medicinal plants from Taif region, Saudi Arabia. Saudi J. Biol. Sci. 25 (7), 1339−1347.

Al Shammari, H.I., AL-Khazraji, H.I., Falih, S.K., 2018. The effectivity of silver nanoparticles prepared by *Jujube ziziphus* sp. extract against whitefly *Bemisia tabaci* nymphs. Res. J. Pharmaceut. Biol. Chem. Sci. 9 (6), 551−558.

AMCA, 2020. Mosquito-Borne Diseases. The American Mosquito Control Association. https://www.mosquito.org/page/diseases.

Amri, I., Hamrouni, L., Hanana, M., Jamoussi, B., 2013. Reviews on phytotoxic effects of essential oils and their individual components: news approach for weeds management. Int. J. Appl. Biol. Pharmaceut. Technol. 4 (1), 96−114.

Amutha, V., Deepak, P., Kamaraj, C., Balasubramani, G., Aiswarya, D., Arul, D., Santhanam, P., Ballamurugan, A.M., Perumal, P., 2019. Mosquito-larvicidal potential of metal and oxide nanoparticles synthesized from aqueous extract of the seagrass, *Cymodocea serrulata*. J. Cluster Sci. 30, 797−812.

Aragao, F.B., Palmieri, M.J., Ferreira, A., Costa, A.V., Queiroz, V.T., Pinheiro, P.F., Andrade-Vieira, L.F., 2015. Phytotoxic and cytotoxic effects of *Eucalyptus* essential oil on lettuce (*Lactuca sativa* L.). Allelopathy J. 35 (2), 259−272.

Araniti, F., Landi, M., Lupini, A., Sunseri, F., Guidi, L., Abenavoli, M.R., 2018. *Origanum vulgare* essential oils inhibit glutamate and aspartate metabolism altering the photorespiratory pathway in *Arabidopsis thaliana* seedlings. J. Plant Physiol. 231, 297−309.

Arthurs, S., Dara, S.K., 2019. Microbial biopesticides for invertebrate pests and their markets in the United States. J. Invertebr. Pathol. 165, 13−21.

Artiga-Artigas, M., Guerra-Rosas, M.I., Morales-Castro, J., Salvia-Trujillo, L., Martin-Belloso, O., 2018. Influence of essential oils and pectin on nanoemulsion formulation: a ternary phase experimental approach. Food Hydrocoll. 81, 209−219.

Athanassiou, C.G., Kavallieratos, N.G., Benelli, G., Losic, D., Rani, P.U., Desneux, N., 2018. Nanoparticles for pest control: current status and future perspectives. J. Pest. Sci. 91, 1−15.

Auld, B.A., Morin, L., 1995. Constraints in the development of bioherbicides. Weed Technol. 9 (3), 638−652.

Ayvaz, A., Sagdic, O., Karaborklu, S., Ozturk, I., 2010. Insecticidal activity of the essential oils from different plants against three stored-product insects. J. Insect Sci. 10, 21.

Bai, L., Huan, S.Q., Gu, J.Y., McClements, D.J., 2016. Fabrication of oil-in-water nanoemulsions by dual-channel microfluidization using natural emulsifiers: saponins, phospholipids, proteins, and polysaccharides. Food Hydrocoll. 61, 703−711.

Bailey, K.L., 2014. The bioherbicide approach to weed control using plant pathogens. In: Abrol, D.P. (Ed.), Integrated Pest Management, Current Concepts and Ecological Perspective. Elsevier Inc., pp. 245−266

Bajwa, A.A., 2014. Sustainable weed management in conservation agriculture. Crop Protect. 65, 105–113.

Bajwa, A.A., Mahajan, G., Chauhan, B.S., 2015. Nonconventional weed management strategies for modern agriculture. Weed Sci. 63 (4), 723–747.

Balasubramani, S., Rajendhiran, T., Moola, A.K., Diana, R.K.B., 2017. Development of nanoemulsion from *Vitex negundo* L. essential oil and their efficacy of antioxidant, antimicrobial and larvicidal activities (*Aedes aegypti* L.). Environ. Sci. Pollut. Res. 24 (17), 15125–15133.

Bale, J.S., van Lenteren, J.C., Bigler, F., 2008. Biological control and sustainable food production. Philos. Trans. R. Soc. Lond. B Biol. Sci. 363 (1492), 761–776.

Barabadi, H., Alizadeh, Z., Rahimi, M.T., Barac, A., Maraolo, A.E., Robertson, L.J., Masjedi, A., Shahrivar, F., Ahmadpour, E., 2019. Nanobiotechnology as an emerging approach to combat malaria: a systematic review. Nanomedicine 18, 221–233.

Bass, C., Field, L.M., 2011. Gene amplification and insecticide resistance. Pest Manag. Sci. 67 (8), 886–890.

Batta, Y.A., Kavallieratos, N.G., 2018. The use of entomopathogenic fungi for the control of stored-grain insects. Int. J. Pest Manag. 64, 77–87.

Beckie, H.J., Ashworth, M.B., Flower, K.C., 2019. Herbicide resistance management: recent developments and trends. Plants 8, 161.

Benchaa, S., Hazzit, M., Abdelkrim, H., 2018. Allelopathic effect of *Eucalyptus citriodora* essential oil and its potential use as bioherbicide. Chem. Biodivers. 15 (8), e1800202.

Benelli, G., 2016. Plant-mediated biosynthesis of nanoparticles as an emerging tool against mosquitoes of medical and veterinary importance: a review. Parasitol. Res. 115, 23–34.

Benelli, G., Caselli, A., Canale, A., 2017. Nanoparticles for mosquito control: challenges and constraints. J. King Saud Univ. Sci. 29 (4), 424–435.

Benelli, G., 2018a. Mode of action of nanoparticles against insects. Environ. Sci. Pollut. Res. Int. 25 (13), 12329–12341.

Benelli, G., 2018b. Gold nanoparticles - against parasites and insect vectors. Acta Trop. 178, 73–80.

Benelli, G., Pavela, R., 2018. Beyond mosquitoes-essential oil toxicity and repellency against bloodsucking insects. Ind. Crop. Prod. 117, 382–392.

Benelli, G., Kadaikunnan, S., Alharbi, N.S., Govindarajan, M., 2018. Biophysical characterization of *Acacia caesia*-fabricated silver nanoparticles: effectiveness on mosquito vectors of public health relevance and impact on non-target aquatic biocontrol agents. Environ. Sci. Pollut. Res. 25 (11), 10228–10242.

Biondi, A., Mommaerts, V., Smagghe, G., Viñuela, E., Zappalà, L., Desneux, N., 2012. The non-target impact of spinosyns on beneficial arthropods. Pest Manag. Sci. 68 (12), 1523–1536.

Boito, J.P., Da Silva, A.S., dos Reis, J.H., Santos, D.S., Gebert, R.R., Biazus, A.H., Santos, R.C.V., Quatrin, P.M., Ourique, A.F., Boligon, A.A., Baretta, D., Baldissera, M.D., Stefani, L.M., Machado, G., 2018. Insecticidal and repellent effect of cinnamon oil on flies associated with livestock. Rev. MVZ Córdoba 23 (2), 6628–6636.

Bostanian, N.J., Vincent, C., Isaacs, R., 2012. Arthropod Management in Vineyards: Pests, Approaches, and Future Directions. Springer.

Botas, G.D., Cruz, R.A.S., de Almeida, F.B., Duarte, J.L., Araujo, R.S., Souto, R.N.P., Ferreira, R., Carvalho, J.C.T., Santos, M.G., Rocha, L., Pereira, V.L.P., Fernandes, C.P., 2017. *Baccharis reticularia* DC. and limonene nanoemulsions: promising larvicidal agents for *Aedes aegypti* (Diptera: *Culicidae*) control. Molecules 22 (11), 1990.

Boyette, C.D., Hoagland, R.E., Stetina, K.C., 2018. Bioherbicidal enhancement and host range expansion of a mycoherbicidal fungus via formulation approaches. Biocontrol Sci. Technol. 28 (3), 307–315.

Boyette, C.D., Hoagland, R.E., Stetina, K.C., 2019. Extending the host range of the bioherbicidal fungus *Colletotrichum gloeosporioides* f.sp. *aeschynomene*. Biocontrol Sci. Technol. 29 (7), 720–726.

Brito, R.F., Miranda, E.H., Gomez, V.R.C., 2019. Biological activity of *Trichilia americana* (Meliaceae) on *Copitarsia decolora* Guenee (Lepidoptera: *Noctuidae*). J. Entomol. Sci. 54 (2), 19–37.

Camara, M.C., Campos, E.V.R., Monteiro, R.A., do Espirito Santo Pereira, A., de Freitas Proença, P.L., Fraceto, L.F., 2019. Development of stimuli-responsive nano-based pesticides: emerging opportunities for agriculture. J. Nanobiotechnol. 17 (1), 100.

Campolo, O., Cherif, A., Ricupero, M., Siscaro, G., Grissa-Lebdi, K., Russo, A., Cucci, L.M., Di Pietro, P., Satriano, C., Desneux, N., Biondi, A., Zappala, L., Palmeri, V., 2017. Citrus peel essential oil nanoformulations to control the tomato borer, *Tuta absoluta*: chemical properties and biological activity. Sci. Rep. 7, 13036.

Campos, E.V.R., Proenca, P.L.F., Oliveira, J.L., Pereira, A.E.S., Ribeiro, L.N.D., Fernandes, F.O., Goncalves, K.C., Polanczyk, R.A., Pasquoto-Stigliani, T., Lima, R., Melville, C.C., Della Vechia, J.F., Andrade, D.J., Fraceto, L.F., 2018. Carvacrol and linalool co-loaded in β-cyclodextrin-grafted chitosan nanoparticles as sustainable biopesticide aiming pest control. Sci. Rep. 8, 7623.

Caraballo, H., King, K., 2014. Emergency department management of mosquito-borne illness: malaria, dengue, and west nile virus. Emerg. Med. Pract. 16 (5), 1–23.

Carvalho, M.S.S., Andrade-Vieira, L.F., dos Santos, F.E., Correa, F.F., Cardoso, M.D., Vilela, L.R., 2019. Allelopathic potential and phytochemical screening of ethanolic extracts from five species of *Amaranthus* spp. in the plant model *Lactuca sativa*. Sci. Hortic. 245, 90–98.

Chandra Mohan, A., Divya, S.R., Vijayalakshmi, T., Kavitha, R., 2019. Herbicidal activity against the common weed *Peperomia wightiana* and synthesis of chitosan nanoparticals. Int. J. Recent Technol. Eng. 7 (6S3), 219–222.

Chattopadhyay, P., Banerjee, G., Mukherjee, S., 2017. Recent trends of modern bacterial insecticides for pest control practice in integrated crop management system. 3 Biotech 7 (1), 60.

Cloyd, R.A., Galle, C.L., Keith, S.R., Kalscheur, N.A., Kemp, K.E., 2009. Effect of commercially available plant-derived essential oil products on arthropod pests. J. Econ. Entomol. 102 (4), 1567–1579.

Cordeau, S., Triolet, M., Wayman, S., Steinberg, C., Guillemin, J.P., 2016. Bioherbicides: dead in the water? A review of the existing products for integrated weed management. Crop Protect. 87, 44–49.

Culliney, T., 2014. Crop losses to arthropods. In: Pimentel, D., Peshin, R. (Eds.), Integrated Pest Management. Springer, Dordrecht, pp. 201–225.

Dahiya, A., Chahar, K., Sindhu, S.S., 2019a. The rhizosphere microbiome and biological control of weeds: a review. Span. J. Agric. Res. 17 (4), e1OR01.

Dahiya, A., Sharma, R., Sindhu, S., Sindhu, S.S., 2019b. Resource partitioning in the rhizosphere by inoculated *Bacillus* spp. towards growth stimulation of wheat and suppression of wild oat (*Avena fatua* L.) weed. Physiol. Mol. Biol. Plants 25 (6), 1483–1495.

Dahmana, H., Mediannikov, O., 2020. Mosquito-borne diseases emergence/resurgence and how to effectively control it biologically. Pathogens 9 (4), 310.

Dakhel, W.H., Latchininsky, A.V., Jaronski, S.T., 2019. Efficacy of two entomopathogenic fungi, *Metarhizium brunneum*, strain F52 alone and combined with *Paranosema locustae* against the migratory grasshopper, *Melanoplus sanguinipes*, under laboratory and greenhouse conditions. Insects 10 (4), 94.

Daniel, J.J., Zabot, G.L., Tres, M.V., Harakava, R., Kuhn, R.C., Mazutti, M.A., 2018. *Fusarium fujikuroi*: a novel source of metabolites with herbicidal activity. Biocatal. Agric. Biotechnol. 14, 314–320.

Dayan, F.E., Duke, S.O., 2014. Natural compounds as next-generation herbicides. Plant Physiol. 166, 1090–1105.

Dedjell, A., Cliquet, S., 2019. Media and culturing protocol using a full 2(5) factorial design for the production of submerged aggregates by the potential bio-herbicide *Plectosporium alismatis* against weed species of Alismataceae. Biocontrol Sci. Technol. 29 (4), 308–324.

de Elguea-Culebras, G.O., Sanchez-Vioque, R., Berruga, M.I., Herraiz-Penalver, D., Gonzalez-Coloma, A., Andres, M.F., Santana-Meridas, O., 2018. Biocidal potential and chemical composition of industrial essential oils from *Hyssopus officinalis, Lavandula* x *intermedia* var. Super, and *Santolina chamaecyparissus*. Chem. Biodivers. 15 (1), e1700313.

de Oliveira, P.N., Andrade, R.D.A., 2018. Polymer nanoparticles: adsorption and desorption of the weedkiller tebuthiuron turned to green chemistry. Orbital - Electron. J. Chem. 10 (5), 402–406.

Deutsch, C.A., Tewksbury, J.J., Tigchelaar, M., Battisti, D.S., Merrill, S.C., Huey, R.B., Naylor, R.L., 2018. Increase in crop losses to insect pests in a warming climate. Science 361 (6405), 916–919.

Dias, M.P., Nozari, R.M., Santarem, E.R., 2017. Herbicidal activity of natural compounds from *Baccharis* spp. on the germination and seedlings growth of *Lactuca sativa* and *Bidens pilosa*. Allelopathy J. 42 (1), 21–36.

Dimetry, N.Z., 2014. Different plant families as bioresource for pesticides. In: Singh, D. (Ed.), Advances in Plant Biopesticides. Springer India, pp. 1–20.

Dougoud, J., Toepfer, S., Bateman, M., Jenner, W.H., 2019. Efficacy of homemade botanical insecticides based on traditional knowledge. A review. Agron. Sustain. Dev. 39, 37.

Drexler, E., Pamlin, D., 2013. Nano-solutions for the 21st Century. Oxford Martin School, University of Oxford, Oxford, UK. https://www.oxfordmartin.ox.ac.uk/downloads/academic/201310Nano_Solutions.pdf.

Duarte, J.L., Amado, J.R.R., Oliveira, A.E.M.F.M., Cruz, R.A.S., Ferreira, A.M., Souto, R.N.P., Falcao, D.Q., Carvalho, J.C.T., Fernandes, C.P., 2015. Evaluation of larvicidal activity of a nanoemulsion of *Rosmarinus officinalis* essential oil. Rev. Bras. Farmacogn. 25 (2), 189–192.

Ebadollahi, A., Sendi, J.J., Aliakbar, A., 2017. Efficacy of nanoencapsulated *Thymus eriocalyx* and *Thymus kotschyanus* essential oils by a mesoporous material MCM-41 against *Tetranychus urticae* (Acari: Tetranychidae). J. Econ. Entomol. 110 (6), 2413–2420.

Einhellig, F.A., Leather, G.R., 1988. Potentials for exploiting allelopathy to enhance crop production. J. Chem. Ecol. 14, 1829–1844.

Elabasy, A., Shoaib, A., Waqas, M., Jiang, M., Shi, Z., 2019. Synthesis, characterization, and pesticidal activity of emamectin benzoate nanoformulations against *Phenacoccus solenopsis* Tinsley (Hemiptera: Pseudococcidae). Molecules 24 (15), 2801.

El-Darier, S.M., El-Kenany, E.T., Abdellatif, A.A., Hady, E.N.F.A., 2018. Allelopathic prospective of *Retama raetam* L. against the noxious weed *Phalaris minor* Retz. growing in *Triticum aestivum* L. fields. Rendiconti Lincei. Sci. Fis. Nat. 29 (1), 155–163.

El-Darier, S.M., Abou-Zeid, H.M., Marzouk, R.I., Hatab, A.S.A., 2020. Biosynthesis of silver nanoparticles via *Haplophyllum tuberculatum* (Forssk.) A. Juss. (*Rutaceae*) and its use as bioherbicide. Egypt. J. Bot. 60 (1), 25–40.

El-Khodary, A.S., Ghanem, N.F., Rakha, O.M., Shoghy, N.W., Ueno, T., 2020. Primary screening of German chamomile oil as insecticides, baits, and fumigants in nanoformulations against the health pest *Periplaneta americana* (L.) (Dictyoptera: *Blattidae*). J. Fac. Agric. Kyushu Univ. 65 (1), 103–112.

EPA, 2020. What are biopesticides? www.epa.gov/oppbppd1/biopesticides/whatarebiopesticides.htm. (Accessed 16 May 2020).

Erida, G., Saidi, N., Hasanuddin, S., 2020. Herbicidal potential of methanolic extracts of *Pinus merkusii* Jungh. et de Vriese, *Acacia mangium* Willd., *Jatropha curcas* L., *Tectona grandis* L.f. and *Terminalia catappa* L. on *Amaranthus spinosus* L. Allelopathy J. 49 (2), 201–216.

Fanning, P.D., VanWoerkom, A., Wise, J.C., Isaacs, R., 2018. Assessment of a commercial spider venom peptide against spotted-wing *Drosophila* and interaction with adjuvants. J. Pest. Sci. 91 (4), 1279–1290.

Farina, W.M., Balbuena, M.S., Herbert, L.T., Mengoni Goñalons, C., Vázquez, D.E., 2019. Effects of the herbicide glyphosate on honey bee sensory and cognitive abilities: individual impairments with implications for the hive. Insects 10 (10), 354.

Favaretto, A., Scheffer-Basso, S.M., Perez, N.B., 2018. Allelopathy in *Poaceae* species present in Brazil. A review. Agron. Sustain. Dev. 38 (2), 22.

Ferreira, R.M.A., Duarte, J.L., Cruz, R.A.S., Oliveira, A.E.M.F.M., Araujo, R.S., Carvalho, J.C.T., Mourao, R.H.V., Souto, R.N.P., Fernandes, C.P., 2019. A herbal oil in water nano-emulsion prepared through an ecofriendly approach affects two tropical disease vectors. Rev. Bras. Farmacogn. 29 (6), 778–784.

Feyereisen, R., Dermauw, W., Van Leeuwen, T., 2015. Genotype to phenotype, the molecular and physiological dimensions of resistance in arthropods. Pestic. Biochem. Physiol. 121, 61–77.

Gandhi, P.R., Jayaseelan, C., Vimalkumar, E., Mary, R.R., 2016. Larvicidal and pediculicidal activity of synthesized TiO_2 nanoparticles using *Vitex negundo* leaf extract against blood feeding parasites. J. Asia Pac. Entomol. 19 (4), 1089–1094.

Gandipilli, G., Ratnakumar, P.K., 2017. In vitro screening of a foliar pathogen for biological control of Horse purslane weed. Indian J. Exp. Biol. 55 (6), 389–395.

George, D.R., Finn, R.D., Graham, K.M., Sparagano, O.A.E., 2014. Present and future potential of plant-derived products to control arthropods of veterinary and medical significance. Parasites Vectors 7, 28.

Gillij, Y.G., Gleiser, R.M., Zygadlo, J.A., 2008. Mosquito repellent activity of essential oils of aromatic plants growing in Argentina. Bioresour. Technol. 99 (7), 2507–2515.

Glab, L., Sowinski, J., Bough, R., Dayan, F.E., 2017. Allelopathic potential of sorghum (*Sorghum bicolor* (L.) Moench) in weed control: a comprehensive review. Adv. Agron. 145, 43–95.

Golden, G., Quinn, E., Shaaya, E., Kostyukovsky, M., Poverenov, E., 2018. Coarse and nano emulsions for effective delivery of the natural pest control agent pulegone for stored grain protection. Pest Manag. Sci. 74 (4), 820–827.

Gour, A., Jain, N.K., 2019. Advances in green synthesis of nanoparticles. Artif. Cell. Nanomed. Biotechnol. 47 (1), 844–851.

Grab, H., Danforth, B., Poveda, K., Loeb, G., 2018. Landscape simplification reduces classical biological control and crop yield. Ecol. Appl. 28 (2), 348–355.

Grichi, A., Nasr, Z., Khouja, M.L., 2018. Identification and phytotoxicity of phenolic compounds in *Eucalyptus camaldulensis*. Allelopathy J. 44 (1), 75–88.

Grillo, R., Pereira, A.E.S., Nishisaka, C.S., Lima, R., Oehlke, K., Greiner, R., Fraceto, L.F., 2014. Chitosan/tripolyphosphate nanoparticles loaded with paraquat herbicide: an environmentally safer alternative for weed control. J. Hazard Mater. 278, 163–171.

Hajek, A., Glare, T., O'Callaghan, M., 2009. Use of Microbes for Control and Eradication of Invasive Arthropods. Springer.

Hajra, A., Dutta, S., Mondal, N.K., 2016. Mosquito larvicidal activity of cadmium nanoparticles synthesized from petal extracts of marigold (*Tagetes* sp.) and rose (*Rosa* sp.) flower. J. Parasit. Dis. 40 (4), 1519–1527.

Hameed, R.S., Fayyad, R.J., Nuaman, R.S., Hamdan, N.T., Maliki, S.A.J., 2019. Synthesis and characterization of a novel titanium nanoparticals using banana peel extract and investigate its antibacterial and insecticidal activity. J. Pure Appl. Microbiol. 13 (4), 2241–2249.

Hashem, A.S., Awadalla, S.S., Zayed, G.M., Maggi, F., Benelli, G., 2018. *Pimpinella anisum* essential oil nanoemulsions against *Tribolium castaneum*-insecticidal activity and mode of action. Environ. Sci. Pollut. Res. 25 (19), 18802–18812.

Hassan, E., Gökçe, A., 2014. Production and consumption of biopesticides. In: Singh, D. (Ed.), Advances in Plant Biopesticides. Springer, New Delhi, pp. 361–379.

Hazrati, H., Saharkhiz, M.J., Niakousari, M., Moein, M., 2017. Natural herbicide activity of *Satureja hortensis* L. essential oil nanoemulsion on the seed germination and morphophysiological features of two important weed species. Ecotoxicol. Environ. Saf. 142, 423–430.

Heydari, M., Amirjani, A., Bagheri, M., Sharifian, I., Sabahi, Q., 2020. Eco-friendly pesticide based on peppermint oil nanoemulsion: preparation, physicochemical properties, and its aphicidal activity against cotton aphid. Environ. Sci. Pollut. Res. 27, 6667–6679.

Huang, X.B., Lv, S.J., Zhang, Z.H., Chang, B.H., 2020. Phenotypic and transcriptomic response of the grasshopper *Oedaleus asiaticus* (Orthoptera: Acrididae) to toxic rutin. Front. Physiol. 11, 52.

Ibanez, M.D., Blazquez, M.A., 2019. Ginger and turmeric essential oils for weed control and food crop protection. Plants-Basel 8 (3), 59.

Ibanez, M.D., Blazquez, M.A., 2018. Phytotoxicity of essential oils on selected weeds: potential hazard on food crops. Plants 7, 79.

Ibrahim, S.S., 2019. Essential oil nanoformulations as a novel method for insect pest control in horticulture. In: Baimey, H.K., Hamamouch, N., Kolombia, Y.A. (Eds.), Horticultural Crops. IntechOpen.

Isman, M.B., 2020. Botanical insecticides in the twenty-first century-fulfilling their promise? Annu. Rev. Entomol. 65, 233–249.

Ishwarya, R., Vaseeharan, B., Kalyani, S., Banumathi, B., Govindarajan, M., Alharbi, N.S., Kadaikunnan, S., Al-anbr, M.N., Khaled, J.M., Benelli, G., 2018a. Facile green synthesis of zinc oxide nanoparticles using *Ulva lactuca* seaweed extract and evaluation of their photocatalytic, antibiofilm and insecticidal activity. J. Photochem. Photobiol. B 178, 249–258.

Ishwarya, R., Vaseeharan, B., Subbaiah, S., Nazar, A.K., Govindarajan, M., Alharbi, N.S., Kadaikunnan, S., Khaled, J.M., Al-anbr, M.N., 2018b. *Sargassum wightii*-synthesized ZnO nanoparticles - from antibacterial and insecticidal activity to immunostimulatory effects on the green tiger shrimp *Penaeus semisulcatus*. J. Photochem. Photobiol. B 183, 318–330.

Jaber, L.R., Ownley, B.H., 2018. Can we use entomopathogenic fungi as endophytes for dual biological control of insect pests and plant pathogens? Biol. Contr. 116, 36–45.

Jabran, K., Mahajan, G., Sardana, V., Chauhan, B.S., 2015. Allelopathy for weed control in agricultural systems. Crop Protect. 72, 57–65.

Jampílek, J., Kráľová, K., 2017a. Nanopesticides: preparation, targeting and controlled release. In: Grumezescu, A.M. (Ed.), Nanotechnology in the Agri-Food Industry, Vol. 10 – New Pesticides and Soil Sensors. Elsevier, London, UK, pp. 81–127.

Jampílek, J., Kráľová, K., 2017b. Nano-antimicrobials: activity, benefits and weaknesses. In: Ficai, A., Grumezescu, A.M. (Eds.), Nanostructures for Antimicrobial Therapy, Nanostructures in Therapeutic Medicine, vol. 2. Elsevier, Amsterdam, Netherlands, pp. 23–54.

Jampílek, J., Kráľová, K., 2018a. Benefits and potential risks of nanotechnology applications in crop protection. In: Abd-Elsalam, K., Prasad, R. (Eds.), Nanobiotechnology Applications in Plant Protection. Springer, Cham, Germany, pp. 189–246.

Jampílek, J., Kráľová, K., 2018b. Nanomaterials applicable in food protection. In: Rai, R.V., Bai, J.A. (Eds.), Nanotechnology Applications in the Food Industry. Taylor & Francis Group, Boca Raton, FL, USA, pp. 75–96.

Jampílek, J., Kráľová, K., 2019a. Nano-biopesticides in agriculture: state of art and future opportunities. In: Koul, O. (Ed.), Nano-Biopesticides Today and Future Perspectives. Elsevier, Amsterdam, pp. 397–447.

Jampílek, J., Kráľová, K., 2019b. Nanoformulations – valuable tool in therapy of viral diseases attacking humans and animals. In: Rai, M., Jamil, B. (Eds.), Nanotheranostics – Applications and Limitations. Springer Nature, Cham, Switzerland, pp. 137–178.

Jampílek, J., Kráľová, K., 2019c. Impact of nanoparticles on photosynthesizing organisms and their use in hybrid structures with some components of photosynthetic apparatus. In: Prasad, R. (Ed.), Plant Nanobionics, Nanotechnology in the Life Sciences. Springer Nature Switzerland AG, pp. 255–332.

Jampílek, J., Kráľová, K., Campos, E.V.R., Fraceto, L.F., 2019. Bio-based nanoemulsion formulations applicable in agriculture, medicine and food industry. In: Prasad, R., Kumar, V., Kumar, M., Choudhary, D.K. (Eds.), Nanobiotechnology in Bioformulations. Springer, Cham, Germany, pp. 33–84.

Jampílek, J., Kráľová, K., 2020a. Impact of nanoparticles on toxigenic fungi. In: Rai, M., Abd-Elsalam, K.A. (Eds.), Nanomycotoxicology − Treating Mycotoxins in the Nano Way. Academic Press & Elsevier, London, UK, pp. 309−348.

Jampílek, J., Kráľová, K., 2020b. Nanocomposites: synergistic nanotools for management of mycotoxigenic fungi. In: Rai, M., Abd-Elsalam, K.A. (Eds.), Nanomycotoxicology − Treating Mycotoxins in the Nano Way. Academic Press & Elsevier, London, UK, pp. 349−383.

Jampílek, J., Kráľová, K., 2020c. Nanoweapons against tuberculosis. In: Talegaonkar, S., Rai, M. (Eds.), Nanoformulations in Human Health − Challenges and Approaches. Springer International Publishing, Cham, Switzerland, pp. 469−502.

Jampílek, J., Kráľová, K., Fedor, P., 2020. Bioactivity of nanoformulated synthetic and natural insecticides and their impact on the environment. In: Fraceto, L.F., de Castro, V.L., Grillo, R., Ávila, D., Oliveira, H.C., de Lima, R. (Eds.), Nanopesticides - from Research and Development to Mechanisms of Action and Sustainable Use in Agriculture. Springer, Cham, Switzerland, pp. 165−225.

Jesser, E., Lorenzetti, A.S., Yeguerman, C., Murray, A.P., Domini, C., Werdin-Gonzalez, J.O., 2020. Ultrasound assisted formation of essential oil nanoemulsions: emerging alternative for *Culex pipiens pipiens* Say (Diptera: Culicidae) and *Plodia interpunctella* Hubner (Lepidoptera: *Pyralidae*) management. Ultrason. Sonochem. 61, 104832.

Jouzani, G.S., Valijanian, E., Sharafi, R., 2017. *Bacillus thuringiensis*: a successful insecticide with new environmental features and tidings. Appl. Microbiol. Biotechnol. 101, 2691−2711.

Kaab, S.B., Rebey, I.B., Hanafi, M., Hammi, K.M., Smaoui, A., Fauconnier, M.L., De Clerck, C., Jijakli, M.H., Ksouri, R., 2020. Screening of Tunisian plant extracts for herbicidal activity and formulation of a bioherbicide based on *Cynara cardunculus*. South Afr. J. Bot. 128, 67−76.

Kachhawa, D., 2017. Microorganisms as a biopesticides. J. Entomol. Zool. Stud. 5 (3), 468−473.

Kalaimurugan, D., Sivasankar, P., Lavanya, K., Shivakumar, M.S., Venkatesan, S., 2019. Antibacterial and larvicidal activity of *Fusarium proliferatum* (YNS2) whole cell biomass mediated copper nanoparticles. J. Cluster Sci. 30 (4), 1071−1080.

Kalaitzaki, A., Papanikolaou, N.E., Karamaouna, F., Dourtoglou, V., Xenakis, A., Papadimitriou, V., 2015. Biocompatible colloidal dispersions as potential formulations of natural pyrethrins: a structural and efficacy study. Langmuir 31 (21), 5722−5730.

Kalpana, V.N., Rajeswari, V.D., 2018. Synthesis of palladium nanoparticles via a green route using *Lagenaria siceraria*: assessment of their innate antidandruff, insecticidal and degradation activities. Mater. Res. Express 5 (11), 115406.

Kamil, D., Prameeladevi, T., Ganesh, S., Prabhakaran, N., Nareshkumar, R., Thomas, S.P., 2017. Green synthesis of silver nanoparticles by entomopathogenic fungus *Beauveria bassiana* and their bioefficacy against mustard aphid (*Lipaphis erysimi* Kalt.). Indian J. Exp. Biol. 55 (8), 555−561.

Karimi, J., Dara, S.K., Arthurs, S., 2019. Microbial insecticides in Iran: history, current status, challenges and perspective. J. Invertebr. Pathol. 165, 67−73.

Karthiga, P., Rajeshkumar, S., Annadurai, G., 2018. Mechanism of larvicidal activity of antimicrobial silver nanoparticles synthesized using *Garcinia mangostana* bark extract. J. Cluster Sci. 29 (6), 1233−1241.

Kaushik, B.D., Kumar, D., Shamim, M., 2020. Biofertilizers and Biopesticides in Sustainable Agriculture. Apple Academic Press.

Khan, A., Nasreen, N., Niaz, S., Ayaz, S., Naeem, H., Muhammad, I., Said, F., Mitchell, R.D., de Leon, A.A.P., Gupta, S., Kumar, S., 2019. Acaricidal efficacy of *Calotropis procera* (*Asclepiadaceae*) and *Taraxacum officinale* (*Asteraceae*) against *Rhipicephalus microplus* from Mardan, Pakistan. Exp. Appl. Acarol. 78 (4), 595−608.

Khatami, M., Iravani, S., Varma, R.S., Mosazade, F., Darroudi, M., Borhani, F., 2019. Cockroach wings-promoted safe and greener synthesis of silver nanoparticles and their insecticidal activity. Bioproc. Biosyst. Eng. 42 (12), 2007−2014.

King, G.F., Hardy, M.C., 2013. Spider-venom peptides: structure, pharmacology, and potential for control of insect pests. Annu. Rev. Entomol. 58, 475−496.

Kitherian, S., 2017. Nano and bio-nanoparticles for insect control. Res. J. Nanosci. Nanotechnol. 7, 1−9.

Kong, M., Zuo, H., Zhu, F.F., Hu, Z.Y., Chen, L., Yang, Y.H., Lv, P., Yao, Q., Chen, K.P., 2018. The interaction between baculoviruses and their insect hosts. Dev. Comp. Immunol. 83, 114−123.

Kong, C.H., Xuan, T.D., Khanh, T.D., Tran, H.D., Trung, N.T., 2019. Allelochemicals and signaling chemicals in plants. Molecules 24 (15), 2737.

Korres, N.E., Norsworthy, J.K., Tehranchian, P., Gitsopoulos, T.K., Loka, D.A., Oosterhuis, D.M., Gealy, D.R., Moss, S.R., Burgos, N.R., Miller, M.R., Palhano, M., 2016. Cultivars to face climate change effects on crops and weeds: a review. Agron. Sustain. Dev. 36 (1), 12.

Kráľová, K., Masarovičová, E., Jampílek, J., 2019. Plant responses to stress induced by toxic metals and their nanoforms. In: Pessarakli, M. (Ed.), Handbook of Plant and Crop Stress, fourth ed. Taylor and Francis, CRC, Boca Raton, pp. 479−522.

Kumar, S., Bhanjana, G., Sharma, A., Dilbaghi, N., Sidhu, M.C., Kim, K.H., 2017. Development of nanoformulation approaches for the control of weeds. Sci. Total Environ. 586, 1272−1278.

Kumar, S., Nehra, M., Dilbaghi, N., Marrazza, G., Hassan, A.A., Kim, K.H., 2019a. Nano-based smart pesticide formulations: emerging opportunities for agriculture. J. Contr. Release 294, 131−153.

Kumar, K.K., Sridhar, J., Murali-Baskaran, R.K., Senthil-Nathan, S., Kaushal, P., Dara, S.K., Arthurs, S., 2019b. Microbial biopesticides for insect pest management in India: current status and future prospects. J. Invertebr. Pathol. 165, 74−81.

Kumaraswamy, R.V., Kumari, S., Choudhary, R.C., Pal, A., Raliya, R., Biswas, P., Saharan, V., 2018. Engineered chitosan based nanomaterials: bioactivities, mechanisms and perspectives in plant protection and growth. Int. J. Biol. Macromol. 113, 494−506.

Kumari, S., Kumaraswamy, R.V., Choudhary, R.C., Sharma, S.S., Pal, A., Raliya, R., Biswas, P., Saharan, V., 2018. Thymol nanoemulsion exhibits potential antibacterial activity against bacterial pustule disease and growth promotory effect on soybean. Sci. Rep. 8, 6650.

Lacey, L.A., 2017. Microbial Control of Insect and Mite Pests: From Theory to Practice. Academic Press & Elsevier, London.

Lacey, L.A., Grzywacz, D., Shapiro-Ilan, D.I., Frutos, R., Brownbridge, M., Goettel, M.S., 2015. Insect pathogens as biological control agents: back to the future. J. Invertebr. Pathol. 132, 1−41.

Lawrance, S., Varghese, S., Varghese, E.M., Asok, A.K., Jisha, M.S., 2019. Quinoline derivatives producing *Pseudomonas aeruginosa* H6 as an efficient bioherbicide for weed management. Biocatal. Agric. Biotechnol. 18. UNSP 101096.

Lee, M.Y., 2018. Essential oils as repellents against arthropods. BioMed Res. Int. 2018, 6860271.

Lee, M.R., Li, D., Lee, S.J., Kim, J.C., Kim, S., Park, S.E., Baek, S., Shin, T.Y., Lee, D.H., Kim, J.S., 2019. Use of *Metarhizum anioplíae* s.l. to control soil-dwelling longhorned tick, *Haemaphysalis longicornis*. J. Invertebr. Pathol. 166, 107230.

Li, Z.R., Amist, N., Bai, L.Y., 2019. Allelopathy in sustainable weeds management. Allelopathy J. 48 (2), 109–138.

Lins, L., Dal Maso, S., Foncoux, B., Kamili, A., Laurin, Y., Genva, M., Jijakli, M.H., De Clerck, C., Fauconnier, M.L., Deleu, M., 2019. Insights into the relationships between herbicide activities, molecular structure and membrane interaction of cinnamon and citronella essential oils components. Int. J. Mol. Sci. 20 (16), 4007.

Lorenzo, P., Reboredo-Duran, J., Munoz, L., Freitas, H., Gonzalez, L., 2020. Herbicidal properties of the commercial formulation of methyl cinnamate, a natural compound in the invasive silver wattle (*Acacia dealbata*). Weed Sci. 68 (1), 69–78.

Luft, L., Confortin, T.C., Todero, I., Neto, J.R.C., Tonato, D., Felimberti, P.Z., Zabot, G.L., Mazutti, M.A., 2021. Different techniques for concentration of extracellular biopolymers with herbicidal activity produced by *Phoma* sp. Environ. Technol. 42, 1392–1401.

Macias, F.A., Mejias, F.J.R., Molinillo, J.M.G., 2019. Recent advances in allelopathy for weed control: from knowledge to applications. Pest Manag. Sci. 75 (9), 2413–2436.

Majeed, M.Z., Naveed, M., Riaz, M.A., Ma, C.S., Afzal, M., 2018. Differential impact of pesticides and biopesticides on edaphic invertebrate communities in a citrus agroecosystem. Invertebr. Surviv. J. 15, 31–38.

Malaikozhundan, B., Vaseeharan, B., Vijayakumar, S., Thangaraj, M.P., 2017. *Bacillus thuringiensis* coated zinc oxide nanoparticle and its biopesticidal effects on the pulse beetle, *Callosobruchus maculatus*. J. Photochem. Photobiol. B 174, 306–314.

Malaikozhundan, B., Vinodhini, J., 2018. Nanopesticidal effects of *Pongamia pinnata* leaf extract coated zinc oxide nanoparticle against the pulse beetle, *Callosobruchus maculatus*. Mater. Today Commun. 14, 106–115.

Malathi, S., Rameshkumar, G., Rengarajan, R.L., Rajagopar, T., Muniasamy, S., Ponmanickam, P., 2019. Phytofabrication of silver nanoparticles using *Annona reticulata* and assessment of insecticidal and bactericidal activities. J. Environ. Biol. 40 (4), 626–633.

Maluin, F.N., Hussein, M.Z., 2020. Chitosan-based agronanochemicals as a sustainable alternative in crop protection. Molecules 25 (7), 1611.

Mangao, A.M., Arreola, S.L.B., San Gabriel, E.V., Salamanez, K.C., 2020. Aqueous extract from leaves of *Ludwigia hyssopifolia* (G. Don) Exell as potential bioherbicide. J. Sci. Food Agric. 100 (3), 1185–1194.

Maranho, L.A., Botelho, R.G., Inafuku, M.M., Nogueira, L.D.R., de Olinda, R.A., de Sousa, B.A.I., Tornisielo, V.L., 2014. Testing the neem biopesticide (*Azadirachta indica* A. Juss) for acute toxicity with *Danio rerio* and for chronic toxicity with *Daphnia magna*. J. Appl. Sci. Technol. 16 (1), 105–111.

Marslin, G., Siram, K., Maqbool, Q., Selvakesavan, R.K., Kruszka, D., Kachlicki, P., Franklin, G., 2018. Secondary metabolites in the green synthesis of metallic nanoparticles. Materials 11 (6), 940.

Martendal, C.O., Mantovanelli, G.C., Reis, B., Cavaleiro, C., Iwamoto, E.L.I., Bonato, C.M., 2018. Effects of *Ocimum gratissimum* L. extract on the germination, respiration and growth of *Euphorbia heterophylla* L. Allelopathy J. 45 (1), 29–44.

Maruyama, C.R., Guilger, M., Pascoli, M., Bileshy-Jose, N., Abhilash, P.C., Fraceto, L.F., de Lima, R., 2016. Nanoparticles based on chitosan as carriers for the combined herbicides imazapic and imazapyr. Sci. Rep. 6, 19768.

Masarovičová, E., Králová, K., 2013. Metal nanoparticles and plants. Ecol. Chem. Eng. S 20 (1), 9–22.

Masarovičová, E., Králová, K., Zinjarde, S.S., 2014. Metal nanoparticles in plants. Formation and action. In: Pessarakli, M. (Ed.), Handbook of Plant and Crop Physiology, third ed. CRC Press, Boca Raton, pp. 683–731.

Mascarin, G.M., Lopes, R.B., Delalibera, I., Fernandes, E.K.K., Luz, C., Faria, M., 2019. Current status and perspectives of fungal entomopathogens used for microbial control of arthropod pests in Brazil. J. Invertebr. Pathol. 165, 46–53.

Masi, M., Freda, F., Sangermano, F., Calabro, V., Cimmino, A., Cristofaro, M., Meyer, S., Evidente, A., 2019a. Radicinin, a fungal phytotoxin as a target-specific bioherbicide for invasive buffelgrass (*Cenchrus ciliaris*) control. Molecules 24 (6), 1086.

Masi, M., Freda, F., Clement, S., Cimmino, A., Cristofaro, M., Meyer, S., Evidente, A., 2019b. Phytotoxic activity and structure-activity relationships of radicinin derivatives against the invasive weed buffelgrass (*Cenchrus ciliaris*). Molecules 24 (15), 2793.

McClements, D.J., Gumus, C.E., 2016. Natural emulsifiers - biosurfactants, phospholipids, biopolymers, and colloidal particles: molecular and physicochemical basis of functional performance. Adv. Colloid Interface Sci. 234, 3–26.

Medo, I., Stojnic, B., Marcic, D., 2017. Acaricidal activity and sublethal effects of the microbial pesticide spinosad on *Tetranychus urticae* (Acari: Tetranychidae). Syst. Appl. Acarol. 22 (10), 1748–1762.

Mejias, F.J.R., Trasobares, S., Lopez-Haro, M., Varela, R.M., Molinillo, J.M.G., Calvino, J.J., Macias, F.A., 2019. In situ eco encapsulation of bioactive agrochemicals within fully organic nanotubes. ACS Appl. Mater. Interfaces 11 (45), 41925–41934.

Meyer, W.L., Gurman, P., Stelinski, L.L., Elman, N.M., 2015. Functional nano-dispensers (FNDs) for delivery of insecticides against phytopathogen vectors. Green Chem. 17, 4173–4177.

Mishra, P., Balaji, A.P.B., Mukherjee, A., Chandrasekaran, N., 2018. Bio-based nanoemulsions: an eco-safe approach towards the eco-toxicity problem. In: Martínez, L.M.T., Kharissova, O.V., Kharisov, B.I. (Eds.), Handbook of Ecomaterials. Springer, pp. 1–23.

Mitchell, C., Brennan, R.M., Graham, J., Karley, A.J., 2016. Herbivorous pests: exploiting resistance and tolerance traits for sustainable crop protection. Front. Plant Sci. 7, 1132.

Mkindi, A.G., Tembo, Y.L.B., Mbega, E.R., Smith, A.K., Farrell, I.W., Ndakidemi, P.A., Stevenson, P.C., Belmain, S.R., 2020. Extracts of common pesticidal plants increase plant growth and yield in common bean plants. Plants 9 (2), 149.

Mnif, I., Ghribi, D., 2015. Potential of bacterial derived biopesticides in pest management. Crop Protect. 77, 52–64.

Morra, M.J., Popova, I.E., Boydston, R.A., 2018. Bioherbicidal activity of *Sinapis alba* seed meal extracts. Ind. Crop. Prod. 115, 174–181.

Mosquitoborne Diseases, 2020. Diseases that Can Be Transmitted by Mosquitoes. Minnesota Department of Health. https://www.health.state.mn.us/diseases/mosquitoborne/diseases.html.

Mosquito Squad, 2020. Essential Oils & Mosquitoes. https://www.mosquitosquad.com/blog/2019/july/essential-oils-mosquitoes/.

Moura, M.S., Lacerda, J.W.F., Siqueira, K.A., Bellete, B.S., Sousa, P.T., Dall'Oglio, E.L., Soares, M.A., Vieira, L.C.C., Sampaio, O.M., 2020. Endophytic fungal extracts: evaluation as photosynthesis and weed growth inhibitors. J. Environ. Sci. Health B 55, 470–476.

Murugan, K., Dinesh, D., Nataraj, D., Subramaniam, J., Amuthavalli, P., Madhavan, J., Rajasekar, A., Rajan, M., Thiruppathi, K.P., Kumar, S., Higuchi, A., Nicoletti, M., Benelli, G., 2018. Iron and iron oxide nanoparticles are highly toxic to *Culex quinquefasciatus* with little non-target effects on larvivorous fishes. Environ. Sci. Pollut. Res. 25 (11), 10504–10514.

Nakasu, E.Y.T., Edwards, M.G., Fitches, E., Gatehouse, J.A., Gatehouse, A.M.R., 2014. Transgenic plants expressing ω-ACTX-Hv1a and snowdrop lectin (GNA) fusion protein show enhanced resistance to aphids. Front. Plant Sci. 5, 673.

Namasivayam, S.K.R., Aruna, A., 2014. Evaluation of silver nanoparticles-chitosan encapsulated synthetic herbicide paraquate (AgNp–CS–PQ) preparation for the controlled release and improved herbicidal activity against *Eichhornia crassipes*. Res. J. Biotechnol. 9, 19–27.

Namasivayam, S.K.R., Tony, B., Bharani, R.S.A., Raj, F.R., 2015. Herbicidal activity of soil isolate of *Fusarium oxysporum* free and chitosan nanoparticles coated metabolites against economic important weed *Ninidam theejan*. Asian J. Microbiol. Biotechnol. Environ. Sci. 17, 1015–1020.

Namasivayam, S.K.R., Bharani, R.S.A., Karunamoorthy, K., 2018. Insecticidal fungal metabolites fabricated chitosan nanocomposite (IM-CNC) preparation for the enhanced larvicidal activity - an effective strategy for green pesticide against economic important insect pests. Int. J. Biol. Macromol. 120, 921–944.

Nenaah, G.E., Ibrahim, S.I.A., Al-Assiuty, B.A., 2015. Chemical composition, insecticidal activity and persistence of three Asteraceae essential oils and their nanoemulsions against *Callosobruchus maculatus* (F.). J. Stored Prod. Res. 61, 9–16.

Noutcha, M.E.A., Edwin-Wosu, N.I., Ogali, R.E., Okiwelu, S.N., 2016. The role of plant essential oils in mosquito (Diptera: Culicidae) control. Ann. Res. Rev. Biol. 10 (6), 28432. ARRB.

Nuruzzaman, M., Rahman, M.M., Liu, Y., Naidu, R., 2016. Nanoencapsulation, nano-guard for pesticides: a new window for safe application. J. Agric. Food Chem. 64, 1447–1483.

Oliveira, A.E.M.F.M., Duarte, J.L., Amado, J.R.R., Cruz, R.A.S., Rocha, C.F., Souto, R.N.P., Ferreira, R.M.A., Santos, K., da Conceiao, E.C., de Oliveira, L.A.R., Kelecom, A., Fernandes, C.P., Carvalho, J.C.T., 2016. Development of a larvicidal nanoemulsion with *Pterodon emarginatus* Vogel oil. PloS One 11 (1), e0145835.

Ootani, M.A., dos Reis, M.R., Cangussu, A.S.R., Capone, A., Fidelis, R.R., Oliveira, W., Barros, H.B., Portella, A.C.F., Aguiar, R.D., dos Santos, W.F., 2017. Phytotoxic effects of essential oils in controlling weed species *Digitaria horizontalis* and *Cenchrus echinatus*. Biocatal. Agric. Biotechnol. 12, 59–65.

Orton, T.J., 2020. Horticultural Plant Breeding. Acadeic Press and Elsevier, London, UK.

Ozturk, B., Argin, S., Ozilgen, M., McClements, D.J., 2014. Formation and stabilization of nanoemulsion-based vitamin E delivery systems using natural surfactants: *Quillaja* saponin and lecithin. J. Food Eng. 142, 57–63.

Pardo-Muras, M., Puig, C.G., Lopez-Nogueira, A., Cavaleiro, C., Pedrol, N., 2018. On the bioherbicide potential of *Ulex europaeus* and *Cytisus scoparius*: profiles of volatile organic compounds and their phytotoxic effects. PloS One 13 (10), e0205997.

Pascual-Villalobos, M.J., Canto-Tejero, M., Vallejo, R., Guirao, P., Rodriguez-Rojo, S., Cocero, M.J., 2017. Use of nanoemulsions of plant essential oils as aphid repellents. Ind. Crop. Prod. 110, 45–57.

Pavela, R., Benelli, G., 2016. Essential oils as ecofriendly biopesticides? Challenges and constraints. Trends Plant Sci. 21, 1000–1007.

Pavela, R., Murugan, K., Canale, A., Benelli, G., 2017. *Saponaria officinalis*-synthesized silver nanocrystals as effective biopesticides and oviposition inhibitors against *Tetranychus urticae* Koch. Ind. Crop. Prod. 97, 338–344.

Pavela, R., Dall'Acqua, S., Sut, S., Baldan, V., Ngahang Kamte, S.L., Nya, P.C.B., Cappellacci, L., Petrelli, R., Nicoletti, M., Canale, A., Maggi, F., Benelli, G., 2018. Oviposition inhibitory activity of the Mexican sunflower *Tithonia diversifolia* (*Asteraceae*) polar extracts against the two-spotted spider mite *Tetranychus urticae* (*Tetranychidae*). Physiol. Mol. Plant Pathol. 101, 85–92.

Pavela, R., Maggi, F., Iannarelli, R., Benelli, G., 2019a. Plant extracts for developing mosquito larvicides: from laboratory to the field, with insights on the modes of action. Acta Trop. 193, 236–271.

Pavela, R., Benelli, G., Pavoni, L., Bonacucina, G., Cespi, M., Cianfaglione, K., Bajalan, I., Morshedloo, M.R., Lupidi, G., Romano, D., Canale, A., Maggi, F., 2019b. Microemulsions for delivery of *Apiaceae* essential oils-towards highly effective and eco-friendly mosquito larvicides? Ind. Crop. Prod. 129, 631–640.

Pavoni, L., Pavela, R., Cespi, M., Bonacucina, G., Maggi, F., Zeni, V., Canale, A., Lucchi, A., Bruschi, F., Benelli, G., 2019. Green micro- and nanoemulsions for managing parasites, vectors and pests. Nanomaterials 9 (9), 1285.

Pavunraj, M., Baskar, K., Duraipandiyan, V., Al-Dhabi, N.A., Rajendran, V., Benelli, G., 2017. Toxicity of Ag nanoparticles synthesized using stearic acid from *Catharanthus roseus* leaf extract against *Earias vittella* and mosquito vectors (*Culex quinquefasciatus* and *Aedes aegypti*). J. Cluster Sci. 28 (5), 2477–2492.

Perotti, V.E., Larran, A.S., Palmieri, V.E., Martinatto, A.K., Permingeat, H.R., 2020. Herbicide resistant weeds: a call to integrate conventional agricultural practices, molecular biology knowledge and new technologies. Plant Sci. 290, 110255.

Perveen, S., Yousaf, M., Mushtaq, M.N., Sarwar, N., Khan, M.Y., Nadeem, S.M., 2019. Bioherbicidal potential of some allelopathic agroforestry and fruit plant species against *Lepidium sativum*. Soil Environ. 38 (1), 119–126.

Pessoa, L.Z.D., Duarte, J.L., Ferreira, R.M.D., Oliveira, A.E.M.D.M., Cruz, R.A.S., Faustino, S.M.M., Carvalho, J.C.T., Fernandes, C.P., Souto, R.N.P., Araujo, R.S., 2018. Nanosuspension of quercetin: preparation, characterization and effects against *Aedes aegypti* larvae. Rev. Bras. Farmacogn 28 (5), 618–625.

Phour, M., Sindhu, S.S., 2019. Bio-herbicidal effect of 5-aminoleveulinic acid producing rhizobacteria in suppression of *Lathyrus aphaca* weed growth. Biocontrol 64 (2), 221–232.

Pinheiro, A.C., Coimbra, M.A., Vicente, A.A., 2016. In vitro behaviour of curcumin nanoemulsions stabilized by biopolymer emulsifiers - effect of interfacial composition. Food Hydrocoll. 52, 460–467.

Pinto, A.P.R., Seibert, J.B., dos Santos, O.D.H., Vieira Filho, S.A., do Nascimento, A.M., 2018. Chemical constituents and allelopathic activity of the essential oil from leaves of *Eremanthus erythropappus*. Aust. J. Bot. 66 (8), 601–608.

Pinto Zevallos, D.M., Querol, M.P., Ambrogi, B.G., 2018. Cassava wastewater as a natural pesticide: current knowledge and challenges for broader utilisation. Ann. Appl. Biol. 173 (3), 191–201.

Plachá, D., Jampílek, J., 2019. Graphenic materials for biomedical applications. Nanomaterials 9 (12), 1758.

Popovic, Z., Kostic, M., Stankovic, S., Milanovic, S., Sivcev, I., Kostic, I., Kljajic, P., 2013. Ecologically acceptable usage of derivatives of essential oil of sweet basil, *Ocimum basilicum*, as antifeedants against larvae of the gypsy moth, *Lymantria dispar*. J. Insect Sci. 13, 161.

Preisler, A.C., Pereira, A.E.S., Campos, E.V.R., Dalazen, G., Fraceto, L.F., Oliveira, H.C., 2020. Atrazine nanoencapsulation improves pre-emergence herbicidal activity against *Bidens pilosa* without enhancing long-term residual effect on *Glycine max*. Pest. Manag. Sci. 76 (1), 141–149.

Priya, S., Santhi, S., 2014. A review on nanoparticles in mosquito control - a green revolution in future. Int. J. Res. Appl. Sci. Eng. Technol. 2 (XII), 378–387.

Puig, C.G., Goncalves, R.F., Valentao, P., Andrade, P.B., Reigosa, M.J., Pedrol, N., 2018a. The consistency between phytotoxic effects and the dynamics of allelochemicals release from *Eucalyptus globulus* leaves used as bioherbicide green manure. J. Chem. Ecol. 44 (7–8), 658–670.

Puig, C.G., Reigosa, M.J., Valentao, P., Andrade, P.B., Pedrol, N., 2018b. Unravelling the bioherbicide potential of *Eucalyptus globulus* Labill: biochemistry and effects of its aqueous extract. PloS One 13 (2), e0192872.

Pyati, P., Fitches, E., Gatehouse, J.A., 2014. Optimising expression of the recombinant fusion protein biopesticide omega-hexatoxin-Hv1a/GNA in *Pichia pastoris*: sequence modifications and a simple method for the generation of multi-copy strains. J. Ind. Microbiol. Biotechnol. 41 (8), 1237–1247.

Radhakrishnan, R., Alqarawi, A.A., Abd Allah, E.F., 2018. Bioherbicides: current knowledge on weed control mechanism. Ecotoxicol. Environ. Saf. 158, 131–138.

Rajakumar, G., Rahuman, A.A., Jayaseelan, C., Santhoshkumar, T., Marimuthu, S., Kamaraj, C., Bagavan, A., Zahir, A.A., Kirthi, A.V., Elango, G., Arora, P., Karthikeyan, R., Manikandan, S., Jose, S., 2014. *Solanum trilobatum* extract-mediated synthesis of titanium dioxide nanoparticles to control *Pediculus humanus capitis*, *Hyalomma anatolicum anatolicum* and *Anopheles subpictus*. Parasitol. Res. 113 (2), 469–479.

Rajkumar, R., Shivakumar, M.S., Nathan, S.S., Selvam, K., 2018. Pharmacological and larvicidal potential of green synthesized silver nanoparticles using *Carmona retusa* (Vahl) Masam leaf extract. J. Cluster Sci. 29 (6), 1243–1253.

Rajkumar, R., Shivakumar, M.S., Nathan, S.S., Selvam, K., 2019. Preparation and characterization of chitosan nanocomposites material using silver nanoparticle synthesized *Carmona retusa* (Vahl) Masam leaf extract for antioxidant, anti-cancerous and insecticidal application. J. Cluster Sci. 30 (4), 1145–1155.

Rajmohan, K.S., Chandrasekaran, R., Varjani, S., 2020. A review on occurrence of pesticides in environment and current technologies for their remediation and management. Indian J. Microbiol. 60, 125–138.

Rakian, T.C., Muhidin, Sutariati, G., Gusnawaty, H.S., Asniah, Fermin, U., 2018a. Selection of deleterious rhizobacterial isolate as bioherbicide to control of weed *Paspalum conjugatum* and *Ageratum conyzoides* on soybean cropland. Biosci. Res. 15 (3), 1695–1702.

Rakian, T.C., Karimuna, L., Taufik, M., Sutariati, G.A.K., Muhidin, Fermin, U., 2018b. The effectiveness of various *Rhizobacteria* carriers to improve the shelf life and the stability of Rhizobacteria as bioherbicide. IOP Conf. Ser. Earth Environ. Sci. 122, 012032. UNSP.

Reichert, F.W., Scariot, M.A., Forte, C.T., Pandolfi, L., Dil, J.M., Weirich, S., Carezia, C., Molinari, J., Mazutti, M.A., Fongaro, G., Galon, L., Treichel, H., Mossi, A.J., 2019. New perspectives for weeds control using autochthonous fungi with selective bioherbicide potential. Heliyon 5 (5), e01676.

Reynolds, O.L., Padula, M.P., Zeng, R.S., Gurr, G.M., 2016. Plant defense against silicon: potential to promote direct and indirect effects on plant defense against arthropod pests in agriculture. Front. Plant Sci. 7, 744.

Romeis, J., Naranjo, S.E., Meissle, M., Shelton, A.M., 2019. Genetically engineered crops help support conservation biological control. Biol. Contr. 130, 136–154.

Rondot, Y., Reineke, A., 2018. Endophytic *Beauveria bassiana* in grapevine *Vitis vinifera* (L.) reduces infestation with piercing-sucking insects. Biol. Contr. 116, 82–89.

Roy, S., Mukhopadhyay, A., Gurusubramanian, G., 2010. Field efficacy of a biopesticide prepared from *Clerodendrum viscosum* Vent. (Verbenaceae) against two major tea pests in the sub Himalayan tea plantation of North Bengal, India. J. Pest. Sci. 83 (4), 371–377.

Ruiu, L., 2018. Microbial biopesticides in agroecosystems. Agronomy 8, 235.

Rusch, A., Chaplin-Kramer, R., Gardiner, M.M., Hawro, V., Holland, J., Landis, D., Thies, C., Tscharntke, T., Weisser, W.W., Winqvist, C., Woltz, M., Bommarco, R., 2016. Agricultural landscape simplification reduces natural pest control: a quantitative synthesis. Agric. Ecosyst. Environ. 221, 198–204.

Saad, M.M.G., Gouda, N.A.A., Abdelgaleil, S.A.M., 2019. Bioherbicidal activity of terpenes and phenylpropenes against *Echinochloa crus-galli*. J. Environ. Sci. Health B 54 (12), 954–963.

Sabry, A.H., Ragaei, M., 2018. Nanotechnology and their applications in insect's pest control. In: Abd-Elsalam, K.A., Prasad, R. (Eds.), Nanobiotechnology Applications in Plant Protection, Nanotechnology in the Life Sciences. Springer International Publishing, pp. 1–28.

Saez, N.J., Herzig, V., 2019. Versatile spider venom peptides and their medical and agricultural applications. Toxicon 158, 109–126.

Said-Al Ahl, H.A.H., Hikal, W.M., Tkachenko, K.G., 2017. Essential oils with potential as insecticidal agents: a review. Int. J. Environ. Plan. Manag. 3 (4), 23–33.

Saini, R., Singh, S., 2019. Use of natural products for weed management in high-value crops: an overview. Am. J. Agric. Res. 4, 25.

Salem, S.S., Fouda, A., 2021. Green synthesis of metallic nanoparticles and their prospective biotechnological applications: an overview. Biol. Trace Elem. Res. 199, 344–370.

Sallam, M.N., 2013. Insect Damage – Post Harvest Operations. Food and Agriculture Organization of the United Nations. http://www.fao.org/3/a-av013e.pdf.

San Aw, K.M., Hue, S.M., 2017. Mode of infection of *Metarhizium* spp. fungus and their potential as biological control agents. J. Fungi 3 (2). UNSP 30.

Saric-Krsmanovic, M., Umiljendic, J.G., Radivojevic, L., Santric, L., Potocnik, I., Durovic-Pejcev, R., 2019. Bio-herbicidal effects of five essential oils on germination and early seedling growth of velvetleaf (*Abutilon theophrasti* Medik.). J. Environ. Sci. Health B 54 (4), 247–251.

Scavo, A., Restuccia, A., Mauromicale, G., 2018. In: Gaba, S., Smith, B., Lichtfouse, E. (Eds.), Allelopathy: Principles and Basic Aspects for Agroecosystem Control, Sustainable Agriculture Reviews, vol. 28. Springer, Cham, pp. 47–101.

Scavo, A., Pandino, G., Restuccia, A., Lombardo, S., Pesce, G.R., Mauromicale, G., 2019a. Allelopathic potential of leaf aqueous extracts from *Cynara cardunculus* L. on the seedling growth of two cosmopolitan weed species. Ital. J. Agron. 14 (2), 78–83.

Scavo, A., Rial, C., Varela, R.M., Molinillo, J.M.G., Mauromicale, G., Macias, F.A., 2019b. Influence of genotype and harvest time on the *Cynara cardunculus* L. sesquiterpene lactone profile. J. Agric. Food Chem. 67 (23), 6487–6496.

Scavo, A., Pandino, G., Restuccia, A., Mauromicale, G., 2020. Leaf extracts of cultivated cardoon as potential bioherbicide. Sci. Hortic. 261, 109024.

Schütte, G., Eckerstorfer, M., Rastelli, V., Reichenbecher, W., Restrepo-Vassalli, S., Ruohonen-Lehto, M., Wuest Saucy, A.G., Mertens, M., 2017. Herbicide resistance and biodiversity: agronomic and environmental aspects of genetically modified herbicide-resistant plants. Environ. Sci. Eur. 29 (1), 5.

Shah, M.A., Wani, S.H., Khan, A.A., 2016. Nanotechnology and insecticidal formulations. J. Food Bioengin. Nanoproc. 1 (3), 285–310.

Shah, F.M., Razaq, M., Ali, A., Han, P., Chen, J.L., 2017. Comparative role of neem seed extract, moringa leaf extract and imidacloprid in the management of wheat aphids in relation to yield losses in Pakistan. PloS One 12 (9), e0184639.

Shang, Y., Hasan, M.K., Ahammed, G.J., Li, M., Yin, H., Zhou, J., 2019. Applications of nanotechnology in plant growth and crop protection: a review. Molecules 24 (14), 2558.

Sharma, H.C., Dhillon, M.K., Hatfield, J.L., Sivakumar, M.V.K., Prueger, J.H., 2018. Climate change effects on arthropod diversity and its implications for pest management and sustainable crop production. In: Hatfield, J.L., Sivakumar, M.V.K., Prueger, J.H. (Eds.), Agroclimatology: Linking Agriculture to Climate, 60. American Society of Agronomy, Crop Science Society of America, and Soil Science Society of America, Inc, pp. 595–619.

Sharma, A., Tripathi, P., Kumar, S., 2020. One-pot synthesis of silver nanocomposites from *Achyranthes aspera*: an eco-friendly larvicide against *Aedes aegypti* L. Asian Pac. J. Trop. Biomed. 10 (2), 54–64.

Shi, W.P., Wang, X.Y., Yin, Y., Zhang, Y.X., Rizvi, U.E.H., Tan, S.Q., Cao, C., Yu, H.Y., Ji, R., 2019. Dynamics of aboveground natural enemies of grasshoppers, and biodiversity after application of *Paranosema locustae* in rangeland. Insects 10 (8), 224.

Shi, L.Q., Wu, Z.Y., Zhang, Y.N., Zhang, Z.G., Fang, W., Wang, Y.Y., Wan, Z.Y., Wang, K.M., Ke, S.Y., 2020. Herbicidal secondary metabolites from actinomycetes: structure diversity, modes of action, and their roles in the development of herbicides. J. Agric. Food Chem. 68 (1), 17–32.

Siegwart, M., Graillot, B., Lopez, C.B., Besse, S., Bardin, M., Nicot, P.C., Lopez-Ferber, M., 2015. Resistance to bio-insecticides or how to enhance their sustainability: a review. Front. Plant Sci. 6, 381.

Silva, M.D., Cocenza, D.S., de Melo, N.F.S., Grillo, R., Rosa, A.H., Fraceto, L.F., 2010. Alginate nanoparticles as a controlled release system for clomazone herbicide. Quim. Nova 33, 1868–1873.

Silva, M.D., Cocenza, D.S., Grillo, R., de Melo, N.F.S., Tonello, P.S., de Oliveira, L.C., Cassimiro, D.L., Rosa, A.H., Fraceto, L.F., 2011. Paraquat-loaded alginate/chitosan nanoparticles: preparation, characterization and soil sorption studies. J. Hazard Mater. 190, 366–374.

Sindhu, S.S., Sehrawat, A., 2017. Rhizosphere microorganisms: application of plant beneficial microbes in biological control of weeds. In: Panpatte, D.G., Jhala, Y.K., Vyas, R.V., Shekat, H.N. (Eds.), Microorganisms for Green Revolution, Vol. 1: Microbes for Sustainable Crop Production, Microorganisms for Sustainability, vol. 6. Springer, Singapore, pp. 391–430.

Singh, J., Dutta, T., Kim, K.H., Rawat, M., Samddar, P., Kumar, P., 2018. 'Green' synthesis of metals and their oxide nanoparticles: applications for environmental remediation. J. Nanobiotechnol. 16 (1), 84.

Singhand, A., Gandhi, S., 2012. Agricultural insect pest: occurrence and infestation level in agricultural fields of Vadodara, Gujarat. Int. J. Sci. Res. Publ. 2 (4), 1–5.

Smith, C.M., Clement, S.L., 2012. Molecular bases of plant resistance to arthropods. Annu. Rev. Entomol. 57, 309–328.

Soni, N., Prakash, S., 2014. Green nanoparticles for mosquito control. Nanomat. Nanodev. 2014, 496362.

Spiassi, A., Fortes, A.M.T., Guedes, L.P.C., de Lima, G.P., Meira, R.O., Valmorbida, R., de Mendonca, L.C., 2019. Phytochemical screening and toxicity of *Crambe abyssinica* Hochst extracts on *Solanum lycopersicum* L., *Euphorbia heterophylla* L., *Bidens pilosa* L. and *Glycine max* (L.) Merril. Biosci. J. 35 (5), 1408–1421.

Stange, E.E., Ayres, M.P., 2010. Climate Change Impacts: Insects. Encyclopedia of Life Sciences. https://doi.org/10.1002/9780470015902.a0022555.

Stevenson, P.C., Green, P.W.C., Farrell, I.W., Brankin, A., Mvumi, B.M., Belmain, S.R., 2018. Novel agmatine derivatives in *Maerua edulis* with bioactivity against *Callosobruchus maculatus*, a cosmopolitan storage insect pest. Front. Plant Sci. 9, 1506.

Sultana, N., Raul, P., Goswami, D., Das, D., Islam, S., Tyagi, V., Das, B., Gogoi, H.K., Chattopadhyay, P., Raju, P.S., 2020. Bio-nanoparticle assembly: a potent on-site biolarvicidal agent against mosquito vectors. RSC Adv. 10 (16), 9356–9368.

Sytykiewicz, H., Kozak, A., Lukasik, I., Sempruch, C., Golawska, S., Mitrus, J., Kurowska, M., Kmiec, K., Chrzanowski, G., Leszczynski, B., 2019. Juglone-triggered oxidative responses in seeds of selected cereal agrosystem plant species. Pol. J. Environ. Stud. 28 (4), 2389–2397.

Taban, A., Saharkhiz, M.J., Khorram, M., 2020. Formulation and assessment of nano-encapsulated bioherbicides based on biopolymers and essential oil. Ind. Crop. Prod. 149, 112348.

Tabari, M.A., Youssefi, M.R., Maggi, F., Benelli, G., 2017. Toxic and repellent activity of selected monoterpenoids (thymol, carvacrol and linalool) against the castor bean tick, *Ixodes ricinus* (Acari: *Ixodidae*). Vet. Parasitol. 245, 86–91.

Tabashnik, B.E., Mota-Sanchez, D., Whalon, M.E., Hollingworth, R.M., Carriere, Y., 2014. Defining terms for proactive management of resistance to Bt crops and pesticides. J. Econ. Entomol. 107 (2), 496–507.

Tahir, H.M., Zahra, K., Khan, A.A., 2018a. Insect-specific peptides in the venom of wolf spiders (Araneae: *Lycosidae*). Turk. J. Zool. 42 (5), 614–616.

Tahir, H.M., Zahra, K., Yagoob, R., Khan, A.A., Butt, A., Hassan, Z., Khan, S.Y., 2018b. Evaluation of venom peptides of two jumping spider species (Araneae: Salticidae) as insecticide potential. Int. J. Agric. Biol. 20 (11), 2423–2427.

Tavares, M., da Silva, M.R.M., de Siqueira, L.B.D., Rodrigues, R.A.S., Bodjolle-d'Almeida, L., dos Santos, E.P., Ricci, E., 2018. Trends in insect repellent formulations: a review. Int. J. Pharm. 539 (1–2), 190–209.

Tawfik, M.M., Ibrahim, N.A., Balah, M.A., Abouzeid, M.M., 2019. Evaluation of bacteria from soil and rhizosphere as herbicidal candidates of some broadleaf weeds. Egypt. J. Bot. 59 (2), 283–291.

Tembo, Y., Mkindi, A.G., Mkenda, P.A., Mpumi, N., Mwanauta, R., Stevenson, P.C., Ndakidemi, P.A., Belmain, S.R., 2018. Pesticidal plant extracts improve yield and reduce insect pests on legume crops without harming beneficial arthropods. Front. Plant Sci. 9, 1425.

Tinio, J.C.P., Rayos, A.L., Aguila, M.J.B., Salamanez, K.C., 2019. Bioherbicidal activity of *Medinilla magnifica* Lindl. leaf extract. Philipp. Agric. Sci. 102 (3), 270–275.

Todero, I., Confortin, T.C., Soares, J.F., Brun, T., Luft, L., Rabuske, J.E., Kuhn, R.C., Tres, M.V., Zabot, G.L., Mazutti, M.A., 2019. Concentration of metabolites from *Phoma* sp. using microfiltration membrane for increasing bioherbicidal activity. Environ. Technol. 40 (18), 2364–2372.

Tomer, H., Blum, T., Arye, I., Faigenboim, A., Gottlieb, Y., Ment, D., 2018. Activity of native and commercial strains of *Metarhizium* spp. against the poultry red mite *Dermanyssus gallinae* under different environmental conditions. Vet. Parasitol. 262, 20–25.

Unal, B.T., Bayram, M., 2019. The allelopathic effects of sunflower and wheat root exudates on *Sinapis arvensis* and *Sinapis alba*. Phyton-Int. J. Exp. Bot. 88 (4), 413–423.

Vinoth, S., Shankar, S.G., Gurusaravanan, P., Janani, B., Devi, J.K., 2019. Anti-larvicidal activity of silver nanoparticles synthesized from *Sargassum polycystum* against mosquito vectors. J. Cluster Sci. 30 (1), 171–180.

Volpato, A., Baretta, D., Zortea, T., Campigotto, G., Galli, G.M., Glombowsky, P., Santos, R.C.V., Quatrin, P.M., Ourique, A.F., Baldissera, M.D., Stefani, L.M., Da Silva, A.S., 2016. Larvicidal and insecticidal effect of *Cinnamomum zeylanicum* oil (pure and nanostructured) against mealworm (*Alphitobius diaperinus*) and its possible environmental effects. J. Asia Pac. Entomol. 19 (4), 1159–1165.

Warsaba, R., Sadasivan, J., Jan, E., 2019. Dicistrovirus-host molecular interactions. Curr. Issues Mol. Biol. 34, 83–112.

Wennersten, R., Fidler, J., Spitsyna, A., 2008. Nanotechnology: a new technological revolution in the 21st century. In: Misra, K.B. (Ed.), Handbook of Performability Engineering. Springer, London, pp. 943–952.

Windley, M.J., Herzig, V., Dziemborowicz, S.A., Hardy, M.C., King, G.F., Nicholson, G.M., 2012. Spider-Venom peptides as bioinsecticides. Toxins 4 (3), 191–227.

Xu, Z.L., Shi, M.Y., Tian, Y.Q., Zhao, P.F., Niu, Y.F., Liao, M.D., 2019. Dirhamnolipid produced by the pathogenic fungus *Colletotrichum gloeosporioides* BWH-1 and its herbicidal activity. Molecules 24 (16), 2969.

Zhang, Y., Yang, X.H., Zhu, Y.B., Li, L.Y., Zhang, Y.L., Li, J.P., Song, X.L., Qiang, S., 2019. Biological control of *Solidago canadensis* using a bioherbicide isolate of *Sclerotium rolfsii* SC64 increased the biodiversity in invaded habitats. Biol. Contr. 139. UNSP 104093.

Zhao, C.H., Shen, X., Guo, M.R., 2018. Stability of lutein encapsulated whey protein nano-emulsion during storage. PloS One 13 (2), e0192511.

Zhou, J., Xiao, K.F., Wei, B.Y., Wang, Z., Tian, Y., Tian, Y.X., Song, Q.S., 2014. Bioaccumulation of Cry1Ab protein from an herbivore reduces anti-oxidant enzyme activities in two spider species. PloS One 9 (1), e84724.

Zohra, Z.F., Houari, B., Malika, B., Abderrazak, M., 2019. Allelopathic effect of *Tetraclinis articulata* (Vahl) mast (from Algeria) on germination and growth of *Lactuca sativa* L. Biosci. Res. 16 (1), 493–499.

Chapter 11

Salvia leucantha essential oil encapsulated in chitosan nanoparticles with toxicity and feeding physiology of cotton bollworm *Helicoverpa armigera*

Devakumar Dinesh[a], Kadarkarai Murugan[a], Jayapal Subramaniam[a,b], Manickam Paulpandi[a], Balamurugan Chandramohan[a], Krishnasamy Pavithra[a], Jaganathan Anitha[a], Murugan Vasanthakumaran[c], Leonardo Fernandes Fraceto[d], Lan Wang[e], Jiang Shoiu-Hwang[f,g] and Hans-Uwe Dahms[h]

[a]*Department of Zoology, School of Life Sciences, Bharathiar University, Coimbatore, Tamil Nadu, India;* [b]*Division of Vector Biology and Control, Department of Zoology, Faculty of Science, Annamalai University, Chidambaram, Tamil Nadu, India;* [c]*Department of Zoology, Kongunadu Arts and Science College, Coimbatore, Tamil Nadu, India;* [d]*Institute of Science and Technology of Sorocaba, São Paulo State University – Unesp, São Paulo, São Paulo, Brazil;* [e]*School of Life Science, Shanxi University, Taiyuan, Shanxi, China;* [f]*Institute of Marine Biology, National Taiwan Ocean University, Keelung, Keelung, Taiwan;* [g]*Center of Excellence for the Oceans, National Taiwan Ocean University, Keelung, Keelung, Taiwan;* [h]*Department of Biomedical Science and Environmental Biology, Kaohsiung Medical University, Kaohsiung, Kaohsiung, Taiwan*

11.1 Introduction

Insect pests inflict important damages to humans, farm animals, crops, stored products either directly or indirectly and impacts on agricultural production and market access, the natural environment, and our lifestyle (Perlatti et al., 2013). The agri-food production is of vital importance as it has been one of the primary drivers of the economy. In addition, it can offer routes to value-added crops. Agricultural practices are often in the public eye because climate change energy, resource constraints and rapidly growing global population are placing unprecedented pressure on food and water resources. The insects such as *Helicoverpa armigera, S. litura* and *P. xylostella* agricultural insect pests are distributed worldwide and most serious and economically potential insect pests of crops throughout the world (Kandaga and Khetagoudar, 2013; Zada et al., 2018). This noctuid pest is distributed eastwards from southern Europe and Africa through the Indian subcontinent to Southeast Asia, and hence to China, Japan, Australia and the Pacific Islands (Murugan et al., 2018). In this situation, protection of agriculture crops from insect damage often requires eco-friendly sound insecticides. The use of synthetic insecticides for controlling insect pests led to huge problems such as their persistent toxicity in food, the development of resistance in pest populations, and effects on non-target organisms (Patel et al., 1992). This has boosted the interest of consumers and growers in application of plant-based insecticides and their use has been increased constantly in recent years (Murugan et al., 2018). Hematpoor et al. (2017) reported the botanical insecticides and combined with nanobase materials are the future alternative for the insect pest management. The Plant secondary natural products are natural chemicals extracted from plants and used as an excellent alternative to synthetic pesticides (Suthisut et al., 2011; Hikal et al., 2017). Plant allelochemicals are the excellent alternative to synthetic chemicals (Suthisut et al., 2011; Hikal et al., 2017) and further, *M. alternifolia* produced several essential compounds and are feeding suppressing compounds to *Helicoverpa armigera* (Liao et al., 2017). To evaluate the toxic activities of Salvia plant oil on different stages of the black cutworm, *Agrotis ipsilon* controlled for possible use as a safe biological method and alternative to chemical pesticides (Sharaby and El-Nujiban, 2013). Natural oil and fungal elements are the good combination for the control stored product insects (Mohammadi et al., 2015). also maintain the quality of seed grains (Aloui et al., 2014). Nowadays, more attention

Biopesticides. https://doi.org/10.1016/B978-0-12-823355-9.00022-5

has been diverted to develop from essential and it will be utilized variously to curtail insect feeding, growth and volatile compounds (Jaya et al., 2014; Kedia et al., 2014). and above are environmentally friendly.

The essential oils from *S. leucantha* have been reported to contain 1, 8-cineole, camphor, borneol, β-pinene, αpinene, and are responsible for therapeutic properties and no-side effects (Zhiming et al., 2013). Nowadays, the synthesis of nanoparticles using biological products received a great interest due to the nanoparticles' unusual optical, chemical, photo electrochemical and electronic properties. Nanoparticles also help to produce new pesticides, insecticides and insect repellents (Owolade et al., 2008; Murugan et al., 2018).

Bhattacharyya et al. (2010) emphasized that future integrated insect pest management nanotechnology is an opt field. Formulations and field evaluations are essential items for the development of nano oriented insecticides. Among which chitosan sandwiches with biological materials will give more eco-friendly, biocompatibility, biodegradability and several drug delivery systems in medicine (Prasanth Koppolu et al., 2014). Moreover, oil and chitosan encapsulations from several plant oils are the best examples. (Ferreira et al., 2019).

Earlier, Murugan et al. (2016) have proposed a novel method of biofabrication of silver nanoparticles (AgNP) using chitosan (Ch) from crab shells and was tested against larvae and pupae of the malaria vector *Anopheles stephensi*. Murugan et al. (2017a,b,c) used spray consisting of a mix of chitosan, extracted from crushed crab shells, and silver nanoparticles can reduce mosquito populations and help to control malaria and other diseases spread by the insect vector. Field studies of the CH−AgNP spray, done in the breeding sites in India, found it lethal to larvae and pupae of the *Anopheles sundaicus* mosquito that spreads coastal malaria. The nanoparticle solution did not have any effect on the fish, indicating that it is an environmentally friendly and non-toxic product. Murugan et al. (2017a) done with Bio-encapsulated chitosan-Ag nanocomplex and bio-encapsulated chitosan-Ag nanocomplex was evaluated in comparison with classic chitosan-synthesized Ag nanoparticles were tested on the toxicity against the malaria vector *A. stephensi*, both in laboratory and in the field. They have also done on the free radical scavenging potential on DPPH and ABT, as well as the cytotoxicity via apoptosis-triggering mechanisms on breast cancer cells (MCF-7) and further the impact of sub-lethal doses of nano-products on the predation efficiency of non-target larvivorous fishes *Poecilia reticulata* predating anopheline larvae. Murugan et al. (2021) investigated the Insecticidal effect of chitosan reduced silver nanocrystals against filarial vector, *Culex quinquefasciatus* and Cotton Bollworm, *Helicoverpa armigera*.

Present study, we shed light on the phyotochemicals present in the *S. leucantha* essential oil using GC/MS analyses and describes a rapid and eco-friendly method for synthesis of oil loaded encapsulation chitosan nanoparticles using essential oil and characterized through the following biophysical methods including UV−vis spectroscopy, Scanning electron microscope (SEM), energy dispersive X-ray spectroscopy (EDX), Fourier transform infrared spectroscopy (FTIR) and Zeta potential. The main purpose of this research reports the larvicidal, pupicidal activity and longevity and fecundity of *S. leucantha* essential oil (SlEO), essential oil encapsulated chitosan nanoparticles (SlEO-ChNPs) against *H. armigera, S. litura* and *P. xylostella.* Furthermore, we evaluated the food utilization measures and digestive enzymes of insect pests with post treatment of encapsulation chitosan nanoparticles. In addition, that explores the polyphagous insect's perspective when exposed to SlEO-ChNPs will help in better pest control strategy design.

11.2 Materials and methods

11.2.1 Plant material

Fresh leaves of *Salvia leucantha* were collected from Kothagiri, Nilgiri district, Tamilnadu. The plant was identified by authorities of Botanical Survey of India, Agricultural University, Coimbatore (BSI/SRC/5/23/207/Tech. 1905). The plant materials were cute small pieces and dried naturally on laboratory condition at room temperature (23−27°C) until they were crisp. The dried aerial parts were stored at 24°C until they were hydro-distilled to extract their essential oils.

11.2.2 Extraction of *S. leucantha* essential oil

Essential oil was obtained by hydro distillation method which is commonly used as an approved reference for the quantification of oils. Dried leaves of *S. leucantha* were subjected to hydro distillation using a Clevenger-type apparatus in order to obtain essential oil. Condition of extraction was: 40 g of leaves, powder and 600 mL distilled water and 4 h distillation. Extracted oil was stored in a refrigerator at 4°C. Anhydrous sodium sulfate was used to remove water after extraction.

11.3 Qualitative analysis

11.3.1 Phytochemical analysis

The essential oil of *S. leucantha* were subjected to chemical tests for the detection of different phytoconstituents using standard procedures as described by Harborne (1984), Trease and Evans (1979). The following methods were used for detection of various phytochemicals by qualitative chemical analysis to give general idea regarding the nature of Phyto-constituents present in essential oil.

11.4 Test for flavonoids

Lead Acetate: Lead acetate solution was added to the small quantity of essential oil, leads to the formation of yellow precipitate indicated the presence of flavonoids.

11.5 Test for alkaloids

a. **Wagner's Reagent:** To 2–3 mL of filtrate, 1 mL of dil. HCl and Wagner's reagent were added and shake well. Formation of redish-brown precipitate showed the presence of alkaloids.
b. **Mayer's Reagent:** To 2–3 mL of filtrate, 1 mL of dil. HCl and Mayer's reagent was added and shake well. Formation of yellow precipitate showed the presence of alkaloids.

11.6 Test for tannins

a. **$FeCl_3$ Solution:** On addition of 5% $FeCl_3$ solution to the essential oil, deep blue black color appeared.
b. **Lead Acetate:** On addition of lead acetate solution to the essential oil, white precipitate appeared.

11.7 Test for phenolics

Phenolics Tests: 0.5 mL of $FeCl_3$ (w/v) solution was added to 2 mL of essential oil, formation of an intense color indicated the presence of phenolics (Gibbs, 1974).

11.8 Test for terpenoids

2 mL of the essential oil was dissolved in 2 mL of $CHCl_3$ and evaporated to dryness. 2 mL of concentrate H_2SO_4 was then added and heated for about 2 min. Finally, the development of a grayish color indicates the presence of terpenoids. Above phytochemicals analysis was carried out using standard procedure (Kokate et al., 2005; Sadashivan and Manickam, 2005).

11.9 Test for saponins

Few drops of Na_2HCO_3 were added to 0.5 mL of essential oil and shaken for 5 min. Formation of froth or lather indicated the presence of saponins.

11.10 Test for glycosides

Dilute NaOH were added to 0.5 mL of essential oil. Yellow colored solution indicated presence of glycosides.

11.11 GCMS analysis of essential oil of *S. leucantha*

11.11.1 GC–MS specification

PerkinElmer (clarus 680), clarus 600 (EI), TurboMass ver 5.4.2 (NIST-2008) with the acquisition parameters Initial temp 60°C for 2 min, ramp 10°C/min to 300°C, hold 6 min, Total Run Time: 32.00 mint, InjAauto = 260°C, Volume = 1 μL, Split = 10:1, Flow Rate: 1 mL/mint, Carrier Gas = He, Column = Elite-5MS (30.0 m, 0.25mm ID, 250 μm df). The mass conditions (EI) were Solvent Delay = 2.00 min, Transfer Temp = 230°C, Source Temp = 230°C, Scan: 50–600 Da.

11.11.2 GC–MS analysis

The Clarus 680 GC was used in the analysis employed a fused silica column, packed with Elite-5MS (5% biphenyl 95% dimethylpolysiloxane, 30 m × 0.25 mm ID × 250 μm df) and the components were separated using Helium as carrier gas at a constant flow of 1 mL/min. The injector temperature was set at 260°C during the chromatographic run. The 1 μL of essential oil injected into the instrument the oven temperature was as follows: 60°C (2 min) followed by 300°C at the rate of 10°C/min and 300°C, where it was held for 6 min. The mass detector conditions were transfer in temperature 230°C ion source temperature 230°C and ionization mode electron impact at 70 eV, a scan time 0.2 s and scan interval of 0.1 s. The fragments from 40 to 600 Da. The spectrums of the components were compared with the database of spectrum of known components stored in the GC–MS NIST (2008) library.

11.12 Collection and processing of crab shells

The Mud crab shells (*Scylla serrata*) were collected from local market in Coimbatore, Tamilnadu, India. Samples were kept chilled in ice during transportation to the laboratory. Using a meat tenderizer the exoskeletons of the crab were cut into smaller pieces. 20 g of crushed crab exoskeletons was measured using a Mettle balance, then labeled and oven-dried at 65°C for four consecutive days to obtain constant weight. The dry weight of the samples was determined and the moisture content measured was based on the differences between the wet and the dry weight. The average moisture content of the crab exoskeletons was 15.66%.

11.13 Isolation and extraction of chitosan from crab shell

The chitosan recovering sequence involved washing of crushed crab exoskeletons in distilled water. Crushed crab exoskeletons were placed in 1000 mL beakers and soaked in boiling sodium hydroxide (2% and 4% w/v) for 1 h, in order to dissolve the proteins and sugars, thus isolating the crude chitin. 4% NaOH is used for chitin preparation, concentration used at the Sonat Corporation (Lertsutthiwong et al., 2002). Then, the samples were boiled in the sodium hydroxide, the beakers containing the shell samples were removed from the hot plate, and allowed to cool for 30 min at room temperature (Lamarque, 2005). The exoskeletons were then crushed to pieces of 0.5–5.0 mm using a meat tenderizer. The obtained pure chitosan was demineralized using 1% HCl (v: v) and 2% NaOH according to the method of Huang et al. (2004). 50% NaOH was added to demineralized chitosan and boiled at 100°C for 2 h on a hot plate then cooled for 30 min at room temperature. The process was repeated to deacetylate the chitosan powder and finally creamy-white form of chitosan was obtained (Muzzarelli and Rochetti, 1985).

11.13.1 Structure of chitosan

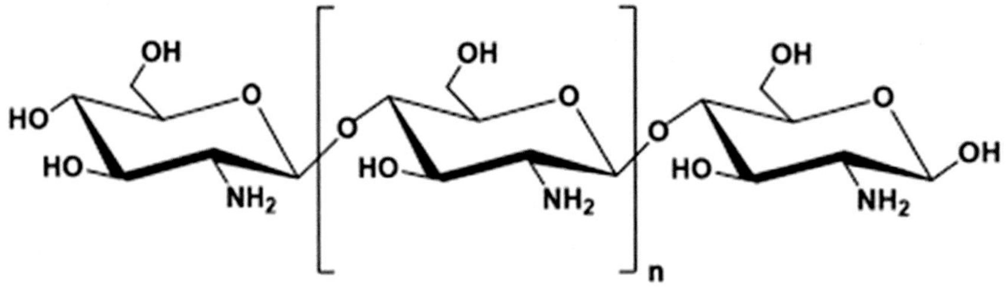

11.14 Chitosan nanoparticles preparation with essential oil

Essential Oil - loaded Chitosan (CS) nanoparticle were prepared according to a method modified from the ones described by Yoksan et al. (2010). Chitosan solutions (1% (w/v)) were prepared by agitating Chitosan in an aqueous (1% 40 mL) acetic acid solution at ambient temperature (23–25°C) overnight. The mixture was then centrifuged using a laboratory centrifuges (Remi C-24BL, India) for 30 min at 9000 rpm; the supernatant was removed then and filtered through 1 μm pore size filters. Tween 80 (HLB 15.0, 0.56 g) was then added as a surfactant to the solution (40 mL) and stirred at 45°C for 2 h to obtain a homogeneous mixture. *S. leucantha* essential oil (0.81 g) was dissolved separately in CH_2Cl_2 (4 mL) and then this oil phase is gradually dropped into the aqueous Chitosan solution (40 mL) during homogenization (Remi RQ-127A/D, India) at a speed of 10,000 rpm for 15 min under an ice-bath condition to obtain an oil-in water emulsion. Sodium Tripolyphosphate (TPP) (0.4% (W/V), 40 mL) was then added dropwise into the agitated emulsion. Agitation was continuously performed for 40 min, the formed particles were collected by centrifugation at 10,000 rpm for 30 min at 4°C and subsequently washed several times with distilled water. Finally, ultra-sonication was performed by a sonication (Digital ultrasonic cleaner LMUC − 2, L6169, India). The suspensions were immediately freeze-dried at −35°C for 72 h using Freeze-dryer.

11.15 Characterization of essential oil loaded chitosan nanomaterials

The morphological and physio-chemical features of the *S. leucantha* essential oil loaded chitosan was characterized by different techniques. The formation of *S. leucantha* essential oil (SlEO)-loaded chitosan nanoparticles (ChNPs) were monitored by UV−vis absorption spectra in a wavelength range from 220 to 400 nm using JASCO spectrophotometer (Model v650). *S. leucantha* essential oil loaded chitosan nanoparticle were digested in HCL solution (2 M) at 95°C to break up the NPs and allow the release of encapsulated EO. The structure and composition of encapsulated nanoparticles was analyzed by using a 10-kV ultrahigh-resolution SEM. In addition, Particle size and its relative functional groups were observed using FTIR spectroscopy with spectra recorded by a PerkinElmer Spectrum 2000, FTIR spectrophotometer (Stuart, 2008). The size of SlEO - loaded ChNPs was determined by the particle analyzer Malveern Zetasizer nanosizer, measuring the size-dependent fluctuation of scattering of laser light on encapsulated ChNPs. Finally, the phase purity of SlEO - loaded ChNPs was studied by XRD and EDAX analysis (Murugan et al., 2016a).

11.16 *H. armigera* and *S. litura* rearing

H. armigera larvae were collected from the Central Institute of Cotton Research, Coimbatore, India. They were cultured in laboratory and fed ad libitum with *Gossypium hirsutum* (*H. armigera*) and *Ricinius communis* (*S. litura*) leaves at 27 ± 2C, 75 85% R.H., and 14:10 (L:D) photoperiod. Pre-pupae of *H. armigera* and *S. litura* were separated and provided with vermiculite clay, which is a good medium for pupation. Pupae of *H. armigera* and *S. litura* were kept on cotton in Petri dishes inside an adult emergence cage. The emerging moths were fed with 10% sucrose solution fortified with a few drops of vitamin mixture (MULTDEC drops) to enhance oviposition. Moths in the ratio of one male to one female were stored inside oviposition cages containing the adult food mentioned above. The egg cage of *H. armigera* and *S. litura* was covered with white muslin cloth for egg laying. The egg clothes were removed daily and surface sterilized using 10% formaldehyde solution to prevent virus infection. The egg clothes were moistened and kept in a plastic container for the eggs to hatch. This process facilitated uninterrupted supply of test insects (Murugan et al., 2000).

11.17 Rearing of *P. xylostella*

P. xylostella larvae and adults were collected from cabbage crops at Thondamuthur (10°59′24.9″N, 76°50′46.3″E), Coimbatore, Tamil Nadu, (India). For egg laying, 500 adults of adults were stored in a plastic cage (50 30 30 cm^3) and provided with leaves of cabbage, *Brassica oleracea* L. var. botrytis (Brassicaceae). Eggs were allowed to hatch on *B. oleracea* leaves, and larvae were maintained at 2571°C; and 6575% relative humidity (RH) under a 16L:8D photoperiod in a growth chamber. Larval instars I−IV and pupae were used for acute toxicity experiments. Newly emerged male and female adults were tested for longevity and fecundity experiments.

11.18 Toxicity against the *H. armigera, S. litura* and *P. xylostella*

Toxicity against *H. armigera, S. litura* and *P. xylostella* larvae and pupae was studied using the leaf disk no-choice method. F2 generation larvae were fed with selected leaf disks treated with different concentrations of essential oil and encapsulation chitosan nanoparticles using the dipping method. After 24 h, the individuals were transferred to untreated fresh selected leaves. The leaves were changed every 24 h. Mortality was recorded after 96 h of treatment. Five replicates were maintained for each treatment with 10 larvae per replicate (total, n = 50). Percentage mortality was calculated using the formula by Hardstone et al. (2009). The survived larvae were fed with untreated cotton leaves until pupation. Pupal mortality was calculated by subtracting the number of emerging adults from the total number of pupae.

11.19 Impact on longevity and fecundity of *H. armigera, S. litura* and *P. xylostella*

Newly emerged males and females of *H. armigera*, *S. litura* and *P. xylostella* were stored into wooden cages at 1:1 sex ratio (n = 20). Ten hours after emergence, they were provided with an artificial diet containing 20 mg sucrose, 1 mL honey in 1 mL sterile distilled water, and 1 mL of aqueous solution containing 100, 200, 300, 400 and 500 ppm of *S. leucantha* essential oil (SlEO) or to 10, 20, 30, 40 and 50 ppm of encapsulated nanoparticles. Control was diet without essential oil and encapsulated nanoparticles. Mean fecundity was calculated checking daily the number of eggs laid on five fresh selected leaves for 4 consecutive days divided by number of females let to mate (n = 20). Mortality was checked daily, and the mean lifespan of each adult was calculated.

11.20 Quantitative food utilization efficiency measures

A gravimetric technique was used to determine weight gain, food consumption and faces produced by *H. armigera, S. litura* and *P. xylostella*. All weights were measured using a monopam balance, accurately to 0.1 mg newly molted only IV instar larvae were starved for 3 h. After measuring the initial weight of the larvae, they were individually introduced into separate containers. The larvae (n = 20 per concentration, five replicates each) were allowed to feed on weighed quantities of treated and untreated selected leaves, for a period of 24 h. Then, larvae were weighed again. The difference in weight of the larvae gives the fresh weight gained during the period of study. Sample caterpillars were weighed, oven dried (48 h at 60°C) reweighed to establish a percentage dry weight of the experimental caterpillars. The leaves remained all the end of each day were oven dried and re-weighed to establish a percentage dry weight of the diet (dry weight) remaining at the end of each experiment from the total dry weight of diet provided. Faces were collected daily and weighed, oven dried and reweighed to estimate the dry weight of excreta. The experiment was continued for four days and observations were recorded every 24 h. Consumption, growth rates and post digestive food utilization efficiencies (all based on dry weight) were calculated following the method by Waldbauer (1968) and Slansky and Scriber (1985); with minor modifications (Murugan and Ancy George, 1992).

Consumption index (CI) = E/TA
Relative growth rate (RGR) = P/TA
Approximate digestibility (AD) = 100 (E-F)
Efficiency of conversion of ingested food (ECI) = 100 P/E
Efficiency of conversion of digested food (ECD) = 100 P/(E-F)

Where,

A = mean dry weight of animal during experiment
E = dry weight of food eaten
F = dry weight of f food eaten
P = dry weight gained by the insect
T = duration of experimental period

11.21 Amylase, protease, proteinase, and lipase assay

Amylase activity in the gut of *H. armigera, S. litura* and *P. xylostella* larvae fed with leaves treated with encapsulated nanomatierals and midgut tissues was analyzed by the dinitrosalycylic acid (DNSA; Sigma-Aldrich, www.sigma-aldrich.com) method (Bernfeld, 1955) as described by Kotkar et al. (2009). One amylase unit was defined as the amount of enzyme required to release 1 μM maltose/min at 37°C under the given assay conditions.

Azocasein (Brock et al., 1982) and trypsin assays of gut homogenates was carried out as described by Tamhane et al. (2005). One protease/proteinase unit was defined as the amount of enzyme in the assay that causes an increase in absorbance by one OD under the given assay conditions.

Lipase activity from gut homogenates was estimated using the p-nitrophenyl palmitate (pNPP; Sigma Aldrich) assay (Winkler and Stuckmann, 1979). The substrate was comprised of solution A and solution B. Solution A contained 0.1 g gum arabica and 0.4 mL Triton X-100 (Sigma-Aldrich) dissolved in 90 mL of distilled water. Solution B contained 30 mg pNPP dissolved in 10 mL isopropanol. The substrate solution was prepared by adding 9.5 mL of solution A to 0.5 mL of solution B drop−wise with constant stirring to obtain an emulsion that was stable for 2 h. The assay mixture, containing 0.9 mL of the substrate, 0.05 mL of buffer (0.02 M sodium− phosphate buffer (pH 6.8) containing 10 mM NaCl and

0.05 mL of gut extract), was incubated at 37°C for 30 min using a thermo-mixer (Eppendorf, www.eppendorf.com). Enzyme activity was stopped by adding 2 mL of 1% sodium carbonate (Na_2CO_3). Absorbance of the samples was measured at 410 nm against a substrate–free blank.

All the assays were performed in duplicate and repeated three times. One unit of lipase activity was defined as the amount of enzyme that causes an increase of one OD under the given assay conditions.

11.22 Statistical analysis

All data were subjected to analysis of variance (ANOVA). LC_{50} and LC_{90} values and their 95% confidence limits were estimated by getting a probit regression model to the observed relationship between percentage mortality of larvae and logarithmic concentration of the substance. The goodness of fitness of the model was tested using Chi-Square test AP value of less than 0.05 was considered as a significant departure of the model from the observations. In case of significant departure a heterogeneity factor were used to calculate the 90% confidence limit for LC_{50} and LC_{90}. All analysis was carried out using SPSS Software version 16.0. Critical differences (CDs) between HSD values were calculated by subtracting subsequent values of the averages and comparing with the calculated CD at $p < 0.05$ and $p < 0.01$ (Kotkar et al., 2009).

11.23 Results and discussion

11.23.1 Phytochemical screening for essential oil of *Salvia leucantha*

In the present study, the essential oil of *S. leucantha* was investigated for its phytochemicals. The results of the phytochemical screening was indicated the presence of different types of active constituents such as alkaloid, flavonoids, phenolics, saponins, tannins, terpenoids, glycosides (Table 11.1). Recently, Noshad et al. (2018) reported that the phytochemical screening of *Black Zira* essential oil showed the existence of phenolic, flavonoids, saponins, alkaloids and tannins. Abdelkader et al. (2014) reported that Phytochemical screening by simple chemical tests and showed a presence of flavonoid, triterpenoids and steroids cinnamic derivatives, chlorogenic acid in *S. officinalis.*

11.24 GC–MS analysis

The composition of essential oil extracted from leaf of *S. leucantha* was analyzed using GC–MS and the results are shown in Fig. 11.1 individual components were identified according to their relative retention indices and their relative percentages calculated (Table 11.2). The main components of the essential oils of *S. leucantha* were bornyl acetate (11.93%), 1-Octadecanesulphonyl Chloride (11.93%), spathulenol (8.28%), caryophyllene oxide (11.93%), 4-Pentadecyne, 15-Chloro (11.24 %) and cis-Muurola-3,5-diene (5.84 %). These compounds were dominanted by the presence of bornyl acetate is a class of terpenoid compound which is efficient in repellent activity reported by Feng et al. (2019). Upadhyaya et al. (2009) identification of sesquiterpenoids (57.6 %) was the major constituents of the oil represented mainly by spathulenol (12.1%), β-caryophyllene (10.7 %), α-himachalene (10.5 %) and γ-cadinene (6.5 %). Monoterpenoids

TABLE 11.1 Phytochemical analysis of *Salvia leucantha* essential oil.

S.No	Phytochemicals	*Salvia leucantha* essential oil
1	Alkaloids	−
2	Terpenoids	+
3	Flavonoids	+
4	Tannins	+
5	Phenolic	−
6	Saponins	+
7	Cardiac glycosides	−

+, present; −, not present.

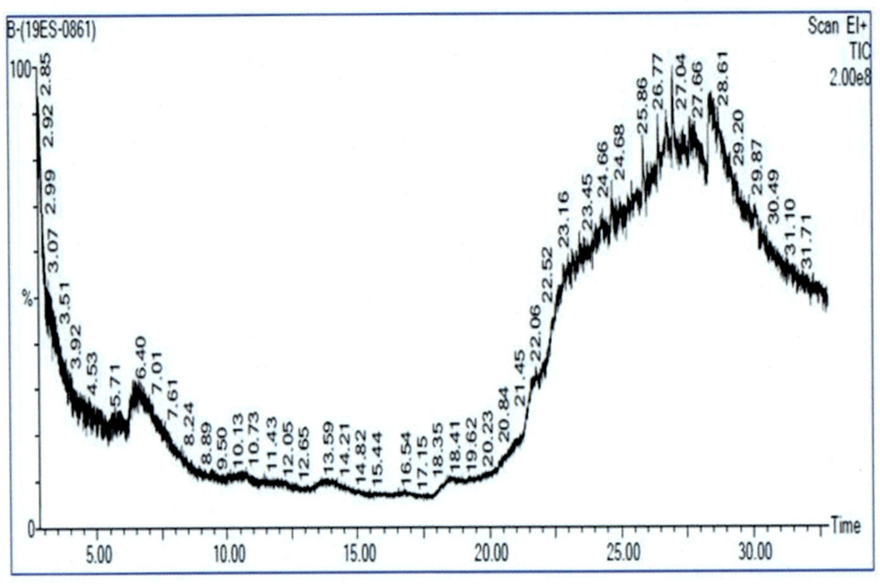

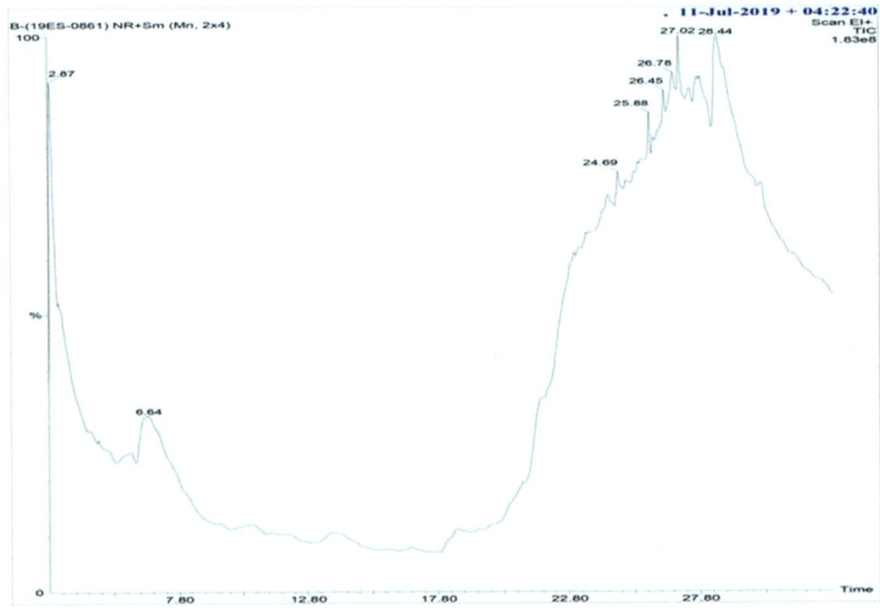

FIG. 11.1 GC–MS chromatogram of the essential oil of *Salvia leucantha*.

constituted 35.7 % of the total oil in *S. leucantha.* The chemical composition of essential oils from aerial parts of *S. leucantha* have been studied and found to be rich in bornyl acetate and sesquiterpene hydrocarbons (Rondon et al., 2005; Negi et al., 2007) suggesting that environmental factors may strongly influence the chemical composition.

11.25 Characterization of essential oil loaded chitosan nanoparticles

11.25.1 UV-VIS spectral analysis of essential oil loaded chitosan nanoparticles

The nanoparticles were primarily characterized by UV—visible spectroscopy, which proved to be a very useful technique for the analysis of nanoparticles (Sastry et al., 1997). *S. leucantha* essential oil encapsulated chitosan nanoparticles in the form of colloidal suspensions were studied using the UV—Visible Spectrophotometer by measuring their absorbance over a predetermined range of wavelength 342 nm (Fig. 11.2). Similarly, Hosseini et al. (2013) found out that the encapsulation of oregano essential oil in chitosan nanoparticles showed UV—vis spectrophotometry from the absorbance at 275 nm. Earlier studies, The formation of Copper chitosan nanoparticle was observed by the peak at 536 nm using UV—Vis spectroscopy

TABLE 11.2 Chemical profile of *Salvia leucantha* essential oil.

Compounds	RT	MW	MF	Area (%)
2-Butanol, 3-methyl-	6.615	88	$C_5H_{12}O$	1.456
Octadecanal	22.816	268	$C_{18}H_{36}O$	2.466
1-Hexyl-2-nitrocyclohexane	25.432	213	$C_{12}H_{23}O_2N$	2.775
2-Octadecyl-propane-1,3-diol	25.863	328	$C_{21}H_{44}O_2$	7.349
Bornyl acetate	26.443	196.29	$C_{12}H_{20}O_2$	11.131
1-Octadecanesulphonyl chloride	26.773	352	$C_{18}H_{37}ClO_2S$	11.932
2-Piperidinone, N-[4-Bromo-N-Butyl]-	27.008	233	$C_9H_{16}ONBr$	8.555
Spathulenol	27.403	220.35	$C_{15}H_{24}O$	8.280
cis-Muurola-3,5-diene	27.628	204	$C_{15}H_{24}$	5.846
4-Pentadecyne, 15-chloro	27.838	242	$C_{15}H_{27}Cl$	11.245
9,12-Tetradecadien-1-Ol, acetate, (Z,E)-	28.474	252	$C_{16}H_{28}O_2$	14.281
26-Hydroxycholesterol	28.709	402	$C_{27}H_{46}O_2$	14.140
Total				99.456

MF, molecular formula; *MW*, molecular weight; *RT*, retention time.

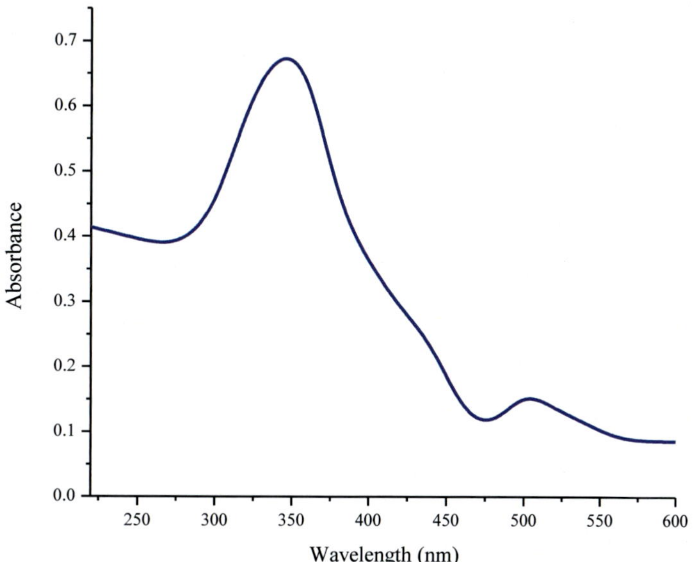

FIG. 11.2 UV−visible spectrum of *S. leucantha* essential oil loaded chitosan nanoparticles.

(Manikandan and Sathiyabama, 2015) and also Murugan et al. (2017a) noticed that UV−Vis absorption spectrum of chitosan synthesized silver nanoparticles obtained the strong peak 426 nm. Further, Oh et al. (2019) illustrates the UV−Visible spectrum of chitosan nanoparticles obtained (sharp intensity) main peak 320 nm.

11.26 SEM analysis

The morphological structure of the SlEO encapsulated ChNPs were observed by FE-SEM (Fig. 11.3). The *S. leucantha* essential oil encapsulated chitosan nanoparticle showed the spherical shapes and smooth surfaces. The size distribution obtained by FE-SEM indicated that possibly due to the loading of SlEO into ChNPs. Recently, Chandirika et al. (2018)

FIG. 11.3 Scanning electron microscopy of *S. leucantha* essential oil loaded chitosan nanoparticles.

reported that SEM analysis showed the *Aerva lanata* plant extract loaded chitosan nanoparticle are homogenous and mostly spherical in shape. Earlier report, Sotelo Boyás et al. (2015) reported chitosan nanoparticles added with essential oils observed SEM result was found in spherical shape nanoparticles. Another study was carried out the chitosan encapsulated *C. martinii* essential oil (Ce-CMEO) nanoparticles of SEM analysis showed spherical surface morphology (Kalagatur et al., 2018).

11.27 Energy-dispersive X-ray spectroscopy analysis

EDX spectrum of the chitosan NPs encapsulated essential oil of *S. leucantha* which indicated the presences of O 39.15 % and Ca 7.23% (Fig. 11.4). Keawchaoon and Yoksan (2011) reported that were concurrent to the findings.

11.28 FTIR analysis of essential oil chitosan nanoparticles

A representative FTIR spectrum is reported in Fig. 11.5. The major peaks in the FTIR spectrum of chitosan NPs encapsulated essential oil of *S. leucantha were* observed at, 3905.15, 3752.8, 3436.53, 2884.99, 2811.70, 2620.79, 2322.01, 2343.09, 2092.39, 1851.33, 1637.27, 1403.92, 1099.23, 989.30, 715.46, 669.17, and 480.18 cm^{-1}, implying the

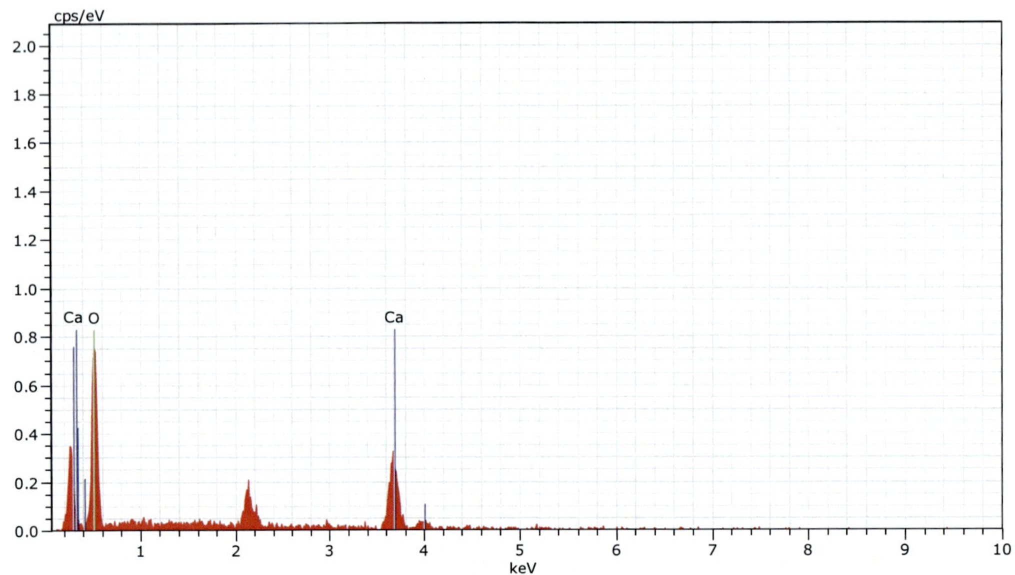

FIG. 11.4 Energy dispersive X-ray analysis of *S. leucantha* essential oil loaded chitosan nanoparticles.

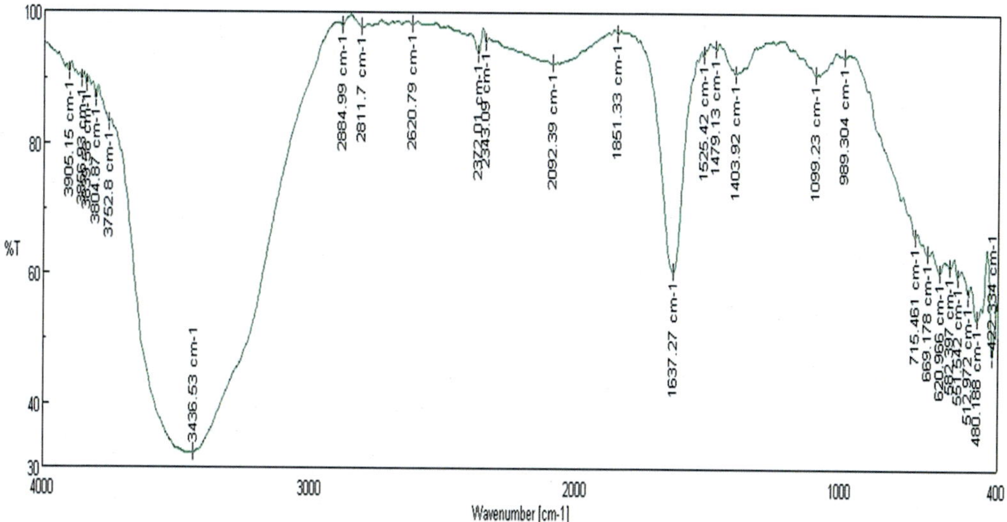

FIG. 11.5 Fourier transform-infrared spectroscopy analysis of *S. leucantha* essential oil loaded chitosan nanoparticles.

complex formation via electrostatic interaction between NH_{3+} groups of chitosan and phosphoric groups of TPP within the nanoparticles (Yoksan et al., 2010). Huang et al. (2009) reported that characteristic peaks at 1080 to 1158 cm-1 due to C—O—H bond in modified starch. These results are consistent with earlier findings of chitosan nanoparticles (Keawchaoon and Yoksan, 2011).

11.29 Zeta potential measurements

Zeta potential (ZP) observed a positive value of +24.12 mV for *S. leucantha* essential oil encapsulated ChNPs given in Fig. 11.6 which resulted from the spontaneous nanocomplex that was formed between CS and TPP with an overall positive surface charge which was in agreement with a previous study (Nallamuthu et al., 2015). Furthermore, summary savory leaves EO loaded chitosan nanoparticles measured negative ZP values ranging from −7.54 to −21.12 mV

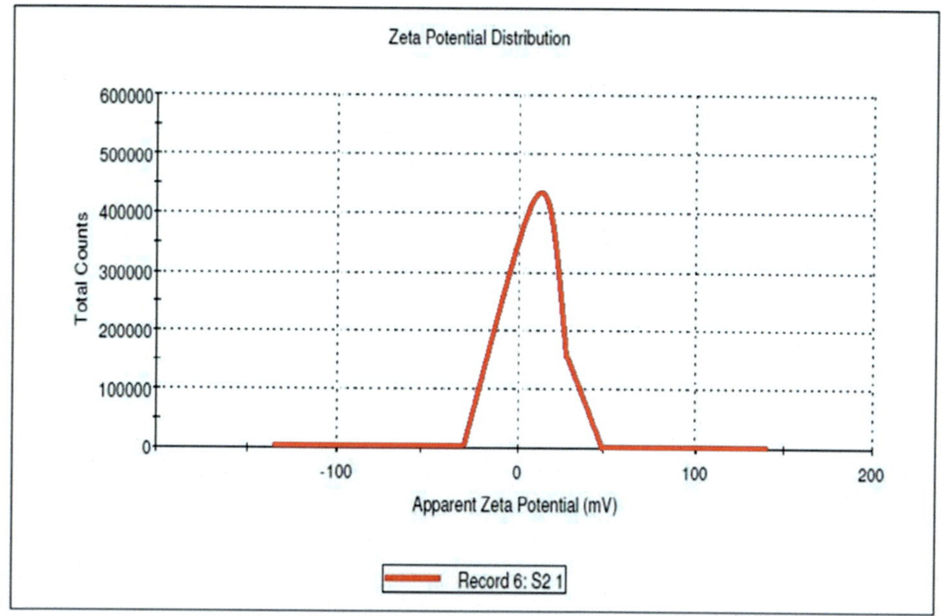

FIG. 11.6 Zeta potential of *S. leucantha* essential oil loaded chitosan nanoparticles.

(Feyzioglu and Tornuk, 2016). Also, Lertsutthiwong et al. (2009) measured negative ZP values ranging from −21.8 to −23.1 mV for turmeric oil loaded CSNPs. ZP values depends on the increasing content of the loaded essential oil concentration.

11.30 Larvicidal and pupicidal toxicity against *H. armigera, S. litura* and *P. xylostella*

The *S. leucantha* essential oil showed moderate toxicity against the larvae and pupae of *H. armigera, S. litura* and *P. xylostella.* LC_{50} ranged from 175.392 ppm (I larvae) to 542.092 ppm (pupa) *H. armigera* (Table 11.3), 200.584 ppm (I larvae) to 559.582 ppm (pupa) *S. litura* (Table 11.4) and 258.412 ppm (I larvae) to 505.061 ppm (pupa) *P. xylostella* (Table 11.5). Alaguchamy and Jayakumararaj (2015) reported that the aqueous extract of *C. roseus* was highly effective on the larvae of *H. armigera*, with LD_{50} ranging from 2.6 mg/cm^2 (I) to 63.34 mg/cm^2 (VI). Ahmad et al. (2015) investigated on the toxic effects of neem-based insecticide products Neemarin (0.15% EC azadirachtin), Neemazal (1% EC azadirachtin), Neemix (0.25% EC azadirachtin) and Neem oil (1% EC azadirachtin) against *H. armigera.* Recently, Murugan et al. (2018) studied that high larvicidal activity of seaweed leaf extract of *Sargassum wightii* against the larvae and pupae of *H. armigera* with LC_{50} of 143.071(I larva) and 261.838 ppm (pupa). The contact toxicity of two essential oils such as *Carum copticum (L.)* and *C. cyminum* against *S. granarius* and *T. confusum* was studied by Ziaee et al. (2014). Recently, Suresh et al. (2017) proposed the potential of *S. maritima*-based herbal coils and green nanoparticles as biopesticides in the fight against the tobacco cutworm *S. litura.* Further studies, *H. musciformis* aqueous extract showed larvicidal and pupicidal toxicity against the cabbage pest *Plutella xylostella* (Roni et al., 2015).

The *S. leucantha* essential oil encapsulated chitosan nanoparticles were more effective than the *S. leucantha* essential oil on *H. armigera, S. litura* and *P. xylostella* LC_{50} ranged from 13.508 ppm (I larvae) to 60.420 ppm (pupa) *H. armigera* (Table 11.6), 17.289 ppm (I larvae) to 68.958 ppm (pupa) *S. litura* (Table 11.7) and 14.566 ppm (I larvae) to 63.283 ppm (pupa) *P. xylostella* (Table 11.8). Roni et al. (2015) reported the *H. musciformis* synthesized silver nanoparticles were highly toxic against *P. xylostella.* LC_{50} were 24.51 ppm (I), 26.47 ppm (II), 28.35 ppm (III), 32.55 ppm (IV) and 38.23 ppm (pupa). Recently, Paulraj et al. (2017) reported that nanoformulation of chitosan tripolyphosphate PONNEEM exhibited highest larvicidal activity 90.2% at 0.3% concentration and growth regulating properties against *H. armigera* larvae at very low concentrations. Rabea et al. (2005) insecticidal activity of synthesized 24 new chitosan derivatives and

TABLE 11.3 Larvicidal and pupicidal effect of *Salvia leucantha* essential oil (SIEO) *against* the cotton bollworm, *Helicoverpa armigera.*

Larval and pupal stage	LC_{50} (LC_{90})	95% confidence limit LC_{50} (LC_{90})		Regression equation	χ^2 (*d.f.* = 4)
		Lower	Upper		
Larva I	175.392 (481.410)	133.652 (435.336)	207.580 (548.440)	$x = 0.004$ $y = -0.735$	1.570 n.s
Larva II	216.680 (575.378)	174.515 (512.394)	250.427 (673.201)	$x = 0.004$ $y = -0.774$	3.174 n.s
Larva III	265.229 (653.816)	226.381 (575.234)	299.950 (781.039)	$x = 0.033$ $y = -0.875$	1.776 n.s
Larva IV	354.069 (844.958)	310.776 (711.281)	407.775 (1094.850)	$x = 0.003$ $y = -0.924$	0.925 n.s
Larva V	415.021 (927.169)	366.008 (769.711)	489.245 (1233.682)	$x = 0.003$ $y = -1.039$	0.803 n.s
Larva VI	478.055 (981.951)	420.787 (811.657)	575.122 (1316.524)	$x = 0.003$ $y = -1.216$	0.689 n.s
Pupa	542.092 (995.129)	477.850 (829.868)	654.450 (1310.482)	$x = 0.003$ $y = -1.533$	1.196 n.s

Mortality rates are means ± SD of three replicates LC_{50}, lethal concentration that kills 50% of the exposed organisms; LC_{90}, lethal concentration that kills 90% of the exposed organisms; *LCL*, lower confidence limit; *UCL*, upper confidence limit; χ^2, chi-square; *n.s.*, not significant ($\alpha = 0.05$). Values followed by the same letter(s) are not significantly different (DMRT, $\alpha = 0.05$).

TABLE 11.4 Larvicidal and pupicidal effect of *Salvia leucantha* essential oil (SlEO) *against* the *Spodoptera litura*.

Larval and pupal stage	LC_{50} (LC_{90})	95% confidence limit LC_{50} (LC_{90}) Lower	Upper	Regression equation	χ^2 (*d.f.* = 4)
Larva I	200.584 (518.421)	161.369 (467.446)	231.874 (593.769)	$x = 0.004$ $y = -0.809$	1.483 n.s
Larva II	232.809 (616.146)	190.102 (543.869)	267.759 (732.130)	$x = 0.003$ $y = -0.778$	2.576 n.s
Larva III	294.258 (713.452)	255.179 (619.792)	332.416 (871.579)	$x = 0.003$ $y = -0.900$	1.808 n.s
Larva IV	397.999 (903.318)	351.217 (753.188)	465.405 (1191.696)	$x = 0.003$ $y = -1.009$	0.673 n.s
Larva V	445.930 (977.881)	391.259 (803.077)	536.147 (1330.033)	$x = 0.002$ $y = -1.074$	0.783 n.s
Larva VI	492.652 (983.365)	434.089 (815.165)	592.682 (1310.450)	$x = 0.003$ $y = -1.287$	0.909 n.s
Pupa	559.582 (990.439)	494.090 (829.868)	674.127 (1293.817)	$x = 0.003$ $y = -1.664$	2.065 n.s

Mortality rates are means ± SD of three replicates LC_{50}, lethal concentration that kills 50% of the exposed organisms; LC_{90}, lethal concentration that kills 90% of the exposed organisms; *LCL*, lower confidence limit; *UCL*, upper confidence limit; χ^2, chi-square; *n.s.*, not significant ($\alpha = 0.05$). Values followed by the same letter(s) are not significantly different (DMRT, $\alpha = 0.05$).

TABLE 11.5 Larvicidal and pupicidal effect of *Salvia leucantha* essential oil (SlEO) *against* the *Plutella xylostella*.

Larval and pupal stage	LC_{50} (LC_{90})	95% confidence limit LC_{50} (LC_{90}) Lower	Upper	Regression equation	χ^2 (*d.f.* = 4)
Larva I	258.412 (521.344)	231.729 (478.037)	283.037 (581.541)	$x = 0.005$ $y = -1.260$	0.229 n.s
Larva II	296.678 (608.365)	267.538 (548.816)	325.563 (696.410)	$x = 0.004$ $y = -1.220$	0.67 n.s
Larva III	367.737 (731.141)	334.433 (644.554)	407.933 (869.313)	$x = 0.004$ $y = -1.297$	0.398 n.s
Larva IV	434.287 (855.378)	390.838 (733.046)	497.520 (1068.991)	$x = 0.003$ $y = -1.332$	0.371 n.s
Pupae	505.061 (979.183)	445.816 (815.348)	606.078 (1292.618)	$x = 0.003$ $y = -1.365$	0.406 n.s

The larval mortalities are expressed as mean ± SD of five replicates. Nil mortality was observed in the control. Within a column means followed by the same letter(s) are not significantly different at 5% level by Duncan's multiple range test. *LFL*, lower fiducidal limit; *UFL*, upper fiducidal limit. x^2, Chi-square value. *Significant at $P < 0.05$ level, *n.s.*, not significant ($\alpha = 0.05$).

tested them against *Spodoptra littoralis*. Goswami et al. (2010) studied the insecticidal properties of different kind of nanoparticles i.e. silver nanoparticles (SNPs), aluminum oxide, zinc oxide and titanium dioxide nanoparticles to control the rice weevil, *S. oryzae*.

After treatment of *Salvia leucantha* essential oil encapsulated in chitosan nanoparticles a decrease in enzymatic activity denotes reduced metabolism in the insect and may be due to the toxic effects nanoparticles and associated plant compounds on membrane permeability, especially on the gut epithelium (Senthil-Nathan et al., 2006).

The complex phytochemical screening was indicated the presence of different types of active constituents such as alkaloid, flavonoids, phenolics, saponins, tannins, terpenoids, glycosides and presence of spathulenol and caryophyllene oxide (Bossou et al., 2013; Gul et al., 2017), which not only provided the toxicity against *H. armigera* larvae and also

TABLE 11.6 **Larvicidal and pupicidal effect of *Salvia leucantha* essential oil encapsulated chitosan nanoparticles against cotton bollworm, *H. armigera*.**

Larval and pupal stage	LC_{50} (LC_{90})	95% confidence limit LC_{50} (LC_{90})		Regression equation	χ^2 (*d.f.* = 4)
		Lower	Upper		
Larva I	13.508 (34.689)	10.116 (31.781)	16.148 (38.569)	$x = 0.061$ $y = -0.817$	2.867 n.s
Larva II	16.440 (46.199)	12.207 (41.851)	19.676 (52.452)	$x = 0.043$ $y = -0.708$	1.828 n.s
Larva III	20.945 (55.686)	16.769 (49.758)	24.265 (64.770)	$x = 0.037$ $y = -0.773$	1.239 n.s
Larva IV	28.113 (64.566)	24.620 (57.246)	31.403 (76.082)	$x = 0.035$ $y = -0.988$	4.023 n.s
Larva V	45.413 (109.073)	38.930 (86.072)	57.552 (162.695)	$x = 0.020$ $y = -0.914$	4.135 n.s
Larva VI	53.474 (122.349)	44.832 (93.846)	73.061 (195.705)	$x = 0.019$ $y = -0.995$	2.776 n.s
Pupae	60.420 (120.791)	50.854 (94.523)	81.494 (183.066)	$x = 0.021$ $y = -1.283$	2.995 n.s

Mortality rates are means ± SD of three replicates *LC50,* lethal concentration that kills 50% of the exposed organisms; LC_{90}, lethal concentration that kills 90% of the exposed organisms; *LCL,* lower confidence limit; *UCL,* upper confidence limit; χ^2, chi-square; *n.s.,* not significant ($\alpha = 0.05$). Values followed by the same letter(s) are not significantly different (DMRT, $\alpha = 0.05$).

TABLE 11.7 **Larvicidal and pupicidal effect of *Salvia leucantha* essential oil encapsulated chitosan nanoparticles against *Spodoptera litura*.**

Larval and pupal stage	LC_{50} (LC_{90})	95% confidence limit LC_{50} (LC_{90})		Regression equation	χ^2 (*d.f.* = 4)
		Lower	Upper		
Larva I	17.289 (41.497)	14.031 (38.061)	19.925 (46.173)	$x = 0.053$ $y = -0.915$	0.510 n.s
Larva II	20.667 (45.476)	17.739 (41.778)	23.159 (50.529)	$x = 0.052$ $y = -1.068$	0.726 n.s
Larva III	23.467 (50.288)	20.575 (46.006)	26.016 (56.275)	$x = 0.048$ $y = -1.121$	2.057 n.s
Larva IV	29.317 (58.619)	26.560 (53.189)	32.045 (66.475)	$x = 0.044$ $y = -1.282$	3.29. n.s
Larva V	53.382 (115.981)	45.366 (91.046)	70.130 (174.573)	$x = 0.020$ $y = -1.093$	0.905 n.s
Larva VI	67.936 (145.726)	54.171 (106.332)	108.364 (268.768)	$x = 0.016$ $y = -1.119$	0.683 n.s
Pupae	68.958 (130.595)	56.773 (100.372)	98.290 (206.945)	$x = 0.021$ $y = -1.1434$	0.266 n.s

Mortality rates are means ± SD of three replicates LC_{50}, lethal concentration that kills 50% of the exposed organisms; LC_{90}, lethal concentration that kills 90% of the exposed organisms; *LCL,* lower confidence limit; *UCL,* upper confidence limit; χ^2, chi-square; *n.s.,* not significant ($\alpha = 0.05$). Values followed by the same letter(s) are not significantly different (DMRT, $\alpha = 0.05$).

TABLE 11.8 Larvicidal and pupicidal effect of *Salvia leucantha* essential oil encapsulated chitosan nanoparticles against, *P. xylostella*.

Larval and pupal stage	LC_{50} (LC_{90})	95% confidence limit LC_{50} (LC_{90})		Regression equation	χ^2 (*d.f.*=4)
		Lower	Upper		
Larva I	14.566 (41.576)	10.415 (37.837)	17.714 (46.794)	$x=0.047$ $y=-0.691$	0.924 n.s
Larva II	22.661 (59.771)	18.447 (52.988)	26.083 (70.484)	$x=0.035$ $y=-0.783$	2.510 n.s
Larva III	29.638 (66.444)	26.215 (58.791)	33.007 (78.565)	$x=0.035$ $y=-1.032$	4.000 n.s
Larva IV	53.691 (115.525)	45.671 (90.836)	70.427 (173.260)	$x=0.021$ $y=-1.113$	3.275 n.s
Pupa	63.283 (120.545)	53.286 (94.896)	85.111 (180.022)	$x=0.022$ $y=-1.416$	2.325 n.s

Mortality rates are means ± SD of three replicates, *LC_{50}*, lethal concentration that kills 50% of the exposed organisms; *LC_{90}*, lethal concentration that kills 90% of the exposed organisms; *LCL*, lower confidence limit; *UCL*, upper confidence limit; χ^2, chi-square; *n.s.*, not significant ($\alpha = 0.05$). Values followed by the same letter(s) are not significantly different (DMRT, $\alpha = 0.05$).

made confusion in the reproductive parameters of insects and the chemicals present in the essential oil and its nanoparticles affected the hormonal control of metamorphosis and particularly post-embryonic growth and development (Bernys and Chapman, 1994). Alternatively, physio-metabolic processes affect the egg maturation and hatching process of a decrease of the rate of the metabolic processes may occur, which can affect the maturation of eggs and related embryonic development and weakens the developing larvae.

11.31 Impact of *S. leucantha* essential oil and encapsulated chitosan nanoparticles on insect longevity and fecundity

The current work after the treatment of nanoparticles on the leaves fed insects showed lesser fecundity and life history performances. Since, feeding and reproduction in insects are very closely related to nutritional factors; the qualitative and quantitative aspects of which have an impact not only on the rates of growth and development but also on fecundity. Moreover, treated insects might have received lesser protein content and it might affect the protein metabolism of insects, hence, treated insects also taken lesser food. Since, the amount, rate and quality of food consumed by a larva influence its performance: growth rate, developmental time, final body weight and probability of survival (Slansky and Scriber, 1985), an understanding of the dietary manifestations and growth of the larvae would be useful.

Protein content in the diets is of paramount importance and it promotes the successful growth of the insects (Mattson, 1980; Scriber, 1984a,b). Hence, in the present study *H. armigera* fed on treated leaves showed less fecundity and poor survival. Engelmann (1970) stressed that the reproductive performance of insects is influenced by different factors and among them protein-rich nutrients are the most single vital factor affecting the total egg output in the majority of species.

Treatments with essential oil and encapsulated chitosan nanoparticles greatly affected adult longevity and fecundity the crop pest *H. armigera, S. litura* and *P. xylostella*. Longevity of *H. armigera* was reduced to 3.4 (male) and 4.2 (female) days after treatment with 50 ppm concentration of encapsulated chitosan nanoparticles, while control was 14.2 (male) and 17.6 (female) days. Fecundity was highly reduced after the treatment of encapsulated chitosan nanoparticles; 856 eggs were recorded in control, while the numbers of eggs recorded in treatments were 74.1 eggs at 50 ppm, respectively (Table 11.9). Longevity of *S. litura* was reduced to 3.9 (male) and 4.7 (female) days after treatment with 50 ppm concentration of encapsulated chitosan nanoparticles, while control was 22.3 (male) and 28.4 (female) days. Fecundity was highly reduced after the treatment of encapsulated chitosan nanoparticles; 730 eggs were recorded in control, while the numbers of eggs recorded in treatments were 99.5 eggs at 50 ppm, respectively (Table 11.10). Longevity in *P. xylostella* was reduced to 4.6 (male) and 5.0 (female) days after treatment with 50 ppm concentration of encapsulated chitosan nanoparticles, while in control it was 17.5 (male) and 20.6 (female) days. Fecundity was highly

TABLE 11.9 Impact of *Salvia leucantha* essential oil and encapsulated chitosan nanoparticles on longevity and fecundity of the crop pest *Helicoverpa armigera.*

		Adult longevity (days)		
Treament	(ppm)	Male	Female	Fecundity (Nos.)
S. leucantha essential oil	Control	14.2 ± 0.35	17.6 ± 0.30	856.7 ± 0.5
	100	12.4 ± 1.1	15.4 ± 1.1	690.5 ± 1.4
	200	10.3 ± 2.4	13.1 ± 1.4	480.4 ± 1.1
	300	8.1 ± 1.6	10.7 ± 2.6	345.4 ± 2.1
	400	6.2 ± 1.7	7.8 ± 2.9	220.4 ± 0.7
	500	5.5 ± 1.6	5.9 ± 1.9	135.9 ± 1.9
Encapsulated chitosan nanoparticles	10	11.6 ± 1.9	13.8 ± 1.4	590.2 ± 1.2
	20	8.7 ± 1.4	10.6 ± 1.7	433.6 ± 1.0
	30	6.4 ± 2.7	8.1 ± 1.7	282.9 ± 1.1
	40	5.2 ± 0.9	5.9 ± 0.9	175.8 ± 0.9
	50	3.4 ± 0.7	4.2 ± 1.9	74.1 ± 0.6

TABLE 11.10 Impact of *Salvia leucantha* essential oil and encapsulated chitosan nanoparticles on longevity and fecundity of the crop pest *spodoptera litura.*

		Adult longevity (days)		
Treament	(ppm)	Male	Female	Fecundity (Nos.)
S. leucantha essential oil	Control	22.3 ± 0.5	28.4 ± 0.30	730.2 ± 0.8
	100	19.6 ± 1.2	24.8 ± 2.4	661.5 ± 1.9
	200	17.2 ± 1.4	22.9 ± 1.1	578.3 ± 1.4
	300	15.4 ± 2.6	20.7 ± 1.0	486.1 ± 2.5
	400	13.9 ± 0.8	17.4 ± 1.7	360.4 ± 1.4
	500	12.1 ± 0.7	13.9 ± 2.3	287.4 ± 2.1
Encapsulated chitosan nanoparticles	10	12.1 ± 2.4	14.2 ± 1.1	610.6 ± 1.04
	20	9.5 ± 1.4	11.3 ± 0.9	450.1 ± 1.1
	30	7.2 ± 2.0	9.3 ± 1.0	304.2 ± 0.57
	40	5.7 ± 1.4	6.5 ± 0.9	190.9 ± 0.8
	50	3.9 ± 0.9	4.7 ± 1.4	99.5 ± 1.0

reduced after the treatment of encapsulated chitosan nanoparticles; 906 eggs were recorded in control, while the numbers of eggs recorded in treatments were 115 eggs at 50 ppm, respectively (Table 11.11). To the best of our knowledge, this is the first study evaluating the impact of encapsulated chitosan nanoparticles on longevity and fecundity of moth pests. More evidence is available about the toxicity of plant extracts against crop pests. For instance, Kumar et al. (2013) recently showed a significant reduction in *S. litura* fecundity and longevity post-treatment with methanolic extract of *S. alata*.

TABLE 11.11 Impact of *Salvia leucantha* essential oil and encapsulated chitosan nanoparticles on longevity and fecundity of the crop pest *plutella xylostella*.

		Adult longevity (days)		
Treament	(ppm)	Male	Female	Fecundity (Nos.)
S. leucantha essential oil	Control	17.5 ± 0.35	20.6 ± 0.30	906.1 ± 0.5
	100	15.2 ± 2.3	18.5 ± 1.0	850.2 ± 1.8
	200	13.6 ± 1.5	16.3 ± 1.9	661.3 ± 1.9
	300	11.8 ± 2.3	15.5 ± 2.1	450.2 ± 1.1
	400	10.6 ± 1.0	13.4 ± 1.1	352.1 ± 1.3
	500	8.1 ± 1.9	11.1 ± 1.0	220.9 ± 0.9
Encapsulated chitosan nanoparticles	10	12.9 ± 2.1	14.9 ± 0.9	648.9 ± 1.1
	20	10.1 ± 1.4	12.0 ± 1.2	472.8 ± 1.3
	30	7.9 ± 2.0	9.8 ± 1.7	328.1 ± 0.7
	40	6.1 ± 1.1	7.2 ± 1.0	214.7 ± 0.5
	50	4.6 ± 1.2	5.0 ± 0.7	115.6 ± 1.1

The endocrine system uses hormones to control and coordinate your body's internal metabolism (or homeostasis) energy level, reproduction, growth and development, and response to injury, stress, and environmental factors. Bernys and Chapman (1994) also reported that enzyme production is clearly related to the feeding behavior amount of food that passes through the alimentary canal It is essential to note that feeding and reproduction. Feeding and reproductive and particularly egg laying is controlled by the endocrine system in insects. Food consumption and fecundity of insects vary with lifestyle and feeding pattern. In general, species which feed during the larval and adult stages and maintain a smaller biomass, allocate a higher percentage of the ingested energy to egg production Moreover, **larval food** quality is the determinant for the fitness of **phytophagous insects** with a non-feeding adult stage. In those species, reproductive potential critically depends upon resource accumulation during the **larval** stage. Moreau and Thiery (2006) by considering the achieved fecundity alone we would have drawn the conclusion that for *L. botrana*, Pinot was more suitable than Riesling. This would have been wrong because there is a trade-off between the number of eggs and their size. Therefore, reproductive effort (fecundity × egg size) that estimates a female's resource allocation to reproduction is more meaningful than egg size or egg number alone. We found that achieved fecundity depends on the larval diet and that there is a positive correlation, although weak, between achieved fecundity and pupal mass. However, we did not find any effect of the diets on pupal mass. This may suggest that, despite body size usually enhancing fitness value (Stearns, 1992).

11.32 Food utilization measures

Food Utilization measures of VI instar larvae of *H. armigera*, *S. litura* and *P. xylostella* after the treatment of *S. leucantha* essential oil and encapsulation chitosan nanoparticles is given in Table 11.12. The essential oil alone treatment does not affect the feeding and growth of the insects studied even at higher concentrations (500 ppm) But synthesized encapsulation of chitosan nanoparticles treated insects showed less feeding and growth of the insect.

In the present study the food utilization parameters of *H. armigera, S. litura* and *P. xylostella* are considerably affected after the treatment of essential oil chitosan encapsulated nanoparticles, which are very similar to Murugan et al. (2018). Devi et al. (2014) reported that *H. armigera* treated with synthesized AgNPs using *E. hirta* also showed a significant reduction in Consumption index (CI), efficiency of ingested (ECI), digested (ECD) relative growth rate in (RGR) 4.86 to 2.00 mg/mg day. Shukla and Patel (2011) investigated that ECI and ECD varied on Spodoptera fed with different host plants.

TABLE 11.12 **Impact of *Salvia leucantha* essential oil and essential oil encapsulated chitosan nanoparticles on food utilization of the, *H. armigera. S. litura* and *P. xylostella*.**

Treatment (ppm)			CI	RGR	ECI (%)	ECD (%)	AD (%)
Control			26.00 ± 1.18^{a}	9.98 ± 0.23^{a}	31.02 ± 1.18^{a}	35.01 ± 1.40^{a}	62.70 ± 1.85^{a}
A.	*S.leucatha* essential oil	100 ppm	23.04 ± 1.23^{b}	7.84 ± 0.56^{b}	28.56 ± 1.65^{b}	32.56 ± 1.30^{b}	58.64 ± 1.36^{b}
		200 ppm	20.62 ± 0.32^{c}	6.50 ± 0.98^{c}	23.28 ± 1.24^{c}	28.48 ± 1.64^{c}	53.28 ± 0.96^{c}
		300 ppm	17.45 ± 0.86^{d}	6.22 ± 0.45^{d}	20.86 ± 0.24^{d}	25.42 ± 0.29^{d}	50.14 ± 0.86^{d}
		400 ppm	16.28 ± 1.24^{e}	5.74 ± 1.24^{e}	18.24 ± 0.76^{e}	22.04 ± 1.45^{e}	48.62 ± 1.09^{e}
		500 ppm	14.14 ± 1.65^{f}	5.18 ± 0.19^{f}	15.52 ± 0.92^{f}	19.26 ± 0.21^{f}	45.84 ± 1.05^{f}
	Essential oil loaded ChNPs	10 ppm	21.52 ± 1.87^{a}	6.89 ± 2.16^{a}	25.40 ± 0.19^{a}	30.32 ± 0.85^{a}	56.06 ± 1.24^{a}
		20 ppm	17.36 ± 0.98^{b}	5.02 ± 1.45^{b}	20.34 ± 1.26^{b}	24.58 ± 1.49^{b}	51.42 ± 1.64^{b}
		30 ppm	13.62 ± 0.64^{c}	4.89 ± 1.89^{c}	16.26 ± 1.82^{c}	19.34 ± 1.08^{c}	45.20 ± 0.28^{c}
		40 ppm	10.25 ± 0.28^{d}	4.12 ± 0.85^{d}	12.02 ± 1.02^{d}	15.28 ± 2.01^{d}	39.26 ± 0.49^{d}
		50 ppm	7.02 ± 1.48^{e}	3.98 ± 0.46^{e}	7.98 ± 1.00^{e}	12.62 ± 1.48^{e}	37.54 ± 0.82^{e}
Control			28.32 ± 0.23^{a}	12.36 ± 0.23^{a}	35.84 ± 0.10^{a}	40.21 ± 0.56^{a}	66.78 ± 0.98^{a}
B.	*S.leucatha* essential oil	100 ppm	22.31 ± 1.75^{b}	11.35 ± 0.98^{b}	33.65 ± 1.23^{b}	37.32 ± 0.36^{b}	63.26 ± 0.41^{b}
		200 ppm	18.56 ± 1.25^{c}	9.65 ± 1.26^{c}	30.21 ± 0.45^{c}	34.19 ± 1.45^{c}	58.12 ± 1.23^{c}
		300 ppm	15.4 ± 0.65^{d}	8.12 ± 1.45^{d}	26.45 ± 0.89^{d}	30.15 ± 1.23^{d}	55.46 ± 2.45^{d}
		400 ppm	13.24 ± 1.36^{e}	7.93 ± 2.03^{e}	22.16 ± 1.20^{e}	26.4 ± 1.87^{e}	52.87 ± 1.21^{e}
		500 ppm	10.3 ± 1.48^{f}	7.02 ± 2.1^{f}	16.4 ± 0.98^{f}	20.4 ± 2.14^{f}	49.46 ± 0.36^{f}
	Essential oil loaded ChNPs	10 ppm	21.09 ± 0.68^{a}	7.12 ± 1.94^{a}	26.38 ± 2.13^{a}	31.24 ± 0.98^{a}	57.08 ± 0.31^{a}
		20 ppm	17.80 ± 0.42^{b}	6.08 ± 1.38^{b}	21.64 ± 1.46^{b}	25.42 ± 0.46^{b}	52.32 ± 1.25^{b}
		30 ppm	14.02 ± 1.07^{c}	5.96 ± 0.28^{c}	17.04 ± 0.93^{c}	20.29 ± 1.25^{c}	46.38 ± 1.54^{c}
		40 ppm	11.10 ± 1.23^{d}	5.24 ± 0.43^{d}	12.86 ± 0.54^{d}	16.82 ± 0.58^{d}	40.64 ± 1.86^{d}
		50 ppm	7.90 ± 1.56^{e}	4.62 ± 0.96^{e}	8.82 ± 1.06^{e}	13.96 ± 0.19^{e}	38.60 ± 0.75^{e}
Control			27.65 ± 1.02^{a}	11.23 ± 1.23^{a}	30.27 ± 0.65^{a}	37.42 ± 0.96^{a}	69.32 ± 0.23^{a}
C.	*S.leucatha* essential oil	100 ppm	24.10 ± 1.31^{b}	10.56 ± 0.23b	28.54 ± 0.51^{b}	34.21 ± 0.21^{b}	66.28 ± 2.36^{b}
		200 ppm	22.69 ± 1.84^{c}	8.13 ± 1.6^{c}	25.83 ± 1.23	30.16 ± 1.45^{c}	63.12 ± 1.85^{c}
		300 ppm	20.56 ± 0.26^{d}	7.10 ± 1.45^{d}	22.65 ± 1.54^{d}	25.47 ± 0.89^{d}	58.41 ± 1.45^{d}
		400 ppm	17.44 ± 1.45^{e}	7.12 ± 0.36^{e}	19.56 ± 1.87^{e}	23.98 ± 0.65^{e}	53.42 ± 2.49^{e}
		500 ppm	15.63 ± 0.98^{f}	6.94 ± 1.54^{f}	16.47 ± 0.28^{f}	18.29 ± 0.45^{f}	48.23 ± 1.05^{f}
	Essential oil loaded ChNPs	10 ppm	22.14 ± 1.29^{a}	7.36 ± 0.73^{a}	26.46 ± 2.09^{a}	31.62 ± 0.17^{a}	58.02 ± 0.36^{a}
		20 ppm	18.26 ± 0.34^{b}	6.56 ± 1.08^{b}	22.02 ± 0.48^{b}	26.21 ± 0.24^{b}	53.20 ± 0.86^{b}
		30 ppm	14.28 ± 0.67^{c}	6.12 ± 1.39^{c}	17.90 ± 1.34^{c}	20.98 ± 1.04^{c}	46.90 ± 0.49^{c}
		40 ppm	11.27 ± 1.24^{d}	5.86 ± 0.51^{d}	13.64 ± 1.72^{d}	17.21 ± 0.93^{d}	41.30 ± 1.24^{d}
		50 ppm	8.34 ± 0.98^{e}	4.10 ± 0.65^{e}	9.64 ± 1.04^{e}	14.02 ± 1.09^{e}	38.64 ± 0.87^{e}

Within each tested product, values followed by the same letter(s) are not significantly different (ANOVA, Tukey's HSD, $\alpha = 0.05$). A. *Helicoverpa armigera*; B. *Spodptera litura*; C. *Plutella xylostella*.

11.33 Gut digestive enzymes of *H. armigera, S. litura* and *P. xylostella* larvae

The present investigations the consumption of *Salvia leucantha* essential oil encapsulated in chitosan nanoparticles treated leaves fed insects, showed decreased food consumption, relative growth rate, efficiency of conversion of ingested and digested food materials and concurrent decrease in the activities of digestive enzymes such protease, amylase and lipases. In general, an increased digestive enzyme activity in the midgut regions promotes the utilization of higher amount food for the growth and development of insects (Senthil-Nathan et al., 2006).

Statistical analysis showed that there was a significant differences between gut amylase, protease, trypsin–like, and lipase activities of larvae fed on treated diets at $P < .01$ and $P < .05$, the *H. armigera gut* amylase activity was significantly high ($P < .01$) in gut extract of fourth instar larvae fed on *Gossypium hirsutum* leaves treated nanoparticles (14.6 U/mg gut tissue) and it was low in after the post treatment *Gossypium hirsutum* –fed *H. armigera* (6.2 U/mg gut tissue). *S. litura gut* amylase activity was significantly high ($P < .01$) in gut extract of fourth instar larvae fed on treated *Ricinius communis* leaves (10.5 U/mg gut tissue) and was low in after the post treatment *Ricinius communis* –fed *S. litura* (5.4 U/mg gut tissue).

For *P. xylostella gut* amylase activity was significantly high ($P < .01$) in gut extract of fourth instar larvae fed on *Brassica oleracea* leaves (12.3 U/mg gut tissue) and it was low in after the post treatment *Brassica oleracea* –fed *P. xylostella* (6.1 U/mg gut tissue) (Table 11.13 and Fig. 11.7). Comparison of lipase activity in the gut extracts of larvae fed on treated leaves that provided the lowest lipid content (selected leaves) revealed significantly high lipase activity in selected leaves–fed larvae, yet in contrast, low lipase activity in after the post treatment with selected leaves–fed larvae lipase activities of larvae fed on diets, while all the remaining diets showed low lipase activities (Table 11.13). Similar results α-amylase interactions of AgNPs were reported earlier the effect of rapid degradation of starch hydrolysis (Ernest et al., 2012). In the present study less feeding and decrease in the gut enzyme activities may due to the secondary anti-feedant effect of plant allelochemicals present in the oil blended with chitosan might affected the resulted from the from the hindrance in the hormonal and further suppression of digestive enzyme secretion (Timmins and Reynolds, 1992).

TABLE 11.13 ***H. armigera, S. litura*** **and** ***P. xylostella*** **gut amylase and trypsin–like enzyme activities of after the treatment larvae fed on selected leaves and diets.**

Insect pest	Treatment	Fed on leaves	Gut amylase (Units/mg)	Gut tripsine (Units/mg)	Gut lipase (Units/mg)
Helicoverpa armigera	Control	*Gossypium hirsutum* leaves	14.64 ± 0.3^{a}	0.153 ± 0.001^{c}	6.56 ± 1.3^{b}
	Essential oil	*Gossypium hirsutum* leaves	09.35 ± 0.1^{c}	0.194 ± 0.004^{b}	2.37 ± 2.36^{d}
	Encapulation ChNPs	*Gossypium hirsutum* leaves	06.26 ± 0.0^{d}	0.215 ± 0.004^{a}	0.00225 ± 0.15^{f}
Spodptera litura	Control	*Ricinius communis* leaves	10.58 ± 0.21^{c}	0.119 ± 0.005^{e}	7.81 ± 1.01^{a}
	Essential oil	*Ricinius communis* leaves	$8.33 \pm 0.04^{c,d}$	0.103 ± 0.001^{f}	1.69 ± 2.56^{e}
	Encapulation ChNPs	*Ricinius communis* leaves	5.48 ± 0.12^{e}	0.98 ± 0.001^{g}	0.004 ± 0.73^{g}
Plutella xylostella	Control	*Brassica oleracea* leaves	12.36 ± 0.08^{b}	0.135 ± 0.003^{d}	3.23 ± 3.54^{c}
	Essential oil	*Brassica oleracea* leaves	9.06 ± 0.1^{c}	0.108 ± 0.001^{f}	0.00513 ± 0.23^{f}
	EncapulationChNPs	*Brassica oleracea* leaves	6.13 ± 0.02^{d}	0.82 ± 0.004^{h}	0.0021 ± 0.21^{g}

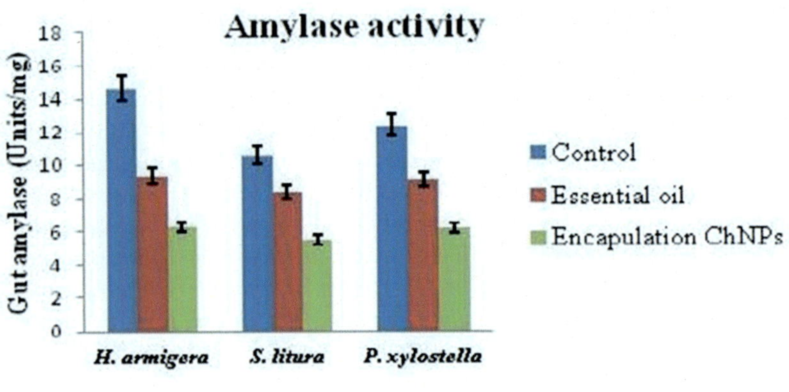

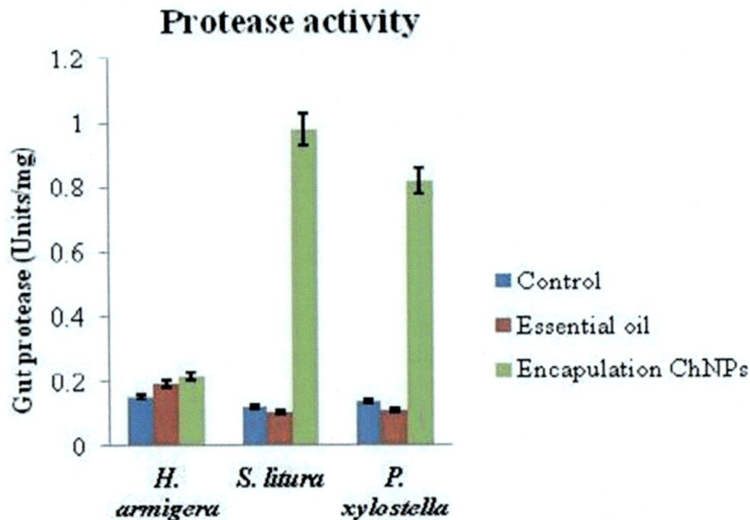

FIG. 11.7 Gut amylase and protease activity of *H. armigera, S. litura* and *P. xylostella* fed on after the treatment selected leaves diets.

11.34 Conclusion

The *salvia leucantha* essential oil and encapsulated chitosan nanoparticles consistently affected the growth, development and insecticidal activity toward *H. armigera, S. litura* and *P. xylostella* larvae by inhibiting midgut proteases. The encapsulated chitosan nanoparticles as insecticidal activity may be an effective strategy to control insect pests that carries a lower toxicological burden on the environment. Further this information will provide for the design of new novel methods to control biological insecticide resistance.

References

Abdelkader, M., Ahcen, B., Rachid, D., Hakim, H., 2014. Phytochemical study and biological activity of sage (*Salvia officinalis L.*). Inter. J. Biolog. Biomole Agri. Food Biotech. Engin. 8 (11), 1231–1235.

Ahmad, S., Ansari, M.S., Muslim, M., 2015. Toxic effects of neem based insecticides on the fitness of *Helicoverpa armigera* (Hübner). Crop Protect. 68, 72–78.

Alaguchamy, N., Jayakumararaj, R., 2015. Larvicidal effect of *Catharanthus roseus L* (G) Don. aqueous leaf extracts on the larvae of *Helicoverpa armigera* (Hübner). Int. J. Life Sci. Educ. Res. 3 (1), 10e14.

Aloui, H., Khwaldia, K., Licciardello, F., Mazzaglia, A., Muratore, G., Hamdi, M., Restuccia, C., 2014. Efficacy of the combined application of chitosan and Locust Bean Gum with different citrus essential oils to control postharvest spoilage caused by *Aspergillus flavus* in dates. Inter. J. Food Microbio. 17 (170), 21–28.

Bernfeld, P., 1955. Amylase, a and b. method. Enzymologia 1, 149–154.

Bernys, E.A., Chapman, R.F., 1994. Host-Plant Selection by Phytophagous Insects. Chapman and Hall.

Bhattacharyya, A., Bhaumik, A., Rani, P.U., Mandal, S., Epidi, T.T., 2010. Nano-particles - a recent approach to insect pest control. Afr. J. Biotechnol. 9 (24), 3489e3493.

Bossou, A.D., Mangelinckx, S., Yedomonhan, H., Pelagie, M.B., Akogbeto, M.C., De Kimpe, N., Avlessi, F., Sohounhloue, D.C.K., 2013. Chemical composition and insecticidal activity of plant essential oils from Benin against *Anopheles gambiae* (Giles). Parasit. Vectors 6, 337.

Brock, F.M., Forsberg, C.W., Buchanan-Smith, J.G., 1982. Proteolytic activity of rumen microorganisms and effects of proteinase inhibitors. Appl. Environ. Microbiol. 44 (3), 561–569.

Chandirika, J.U., Sindhu, R., Selvakumar, S., Annadurai, G., 2018. Herbal extract encapsulated in chitosan nanoparticle: a novel strategy for the treatment of urolithiasis. INDO Amer. J. Pharm. Sci. 5 (3), 1955–1961.

Devi, G.D., Murugan, K., Selvam, C.P., 2014. Green synthesis of silver nanoparticles using *Euphorbia hirta* (Euphorbiaceae) leaf extract against crop pest of cotton bollworm, *Helicoverpa armigera* (Lepidoptera: Noctuidae). J. Biopestic. 7, 54.

Engelmann, F., 1970. The Physiology of Insect Reproduction. Academic Press, New York.

Ernest, V., Shiny, P.J., Mukherjee, A., Chandrasekaran, N., 2012. Silver nanoparticles: a potential nanocatalyst for the rapid degradation of starch hydrolysis by α-amylase. Carbohydr. Res. 352, 60–64.

Feng, Y.-X., Wang, Y., Chen, Z.-Y., Du, S.-S., 2019. Efficacy of bornyl acetate and camphene from Valeriana officinalis essential oil against two storage insects. Environ. Sci. Pollut. Control Ser. 26 (90). https://doi.org/10.1007/s11356-019-05035-y.

Ferreira, T.P., Haddi, K., Correa, R.F., Zapata, V.L., Piau, T.B., Souza, L.F., Santos, S.M., Oliveira, E.E., Jumbo, L.O., Ribeiro, B.M., Grisolia, C.K., 2019. Prolonged mosquitocidal activity of *Siparuna guianensis* essential oil encapsulated in chitosan nanoparticles. PLoS Neglected Trop. Dis. (8), e0007624.

Feyzioglu, G.C., Tornuk, F., 2016. Development of chitosan nanoparticles loaded with summer savory (*Satureja hortensis L.*) essential oil for antimicrobial and antioxidant delivery applications. LWT - Food Sci. Technol. (Lebensmittel-Wissenschaft -Technol.) 70, 104–110.

Gibbs, R.D., 1974. Chemotaxonomy of Flowering Plants, vols. I and III. McGill Queen's University Press, Montreal and London.

Goswami, A., Roy, I., Sengupta, S., Debnath, N., 2010. Novel applications of solid and liquid formulations of nanoparticles against insect pests and pathogens. Thin Solid Films 519, 1252e1257.

Gul, R., Jan, S.U., Faridullah, S., Sherani, S., Jahan, N., 2017. Preliminary phytochemical screening, quantitative analysis of alkaloids, and antioxidant activity of crude plant extracts from *Ephedra intermedia* indigenous to balochistan. Sci. World J. https://doi.org/10.1155/2017/5873648.

Harborne, J.B., 1984. Phytochemical Methods to Modern Techniques of Plant Analysis. Chapman and Hall, London.

Hardstone, M.C., Leichter, C.A., Scott, J.G., 2009. Multiplicative interaction between the two major mechanisms of permethrin resistance, KDR and cytochrome P450-monooxygenase detoxification, in mosquitoes. J. Evol. Biol. 22, 416e423.

Hematpoor, A., Liew, S.Y., Azirun, M.S., Awang, K., 2017. Insecticidal activity and the mechanism of action of three phenylpropanoids isolated from the roots of *Piper sarmentosum* Roxb. Sci. Rep. 7 (1), 1–3.

Hikal, W.M., Baeshen, R.S., Said-Al Ahl, H.A., 2017. Botanical insecticide as simple extractives for pest control. Cogent Biol. 3 (1), 1404274.

Hosseini, S.F., Zandi, M., Rezaei, M., Farahmandghavi, F., 2013. Two-step method for encapsulation of oregano essential oil in chitosan nanoparticles: preparation, characterization and in vitro release study. Carbohydr. Polym. 95 (1), 50–56.

Huang, K.S., Sheu, Y.R., Chao, I.C., 2009. Preparation and properties of nanochitosan. Polym. Plast. Technol. Eng. 48, 1239–1243.

Huang, M., Khor, E., Lim, L.Y., 2004. Uptake and cytotoxicity of chitosan molecules and nanoparticles: effects of molecular weight and degree of deacetylation. Pharma Res. 21 (2), 344–353.

Jaya, P.S., Prakash, B., Dubey, N.K., 2014. Insecticidal activity of *Ageratum conyzoides* L., *Coleus aromaticus* Benth. and *Hyptis suaveolens* (*L.*) Poit essential oils as fumigant against storage grain insect *Tribolium castaneum* Herbst. J. Food Sci. Technol. 51 (9), 2210.

Kalagatur, N.K., Ghosh, N., Oriparambil, S., Sundararaj, N., Mudili, V., 2018. Antifungal activity of chitosan nanoparticles encapsulated with *Cymbopogon martinii* essential oil on plant pathogenic fungi *Fusarium graminearum*. Front. Pharmacol. 6 (9), 610.

Kandaga, A.S., Khetagoudar, M.C., 2013. Stud on larvicidal activit of weed e tracts against *Spodoptera litura*. J. Environ. Biol. 34, 253–257.

Keawchaoon, L., Yoksan, R., 2011. Preparation, characterization and in vitro release study of carvacrol-loaded chitosan nanoparticles. Colloids Surf. B Biointerf. 84 (1), 163–171.

Kedia, A., Prakash, B., Mishra, P.K., Dubey, N.K., 2014. Antifungal and antiaflatoxigenic properties of *Cuminum cyminum (L.)* seed essential oil and its efficacy as a preservative in stored commodities. Inter. J. Food Microbio. 3 (168), 1–7.

Kokate, C.K., Purohit, A.P., Gokhale, S.B., 2005. Pharmacognosy. Nirali Prakashan, Pune.

Kotkar, H.M., Sarate, P.J., Tamhane, V.A., Gupta, V.S., Giri, A.P., 2009. Responses of midgut amylases of *Helicoverpa armigera* to feeding on various host plants. J. Insect Physiol. 55, 663–670.

Kumar, S.S., Venkateswarlu, P., Rao, V.R., Rao, G.N., 2013. Synthesis, characterization and optical properties of zinc oxide nanoparticles. Int. Nano Lett. 3, 30.

Lamarque, G., 2005. Physicochemical behavior of homogeneous series of acetylated chitosans in aqueous solution: role of various structural parameters. Biomacromolecules 6 (1), 131–142.

Lertsutthiwong, P., How, N.C., Chandrkrachan, S., Stevens, W.F., 2002. Effect of chemical treatment on the characteristics of shrimp chitosan. J. Metals Mater. & Miner. 12 (1), 11–18.

Lertsutthiwong, P., Rojsitthisak, P., Nimmannit, U., 2009. Preparation of turmeric oil-loaded chitosan-alginate biopolymeric nanocapsules. Mater. Sci. Eng. C Mater. 29, 856–860.

Liao, M., Xiao, J.J., Zhou, L.J., Yao, X., Tang, F., Hua, R.M., Wu, X.W., Cao, H.Q., 2017. Chemical composition, insecticidal and biochemical effects of *Melaleuca alternifolia* essential oil on the *Helicoverpa armigera*. J. Appl. Entomol. 1 (9), 721–728.

Manikandan, A., Sathiyabama, M., 2015. Green synthesis of copper-chitosan nanoparticles and study of its antibacterial activity. J. Nanomed. Nanotech. 6 (1), 1.

Mattson, W.J., 1980. Herbivory in relation to plant nitrogen content. A. Rev. Ecol. Sysr. 11, 119–161.

Mohammadi, A., Hashemi, M., Hosseini, S.M., 2015. Nanoencapsulation of *Zataria multiflora* essential oil preparation and characterization with enhanced antifungal activity for controlling *Botrytis cinerea*, the causal agent of gray mould disease. Innovat. Food Sci. Emerg. Technol. 1 (28), 73–80.

Moreau, B.B., Thiery, D., 2006. Assessing larval food quality for phytophagous insects: are the facts as simple as they appear? J. Funct. Ecol. 20, 592–600, 592.

Murugan, K., Ancy George, S., 1992. Feeding and nutritional influence on growth and reproduction of *Daphnis nerii* (Linn.) (Lepidoptera: Sphingidae). J. Insect Physiol. 38, 961–968.

Murugan, K., Anitha, J., Dinesh, D., Suresh, U., Rajaganesh, R., Chandramohan, B., Subramaniam, J., Paulpandi, M., Vadivalagan, C., Amuthavalli, P., Wang, L., Hwang, J.-S., Hui, W., Saleh Alsalhi, M., Devanesan, S., Kumar, S., Pugazhendy, K., Higuchi, A., Nicoletti, M., Benelli, G., October 2016. Fabrication of nano-mosquitocides using chitosan from crab shells: impact on nontarget organisms in the aquatic environment. Ecotoxicol. Environ. Saf. 132, 318–328. https://doi.org/10.1016/j.ecoenv.2016.06.021. Epub 2016 Jun 24.

Murugan, K., Anitha, J., Suresh, U., Rajaganesh, R., Panneerselvam, C., Tseng, L.C., Kalimuthu, K., Alsalhi, M.S., Devanesan, S., Nicoletti, M., Sarkar, S.K., 2017a. Chitosan-fabricated Ag nanoparticles and larvivorous fishes: a novel route to control the coastal malaria vector *Anopheles sundaicus*? Hydrobiologia 797 (1), 335–350.

Murugan, K., Jaganathan, A., Suresh, U., Rajaganesh, R., Jayasanthini, S., Higuchi, A., Kumar, S., Benelli, G., 2017b. Towards bio-encapsulation of chitosan-silver nanocomplex? Impact on malaria mosquito vectors, human breast adenocarcinoma cells (MCF-7) and behavioral traits of non-target fishes. J. Cluster Sci. 28 (1), 529–550.

Murugan, K., Jaganathan, A., Suresh, U., et al., 2017c. Towards bio-encapsulation of chitosan-silver nanocomplex? Impact on malaria mosquito vectors, human breast adenocarcinoma cells (MCF-7) and behavioral traits of non-target fishes. J. Cluster Sci. 28, 529–550. https://doi.org/10.1007/s10876-016-1129-1.

Murugan, K., Panneerselvam, C., Subramaniam, J., Madhiyazhagan, P., Hwang, J.S., Wang, L., Dinesh, D., Suresh, U., Roni, M., Higuchi, A., Nicoletti, M., 2016a. Eco-friendly drugs from the marine environment: spongeweed-synthesized silver nanoparticles are highly effective on *Plasmodium falciparum* and its vector *Anopheles stephensi*, with little non-target effects on predatory copepods. Environ. Sci. Pollut. Control Ser. 23 (16), 16671.

Murugan, K., Roni, M., Panneerselvam, C., Suresh, U., Rajaganesh, R., Aruliah, R., Mahyoub, J.A., Trivedi, S., Rehman, H., Al-Aoh, H.A., Kumar, S., 2018. Sargassum wightii-synthesized ZnO nanoparticles reduce the fitness and reproduction of the malaria vector *Anopheles stephensi* and cotton bollworm *Helicoverpa armigera*. Physiol. Mol. Plant Pathol. 101, 202–213.

Murugan, K., Senthil Kumar, N., Jeyabalan, D., Senthil-Nathan, N., Swamiappan, M., 2000. Influence of *Helicoverpa armigera* (hubner) diet on its parasitoid, *Compoletis chlorideae uchida*. Insect Sci. & Appl. 20 (1), 23–31.

Murugan, K., Wang, L., Anitha, J., Dinesh, D., Amuthavalli, P., Vasanthakumaran, M., Paulpandi, M., Hwang, J.S., 2021. Insecticidal effect of chitosan reduced silver nanocrystals against filarial vector, Culex quinquefasciatus and cotton bollworm, Helicoverpa armigera. In: Advances in Nano-Fertilizers and Nano-Pesticides in Agriculture. Woodhead Publishing, pp. 469–486.

Muzzarelli, R.A.A., Rochetti, R., 1985. Determination of the degree of deacetylation of chitosan by first derivative ultraviolet spectrophotometry. J. Carbohydr. Polym. 5, 461–472.

Nallamuthu, I., Devi, A., Khanum, F., 2015. Chlorogenic acid loaded chitosan nanoparticles with sustained release property, retained antioxidant activity and enhanced bioavailability. Asian J. Pharm. Sci. 10 (2015), 203–211. https://doi.org/10.1016/j.ajps.2014.09.005.

Negi, A., Javed, Mohammad, S., Melkani, A.B., Dev, V., Beauchamp, P.S., 2007. Steam volatile terpenoids from Salvia leucantha. J. Essent. Oil Res. 19, 463–465.

Noshad, M., Hojjati, M., Behbahani, B.A., 2018. Black Zira essential oil: chemical compositions and antimicrobial activity against the growth of some pathogenic strain causing infection. Microb. Pathog. 116, 153–157. https://doi.org/10.1016/j.micpath.2018.01.026.

Oh, J.W., Chun, S.C., Chandrasekaran, M., 2019. Preparation and in vitro characterization of chitosan nanoparticles and their broad-spectrum antifungal action compared to antibacterial activities against phytopathogens of tomato. Agronomy 9 (1), 21.

Owolade, O.F., Ogunleti, D.O., Adenekan, M.O., 2008. Titanium dioxide affects disease development and yield of edible cowpea. Electron. J. Environ. Agric. Food Chem. 7 (50), 2942–2947.

Patel, C.C., Mehta, D.M., Patel, J.R., Patel, N.M., 1992. Resistance of gram-podborer (*Helicoverpa armigera*) to insecticides in Gujarat. Indian J. Agric. Sci. 62 (6), 421–423.

Paulraj, G.M., Ignacimuthu, S., Gandhi, M.R., Shajahan, A., Ganesan, P., Packiam, S.M., Al-Dhabi, N.A., 2017. Comparative studies of tripolyphosphate and glutaraldehyde cross-linked chitosan-botanical pesticide nanoparticles and their agricultural applications. Int. J. Biol. Macromol. 104, 1813–1819. https://doi.org/10.1016/j.ijbiomac.2017.06.043.

Perlatti, B., de Souza Bergo, P.L., Fernandes da Silva, M.F.G., Fernandes, J.B., Forim, M.R., 2013. Polymeric nanoparticle-based insecticides: a controlled release purpose for agrochemicals. In: Insecticides-Development of Safer and More Effective Technologies. IntechOpen. https://doi.org/10.5772/53355.

Prasanth Koppolu, B., Smith, S.G., Ravindranathan, S., Jayanthi, S., Kumar, T.K., Zaharoff, D.A., 2014. Controlling chitosan-based encapsulation for protein and vaccine delivery. Biomaterials 35 (14), 4382–4389.

Rabea, E.I., Badawy, M.E., Rogge, T.M., Stevens, C.V., Höfte, M., Steurbaut, W., Smagghe, G., 2005. Insecticidal and fungicidal activity of new synthesized chitosan derivatives. Pest Manag. Sci. 1 (10), 951–960.

Rondon, M., Velasco, J., Morales, A., Rojas, J., Carmona, J., Gualtieri, M., Hernandez, V., 2005. Composition and antibacterial activity of the essential oil of *Salvia leucantha* Cav. cultivated in *Venezuela Andes*. Rev. Latinoam. Quim. 33, 55–59.

Roni, M., Murugan, K., Panneerselvam, C., Subramaniam, J., Nicoletti, M., Madhiyazhagan, P., Dinesh, D., Suresh, U., Khater, H.F., Wei, H., Canale, A., November 1, 2015. Characterization and biotoxicity of *Hypnea musciformis*-synthesized silver nanoparticles as potential eco-friendly control tool against *Aedes aegypti* and *Plutella xylostella*. Ecotoxicol. Environ. Saf. 121, 31–38.

Sadashivan, S., Manickam, A., 2005. Biochemical Methods, second ed. New Age International (P) Ltd. Publisher, New Delhi.

Sastry, M., Mayya, K.S., Bandyopadhyay, K., 1997. pH dependent changes in the optical properties of carboxylic acid derivatized silver colloidal particles. Colloids Surf., A 127, 221–228.

Scriber, J.M., 1984a. Host-plant suitability. In: Bell, W.J., Carde, R.T. (Eds.), Chemical Ecology of Insects. Chapman and Hall, London, pp. 159–202.

Scriber, J.M., 1984b. Nitrogen nutrition of plants and insect invasion. In: Hauck, R.D. (Ed.), Nitrogen in Crop Production, p. 4414 (Am. Sk. Agron, Madison, WI).

Senthil-Nathan, S., Chung, P.G., Murugan, K., 2006. Combined effect of biopesticides on the digestive enzymatic profiles of *Cnaphalocrocis medinalis* (Guenee) (the rice leaffolder) (Insecta: Lepidoptera: Pyralidae). Ecotoxicol. Environ. Saf. 64 (3), 382–389.

Sharaby, A., El-Nujiban, A., 2013. Biological activity of essential oil of sage plan leaves *Salvia offecinalis* L. against the black cutworm *Agrotis ipsilon* (Hubn.). Int. J. Sci. Res. 4, 737–741.

Shukla, A., Patel, P.R., 2011. Biology, food utilization and seasonal incidence of *Spodoptera litura* (Fab.) on Banana cv. Grand Naine. Res. J. Agric. Sci. 2 (1), 49–51.

Slansky, F., Scriber, J.M., 1985. Food consumption and utilization. In: Kerkut, G.A., Gilbert, L.I. (Eds.), Comprehensive Insect Physiology, Biochemistry and Pharmacology, vol. 4. Pergamon Press, New York, pp. 87–163.

Sotelo-Boyás, M.E., Valverde-Aguilar, G., Plascencia-Jatomea, M., Correa-Pacheco, Z.N., Jiménez-Aparicio, A., Solorza-Feria, J., Barrera-Necha, L., Bautista-Baños, S., 2015. Characterization of chitosan nanoparticles added with essential oils. in vitro effect on pectobacterium carotovorum. Revista Mexicana de Ingeniería Química vol. 14 (3), 589–599.

Stearns, S.C., 1992. The Evolution of Life Histories. Oxford University Press, Oxford.

Stuart, B.H., 2008. Polymer Analysis, vol. 30. John Wiley & Sons.

Suresh, U., Murugan, K., Panneerselvam, C., Rajaganesh, R., Roni, M., Al-Aoh, H.A., Trivedi, S., Rehman, H., Kumar, S., Higuchi, A., Canale, A., 2017. Suaeda *maritima*-based herbal coils and green nanoparticles as potential biopesticides against the dengue vector *Aedes aegypti* and the tobacco cutworm *Spodoptera litura*. Physiol. Mol. Plant Pathol. 101, 225–235.

Suthisut, D., Fields, P.G., Chandrapatya, A., 2011. Contact toxicity, feeding reduction, and repellency of essential oils from three plants from the ginger family (Zingiberaceae) and their major components against *Sitophilus zeamais* and *Tribolium castaneum*. J. Econ. Entomol. 104 (4), 1445–1454.

Tamhane, V.A., Chougule, N.P., Giri, A.P., Dixit, A.R., Sainani, M.N., Gupta, V.S., 2005. In vivo and in vitro effect of *Capsicum annum* proteinase inhibitors on *Helicoverpa armigera* gut proteinases. Biochim. Biophys. Acta Gen. Subj. 1722 (2), 156–167.

Timmins, W.A., Reynolds, S.E., 1992. Azadirachtin inhibits secretion of trypsin in midgut of *Manduca sexta* caterpillars: reduced growth due to impaired protein digestion. Entomol. Exp. Appl. 63, 47–54.

Trease, G.E., Evans, W.C., 1979. Textbook of Pharmacognosy, twelfth ed. Balliere-Tindal, London, p. 343.

Upadhyaya, K., Dixit, V., Padalia, R.C., Mathela, C.S., 2009. Terpenoid composition and antioxidant activity of essential oil from leaves of salvia leucantha cav. J. Essent. Oil-Bear. Plants JEOP 12 (5), 551–556.

Waldbauer, G.P., 1968. The consumption and utilization of food by insects. Adv. Insect Physiol. 5, 229e288.

Winkler, U.K., Stuckmann, M., 1979. Glycogen, hyaluronate, and some other polysaccharides greatly enhance the formation of exolipase by *Serratia marcescens*. J. Bacteriol. 138 (3), 663–670.

Yoksan, R., Jirawutthiwongchai, J., Arpo, K., 2010. Encapsulation of ascorbyl palmitate in chitosan nanoparticles by oil-in-water emulsion and ionic gelation processes. Colloids Surf. B Biointerf. 76 (1), 292–297.

Zada, H., Ahmad, B., Hassan, E., Ur Rehman Saljoqi, A., Naheed, H., 2018. Toxicity potential of different azadirachtin against *plutella* Xy-*lostella* (Lepidoptera; Plutellidae) and its natural enemy, diadegma species. J. Agron. Agri. Sci. 1 (003).

Zhiming, F., Wang, H., Hu, X., Sun, Z., Han, C., 2013. The pharmacological properties of salvia essential oils. J. Appl. Pharmaceut. Sci. 3 (07), 122–127.

Ziaee, M., Moharramipour, S., Mohsenifar, A., 2014. MA-chitosan nanogel loaded with *Cuminum cyminum* essential oil for efficient management of two stored product beetle pests. J. Pest. Sci. 87, 691–699.

Chapter 12

Microbial bio-pesticide as sustainable solution for management of pests: achievements and prospects

Udayashankar C. Arakere[a,1], Shubha Jagannath[c], Soumya Krishnamurthy[d], Srinivas Chowdappa[b] and Narasimhamurthy Konappa[b,1]

[a]*Department of Studies in Biotechnology, University of Mysore, Mysore, Karnataka, India;* [b]*Department of Microbiology and Biotechnology, Bangalore University, Bengaluru, Karnataka, India;* [c]*Department of Botany, Molecular Biology division, Jnana Bharathi Campus, Bangalore University, Bengaluru, Karnataka, India;* [d]*Department of Microbiology, Field Marshal K. M. Cariappa College, A Constituent College of Mangalore University, Madikeri, Karnataka, India*

12.1 Introduction

Agriculture plays a pivotal role in the upliftment of our country's economy by meeting the necessary demands of the drastically increasing population. India is considered to have largest agrarian economy as its millions of citizens are dependent on farming and its allied activities for their livelihood. Agricultural has to meet up to the changes of providing immunity boosting, nutritious food to the growing population by inventing easily cultivable and fast-growing crops. The scenario is not different globally, as the global population presently is approximately 7 billion and is anticipated to escalate up to 9.2–10 billion by 2050, an upsurge of 30% (World Population Prospects, 2019). The quick-tempered growth of population has predicted the mandate for increased production of agricultural produce but, meeting this demand is challenging because of decreased availability of cultivable lands due to urbanization, growing industry, lack of skilled man power, water crisis, increased destruction of crops by pests, post-harvest losses etc. (FAO, 2009). The major cause for massive destruction of crops includes bacteria, fungi, viruses, insects and weeds. Globally almost one-third of the crops are lost annually owed to the illness caused by various phytopathogens (Al-Sadi, 2017).

The use of pesticides has positively contributed in uplifting the agricultural produce both in terms of quality and quantity thus, leading to swelling of Agri-based income. The chemical pesticides are profusely used in integrated pest management to protect crops from pests, insects and pathogenic microorganisms. It is assessed that annually two million tons of chemical pesticides are used worldwide; Europe (45% of total consumed), United States (25%) and rest of the globe (25%). The chemical pesticides are not only reported to cause various environmental hazards, health hazards and toxicity but also, development of resistance in pests due to repeated use resulting in pest resurgence (De Oliveira et al., 2014; Carvalho, 2017). The chemical pesticides have proved to have hostile impact on soil fertility, structure, soil microbial flora and mineral recycle (Randall et al., 2013; Lade et al., 2017). It also is reported to cause major health issues like breast cancer, endocrine disruption, reproductive toxicity and cytotoxicity (Nicolopoulou-Stamati et al., 2016).

Presently, with changing time and environmental concern, the agrarian sector is moving toward natural pesticides or biopesticides as an alternative to chemicals for the management of pests. The United State Environmental Protection Agency (EPA) has defined biopesticides as, "Types of pesticides derived from natural materials such as animals, plants and microorganisms." The biocontrol agents combat the plant pathogens by competing for space, nutrition, production of antimicrobial compounds, direct parasitism and induced resistance. The biopesticide trade is predicted to elevated from $ 3.22 billion in 2017 up to $ 6.6 billion by 2022 at 15.43% compound annual growth. Hence, need of the hour is to develop

1. contributed equally.

Biopesticides. https://doi.org/10.1016/B978-0-12-823355-9.00016-X

biopesticides which are harmless to farmers, easily biodegradable, ecologically friendly, economically viable and target specific (Mazid et al., 2011).

The biopesticides can be broadly classified into three types:

12.1.1 Biochemical pesticides

Insect Pheromones, Chitosan, Plant extracts.

12.1.2 Microbial pesticides

Bacteria, Fungi, Viruses, Protozoa.

12.1.3 Plant incorporated protectants

Genetically Modified crops.

12.2 Biochemical pesticides

12.2.1 Insect pheromones

The pheromones are chemicals secreted by animals and insects to interact with the other members of the similar species. The pheromones are used in the integrated management of pests. The pheromones are used to monitor presence or absence of insects in a specific locality. It is also used to find the significant load of insects present in an area to implement costly treatments. The pheromones are widely used in urban areas to regulate pests like cockroaches as it is useful in maintenance of stored food grains in distribution centers. It is also used to track the countrywide blowout of significant agricultural pests like Japanese beetle, carob moth, armyworm, med fly, fruit flies, mountain pine beetle, Asian citrus psyllid, red palm weevil and gypsy moth. The pheromones are used to trap and remove the abundant number of insects from the breeding and feeding population. It is also used to disrupt the mating of insect population.

The pheromones commonly used in pest management are: Sex pheromones: released by either male or female insects to attract its mate. Trail pheromones: commonly released by social insects which helps in tracking its feed and nest. Alarm pheromones: commonly released by gregarious insects as a response to predation giving alarm for quick group dispersal of fellow insects. Host marking pheromones: released to reduce competition within the members of the identical species. Aggregation pheromones: causes insects of both male and female sex to crowd for reproduction and feeding.

12.3 The few examples of pheromones used in agricultural pest management are as follows

The sex pheromone traps of tobacco caterpillar (*Spodoptera litura*) are used to manage pests associated with groundnut, castor and sunflower. The gram pod borer (*Helicoverpa armigera*) sex pheromones are used to monitor pests allied with groundnut, sunflower and soybean.

Sex pheromone trap of leaf miner (*Aproaerema modicella*) is operative in trapping of moths in large numbers and thus reducing the loss of groundnut. The methoxy benzene (anisole) from abdominal glands of female adults of *Holotrichia consanguinea* is an aggregation pheromone which attracts both the sexes of beetle thus can be successfully used to attract and trap beetles. The sex pheromones are used to trigger sexual confusion in the target pests and thus prevent them from laying eggs in the crops hence, proving efficient in pest control and management.

12.3.1 Chitosan

Chitosan is obtained by deacetylation of chitin to varying degrees. It is made by the alkali treatment of shells of crustaceans like shrimp, lobsters and crabs. It is structurally a linear polysaccharide made up of arbitrarily distributed β-(1−4)-linked D-glucosamine (deacetylated) and N-acetyl-D-glucosamine (acetylated). Chitosan is reported to be beneficial in the control and management of plant pathogens. The chitosan chelates minerals (Fe, Cu) and nutrients thus when applied on plants, protect them from diseases. The chitosan also is known to bind mycotoxins thus protecting the plant tissues from getting damaged due to toxins. The chitosan inhibits the spore germination and hyphal growth of frequent fungal pathogens like *Botrytis cinerea*, *Monilina laxa*, *Pythium aphanidermatum*, *Fusarium oxysporum*, and *Alternaria alternata*. The success of chitosan as biopesticide depends on its concentration, pH, viscosity, grade of diacylation and the formula used in application like chitosan alone, foliar application, soil amendment etc.

The root rot disease of tomato caused by *Fusarium oxysporum* f. sp. *radicis-lycopersici* was suppressed by chitosan derivatives. The modified chitosan derivatives like, N-phosphonomethyl chitosan, methylpyrrolidinone chitosan, N-carboxymethyl chitosan inhibited the growth of *Saprolegnia parasitica*. The chitosan is widely used in regulation of post -harvest diseases, its coating on tomatoes delays ripening thus reduce decay.

12.3.2 Plant extract biopesticides

These pesticides are derived from plants and its extracts. The most commonly used sources of plant based biopesticides are leaves, bark, seeds of Neem (*Azadiracha indica*), Tulsi (*Ocimum basilicum*), *Lavandula officinalis* and Chinaberry (*Melia azedarach*); flowers of pyrethrum daisy, Chrysanthemum; roots of Derris, Lonchocarpus; seeds of tropical lily, *Schoenocaulon officinale*, European *Veratrum album*; citrus oil and many more. More than two thousand plant species are beneficial in effective pest management. The modes of action of these pesticides are contact poison or stomach poison. They may exhibit pesticidal activity, act as antifeedants, repellents, attractants and growth inhibitors of common plant pathogens.

12.4 Microbial biopesticides

Microbial biopesticides constitutes the various microorganisms like bacteria, fungi, viruses, protozoa and algae which are used to combat variety of pests causing destruction of agricultural economy. The microbes as biopesticides contribute a major share to the group of existing broad-spectrum biopesticides. The microbial biopesticides target the pests precisely and will not cause any harm to the non-target and environmentally friendly organisms. There are thousands of microorganisms reported to be used successfully used as biopesticides and around two hundred microorganisms are successfully available as to be used as biopesticides in the 30 countries affiliated to OECD (organization for Economic Co-operation and Development). The different microbial biopesticides have been officially registered to be used in USA, Canada, European Union and Asian counties; the microbe-based formulations of biopesticides are also on the upsurge due to the innovative technologies used.

In India, the use of microbe based biopesticides is increasing at a quick stride as compared to the chemical pesticides due to awareness booting among the farming community. The India biopesticide market promotes the microbial pesticides as bioherbicides, bioinsecticides, bio fungicides etc in either dry or liquid formulations. The different microorganism used as biofungicides are *Trichoderma viridae* and *T. harzianum*, *Fusarium proliferatum*, *Pseudomonas fluorescens*, *Bacillus subtilis*, *B. pumilus* and *Ampelomyces quisqualis*. The microbes commonly used as bioinsecticides in the market are *Bacillus thuringiensis* var. *kurstaki*, *Beauveria bassiana*, *Verticillium lecanii*, *Metarhizium anisopliae*, *Paecilomyces lilacinus* and others. These microbes based biopesticides are widely used to manage and control the common pests affecting commercial crops, ornamental crops, grains, cereals, pulses, oil seeds, fruits and vegetables (Table 12.1).

12.5 Bacteria as biopesticides

The bacteria that are used as biopesticides are broadly classified into spore forming bacteria and non-spore forming bacteria. Spore forming bacteria used as biopesticides mainly belong to the family of *Bacilliaceae*. Non-spore forming bacteria used as biopesticides mainly belong to the families of *Pseudomonadaceae* and *Enterobacteriaceae*.

12.6 Members of *Bacilliaceae* as biopesticides (spore formers)

The bacteria belonging to the family *Bacillaceae* are rod shaped, Gram-positive bacteria producing endospores. They are heterotrophic bacteria capable of utilizing inorganic and organic materials from its surroundings. They are mostly aerobic while some are facultative and strict anaerobes. The members of *Bacillaceae* family are capable of producing varied array of bioactive/biologically significant molecules capable of successfully inhibiting the pathogens affecting economically important crops. The bacteria belonging to *Bacillus* sp., *Paenibacillus* sp., *Halobacillus* sp., *Aneurinibacillus* sp., and *Brevibacillus* spp. produce peptides and its derivatives having exclusive structures that augment resistant to protease hydrolysis thus display strong and swift destruction of wide range of pathogens. These peptides exhibit similar activities as compared to the presently available antibiotics in the market. The novel peptide compounds produced are surfactin-like lipopetides, fengycin c, cyclic lipopetides, linear lipopetides, 2,5-diketopiperazine, iturinic lipopetides; 119 ingenious peptide compounds have been reported since 2000 from *Bacillaceae* members (Zhao et al., 2018).

TABLE 12.1 Microbial biopesticides for different target pathogens.

Microbial biopesticides	Target pest/mechanism
A. radiobacter	Crown galls
Adoxophyes orana granulovirus (AoGVs)	Summer fruit tortrix moth (*Adoxophyes orana*)
Altemaria destruens	Cuscuta control
Ampelomyces quisqualis	Hyperparasitic, powdery mildew control and damping off disease control
Aspergillus flavus strain AF36	Fungicide for cotton Pistachio orchards
B. bassiana	Effective against variety of insects such as crickets, white grubs, fire ants, flea beetles, whiteflies, plant bugs, grasshoppers, thrips, aphids, mites, mosquito larvae and many others
B. firmus	Diamondback moths
B. popilliae	Stomach poison, larvae of various beetles
B. pumilus	Effective against rust, downy and powdery mildews
B. sphaericus	Stomach poison mosquitoes
B. subtilis	Effective against root rot caused by *Rhizoctonia, Fusarium, Alternaria, Aspergillus* and *Pythium.* Also effective against some foliar diseases
B. subtilis FZB24	Effective against *Rhizoctonia, Fusarium, Alternaria, Verticillium* and *Streptomyces* on vegetables and ornamental plants
B. thuringiensis aizawai 7.29	Lepidoptera
B. thuringiensis aizawai IC 1	*Lepidoptera, Diptera*
B. thuringiensis Berliner	*Lepidoptera*
B. thuringiensisentomocidus 6.01	*Lepidoptera*
B. thuringiensis israelensis	*Diptera*
B. thuringiensis japonensis Strain Buibui	Activity against soil-inhabiting beetles from cotton
B. thuringiensis kurstaki HD-1	*Lepidoptera, Diptera*
B. thuringiensiskurstaki KTO, HD-1	*Lepidoptera*
B. thuringiensis morrisoni PG14	*Diptera*
B. thuringiensis ssp. Kurstakiandaizawai	Act as insecticides against lepidopteran larval species from Soyabean castor, tobacco
B. thuringiensis tenebrionis	Activity against coleopteran adults and larvae, most notably the Colorado potato beetle (*Leptinotarsa decemlineata*) From tomato, Chilies, rice red gram,
B. thuringiensis tenebrionis (san diego)	*Coleoptera*
B. thuringiensis var. aizawai	Stomach poison, effective against lepidopterans in vegetables and maize
B. thuringiensis var. galleriae	Stomach poison, effective against American bollworm on cotton, tobacco caterpillar on chillies, leaf folder of rice and diamond-back moth on vegetables
B. thuringiensis var. israelensis	Stomach poison, effective against mosquitoes and black fly larvae and midges
B. thuringiensis var. kurstaki	Stomach poison, effective against most lepidopteran larvae and some leaf beetles
B. thuringiensis var. tenebrionis	Stomach poison, effective against coleopteran beetles on vegetables
Bacillus firmus	Nematodes

TABLE 12.1 Microbial biopesticides for different target pathogens.—cont'd

Microbial biopesticides	Target pest/mechanism
Bacillus sphaericus	Diamondback moths
Bacillus subtilis	Pathogens such as *Rhizoctonia, Fusarium, Aspergillus,* banana Sigatoka (caused by *Mycosphaerella musicola*), Paddy (bacterial leaf blight)
Bacillus thuringiensis aizawai	Armyworms, diamondback moth
Bacillus thuringiensis and *Bacillus sphaericus 2362 (Bs)*	Effective against mosquito and other dipteran larvae black fly (simuliidae), and fungus gnats from cabbage and Cauliflower
Bacillus thuringiensis israelensis	Mosquitoes and black flies
Bacillus thuringiensis kurstaki	Lepidoptera
Bacillus thuringiensis sphaericus	Mosquitoes
Bacillus thuringiensis subsp. israelensis	Lepidopteran pests
Bacillus thuringiensis subsp. Kurstaki	Lepidopteran pests
Bacillus thuringiensis tenebrionis	Colorado potato beetle
Bacillus thuringiensis var. galleriae	*Helicoverpa armigera*
Bacillus thuringiensis var. israelensis	Diamondback moths
Bacillus thuringiensis var. kurstaki	Diamondback moths
Beauveria bassiana	Mango hoppers and mealy bugs and coffee pod borer
Beauveria bassiana	Wide range of insects and mites
Beauveria bassiana	Many insect species such as aphids, whiteflies, thrips
Beauveria brongniartii	*Helicoverpa armigera,* berry borer, root grubs
Burkholderia cepacia	Effective against soil fungal pathogens
Burkholderia spp.	Chewing and sucking insects and mites; nematodes
Candida oleophila	Effective against postharvest pathogens like botrytis and Penicillium
Chromobacterium subtsugae	Chewing and sucking insects and mites
Close analog of the fumonisin mycotoxins	Highly toxic used to kill weed species. Highly toxic to mammals by inhibition of ceramide synthase. Broadleaf plants (e.g. jimsonweed, prickly sida and black nightshade)
Coniothyrium minitans	Effective against Sclerotinia species on canola, sunflower, peanut, soyabean and vegetables
Cryptolaemusm ontrouzieri	To control mealy bugs especially in fruits like citrus, coffee, grapes and several other fruit crops and ornamental
Cryptophlebia leucotreta granulovirus	False codling moth (*Thaumatotibia leucotreta*)
Cydia pomonella granulovirus (CpGVs)	*Cydia pomonella*
Entomopathogenic nematodes such as Steinernema carpocapsae	Thrips, fungus gnats (*Bradysia* spp.), mole crickets (*Scapteriscus* spp.) and other insects
Erwinia amylovora (HrpN harpin protein)	Insecticide, fungicide and nematicide. Multi-spectrum
Fusarium oxysporum (non-pathogenic)	Effective against pathogenic *Fusarium* on basil, carnation, cyclamen, tomato
Gliocladium catenulatum *Gliocladium spp.*	Effective against *Pythium, Rhizoctona, Botrytis* and *Didymella* species on greenhouse crops
Gliocladium virens	Effective against soil pathogens causing damping off and root rot

Continued

TABLE 12.1 Microbial biopesticides for different target pathogens.—cont'd

Microbial biopesticides	Target pest/mechanism
Granulosis virus	Effective against leafroller and codling moth and other lepidopterans
H. bacteriophora	Effective against many lepidopteran larvae, turf and Japanese beetles and soil insects
H. bacteriophora	Borers
H. megidis	Effective against black vine weevils and soil insects
Helicoverpa armigera NPV	*H. armigera*
Helicoverpa armigera Nucleopolyhedrosis virus and (nucleopolyhedrosis) NPV for other *Lepidoptera*	*Helicoverpa* spp. (for HNPV) and other *Lepidoptera* species
Helicoverpa armigera nucleopolyhedrovirus (HearNPV)	African cotton bollworm (*Helicoverpa armigera*), corn earworm (*H. zea*) and other *Helicoverpa* species (*H. virescens, H. punctigera*)
Helicoverpa zea Nuclear Polyhedrosis Virus	*Heliothis* and *Helicoverpa* species
Helicoverpa zea nucleopolyhedrovirus	*Helicoverpa* spp. and *Heliothis virescens*
Helicoverpaarmigera (Hubner) and *Spodopteralitura (Fabricius)*	Used against bollworms in cotton and pod borers species specific for *Lepidoptera, Hymenoptera* and *Diptera* from cotton, flax, groundnuts, jute, maize, soya beans, tea, tobacco, vegetables, lucerne
Heterorhabditis bacteriophora	*Otiorhynchus* spp., chestnut, moths, black vine weevil and soil-dwelling beetle larvae, *Melolontha*, caterpillars, cutworms, leafminers
Heterorhabditis downesi	Black Vine weevil, *Otiorhynchus sulcatus*
Hirsutella thompsonii	Stimulates premature fungal epizootics, effective against mites
Hirsutella thompsonii *Isaria fumosorosea*	Spider mites
Lagenidium giganteum	Effective against mosquito larvae and related dipterans. Kill through zoospores
Lecanicillium lecanii	Aphids, leafminers, mealybugs, scale insects, thrips, whiteflies
Lymantria dispar multiple nucleopolyhedrovirus (LdMNPV)	*Lymantria dispar*
M. anisopilae	Effective against range of pests. Green muscle is specific for locusts and grasshoppers
Metarhizium anisopliae	Beetles & caterpillar pests; grasshoppers, termites
Metarhizium anisopliae	Insects such as crickets and grasshoppers
Metarhizium anisopliae Entomopathogenic fungus	Insect pest control used as a biocontrol agent, particularly for malaria vector species from cabbage, canola (*Brassica napus* L.)
Myrothecium vaerrcaria *Paecilomyces fumosoroseus*	Black vine weevil, effective against many nematodes effective against whiteflies in greenhouse
Myrothecium verrucaria	Nematodes
Neodiprion abietis nucleopolyhedrovirus (NeabNPV)	*Neodiprion abietis*
NPV for *A. gemmatalis*	Effective against velvetbean caterpillar and sugarcane borer
NPV for *Anagrapha falcifera*	Indian meal moth, effective against lepidopterans
NPV for *Autographa calofornica*	Effective against Alfalfa looper
NPV for *H. zea and H. virescens*	Effective against bollworms
NPV for *L. dispar*	Effective against gypsy moth

TABLE 12.1 Microbial biopesticides for different target pathogens.—cont'd

Microbial biopesticides	Target pest/mechanism
NPV for *Mamestra brassicae*	Effective against lepidopterans
NPV for *Neodiprion sertifer, N. lecontei and N. abietis*	Effective against sawfly larvae
NPV for *Orgyia pseudotsugata*	Effective against Douglas-fir tussock moth
NPV for *Spodoptera exigua*	Effective against beet and lesser army worms, pig weed caterpillar and mottled willow moth
NPV for *Syngrapha falcif*	Effective against *Helicoverpa* and *Cydia* spp.
NPV of Spodoptera litura	*Spodoptera litura*
P. hermaphrodita	Effective against slugs eating nematode
Paecilomyces fumosoroseus	Insects, mites, nematodes, thrips
Paecilomyces lilacinus	Effective against nematodes
Paecilomyces lilacinus	Plant pathogenic nematodes
Phasmarhabditis hermaphrodita	Molluscs
Phelbia gigantea	Management of fungal pathogens in various crops
Phromone lures for *Helicoverpaarmigera (Hubner)* and *Spodopteralitura*	To trap productive male of gram borer and tobacco caterpillar, *Spodopteralitura* from tomato, potato: Cabbage: *Brassica oleracea*, brinjal
Plutella xylostella granulovirus	*Plutella xylostella*
Pseudomonas fluorescens	Whitefly
Pseudomonas fluorescens,	Several fungal, viral, and bacterial diseases such as frost forming bacteria. Wheat (Loose Smut) Paddy (bacterial leaf blight) Ground nut (late leaf spot) rice (leaf and neck blast) chili seedlings (damping off) tomato (wilt)
S. carpocapsae	Effective against black vine weevils, strawberry root weevils, cutworms, cranberry girdler and termites
S. glaseri	Effective against root weevils, cutworms, fleas, borers and fungal gnats
S. viridochromogenes Streptomyces hygroscopicus	Anntnual and perennial broadleaf weeds and grasses. L-phosphinothricin inhibitor of glutamine synthetase (GS)
Saccharopolyspora spinosa (Actinomycete bacteria)	Used for the control of a very wide range of caterpillars, leaf miners, thrips and foliage feeding beetles
Saccharopolyspora spinosa	Insects
Serratia entomophila	Grass grub
Spodoptera exigua nucleopolyhedrovirus (SeNPVs)	*Spodoptera exigua*
Spodoptera littoralis nucleopolyhedrovirus (SpliNPV)	African cotton leaf worm (*Spodoptera littoralis*)
Spodoptera litura NPV	*S. litura*
Spodoptera litura nucleopolyhedrovirus	*Spodoptera litura*
Steinernema carpocapsae	Borer beetles, caterpillars, cranefly, moth larvae, *Rhynchophorus ferrugineus,* Tipulidae.
Steinernema feltiae	Bradysia spp., *Chromatomyia syngenesiae, Phytomyza vitalbae,* soil dwelling pests, codling moth larvae, sciarids, thrips
Steinernema kraussei	Vine weevil larvae
Streptomyces avermitilis (soil bacterium)	Ornamental plants, citrus, cotton, pears and vegetable crops to control phytophagous mites

Continued

TABLE 12.1 Microbial biopesticides for different target pathogens.—cont'd

Microbial biopesticides	Target pest/mechanism
Streptomyces griseoviridis	Effective against wilt, seed rot and stem rot
Streptomyces hygroscopicus	Morning glories, hemp (*Sesbaniabispinosa*), Pennsylvania smartweed (*Polygonum pensylvanicum*) and yellow nutsedge. Natural GS inhibitors, used for weed control in Japan.
Streptomyces lydicus	Effective against soil borne diseases of turf, nursery crops
Streptomyces sp. culture broths	Useful in glutamate and GABA gated chloride-channel opening. From fruit, vegetables and ornamentals plants
T. harzianum + *T. viride*	Effective against Armillaria and botryoshaeria and others
Talaromyces flavus V117b	Effective against fungal pathogens of tomato, cucumber, strawberry and rape oilseeds
Trichoderma viride/harzanum	Control root rot and wilt diseases especially on pulses and Paddy (Foot rot)
Trichoderma harzianum	Effective against variety of soil pathogens and wound pathogens
Trichoderma sp.	Suppresses root pathogens
Trichoderma sp. (egg parasite)	Used for control of sugarcane, early shoot borer, cotton bollworms, sorghum stem borer
Trichoderma viride	Effective against rot diseases
Trichoderma viride	Root rots and wilts
Trichogramma parasitoid	Sugarcane borers
Trichogramma spp.	Eggs of Lepidoptera
Verticillium lecanii	Mealy bugs and sucking insects

12.6.1 *Paenibacillus popilliae* (*Bacillus popillae*) and *B. lentimorbus*

They are soil-dwelling bacteria solely used for the devastation of *Popillia japonica* (Coleoptera: *Scarabaeidae*) commonly called as Japanese beetle by causing milky disease. These beetles cause grave harm to the vast range of economically significant crops by eating the plant tissues. These bacteria naturally dwell in soil and can be mass produced and used as biocontrol agent. The bacterial formulation is applied to the soil and the spore ingested by the larvae germinates in its gut mass-producing vegetative bacteria which penetrates the walls of the gut and enter into haemolymph causing death of larvae. The spores get into the soil by decomposition of dead larvae and are invulnerable to adverse climatic conditions. This bacterium is credited to be the first officially registered microbial biocontrol agent in United States.

12.6.2 *Lysinibacills sphaericus* (*Bacillus sphaericus*)

This bacterium is used to control and inhibit the growth of blackflies, grasshoppers and mosquitoes by synthesizing mosquitocidal toxins - Mtx, crystal proteins: Bin A (41.9 kDa), Bin B (51.4 kDa). These toxins exhibit resemblance with Bt cry toxins. The *L. sphaericus* isolated as a diazotrophic endophytic bacteria from rice plants, produced volatile organic compounds which efficiently exhibited 100% growth inhibition of *Rhizoctonia solani*, a causative agent of sheath blight in rice; thus, proved superior to chemical fungicide treatment (Shabanamol et al., 2017). The culture filtrate of *L. sphaericus* (ZA9) displayed significant antagonistic activity against *Bipolaris spicifera*, *Scerotinia* sp., *Trichophyton* sp., *Aspergillus* sp., *Curvularia lunata* and *Alternaria alternata*. The active compounds exhibiting antifungal activity was identified to be 2-pentyl-4-quinoline carboxylic acid, a quinoline alkaloid which was analogous in activity with the Benlate fungicide (Naureen et al., 2017).

12.6.3 Bacillus subtilis

This bacterial is globally used as a successful biocontrol agent. The different strains of *B. subtilis* are reported to exhibit broad spectrum action to counter Phyto nematodes. The various hydrolytic enzymes like cellulase, β-gluconase, lipase and protease produced by the bacteria restricts the replication cycle and growth of nematode larvae infesting the plants. *Bacillus subtilis* QST 713 played a significant role in controlling *Phakopsora pachyrhizi* (causing Soybean Rust) by reducing fungal sporulation; this action was compared to be similar to acibenzolar S methyl, a chemical pesticide (Twizeyimana and Hartman, 2019). The *B. subtilis* TM4 biopesticide formulation played a significant role in controlling maydis leaf blight (MLB) disease caused by *Bipolaris maydis* in maize by 21% (Djaenuddin and Muis, 2020). The *B. subtilits* based biopesticide (Biocure-X) was found to be effective against Wilt disease caused by *Fusarium oxysporum* f. sp. *ciceri*; root-knot disease caused by *Meloidogyne incognita* and wilt disease complex (*F. oxysporum* f. sp. *cicero*and *M. incognita*) in chickpea (*Cicer arietinum* L.) (Mohiddin and Khan, 2019). *Bacillus subtilis* based preparation suppressed greater than 40% growth of *Cryphonectria parasitica*, a causative agent of chestnut blight (Murolo et al., 2019). The various bioactive compounds produced by *B. subtilis* are Bacteriocin, Protein E2, Iturin A, surfactin, Gageotetrins A − C, lipopeptides, mycolytic enzymes −chitinase, protease, cellulase, glucanase. These bioactive compounds efficiently exhibit mosquitocidal, nematocidal and antimicrobial activities.

12.6.4 Bacillus firmus

It is a soil bacterium rarely present in the environment. It is proved to have significant nematocidal activity and hence is of interest to the scientist in recent times. *Bacillus firmus* has demonstrated fatal effect against nematodes like *Rodopholus similis*, *Xiphinema index*, *Heterodera* sp., *Ditylenchus* sp., *Tylenchulus semipenetrans* and *Meloidogyne* sp. The toxins produced by *B. firmus* causes damage to the peripheral egg pellicle of gall forming nematodes thus preventing the hatching of nematode eggs.

Bacillus firmus based product when used as a biopesticide played a significant role in the management of *Meloidogyne incognita* (southern root-knot nematode) and *Pseudopyrenochaeta lycopersici* causing significant damage to tomato crops (d'Errico et al., 2019).

12.6.5 *Bacillus thuringiensis* (Bt)

It is an aerobic spore-forming bacterium producing Bt toxin or Cry toxin (65−145 kDa, based on the strain)encoded by cry genes. It is produced by various strains of *B. thuringiensis* during sporulation, exists in the parasporal crystal and has proven to possess potent insecticidal activity. This bacterium was first reported by Ishiwata of Japan in 1901 as the causative agent of sotto disease, infecting silk worm (*Bombyx mori*) larvae but, named as *Bacillus thuringiensis* by Berliner, German Biology in 1911. The *Bt* produced crystalline proteins are target specific infecting only invertebrates but not vertebrates and thus proved to be an ecofriendly, popular biocontrol agent acquiring 95% market of the total biocontrol products. The Bt crystalline protein (δ - endotoxin) when ingested by the larvae binds to the insect gut receptors and gets active due to its slightly alkaline pH and protease action, causes damage to the gut tissues resulting in leakage followed by paralysis thus the infected larvae will stop over feeding and finally perishes as a result of both starvation and mid-gut epithelium injury. The activation of midgut proteases varies with different insects due to the variation in toxin structure, affecting its binding to the gut receptors thus responsible for their host specificity.

The various stains of *B. thuringiensis* produce different crystal toxins (Cry) which constitutes around 30% of the bacterium dry weight and causes death of variety of insects belonging to *Lepidoptera*, *Coleoptera* and *Diptera*. These toxins were grouped into four major classes (CryI, CryII, CryII and CryIV) based on activity and further classified into subclasses (A, B, C, D etc.) and subgroups (a, b, c, d etc.) The Cry toxins are classified into approximately 40 different families named from Cry1 to Cry40 based on its sequence similarity and molecular size which ranges approximately between 65 kDa and 145 kDa. The *Bt* var. *berliner* is reported to produce Cry1Aa toxin with molecular size 130−140 kDa targeting *Lepidoptera* species; *Bt* var. *kurstaki* KTD, HD1 is reported to produce Cry1Ab toxin with molecular size 130−140 kDa targeting *Lepidoptera* species; *Bt* var. *entomocidus* 6.01 is reported to produce Cry1Ba toxin with molecular size 130−140 kDa targeting *Lepidoptera* species; *Bt* var. *aizawai* 7.29is reported to produce Cry1Ca toxin with molecular size 130−140 kDa targeting *Lepidoptera* species; *Bt* var. *aizawai* IC 1is reported to produce CryII (Cry1Da) toxin with molecular size 135 kDa targeting *Lepidoptera* and *Diptera* species; *Bt* var. *kurstaki* HD-1is reported to produce CryII (Cry2Ab) toxin with molecular size 71 kDa targeting *Lepidoptera* and *Diptera* species; *Bt* var. *tenebrionis* (sd) is reported to produce CryIII (Cry3Aa) toxin with molecular size 66–73 kDa targeting *Coleoptera* species; *Bt* var. *morrisoni* PG14is reported to produce CryIV (Cry4Aa) toxin with molecular size 125–145 kDa targeting *Diptera* species; *Bt* var. *israelensis*is reported to produce CryIV (Cry4Ba) toxin with molecular size 68 kDa targeting *Diptera* species.

More than 150 subspecies of *B. thuringiensis* are reported till date and is of great economic significance due to its effective application as biocontrol agent in agricultural sector. The first *B. thuringiensis* based commercial biopesticide "Sporeine" was used in France (1938) but, the effective utilization started with "Thuricide" as the first registered *B. thuringiensis* biopesticide in 1961. The different strains of *B. thuringiensis* are available in various trade names are targeted against variety of insect pathogens, few examples to mention are *Bt* var. *aizurai* is sold under trade name Florback and Centari, targeted against Diamondback moth; *Bt* var. *galleriae* is sold under trade name Certan, target against wax moth larvae in honey combs; *Bt* var. *israelensis* is sold under trade name Thurimos, Vectobac, Bactis, Bactimos, BMP, Gnatrol, Aquabac target against mosquito larvae, Sciarids, fungus gnat larvae and black flies; *Bt* var. *kurstaki* is sold under trade name Biobit, Halt, Javelin, Dipel, Delfin, Foray, Batik, Wilbur-Ellis BT 320, Javelin WG, Green light, Hi-yield target against lepidopterous larvae, butterflies, moths, skippers, gypsy moths, cabbage loopers, tobacco horn worms, cabbage worms, spruce bud worms, variety of caterpillars; *Bt* var. *sandiego* is sold under trade name M– one plus, Diterra, target against weevils and beetles; *Bt* var. *thuringiensis* is sold under trade name Thuricide, Muscubac, target against flies, lepidopterous larvae. *B. thuringiensis* var. *tenebrionis*sold under trade name Novodor is successfully used against Colorado potato beetle, boll weevil, elm leaf beetle larvae, *Agelastica alni* - Alder leaf beetle (Coleoptera: Chrysomelidae); *Bt* var. *aizawai* is sold under trade name Florbac, M-Trak, Agree WG, Turex, targeted against Plutella, diamondback moth caterpillar, wax moth larvae.

12.6.6 Antimicrobial activity of *B. thuringiensis* based biopesticides

The *Bt* derived biopesticides also act as both antifungal and antibacterial agents thus controlling the microbial phytopathogens. The growth of *Guignardia citricarpa*, a fungal pathogen infesting citrus fruit was successfully inhibited by *Bt* isolates. The *Erwinia carotovora* and potato soft rot pathogen was inhibited by Bt due to production of acyl-homoserine lactone-lactonase. The Bt δ-endotoxin exhibited antifungal activity against *Phytophthora* and *Fusarium*, potent phytopathogens; *Fusarium oxysporum* f. sp. *lycopersici*, causing late blight and wilt of tomato was inhibited with Bt treatment. The exochitinase enzyme produced by different strains of Bt acted on the fungal chitin layer thus inhibiting phytopathogenic fungi like *Sclerotium rolfsii*, *Aspergillus* sp., *Nigrospora* sp., *Fusarium* sp., *Helminthosporium* sp., *Curvularia* sp., *Absidia* sp., *Penicillium* sp., *Rhizoctonia* sp. etc. The bioactive compounds produced by different *Bt* strains such as aminopolyol antibiotic, zqittermicin A, Bacthuricin F103, Thuricin Bn1, entomocin 110 inhibited bacterial pathogenic like *Agrobacterium tumefaciens*, *Pseudomonas syringae*, *Pseudomonas savastanoi*. *Paucimonas lemoignei* and *Paenibacillus* sp. The various lipoproteins produced by different strains of *Bt* like fengycin inhibited growth of phytopathogenic fungi like *Colletotrichum gloeosporioides*, *Stachybotrys charatum* etc. (Table 12.1).

12.6.7 *Bacillus thuringiens* is used as nano pesticides

Nanoparticles being very small in size exhibit various exclusive properties like ion adsorption, cation exchange capacity, and enhanced surface area. The technology of nanoparticles can be used with tiny constituents of active ingredients in pesticides or other tiny engineered particles exhibiting pesticidal activities to make nano pesticides. Nano pesticides are more efficient due to its specific surface area which exhibits greater target specificity; it also increases the dampness and diffusion in agricultural fields.

Bacillus thuringiensis can be formulated into nanoform by utilizing techniques such as ball milling jet milling, and high-pressure blending which converts coarse-grained powder to extremely small (ultrafine) powder ranging from less than 1 μm up to 5 μm. The ultrafine powder of *Bt* var. *aizawai* NT0423 (1.9 μm) have proved to enhance the mortality of diamondback moth larvae because, nanoparticles increases solubility of crystal toxin in midgut of larvae having alkaline pH resulting in release of larger quantity of toxin finally leading to rapid gut paralysis.

The active ingredient of *Bacillus thuringiensis* like crystal toxins can be combined with nanoparticles of organic and inorganic nanoparticles, metal oxides which have many advantages like enhanced shelf life, more efficient and reduced dosage. The *Bt* combined with nanoparticles in various formulations like nanosuspension, nanocapsule, nanogel, nanoemulsion increases the efficacy when used in agricultural fields by increased stability and action of the residues. The combination of *Bt* with nanosheets of graphene oxide and olive oil protects *Bt* from UV radiation. The *Bt* combined with zinc oxide nanoparticles were efficient in combating pulse beetle by causing 100% mortality (Malaikozhundan et al., 2017). The silica nanoparticles with very tiny *Bt* chitinase turned out to be highly efficient with broader pH tolerance, UV resistant, high thermostability, enhanced nematocidal activity (100%) with lesser dose and time against *Caenorhabditis elegans* (Qin et al., 2016). The *Bt* crystal protein laden with nano-Mg $(OH)_2$ exhibit structural stability with enhanced UV resistance and insecticidal action.

12.7 Members of *Pseudomonadaceae* and *Enterobacteriaceae* as biopesticides (non-spore formers)

12.7.1 Pseudomonadaceae

The members of *Pseudomonas* are Gram-negative, aerobic, motile rods, wide spread in environment. It commonly resides in the rhizosphere regions of plants and plays a vital role in combating soil borne plant pathogens. The various bioactive compounds produced by different species of *Pseudomonas* are hydrogen cyanide, dialkyl resorcinols, siderophores, antibiotics-bacteriocins, pyocyanin, phloroglucinol, rhamnolipid, masetolid, sclerosin, amphisin lipopeptide, pyoluteorin, pyrrolnitrin, phenazines; degradative enzymes - cellulase, chitinase, glucanase, protease. The metabolites produced by *Pseudomonas* sp. are contributing these bacteria as a diverse biopesticide to combat phytopathogenic fungi, larvae and nematodes.

The *Pseudomonas* sp. based biopesticides is commercially available under various trade names for Example: The phenazine-1-carboxylic acid derived from *Pseudomonas* sp., sold under the trade name "Shenginmycin" is used to inhibit fungal phytopathogens. The *Pseudomonas chlororaphis* strains, TX-1 and 63−28 are registered as biopesticides with EPA for combating wilts, root rots and turf diseases. The viable *P. chlororaphis* strains are marketed under several trade names -Cedomon, Cerall, Cedress, AteZe, VioJect, Nematokill, ItaEpi, Helper Plus, Bastapa to combat various pathogens like Pythium, Dollar spot, pink snow mold in turf, seed borne pathogens in barley, oat, wheat, rye, triticale, corn worm larvae, Ascochytain pea, *Acrothecium carotae* on carrots, *Gaeumannomyces graminis*, root-knot nematodes and aphids. *Pseudomonas syringae* strains ESC-10 and ESC-11 are registered by U.S. Environmental Protection Agency (EPA) for suppression of post-harvest fungal disease of pome, stone fruits, potatoes and citrus and are sold by Jet Harvest solutions, Longwood, FL under the trade names Bio-Save 11 LP, Bio-Save 10 LP. *Pseudomonas fluorescens* strain A506 is marketed under the trade name BlightBan A506, by NuFarm Americas, Burr Ridge, IL and used to combat bacterial fire blight disease, fruit russet and frost injury. *Pseudomonas fluorescens* strain CL154A and CL145A is a biopesticide, efficiently used to combat quagga, dreissenid and zebra mussels (Bivalvia: Dreissenidae).

12.8 Enterobacteriaceae

Xenorhabus and *Photorhabdus* are motile, Gram-negative bacteria, belonging to *Enterobacteriaceae* family. Some strains of *Xenorhabus* and *Photorhabdus* are reported to be extremely virulent against insect pests like *Gelleria mellonella* and *Mandula sexta*. *Xenorhabus* sp. secretes toxins and hydrolytic exoenzymes which help to act as antifungal, antibacterial and insecticidal agent. *Xenorhabus* is reported to exist as a symbiont with *Steinernema* and *Photorhabdus* with *Heterorhabditis*; no evidence of this bacteria in the natural environment including soil. The *X. indica*, *X. nematophila*, *X. bovienii* secretes toxins which are fatal to pest insects. The toxins reported to be produced by *X. nematophila* are xenocin, 42-kDa protein, Txp 40 (reported in 59 strains of *Xenorhabdus* and *Photorhabdus* bacteria), 17-kDa pilinactive against Leptidopteran pests. The toxins produced by *Photorhabdus* sp. are*Photorhabdus* insect-related (Pir) proteins, *Photorhabdus* virulence cassettes (PVCs), makes caterpillar floppy (Mcf) toxins, toxin complexes (Tcs - TcA, TcB, TcC and TcD), PaTox and photox toxins, transcinnamic acid (TCA) targeted against olive moth, Mediterranean flour moth, mushroom mite, beet armyworm, oriental leaf worm moth, onion thrips, western flower thrips, sugarcane stalk borer, desert locust and red flour beetle (Javed et al., 2017).

The bioformulations of *Xenorhabdus stockiae* PB09 exhibited miticidal activity and is used against *Luciaphorus perniciosus* (mushroom mite) (Namsena et al., 2016). The Txp40 toxin from *X. nematophila* is successfully used against *Plutella xylostella* larvae (Park et al., 2012). The xenocoumacin 1 antifungal produced by *Xenorhabdus nemotophilus* var. *pekingensis* is used to combat *Phytophthora infestans*. The Cabanillasin produced by *Xenorhabdus cabanillasii* JM26 was found to combat filamentous fungi and yeast causing opportunistic infections (Houard et al., 2013). The antimicrobial dipeptide compound nematophin produced from *X. nematophila* was used as a potent biopesticide against *Rhizoctonia solani* (Zhang et al., 2019).

12.8.1 Fungi as biopesticides

The entomopathogenic fungi are the group of pathogenic fungi which exist as obligate/facultative, symbionts or commensals of insects like mosquitoes, mites, thrips, aphids, bugs, scale insects etc. These fungi act as significant natural controller of insect inhabitants as they exhibit myco-insecticidal activity against vast range of insect pests causing massive destruction to agricultural crops. The entomopathogenic fungi causes infections to its host insects by piercing the cuticle, enter into haemolymph, grow by using nourishment in haemocoel, produce toxins leading to death thus, proving as a gifted bio-pesticide. The most commonly used entomopathogenic fungi as biopesticides are *Beauveria bassiana, Paecilomyces farinosus, Metarhizium anisopilae, Verticillium lecanii, Nomurea rileyi, Lecanicillium lecanii.*

The various metabolites are produced by entomopathogenic fungi which plays significant role in fighting against the pathogens thus proving it as a successful biopesticide in the market. Some of the important metabolites produced by selected fungi are mentioned here. The destruxins more than 27 types and cytochalasin metabolites are produced by *Metarhizium anisopliae*; bassianin, bassianolide, beauvericin and tenellin metabolites are produced by *Beauveria bassiana*; Oosporein metabolite is secreted by *Beauveria brogniartii*; beauvericin, pyridine-2, 6- dicarboxylic acid, beauverolies are reported from *Paecilomyces fumosoroseus*; bassianolide, vertilecannins, hydroxycarboxylic acid, dipcolonic acid are produced in *Verticillium lecanii*; efrapeptins (5 types) and cyclosporin are reported in *Tolypocladium* spp.; hirsutellin A and B, phomalatone is produced in *Hirustella thompsonii.*

The entomopathogenic fungi based biopesticides is commercially available under various trade names. The *Verticillium lecanii* fungus is commercially available under the trade names Mycotol and Vertalec to target whitefly, aphids and thrips. *Metazhizium anisopliae* is used against termites with the trade name Meta guard; against locusts with trade name Biogreen; against cockroaches with trade name Bio-path; against termites with trade name Bio-Blast; against sugarcane spittle bug with trade name Cobicant; against Cane grubs with trade name Bio-Cane; against locusts, grasshoppers with trade name Green muscle. The *Beauveria bassiana* fungus is used against coffee berry borer with trade name Conidia; against European corn borer with trade name Corn guard; against cotton pests with trade name Naturalis-L; against Locusts and grass hoppers under trade name Mycotrol GH; against white fly, aphids and thrips under trade names Mycotrol WH and Botanigard. The *Paecilomyces fumosoroseus* fungi are used to combat white fly under the trade name PFR-97 ang Pae-sin.

12.8.2 *Trichoderma* spp. as biopesticide

These fungi has dual activity as it possesses both biocontrol and plant growth promoting properties and hence used in many crop varieties as it grows luxuriantly in different soil types and persist for long period of time (Narasimha Murthy et al., 2013b, 2018). The various metabolites produced by *Trichoderma* spp. are hydrolytic enzymes like - cellulase, glucanase, glucosidase, xylanase, chitinase, peptidase, lipase, glucose oxidase; secondary metabolites like − pyrones, lactones, koninginis, trichodermamides, viridins, harzianopyridone, harzianic acid, diketopiperazines − gliotoxin and gliovirin, azaphilones, cerinoactone, trichosordarin A, harzianol A, hydrophobins, sesquiterpenoids, trichocitrin, trochosardarin A, harzianone, harzianic acid and harzianopyridone. These metabolites exhibit antimicrobial activities thus making *Trichoderma* spp. an efficient biocontrol agent (Narasimha Murthy et al., 2013a; Sood et al., 2020, Table 12.1).

The different species of *Trichoderma* are commercially available under different trade names; various strains of *Trichoderma asperellum* is available under the trade names -T34 Biocontrol, Remedier WP, VIRISAN, TUSAL; *T. atroviride* is available with trade names − Trichopel, Trichodry, Trichospray, Vinevax, Biodowel, Sentinel, Tenet, Esquive WP (Agrauxine), BINAB TF WP, Binab T vector, Ecohope, Trichodermax EC, Trichotech; *T. harzianum* strains are available as commercial products under the labels − T-Gro, Binab, Rootshield WP, T-22 technical, T-22 wp biological fungicide, Bio-trek hb, Antagon WP, Fitotripen SP, Bio Traz, BioFit, Trichobiol WP, Agroguard WG, ROOTgard, SF Bio-Tricho, Supresivit, 'Ecosom-TH, TRIANUM-P, TRIANUM-G and Trichodex; *T. polysporum* with trade names BINAB TF WP, Binab T vector; *T. virens* is available with trade names G-41, BW240 WP biological fungicide; *T. viride* is commercially available as Biocure, Bio-shield, Bioveer, Mycofungicyd, Trichodermin (Sheridan et al., 2014).

Trichoderma strains are the only conceivable solution to counter some of the plant disease like wood disease of grapes (as no chemicals are effective), gray mold (*Botrytis cinerea*). The endophytic *Trichoderma citrinoviridae* PG87 isolated from mountain-cultivated ginseng produced bioactive metabolite ginsenoside which significantly inhibited the disease caused by *Botrytis cinerea* and *Cylindrocarpon destructans* (Park et al., 2019). *Trichoderma asperellum* BCC1 is proved to be an efficient biocontrol agent against *Sclerotium cepivorum* (necrotrophic ascomycete), a causative agent of onion white rots (Rivera-Mendez et al., 2020); *T. asperellum* is used as a potential biocontrol agent against *Rigidoporus microporus* (*Hevea brasiliensis*), causing devastating disease in rubber plants (Go et al., 2019). The endophytic

Trichoderma atroviridae BC0584 is used against *Fusarium avenaceum* and *Fusarium culmorum* causing *Fusarium* damping-off on maize (Coninck et al., 2020). *Trichoderma* spp. is used as a biocontrol agent against Sclerotinia stem rot or white mold infecting soybeans (Juliati et al., 2019).

12.9 *Coniothyrium minitans* as biopesticide

Coniothyrium minitans is a coelomycete distributed worldwide, occurs in sclerotia of *Sclerotinia trifoliorum* and *S. sclerotiorum*. This fungus is an extremely specific antifungal representative targeting sclerotia of *Deuteromycotina* and *Ascomycotina* (ex: *S. trifoliorum* and *S. sclerotiorum*). It penetrates into the target pest by making small pores or lacerate the surface of target pest by producing enzymes like chitinase and glucanase then, breaches into the sub-cortex and medulla producing fruiting bodies and causes the pest cells to shrink due to osmosis. This fungus is used to protect economically viable crops like oilseeds, beans, peas, lettuce etc.

12.10 *Gliocladium catenulatum* as biopesticide

This is a saprophytic fungus which is widely present in the atmosphere. It is reported to produce antibiotics, toxins and enzymes like pectinase and glycosidases. It is available in the trade name JI446 Prestop™ Verdera, ArgaQuest, effectively used against soil borne pathogens causing gray molds, seed, root and stem rots, wilt diseases by *Pythium, Fusarium, Rhizoctonia, Botrytis, Cladosporium, Penicillium, Alternaria* and *Plicaria* species. It is used to protect vegetables, fruits, ornamental plants, herbs and spices against the pathogens causing devastation. *Gliocladium catenulatum* is reported to suppress and control the pathogen *Plasmodiophora brassica* (>80% control) infecting Chinese cabbage (Peng et al., 2011).

12.10.1 *Purpureocillium lilacinum* as biopesticide

This is a common fungus found in soil. The *P. lilacinum* 251 is used as an active ingredient under the trade name BioAct WG, Bio-Nematon, Nematofree, Bionemat, Bioniconema, Ecogreen, Krishi Bio Vikalp, Paceilo, Mysis, Nemato Gurard, Nematolin-LF, Nemator, is used effectively against *Meloidogyne* sp. (root knot nematodes); *Heterodera* spp. and *Globodera* spp. (cyst nematode); *Pratylenchus* sp. (root lesion nematodes); *Radopholus similis* (burrowing nematodes). This fungus acts by contaminating eggs, juveniles and adult females of various pathogenic nematodes infecting food crops. The *P. licacinum* CKPL-053 inhibited *Thrips palmi* (Thysanoptera: Thripidae) in orchid farms, Thailand by producing paecilotoxins A, B and C (Hotaka et al., 2015); *P. lilacinum* reduced root-knot nematode galls causing infection in tomato plants (Kepenekci et al., 2018); *P. lilacinum* exhibited the entomopathogenic potential against *Galleria mellonella* (Lepidoptera: Pyralidae), a model insect thus proving as a potential biopesticide (Duman and Altuntas, 2018).

12.10.2 *Beauveria bassiana* as biopesticide

The various strains of *B. bassiana* used as biopesticides are BB-IARI-RJP, HaBa, AAI, IIHR available under different trade names Beauva Grip, Beauveria, Beauvo Royal, Bio Be Ba, Biojaal, Biolarvex, Biorin, Bio-wonder, Green Beauveria, Metabeave Beauveria, Sun Agro Beviguard, Beaulife, ABTEC Beauveria and Basicon. It is effectively used against beetles, thrips, white grubs, aphids, coffee berry borer, white flies, grasshoppers and many Lepidopteran pests. This fungus causes white muscardine disease in insect pests, it grows as a white mold, produce powdery, dry conidia in characteristic white spore balls which is made up of cluster of conidiogenous cells.

The *B. bassiana* increased the effectiveness of *Nicotiana tabacum* extract as a biopesticide against coffee berry borer (*Hypothenemus hampei*) (Haryuni et al., 2019); *B. bassiana* - Myco jaal and Bba5653 is used against *Plutella xylostella* and *Pieris brassicae* infecting crucifers; *B. bassiana* CPD9 and *Metarhizium anisopliae* CPD 5 and 12 is targeted against pod sucking bug and pod borer infecting cowpea (Srinivasan et al., 2019).

12.10.3 *Lecanicillium* (*Verticillium*) *lecanii* as biopesticide

The different strains of *L. lecanii* (AAI, As-MEGHVL, ICAR, NCIM 1312) is commercially available with the trade names - WP ABTEC Verticillium, Biosar, Biovert Rich, AvishiVerti, Bio-catcg, Kavach *Verticillium, Krishi* Bio Yug, Mealikil VL, Phule Bugicide, Sun Agro Vetri, Mealikil Plus, Varunastra, *Verticillium lecanii*, Verticim, Vertisoft, Verti-star and Vertisweep. It is targeted against scale insects, thrips, whiteflies, mites, mealybugs, Aphids and other Homopterans. *Lecanicillium lecanii* (Zimm.) is used as a biocontrol agent against green peach aphid (Hemiptera: Aphididae) causing infection in many economically significant agricultural crops (Hanan et al., 2020). The crude toxins - toxinv3450 and toxinVp28 from *Lecanicillium* (*Verticillium*) *lecanii* Gams and Zare strains exhibited toxicity and antifeedant property on sweet potato whitefly (*Bemisia tabaci*) (Homoptera: Aleyrodidae).

12.10.4 Endophytic fungi as biocontrol agents

Endophytic fungi grow with in the plant tissues without causing any infections or disease rather; it lives a symbiotic life with its host plant. The endophytic fungi associated with different plants have proved as a biocontrol agent by production of secondary metabolites which successfully helps in combating pathogens. The endophytic fungi *Curvularia lunata* (MF113056) isolated from *Melia azedarach*, a medicinal plant; *Aspergillus solani* (MG7865453) isolated from *Pelargonium graveolens*, an aromatic plant and *Alternaria alternata* (MG786545) isolated from *Chenopodium album* was successful in inhibiting cotton leaf worm larvae (*Spodoptera littoralis*), Lepidoptera: Noctuidae (Saad et al., 2019). The methanolic extract of *Aspergillus flavus* isolated from *Tectona grandis* caused mortality of *Hyblea purea*, *Eligma narcissus* and *Atteva fabriciella*. The endophytic fungi Aspergillus *niger*, *A. terreus* and *Penicillium chrysogenum* isolated from various parts of *Alhagi maurorum* plant successfully inhibited *Fusarium oxysporoum* f. sp. *seseame*, the causative agent of wilt disease in sesame (Hegazy et al., 2019). The endophytic fungus *Epicoccum nigrum* (ASU11) isolated from healthy potato plant was used as a biocontrol agent to combat *Pectobacterium carotovora* subsp. *atrosepticum* PHY7, causative agent of blackleg disease in potato (Bagy et al., 2019). The dual strains of *Trichoderma harzianum* and *T. lentiforme*, endophytic fungi from healthy watermelon plants was effectively used as biocontrol agents against the significant fungal pathogens causing wilt (*Fusarium oxysporum* f. sp. *neveum*), carbonaceous rot (*Macrophomina phaseolina*) and collapse (*Monosporascus cannonballus*) affecting watermelon and melon crops (Ganozalez et al., 2020).

12.11 Yeast as biocontrol agents

The *Saccharomyces* and *Zygosaccharomyces* yeast used in wine preparation are reported to inhibit soil borne pathogenic fungi like *Sclerotinia sclerotiorum*, *Phomopsis longicolla*, *Macrophomina phaseolina*, *Rhizoctonia fragariae* and *Botrytis squamosa*. *Candia* (*Pichia*) *guilliermondii* is targeted against the phytopathogenic fungi *Colletotrichum capsici*, *Rhizopus nigricans*, *Rhizopus stolonifera*, *Botrytis cinereal*, *Colletotrichum gloeosporioides* causing infections in chili, tomato, kiwifruit and papaya. The *Torulaspora globosa* yeast exhibited antagonistic effect against *Colletotrichum graminicola* causing anthracnose disease in maize (Rosa-Magri et al., 2011). *Saccharomyces cerevisiae* and *Pichia anomala* (*Wickerhamomyces anomalus*) inhibited Basidiomycetes pathogens − *Lentinus lepideus*, *Serpula lacrymans*, *Postia placenta* and *Ophiostoma ulmi* and other phytopathogens − *Fusarium equiseti*, *Botrytis fabae*, *Phytophthora infestans* and *Rhizoctonia solani* (Ferraz et al., 2019). The *Aureobasidium pullulans* is commercially available under the trade name Boni Protect, Bio-Ferm, AT to combat pathogenic *Penicillium*, *Monilinia* and *Botrytis* causing infection in Pome; *Candida oleophila* is available under the trade names Nexy, Lesaffre, BE to fight against *Penicillium*, *Botrytis* infecting Pome; *Metschnikowia fructicola* is available with trade name Shemer, Bayer/Kopppert, NL against *Penicillium*, *Botrytis*, *Rhizopus* and *Aspergillus* causing infection to pome, stone fruits, table grape, strawberry and sweet potato (Wisniewski et al., 2016).

12.11.1 Insect viruses as biopesticides

Insect viruses are the living entities which causes infection to a particular host insect by exclusively multiplying with in the living hosts. The different insect viruses belonging to the families *Entomopoxvirinae*, *Densovirinae*, *Reoviridae*, *Polydnaviridae*, *Ascoviridae*and *Baculoviridae* causes infection in insect population. The baculoviruses (granulovirus and nucleopolyhedrovirus) are extensively chosen and used viruses as biocontrol agents followed by cypoviruses, nudiviruses and densoviruses which are also reported as biocontrol agents. The codling moth granulovirus, CpGVs (*Cydia pomonella* Granulovirus) is used as a biocontrol agent against apple pests and codling moth caterpillar. The Baculoviruses are rod shaped, double stranded DNA virus which causes infection through the mouth of insects, rapidly replicate and causes destruction of host cells and tissues; extremely host specific mostly causing infection in insects of order *Lepidoptera* also, can infect insects belonging to *Hymenoptera*, *Coleoptera*, *Trichoptera* and *Diptera*. The phenotypes generated during replication of Baculoviruses are occlusion derived virus (ODVs) which spread infection among larvae and budded virus (BVs) which Disseminate infection within the host (Table 12.1).

The Nucleo polyhedro viruses (NPVs) are used in India to control army worms belonging to *Spodoptera litura* and *S. exigua*; boll worms belonging to *Helicoverpa armigera* and *H. zea* causing major damage to commercially significant crops like cocoa, groundnut, legumes, sorghum, potato, rice, maize, rubber, castor, citrus, tomato and vegetables. The

HzNPV is targeted against *Helicoverpa zea*; SLNPV is targeted against *Spodoptera litura*; HaNPV is targeted against *Helicoverpa armigera*; AgNPV is targeted against *Anticarsia gemmatalis*; SfNPV is targeted against *Spodoptera frugiperda*. The NPV isolated from *Spodoptera littoris* (SpliNPV) is used against cotton leaf worm (Lepidoptera: Noctuidae) causing infections in cotton, vegetable crops, ornamental and orchard trees (Elmenofy et al., 2020). Granulo viruses (GVs) are reported to be used against sugarcane shoot borer (*Chilo infuscatellus*), *Cnaphalocrosis medinalis*, *Pericallia ricini*, *Achaea janata*, *Phthorimaea operculella*; Cytoplasmic polyhedrosis virus is used against *Helicoverpa armigera*; Pox virus is used against *Amsacta moore*.

Some of these insect viruses are commercially produced and marketed under registered trade names. The *Cydia pomonella* (CpGV) virus is commercially produced with the trade names Cyd-X, Virosoft CP4, Carpovirusine, Madex, Granupom, Granusal, Virin-CyAP, used against codling moth causing infection in pear and apple. *Helicoverpa armigera* NPV is available under the trade name Gemstar LC against cotton bollworm, corn earworm, tobacco budworm infecting cotton and row crops. *Spodoptera exigua* NPV is marketed under the trade name Spod-X LC against beet armyworm infecting vegetables (Abd-Alla et al., 2020).

12.11.2 Protozoans as biopesticides

Approximately 1000 species of Entomopathogenic protozoans belonging to the Phylum: Sarcomastigophora, Microspora, Apicomplexa and Ciliophora are reported to be pathogenic to insects and are known as microsporidians. The protozoans are highly specific to the host causing chronic infection leading to exhaustion of the host; used as potential pest control in aquatic ecosystem, stored products, soil, forest and grasslands. The infectious protozoan spores ingested by the host insect germinates in the mid gut releasing sporoplasm which invades the target cells causing disease to the host which results in reduced foodintake and fertility, increased mortality. *Nosema locustae* is the only microsporidian available in the market, used to combat grass hoppers and crickets; *N. pyrausta* causes death of European corn borer larvae, reduces the fertility and life span of adult insects. The *Paecilomyces fumosoroseus* causes greater infection to the *Plutella xylostella* starved larvae when compared to fed larvae (Table 12.1).

The various protozoans are reported to target different insects — *Nosema acridophagous*, *N. cuneatum* and *N. locustae* are targeted against grasshoppers; *N. algerae* is used against *Anopheles albimanus*, *Culex tritaeniorhynchus* mosquitoes; *N. fumifueranae* is used against Spruce budworm; *N. heliothidis* is used against *Helicoverpa zea*; *N. pyrausta* is used against *Ostrinia nubialis*; *N. whitei* is used against *Tribolium castaneum*; *Nosema* spp. is used against *Helicoverpa armigera*, *Spodoptera litura*; *Vairimorpha necatrix* is used against *Agrotis ipsilon*, *Helicoverpa zea*; *Gregarina garnhani* is targeted against *Schistocera gregaria*; *G. polymorpha* is used against *Tenebrio molitor*; *Ascogregarina culicis* and *A. geniculate* is used against *Aedes aegypti* mosquito; *Mattesia trogodermae* is used against *Trogoderma granarium*; *Farinocystis tribolii* is used against *Tribolium castaneum*.

12.12 Plant incorporated protectants: genetically modified (GM) crops

The genetically modified (GM) crops are the crops whose genetic material (DNA) is manipulated either through change of gene structure, gene doubling, deletion and insertion of selected genes by using techniques like virus induced gene silencing (VIGS), RNA interference (RNAi) and genome editing to alter the phenotypic traits of crops. Depending on the gene manipulation, there are different types of transgenic plants: cisgenic -the gene is transferred between members of same species (Ex: gene from rice origin is inserted into a rice variety), intragenic - gene is transferred from different members of closely related species (Ex: gene transferred from wheat to barley), transgenic - gene is inserted from different species (Ex: gene from a bacteria is inserted into plants -*Bacillus thuringiensis* toxin gene into cotton plants). The transgenic plants help us to overcome the present problems faced by the developing nations like malnourishment, biotic stress (pests and diseases), abiotic stress (drought and salinity), and post-harvest challenges (shelf life and pests).

The various transgenic crops developed exhibit several advantages: Bt cotton - the crop losses will be minimized resulting in economic gains; Bt maize - reduced mycotoxin (fumonisin, a carcinogenic toxin produced by *Fusarium* spp.) contamination in food; Rotavirus vaccine completed rice - reduces infant death due to rotavirus infections; Golden rice - increased nutritive value with enhanced Vitamin A; Virus resistant sweet potato — High tuber yield, more profit; Biofortified cassava — increased nutritive value with enhanced iron, zinc, vitamins and protein; Golden mustard — high in beta-carotene; herbicide tolerant canola — more profitable, transgenic potato — enhance tolerance to freezing, resistant to aphids (Kim et al., 2016); transgenic carrot — resistant to fungal pathogens. The transgenic crops like papaya, soybean, tomato, maize, grapes are with insect resistant and herbicide resistant genes grown in large areas throughout the world (Gosal and Wani, 2018).

The GM crops have many possible risks: specific gene products may be allergic and cause harm to beneficial insects, wild life and human health, insects may develop resistant to toxins expressed by genes integrated with GM crops, herbicide resistant crops may cross-pollinate with other weeds in its vicinity and may result in "superweeds" which turn out to be highly resistant.

12.13 Advantages of microbial biopesticides

The microbial biopesticides used are highly host specific and generally ecofriendly being nonpathogenic, non-toxic to other organisms in the environment. Microbial biopesticides can be used along with synthetic chemical pesticides and has proven to have more effective activity against the pests during integrated pest management. The residue of microbial biopesticides does not have ill effects on any other organisms including humans unlike chemical pesticides. The microbial biopesticides gets established among the pest habitat and control the pests during subsequent crops.

12.14 Disadvantages of microbial biopesticides

Microbial biopesticides are highly specific to target species and hence cannot be effectively used when a crop is infested by more than one pest. Microbial biopesticides are highly sensitive to external environmental conditions like temperature, pH desiccation, and exposure to ultra-violet light. The carrier materials and storage procedures are stringent for some microbial pesticides. The microbial biopesticides do not exhibit broad spectrum pesticidal activity hence, economically not viable. Microbial biopesticides exhibit delayed action against the pest thus may not be able to protect the crop at the right time.

References

Abd-Alla, A.M., Meki, I.K., Demirbas-Uzel, G., 2020. Insect viruses as biocontrol agents: challenges and opportunities. In: Cottage Industry of Biocontrol Agents and Their Applications. Springer, Cham, pp. 277–295.

Al-Sadi, A.M., 2017. Impact of plant diseases on human health. Int. J. Nutr. Pharmacol. Neurol. Dis. 7, 21–22.

Bagy, H.M.K., Hassan, E.A., Nafady, N.A., Dawood, M.F., 2019. Efficacy of arbuscular mycorrhizal fungi and endophytic strain *Epicoccum nigrum* ASU11 as biocontrol agents against blackleg disease of potato caused by bacterial strain *Pectobacterium carotovora* subsp. *atrosepticum* PHY7. Biol. Contr. 134, 103–113.

Carvalho, F.P., 2017. Pesticides, environment, and food safety. Food Energy Secur. 6, 48–60.

Coninck, E., Scauflaire, J., Gollier, M., Liénard, C., Foucart, G., Manssens, G., Munaut, F., Legrève, A., 2020. *Trichoderma atroviride* as a promising biocontrol agent in seed coating for reducing *Fusarium* damping-off on maize. J. Appl. Microbiol. https://doi.org/10.1111/jam.14641.

d'Errico, G., Roberta, M., Aniello, C., Davino, S.W., Fanigliulo, A., Woo, S.L., Lorito, M., 2019. Integrated management strategies of *Meloidogyne incognita* and *Pseudopyrenochaeta lycopersici* on tomato using a *Bacillus* firmus-based product and two synthetic nematicides in two consecutive crop cycles in greenhouse. Crop Protect. 122, 159–164.

De Oliveira, J.L., Campos, E.V.R., Bakshi, M., Abhilash, P.C., Fraceto, L.F., 2014. Application of nanotechnology for the encapsulation of botanical insecticides for sustainable agriculture: prospects and promises. Biotechnol. Adv. 32, 1550–1561.

Djaenuddin, N., Muis, A., April 2020. Effectiveness of *Bacillus* subtilis TM4 biopesticide formulation as biocontrol agent against maydis leaf blight disease on corn. In: IOP Conference Series: Earth and Environmental Science, vol. 484, p. 012096 (1).

Duman, E., Altuntaş, H., 2018. Genotoxicity of azadirachtin on *Galleria mellonella* L. (Lepidoptera: Pyralidae). Biol. Divers. Conserv. 24–30.

Elmenofy, W., Salem, R., Osman, E., Yasser, N., Abdelmawgod, A., Saleh, M., Zaki, A., Hanafy, E., Tamim, S., Amin, S., El-Bakry, A., 2020. Evaluation of two viral isolates as a potential biocontrol agent against the Egyptian cotton leafworm, *Spodoptera littoralis* (Boisd.) (Lepidoptera: Noctuidae). Egyp. J. Biolog. Pest Contr. 30 (1), 1–8.

FAO, 2009. FAO's director-general on how to feed the world in 2050. Popul. Dev. Rev. 35, 837–839. https://doi.org/10.1111/j.1728-4457.2009.00312.x.

Ferraz, P., Cássio, F., Lucas, C., 2019. Potential of yeasts as biocontrol agents of the phytopathogen causing cacao Witches' Broom Disease: is microbial warfare a solution? Front. Microbiol. 10, 1766.

Go, W.Z., H'ng, P.S., Wong, M.Y., Chin, K.L., Ujang, S., Noran, A.S., 2019. Evaluation of *Trichoderma asperellum* as a potential biocontrol agent against *Rigidoporus microporus Hevea brasiliensis*. Arch. Phytopathol. Plant Protect. 52 (7–8), 639–666.

Gonzalez, V., Armijos, E., Garcés-Claver, A., 2020. Fungal endophytes as biocontrol agents against the main soil-borne diseases of melon and watermelon in Spain. Agronomy 10 (6), 820.

Gosal, S.S., Wani, S.H., 2018. Plant genetic transformation and transgenic crops: methods and applications. In: Biotechnologies of Crop Improvement, vol. 2. Springer, Cham, pp. 1–23.

Hanan, A., Nazir, T., Basit, A., Ahmad, S., Qiu, D., 2020. Potential of *Lecanicillium lecanii* (Zimm.) as a microbial control agent for green peach aphid, *Myzus persicae* (Sulzer)(Hemiptera: Aphididae). Pakistan J. Zool. 52 (1), 131.

Haryuni, H., Dewi, T.S.K., Dewi, E., Rahman, S.F., Gozan, M., 2019. Effect of *Beauveria bassiana* on the effectiveness of *Nicotiana tabacum* extract as biopesticide against *Hypothenemus hampei* to robusta coffee. Int. J. Technol. 10 (1), 159–166.

Hegazy, M.G.A., El Shazly, A.M., Mohamed, A.A., Hassan, M.H.A., 2019. Impact of certain endophytic fungi as biocontrol agents against sesame wilt disease. Arch. Agric. Sci. J. 2 (2), 55–68.

Hotaka, D., Amnuaykanjanasin, A., Maketon, C., Siritutsoontorn, S., Maketon, M., 2015. Efficacy of *Purpureocillium lilacinum* CKPL-053 in controlling *Thrips palmi* (Thysanoptera: Thripidae) in orchid farms in Thailand. Appl. Entomol. Zool. 50 (3), 317–329.

Houard, J., Aumelas, A., Noël, T., Pages, S., Givaudan, A., Fitton-Ouhabi, V., Villain-Guillot, P., Gualtieri, M., 2013. Cabanillasin, a new antifungal metabolite, produced by entomopathogenic *Xenorhabdus cabanillasii* JM26. J. Antibiot. 66 (10), 617–620.

Javed, N., Kamran, M., Abbas, H., 2017. Toxic secretions of *Xenorhabdus* and their efficacy against crop insect pests. In: Biocontrol Agents: Entomopathogenic and Slug Parasitic Nematodes, p. 223.

Juliatti, F.C., Rezende, A.A., Juliatti, B.C.M., Morais, T.P., 2019. *Trichoderma* as a biocontrol agent against *Sclerotinia* stem rot or white mold on soybeans in Brazil: usage and technology. In: *Trichoderma*-The Most Widely Used Fungicide. Intech Open. https://doi.org/10.5772/intechopen.84544.

Kepenekci, I., Hazir, S., Oksal, E., Lewis, E.E., 2018. Application methods of *Steinernema feltiae*, *Xenorhabdus bovienii* and *Purpureocillium lilacinum* to control root-knot nematodes in greenhouse tomato systems. Crop Protect. 108, 31–38.

Kim, S.W., Park, J.K., Lee, C.H., Hahn, B., Koo, J.C., 2016. Comparison of the antimicrobial properties of chitosan oligosaccharides (COS) and EDTA against *Fusarium fujikuroi* causing rice bakanae disease. Curr. Microbiol. 72 (4), 496–502.

Lade, B.D., Gogle, D.P., Nandeshwar, S.B., 2017. Nano-biopesticide to constraint plant destructive pests. J. Nano Res. 6, 1–9.

Malaikozhundan, B., Vaseeharan, B., Vijayakumar, S., Thangaraj, M.P., 2017. *Bacillus thuringiensis* coated zinc oxide nanoparticle and its biopesticidal effects on the pulse beetle, *Callosobruchus maculatus*. J. Photochem. Photobiol., B 174, 306–314.

Mazid, S., Kalida, J.C., Rajkhowa, R.C., 2011. A review on the use of biopesticides in insect pest management. Int. J. Sci. Adv. Technol. 1, 169–178.

Mohiddin, F.A., Khan, M.R., 2019. Efficacy of newly developed biopesticides for the management of wilt disease complex of chickpea (*Cicer arietinum* L.). Legume Res. Int. J. 42 (4).

Murolo, S., Concas, J., Romanazzi, G., 2019. Use of biocontrol agents as potential tools in the management of chestnut blight. Biol. Contr. 132, 102–109.

Namsena, P., Bussaman, P., Rattanasena, P., 2016. Bioformulation of *Xenorhabdus stockiae* PB09 for controlling mushroom mite, *Luciaphorus perniciosus* Rack. Bioresour. & Bioprocess. 3 (1), 19.

Narasimha Murthy, K., Nirmala Devi, D., Srinivas, C., 2013a. Efficacy of *Trichoderma asperellum* against *Ralstonia solanacearum* under greenhouse conditions. Ann. Plant Sci 2, 342–350.

Narasimha Murthy, K., Uzma, F., Srinivas, C., 2013b. Induction of systemic resistance by *Trichoderma asperellum* against bacterial wilt of tomato caused by *Ralstonia solanacearum*. Int. J. Adv. Res. 1, 181–194.

Narasimha Murthy, K., Soumya, K., Chandranayak, S., Niranjana, S.R., Srinivas, C., 2018. Evaluation of biological efficacy of *Trichoderma asperellum* against tomato bacterial wilt caused by *Ralstonia solanacearum*. Egyp. J. Biol. Pest Contr. 28 (63).

Naureen, Z., Rehman, N.U., Hussain, H., Hussain, J., Gilani, S.A., Al Housni, S.K., Mabood, F., Khan, A.L., Farooq, S., Abbas, G., Harrasi, A.A., 2017. Exploring the potentials of *Lysinibacillus sphaericus* ZA9 for plant growth promotion and biocontrol activities against phytopathogenic fungi. Front. Microbiol. 8, 1477.

Nicolopoulou-Stamati, P., Maipas, S., Kotampasi, C., Stamatis, P., Hens, L., 2016. Chemical pesticides and human health: the urgent need for a new concept in agriculture. Front. Public Health 4 (148), 1–8.

Park, J.M., Kim, M., Min, J., Lee, S.M., Shin, K.S., Oh, S.D., Oh, S.J., Kim, Y.H., 2012. Proteomic identification of a novel toxin protein (Txp40) from *Xenorhabdus nematophila* and its insecticidal activity against larvae of *Plutella xylostella*. J. Agric. Food Chem. 60 (16), 4053–4059.

Park, Y.H., Mishra, R.C., Yoon, S., Kim, H., Park, C., Seo, S.T., Bae, H., 2019. Endophytic *Trichoderma citrinoviride* isolated from mountain-cultivated ginseng (*Panax ginseng*) has great potential as a biocontrol agent against ginseng pathogens. J. Ginseng Res. 43 (3), 408–420.

Peng, D., Chai, L., Wang, F., Zhang, F., Ruan, L., Sun, M., 2011. Synergistic activity between *Bacillus thuringiensis* Cry6Aa and Cry55Aa toxins against *Meloidogyne incognita*. Microb. Biotechnol. 4, 794–798.

Qin, X., Xiang, X., Sun, X., Ni, H., Li, L., 2016. Preparation of nanoscale *Bacillus thuringiensis* chitinases using silica nanoparticles for nematicide delivery. Int. J. Biol. Macromol. 82, 13–21.

Randall, C., Hock, W., Crow, E., Hudak-Wise, C., Kasai, J., 2013. National Pesticide Applicator Certification Core Manual. National Association of State Departments of Agriculture Research Foundation, Washington, DC, pp. 1–17.

Rivera-Méndez, W., Obregón, M., Morán-Diez, M.E., Hermosa, R., Monte, E., 2020. *Trichoderma asperellum* biocontrol activity and induction of systemic defenses against *Sclerotium cepivorum* in onion plants under tropical climate conditions. Biol. Contr. 141, 104145.

Rosa-Magri, M.M., Tauk-Tornisielo, S.M., Ceccato-Antonini, S.R., 2011. Bioprospection of yeasts as biocontrol agents against phytopathogenic molds. Braz. Arch. Biol. Technol. 54, 1–5.

Saad, M.M., Ghareeb, R.Y., Saeed, A.A., 2019. The potential of endophytic fungi as bio-control agents against the cotton leafworm, *Spodoptera littoralis* (Boisd.)(Lepidoptera: Noctuidae). Egyp. J. Biol. Pest Contr. 29 (1), 1–7.

Shabanamol, S., Sreekumar, J., Jisha, M.S., 2017. Bioprospecting endophytic diazotrophic *Lysinibacillus sphaericus* as biocontrol agents of rice sheath blight disease. 3 Biotech 7 (5), 337.

Sheridan, L.W., Michelina, R., Francesco, V., Marco, N., Roberta, M., Nadia, L., Alberto, P., Stefania, L., Gelsomina, M., Matteo, L., 2014. *Trichoderma*-based products and their widespread use in agriculture. Open Mycol. J. 8 (Suppl. 1, M4), 71–126.

Sood, M., Kapoor, D., Kumar, V., Sheteiwy, M.S., Ramakrishnan, M., Landi, M., Araniti, F., Sharma, A., 2020. *Trichoderma*: the secrets of a multi-talented biocontrol agent. Plants 9 (6), 762.

Srinivasan, R., Sevgan, S., Ekesi, S., Tamò, M., 2019. Biopesticide based sustainable pest management for safer production of vegetable legumes and brassicas in Asia and Africa. Pest Manag. Sci. 1–25. https://doi.org/10.1002/ps.5480.

Twizeyimana, M., Hartman, G.L., 2019. Effect of selected biopesticides in reducing soybean rust (Phakopsora pachyrhizi) development. Plant Dis. 103 (9), 2460–2466.

World Population Prospects, 2019. United Nations, Department of Economic and Social Affairs. Population Division Highlights (ST/ESA/SER.A/423).

Wisniewski, M., Droby, S., Norelli, J., et al., 2016. Alternative management technologies for postharvest disease control: the journey from simplicity to complexity. Postharvest Biol. Technol. 122, 3–10.

Zhang, S., Liu, Q., Han, Y., Han, J., Yan, Z., Wang, Y., Zhang, X., 2019. Nematophin, an antimicrobial dipeptide compound from *Xenorhabdus nematophila* YL001 as a potent biopesticide for *Rhizoctonia solani* control. Front. Microbiol. 10, 1765.

Zhao, P., Xue, Y., Gao, W., Li, J., Zu, X., Fu, D., Bai, X., Zuo, Y., Hu, Z., Zhang, F., 2018. Bacillaceae-derived peptide antibiotics since 2000. Peptides 101, 10–16.

Chapter 13

Nano bio pesticide: today and future perspectives

Camelia Ungureanu
University POLITEHNICA of Bucharest, Bucharest, Romania

The explosive increase of the global population at an average of 1.1% a year up to 2030 generate a demand for food production (Feeding the world in 2050, 2009). The decrease of current output losses caused by plant disease is a great challenge to agricultural production and it is very important to understand the plant disease and the pesticide scenario at the same time.

In recent years, the diseases that appear in the fruit trees and the vines are becoming more frequent and very virulent. Explanations would be the climate change but also the excessive or unilateral application of the nitrogen fertilizers. Also, the resistance of pathogenic germs (pest resistance) to the multitude of synthetic treatments is a cause. Synthetic treatments use has also shown adverse effects on the soil microbial flora, and on the soil structure, fertility and mineral cycles (Lade et al., 2019; Ungureanu et al., 2019).

One solution is to replace the treatments obtained by chemical synthesis with pesticides obtained by green and eco-friendly ways (Zgura et al., 2020; Barbinta-Patrascu et al., 2019a,b) (e.g. with the help of plants, the use of plant extracts like natural bio-pesticide has received extensive acceptance from consumers because that plant extracts can be non-toxic at higher concentrations). The nano bio pesticides could be fabricated using any metal such as Cu, Ag, ZnO, SiO_2, with broad-spectrum pest protection efficiency.

The term "pesticide" comes from pest (disease) and "cide" from the Latin word that means "to kill" (Lade et al., 2017). The plant pesticides include fungicides (yeast, mold, mildew, mushrooms, etc.), herbicides (tree killers, weed killers, and defoliants) and algicides for algae disease.

Pesticides or plant protection products contain at last one active substance and are used for (Koul, 2019):

- plant protection against pests and diseases;
- influencing the degree of plant growth;
- plant based products preservation;
- eradication or impediment of the growth of unwanted plants.

The UE is controlling the use of pesticides in order to minimize health and environmental risks.

All pesticides are composed of active substances acting on pests or unwanted plants. Before being placed within a certain product on the market, all active substances used for plant protection products in the EU must be approved by the European Commission, in order to ensure that they are not dangerous. Substances must be safe for both human and animal health, as well as for the environment (Pahun et al., 2018).

The European Food Safety Authority (EFSA) assesses pesticides from their risk point of view and provides scientific support to the European Commission and Member States in the decision-making process. Approvals are given after such intensive evaluations. EFSA is the EU agency responsible for this domain and provides detailed information on its official website, which you can access via the link below.

EFSA manages the legislation on chemical products' classification and labeling. The evaluation and labeling of a chemical product is based on the hazardous properties that product may have (Herman et al., 2019).

Another method for limiting the pesticides emission in the environment and human's exposure to them consists in environmentally friendly products consists in manufacturing environmentally. The use of chemical pesticides, as well as of

Biopesticides. https://doi.org/10.1016/B978-0-12-823355-9.00006-7

synthetic fertilizers, antibiotics and other substances is highly restricted. Environmentally friendly farms must comply to strict rules to make possible the "bio products" name for their products.

Some of the most important principles of environmentally friendly products are as following:

- very strict limits regarding the use of agricultural chemical substances, pesticides, fertilizers, antibiotics and food additives;
- avoidance of genetically modified organisms (GMOs);
- local resources use;
- the choice of plant and animal species resistant to diseases and adapted at the local conditions (Bauer-Panskus et al., 2020).

Biopesticides are substances present in nature, such as fungi, bacteria, plant extracts, fatty acids or pheromones. The use of nano bio pesticides is growing rapidly worldwide and thy are more and more present in Integrated Pest Management (IPM), in order to increase the yields and quality of the combat/control programs, and also their low environmental impact. Biopesticides offer additional benefits, such as complex and new modes of action for resistance management and life extension of conventional pesticide products. They also add flexibility to a classic combat/control program, with short intervals from the last harvest treatment, a very good pesticide residue management for exported products and an excellent eco-toxicological profile for humans, animals and useful entomofauna. The Environmental Protection Agency (EPA) defines nano-biopesticides as being certain types of pesticides derived from natural sources - such as animals, plants, fungi, bacteria - and certain minerals along with metallic nanoparticles. For example, the canola oil, the garlic, the peppermint or chrysanthemum oil, have applications in the pesticide field and are considered as biopesticides (Khandelwal et al., 2016).

Biopesticides are already recognized as an effective option for diseases, pests and even weeds combat of the environmentally friendly crops, especially for crops of vegetables, fruit trees, vines, apartment or garden ornamental plants, grass, but also for most field crops where the objective is to obtain bio-crops. Nowadays, there are quite numerous active substances with a biopesticide profile registered at European level, which are used in various products, as deer repellents in forestry, mosquito or tick repellents, but also biopesticides for commercial agriculture, especially for pest control. As more natural sources of biopesticides are identified and synthesized, the number of such registered products will increase. The EPA currently recognizes three major classes of biopesticides: microbial, biochemical and plant incorporated pesticides (PIPs). The demand for nano bio pesticides is constantly increasing in the worldwide countries, and at European level it represents a safe alternative for replacement of many active substances on the rather wide list of the ones that will be excluded in the medium- and long-term periods of time. More often, the public opinion and the users of conventional pesticides are asking pertinent questions about their impact on the environment and possible risks to human health but also to the workers' safety, their toxicity to birds, the surface water and groundwater contamination (Goszczyński, 2020; Oprea et al., 2020).

One thing that is certain and of which the farmers and final consumers of agricultural products are well aware is that without the classic products, chemically synthesized (pesticides) for plants protection it cannot be designed as an efficient technology for diseases, pests and weeds control. These technologies must ensure a reasonable global production, which in the medium- and long-term periods of time must cover the food needs of an exponentially growing humanity. The large multinational companies, which invest hundreds of millions of Euros per year in research, provide to the officials, farmers and consumers detailed studies on the ecotoxicological profile of active substances placed on the market and take on major responsibilities regarding their management. Thus, they offer clear information on recommendations for their use and the protection of the user. The plant protection products are in fact plant medicines, that are indispensable for preserving their health. The condition to this is a strict compliance with the label (as with medicines), where all the details of use are presented: the approved dose per hectare, the optimal time for application, the recommended number of treatments and alternation of active substances, the period of time from the last treatment until harvest, etc. At European level, ECPA (European Crop Protection Association) and, at national level (e.g., AIPROM - Romanian Plant Protection Industry Association) offer extensive studies on the responsible use of plant protection products, but also on the major impact on the national and European agriculture that the removal of several active substances from the market, could have. Some of them have practically no substitute (Dobe et al., 2020; van Asseldonk et al., 2019).

Soil Association in the United Kingdom and the Biological Farmers of Australia (BFA) have assessed numerous biopesticides considering nano bio pesticides as of organic standards (Scrinis and Lyons, 2013).

Both categories of plant protection products, nano bio pesticides and the classical products that have been chemically synthesized, find their place for sure in the field of technologies for integrated diseases, pests and weeds control at farm level, when applied alternately and strictly complying with the Good Agricultural Practices. Conventional pesticides are obtained as chemically produced synthetically that kill a pest (United States Environmental Protection Agency, 2018).

Nano bio pesticides can be successfully used as a good alternative to synthetic pesticides. For example, the nanoporous zeolites, nanocapsules, and nano bio pesticides obtained from different non-dangerous biostructures, are releasing slowly and efficiently active agents that function as pesticides.

Micro-biopesticides includes microorganisms such as fungi or bacteria that kill the pest. The best-known microorganism is bacterium *Bacillus thuringiensis* (Bt) with different strains which produce various toxins (a different mix of proteins) that kills insects or larvae (Crampton, 2017; Jalali et al., 2020).

The use of different living organisms as "nanoparticles factories" involves environmentally friendly and safer processes. Various biological entities such as bacteria, fungi, diatomea, plants, actinomycetes and viruses have been used for this purpose. Metallic salts reduction to the corresponding nanoparticles (Au, Ag, Hg, Zn, Pt, etc.) is made possible by the new biosynthetic methods. Nature has developed different processes for the synthesis of scaled inorganic nano- and micromaterials. Moreover, biologically manufactured nanostructures offer substantial different properties such as a good adhesion and reduced toxicity, which makes them more valuable as nano bio fertilizers (Murugan et al., 2016; Barbinta-Patrascu et al., 2016).

Currently, the nanotechnology applications are found in different fields such as alimentary, textile, pesticides and fertilizers, plant protection, nutrition and agrochemical industry.

The most used nanoparticles used in nanotechnology for plant disease control, environment pollution control but also for biopesticides production are those based on Cu, Ag, ZnO and SiO_2 (Gupta et al., 2013; Worrall et al., 2018). The nanoparticles present many advantages, of which we list the most important: they are stable and present slow kinetics, which makes room temperature reactions possible.

Soil treatment with Zn nanoparticles leads to multiple benefits such as: soil zinc amount increase, control over a wide range of plants diseases but also the fact that vegetable crops grown in such treated fields become a zinc source for the consumers (Kim et al., 2011).

Nano bio pesticides are chemical complexes biologically derived using nanoparticles intended for pesticide use. The efficiency and effectiveness of nano-biopesticides are improved by the use of several metallic oxides, polymers but also of several active particles combined with mycéliums etc.

Nanoformulations based on nanoparticles and their synthesis have been widely studied in the last decade. For example, biological synthesis of silver nanoparticles includes room temperature synthesized microorganisms and plant extracts (Narayanan and Sakthivel, 2010).

Numerous extracts obtained from plants present pesticidal activity when prepared using stable metallic nanoparticles. Among the plants from which these extracts are obtained we can mention the following: ivy, ferns, *Vitex negundo*, *Toddalia asiatica*, and *Acorus calamus* (Benelli et al., 2017; Cîrstea et al., 2019). These methods are relatively inexpensive, and most importantly, they can be of real help for the recycling goals (e.g. leaf of *Raphanus sativus* L.). It is highly important in this case to optimize the reaction conditions and the composition-based plant selection; these nano-factors can be used furtherly for the synthesis of stable nanoparticles with well-defined dimensions, morphology and composition (Iravani, 2011).

These silver nanoparticles products obtained from radish leaves but also several ferns types (Călinescu et al., 2020) have been successfully tested against the *Podosphaera leucotricha* and *Venturia inaequalis* (Călinescu et al., 2019).

Another use of nanoparticles in agriculture is also their utility for ensuring plants growth and physiology. For example, the applying of TiO_2 nanoparticles produces chlorophylle amount increase by 45% There is another advantage of using metallic nanoparticles, namely the fact that no soil microorganisms damage has been reported (Mishram et al., 2014).

The future strategies are based on the development of nano bio pesticides with rapid biodegradation and phytotoxic effects as low as possible, without affecting the seed germination but also neither the human health.

The nano-bioimpact of *Citrus*-based Ag nanoparticles on four invasive wetland plants: Cattail (*Typha latifolia*), Flowering-rush (*Butomus umbellatus*), Duckweed (*Lemna minor*) and Water-pepper (*Polygonum hydropiper*) was demonstrated by Patrascu et all (Barbinta-Patrascu et al., 2017).

Our results pointed out the potential use of *Citrus*-based Ag nanoparticles as a solution in controlling invasive wetland plants in aqueous media and in management of aquatic weeds growth. Uncontrolled growth of these pathogens can be stopped with chemicals pesticides, but these could be harmful for human health and the environment, too.

For example, in Fig. 13.1 is given the diagrammatic representation of silver nanoparticle synthesis using aqueous extract of *Citrus reticulata* peels under specific conditions (personal studies Barbinta-Patrascu et al., 2017).

The nanoemulsions, nanocapsules, nanocontainers and the nanocages are several of the last decade nano bio pesticide administration techniques (Bergeson, 2010).

There is a lot of investment in obtaining the nano bio pesticides due to the above-mentioned advantages, since crops productivity increase is needed.

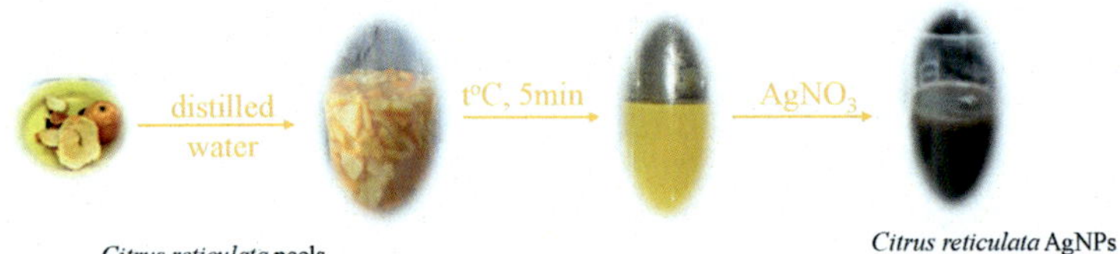

FIG. 13.1 Diagrammatic representation of silver nanoparticle synthesis using aqueous extract of *Citrus reticulata* peels under specific conditions.

The future strategies must take into account the following key points:

(a) Bio-nanopesticides natural degradation but also residues biosecurity;
(b) Targeted biodeposition of nano-biopesticides on plant leaves;
(c) Determination of the mechanism for establishing the applied dose;
(d) Development of dispersion systems based on water.

The polymers is also used for obtaining the nano bio pesticides and can be both natural or synthesis ones and they act as transporters, helping the active nano-biopesticides transport. Polymers such as starch, gelatin and chitosan are used as binders, disintegrants or dilutents and additionally are used as transporters. There exist several thermosensitive polymers, such as poly-N-isopropyl, acrylamide, etc. (Mutharani et al., 2019).

In the polymer-based formulations, an important role is taken by the polymer type, molecular size, drug loading, particle size but also the interactions between the nanoparticles and the polymer. In conclusion, on must consider the obtaining of nano bio pesticide from polymers, using acceptable ways from the environmental protection point of view, including a low consumption of energy and raw materials, maintaining an ecological balance as favorable as possible (Horowitz and Ishaaya, 2013).

The most known and used nanostructures for obtaining nano bio pesticides using plant secondary metabolites comprise nanospheres, nanocapsules, nanogels and myceliums (Qin et al., 2018). These nano bio pesticides structures have been studied in different countries. For example, in the USA in 2015, Forim and coworkers (Forim et al., 2015) patented the obtaining of biopolymeric nanoparticles containing *Azadirachta indica* oil and extracts, biopolymeric nanoparticles, and powder microparticles.

Nanoformulations distribution is an important part for nanobiopesticides use. Normally, these products contain organic compounds (e.g., polymers) or/and an inorganic compound (e.g., metallic oxides) in various forms that increase the apparent solubility of the low soluble active ingredient and therefore, its protection from a premature degradation.

Once a nanoformulation is developed, its use must be controlled so that the interaction between the nanoparticles and the targeted disease can be seen. In other words, an efficient controlled release formulation must remain inactive until the active compound is released. Nanopolymers such as chitosan, polyethylene glycol, and sodium alginate are controlling the quantity and rate of nanoparticle release. This latter depends on the polymers chemical structure, the chemical bond strength, and the size and structure of formed macromolecules (Shakiba et al., 2020).

Various studies shows that nanoparticles are toxic to insect pest such as *Ragmus morosus*, Termites, *Episomus lacerta*, etc. (Lade et al., 2017). The impact on mosquitoes (Sivapriyajothi et al., 2014), on domestic fly *Musca domestica* (Kamaraj et al., 2012) and on silkworms (Routray et al., 2016) has been also studied. Stadler et al. (2018) proved the efficacy of nanoparticles on *Sitophilus oryzae* and *Rhyzoperta dominica* while Routray et al. (2016), proved in a study from 2016 the efficacy of a pheromone nanogel on *Bactrocera dorsalis* saving a very large amount of guava fruits. At the same time, there are products containing CdS, TiO_2 and Ag showing activity against some lepidopterans (El-bendary and El-Helaly, 2013; Zgura et al., 2020) but also efficacy of the Zn and Ag nanoparticles on aphids (Mohammad et al., 2012).

A special type of the future nano bio pesticides is the class of pesticides containing magnetic nanoparticles. Magnetite octadecylsilane nanoparticles were synthesized and used like also as pesticides (Maddah and Shamsi, 2012).

The use of nano bio pesticides in the coming decade will be influenced by the nano formulation research and will include the involvement of all entities (e.g. universities, industry, governments and non-governmental entities).

Acknowledgments

This work was supported by a grant of the Romanian Ministery of Research and Innovation, CCCDI-UEFISCDI, project number PN-III-P1-1.2-PCCDI-2017-0332, contract 6PCCDI/2018, within PNCDI III.

References

Barbinta-Patrascu, M.E., Badea, N., Pirvu, C., Bacalum, M., Ungureanu, C., Nadejde, P.L., Ion, C., Rau, I., 2016. Multifunctional soft hybrid bio-platforms based on nano-silver and natural compounds. Mater. Sci. & Eng. C 69, 922–932.

Barbinta-Patrascu, M.E., Badea, N., Ungureanu, C., Iordache, S.M., Constantin, M., Purcar, V., Rau, I., Pirvu, C., 2017. Ecobiophysical aspects on nanosilver biogenerated from *Citrus reticulata* peels, as potential biopesticide for controlling pathogens and wetland plants in aquatic media. J. Nanomater. Art. no. 4214017.

Barbinta-Patrascu, M.E., Badea, N., Bacalum, M., Ungureanu, C., Suica-Bunghez, I.R., Iordache, S.M., Pirvu, C., Zgura, I., Maraloiu, V.A., 2019a. 3D hybrid structures based on biomimetic membranes and *Caryophyllus aromaticus* - "green" synthesized nano-silver with improved bioperformances. Mater. Sci. & Eng. C 101, 120–137.

Barbinta-Patrascu, M.E., Constantin, M., Badea, N., Ungureanu, C., Iordache, S.M., Purcar, V., Antohe, S., 2019b. Tangerine-generated silver - silica bioactive materials. Roman. J. Phys. 64 (3–4). Art. no. 701.

Bauer-Panskus, A., Miyazaki, J., Kawall, K., Then, C., 2020. Risk assessment of genetically engineered plants that can persist and propagate in the environment. Environ. Sci. Eur. 32 (1). Art. no. 32.

Benelli, G., Canale, A., Toniolo, C., Higuchi, A., Murugan, K., Pavela, R., Nicoletti, M., 2017. Neem (*Azadirachta indica*): towards the ideal insecticide? Nat. Prod. Res. 31 (4), 369–386.

Bergeson, L.L., 2010. Nanosilver pesticide products: what does the future hold? Environ. Qual. Manag. 19 (4), 73–82.

Călinescu, M., Ungureanu, C., Marin, F.C., Militaru, M., Soare, C., Fierăscu, R.C., Fierăscu, I., 2019. Antifungal activities of vegetal extract obtained from *Dryopteris* filix-mas (l.) fern. Fruit Grow. Res. XXXV.

Călinescu, M., Ungureanu, C., Soare, C., Fierascu, R.C., Fierăscu, I., Marin, F.C., 2020. Green matrix solution for growth inhibition of Venturia inaequalis and Podosphaera leucotricha. Acta Hortic. 1289, 61–66. https://doi.org/10.17660/ActaHortic.2020.1289.9.

Cîrstea, G., Călinescu, M., Ducu, C., Moga, S., Mihăescu, C., Sumedrea, D., Ungureanu, C., Butac, M., Vălu, M.-V., 2019. Bioformulations of plant protection products to control *Podosphaera leucotricha* and *Venturia inaequalis phytopathogens*. Fruit Grow. Res. XXXV.

Crampton, L., 2017. Biological vs. Chemical Pest Control: Benefits and Disadvantages. https://owlcation.com/agriculture/Biological-vs-Chemical-Pest-Control (Accessed August 29, 2018).

Dobe, C., Bonifay, S., Krass, J.D., McMillan, C., Terry, A., Wormuth, M., 2020. REACH Specific Environmental Release Categories for Plant Protection Product Applications (Integrated Environmental Assessment and Management).

El-bendary, H.M., El-Helaly, A.A., 2013. First record nanotechnology in agricultural: silica nanoparticles a potential new insecticide for pest control. Appl. Scientif. Rep. 4, 241–246.

FAO, 2009. Feeding the world in 2050. In: World Agricultural Summit on Food Security 16–18 November 2009. Food and Agriculture Organization of the United Nations, Rome.

Forim, M.R., Fernandes Da Silva, M.F.F.D.G., Fernandes, J.B., Vieira, P.C., 2015. Process for Obtaining Biopolymeric Nanoparticles Containing *Azadirachta indica* A. Juss. (neem) Oiland Extracts, Biopolymeric Nanoparticles, and Powder Microparticles. United States, patent application publication, pub. no.: US 2015/0320036A1, pp. 1–23.

Goszczyński, W., 2020. Search of the vocabulary for Eastern European food studies. Conceptual remarks after the workshop: alternative food supply networks in Central and Eastern Europe. East. Eur. Countryside 25 (1), 273–279.

Gupta, N., Fischer, A.R.H., George, S., Frewer, L.J., 2013. Expert views on societal responses to different applications of nanotechnology: a comparative analysis of experts in countries with different economic and regulatory environments. J. Nanopart. Res. 15 (8). Art. no. 1838.

Herman, R.A., Storer, N.P., Walker, C.E., 2019. Clarification on "EFSA genetically engineered crop composition equivalence approach: performance and consistency". J. Agric. & Food Chem. 67 (14), 4080–4088 (2020) J. Agric. & Food Chem., 68 (21), pp. 5787–5789.

Horowitz, A.R., Ishaaya, I., 2013. Advanced technologies for managing insect pests: an overview. Adv. Technol. Manag. Insect Pests 1–11.

Iravani, S., 2011. Green synthesis of metal nanoparticles using plants. Green Chem. 13 (10), 2638–2650.

Jalali, E., Maghsoudi, S., Noroozian, E., 2020. A novel method for biosynthesis of different polymorphs of TiO_2 nanoparticles as a protector for *Bacillus thuringiensis* from ultraviolet. Sci. Rep. 10 (1). Art. no. 426.

Kamaraj, C., Rajakumar, G., Rahuman, A.A., Velayutham, K., Bagavan, A., Zahir, A.A., Elango, G., 2012. Feeding deterrent activity of synthesized silver nanoparticles using *Manilkara zapota* leaf extract against the house fly, *Musca domestica* (Diptera: Muscidae). Parasitol. Res. 111 (6), 2439–2448.

Khandelwal, N., Barbole, R.S., Banerjee, S.S., Chate, G.P., Biradar, A.V., Khandare, J.J., Giri, A.P., 2016. Budding trends in integrated pest management using advanced micro- and nano-materials: challenges and perspectives. J. Environ. Manag. 184, 157–169.

Kim, S., Kim, J., Lee, I., 2011. Effects of Zn and ZnO nanoparticles and Zn^{2+} on soil enzyme activity and bioaccumulation of Zn in *Cucumis sativus*. Chem. & Ecol. 27 (1), 49–55.

Koul, O., 2019. Nano-biopesticides Today and Future Perspectives, pp. 1–485.

Lade, B.D., Gogle, D.P., Nandeshwar, S.B., 2017. Nano-biopesticide to constraint plant destructive pests. J. Nanomed. Res. 6, 1–9.

Lade, B.D., Gogle, D.P., Lade, D.B., Moon, G.M., Nandeshwar, S.B., Kumbhare, S.D., 2019. Nanobiopesticide Formulations: Application Strategies Today and Future Perspectives, pp. 179–206. Nano-Biopesticides Today and Future Perspectives.

Maddah, B., Shamsi, J., 2012. Extraction and preconcentration of trace amounts of diazinon and fenitrothion from environmental water by magnetite octadecylsilane nanoparticles. J. Chromatogr. A 1256, 40–45.

Mishram, V., Mishra, R.K., Dikshit, A., Pandey, A.C., 2014. Chapter 8 - interactions of nanoparticles with plants. In: An Emerging Prospective in the Agriculture Industry, Emerging Technologies and Management of Crop Stress Tolerance. Biological Techniques, vol. 1, pp. 159–180.

Mohammad, R., Mohammad Amin, S., Salma, K., 2012. Insecticied effect of silver and zinc nanoparticles against *Aphis nerii* Boyer of fonscolombe (Hemiptera: Aphididae). Chilean J. Agric. Res. 72 (4).

Murugan, K., Raman, C., Panneerselvam, C., Madhiyazhagan, P., Subramanium, J., et al., 2016. Nano-insecticides for the control of human and crop pests. In: Raman, C., Goldsmith, M.R., Agunbiade, T.A. (Eds.), Short Views on Insect Genomics and Proteomics: Insect Proteomics, vol. 2. Springer International Publishing, Berlin, Germany, pp. 229–251. ISBN-13: 9783319242446.

Mutharani, B., Ranganathan, P., Chen, S.-M., Karuppiah, C., 2019. Enzyme-free electrochemical detection of nanomolar levels of the organophosphorus pesticide paraoxon-ethyl by using a poly(N-isopropyl acrylamide)-chitosan microgel decorated with palladium nanoparticles. Microchimica Acta 186 (3). Art. no. 167.

Narayanan, K.B., Sakthivel, N., 2010. Biological synthesis of metal nanoparticles by microbes. Adv. Coll. & Interf. Sci. 156 (1–2), 1–13.

Oprea, F., Onofrei, M., Lupu, D., Vintila, G., Paraschiv, G., 2020. The determinants of economic resilience. The case of Eastern European regions. Sustainability (Switzerland) 12 (10). Art. no. 4228.

Pahun, J., Fouilleux, E., Daviron, B., 2018. Bioeconomy: from early meanings to the emergence of a new framework for public action [Article@De quoi la bioéconomie est-elle le nom? Genèse d'un nouveau référentiel d'action publique]. Nat. Sci. Soc. 26 (1), 3–16.

Qin, Y., Xiong, L., Li, M., Liu, J., Wu, H., Qiu, H., Mu, H., Xu, X., Sun, Q., 2018. Preparation of bioactive polysaccharide nanoparticles with enhanced radical scavenging activity and antimicrobial activity. J. Agric. & Food Chem. 66 (17), 4373–4383.

Routray, S., Dey, D., Baral, S., Das, A.P., Patil, V., 2016. Potential of nanotechnology in insect pest control. Progress. Res. Int. J. 11 (Special-2), 903–906.

Scrinis, G., Lyons, K., 2013. Nanotechnology, and the techno-corporate agri-food paradigm. Food Secur. Nutr. & Sustain. 252–270.

Shakiba, S., Astete, C.E., Paudel, S., Sabliov, C.M., Rodrigues, D.F., Louie, S.M., 2020. Emerging investigator series: polymeric nanocarriers for agricultural applications: synthesis, characterization, and environmental and biological interactions. Environ. Sci.: Nano 7 (1), 37–67.

Sivapriyajothi, S., Mahesh Kumar, P., Kovendan, K., Subramaniam, J., Murugan, K., 2014. Larvicidal and pupicidal activity of synthesized silver nanoparticles using *Leucas aspera* leaf extract against mosquito vectors, *Aedes aegypti* and *Anopheles stephensi*. J. Entomol. & Acarol. Res. 46, 1787.

Stadler, T., Buteler, M., Valdez, S.R., Gitto, J.G., 2018. Particulate nanoinsecticides: a newconcept in insect pest management. In: Begum, G. (Ed.), Insecticides Agriculture and Toxicology. IntechOpen, London, pp. 83–105.

Ungureanu, C., Calinescu, M., Ferdes, M., Soare, L., Vizitiu, D., Fierascu, I., Fierascu, R.C., Raileanu, S., 2019. Isolation and cultivation of some pathogen fungi from apple and grapevines grown in Arges county. Revista de Chimie 70 (11), 3913–3916.

EPA, 2018. United States Environmental Protection Agency. Available from: https://www.epa.gov/home/forms/contact-epa (Accessed August 25, 2018).

van Asseldonk, M., Jongeneel, R., van Kooten, G.C., Cordier, J., 2019. Agricultural risk management in the European union: a proposal to facilitate precautionary savings [Article@La gestion des risques agricoles dans l'Union européenne: proposition pour faciliter l'épargne de précaution]. EuroChoices 18 (2), 40–46.

Worrall, E.A., Hamid, A., Mody, K.T., Mitter, N., Pappu, H.R., 2018. Nanotechnology for plant disease management. Agronomy 8 (12).

Zgura, I., Preda, N., Enculescu, M., Diamandescu, L., Negrila, C., Bacalum, M., Ungureanu, C., Barbinta-Patrascu, M.E., 2020. Cytotoxicity, antioxidant, antibacterial, and photocatalytic activities of ZnO-CdS powders. Materials 13 (1), 182.

Chapter 14

Current development, application and constraints of biopesticides in plant disease management

Shweta Meshram[a], Sunaina Bisht[b] and Robin Gogoi[a]
[a]*Division of Plant Pathology, ICAR-Indian Agricultural Research Institute, New Delhi, Delhi, India;* [b]*Rani Lakshmi Bai Central Agricultural University, Jhansi, Uttar Pradesh, India*

14.1 Introduction

Pesticides are any substances that kill or control pests like insects, microorganism, weeds and rodents. Pesticide term includes insecticides, fungicides, bactericides, nematicides, and herbicides (Gevao and Jones 2002). Pesticides are classified on the basis of chemical nature (as inorganic and organic); on the basis of target pest (as fungicides, bactericide, nematicide, herbicide); on the basis of mode of action (as non-systemic and systemic); on the basis of function (as protectant, therapeutant and eradicant) and on the basis of method of application (as foliar spray, seed treatment, fumigants and wound dresser). Pesticides are broadly categorized in synthetic and bio pesticides. According to Food and Agriculture organization (FAO) "pesticides are any substance or mixture of substances intended for preventing, destroying or controlling any pest, including vectors of human or animal disease, unwanted species of plants or animals causing harm during or otherwise interfering with the production, processing, storage, transport or marketing of food, agricultural commodities, wood and wood products or animal feedstuffs, or substances which may be administered to animals for the control of insects or other pests in or on their bodies." In agriculture most of the farmers rely on the synthetic pesticides because they give quick results has increased 20%−50% food production over the last 40 years (Warren, 1998; Webster et al., 1999) and some benefits over conventional agricultural practices such as pesticides reduced deforestation by producing better crop yield in less land needed; reduced the use of fertilizer; pesticides enable farmers to till less (less weeds) in turn causing less soil erosion; pesticides helps in controlling insects, pathogen and toxic weeds; pesticides helps in protecting livestock from insects and rodents and pesticides are cost effective-more money for the farmers results in better prices in the grocery store (Lamichhane et al., 2016). In agriculture most of the farmers rely on the synthetic pesticides because they give quick results. Synthetic pesticides in one hand have given boost to the production by its great capacity to manage pests and diseases but on other hand its indiscriminate use for long term is a challenge for health and environment. However, financial benefits arising from the use of pesticides are often outweighed by negative environmental and health effects (Ministry of the Environment, 2002). The main concerns are the risk of poisoning humans or animals, contamination of livestock products, harm to beneficial insects by its residual effect which will contaminate food products, waterways and soil. Development resistant strain of target fungi is also one of the major problems with selective fungicides. Although the use of chemicals in modern agriculture has significantly increased productivity. There has been an increase in the concentration of pesticides in food and in our environment, with associated negative effects on human health and the environment (Fig. 14.1). Unfortunately, problems are encountered with some pesticides, due to their build-up in the aquatic environment and subsequently the food chain, and pressure has therefore increased in many parts of the world to regulate and monitor them more closely (Sarmah et al., 2004).

There is now overwhelming evidence that some of these chemicals do pose a potential risk to humans and other life forms and unwanted side effects to the environment hence, there is a need of use safer, non-chemical pest control (including weed control) methods. Biopesticides play a very important role in management of pest and pathogens and they are safe for the

Biopesticides. https://doi.org/10.1016/B978-0-12-823355-9.00004-3

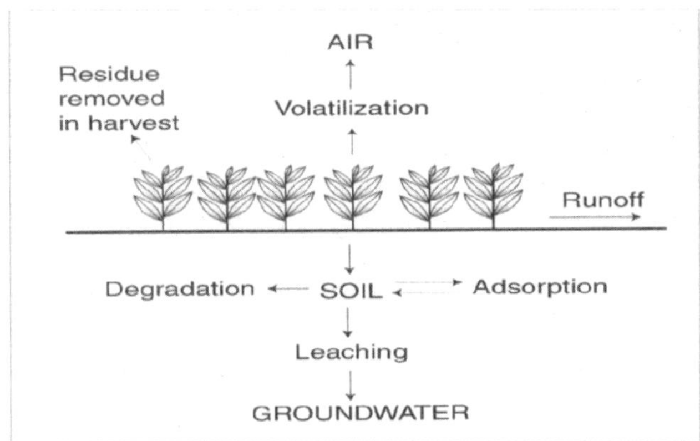

FIG. 14.1 Processes responsible for the fate of applied pesticides in the environment (Sarmah et al., 2004).

environment. They offer an ecologically beneficial and effective control to pest problems. They are safer to the environment and to human health as they are degradable, and there is no residue effect on human beings (Pathak et al., 2017).

14.2 History of synthetic pesticides used in plant disease evolution

"Development" of the first fungicide was the result of good observations. The first use of brining of grain with salt water followed by liming took place in the middle of the 17th century to control bunt, and followed the observation that seed wheat salvaged from the sea was free of bunt. This had occurred long before Tillet (1755) established that seed borne fungi (*Tilletia tritici*, *T. laevis*) caused bunt of wheat and that it could be controlled by seed treatments of lime, or lime and salt. Another important discovery was made in France in 1882 by Millardet, who noticed that grape vines that had been sprayed with a bluish white mixture of copper sulfate and lime to deter pilferers retained their leaves through the season, whereas the unsprayed vines lost their leaves. After numerous spraying experiments Millardet concluded that a mixture of copper sulfate and hydrated lime could effectively control downy mildew of grape. Hooker in 1923 stated in his paper on colloidal copper hydroxide that there were no entirely satisfactory fungicides available. Then he went on to describe the shortcomings of the Bordeaux mixture and lime sulfur, the latter also being "most disagreeable to handle." Up until the 1940s chemical disease control relied upon inorganic chemical preparations, frequently prepared by the user.

Many of the early efforts to produce healthy crops involved diseases that had newly been introduced and left the growers quite helpless. In the following sections we highlight trends in fungicide development and use over the last century in light of the ever changing spectra and intensities of fungal pathogens, which have often occurred as a consequence of changing cropping systems. Morpholine as dodemorph (1965) and tridemorph (1969), Carboxanilide as carboxin (1966), oxycarboxin (1966), Benzimidazoles in thiabendazole (1964), benomyl (1968), and thiophanate methyl (1970) were released for use. Phenyl amides came in 1977 which was adapted worldwide because it's high potency, excellent curative activity, protectant activity, excellent redistribution and protection of new growth.

14.3 Current global scenario

The use of pesticides in agriculture is increasing rapidly in developing countries, especially in Southeast Asia (Kunstadter, 2007; Schreinemachers and Tipraqsa, 2012). WHO has reported that approximately 20% of pesticides are used in developing countries with increasing rate of usage. The manufacturing of pesticides in India started in 1952, with the production of benzene hexachloride, followed by DDT. The synthesis of pesticides increased enormously. India is one of the major pesticides producing countries in Asia with annual production of 90,000 tonnes, and it stands at 12th position in the world in the manufacturing of pesticides (Abhilash and singh, 2009; Khan et al., 2010). India is the largest manufacture of mancozeb (500,000 MTPA) and chlorpyrifos for both the domestic and international markets (Agribusiness Global, 2020). Recently, India has moved to has moved to ban 27 pesticides including mancozeb and chlorpyrifos (DPPQS, 2020). India is also major consumer of pesticides, in accord to consumption of pesticides India rank second globally (Table 14.1). Pesticides storage in 85% countries are subjected to legislation among them 27% restricted their storage specifically for agricultural related pesticides, rest 15% do not have pesticides storage legislation. Regarding disposal 51% countries impose legislation where 18% countries have disposal of legislation for agricultural pesticides. In terms of overall agricultural policy, 83% of countries consider IPM of high priority. The approaches of agro ecology and organic farming are considered a high priority in 54% and 56% of countries, respectively, particularly in the European Region (Source: FAO).

TABLE 14.1 Annual pesticide consumption in different Asian countries (FAO, 2017).

S. No.	Country	Tonnes pesticides Use
1	China	1,807,000
2	India	56,120
3	Malaysia	49,199
4	Pakistan	27,885
5	Thailand	21,800
6	Vietnam	19,154
7	South Korea	19,788
8	Bangladesh	15,833
9	Myanmar	5,583
10	Nepal	454
11	Bhutan	12

Although the credits of pesticides include enhanced economic potential in terms of increased production of food and fiber, and amelioration of vector-borne diseases, then their debits have resulted in serious health implications to man and his environment. Due to ill-effects of chemical many chemical are banned. Keeping in mind the environmental hazards due to pesticides, biopesticides comes into picture as they play a very important role in management of pest and pathogens in eco-friendly manner (Pathak et al., 2017). Among nations The USA, China, Russia, and India are the leading producers of microbial pesticides (Leng et al., 2011). In India up to 2012 there were 410 biopesticides manufacturing companies were registered (Hassan and Gökçe, 2014).

14.4 Biopesticides

The biopesticides are certain types of pesticides derived from natural materials such as microorganisms, plants, animals and certain minerals (Pathak et al., 2017). They are living organisms (natural enemies) or their products (phytochemicals/ botanical pesticides, microbial products) or by-products (Fig. 14.2), which can be used for the management of pests that are injurious to crop plants. Bio pesticides are obtained via natural resources and used against targeted pest. They are very selective to their targets than synthetic pesticides and are used broadly with intention to manage disease rather eradicating. There are rare chances of development of resistance in pests against these natural pesticides (Leng et al., 2011).

Bio pesticides are not new thing for the world, before discovery of synthetic pesticides people used to relay on natural or herbal substances for control of pest and diseases. The first known documentation of plant-based pesticides use was as 2000 BCE (Ignacimuthu, 2012) from India. Farmers in 17th century used nicotine to control plum beetle (Lopez et al., 2011). Pyrethrins were used as house hold and industrial spray, rotenone used as garden whereas nicotine used for green house (Henn and Weinzierl, 1989). Book written by Rachel Carson highlighted the issue of pesticides harmful effect in her book "Silent Spring" (1962) which lead to ban on DDT pesticides (Hassan and Gökçe, 2014). Worse effects of synthetic pesticides on fauna and flora urge the need to promote of bio pesticides for management of pest and diseases in integrated pest management strategies (IPMs). Biological control, despite being known and studied for more than 100 years, still presents low diversity and availability of products in the market. However, the interest in biopesticides has increased, because the use of biological control agents as part of the IPM program represents a change in the model of agriculture being practiced in the country (Panizzi et al., 2012; Faria, 2017; Tripathi et al., 2020). By 2001, there were about 195 registered biopesticide active ingredients and 780 products. These are biochemical pesticides, which are naturally occurring sub-stances for control pests by nontoxic mechanisms (Pathak et al., 2017).

14.5 Classification of biopesticides

The three different classes of biopesticides that the US Environmental Protection Agency (EPA) has identified are microbial, biochemical and plant-incorporated protectants (PIPs).

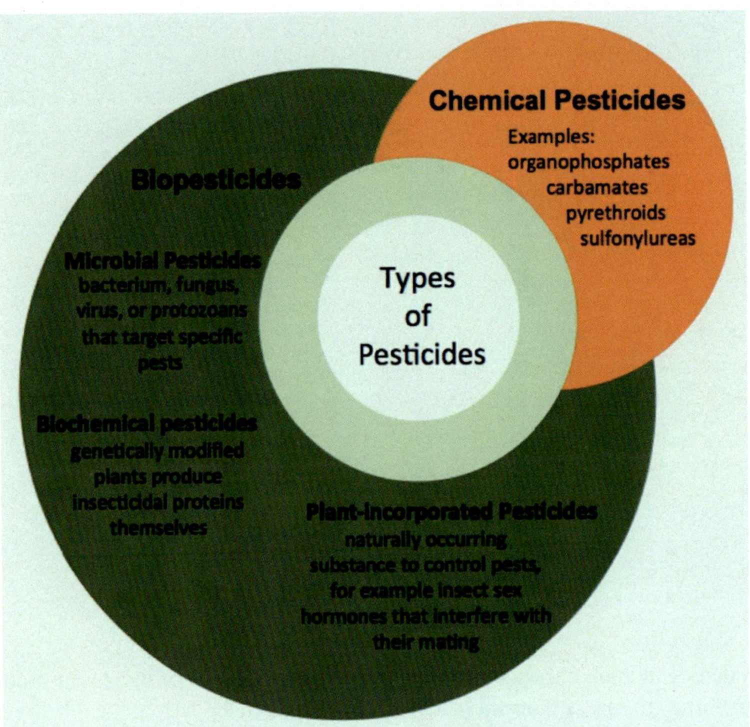

FIG. 14.2 Classification of pesticides (Hsaio, 2017).

1. **Biochemical pesticides** are naturally occurring substances that control pests by non-toxic mechanisms. Conventional pesticides, by contrast, are generally synthetic materials that directly kill or inactivate the pest. Biochemical pesticides include substances that interfere with mating, such as insect sex pheromones, as well as various scented plant extracts that attract insect pests to traps. Because it is sometimes difficult to determine whether a substance meets the criteria for classification as a biochemical pesticide, EPA has established a special committee to make such decisions. For a substance or a mixture of substances to be classified as a biochemical pesticide active ingredient, three criteria must be met (Jones, 2010)
   i. It must be naturally occurring or, if not naturally occurring, it must be structurally similar to and functionally identical with a naturally occurring substance.
   ii. It must have a nontoxic mode of action against the target pest.
   iii. It must have a history of nontoxic exposure to humans and the environment.
2. **Microbial pesticides** consist of a microorganism (e.g., a bacterium, fungus, virus or protozoan) as the active ingredient. Microbial pesticides can control many different kinds of pests, although each separate active ingredient is relatively specific for its target pest [s]. Microbial pesticides are living microorganisms that control pests via pathogenicity, competition, or the synthesis of myco- or bacterial toxins. The microbial pesticides, however, are safe for humans and the environment because their pesticidal effects are usually highly specific to the target organism, exposure to non target organisms is low, and they are non-persistent in their biologically active stat (Jones, 2010). Typically, the microbial pesticide organism population declines as its host organism (the pest) declines. Dormant spores may remain in the environment, but germinate only when its host organism returns. For example, there are fungi that control certain weeds and other fungi that kill specific insects. The most widely used microbial pesticides are subspecies and strains of *Bacillus thuringiensis*, or Bt. Each strain of this bacterium produces a different mix of proteins and specifically kills one or a few related species of insect larvae. While some Bt ingredients control moth larvae found on plants, other Bt ingredients are specific for larvae of flies and mosquitoes. The target insect species are determined by whether the particular Bt produces a protein that can bind to a larval gut receptor, thereby causing the insect larvae to starve.
3. **Plant-incorporated-protectants (PIPs)** are pesticidal substances that plants produce from genetic material that has been added to the plant. Similar to the microbial pesticides, PIPs may not be considered green chemistry alternatives by many. The PIPs registered to date encompass plants genetically engineered to produce a *B. thuringiensis* toxin (Bt-PIPs), or plants that have been engineered to produce plant viral coat proteins (PVCP-PIPs). The Bt-PIPs and

PVCPPIPs, respectively, protect plants from insect pests and pathogenic viruses. The gene for the Bt pesticidal protein are introduce into the plant's own genetic material. Then the plant, instead of the Bt bacterium, manufactures the substance that destroys the pest. All the prototype biopesticides have evolved from the bacteria *B. thuringiensis* (Bt) that produce a toxin (Bt toxin) which binds to insect gut receptor protein and disrupt the gut upon ingestion. They are already present in the environment and pose no risk to human health.

All the three varieties of biopesticides (microbial, biochemical and PIP varieties) are derived from this bacteria and its toxin.

14.6 Microbial biopesticides

Plants have fungi, Bacteria, oomycetes, viruses and protozoa as their natural pathogen, likewise they have natural antagonistic microbe against these pathogens. Many species of microbes are reported to be used as biopesticides by many workers (Table 14.3). Out of the total global biopesticide present in market for all crop types, the contribution of bacterial biopesticides is about 74%, fungal biopesticides 10%, viral biopesticides 5%, predator biopesticides 8% and other biopesticides 3% (Mishra et al., 2015). Most popular bioinsecticide is *B. thuringiensis* (Bt) specially against insects of cotton, similarly *Trichoderma* spp. Is most popular and broadly adopted against plant pathogen. *Trichoderma* has immense potential as biocontrol agent with insuring ecological safety which is the most desirable criteria for any BCAs (Hassan and Gökçe, 2014). Microbial pesticides accounted for approximately 63% of the global biopesticide market with *B. thuringiensis* and its subspecies being the most widely used (Hassan and Gökçe, 2014). The major species of *B. thuringiensis* species are used for suppression of different types of lepidopteran pests, forest pests, mosquitoes and black flies represented in following Table 14.3. Another fine example of fungus based microbial pesticides is *Chaetomium* spp., since 1989 many biological products in the form of pellet and formulations has been developed, which used 22 strains of various *Chaetomium* spp., and still counting (Soytong et al., 2001). Most preferred species are *C. globosum* and *C. cupreum* (Table 14.2).

TABLE 14.2 Some microbial biopesticides used against pathogen and plant diseases.

Pathogen	Disease	Active BCAs
Rhizoctonia solani	Damping-off of bean	*T. lignorum, T. virens, T. hamatum, T. harzianum*
Fusarium oxysporum	Fusarium wilt	*T. viride, T. harzianum*
Aspergillus flavus	Afla toxin, seed toxicity	*T. viride, T. harzianum*
Sclerotinia sclerotiorum	Fungal-soybean plant	*T. harzianum*, BAFC 742
Rhizoctonia solani, Pythium ultimum,	Damping-off of cucumber	*T. virens* isolates GL3
Magnaporthe grisea	Rice blast	*Chaetomium globosum* strain F0142 Park et al. (2005)
Phytophthora capsici	foot rot and stem blight of Pepper	*Chaetomium globosum* CGOI-CG 12
Fusarium oxysporum f. sp. *lycopersici*	Wilt of Tomato	*Chaetomium globosum* and *Trichoderma* sp.
Phytophthora palmivora	Tangerine	*Chaetomium cupreum* CCOI-CC I0
Hemileia vastatrix,	Rust disease of coffee, wheat leaf rust, carnation rust	*Verticillium lecanii*
Broad range of pathogens, *Macrophomina, Fusarium, Alternaria, Sclerotium, Rhizoctonia,* bacterial wilt and leaf spots	Broad range of disease, wilt, leaf spots, rotting etc.	*Pseudomonas fluorescens*
Fusarium, Aspergillus, P. savastanoi	Several bacterial and fungal diseases.	*Bacillus subtilis, B. amyloliquefaciens, B. thuringiensis* etc.

TABLE 14.3 Bacterial species used as biopesticides.

Bt variety	Target pest	References
Serratia entomolphila	Grass grub	Johnson et al. (2001)
Bacillus thuringiensis subsp. *israelensis*	Mosquito and blackflies	Kabaluk et al. (2010)
B. thuringiensis subsp. *kurstaki*	Lepidopteran larvae	
B. thuringiensis subsp. *galleriae*	Colorado potato beetle	
Lysisnibacillus sphaericus	Mosquito larvae	Berry (2012)
Bacillus moritai	*Diptera*	Kunimi (2007)
Burkholderia spp.	Chewing and sucking insects and mites; nematodes	Ruiu (2018)
Saccharopolyspora spinose	Insects	

14.7 Insight into popular fungal and bacterial biopesticides used in plant disease management

14.7.1 *Trichoderma* spp

Trichoderma is naturally occurring fungus, has tremendous capacity to antagonize plant pathogen. *Trichoderma* formulations are highly effective on roots and root stems which makes it quite potential biopesticides against soil borne pathogens. It also manages foliar pathogen such as gray mold and seed borne pathogen. Common methods of application of *Trichoderma* are soil application, seed priming and uniform foliar spray (*Trichoderma* Fact sheet, 2000). In India many agrochemical companies offer *Trichoderma* formulations with various trade names. Common commercial names of *Trichoderma* formulations are Bio Protectore, Bio-Shield, Basderma, Biocure F, Bioderma, Bioveer, Krishi Bio, and Sanjeevni for *T. virde*. Bioderma H, Bioharz, Commander, Sardar Eco Green, Ecosom-TH for *T. harzianum*. Tricone V and Neemoderm A are common name for *T. viride/T. harzianium* wettable powder (WP) formulations (Kumar et al., 2017).

Popular formulations for *Trichoderma*

1. Talc based

Grown in the liquid medium is mixed with talc powder in the ratio of 1:2 and dried to 8% moisture under shade.

Shelf life: 3-4 months

2. Vermiculite-wheat bran

100 g vermiculite + 33 g wheat bran, sterilized in an oven at 70°C for 3 days.

Medium: Molasses-Yeast, 0.05 N medium + HCL

3. Pesta granules based

52 ml Fermentor biomass + wheat flour (100 g) +one mm thick sheets.

Pass through an eighteen mesh and collect the granules.

4. Alginate prills based

1 part sodium alginate + 1 portion food base in distilled water. Autoclaved and blend together.

5. Press mud based

Press mud (by product of the sugar) mix with *Trichoderma* culture, sprinkle water, cover with gunny bag, and keep in shed for 25 days

6. Banana waste based

Banana waste + urea + rock + phosphate, culture *T. viride* are used, buried in the pit for 45 days.

7. Coffee husk based

First time developed in Karnataka.

Waste substances of coffee industries are used as substrate

8. Oil-based

Harvested conidia + vegetable or mineral oil + surface active agents.

Mostly suitable for foliar spray

14.8 Mass production of *Trichoderma* for commercial purpose

In biopesticides production important aspect is to obtain culture in large scale insuring cost effectiveness of production. Successful mass production requires adequate growth conditions such as temperature and nutrition (Kumar et al., 2014). Before releasing of formulation efficacy, compatibility and filed performance is mandatory to test. Following popular methods are used for production of efficacy.

Fermentation

Liquid fermentation

- Water and soluble materials are used as broth.
- Requirement : Readily available, inexpensive
- Example: V-8 juice, Molasses-yeast medium, Wheat bran

Solid fermentation

- Insoluble materials, water is required less than the saturation capacity.
- Microbial propagation solid state fermentation gives high amount of conidia Therefore more preferable.
- Grains are moistened, sterilized and inoculated with *Trichoderma* and incubated for 10 to 15 days
- Medium: Sorghum, Millets, Ragi.

14.8.1 Pseudomonas fluorescens

Pseudomonas fluorescens is a potential bacterium that has tremendous capacity to control plant pathogen. This rhizosphere bacterium is highly effective against soil and root pathogen. The Pseudomonad has been well recognized biocontrol agent as well as plant growth promoter. It is also manages diseases like molds and rots. Biopesticides formulation is most important aspect for successful performance of *Pseudomonas* under field and green house condition. The effective methods of application are spraying, dipping plant parts or seed treatment.

14.9 Formulations for *P. fluorescens*

In formulations active ingredient is mixed with carrier material which ensures efficient delivery and stability of formulation.

14.9.1 Organic carriers

For mass production of *P. fluorescens* organic substrates are used such as starch, peat, compost, talk and manure.

14.9.2 Inorganic carriers

Alginate, $CaCO_3$, talk, clay-based carriers etc. are used as inorganic carrier for formulation preparation.

14.10 Methods

14.10.1 Powder formulations

Al-Waily et al. (2018) suggested method for preparation for powder formulations with the following steps:

Culture *P. fluorescens* on Kings B medium at 28°C ⟶ Autoclave 1 kg of any organic/inorganic carrier at 121°C and 15 pound/inch^2 for 30 min ⟶ Mix 400 mL of previously grown one day old culture of *P. fluorescens*, 306.7×10^{11} CFU/mL⟶

Pack mixer in polythene bag, moisten it with water up to 35% ⟶ Keep the bag in room temperature.

14.11 Liquid formulation

Liquid formulation utilizes the substrates like KNO_3, phosphate buffer and glycerol. One of the standardized methods for preparation of liquid formulations of *P. fluorescens* against tomato wilt pathogen *Fusarium solani* was given by

Manikandan et al. (2010). Viability of *P. fluorescens* Pf1 was enhanced in nutrient broth (NB) amended with glycerol which leads to the development of liquid-based bioformulation of *P. fluorescens*. The NB was prepared with addition of 2% glycerol + 1 mL (3×10^{10} CFU/mL) to enrich the bacterial culture was inoculated and incubated at 25 ± 2°C. The formulation should be sealed in plastic containers for further use.

14.11.1 *Bacillus* sp.

*Bacillus thuringensis*is well known biocontrol agent against insect pest of crops, in case of plant pathogen *B. subtilis* has gained popularity. *Bacillus* based biopesticides has shown efficient capacity to control plant diseases against broad range of pathogen. Bacillus like *Pseudomonas* and *Trichoderma* has the capacity to take part in growth promotion and inducing systemic resistance as well. *Bacillus* is effective against fungal pathogen of cotton, vegetables, peanuts, and soybeans suppress plant disease fungal organisms such as *Rhizoctonia, Fusarium, Aspergillus*, and others (EPA fact sheet, 1999). Application method as seed treatment or slurry mix (Water + *Bacillus* fungicide + seed) should be used within 72 h.

14.11.1.1 Carriers used

Liquid carriers- Vegetable oils, Mineral carriers - Kaolinite clay, diatomaceous earth, Organic carriers -Grain flours, Stabilizers - Lactose, sodium benzoate, Nutrients- Molasses, peptone, Binders- Gum arabic, carboxymethylcellulose (CMC), Desiccants- Silica gel, anhydrous salts, Thickeners- Xanthan gum, Stickers- Pregelatinized corn flour.

14.11.1.2 Liquid formulations

This formulations utilizes broth culture of bacteria, include nutrients, cell protectants and inducers responsible for cell/spore/cyst formation (Rao et al., 2015).

14.11.1.3 Dry formulations

Dry formulation products include wettable powders, dusts, and granules. In case of *Bacillus* WP are mostly used. WP consist of dry inactive and active ingredients (biomass) intended to be applied as a suspension in liquid (Schisler et al., 2004).

14.11.1.4 Some commercial names of the products

Serenade, EcoGuard, Kodiak, Yield Shield, BioYield, Subtilex, Hi Stick L + Subtilex.

14.12 Improvement of formulation efficacy

Most approaches for biological control of plant diseases have focused on some basic principles, some of them are:

i. Modifying genetics of the bio-control agent to add mechanism of disease suppression that are operable against more 14.4 than one pathogen,
ii. Alteration of environment to favor the biological control agent and inhibit other competitive microflora
iii. Development of microbial mixtures or consortia with superior bio-control activity.

14.13 Molecular approach for improvement of formulation efficacy

The genetic improvement of formulation aims to make them more effective, it can be achieved by following ways:

14.13.1 Protoplast fusion

It is a quick and easy method in strain development for bringing genetic recombination and developing hybrid strains in filamentous fungi (Kumar et al., 2014). Protoplasts are the cells of which cell walls are removed and cytoplasmic membrane is the outermost layer in such cells. During fusion two or more protoplasts come in contact and adhere with one another either spontaneously or in presence of fusion inducing agents. By protoplast fusion it is possible to transfer some useful genes from one species to another. Protoplast fusion is an important tools in strain improvement. These are the powerful techniques for engineering of microbial strains for desirable industrial properties which helps in combining the advantageous properties of distinct promising agents. Protoplast fusion technique has a great potential for strain improvement (specially for industrially microorganisms). Various enzymes like cellulas, pectinase, xylanase, lysozyme, novozyme are used for breaking the cells. Protoplast fusion can be broadly classified into two categories: Spontaneous fusion (fuse through their plasmodesmata) and Induced fusion (needs a fusion inducing chemicals). Several chemicals has been used to induce protoplast fusion such as sodium nitrate, polyethylene glycol, calcium ions. Polyethylycol and Calcium chloride (Penttilä et al., 1987) chemicals are used to increase efficiency by enhancing cellulose production of *T. reesei* by inter-specific protoplast fusion. Self-fusion of protoplasts from *T. harzianum* strain PTh18 showed two fold increases in chitinase and biocontrol activity as compared to the parent strain against *Rhizoctonia solani* (Prabavathy et al., 2006).

14.13.2 Genetic recombination

Genetic recombination (genetic reshuffling) is the exchange of genetic material between different organisms which leads to production of offspring with combinations of traits that differ from those found in either parent. The process occurs naturally and can also be carried out in the lab. Recombination increases the genetic diversity in sexually reproducing organisms and can allow an organism to function in new ways. Integrating foreign DNA into the genome of a biocontrol helps in developing new efficient strain. This involves the construction of strains that produce increased levels of lytic enzymes and antibiotics. Biocontrol Plant Growth Promoting Bacteria (PGPB) may be improved by genetically engineering them to overexpress one or more of traits so that strains with several different anti-phytopathogen traits which can act synergistically are created. The suppression or deletion of gene from the biocontrol strains could also enhance the sustained biocontrol activity of the strains (Glick and Bashan, 1997). The biosynthesis of cell wall degrading enzymes like glucanases, chitinase and protease etc. produced by *Trichoderma* spp. which are involved in mycoparasitism is controlled mainly at transcriptional level and responsible genes are present as single copy genes. To overproduce these enzymes, their gene copy number has been increased by transformation, transformants overproducing these enzymes are more efficient as biocontrol agents (Kumar, 2013).

14.13.3 Mutation

A mutation is a permanent alteration in the sequence of nitrogenous bases of a DNA molecule. Mutation is employed to generate variability in populations, which give an opportunity to select the desirable type. They are spontaneous heritable changes. The mutants differed from the wild type strains in phenotype, growth rate, sporulation and antagonistic potential N– methyl-N– nitrosoguanidine (NTG) has been the most widely used chemical for inducing mutations in fungal *Trichoderma*. By exposing the conidia of *Trichoderma* spp., to NTG, Ahmad and Baker (1987) generated mutants that were rhizosphere competent and superior to wild type in respect of controlling *Pythium ultimum*. These mutants were insensitive to up to 100 μg/mL of benomyl. Mutants of *T. virens* by exposing the cultures to 125 k rad of gamma radiation (Mukherjee et al., 2003), carboxin tolerant mutant *T. viride* by exposing the culture to UV-light and ethylmethane sulfonate (Selvakumar et al., 2000), Triazole tolerant biotypes of *T. viride* developed by UV-irradiation (Kumar and Gupta, 1999) showed wide variation in enzyme activity especially β-1,3-glucanase and chitinases. Biocontrol efficiency of Trichoderma has been improved by transformation with genes prbl (protease), egll (β-1,4-glucanase) and chit 33 (chitinase). Lorito et al. (1998) developed transformant of *T. harzianum* strain 1295-22 by integrating β–glucuronidase (GUS) and hygromycin B (hyg B) phosphor-transferase genes that exhibited increased biocontrol activity *R. solani* as compared with the wild type.

14.14 Development of compatible consortia for improvement of formulation efficiency

Mixtures of two or more antagonists may increase the efficacy or decrease the variability associated with biocontrol treatments (Haggag and Nofal, 2006). A microbial consortium is a group of different species of microorganisms that act together as a community. For developing a consortium one can choose microorganisms that are resistant to environmental shock, fast acting, synergistically active, producing natural enzymatic activity, easy to handle, having long self-life, good sustainability, nonpathogenic, noncorrosive of consistent quality and economical. Commonly, control is based on the use of single biocontrol agents. This strategy must be changed because, from the ecological point of view, the disease is part of a complex agro-ecosystem. In recent years, more emphasis is laid on the combined use of biocontrol agents with different mechanisms of disease control, to improved disease control and also to overcome the inconsistent performance of the introduced biocontrol agents. It can be achieved by following ways:

14.14.1 Combining various microbes

Combinations with other bacteria or fungi often provided more effective disease control than the application of an individual biocontrol of *Trichoderma* alone (de Boer et al., 2003). Jetiyanon and Kloepper (2002) discovered that the use of mixtures of PGPR strains has a high potential for inducing systemic resistance against diseases of several different plant hosts in the greenhouse. Combined application of *Pichia guilermondii* and *Bacillus mycoides* (B16) reduced the infection of *Botrytis cinerea* by 75% on fruits in strawberry plants. Microbial consortia of *T. harzianum* TNHU27 and *B. subtilis* BHHU100 helps in management of soft rot pathogen *Sclerotinia sclerotiorum* (Jain et al., 2012).

14.14.2 Combining different mode of action

Application of mixtures of biocontrol agents combinations of different disease suppressive mechanisms that are complementary to each other. Cocktail of various *Trichoderma* strains provided enhanced protection than the single organism (Kumar et al., 2014). Integration of seed treatment with *T. viride* with *Rhizobium* significantly improved root and shoot length, number of nodules per plant, their fresh and dry weight in chickpea (Dubey and Patel, 2012).

14.14.3 Development of strain mixtures

Non-competitive nature of these fungal and bacterial strains is required to develop consortia which will have an additive effect in increasing the yield and plant growth apart from effective management of the diseases.

14.15 General mode of actions of microbial pesticides against plant pathogens

Fungal biopesticides compete with pathogen for nutrient and space, therefore exclude by competition. Mycoparasitism is another mode by which it parasitizes pathogenic fungus. Another way is production of secondary metabolites which is toxic to pathogens. *Trichoderma* produce enzymes like cellulases, chitinases which are cell wall degrading enzymes which help parasitizing pathogenic fungus. ATP-ABC (Adenosine triphosphate and binding cassette) transporter proteins complex assists *Trichoderma* in mycoparacitism as well as nutrient uptake (Sarma et al., 2014). As far as antibiosis is concern it produces volatile antibiotics such as 6-pentyl-a-pyrone, water-soluble acids like heptelidic acid and peptaibols which are responsible for imposing antibiosis character in *Trichoderma* (Kumar et al., 2017).

In similar way *Chaetomium* spp. as biopesticides antagonizes soil and foliar pathogens of rice, maize and wheat. It produces metabolites such as chaetomin, chaetoglobosin, cochliodinol, chaetosin and prenisatin and cellulolytic enzyme chitinase (Moya et al., 2016).

Bacterial biocontrol *P. fluorescens* anti-fungal metabolite 2,4-diacetyl phloroglucinol which act against pathogen very effectively. Antibiotic compound pyrollnitrinand phenazine-1-carboxylic acid also produced by *P. fluorescens* (Ganeshan and Manoj Kumar, 2005). It is also reported that *P. fluorescens* inactivates cell wall degrading enzymes of plant pathogenic fungi (Borowicz et al. 1992). Another gram positive bacteria *B. subtilis* which forms biofilm produces cellulases, proteases, and beta-glucanases which act against pathogens. Bacteria also produce the cell-wall-degrading enzymes and various metabolites that can limit the growth or activity of other harmful pathogen (Hashem et al., 2019).

14.16 Nanobiopesticides

Nanotechnology is an emerging science and has lot of potential in various fields. Term "nano" means 10^{-9}. Thus in case of biopesticides term nano differentiates nanobiopesticides from conventional biopesticides. Nanobiopesticides are biologically derived chemical complexes along with nanoparticles, which are projected to be utilized as pesticides (Figs. 14.3 and 14.4). The efficiency and effectiveness of nanobiopesticides are enhanced by using polymers, metal oxides, active particles combined with micelles, etc. (Lade et al., 2019). That way, nanobiopesticides is broad term and when it is related to nanobipesticides it is completely depended on purpose served by the nanomaterial. Thus nanobiopesticides is a sub class classified under nano pesticides. According to definition of Bergeson (2010) "very small particles of active ingredients or other small-engineered structures with the properties of preventing, destroying, repelling or mitigating pests will be deemed nanopesticides." As biopesticides, use of nano-organic and inorganic carriers are gaining interest along with ensuring eco-friendly nature. Nano carriers with active ingredient of biocontrol agent enables the efficient delivery system as it is prepared by the following the procedure such as adsorption, attachment, entrapment, and encapsulation techniques (Nuruzzaman et al., 2019). Components which are nanocomposite matrix + emulsion-based + nanodroplet, these provides nanobiopesticides good solubility and stability. Further, nanomaterial such as nanoencapsulates, nanocontainers, and nanocagesare used for pesticide delivery, utilization and prevention from premature degradation. Preparation of nanocomposite materials is gaining much attention in current time. Nuruzzaman et al. (2019) described preparation detailed method of nanobiopestcides in their review. Broadly nanobiopesticides are combination of biopesticides (botanical/microbial) along with nano carriers. Some examples of nanobiopesticides are cited in Tables 14.4–14.6.

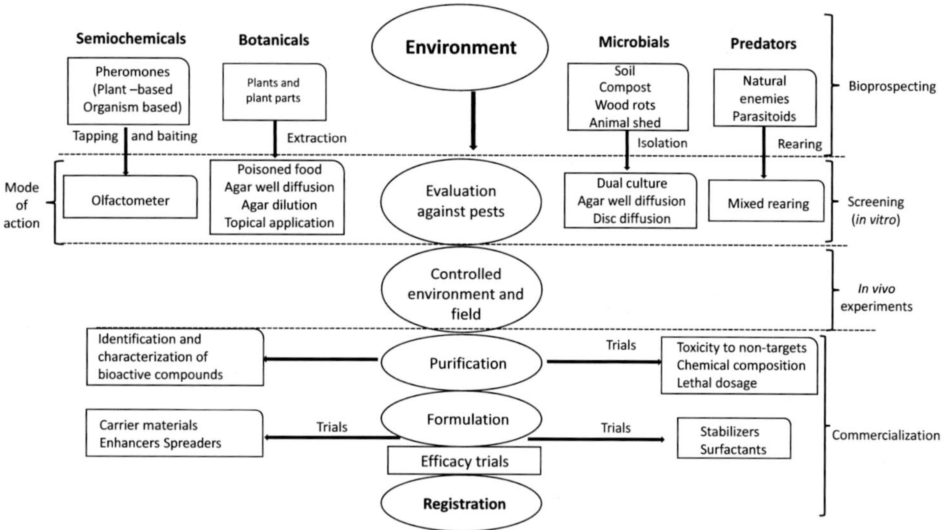

FIG. 14.3 Steps in production and evaluation of different types of biopesticides (Lengai and Muthomi, 2018).

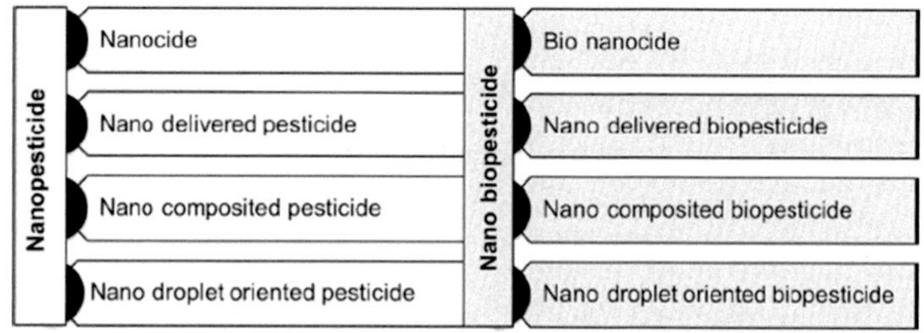

FIG. 14.4 Categories of nano-pesticides and nano-biopesticides (Nuruzzaman et al., 2019).

TABLE 14.4 Most commonly used biopesticides active ingredient in India as per biopesticides registrants (Directorate of Plant Protection, and Quarantine & Storage India (DPPQs)).

Biopesticides	Active ingredient
Trichoderma viridae	1.15% WP
Trichoderma viridae	1% WP
Trichoderma harzianum	0.5% WS
Trichoderma harzianum	2% WP
Verticiliumlecanii	1.15% WP
Pseudomonas flourescens	1.75% WP
Pseudomonas fluorescens	0.5% WP
Pseudomonas fluorescens	0.5% WP (VFU 2 × 106 gm/min)
Bacillus thuringiensis kurstaki 3a, 3b, 3c (Serotype)	0.5% WP

14.17 Biopesticides and their association with growth promoter

Microbial based biopesticides which includes fungi and bacteria also influences germination, growth and development of plants. They acts as biofertilizers and hence termed as plant growth-promoting microbes (PGPMs). This PGPM further divided into two main groups: plant growth-promoting rhizobacteria (PGPR) and plant growth-promoting fungi (PGPF), they interact in mutualistic manner with plants and enhances plant growth (Mishra et al., 2017). Among PGPM *Trichoderma* is reported extensively for growth promotion with varying degree. Stewart and Hill (2014) addressed that *Trichoderma* mainly promotes growth in terms of root and shoot biomass, but in some cases it also observed that it

TABLE 14.5 Recent examples of biopesticides recommendations in India (Directorate of Plant Protection, and Quarantine & Storage India (DPPQs, 2020)).

Bio-pesticide	Type of formulations	Formulation (Days) (g/mL)/%	Disease
Pseudomonas fluorescens	1.75% WP Accession No. MTCC 5176	05 g/kg seed (Seed treatment) 05 g/L (Foliar spray)	Loose smut of wheat
	2.0% AS Accession No. MTCC 5727	10 mL/L of water 10 mL/L of water	Seedling root, Bacterial leaf blight of rice
	0.5% WP Accession No. ITCC BE 0005	10 g/kg seed	Groundnut late leaf spot Damping off (*Pythium aphanidermatum*) Wilt (*Fusarium oxysporum*)
	1.5% Accession No. MTCC 5866	5 gm/kg of seed	Bacterial leaf blight of rice
	1.0% Accession No.MTCC5727	5 gm/kg of seed	Wilt (*Fusarium oxysporum*), Damping off (*Pythium aphanidermatum*) Root rot (*Rhizoctonia* spp.)
Bacillus subtilis	1.50% L.F Strain MTCC 25072	5 L/ha (Foliar spray)	Banana sigatoka
	2.0% AS Accession No. MTCC 5728	10 mL/L of water	Bacterial leaf blight of rice
Trichoderma harzianum/ viridae 50% WS	50% WS	100 gm/plant	Capsule rot (*Phytophthora meadii*)
	1.0% WP Accessions No. ITCC6888	20 gm/kg seed	*F. oxysporum, F. solani,* Root rot (*Athelia rolfsii*)
	1.0% WP	4 gm/kg seed	Stalk rot (*Sclerotinia sclerotiorum*)
	2.0% WP	20 gm/kg seed	Root rot (*Fusarium verticillioides*) in maize
1.0% WP	8 gm/kg seed 5 gm/kg seed	Wilt, root rot of pulses, root rot cowpea	
	1.5% WP Accession No. ITCC 6889	20 gm/kg	*F. oxysporum, F. solani,* root rot (*Sclerotium rolfsii*)
Ampelomyces quisqualis	2.0% WP MTCC-5683	2.5 kg	Powdery mildew (*Erysiphe cichoracearum*)
Azadirachtin	0.15% EC	Neem seed kernel based	As protectant fungicides which helps to prevent *F. oxysporum, Rhizoctonia solani, Alternaria solani* and *S. sclerotiorum*

positively affects morphology and development of plants. The evidence which supports the fact that fungal based biopesticides especially *Trichoderma* are increased nutrient uptake, carbohydrate metabolism, photosynthesis, and phytohormone synthesis. Among PGPR most popular bacteria is *P. fluorescence,* there are several PGPR based formulations available commercially which are not only useful for managing plant disease but also as biofertilizers for use in agriculture microbes as biofertilizers interacts with plants in symbiosis and assists them in nitrogen fixation, P-solubilization, siderophore production and HCN production especially in case of legumes it has reported remarkably significant (Backer et al., 2018).

14.18 Inducer of systemic resistance in plant against plant pathogen

Plants responds to various stimuli such as physical, chemical and physiological, in similar way it also responds to stress such as biotic and abiotic. These stimuli can trigger induced defense response inside plants against stress. It can be local or systemic *viz.*, salicylic acid mediated systemic acquired resistance (SAR) which subsequently leads to the expression of PR proteins. Jasmonic acid can also mediate induced systemic resistance (ISR) in the host plants. Microbial pesticides are capable of inducing either one or both types of defense responses.

TABLE 14.6 List of nanobiopesticides formulations.

Biopesticides	Nanocarrier	References
Garlic essential oil	Poly-ethyleneglycol	Yang et al. (2009)
Validamycin	Poroushollow, silicanano particles	Liu et al. (2006)
Avermectin	Poroushollow, silicananoparticles	Li et al. (2007)
Azadirachtin-A	Nanomicelles	Kumar et al. (1989)
Neem oil	Poly-ε-caprolactone	Carvalho et al. (2012)
Neem seed kernel	PCL poly-ε-caprolactone	Forim et al. (2015)
Neem oil	Silicanano particles	El-Samahy et al. (2014)

Bacterial biopesticides especially *P. fluorescence* mainly induces ISR response whereas *P. fluorescence* strain CHA0 in tobacco induces SAR (Notz et al., 2001). Bacterial determinants which induces systemic defense response are peroxidase, chitinase, β-1,3-glucanase, 2,3-butanediol, siderophore, lipopolysaccharide, lipopolysaccharide Z, 3-hexenal and iron regulated factor. Apart from *P. fluorescence, Serratia marcescens* strain 90-166 also induces ISR in cucumber due to siderophore activity (Press et al., 2001). Fungal biopesticides especially *Trichoderma* sp. and its associated strains are widely reported strong inducer of systemic response. *Trichoderma* spp. capable of inducing SAR in plants by various means such as generation of reactive oxygen species (ROSs), PR protein accumulation, phytoalexin production and interestingly induction of lignifications against soil borne pathogens to prevent them from invading host tissues (Meshram et al., 2019). Various species of the biocontrol agent *Trichoderma* produce glucan, chitin, xylan which also play the role of microbial associated molecular patterns (MAMPs) there by act as inducer of SAR in the plants (Köhl et al., 2019).

14.19 Botanical biopesticides usage against plant pathogen

Plants produce many secondary metabolites and some of these metabolites have the property to be used as pesticides. These secondary metabolites affect indirectly such as repellent or directly as toxicant. There are number of plant based pesticides available commercially for application against pest and diseases of plants (Cloyd et al., 2009). These compounds include terpenes, phenolics, and alkaloids. Azadirachtins are compounds that are obtained from neem extracts and contain triterpenoids as principal component. Active ingredient of this biopesticides affects reproductive and digestive processes of insect pests. Important pathogenic fungi namely *Fusarium oxysporum*, *Rhizoctonia solani*, *Alternaria solani* and *Sclerotinia sclerotiorum* were reported to inhibit by neem seed extract (Moslem and El-Kholie, 2009). In similar way *Nicotiana tabacum* extract also exhibits antifungal activities against fungal pathogens such as *Aspergillus viridae, Penicillium digitatum* and *Rhizopus* sp. (Suleiman, 2011).

14.20 Essential oils

Essential oils (EOs) are more recently added as biopesticides and their demands are increasing with time. Examples of some plants which are endowed with EOs are mentha, geranium, cinnamon, cedar, rose, ginger, fennal, balsam, orange, lemon etc. These EOs can be obtained from every plants leaves, bark, flower, roots seeds and gum extracts. EOs are found in various forms which act as biopesticides like limonene, citral, geraniol, esters like benzoates acetates, camphor, cinol, anethol, eugenol, cymene, menthone etc. (Handa, 2008). Principal compounds in EOs are monoterpenoids and sesquiterpenoids. EOs are plant products usually obtained by distillation or other extraction methods, most simple is hydro-distillation and steam distillation for volatile compounds and for volatile compounds cold fat extraction, expression, maceration, and solvent extraction methods are popular. A wide number of plant pathogens are reported to be sensitive to EOs (Table 14.7).

14.21 Advantages and limitations of biopesticides

14.21.1 Advantages

- Biopesticides are less likely to contaminate soil and groundwater.
- Less chances of pesticides resistance in target organisms because of complexity of mixtures.

TABLE 14.7 Essential oils used against plant diseases.

Pathogen	Disease	Plant source	References
Aspergillus flavus	Aflatoxins production	*Mentha x piperita*	Camiletti et al. (2014)
Aspergillus niger	Mold	*Ocimum basilicum*	Hanif et al. (2017)
Bipolaris oryzae	Brown spot	*Piper sarmentosum*	Irshad et al. (2012)
Bipolaris sorokiniana	Leaf blight/spot	*Pinus pinea*	Amri et al. (2012)
Botrytis cinerea	Gray mold	*Foeniculum vulgare*	Aminifard and Mohammadi (2012)
Colletotrichum capsici	Leaf spot	*Cestrum nocturnum*	Al-Reza et al. (2010)
Eurotium sp.	Mold	*Citrus x limon*	Dimi'c et al. (2015)
Fusarium oxysporum	Fusarium wilt	18 Egyptian plant species	Badawy and Abdelgaleil (2014)
Rhizoctonia solani	Damping-off, root and stems rot	*Mentha* spp., *Ocimum, Cestrum, Piper chaba*	Khaledi et al. (2015)

- Multiple modes of action also an important factor which makes it difficult to detoxify biopesticides, especially the botanical biopesticides.
- Most biopesticides are fast acting, therefore reduces chances of multiplication of pathogen.
- Mostly biopesticides especially microbial biopesticides are target specific and degrade quickly in environment, so less chances of human and animal toxicity.
- Help maintain beneficial insect populations, their non-hazardous application, and their effective use in resistance management programs.
- Most biopesticides worldwide are exempted from residue limits on fresh and processed foods

14.21.2 Limitations

- When exposed to sunlight, plant-based compounds breakdown easily into nontoxic substances.
- More frequent applications and dosage are required.
- Commercial formulations are more expensive than synthetic pesticides due to the high cost of raw materials.
- Takes more time for disease control compared to synthetic pesticides which gives quick results.
- Quantity of plant secondary metabolites varies season to season.
- Performance of microbial biopesticides influenced by environmental fluctuations.

14.22 Factors affecting biopesticides marketing

In current time pesticide preference of the farming community is changing and this change is influenced by factors like increasing demand of organic or chemical free commodity. Another important factor that influences biopesticides market positively is integration of biological crop protection techniques, which eventually increases the demand. The factors which offer challenges for biopesticides markets are limited known biological control technique and registration of pesticides (Anonymous, 2020). These challenges can be overcome by promoting extensive research, extending partnership globally targeting R & D and fast-tracking the registration of products. Rapid promotion of IPM strategies among farmers will strengthen the demand of biopesticides. Some important factors which impact biopesticides market are shown below (Fig. 14.5).

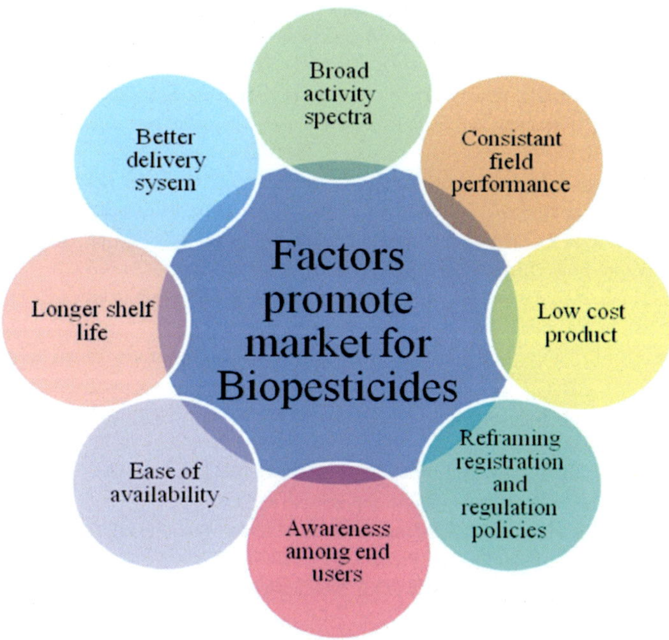

FIG. 14.5 Diagrammatic representation of factors affecting market of biopesticides (Sachdev and Singh, 2016).

14.23 Conclusion

Use of microbes as biofertilizers and biopesticides is providing a solution to reduce the reliance on synthetic pesticides. However, implementation of biopesticides strategies at commercial level is still challenging because it requires high establishment cost and sometimes for management of single pathogen, multiple biocontrol strategies needs to be adopted. Intensive use of synthetic pesticides in activities can cause pesticide resistance development and therefore resurgence of target pests occur. When we talk about the application of biopesticides we should also consider the compatibility between two groups *viz.*, microbial biopesticides and botanical biopesticides, because botanical biopesticides are non-selective in action and they may affect beneficial microbes negatively. Biopesticides although have less effectiveness compared to synthetic pesticides but it is gaining popularity for securing environmental health intact which otherwise corroded by indiscriminate usage of synthetic pesticides. More research toward enhancement of stability and efficacy will improve the wide adaption of biopesticides application in agriculture disease management such as focus on improving quality of nano-carriers in nano-biopesticide, efficient encapsulation of biocontrol agent, essential oils and biopesticides. Further, authorizations for biopesticides commercialization and production must be promoted.

References

Abhilash, P., Singh, N., 2009. Pesticide use and application: an Indian scenario. J. Hazard Mater. 165 (1–3), 1–12.

Agribusiness Global, 2020. India Moves to Ban 27 Pesticides. https://www.agribusinessglobal.com/markets/india-moves-to-ban-27-pesticides/.

Ahmad, J.S., Baker, R., 1987. Competitive saprophytic ability and cellulolytic activity of rhizosphere-competent mutants of *Trichoderma harzianum*. Phytopathology 77 (2), 358–362.

Al-Reza, S.M., Rahman, A., Ahmed, Y., Kang, S.C., 2010. Inhibition of plant pathogens in vitro and in vivo with essential oil and organic extracts of *Cestrum nocturnum* L. Pestic. Biochem. & Phys. 96, 86–92.

Al-Waily, D.S., Al-Saad, L.A., Al-Dery, S.S., 2018. Formulation of *Pseudomonas fluorescens* as a biopesticide against soil borne root pathogens. Iraqi J. Agric. Sci. 49 (2), 235–242.

Aminifard, M.H., Mohammadi, S., 2012. Essential oils to control *Botrytis cinerea* in vitro and in vivo on plum fruits. J. Sci. Food Agric. 93, 348–353.

Amri, I., Gargouri, S., Hamrouni, L., Hanana, M., Fezzani, T., Jamoussi, B., 2012. Chemical composition, phytotoxic and antifungal activities of *Pinus pinea* essential oil. J. Pestic. Sci. 85, 199–207.

Anonymous, 2020. Global market study on biopesticides: the growing focus on sustainable and organic agriculture. In: Persistance Market Research Report. https://www.persistencemarketresearch.com/market-research/biopesticides-.

Backer, R., Rokem, J.S., Ilangumaran, G., Lamont, J., Praslickova, D., Ricci, E., Smith, D.L., 2018. Plant growth-promoting rhizobacteria: context, mechanisms of action, and roadmap to commercialization of biostimulants for sustainable agriculture. Front. Plant Sci. 9, 14–73.

Badawy, M.E.I., Abdelgaleil, S.A.M., 2014. Composition and antimicrobial activity of essential oils isolated from Egyptian plants against plant pathogenic bacteria and fungi. Ind. Crop. Prod. 52, 776–782.

Bergeson, L.L., 2010. Nanosilver: US EPA's pesticide office considers how best to proceed. Environ. Qual. Manag. 19, 79–85.

Berry, C., 2012. The bacterium, *Lysinibacillus sphaericus*, as an insect pathogen. J. Invertebr. Pathol. 109 (1), 1–10.

Borowicz, J., Pietr, S.J., Stankiewcz, M., Lewicka, T., Zukowska, Z., 1992. Inhibition of fungal cellulase, pectinase and xylanase activity by plant growth-promoting fluorescent *Pseudomonas* spp. Bull. OILB SROP 15, 103106.

Camiletti, B.X., Asensio, C.M., Pecci, M.D.L.P.G., Lucini, E.I., 2014. Natural control of corn postharvest fungi *Aspergillus flavus* and *Penicillium* sp. using essential oils from plants grown in Argentina. J. Food Sci. 79, M2499–M2506.

Carvalho, S.S., Vendramim, J.D., Pitta, R.M., Forim, M.R., 2012. Efficiency of neem oil nanoformulations to *Bemisia tabaci* (GENN.) Biotype B (Hemiptera: *Aleyrodidae*). Semin: Ciênc. Agrár. 33 (1), 193–201.

Cloyd, R.A., Galle, C.L., Keith, S.R., Kalscheur, N.A., Kemp, K.E., 2009. Effect of commercially available plant-derived essential oil products on arthropod pests. J. Econ. Entomol. 102 (4), 1567–1579.

de Boer, M., Bom, P., Kindt, F., Keurentjes, J.J.B., van dex Sluism, I., van Loon, L.C., Bakker, P.A.H.M., 2003. Control of *Fusarium* wilt of radish by combining *Pseudomonas putida* strains that have different disease suppressive mechanisms. Phytopath 93, 626–632.

Dimi'c, G., Koci'c-Tanackov, S., Mojovi, L., Pejin, J., 2015. Antifungal activity of lemon essential oil, coriander and cinnamon extracts on foodborne molds in direct contact and the vapor phase. J. Food Process. Preserv. 39, 1778–1787.

Directorate of Plant Protection, and Quarantine & Storage India (DPPQS). Notification. https://d6kq167ddwbdq.cloudfront.net/farmchemint/wpcontent/uploads/2020/05/Untitled-1.pdf.

Directorate of Plant Protection, Quarantine & Storage India (DPPQS), 2020. Major Uses of Bio-Fungicides. http://ppqs.gov.in/sites/default/files/approved_use_of_bio-pesticides.pdf.

Dubey, S.C., Patel, B., 2012. Determination of tolerance in *Thanatephorus cucumeris*, *Trichoderma viride*, *Gliocladium virens* and *Rhizobium* sp. to fungicides. Indian Phytopathol. ISSN: 0367-973X.

El-Samahy, M.F., El-Ghobary, A.M., Khafagy, I.F., 2014. Using silica nanoparticles and neemoil extract as new approaches to control *Tuta absoluta* (meyrick) in tomato under field conditions. Int. J. Phys. Soc. Sci. 3, 1355–1365.

EPA, 1999. Impact Reports and Fact Sheets. https://www.epa.gov/research/impact-reports-and-fact-sheets.

FAO, 2017. Pesticide Residues in Food. Joint FAO/WHO Meeting on Pesticide Residues. Food and Agriculture Organization of the United Nations, WHO, Rome FAO. http://www.fao.org/3/ca7032en/ca7032en.pdf.

Faria Jr., P., 2017. Biocontrol in Brazil: opportunities and challenges. Agropag. Biopest. Suppl. 12–13.

Forim, M.R., Da Silva, F., Fernandes, J.B., Vieira, P.C., 2015. Process for Obtaining Biopolymeric Nanoparticles Containing *Azadirachta indica* A. Juss. (neem.) Oil and Extracts, Biopolymeric Nanoparticles, and Powder Microparticles. United States, Patent Application Publication, Pub. No.: US 2015/0320036A1, pp. 1–23.

Ganeshan, G., Manoj Kumar, A., 2005. Pseudomonas fluorescens, a potential bacterial antagonist to control plant diseases. J. Plant Interact. 1 (3), 123–134.

Gevao, B., Jones, K.C., 2002. Pesticides and persistent organic pollutants. In: Haygarth, P., Jarvis, S. (Eds.), Agriculture, Hydrology and Water Quality. CABI Publishing, North Wyke, UK, pp. 83–100.

Glick, B.R., Bashan, Y., 1997. Genetic manipulation of plant growth-promoting bacteria to enhance biocontrol of phytopathogens. Biotechnol. Adv. 15 (2), 353–378.

Haggag, W.M., Nofal, M.A., 2006. Improving the biological control of *Botryodiplodia* disease on some Annona cultivars using single or multi-bioagents in Egypt. Biol. Control 38 (3), 341–349.

Handa, S.S., 2008. An overview of extraction techniques for medicinal and aromatic plants. In: Handa, S.S., Khanuja, S.P.S., Longo, G., Rakesh, D.D. (Eds.), Extraction Technologies for Medicinaland Aromatic Plants. UNIDO International Centre for Science and High Technology, Trieste, pp. 21–54.

Hanif, M.A., Nawaz, H., Ayub, M.A., Tabassum, N., Kanwal, N., Rashid, N., Saleem, M., Ahmad, M., 2017. Evaluation of the effects of Zinc on the chemical composition and biological activity of basil essential oil by using Raman spectroscopy. Ind. Crop. Prod. 96, 91–101.

Hashem, A., Tabassum, B., Abd Allah, E.F., 2019. *Bacillus subtilis*: a plant-growth promoting rhizobacterium that also impacts biotic stress. Saudi J. Biol. Sci. 6 (26), 1291–1297.

Hassan, E., Gökçe, A., 2014. Production and consumption of biopesticides. In: Advances in Plant Biopesticides. Springer, New Delhi, pp. 361–379.

Henn, T., Weinzierl, R., 1989. Botanical insecticides and insecticidal soaps. Circular 1296, 1–18.

Hsaio, J., 2017. GMOs and pesticides: helpful or harmful? Sci. News 29.

Ignacimuthu, S., 2012. Insect Pest Control: Using Plant Resources. Alpha Science International, Oxford.

Irshad, M., Aziz, S., Hussain, H., 2012. GC-MS analysis and antifungal activity of essential oils of *Angelica glauca*, *Plectranthus rugosus* and *Valeriana wallichii*. J. Essent. Oil Bear. Plants 15, 15–21.

Jain, A., Singh, S., Kumar Sarma, B., Bahadur Singh, H., 2012. Microbial consortium mediated reprogramming of defence network in pea to enhance tolerance against *Sclerotinia sclerotiorum*. J. Appl. Microbiol. 112 (3), 537–550.

Jetiyanon, K., Kloepper, J.W., 2002. Mixtures of plant growth-promoting rhizobacteria for induction of systemic resistance against multiple plant diseases. Biol. Control 24 (3), 285–291.

Johnson, V.W., Pearson, J.F., Jackson, T.A., 2001. Formulation of *Serratia entomophila* for biological control of grass grub. In: Proceedings of the New Zealand Plant Protection Conference. New Zealand Plant Protection, vol. 54, pp. 125–127.

Jones, R.S., 2010. Biochemical pesticides: green chemistry designs by nature. In: Handbook of Green Chemistry: Online, pp. 329–347.

Kabaluk, J.T., Svircev, A.M., Goettel, M.S., Woo, S.G., 2010. The Use and Regulation of Microbial Pesticides in Representative Jurisdictions Worldwide. International Organization for Biological Control of Noxious Animals and Plants, (IOBC), p. 99.

Khaledi, N., Taheri, P., Tarighi, S., 2015. Antifungal activity of various essential oils against *Rhizoctonia solani* and *Macrophomina phaseolina* as major bean pathogens. J. Appl. Microbiol. 118, 704–717.

Khan, M.J., Zia, M.S., Qasim, M., 2010. Use of pesticides and their role in environmental pollution. World Acad. Sci. Eng. & Technol. 72, 122–128.

Köhl, J., Kolnaar, R., Ravensberg, W.J., 2019. Mode of action of microbial biological control agents against plant diseases: relevance beyond efficacy. Front. Plant Sci. 10, 845.

Kumar, S., 2013. *Trichoderma*: a biological weapon for managing plant diseases and promoting sustainability. Int. J. Agric. Sci. & Med. Veter. 1 (3), 106–121.

Kumar, A., Gupta, J.P., 1999. Variations in enzyme activity of tebuconazole tolerant biotypes of *Trichoderma viride*. Indian Phytopathol. 52 (3), 263–266.

Kumar, J., Shakil, N.A., Singh, M.K., Singh, M.K., Pandey, A., Henn, T., Weinzierl, R., 1989. Botanical insecticides and insecticidal soaps. Circular 1296, 1–18.

Kumar, S., Thakur, M., Rani, A., 2014. *Trichoderma*: mass production, formulation, quality control, delivery and its scope in commercialization in India for the management of plant diseases. Afr. J. Agric. Res. 9 (53), 3838–3852.

Kumar, G., Maharshi, A., Patel, J., Mukherjee, A., Singh, H.B., Sarma, B.K., 2017. *Trichoderma*: a potential fungal antagonist to control plant diseases. SATSA Mukhapatra Ann. Techn. Iss. 21, 206–218.

Kunimi, Y., 2007. Current status and prospects on microbial control in Japan. J. Invertebr. Pathol. 95 (3), 181–186.

Kunstadter, P., 2007. Pesticides in Southeast Asia: Environ-Mental, Biomedical, and Economic Uses and Effects. Silkworm Books, Ms Trasvin, p. 326.

Lade, B.D., Gogle, D.P., Lade, D.B., Moon, G.M., Nandeshwar, S.B., Kumbhare, S.D., 2019. Nanobiopesticide formulations: application strategies today and future perspectives, Chapter 7. In: Koul, O. (Ed.), Nano-Biopesticides Today and Future Perspectives. Elsevier Inc, pp. 179–206. https://doi.org/10.1016/B978-0-12-815829-6.00007-3.

Lamichhane, J.R., Dachbrodt-Saaydeh, S., Kudsk, P., Messéan, A., 2016. Conventional pesticides in agriculture: benefits versus risks. Plant Dis. 100 (1), 10–24.

Leng, P., Zhang, Z., Pan, G., Zhao, M., 2011. Applications and development trends in biopesticides. Afr. J. Biotechnol. 10 (86), 19864–19873.

Lengai, G.M., Muthomi, J.W., 2018. Biopesticides and their role in sustainable agricultural production. J. Biosci. Med. 6 (06), 7.

Li, Z.Z., Chen, J.F., Liu, F., Liu, A.Q., Wang, Q., Sun, H.Y., Wen, L.X., 2007. Study of UV shielding properties of novel porous hollow silica nanoparticle carriers for avermectin. Pest Manag. Sci. 63, 241–246.

Liu, F., Wen, L.-X., Li, Z.-Z., Yu, W., Sun, H.-Y., Chen, J.-F., 2006. Porous hollow silica nanoparticles as controlled delivery system for water-soluble pesticide. Mater. Res. Bull. 41, 2268–2275.

Lopez, Ó., Fernández-Bolaños, J.G., Gil, M.V., 2011. Classical insecticides: past, present and future. In: Green Trends in Insect Control. Royal Society of Chemistry, pp. 53–93.

Lorito, M., Woo, S.L., Fernandez, I.G., Colucci, G., Harman, G.E., Pintor-Toro, J.A., Tuzun, S., 1998. Genes from mycoparasitic fungi as a source for improving plant resistance to fungal pathogens. Proc. Natl. Acad. Sci. U. S. A. 95 (14), 7860–7865.

Manikandan, R., Saravanakumar, D., Rajendran, L., Raguchander, T., Samiyappan, R., 2010. Standardization of liquid formulation of *Pseudomonas fluorescens* Pf1 for its efficacy against *Fusarium* wilt of tomato. Biol. Contr. 54 (2), 83–89.

Meshram, S., Patel, J.S., Yadav, S.K., Kumar, G., Singh, D.P., Singh, H.B., Sarma, B.K., 2019. *Trichoderma* mediate early and enhanced lignifications in chickpea during *Fusarium oxysporum* f. sp. ciceris infection. J. Basic Microbiol. 59 (1), 74–86.

Ministry of the Environment, 2002. Towards a Pesticide Risk Reduction Policy for New Zealand. A Public Discussion Paper, pp. 1–67 (MfE,Wellington).

Mishra, J., Tewari, S., Singh, S., Arora, N.K., 2015. Biopesticides: where we stand?. In: Plant Microbes Symbiosis. Applied Facets. Springer, New Delhi, pp. 37–75.

Mishra, J., Singh, R., Arora, N.K., 2017. Plant growth-promoting microbes: diverse roles in agriculture and environmental sustainability. In: Probiotics and Plant Health. Springer, Singapore, pp. 71–111.

Moslem, M.A., El-Kholie, E.M., 2009. Effect of neem (Azardirachtaindica A. Juss) seeds and leaves extract on some plant pathogenic fungi. Pakistan J. Biol. Sci. 12 (14), 1045.

Moya, P., Pedemonte Roman, D., Susana, A., Franco, M.E.E., Sisterna, M.N., 2016. Antagonism and Modes of Action of *Chaetomium globosum* Species Group, Potential Biocontrol Agent of Barley Foliar Diseases. ISSN: 0373-580X.

Mukherjee, M., Hadar, R., Mukherjee, P.K., Horwitz, B.A., 2003. Homologous expression of a mutated beta-tubulin gene does not confer benomyl resistance on *Trichoderma virens*. J. Appl. Microbiol. 95 (4), 861–867.

Notz, R., Maurhofer, M., Schnider-Keel, U., Duffy, B., Haas, D., Defago, G., 2001. Biotic factors affecting expression of the 2,4-diacetylphloroglucinol biosynthesis gene phlA In: *Pseudomonas fluorescens* biocontrol strain CHA0 in the rhizosphere. Phytopathology 91, 873–881.

Nuruzzaman, M., Liu, Y., Rahman, M.M., Dharmarajan, R., Duan, L., Uddin, A.F.M.J., Naidu, R., 2019. Nanobiopesticides: composition and preparation methods. In: Nano Biopesticides Today and Future Perspectives. Academic Press, pp. 69–131.

Panizzi, A.R., Bueno, A.D.F., Silva, F.D., 2012. Insects that Attack Pods and Grains. Soy: Integrated Management of Insects and Other Pest Arthropods. Embrapa, Brasília, pp. 335–420.

Park, J.H., Choi, G.J., Jang, K.S., Lim, H.K., Kim, H.T., Cho, K.Y., Kim, J.C., 2005. Antifungal activity against plant pathogenic fungi of chaetoviridins isolated from *Chaetomium globosum*. FEMS (Fed. Eur. Microbiol. Soc.) Microbiol. Lett. 252 (2), 309–313.

Pathak, D.V., Yadav, R., Kumar, M., 2017. Microbial pesticides: development, prospects and popularization in India. In: Plant-Microbe Interactions in Agro-Ecological Perspectives. Springer, Singapore, pp. 455–471.

Penttilä, M., Nevalainen, H., Rättö, M., Salminen, E., Knowles, J., 1987. A versatile transformation system for the cellulolytic filamentous fungus *Trichoderma reesei*. Gene 61 (2), 155–164.

Prabavathy, V.R., Mathivanan, N., Sagadevan, E., Murugesan, K., Lalithakumari, D., 2006. Self-fusion of protoplasts enhances chitinase production and biocontrol activity in *Trichoderma harzianum*. Bioresour. Technol. 97 (18), 2330–2334.

Press, C.M., Loper, J.E., Kloepper, J.W., 2001. Role of iron in rhizobacteria mediated induced systemic resistance of cucumber. Phytopathology 91, 593–598.

Rao, M.S., Umamaheswari, R., Chakravarthy, A.K., Grace, G.N., Kamalnath, M., Prabu, P., 2015. A frontier area of research on liquid biopesticides: the way forward for sustainable agriculture in India. Curr. Sci. 108 (9), 1590–1592.

Ruiu, L., 2018. Microbial biopesticides in agro ecosystems. Agronomy 8 (11), 235.

Sachdev, S., Singh, R.P., 2016. Current challenges, constraints and future strategies for development of successful market for biopesticides. Clim. Chang. & Environ. Sustain. 4 (2), 129–136.

Sarma, B.K., Yadav, S.K., Patel, J.S., Singh, H.B., 2014. Molecular mechanisms of interactions of *Trichoderma* with other fungal species. Open Mycol. J. 8, 140–147.

Sarmah, A.K., Müller, K., Ahmad, R., 2004. Fate and behaviour of pesticides in the agroecosystem a review with a New Zealand perspective. Soil Res. 42 (2), 125–154.

Schisler, D.A., Slininger, P.J., Behle, R.W., Jackson, M.A., 2004. Formulation of *Bacillus* spp. for biological control of plant diseases. Phytopathology 94 (11), 1267–1271.

Schreinemachers, P., Tipraqsa, P., 2012. Agricultural pesticides and land use intensification in high, middle and low income countries. Food Pol. 37 (6), 616–626.

Selvakumar, R., Srivastava, K.D., Rashmi, A., Singh, D.V., Prem, D., 2000. Studies on development of Trichoderma viride mutants and their effect on Ustilago segetum tritici. Indian Phytopath. 53 (2), 185–189.

Soytong, K., Kanokmedhakul, S., Kukongviriyapa, V., Isobe, M., 2001. Application of *Chaetomium* species (Ketomium) as a new broad spectrum biological fungicide for plant disease control. Fungal Divers. 7, 1–15.

Stewart, A., Hill, R., 2014. Applications of *Trichoderma* in plant growth promotion. In: Biotechnology and Biology of *Trichoderma*. Elsevier, pp. 415–428.

Suleiman, M.N., 2011. Antifungal properties of leaf extract of neem and tobacco on three fungal pathogens of tomato (*Lycopersicon Esculentum* Mill). Adv. Appl. Sci. Res. 2 (4), 217–220.

Trichoderma Fact Sheet, 2000. https://www3.epa.gov/pesticides/chem_search/reg_actions/registration/fs_PC-119200_01May-00.pdf.

Tripathi, Y.N., Divyanshu, K., Kumar, S., Jaiswal, L.K., Khan, A., Birla, H., Upadhyay, R.S., 2020. Biopesticides: current status and future prospects in India. In: Bioeconomy for Sustainable Development. Springer, Singapore, pp. 79–109.

Warren, G.F., 1998. Spectacular increases in crop yields in the United States in the twentieth century. Weed Technol. 12, 752.

Webster, J.P.G., Bowles, R.G., Williams, N.T., 1999. Estimating the economic benefits of alternative pesticide usage scenarios: wheat production in the United Kingdom. Crop Prod. 18, 83.

Yang, F.-L., Li, X.-G., Zhu, F., Lei, C.-L., 2009. Structural characterization of nanoparticles loaded with garlic essential oil and their insecticidal activity against *Tribolium castaneum* (Herbst) (Coleoptera: Tenebrionidae). J. Agric. Food Chem. 57, 10156–10162.

Chapter 15

Insights into the genomes of microbial biopesticides

A.B. Vedamurthy[a], Sudisha Jogaiah[c] and S.D. Shruthi[b]

[a]*Department of Biotechnology and Microbiology, Karnatak University, Dharwad, Karnataka, India;* [b]*Microbiology and Molecular Biology Lab, BioEdge Solutions, Bangalore, Karnataka, India;* [c]*Laboratory of Plant Healthcare and Diagnostics, PG Department of Biotechnology and Microbiology, Karnatak University, Dharwad, Karnataka, India*

15.1 Introduction

Microbial pesticides can be used effectively as an alternative to conventional chemical insecticides. Biopesticides are employed in agricultural use for the purposes of insect control, disease control, weed control, nematode control and plant physiology and productivity. They usually occur from natural origin or genetically altered bacteria, fungi, algae, viruses or protozoans. Microbial toxin can be defined as biological toxin material derived from any microorganism. But the pathogenic effect of these microorganisms on the target pests is very specific to species. There are at least 1500 naturally occurring insect-specific microorganisms, 100 of which are insecticidal and over 200 microbial biopesticides are available so far in 30 countries affiliated to the Organization for Economic Co-operation and Development (OECD). They represent a wide range of bio-based substances acting against pests with different mechanisms of action and are classified based on occurrence of biochemicals, microbial entomopathogens; and plant-incorporated protectants which are genetically engineered (Jogaiah et al., 2018). The basic concept of employing organisms and natural products leverages the properties of natural ecosystem components to counteract the biotic and reproductive potential of pests (Kenis et al., 2017; Kaya and Vega, 2012; Marrone, 2014; Jogaiah et al., 2016). Microbial pesticides and other entomopathogens generally refer to pesticides that contain microorganisms like bacteria, fungi, or virus, attacking specific pest species. In other way, these pesticides will be having entomopathogenic nematodes as active ingredients. Although most of these agents attack insect species normally called as entomopathogens and the products are referred as bioinsecticides. There are also microorganisms like fungi which control weeds referred as bioherbicides.

15.1.1 Entomopathogenic bacteria

From long time different bacterial species have been the object of studies toward investigating their pathogenic relationship with insects. Coming to bacterial biopesticides, the bacteria that are used as biopesticides can be divided into four categories: crystalliferous spore formers (such as *Bacillus thuringiensis*); obligate pathogens (such as *Bacillus popilliae*); potential pathogens (such as *Serratia marcesens*); and facultative pathogens (such as *Pseudomonas aeruginosa*). Different bacterial species in the family Bacillaceae have been the object of studies investigating their pathogenic relationship with invertebrates, especially insects from long time (Castagnola and Stock, 2014; Ruiu, 2015). Among them spore formers have been most widely adopted for commercial use because of their safety and effectiveness, like *Bacillus thuringiensis, Bacillus sphaericus* etc. Subspecies which are used till now are *B. thuringiensis tenebrionis, B. thuringiensis kurstaki, B. thuringiensis israelensis* and *B. thuringiensis aizawai*. Other species of bacteria have little impact on pest management, still some commercial products based on *Agrobacterium radiobacter*, *Bacillus popilliae*, *Bacillus subtilis*, *Pseudomonas cepacia*, *Pseudomonas chlororaphis*, *Pseudomonas flourescens*, *Pseudomonas solanacearum* and *Pseudomonas syringae* are available (Babu et al., 2015).

The group of bacterial pathogens is well represented by *Bacillus thuringiensis* (Bt) which is most studied and commercially utilized. The biosynthesis of crystaltoxins (Cry, Cyt) associated with parasporal bodies produced during the sporulation phase and other toxins and a virulence factor determines the insecticidal activity of this bacterium. Some of the

Biopesticides. https://doi.org/10.1016/B978-0-12-823355-9.00026-2

toxins and virulence factors are also produced and released by the cell during the vegetative phase of growth (Jurat-Fuentes and Jackson, 2012). Varying gene sequences of toxins result in different affinity with insect midgut receptors due to which different strains are characterized by diverse insecticidal protein toxins and strain-specific insecticidal properties (Pigott and Ellar, 2007). Hence, different Bt strains are effective only against a narrow target range. This leads to the search for new strains and insecticidal toxins at the scientific and industrial level. These toxins specifically bind to insect mid gut receptors after being ingested triggering a pore-forming process that determines the alteration in permeability of epithelial membrane. This consequently disrupts the functions of intestinal barrier and eventual bacterial septicemia leading to insect death (Bravo et al., 2007). Same kind of mechanism is associated with another species *Bacillus sphaericus* (*Lysinibacillus sphaericus*) which act against mosquitoes and blackflies through the production of complementary crystal proteins (BinA, BinB) and mosquitocidal toxins (Mtx) (Charles et al., 2000). There are insecticidal toxins which show high homology with Bt Cry toxins in species belonging to *Paenibacillus* genus. Another bacterium in the same family showing wide spectrum of pesticidal activity is *Brevibacillus laterosporus*. It is characterized by a swollen sporangium containing a spore with a canoe-shaped parasporal body attached to one side of it (Ruiu, 2013; Marche et al., 2017). This bacterium is said to have several virulence factors and its use in integrated management of different pests has been proposed (Marche et al., 2018; Ruiu et al., 2011, 2014; Babu et al., 2015).

The activity of *B. thuringiensis* on crop foliage or applications via soil can be enhanced by genetic manipulation. Genetic transformation of *B. thuringiensis* has produced a strain that displays insecticidal activity against both coleopteran and lepidopteran insects. Crystal proteins (Cry34, Cry35) of *B. thuringiensis* are closely related to each other, are environmentally ubiquitous and share similarity in sequences with activity through disruption of membrane in target organisms. Cry35 proteins can be modified such that their segments, domains and motifs will be exchanged with other proteins to enhance insecticidal activity, which provides excellent control of plant pests and rootworms (Schnepf et al., 2007). Similarly, a toxin polypeptide from *B. thuringiensis* called Cry8Bb1has been engineered to contain a proteolytic protection site, which makes it insensitive to a plant protease. This helps to protect the toxin from any proteolytic inactivation. Modified Cry8Bb1 has been used for controlling corn root worms, wireworms, boll weevils, Colorado potato beetles and the alfalfa weevils (Abad et al., 2008).

Recent study shows that *B. cereus* genomes have a *Bacillus* enhancin-like (bel) gene, which has potential to increase the insecticidal activity of *B. thuringiensis*-based biopesticides and transgenic crops. These bel genes encode peptides having 20%–30% similarity with viral enhancin proteins. Complete annotated genome sequence of *B. thuringiensis* YC-10, which is highly toxic to nematodes is already reported. It consists of a circular chromosome and nine circular plasmids, which harbors six parasporal crystal proteins genes consisting of cry1Aa, cry1Ac, cry1Ia, cry2Aa, cry2Ab and cryB1. The crystals proteins of Cry1Ia and Cry1Aa have high nematicidal activity against *Meloidogyne incognita*. Hence, the availability of the complete genome sequence of YC-10 will strongly contribute to a better exploration of more nematicidal gene resource and for developing nematocidal products. Among all Gammas proteobacteria represents a heterogeneous group of species which includes several entompathogens like *Photorhabdus*, *Xenorhabdus*, and *Serratia* species. The insecticidal action of these strains is a toxin mediated process (French-Constant and Waterfield, 2006; Hurst et al., 2000). In the same group there are non-spore forming species like *Yersinia entomophaga* producing the toxin complex Yen-Tc, containing toxins and chitinases, and *Pseudomonas entomophila* holding a toxin secretion system; and both of them act by ingestion process (Landsberg et al., 2011; Vodovar et al., 2006). Till now research on genome sequence of *B. thuringiensis* strain YBT-1518 with toxicity to nematode is done and reported to have one chromosome and six circular plamid and harbors three nematicidal crystal protein genes. In addition, the availability of complete genome sequence of YC-10 structured will strongly contribute in better exploration of more nematicidal gene resource and developing nematocidal products (Cheng et al., 2015).

Paenibacillus polymyxa (also called *Bacillus polymyxa*) has been extensively studied for agricultural applications as a plant-growth-promoting rhizobacterium and is also an effective biocontrol agent. *P. polymyxa*strain HY96-2 is developed from the tomato rhizosphere as the first microbial biopesticide based on *P. polymyxa* for controlling plant diseases. The complete genome sequence of HY96-2 and the results of comparative genomic analysis between different *P. polymyxa* strains are reported. The complete genome size of HY96-2 was found to be 5.75 Mb and predicted 5207 coding sequences. HY96-2 was compared with seven other *P. polymyxa* strains for which complete genome sequences have been published, using phylogenetic tree, pan-genome, and nucleic acid co-linearity analysis. Other parameters such as genes and gene clusters involved in biofilm formation, antibiotic synthesis and production of systemic resistance inducers were compared between strains of HY96-2, SC2 and E681. The results revealed that all three of the *P. polymyxa* strains have the ability to control plant diseases via the mechanisms of colonization-biofilm formation, antagonism-antibiotic production, and induced resistance-systemic resistance inducer production. However, slight variations of corresponding genes or gene clusters between the three strains may lead to different antimicrobial spectra and efficacy of biocontrol agents. Scientific basis for further optimization of field applications and quality standards of industrial microbial biopesticides based on HY96-2 at the pathway level is needed. This will further serve as reference standard for comparing different biocontrol agents (Luo et al., 2018). Other studies have demonstrated that HY96-2 even inhibits the growth of many fungal and bacterial pathogens, such as *Colletotrichum gloeosporioides*, *Rhizoctonia solani*, and *Erwinia carotovora* (Fan et al., 2012). Although microbial biopesticides derived from *P. polymyxa* HY96-2 have been manufactured and sold, further research at molecular level and understanding its precise biocontrol mechanism is still needed.

Pseudomonas species which colonize in roots of crop plants and produce antifungal metabolites represent a real alternative for chemical fungicides. Betaproteobacteria is another class having species like *Burkholderia rinojensis* having significant potential to be a biocontrol agent and products have been developed against diverse chewing and sucking insects and mites (Cordova-Kreylos et al., 2013). Another strain of *Chromobacterium subtsugae* is commercially successful and its metabolites show a broad spectrum insecticidal activity against different species of *Lepidoptera*, *Hemiptera*, *Colepotera*, and *Diptera* (Martin et al., 2007). A different class called Actinobacteria includes *Streptomyces* species which produce a variety of toxins like macrocyclic lactone derivatives which attacks the peripheral nervous system of insects (Copping and Menn, 2000). *Saccharopolyspora spinosa* belongs to the same phylum which produces potent and broad spectrum insecticidal toxins called spinosins whose natural and semisynthetic derivatives have a good commercial success (Kirst, 2010).

Genomes are a very useful resource for understanding the molecular mechanism of biocontrol agents at the organism level (Kohl et al., 2019). The key genes responsible for the production of antimicrobial agents and volatile organic compounds, indole acetic acid (IAA) synthesis, siderophore secretion, phosphate transporter and phosphonate cluster biosynthesis in *Paenibacillus yonginensis* DCY84T by genome sequencing is documented (Kim et al., 2017) They also confirmed the ability of this strain to induce plant resistance and protect plant growth proved by a combination of physiological experiments. By comparing the genomes of *B. amylolique faciens* FZB42 with *B. subtilis,* Chen et al. (2007) discovered that over 8.5% of the genome of *B. amyloliquefaciens* FZB42 is involved in antibiotic and siderophore synthesis. Research done by Stein (2005) estimated that not more than 4%−5% of the average *B. subtilis* genome is devoted to antibiotic production. This genomic comparison hence demonstrated that.

B. amyloliquefaciens FZB42 is capable of protecting plants from diseases through production of antibiotics at the molecular level. Despite their importance as biocontrol agents, there have been only small numbers of comprehensive studies into the biocontrol mechanism using genomic comparisons or other molecular methods. To understand completely the biocontrol mechanism at molecular level firstly its genome should be sequenced and compared with other strains. So far, the complete genomes of seven *P. polymyxa* strains have been published such as SC2, E681, YC0136, M−1, SQR-21, CR1 and YC0573 (Ma et al., 2011; Kim et al., 2010; Liu et al., 2017a,b; Niu et al., 2011; Li et al., 2014; Eastman et al., 2014). SC2 is isolated from the rhizosphere of pepper plants in Guizhou Province China, and demonstrated to inhibit plant pathogenic fungi, such as *F. oxysporum*, *B. cinerea*, *Pseudoperonospora cubensis* etc. (Zhu et al., 2008). E681 is isolated from the rhizosphere of barley plants in South Korea, and was reported to possess inhibitory activity against plant pathogenic fungi such as *F. oxysporum*, *B. cinerea*, and *R. solani* (Ryu et al., 2006) and plant pathogenic bacterium *P. syringae* (Kwon et al., 2016).

Among all the strains tested HY96-2 and SC2 exhibited relatively high homology, whereas HY96-2 and E681 had relatively low homology in terms of the genes and gene clusters involved in antibiotic production. The variations in these genes and gene clusters between the three strains might be responsible for the differences in the biocontrol targets and efficacies. Gene clusters for fusaricidins were found in *P. polymyxa* as the main antifungal metabolite in all three strains. Till date, at least ten members of the fusaricidin family have been isolated from *P. polymyxa* like A− D, LI-F03, LI-F04, LI-F05, LI-F06, LI-F07 and LI-F08 (Liu et al., 2011). Fusaricidins display excellent antifungal activities against many plant pathogenic fungi, especially *F. oxysporum.* Fusaricidin B is particularly effective against *Candida albicans* and *Saccharomyces cerevisiae*. They also exhibit excellent germicidal activity against G + bacteria such as *Staphylococcus aureus* (Kajimura and Kaneda, 1995, 1996). The gene cluster for fusaricidin synthesis in strain HY96-2 showed maximum identity with SC2 followed by strain E681. Fusaricidin A was mainly found in HY96-2 strain (Liu et al., 2011). Choi et al. (2008) discovered that the fusA gene in E681 plays an important role in biosynthesis of fusaricidin. Inactivation of this gene will lead to complete loss of antifungal activity against *F. oxysporum*, and fusA gene can produce more than one kind of fusaricidin. Studies done by Mikkola et al. (2017) says that fusaricidin A and B possess toxicity to mammalian cells at a certain concentration, whereas *P. polymyxa* HY96-2 showed very slight toxicity on animal and environment in the tests conducted. In addition, the gene clusters for paenilarvins, a class of iturin-like compounds with broad-spectrum antifungal activity were also found in HY96-2 and SC2 strains but not in E681 strain. All three strains are reported to suppress the plant pathogenic fungus *F. oxysporum* and *B. cinerea* with control targets being different in three different strains. Strains HY96-2 and E681 were also found to suppress *R. solani*. The different biocontrol targets of these three strains may be attributable to the variation in their gene clusters responsible for the synthesis of antifungal metabolites.

Earlier research has showed that Spo0A is a key transcriptional regulatory protein in *B. subtilis* that controls the expression of over 100 genes including those which are involved in formation of biofilm and sporulation. DegU is also a global regulator in *B. subtilis* and controls multiple cellular processes such as competence, motility, hydrolase secretion, which are very much essential in formation of biofilm (Kobayashi, 2007). The possible pathway in *P. polymyxa* for biofilm formation is due to response to an environmental stimulus, KinB phosphorylates the Spo0F response regulator, which subsequently transfers the phosphate group directly to Spo0A, and this triggers biofilm formation. The main differences in the pathways explained above between *P. polymyxa* and *B. subtilis*are that the mediator Spo0B transfers the phosphate group between Spo0F and Spo0A, is present in *B. subtilis* but absent in *P. polymyxa*. The phosphate group can be directly transferred from Spo0F to Spo0A in *P. polymyxa* and Spo0B need not act as mediator. Another possible pathway for biofilm formation in *P. polymyxa* was deduced, wherein the histidine kinase sensor DegS phosphorylates the response regulator protein DegU to form the biofilm in response to an external stimulus. Above are the two possible pathways for biofilm formation in *P. polymyxa* predicted so far. Recent research is aimed at understanding the molecular level, the mechanisms that enable these strains to act as efficient biological control agents. This approach is facilitating the development of novel strains with modified traits for enhanced biocontrol efficacy (Jogaiah et al., 2018).

15.1.2 Entomopathogenic fungi

The pathogenic fungi are another group of microbial pest management organisms which grow in both aquatic as well as terrestrial habitats and when specifically associated with insects are called as entomopathogenic fungi. Entomopathogenic fungi are considered to play vital role as biological control agent of insect populations. These fungi may be obligate or facultative, commensals or symbionts of insects. The pathogenic action of fungus depends on contact, either they infect or kill sucking insect pests such as aphids, thrips, mealy bugs, whiteflies, scale insects, mosquitoes and all types of mites. Entomopathogenic fungi are considered as promising microbial biopesticides that have multiple mechanisms for pathogenesis. They belong to 12 classes within six phyla and belong to four major groups; Laboulbeniales, Pyrenomycetes, Hyphomycetes and Zygomycetes. Few of most widely used species are *Beauveria bassiana*, *Metarhizium anisopilae*, *Nomuraea rileyi*, *Paecilomyces farinosus* and *Verticillium lecanii*.

Metarhizium anisopliae and *Beauveria bassiana* are common fungi which have been extensively studied for elucidation of pathogenic processes and manipulation of the genes of the pathogens to improve biocontrol performance. A group of scientists added copies of the gene encoding the regulated cuticle-degrading protease Pr1 were inserted into the genome of *M. anisopliae* and overexpressed (St. Leger and Wang, 2010). The resultant strain reduced survival time in tobacco hornworm (*M. sexta*) by 25% when compared with the parent wild-type strain. The remarkable extent to which virulence can be increased is shown in the case of the scorpion toxin (AaIT) expressed in the *M. anisopliae* strain ARSEF 549. The fungus which was modified gave the same mortality rates in *M. sexta* at lower spore doses and reduced survival rates at same doses when compared to wild type (St. Leger et al., 1996; Wang and Leger, 2007).

The main route of entrance of the entomopathogen is through integument and it will infect the insect by ingestion method or through the wounds or trachea. Entomopathogenic fungi have a great potential as control agents, as they constitute a group with over 750 species and when dispersed in the environment it provokes fungal infections in insect populations. These fungi begin their infective process when spores are retained on the integument surface from where initiation of germinative tube formation occurs. The fungi start to excrete enzymes such as proteases, chitinases, quitobiases, upases and lipoxygenases (Holder and Keyhani, 2005). The infection process of entomopathogenic fungi normally starts with the germination of conidia or spores that have come into contact with the host cuticle. With the combined enzymatic and mechanical action fungus penetrates the host body and allows the mycelium to develop internally. This often produces different types of conidia or spores after colonizing inside the host. During the vegetative growth of fungus it produces and releases a variety of metabolites which favors its growth and act as virulence factors or toxins. New conidia or spores will be produced outside the infected host and completely stops the further spreading in the environment. The host affected by both the biochemical and mechanical action of the fungus normally dies before it reaches this stage. The infection which is triggered by the first conidium or spore germination normally requires specific environmental conditions such as temperature and relative humidity. Most commercial products available are based on suspensions of conidia and these fungal entomopathogens include species from Chytridiomycota, Zygomycota, Oomycota, Ascomycota and Deuteromycota (Vega et al., 2012).

First and the most used fungal bioinsecticide for insect microbial control since 19th century is *Beauveria bassiana*. *Beauveria bassiana* and *Beauveria brongniartii* are strains of same genus which shows varying level of virulence against diverse targets and hence used as active substances in diverse formulations (Zimmermann, 2007). Recent studies have highlighted the potential of these strains as endophytes in biological control applications (McKinnon et al., 2017). Another well exploited fungal species is *Metarhizium anisopliae* which holds diverse strains acting against a wide range of targets (Zimmermann, 2007). *M. anisopliae* strains have been identified for producing a variety of insecticidal toxins and virulence factors (Schrank and Vainstein, 2010). Few other fungal species which are commercially exploited worldwide for pest management includes *Verticillium lecanii*, *Lecanicillium* spp., *Hirsutella* spp., *Paecilomyces*, *Isaria* spp. etc., whose action is associated with the production of insecticidal metabolites (Sugimoto et al., 2003; Kim et al., 2008; Kaya and Koppenhöfer, 1996; Zimmermann, 2008). Several other species are associated with insects and play an important role by spreading their spores to control insects in a natural way (Sawyer et al., 1994). A limitation of fungal based biopesticides is their action by contact through conidia and spore germination, so employing endophitic strains targeting insects after their penetration inside the plant are under development (Vidal and Jaber, 2015).

Another fungus *Piriformospora indica* is a beneficial cultivable phytopromotional, biotrophic mutualistic root endosymbiont belonging to the order Sebacinales (Basidiomycota). It has been reported to mimic the functions and capabilities of typical arbuscular mycorrhizal (AM) fungi. It performs multifarious functions during biological hardening and transplantation of micro-propagated plantlets (Singh et al., 2003). It increases endogenous content of spilanthol after realization of its mutual interaction with medicinal plants *Spilanthes calva* and its infestation in *Helianthus annus* and *Aristolochia elegans* has resulted in stimulation valuable compounds synthesis (Rai et al., 2004; Bagde et al., 2010). With this reason the fungus can be cultured very easily in bioreactor in order to prepare effective biofertilizer formulations (Oelmüller et al., 2009; Qiang et al., 2011). Moreover, *P. indica* root endophyte has been proved to minimize the use of chemical fertilizers by controlling the yield of crop, and also to provide increased resistance and tolerance in plants against biotic and abiotic stresses (Unnikumar et al., 2013). Zuccaro et al. (2011) presented the first in-depth genomic study and unveiled a mutualistic symbionce bearing 25 Mb genome; the authors characterized fungal transcriptional responses associated with the colonization of living and dead *H. vulgare* roots. A biphasic root colonization strategy of *P. indica* was revealed by microarray analysis, where a tightly controlled expression of the lifestyle-associated gene-sets was reported during the onset of the symbiosis. It was also observed that about 10% of the fungal genes induced during the biotrophic colonization encoded putative small secreted proteins (SSPs). SSP included several lectin-like proteins and members of a *P. indica*-specific gene family (DELD) with a conserved novel seven-amino acids motif at the C-terminus. This correlated with the presence of transposable elements in gene-poor repeat rich regions of the genome similar to the effectors found in other filamentous organisms. This information suggested a series of incremental shifts along the continuum from saprotrophy toward biotrophy in the evolution of mycorrhizal association from decomposer fungi (Zuccaro et al., 2011).

Phytohormones and Ca^{2+} release can direct the molecular and physiological processes responsible for *P. indica* colonization with plants. In another study authors concluded that *P. indica* is an efficient biocontrol agent that protects *Arabidopsis* from *Verticillium dahliae* infection. They demonstrated that *Verticillium dahliae* growth is restricted in the presence of *P. indica* and signals generated from *P. indica* participate in regulation of immune response against pathogen (Sun et al., 2014). The research done by Alga et al., shown that *P. indica* possess at least six chromosomes having genome size of about 15.4–24 Mb. Sequences of the genes encoding the elongation factor 1-α (TEF) and glyceraldehyde-3-phosphate dehydrogenase (GAPDH) were used for genome size estimation through real-time PCR analysis. Chromosomal location investigated by Southern blot and expression analysis suggested that TEF and GAPDH are single-copy genes with strong and constitutive promoters. A genetic transformation system was established using a fragment of the TEF promoter region for construction of vectors carrying the selectable marker hygromycin B phosphotransferase. These results demonstrated that *P. indica* can be stably transformed by random genomic integration of foreign DNA as it relatively possess small genome as compared to other members of the Basidiomycota (Alga et al., 2009).

15.1.3 Viral biopesticides

Viral biopesticides are prepared from isolating insect-infecting, likely from *Lepidoptera*, *Hymenoptera*, *Coleoptera*, *Diptera* and *Orthoptera*. The viruses used for insect control are the DNA-containing baculoviruses (BVs), nucleopolyhedrosis viruses (NPVs), granuloviruses.

(GVs), acoviruses, iridoviruses, parvoviruses, polydnaviruses, poxviruses and the RNA-containing reoviruses, cytoplasmic polyhedrosis viruses, nodaviruses, picrona-like viruses and tetraviruses. But the main classes used in pest management are NPVs and GVs, as these viruses are widely used for control of vegetables, field crop pests globally, and are effective against plant-chewing insects. Species belonging to the family Baculoviridae represents DNA viruses which establishes pathogenic relationships with invertebrates and plays potential role in biological control (Clem and Passarelli, 2013; Haase et al., 2015). The virus infectivity is associated with the production of crystalline occlusion bodies inside the host cell which contains infectious particles. Based on the morphology of these occlusion bodies Baculoviruses are divided into nucleo polyhedro viruses (NPVs), in which these bodies are polyhedron-shaped and develop in cell nuclei, and the granuloviruses (GVs), in which these bodies are granular-shaped. A different double-stranded RNA virus family called Reoviridae presents polyhedron-shaped occlusion bodies in the cell cytoplasm (CPVs) (Rohrmann, 2011).

Baculoviruses act orally against insects in which first infection normally takes place after ingestion of contaminated food. Ingested occlusion bodies within the midgut environment release specific types of virions, called occlusion-derived viruses (ODVs). These ODVs interact directly with the membrane of microvillar epithelial cells through the action of their envelop proteins (PIFs). Later a second type of virions called budded viruses (BVs) are produced in the nucleus of infected midgut cells, which spread the virus throughout the host. With the spread of infection the insect dies and body liquefies progressively, favoring the dispersal of virus particles in the environment. Viral infections are also able to induce behavioral changes in the hosts by affecting their gene expression mechanisms (Williams et al., 2017; Katsuma et al., 2012). The first baculovirus genome sequence was published from *Autographa californica* multiple nucleopolyhedrovirus (AcMNPVs) and now the number of complete genomes available have raised to 30 (Ayres et al., 1994). The most successful example of virus as biological pesticideis *Anticarsia gemmatalis* MNPV (AgMNPV) in the control of velvet bean caterpillar (Moscardi, 1999).

The production, commercialization and field application of AgMNPV have increased constantly right after initial isolation from infected larvae in Brazil (Allen and Knell, 1977). Comparisons among AgMNPV temporal isolates have shown that the viral heterogeneity of the commercial preparation has increased and that changes are concentrated at "hot spots" (Maruniak et al., 1999). AgMNPV isolate 2D (AgMNPV-2D) was cloned by plaque purification and chosen as the prototype as it was the major genotype present among wild-type population. Moreover, genetic modification of AgMNPV has allowed studies of its pathology in *A. gemmatalis* larvae. This led to characterization of many *A. gemmatalis* larvae hemocytes and all of which are shown to be susceptible for AgMNPV infection (da Silveira et al., 2003, 2004). Several individual genes of AgMNPV-2D like polh (polyhedrin), gp41, egt (ecdysteroid UDP glucosyl transferase), p10, gp64, v-trex (viral 39 repair exonuclease), p143 (helicase), dnapol (DNA polymerase), iap-3 and p74 have been sequenced and analyzed. In addition, a homologous region (hr4) of AgMNPV-2D has been characterized (Garcia-Maruniak et al., 1996). Phylogenetic analysis of polh indicates that AgMNPV belongs in the group I NPVs and comparisons of some genes have shown that AgMNPV is closely related to *Choristoneura fumiferana* defective MNPV (CfDefNPV) (Lima et al., 2004; Lauzon et al., 2005). Few scientists have modified AgMNPV-2D genetically and it behaves as a viable expression system for heterologous genes (Arana et al., 2001; Ribeiro et al., 2001). In addition, AgMNPV is declared as one of the most important baculovirus systems under study, as it is one of the few viruses that will allow integrating studies from large scale field application to the genomic and post-genomic levels.

In another study, Recombinant baculoviruses (vEV-Tox34) expressing the gene Tox-34 from a mite *Pyemotes tritici*, took less time to kill corn earworm *Helicoverpa zea*. Similarly, two genetically enhanced isolates of the *Autographa califomica* nuclear polyhedrosis virus (AcMNPVs) expressing insect-specific neurotoxin genes from the spiders *Diguetia canities* and *Tegenaria agrestis* (vAcTaITX-1, vAcDTX9.2) have been evaluated for their potency against lepidopteran insects. The genome of *Anticarsia gemmatalis* multiple nucleopolyhedrovirus isolate 2D (AgMNPV-2D), is the most extensively used virus pesticide in the world, was completely sequenced and shown to have 132 239 bp (G + C content 44.5 mol%) and to be capable of encoding 152 non-overlapping open reading frames (ORFs). Given the close relationship and specificity with the host, the name of the entomopathogenic virus includes the initial of the host name. For instance, LdMNPV refers to the *Lymantria dispar* multicapsid nucleopolyhedrovirus. The commercially available baculovirus-based products are active only against chewing insects, especially *Lepidopteran caterpillars* through their mode of action.

Because of the low stability of baculovirus formulations in the environment and their high production costs related to the need to reproduce them within their host, their use in biological pest management is limited to specific niche market segments (Harrison and Hoover, 2012; Sun, 2015). More research is needed to find out the functioning of chitinase and cathepsin present in all other baculoviruses sequenced. This allows field production and harvesting of relatively intact insects full of virus needed for subsequent applications (de Castro Oliveira et al., 2006).

15.1.4 Entomopathogenic nematodes

Another group of microorganisms used as pesticide are entomopathogenic nematodes, which control weevils, gnats, white grubs and various species of family Sesiidae. They are soft bodied, non-segmented roundworms that are obligate or sometimes facultative parasites of insects. Entomopathogenic nematodes occur naturally in soil environments and locate their host in response to carbon dioxide, vibration, and other chemical cues (Kaya and Gaugler, 1993). Commonly used nematodes in pest management belong to the genera *Steinernema* and *Heterorhabditis*, which attack the hosts as infective juveniles.

Entomopathogenic nematode (EPN) species belonging to genera *Heterorhabditis* and *Steinernema* act as obligate parasites. Because of their mutualistic symbiosis with insect pathogenic bacteria in genera *Photorhabdus* and *Xenorhabdus* they possess a significant insecticidal potential (Lewis and Clarke, 2012). Insect pathogenic nematodes easily enter inside the host through its natural openings (oralcavity, anus, and spiracles) and release their symbiotic bacteria in the hemocoel. Followed by proliferation of bacteria it is accompanied by release of toxins and virulence factors that weaken the host, and produce metabolites that favor a suitable environment for nematode reproduction (Poinar, 1990). Among other virulence factors, *Photorhabdus* and *Xenorhabdus* species produce an insecticidal toxins complex (Tc), including different subunits that show toxicity against insects by ingestion (Ffrench-Constant et al., 2007). Nematodes continue to feed on the host tissue after its death, mature and reproduce in host itself. The progeny nematodes develop through four juvenile stages to the adult. Depending on the available resources, one or more generations may occur and large number of infective juveniles are eventually released into environment to infect other hosts and continue their life cycle (Bedding and Molyneux, 1982).

To produce nematodes at small and large-scale, a variety of improved in vivo and in vitro methods have been developed. The quality of the final formulation plays a major role in the efficacy of nematode-based biological control applications against pests. A variety of products based on different nematode species targeting specific pest species and market segments are commercialized worldwide (Shapiro-Ilan et al., 2012). Artificial selection has been successful in increasing infectivity and nematicide resistance in case of entomopathogenic nematodes. The recent discovery stating that maize roots damaged by the western corn rootworm emit a key attractant for insect-killing nematode has opened the way to explore selection strategy can improve the control of root pests. Two commonly used entomopathogenic fungi, *Metarhizium anisopliae* and *Beauveria bassiana* have been extensively studied for elucidation of pathogenic processes and manipulation of the genes of the pathogens to improve biocontrol performance. In recent studies, additional copies of the gene encoding regulated cuticle-degrading protease Pr1 were inserted into the genome of *M. anisopliae* and overexpressed.

15.1.5 Entomopathogenic protozoans

Protozoan pathogens are also used as biopesticide agents as they infect a wide range of pests naturally and induce chronic and debilitating effects that reduce the target pest populations. They are taxonomically subdivided into several phyla, some of which contain entomogenous species. They are extremely diverse group of organisms comprising around 1000 species attacking invertebrates including insect species and are commonly referred as microsporidians (Brooks, 1988). They are generally host specific and act slowly, producing chronic infections with general debilitation of the host. The spore formed by the protozoan is the infectious stage and has to be ingested by the insect host for pathogenicity. The spore germinates in the midgut and sporoplasm is released invading the target cells causing infection of the host. This leads to reduced feeding, vigor, fecundity and longevity of the insect host. Only few species tested so far has been moderately successful (Solter and Becnel, 2000).

Nosema locustae has been used as grasshopper biocontrol agent, but it remains questionable because of the great difficulty in assessing the efficacy in case of a highly mobile insect (Lacey and Goettel, 1995). *Nosema pyrausta* is another beneficial microsporidian that reduces fecundity and longevity of the adults and also causes mortality of the larvae of European corn borer (Siegel et al., 1986). Extensive research is done with microsporan protozoans as possible components of integrated pest management programmes. Few bioassays are conducted to test the presence, viability, or quantity of infective forms of a protozoan. To conduct a quantitative bioassay, spores must be counted and a dilution series should be made so that a known concentration of spores can be fed to the host. To calculate dosage or concentration for any bioassay, the spores must be accurately counted which is difficult. Alternatively spores can be fed to the host on artificial diet or in their natural food. Proper experimentation is needed in order to find out the optimum dosage for best spore production. *Nosema* and *Endoreticulatus* spp. in *Lepidoptera* develop more slowly than *Vairimorpha necatrix*, dictating that infections be initiated earlier in larval life. Further research related to molecular aspects and genes responsible for pathogenecity of protozoan is needed.

15.2 Advantages of genetic manipulation and their commercialization

Microbial pathogens of insects are intensively investigated to develop environmental friendly pest management strategies in agriculture. In the recent years biopesticides are replacing the chemicals pesticides to overcome the harmful effect of the chemicals on non-target organism. In comparison with synthetic chemicals these microorganisms are employed as active substances in pest management and do not harm non-target species. They have very specific mode of action which limits their efficacy against one or a narrow range of pest species (Kaya and Vega, 2012). Microbiological pesticides exhibit a multi-site action which hinders the development of resistant pests. Few technologies are developed by industry which maximizes the effects on the target and try to improve the application features of products (Satinder et al., 2006).

The whole market of biopesticides has significantly grown during the last years, as a result of an increased awareness of their potential and a growing attention to the environmental and health risks associated with conventional chemicals (Lacey et al., 2001; Glare et al., 2012).

Till today several hundreds of commercial products of fungi, bacteria and viruses are available worldwide for the biological control of insect pests in agriculture and forestry. The growth rate of bio-pesticide industry has been forecasted to increase 10%−15% per annum in contrast to 2%−3% for chemical pesticides in next 10 years. Entomopathogenic nematodes fit nicely into integrated pest management as they are considered nontoxic to humans, relatively specific to target pests, and can be applied with standard pesticide equipment (Shapiro-Ilan et al., 2006). There are many challenges that need to be overcomed owing to some of the early successes and continuing growth of biopesticide market. However, Commercialization is the final and most difficult step in the development of a genetically modified microbial product. The most critical factors which decide the fate are developmental cost and time to market. Therefore, to examine all these critical factors in the successful commercialization of product is essential (Ravensberg, 2011).

15.3 Conclusions

Availability of biopesticides acting against diverse crop pests is essential to ensure the management of agro-ecosystems respecting the environment and human health. The growing demand from farmers is accompanied by an increasing market offer of newly introduced and improved products that can be used alone and in rotation or combination with conventional chemicals. Due to the increasing demand in global market academic and industrial investments in the biopesticide sector are experiencing a significant growth and many discoveries are being developed. This includes the development of novel solutions against new targets or the incorporation of new technologies that enhance the efficacy of already available active substances. Advanced molecular studies on insect microbial community diversity are also opening new frontiers for the development of innovative pest management strategies. On the other hand, recent findings are contributing to foster a deeper understanding of the insect-microbial interactions within ecosystem. The modern legislative frameworks requiring in following criteria and principles of integrated pest management are further fueling a significantly expanding market. Added to this are the efforts made by scientists working in the field of genetics and molecular biology, whose studies aim to give light to new and increasingly effective microbial based active substances. Microorganisms as such have natural capability of causing disease at epizootic levels due to their persistence in soil and transmission efficiency. The cost of development and registration of microbial insecticides is much less than that of chemical insecticides is an added advantage. The self-perpetuating nature of most of the pathogens in both space and time would certainly prove to be an asset in sustainable agriculture.

References

Abad, A.R., Flannagan, R.D., McCutchen, B.F., Yu, C.G., 2008. *Bacillus thuringiensis* cry gene and protein. US Patent 20087329736.

Allen, G.E., Knell, J.D., 1977. A nuclear polyhedrosis virus of *Anticarsia gemmatalis*: I. Ultrastructure, replication and pathogenicity. Fla. Entomol. 60, 233−240.

Arana, E.I., Albarin, C.G., O'Reilly, D., Ghiringhelli, P.D., Romanowski, V., 2001. Generation of a recombinant *Anticarsia gemmatalis* multicapsid nucleopolyhedrovirus expressing a foreign gene under the control of a very late promoter. Virus Gene. 22, 363−372.

Ayres, M.D., Howard, S.C., Kuzio, J., Lopez-Ferber, M., Possee, R.D., 1994. The complete DNA sequence of *Autographa californica* nuclear polyhedrosis virus. Virology 202, 586−605.

Babu, A.N., Jogaiah, S., Ito, S-I, Nagaraj, A.K., Tran, L.-S.P., 2015. Improvement of growth, fruit weight and early blight disease protection of tomato plants by rhizosphere bacteria is correlated with their beneficial traits and induced biosynthesis of antioxidant peroxidase and polyphenol oxidase. Plant Sci. 231, 62−73.

Bagde, U.S., Prasad, R., Varma, A., 2010. Interaction of *Piriformospora indica* with medicinal plants and of economic importance. Afr. J. Biotechnol. 9, 9214−9226.

Bedding, R., Molyneux, A., 1982. Penetration of insect cuticle by infective juveniles of *Heterorhabditis* spp. (Heterorhabditidae: Nematoda). Nematologica 28, 354−359.

Bravo, A., Gill, S.S., Soberon, M., 2007. Mode of action of *Bacillus thuringiensis* Cry and Cyt toxins and their potential for insect control. Toxicon 49, 423−435.

Brooks, W.M., 1988. Entomogenous protozoa. In: Ignoffo, C.M., Mandava, N.B. (Eds.), Handbook of Natural Pesticides, Microbial Insecticides, Part A, Entomogenous Protozoa and Fungi, vol. V. CRC Press, Baco Raton, FL, pp. 1−149.

Castagnola, A., Stock, S.P., 2014. Common virulence factors and tissue targets of entomopathogenic bacterial for biological control of *Lepidopteran* pests. Insects 5, 139−166.

Charles, J.F., Silva-Filha, M.H., Nielsen-LeRoux, C., 2000. Mode of action of *Bacillus sphaericus* on mosquito larvae: incidence on resistance. In: Charles, J.F., Delecluse, A., Nielsen-LeRoux, C. (Eds.), Entomopathogenic Bacteria: From Laboratory to Field Application. Kluwer Academic Publishers, London, UK, pp. 237–252.

Chen, X.H., Koumoutsi, A., Scholz, R., Eisenreich, A., Schneider, K., Heinemeyer, I., et al., 2007. Comparative analysis of the complete genome sequence of the plant growth- promoting bacterium *Bacillus amyloliquefaciens* FZB42. Nat. Biotechnol. 25, 1007–1014.

Cheng, F., Wang, J, Song, Z., Cheng, J., Zhang, D., Liu, Y., September 20, 2015. Complete genome sequence of *Bacillus thuringiensis* YC-10, a novel active strain against plant-parasitic nematodes. J. Biotechnol. 210, 17–18.

Choi, S.K., Park, S.Y., Kim, R., Lee, C.H., Kim, J.F., Park, S.H., 2008. Identification and functional analysis of the fusaricidin biosynthetic gene of *Paenibacillus polymyxa* E681. Biochem. Biophys. Res. Commun. 365, 89–95.

Clem, R.J., Passarelli, A.L., 2013. Baculoviruses: sophisticated pathogens of insects. PLoS Pathog. 9, e1003729.

Copping, G.L., Menn, J.J., 2000. Biopesticides: a review of their action, applications and efficacy. Pest Manag. Sci. 56, 651–676.

Cordova-Kreylos, A.L., Fernandez, L.E., Koivunen, M., Yang, A., Flor-Weiler, L., Marrone, P.G., 2013. Isolation and characterization of *Burkholderia rinojensis* sp. nov., a non- *Burkholderia cepacia* complex soil bacterium with insecticidal and miticidal activities. Appl. Environ. Microbiol. 79, 7669–7678.

da Silveira, E.B., Ribeiro, B.M., Báo, S.N., 2003. Characterization of larval haemocytes from the velvetbean caterpillar *Anticarsia gemmatalis* (Hübner) (Lepidoptera: Noctuidae). J. Submicr. Cytol. Pathol. 35, 129–139.

da Silveira, E.B., Cordeiro, B.A., Ribeiro, B.M., Báo, S.N., 2004. Morphological characterization of *Anticarsia gemmatalis* M nucleopolyhedrovirus infection in haemocytes from its natural larval host, the velvet bean caterpillar *Anticarsia gemmatalis* (Hübner) (Lepidoptera: Noctuidae). Tissue Cell 36, 171–180.

de Castro Oliveira, J.V., Wolff, J.L.C., Maruniak, A.G., Morais Ribeiro, B., Batista de Castro, M.E., Lobo de Souza, M., Moscardi, F., Maruniak, J.E., de Andrade Zanotto, P.M., 2006. Genome of the most widely used viral biopesticide: *Anticarsia gemmatalis* multiple nucleopolyhedrovirus. J. Gen. Virol. 87, 3233–3250.

Eastman, A.W., Weselowski, B., Nathoo, N., Yuan, Z.C., 2014. Complete genome sequence of *Paenibacillus polymyxa* CR1, a plant growth promoting bacterium isolated from the corn rhizosphere exhibiting potential for biocontrol, biomass degradation, and biofuel production. Genome Announc. 2. https://doi.org/10.1128/genomeA.01218-13 e01218-13.

Fan, L., Zhang, D.J., Liu, Z.H., Tao, L.M., Luo, Y.C., 2012. Antifungal lipopeptide produced by *Paenibacillus polymyxa* HY96-2. Nat. Prod. Res. Dev. 24, 729–735. https://doi.org/10.1007/s00253-013-5157-6.

Ffrench-Constant, R., Waterfield, N., 2006. An ABC guide to the bacterial toxin complexes. Adv. Appl. Microbiol. 58, 169–183.

Ffrench-Constant, R.H., Dowling, A., Waterfield, N.R., 2007. Insecticidal toxins from *Photorhabdus bacteria* and their potential use in agriculture. Toxicon 49, 436–451.

Garcia-Maruniak, A., Pavan, O.H.O., Maruniak, J.E., 1996. A variable region of *Anticarsia gemmatalis* nuclear polyhedrosis virus contains tandemly repeated DNA sequences. Virus Res. 41, 123–132.

Glare, T., Caradus, J., Gelernter, W., Jackson, T., Keyhani, N., Kohl, J., Marrone, P., Morin, L., Stewart, A., 2012. Have biopesticides come of age? Trends Biotechnol. 30, 250–258.

Haase, S., Sciocco-Cap, A., Romanowski, V., 2015. Baculovirus insecticides in Latin America: historical overview, current status and future perspectives. Viruses 7, 2230–2267.

Harrison, R., Hoover, K., 2012. Baculoviruses and other occluded insect viruses. In: Vega, F., Kaya, H. (Eds.), Insect Pathology, second ed. Academic Press, London, UK, pp. 73–131.

Holder, D.J., Keyhani, N.O., 2005. Adhesion of the entomopathogenic fungus *Beauveria* (Cordyceps) bassiana to substrata. Appl. Environ. Microbiol. 71 (9), 5260–5266.

Hurst, M.R., Glare, T.R., Jackson, T.A., Ronson, C.W., 2000. Plasmid-located pathogenicity determinants of *Serratia entomophila*, the causal agent of amber disease of grass grub, show similarity to the insecticidal toxins of *Photorhabdus luminescens*. J. Bacteriol. 182, 5127–5138.

Jogaiah, S., Shetty, H.S., Ito, S.I., Tran, L.-S.P., 2016. Enhancement of downy mildew disease resistance in pearl millet by the G_app7 bioactive compound produced by *Ganoderma applanatum*. Plant Physiol. Biochem. 105, 109–117.

Jogaiah, S., Abdelrahman, M., Tran, L.-S.P., Ito, S.-I., 2018. Different mechanisms of *Trichoderma* virens-mediated resistance in tomato against *Fusarium* wilt involve the jasmonic and salicylic acid pathways. Mol. Plant Pathol. 19, 870–882.

Jurat-Fuentes, J.L., Jackson, T.A., 2012. Bacterial entomopathogens. In: Vega, F., Kaya, H. (Eds.), Insect Pathology, second ed. Academic Press, London, UK, pp. 265–349.

Kajimura, Y., Kaneda, M., 1995. Fusaricidin A, a new depsipeptide antibiotic produced by *Bacillus polymyxa* KT-8. Taxonomy, fermentation, isolation, structure elucidation and biological activity. J. Antibiot. 49, 129–135. https://doi.org/10.7164/antibiotics.49.129.

Kajimura, Y., Kaneda, M., 1996. Fusaricidins B, C and D, new depsipeptide antibiotics produced by *Bacillus polymyxa* KT-8: isolation, structure elucidation and biological activity. J. Antibiot. 50, 220–228. https://doi.org/10.7164/antibiotics.50.220.

Katsuma, S., Koyano, Y., Kang, W., Kokusho, R., Kamita, S.G., Shimada, T., 2012. The baculovirus uses a captured host phosphatase to induce enhanced locomotory activity in host caterpillars. PLoS Pathog. 8, e1002644.

Kaya, H.K., Gaugler, R., 1993. Entomopathogenic nematodes. Annu. Rev. Entomol. 38, 181–206.

Kaya, H.K., Koppenhöfer, A.M., 1996. Effects of microbial and other antagonistic organism and competition on entomopathogenic nematodes. Biocontrol Sci. Technol. 6, 357–371.

Kaya, H.K., Vega, F.E., 2012. Scope and basic principles of insect pathology. In: Vega, F., Kaya, H. (Eds.), Insect Pathology, second ed. Academic Press, London, UK, pp. 1–12.

Kenis, M., Hurley, B.P., Hajek, A.E., Cock, M.J.W., 2017. Classical biological control of insect pests of trees: facts and figures. Biol. Invasions 19, 3401–3417.

Kim, J.J., Goettel, M.S., Gillespie, D.R., 2008. Evaluation of Lecanicillium longisporum, Vertalec® for simultaneous suppression of cotton aphid, *Aphis gossypii*, and cucumber powdery mildew, *Sphaerotheca fuliginea*, on potted cucumbers. Biol. Contr. 45, 404–409.

Kim, J.F., Jeong, H., Park, S.Y., Kim, S.B., Park, Y.K., Choi, S.K., et al., 2010. Genome sequence of the polymyxin-producing plant-probiotic rhizobacterium *Paenibacillus polymyxa* E681. J. Bacteriol. 192, 6103–6104. https://doi.org/10.1128/JB.00983-10.

Kim, Y.J., Sukweenadhi, J., Seok, J.W., Kang, C.H., Choi, E.S., Subramaniyam, S., et al., 2017. Complete genome sequence of *Paenibacillus yonginensis* DCY84T, a novel plant symbiont that promotes growth via induced systemic resistance. Stand. Genomic Sci. 12, 63. https://doi.org/10.1186/s40793-017-0277-8.

Kirst, H.A., 2010. The spinosyn family of insecticides: realizing the potential of natural products research. J. Antibiot. 63, 101–111.

Kobayashi, K., 2007. *Bacillus subtilis* pellicle formation proceeds through genetically defined morphological changes. J. Bacteriol. 189, 4920–4931. https://doi.org/10.1128/JB.00157-07.

Köhl, J., Kolnaar, R., Ravensberg, W.J., July 19, 2019. Mode of action of microbial biological control agents against plant diseases: relevance beyond efficacy. Front. Plant Sci. 10, 845. https://doi.org/10.3389/fpls.2019.00845.

Kwon, Y.S., Lee, D.Y., Rakwal, R., Baek, S.B., Lee, J.H., Kwak, Y.S., et al., 2016. Proteomic analyses of the interaction between the plant-growth promoting rhizobacterium *Paenibacillus polymyxa* E681 and *Arabidopsis thaliana*. Proteomics 16, 122–135. https://doi.org/10.1002/pmic.201500196.

Lacey, L.A., Goettel, M.S., 1995. Entomophaga 40, 3–27.

Lacey, L.A., Frutos, R., Kaya, H.K., Vail, P., 2001. Insect pathogens as biological control agents: do they have a future? Biol. Contr. 21, 230–248.

Landsberg, M.J., Jones, S.A., Rothnagel, R., Busby, J.N., Marshall, S.D.G., Simpson, R.M., Lott, J.S., Hankamer, B., Hurst, M.R.H., 2011. 3D structure of the *Yersinia entomophaga* toxin complex and implications for insecticidal activity. Proc. Natl. Acad. Sci. U.S.A. 108, 20544–20549.

Lauzon, H.A.M., Jamieson, P.B., Krell, P.J., Arif, B.M., 2005. Gene organization and sequencing of the *Choristoneura fumiferana* defective nucleopolyhedrovirus genome. J. Gen. Virol. 86, 945–961.

Lewis, E.E., Clarke, D.J., 2012. Nematode parasites and entomopathogens. In: Vega, F., Kaya, H. (Eds.), Insect Pathology, second ed. Academic Press, London, UK, pp. 395–424.

Li, S.Q., Yang, D.Q., Qiu, M.H., Shao, J.H., Guo, R., Shen, B., et al., 2014. Complete genome sequence of *Paenibacillus polymyxa* SQR-21, a plant growth-promoting rhizobacterium with antifungal activity and *Rhizosphere colonization* ability. Genome Announc. 2. https://doi.org/10.1128/genomeA.00281-14 e00281–14.

Lima, L., Pinedo, F.J., Ribeiro, B.M., Zanotto, P.M.A., Wolff, J.L., 2004. Identification, expression and phylogenetic analysis of the *Anticarsia gemmatalis* multicapsid nucleopolyhedrovirus (AgMNPV) helicase. Virus Gene. 29, 345–352.

Liu, Z.H., Fan, L., Zhang, D.J., Li, Y.G., 2011. Antifungal depsipeptide compounds from *Paenibacillus polymyxa* HY96-2. Chem. Nat. Compd. 47, 496–497. https://doi.org/10.1007/s10600-011-9978-1.

Liu, H., Liu, K., Li, Y.H., Wang, C.Q., Hou, Q.H., Xu, W.F., et al., 2017a. Complete genome sequence of *Paenibacillus polymyxa* YC0136, a plant growth-promoting rhizobacterium isolated from tobacco rhizosphere. Genome Announc. 5. https://doi.org/10.1128/genomeA.01635-16 e01635–16.

Liu, H., Wang, C.Q., Li, Y.H., Liu, K., Hou, Q.H., Xu, W.F., et al., 2017b. Complete genome sequence of *Paenibacillus polymyxa* YC0573, a plant growth promoting rhizobacterium with antimicrobial activity. Genome Announc. 5. https://doi.org/10.1128/genomeA.01636-16 e01636–16.

Luo, Y., Cheng, Y., Yi, J., Zhang, Z., Luo, Q., Zhang, D., Li, Y., 2018. Complete genome sequence of industrial biocontrol strain *Paenibacillus polymyxa* HY96-2 and further analysis of its biocontrol mechanism. Front. Microbiol. 9, 1520.

Ma, M.C., Wang, C.C., Ding, Y.Q., Li, L., Shen, D.L., Jiang, X., et al., 2011. Complete genome sequence of *Paenibacillus polymyxa* SC2, a strain of plant growth- promoting Rhizobacterium with broad-spectrum antimicrobial activity. J. Bacteriol. 193, 311–312. https://doi.org/10.1128/JB.01234-10.

Marche, M.G., Mura, M.E., Falchi, G., Ruiu, L., 2017. Spore surface proteins of *Brevibacillus laterosporus* are involved in insect pathogenesis. Sci. Rep. 7, 43805.

Marche, M.G., Camiolo, S., Porceddu, A., Ruiu, L., 2018. Survey of *Brevibacillus laterosporus* insecticidal protein genes and virulence factors. J. Invertebr. Pathol. 155, 38–43.

Marrone, P.G., 2014. The market and potential for biopesticides. In: Gross, A.D., Coats, J.R., Duke, S.O., Seiber, J.N. (Eds.), Biopesticides: State of the Art and Future Opportunities. American Chemical Society, Washington, DC, USA, pp. 245–258.

Martin, P.A.W., Gundersen-Rindal, D., Blackburn, M., Buyer, J., 2007. *Chromobacterium subtsugae* sp. nov., a betaproteobacterium toxic to Colorado potato beetle and other insect pests. Int. J. Syst. Evol. Microbiol. 57, 993–999.

Maruniak, J.E., Garcia-Maruniak, A., Souza, M.L., Zanotto, P.M.A., Moscardi, F., 1999. Physical maps and virulence of *Anticarsia gemmatalis* nucleopolyhedrovirus genomic variants. Arch. Virol. 144, 1991–2006.

McKinnon, A.C., Saari, S., Moran-Diez, M.E., Meyling, N.V., Raad, M., Glare, T.R., 2017. *Beauveria bassiana* as an endophyte: a critical review on associated methodology and biocontrol potential. BioControl 62, 1–17.

Mikkola, R., Andersson, M.A., Grigoriev, P., Heinonen, M., SalkinojaSalonen, M.S., 2017. The toxic mode of action of cyclic lipodepsipeptide fusaricidins, produced by *Paenibacillus polymyxa,* toward mammalian cells. J. Appl. Microbiol. 123, 436–449. https://doi.org/10.1111/jam.13498.

Moscardi, F., 1999. Assessment of the application of baculoviruses for control of lepidoptera. Annu. Rev. Entomol. 44, 257–289.

Niu, B., Rueckert, C., Blom, J., Wang, Q., Borriss, R., 2011. The genome of the plant growth-promoting rhizobacterium *Paenibacillus polymyxa* M-1 contains nine sites dedicated to nonribosomal synthesis of lipopeptides and polyketides. J. Bacteriol. 193, 5862–5863. https://doi.org/10.1128/JB.05806-11.

Oelmüller, R., Sherameti, I., Tripathi, S., Varma, A., 2009. *Piriformospora indica*, a cultivable root endophyte with multiple biotechnological applications. Symbiosis 49, 1–17. https://doi.org/10.1007/s13199-009-0009-y.

Pigott, C.R., Ellar, D.J., 2007. Role of receptors in *Bacillus thuringiensis* crystal toxin activity. Microbiol. Mol. Biol. Rev. 71, 255–281.

Poinar, G.O., 1990. Biology and taxonomy of steinernematidae and heterorhabditidae. In: Gaugler, R., Kaya, H.K. (Eds.), Entomopathogenic Nematodes in Biological Control. CRC Press, Boca Raton, FL, USA, pp. 23–62.

Qiang, X., Weiss, M., Kogel, K.H., Schäfer, P., 2011. *Piriformospora indica* - a mutualistic basidiomycete with an exceptionally large plant host range. Mol. Plant Pathol. 13, 508–518. https://doi.org/10.1111/j.1364-3703.2011.00764.x.

Rai, M.K., Varma, A., Pandey, A.K., 2004. Antifungal potential of *Spilanthes calva* after inoculation of *Piriformospora indica*. Mycoses 47, 479–481. https://doi.org/10.1111/j.1439-0507.2004.01045.x.

Ravensberg, W.J., 2011. A Roadmap to the Successful Development and Commercialization of Microbial Pest Control Products for Control of Arthropods. Springer, Dordrecht.

Ribeiro, B.M., Gatti, C.D.C., Costa, M.H., Moscardi, F., Maruniak, J.E., Possee, R.D., Zanotto, P.M.A., 2001. Construction of a recombinant *Anticarsia gemmatalis* nucleopolyhedrovirus (AgMNPV- 2D) harbouring the b-galactosidase gene. Arch. Virol. 146, 1355–1367.

Rohrmann, G.F., 2011. Baculovirus Molecular Biology, second ed. National Library of Medicine (US), National Center for Biotechnology Information, Bethesda, MD, USA.

Ruiu, L., 2013. *Brevibacillus laterosporus*, a pathogen of invertebrates and a broad-spectrum antimicrobial species. Insects 4, 476–492.

Ruiu, L., 2015. Insect pathogenic bacteria in integrated pest management. Insects 6, 352–367.

Ruiu, L., Satta, A., Floris, I., 2011. Comparative applications of azadirachtin- and *Brevibacillus laterosporus*-based formulations for house fly management experiments in dairy farms. J. Med. Entomol. 48, 345–350.

Ruiu, L., Satta, A., Floris, I., 2014. Administration of *Brevibacillus laterosporus* spores as a poultry feed additive to inhibit house fly development in feces: a new eco-sustainable concept. Poultry Sci. 93, 519–526.

Ryu, C.M., Kim, J., Choi, O., Kim, S.H., Park, C.S., 2006. Improvement of biological control capacity of *Paenibacillus polymyxa* E681 by seed pelleting on sesame. Biol. Contr. 39, 282–289.

Satinder, K.B., Verma, M., Tyagi, R.D., Valéro, J.R., 2006. Recent advances in downstream processing and formulations of *Bacillus thuringiensis* based biopesticides. Process Biochem. 41, 323–342.

Sawyer, A.J., Griggs, M.H., Wayne, R., 1994. Dimensions, density, and settling velocity of entomophthoralean conidia: implications for aerial dissemination of spores. J. Invertebr. Pathol. 63, 43–55.

Schnepf, H.E., Narva, K.E., Evans, S.L., 2007. Modified Chimeric Cry35 Proteins. US Patent 20077309785.

Schrank, A., Vainstein, M.H., 2010. Metarhizium anisopliae enzymes and toxins. Toxicon 56, 1267–1274.

Shapiro-Ilan, D.I., Gough, D.H., Piggott, S.J., Patterson Fife, J., 2006. Application technology and environmental considerations for use of entomopathogenic nematodes in biological control. Biol. Contr. 38, 124–133.

Shapiro-Ilan, D.I., Han, R., Dolinksi, C., 2012. Entomopathogenic nematode production and application technology. J. Nematol. 44, 206–217.

Siegel, J.P., Maddox, J.V., Ruesink, W.G., 1986. J. Invertebr. Pathol. 48, 167–173.

Singh, A., Singh, A., Kumari, M., Rai, M.K., Varma, A., 2003. Biotechnological importance of *Priformospora indicia* - a novel symbiotic mycorrhiza-like fungus: an overview. Indian J. Biotechnol. 2, 65–75.

Solter, L.F., Becnel, J.J., 2000. Entomopathogenic microsporodia. Field manual of technique in invertebrate pathology. In: Lacey, L.A., Kaya, H.K. (Eds.), Application and Evaluation of Pathogens for Control of Insects and Other Invertebrate Pests. Kluwer Academic, Dordrecht, pp. 231–254.

St. Leger, R.J., Wang, C., 2010. Genetic engineering of fungal biocontrol agents to achieve greater efficacy against insect pests. Appl. Microbiol. Biotechnol. 85, 901–907.

St. Leger, R.J., Joshi, L., Bidochka, M.J., Roberts, D.W., 1996. Construction of an improved mycoinsecticide overexpressing a toxic protease. Proc. Nat. Acad. Sci. U. S. A. 93, 6349–6354.

Stein, T., 2005. *Bacillus subtilis* antibiotics: structures, syntheses and specific functions. Mol. Microbiol. 56, 845–857.

Sugimoto, M., Koike, M., Hiyama, N., Nagao, H., 2003. Genetic, morphological, and virulence characterization of the entomopathogenic fungus *Verticillium lecanii*. J. Invertebr. Pathol. 82, 176–187.

Sun, X., 2015. History and current status of development and use of viral insecticides in China. Viruses 7, 306–319.

Sun, C., Shao, Y., Vahabi, K., Lu, J., Bhattacharya, S., Dong, S., Yeh, K.-W., Irena Sherameti, Lou, B., Baldwin, I.T., Oelmüller, R., 2014. The beneficial fungus *Piriformospora indica* protects *Arabidopsis* from *Verticillium dahlia* infection by downregulation plant defense responses. BMC Plant Biol. 14, 268. Article number.

Unnikumar, K.R., Sowjanya, S.K., Varma, A., 2013. *Piriformospora indica*: a versatile root endophytic symbiont. Symbiosis 60, 107–113.

Vega, F.E., Meyling, N.V., Luangsa-ard, J.J., Blackwell, M., 2012. Fungal entomopathogens. In: Vega, F., Kaya, H. (Eds.), Insect Pathology, second ed. Academic Press, London, UK, pp. 171–220.

Vidal, S., Jaber, L.R., 2015. Entomopathogenic fungi as endophytes: plant-endophyte- herbivore interactions and prospects for use in biological control. Curr. Sci. 109, 46–54.

Vodovar, N., Vallenet, D., Cruveiller, S., Rouy, Z., Barbe, V., Acosta, C., Cattolico, L., Jubin, C., Lajus, A., Segurens, B., et al., 2006. Complete genome sequence of the entomopathogenic and metabolically versatile soil bacterium *Pseudomonas entomophila*. Nat. Biotechnol. 24, 673–679.

Wang, C.S., St. Leger, R.J., 2007. A scorpion neurotoxin increases the potency of a fungal insecticide. Nat. Biotechnol. 25, 1455–1456.

Williams, T., Virto, C., Murillo, R., Caballero, P., 2017. Covert infection of insects by baculoviruses. Front. Microbiol. 8, 1337.

Zhu, H., Yao, L.T., Tian, F., Du, B.H., Ding, Y.Q., 2008. Screening and study on biological characteristics of antagonistic bacteria against *Fusarium solani*. Biotechnol. Bull. 1, 156–159.

Zimmermann, G., 2007. Review on safety of the entomopathogenic fungi *Beauveria bassiana* and *Beauveria brongniartii*. Biocontrol Sci. Technol. 17, 553–596.

Zimmermann, G., 2008. The entomopathogenic fungi *Isaria farinosa* (Formerly *Paecilomyces farinosus*) and the *Isaria fumosorosea* species complex (Formerly *Paecilomyces fumosoroseus*): biology, ecology and use in biological control. Biocontrol Sci. Technol. 18, 865–901.

Zuccaro, A., Basiewicz, M., Zurawska, M., Biedenkopf, D., Kogel, K.-H., 2009. Karyotype analysis, genome organization, and stable genetic transformation of the root colonizing fungus *Piriformospora indica*. Fungal Genet. Biol. 46 (8), 543–550.

Zuccaro, A., Lahrmann, U., Güldener, U., Langen, G., Pfiffi, S., Biedenkopf, D., et al., 2011. Endophytic life strategies decoded by genome and transcriptome analyses of the mutualistic root symbiont *Piriformospora indica*. PLoS Pathog. 7, e1002290. https://doi.org/10.1371/journal.ppat.1002290.

Chapter 16

Genetic engineering intervention in crop plants for developing biopesticides

Shambhu Krishan Lal[a,b], Sahil Mehta[b], Sudhir Kumar[c], Anil Kumar Singh[a], Madan Kumar[c], Binay Kumar Singh[c], Vijai Pal Bhadana[c] and Arunava Pattanayak[a,c]

[a]*School of Genetic Engineering, ICAR-Indian Institute of Agricultural Biotechnology, Ranchi, Jharkhand, India;* [b]*Crop Improvement Group, International Centre for Genetic Engineering and Biotechnology, New Delhi, Delhi, India;* [c]*School of Genomics and Molecular Breeding, ICAR-Indian Institute of Agricultural Biotechnology, Ranchi, Jharkhand, India*

16.1 Biopesticides

Advancements in scientific technologies have made our lives convenient. Agriculture plays a vital role in human civilization. The application of modern technologies in agriculture will not only boost agricultural production but will also reduce the cost of production and lead to the reclamation of polluted land, thereby increasing the farmer's income. The first green revolution during the 1970s was possible due to the development of high-yielding varieties and application of chemical fertilizers, which enhanced agricultural productivity, and fulfilled the global food demand (Chakravarti, 1973). During the green revolution, great emphasis was laid on infrastructure development and formulation of policies in favor of tackling the food crisis by boosting research and development in agriculture. Use of agricultural inputs such as high yielding varieties, chemical fertilizers and irrigation resulted in a surplus grain production. The darker shadow of the green revolution is the degradation of land, water, and environmental pollution that have been realized later (Pingali, 2012). Green revolution contributed to environmental pollution that severely affects biodiversity and human health (Benbi, 2017). After the green revolution, now the priority is to restore the polluted environment and minimize further deterioration of natural resources. There has been extensive exploitation of natural resources and non-judicious use of chemical fertilizers, pesticides, groundwater, etc., which drastically affected the practiced old age agriculture (Kesavan and Swaminathan, 2008). Chemical pesticides lost their efficacy to control insect pests, namely *Helicoverpa armigera*, *Spodoptera litura* in cotton plants (Kranthi et al., 2002). The alternative to synthetic pesticides are biopesticides that are greener, safer and environmentally friendly; hence there are least health hazards of their use in agriculture. Despite their beneficial effects, the technology is combating with already established markets of chemical pesticides in respect of cost, storability, and availability in time. Seed priming is another potent technology that is widely used in agriculture for getting tolerance of crop plants to pests (Lal et al., 2018). Biopesticides are derived from natural sources such as microbes, plants, animals, and possess the ability to kill pests (EPA, 2020, accessed on August 9, 2020). Microbial biopesticides are mostly made from soil-borne bacteria, *Bacillus thuringiensis* (Kumar and Singh, 2015). Biopesticides contribute 5% to the total pesticide market and globally share the value of $3 billion, worldwide (Marrone, 2014; Olson, 2015). The use of biopesticides increases by approximately 10% every year (Kumar and Singh, 2015). At present, countries like, United States of America (USA), China, India and Brazil use more number of biopesticides as compared to the European Union (Balog et al., 2017). In India, both indigenous and imported pesticides is being used by farmers and there is large cultivated area comes under pesticide usage (Figs. 16.1 and 16.2). The regulation of biopesticides is much more complicated in European countries as compared to India and USA. Application of high doses of chemical-based pesticides over the prolonged period has witnessed the evolution of super-weeds and novel biotypes (Fernandez-Cornejo et al., 2014). Due to repeated cultivation of transgenic crops over the period, pests evolved new biotypes that become resistant to toxins produced by transgenes. In the last decade, there has been a report of five significant pests that become resistant to transgenic cotton and corn (Tabashnik et al., 2013a, Fig. 16.4) Thus, a refined

Biopesticides. https://doi.org/10.1016/B978-0-12-823355-9.00020-1

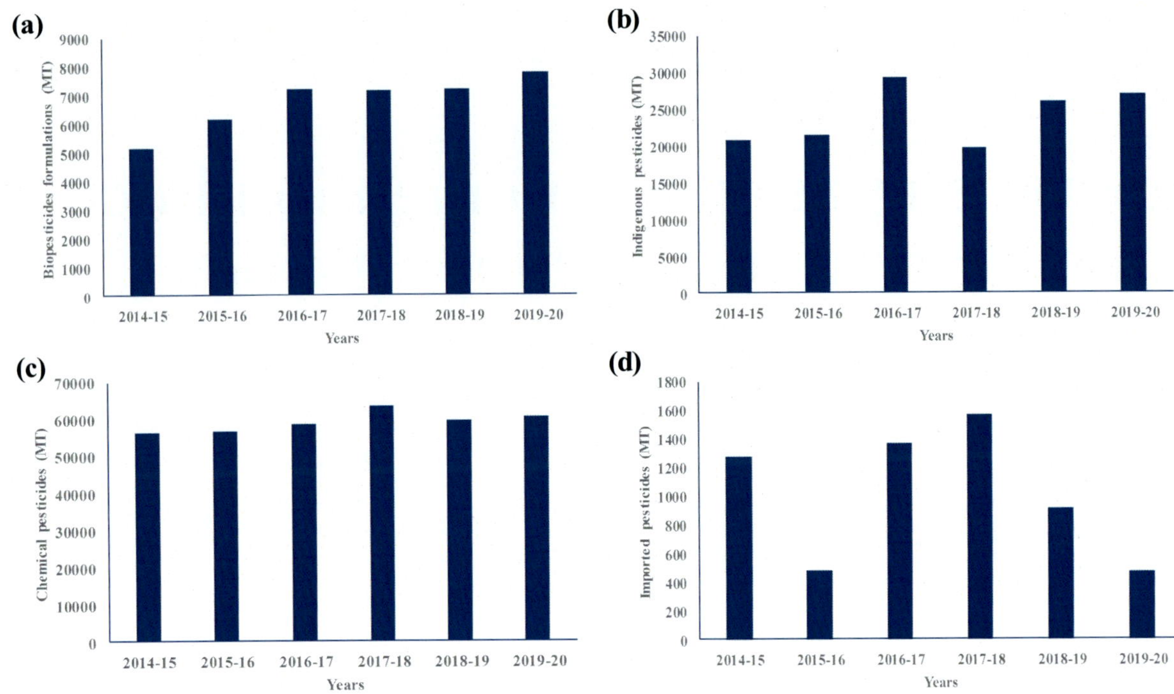

FIGURE 16.1 Consumption of different categories of pesticides (Metric tonnes) in India during years 2014–20. *Data accessed on August 6, 2020 (Pesticide consumption in India, n.d., Government of India, Ministry of Agriculture and Farmers welfare: pqs.gov.in/statistical-database).*

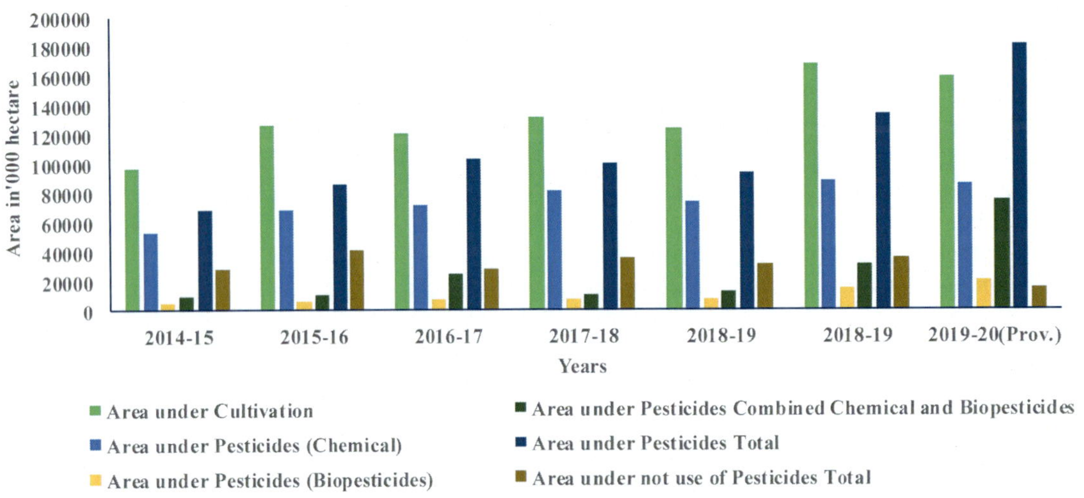

FIGURE 16.2 Area under cultivation (000 ha) with different categories of pesticides application in India during years 2014–20. *Data accessed on August 6, 2020 (Pesticide consumption in India, n.d., Government of India, Ministry of Agriculture and Farmers welfare: pqs.gov.in/statistical-database).*

and sophisticated technology is a prerequisite for controlling the evolution of resistant pests in agriculture. In this context, modern genetic engineering (GE) approaches for improving crop tolerance against pests have been discussed in this article. Recent advancements in genome editing technology could be the future of agriculture, and it will inculcate precise breeding for the development of new traits in existing crop varieties.

16.2 Engineering of Bt genes for insect resistance

A number of crop plants resistant to insect pests have been developed by expressing cry genes of *Bacillus thuriengenesis* (Bt) through transgenic approach (Adang et al., 2014). The Cry proteins get solubilized in the alkaline pH of the midgut of insects, and pro-toxins are formed by cleavage of C or N terminals. The active toxins bind to epithelial cells of midgut and create pores in the membrane of the gut lining, causing death of insects (Vachon et al., 2012). Syngenta is marketing the transgenic maize co-expressing genes encoding Bt Cry1Ab delta-endotoxin and Vip3Aa2 (Lee et al., 2003). Plants secrete protease inhibitors in response to physical injury and insect attack, which retard the function of protease present in the insect gut. Cotton plants expressing cowpea trypsin proteinase (CpTI) got commercialized in China in 1999 (Smigocki et al., 2013). Some Insect species show less susceptibility to Cry protein, and overtime insects develop resistance for it. Cry protein consists of three domains, namely I, II, and III. The splicing modification gives a broad spectrum of toxicity against insects. Fifty-six amino acid cleavage at N-terminus of domain I in Cry1Ab and Cry1Ac toxins produce modified version of toxins that are active against resistant lepidopteran insects for Cry proteins (Tabashnik et al., 2011, 2013b; Deist et al., 2014). Cry1Ab toxin mutant, V171C showed 25 times more toxicity against gypsy moth (Alzate et al., 2010). Introduction of Chymotrypsin G cleavage point in domain I of Cry3A between 3 and 4 helices showed higher toxicity against Western corn rootworm (Walters et al., 2008). A fusion protein was made by stacking delta-endotoxin Cry1Ac and ricin −B chain galactose binding domain (RB), the fusion product, BtRB exhibited an additional binding domain with higher interaction. The transgenic maize and rice expressing this fusion protein exhibited higher tolerance as compared to Bt alone (Mehlo et al., 2005). Hybrid engineered protein made by variable region exchange Cry1Ab and Cry3A exhibited higher toxicity against larvae of western corn rootworm (Walters et al., 2010). The truncated form of Bt toxin could target other beneficial insects and affect biodiversity (Tabashnik et al., 2013b).

16.3 Bt cotton adoption in India

Cotton is a commercial crop as it earns foreign exchange in addition to providing raw materials to the textile industry. This crop is widely adopted by Indian farmers and is being grown in a large area mainly due to remunerative returns. However, yield of cotton in India was far below than the world average, mainly due to the infestation of cotton bollworm that severely affects yield loss. Bt cotton was released for commercial cultivation in India in 2002, and thereafter, the outbreak of bollworm reduced drastically. The total area under cultivation of Biotech cotton crop is 11.6 million hectares that share 95% of the total area under cotton cultivation in India (Fig. 16.3). The Bt cotton was grown in 2002 in six states, namely Maharashtra, Madhya Pradesh, Andhra Pradesh, Tamilnadu, Karnataka, and Gujarat. Studies conducted on the use of Bt cotton revealed 31% enhancement in yield and 39% reduction in insecticide spray that finally gives 88% enhancement in profit (US$250/ha) to farmers (ISAAA, 2020, Accessed on August 10, 2020).

16.4 Genetic engineering approaches for combating aphid infestation in crop plants

Aphids are responsible for severe yield losses in crops, primarily through infestation and indirectly by transmitting plant viruses (Miles, 1989; Hossain et al., 2006; Tagu et al., 2008). Worldwide, massive losses of crop produce are reported due

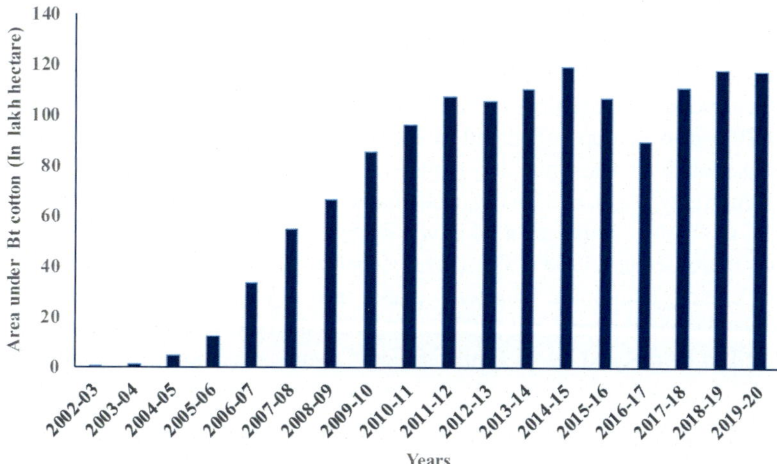

FIGURE 16.3 Year-wise change in area under Bt cotton in India. *Department of Agriculture, Cooperation and Farmers Welfare, State Governments and Directorate of Cotton Development, Nagpur, India 2020.*

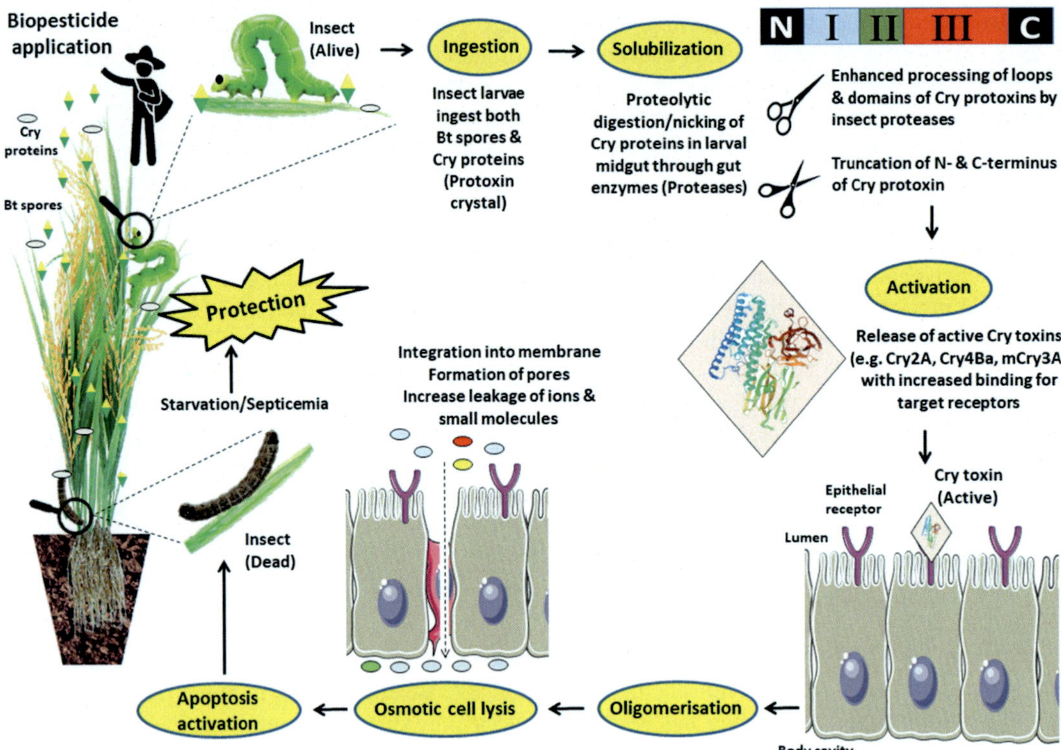

FIGURE 16.4 Schematic representation of mode of action of Bt toxin and genetic engineering in its domain for enhancing specificity and toxicity against insects.

to aphids (Morrison and Peairs, 1998). Long term application of pesticides leads to development of resistance in aphids and chemical pesticides also kill other useful organisms (Sabater-Munoz et al., 2006; Puinean et al., 2010; Yu et al., 2012). In most of the major crops, there is a lack of germplasms with aphid tolerance, therefore breeders are still struggling to develop aphid tolerance in major crops. Thus, genetic engineering is the only option left with the researcher for developing aphid resistance in crop plants (Van Eck et al., 2010; Yu et al., 2012).

Engineering of Potato and maize by expressing *Galanthus nivalis* agglutinin (GNA) related lectins provide aphid tolerance (Down et al., 1996, 2003; Wang et al., 2005).

Pinellia ternate agglutinin (PTA) expressed in tobacco chloroplast lead to accumulation of PTA up to 9.2% of total soluble proteins and provided high level of tolerance against aphids (Jin et al., 2012).

Expression of Concanavalin A (ConA) in potato retarded development of green peach and peach-potato aphids (Gatehouse et al., 1999). Wheat germ agglutinin (WGA) expression in mustard exhibited significant tolerance to mustard aphids (Kanrar et al., 2002). The cysteine proteinase inhibitors (PIs) *Oryza* cystatin (OC−I) administered in diet inhibited aphid growth. The transgenic oilseed rape and eggplant expressing OC-I caused growth inhibition of aphids (Rahbe et al., 2003; Azzouz et al., 2005; Ribeiro et al., 2006). Barley cysteine PIs exhibited toxicity for nymphs of pea aphids. The ectopic expression of *HvCPI-6* in Arabidopsis resulted in delayed aphids development (Carrillo et al., 2011). Plant virus vector made by fusing coat protein of lute virus and peptide ω-hexatoxin- Hv1a (insect toxin) was delivered in insect hemocoel resulted in high mortality. Artificial feeding and fusion protein expression in Arabidopsis caused mortality in aphid species (Bonning et al., 2014). Expression of plant R genes, namely, Mi-1.2 in susceptible tomato conferred tolerance to potato aphids (Rossi et al., 1998). Transgenic tobacco plants possessing suppressed expression of P450 producing more diterpene cembratriene-ol and transgenic plants were found to be less susceptible to aphid colonization (Wang et al., 2001). Transgenic tobacco expressing β-Glucosidase produced enhanced sucrose esters and trichrome density and aphid growth was found to be reduced by 15% in transgenic plants as compared to control plants (Jin et al., 2011).

16.5 Applications of RNA interference (RNAi) to control pests

RNAi relies on the silencing of genes by targeting specific transcript. When dsRNA is fed to insects, it causes gene silencing in an insect that is measured by gene expression and transcript level. Transgenic plants expressing dsRNA results in resistance to insects (Turner et al., 2006). Transgenic maize plants expressing dsRNA having specificity for corn rootworm V-type ATPase showed the reduction in damage by insects and silencing of insect transcript as compared to wild type plants (Baum et al., 2007). Expression of dsRNA in Arabidopsis and tobacco targeting against CytP450 gene, CYP6AE14 rendered insects more susceptible to gossypol (Mao et al., 2007).

RNAi technology has been explored to silence AC1 gene of Bean golden mosaic virus (BGMV). The transgenic common bean plants (*Phaseolus vulgaris*) expressing hairpin structure against AC1 viral gene are tolerant against the virus (Bonfim et al., 2007).

Engineered papaya and plum plants expressing antisense strand for viral coat protein gene exhibited tolerance to *Papaya ringspot* and *Plum pox* virus, respectively (Gonsalves et al., 2004; Scorza et al., 2013). Thus, the RNAi technology has great potential in controlling insect pests in agricultural crops.

16.6 Applications of genome editing to control pests

Genome editing is mediated by Cas9 endonuclease, creating site-specific mutation or insertion of DNA fragment by homology-based recombinations. Nowadays, it is widely used to get the desired mutation in the genome (Mehta et al., 2020). The oligonucleotide-based correction has been used for developing crop plants such as maize, rice, wheat for acetolactate synthase herbicide tolerance (Okuzaki and Toriyama, 2004; Zhu et al., 2000; Dong et al., 2006). Targeted base editing of acetolactate synthase (ALS1, ALS2) gene in maize confered tolerance to sulfonylurea. The edited ALS1 plants tolerated up to 15 fold recommended doses of pesticides, and ALS2 plants tolerate up to 200 mgL^{-1} of chlorsulfuron (Li et al., 2020). Acetolactate synthase and acetyl-coenzyme A carboxylase (ACCase) mutation mediated by CRISPR/Cas9 based editing has been accomplished in wheat. The Co-edited plants for both genes exhibited tolerance to herbicide nicosulfuron and quizalofop (Zhang et al., 2019). Single amino acid mutation in acetohydroxyacid synthase (AHAS) created by template-based homologous recombination through zinc finger nuclease in wheat provided tolerance to imidazolinone herbicides (Ran et al., 2018). Targeted genome editing of Acetolactate synthase using a repair template through CRISPR/Cas9 in maize provides tolerance to chlorsulfuron (Svitashev et al., 2015, 2016). Targeted base editing of CESA3 in Arabidopsis confers tolerance for cellulose biosynthesis inhibitors, namely C17 and isoxaben (Hu et al., 2019, Table 16.1).

TABLE 16.1 List of genetically engineered plants as biopesticides.

Gene	Plant	Trait	Technology	References
Bt Cry1Ab delta-endotoxin and Vip3Aa2	Maize	Insect resistance	Transgenic	Lee et al. (2003)
Trypsin proteinase (CpTI)	Cotton	Insect resistance	Transgenic	Smigocki et al. (2013)
Delta-endotoxin Cry1Ac and ricin −B chain galactose binding domain (RB), BtRB	Maize, rice	Insect resistance	Transgenic	Mehlo et al. (2005)
Galanthus nivalis agglutinin (GNA)	Potato, maize	Aphid resistance	Transgenic	Down et al. (1996), Wang et al. (2005)
Pinellia ternate agglutinin (PTA)	Tobacco	Aphid resistance	Transgenic	Jin et al. (2012)
ConA	Potato	Aphid resistance	Transgenic	Gatehouse et al. (1999)
WGA	Mustard	Aphid resistance	Transgenic	Kanrar et al. (2002)
oryza cystatin (OC−I)	Eggplant and oil seed rape	Aphid resistance	Transgenic	Rahbe et al. (2003), Ribeiro et al. (2006)
HvCPI-6	Arabidopsis	Aphid resistance	Transgenic	Carrillo et al. (2011)
peptide ω-hexatoxin- Hv1a	Arabidopsis	Aphid resistance	Transgenic	Bonning et al. (2014)
Mi-1.2	Tomato	Aphid resistance	Transgenic	Rossi et al. (1998)
β-Glucosidase	Tobacco	Aphid resistance	Transgenic	Jin et al. (2011)
P450	Tobacco	Aphid resistance	Transgenic	Wang et al. (2001)

Continued

TABLE 16.1 List of genetically engineered plants as biopesticides.—cont'd

Gene	Plant	Trait	Technology	References
V-type ATPase	Maize	Insect resistance	RNAi	Baum et al. (2007)
CYP6AE14	Arabidopsis, tobacco	Insect resistance	RNAi	Mao et al. (2007)
AC1	Common bean plants	Virus resistance	RNAi	Bonfim et al. (2007)
PPV-coat protein (CP)	Tobacco	Virus resistance	RNAi	Scorza et al. (2013), Gonsalves et al. (2004)
ALS1, ALS2	Maize	Herbicide tolerance	Genome editing	Li et al. (2020)
ALS	Maize, rice, wheat	Herbicide tolerance	Homology based recombination	Zhu et al. (2000), Okuzaki and Toriyama (2004), Dong et al. (2006)
ALS and ACCase	Wheat	Herbicide tolerance	Genome editing	Zhang et al. (2019)
Acetohydroxy acid synthase	Wheat	Herbicide tolerance	Genome editing	Ran et al. (2018)
ALS	Maize	Herbicide tolerance	Genome editing	Svitashev et al. (2015)
CESA3	Arabidopsis	Herbicide tolerance	Genome editing	Hu et al. (2019)

16.7 Future perspectives

Biopesticides provide a safe and environmental friendly alternative to chemical-based pesticides to acquire genetic yield potential of crops and to fulfill the demand of the exponentially growing population. It is a boon for human society in the prevailing climate change and deteriorating ecosystem scenario. Many biopesticides have been developed, but there is a great challenge in public acceptability and implementation of this technology. Still, refinement and innovation like new formulations, delivery of biopesticides in crop plants, efficient genetically engineered crop plants and effective biocontrol agents are needed to get rid of pests and chemical pesticides. Through precise modification in the genome of crop plants through advanced technology like transgenic, RNAi, Protein engineering, genome editing enables plants for enhanced tolerance to insect pests. The chances of insect pests acquiring resistance are least when more genes are stacked together and introduced inside crop plants. The emphasis would be given to provide this technology at a low cost with significant profit after using biopesticides in the agricultural field. Broadly speaking, there is need of the hour for a public-private partnership in R & D, the policy decision of the government for providing subsidy to farmers and awareness program by extension workers, which may lead to large scale implementation of the technology.

References

Adang, M.J., Crickmore, N., Jurat-Fuentes, J.L., January 1, 2014. Diversity of *Bacillus thuringiensis* crystal toxins and mechanism of action. In: Advances in Insect Physiology, vol. 47. Academic Press, pp. 39–87.

Alzate, O., Osorio, C., Florez, A.M., Dean, D.H., December 1, 2010. Participation of valine 171 in α-helix 5 of *Bacillus thuringiensis* Cry1Ab δ-endotoxin in translocation of toxin into *Lymantria dispar* midgut membranes. Appl. Environ. Microbiol. 76 (23), 7878–7880.

Azzouz, H., Cherqui, A., Campan, E.D., Rahbe, Y., Duport, G., Jouanin, L., Kaiser, L., Giordanengo, P., January 1, 2005. Effects of plant protease inhibitors, oryzacystatin I and soybean Bowman–Birk inhibitor, on the aphid *Macrosiphum euphorbiae* (Homoptera, Aphididae) and its parasitoid *Aphelinus abdominalis* (Hymenoptera, Aphelinidae). J. Insect Physiol. 51 (1), 75–86.

Balog, A., Hartel, T., Loxdale, H.D., Wilson, K., November 2017. Differences in the progress of the biopesticide revolution between the EU and other major crop-growing regions. Pest Manag. Sci. 73 (11), 2203–2208.

Baum, J.A., Bogaert, T., Clinton, W., Heck, G.R., Feldmann, P., Ilagan, O., Johnson, S., Plaetinck, G., Munyikwa, T., Pleau, M., Vaughn, T., November 2007. Control of coleopteran insect pests through RNA interference. Nat. Biotechnol. 25 (11), 1322–1326.

Benbi, D.K., January 2017. Nitrogen balances of intensively cultivated rice—wheat cropping systems in original green revolution states of India. In: The Indian Nitrogen Assessment, vol. 1. Elsevier, pp. 77—93.

Bonfim, K., Faria, J.C., Nogueira, E.O., Mendes, É.A., Aragão, F.J., June 2007. RNAi-mediated resistance to Bean golden mosaic virus in genetically engineered common bean (*Phaseolus vulgaris*). Mol. Plant-Micr. Interac. 20 (6), 717—726.

Bonning, B.C., Pal, N., Liu, S., Wang, Z., Sivakumar, S., Dixon, P.M., King, G.F., Miller, W.A., January 2014. Toxin delivery by the coat protein of an aphid-vectored plant virus provides plant resistance to aphids. Nat. Biotechnol. 32 (1), 102—105.

Carrillo, L., Martinez, M., Alvarez-Alfageme, F., Castanera, P., Smagghe, G., Diaz, I., Ortego, F., April 1, 2011. A barley cysteine-proteinase inhibitor reduces the performance of two aphid species in artificial diets and transgenic *Arabidopsis* plants. Transgenic Res. 20 (2), 305—319.

Chakravarti, A.K., September 1, 1973. Green revolution in India. Ann. Assoc. Am. Geogr. 63 (3), 319—330.

Deist, B.R., Rausch, M.A., Fernandez-Luna, M.T., Adang, M.J., Bonning, B.C., October 2014. Bt toxin modification for enhanced efficacy. Toxins 6 (10), 3005—3027.

Department of Agriculture, Cooperation and Farmers Welfare (DAC&FW), State Governments and Directorate of Cotton Development, 2020, Nagpur, India.

Dong, C., Beetham, P., Vincent, K., Sharp, P., May 1, 2006. Oligonucleotide-directed gene repair in wheat using a transient plasmid gene repair assay system. Plant Cell Rep. 25 (5), 457—465.

Down, R.E., Ford, L., Woodhouse, S.D., Davison, G.M., Majerus, M.E., Gatehouse, J.A., Gatehouse, A.M., April 1, 2003. Tritrophic interactions between transgenic potato expressing snowdrop lectin (GNA), an aphid pest (peach—potato aphid; *Myzus persicae* (Sulz.) and a beneficial predator (2-spot ladybird; *Adalia bipunctata* L.). Transgenic Res. 12 (2), 229—241.

Down, R.E., Gatehouse, A.M., Hamilton, W.D., Gatehouse, J.A., November 1, 1996. Snowdrop lectin inhibits development and decreases fecundity of the glasshouse potato aphid (*Aulacorthum solani*) when administered in vitro and via transgenic plants both in laboratory and glasshouse trials. J. Insect Physiol. 42 (11—12), 1035—1045.

EPA USA, 2020. Biopesticides. Available online: www.epa.gov/pesticides/biopesticides. (Accessed 9 August 2020).

Fernandez-Cornejo, J., Wechsler, S., Livingston, M., Mitchell, L., February 1, 2014. Genetically Engineered Crops in the United States. USDA-ERS Economic Research Report 162.

Gatehouse, A.M., Davison, G.M., Stewart, J.N., Gatehouse, L.N., Kumar, A., Geoghegan, I.E., Birch, A.N., Gatehouse, J.A., May 1, 1999. Concanavalin A inhibits development of tomato moth (*Lacanobia oleracea*) and peach-potato aphid (*Myzus persicae*) when expressed in transgenic potato plants. Mol. Breed. 5 (2), 153—165.

Gonsalves, D., Gonsalves, C., Ferreira, S., Pitz, K., Fitch, M., Manshardt, R., Slightom, J., 2004. Transgenic Virus Resistant Papaya: From Hope to Reality for Controlling *Papaya ringspot* virus in Hawaii. APS net feature story for July.

Hossain, M.A., Ferdous, J., Salim, M.M., 2006. Relative abundance and yield loss assessment of lentil aphid, *Aphis craccivora* Koch in relation to different sowing dates. J. Agric. Rural Dev. 4 (1), 101—106.

Hu, Z., Zhang, T., Rombaut, D., Decaestecker, W., Xing, A., D'Haeyer, S., Höfer, R., Vercauteren, I., Karimi, M., Jacobs, T., De Veylder, L., June 1, 2019. Genome editing-based engineering of CESA3 dual cellulose-inhibitor-resistant plants. Plant Physiol. 180 (2), 827—836.

ISAAA (The International Service for the Acquisition of Agri-biotech Applications), 2020. The Story of Bt Cotton in India. (Accessed 10 August 2020).

Jin, S., Kanagaraj, A., Verma, D., Lange, T., Daniell, H., January 1, 2011. Release of hormones from conjugates: chloroplast expression of β-glucosidase results in elevated phytohormone levels associated with significant increase in biomass and protection from aphids or whiteflies conferred by sucrose esters. Plant Physiol. 155 (1), 222—235.

Jin, S., Zhang, X., Daniell, H., April 2012. *Pinellia ternata* agglutinin expression in chloroplasts confers broad spectrum resistance against aphid, whitefly, lepidopteran insects, bacterial and viral pathogens. Plant Biotechnol. J. 10 (3), 313—327.

Kanrar, S., Venkateswari, J., Kirti, P., Chopra, V., March 1, 2002. Transgenic Indian mustard (*Brassica juncea*) with resistance to the mustard aphid (*Lipaphis erysimi* Kalt.). Plant Cell Rep. 20 (10), 976—981.

Kesavan, P.C., Swaminathan, M.S., February 27, 2008. Strategies and models for agricultural sustainability in developing Asian countries. Phil. Trans. Biol. Sci. 363 (1492), 877—891.

Kranthi, K.R., Russell, D., Wanjari, R., Kherde, M., Munje, S., Lavhe, N., Armes, N., February 1, 2002. In-season changes in resistance to insecticides in *Helicoverpa armigera* (Lepidoptera: Noctuidae) in India. J. Econ. Entomol. 95 (1), 134—142.

Kumar, S., Singh, A., 2015. Biopesticides: present status and the future prospects. J. Fertil. Pestic. 6 (2), 100—129.

Lal, S.K., Kumar, S., Sheri, V., Mehta, S., Varakumar, P., Ram, B., Borphukan, B., James, D., Fartyal, D., Reddy, M.K., 2018. Seed priming: an emerging technology to impart abiotic stress tolerance in crop plants. In: Advances in Seed Priming. Springer, Singapore, pp. 41—50.

Lee, M.K., Walters, F.S., Hart, H., Palekar, N., Chen, J.S., August 1, 2003. The mode of action of the *Bacillus thuringiensis* vegetative insecticidal protein Vip3A differs from that of Cry1Ab δ-endotoxin. Appl. Environ. Microbiol. 69 (8), 4648—4657.

Li, Y., Zhu, J., Wu, H., Liu, C., Huang, C., Lan, J., Zhao, Y., Xie, C., June 1, 2020. Precise base editing of non-allelic acetolactate synthase genes confers sulfonylurea herbicide resistance in maize. The Crop J. 8 (3), 449—456.

Mao, Y.B., Cai, W.J., Wang, J.W., Hong, G.J., Tao, X.Y., Wang, L.J., Huang, Y.P., Chen, X.Y., November 2007. Silencing a cotton bollworm P450 monooxygenase gene by plant-mediated RNAi impairs larval tolerance of gossypol. Nat. Biotechnol. 25 (11), 1307—1313.

Marrone, P.G., 2014. The market and potential for biopesticides. In: Gross, A.D., Coats, J.R., Duke, S.O., Seiber, J.N. (Eds.), Biopesticides: State of the Art and Future Opportunities. American Chemical Society, Washington, DC, USA, pp. 245—258.

Mehlo, L., Gahakwa, D., Nghia, P.T., Loc, N.T., Capell, T., Gatehouse, J.A., Gatehouse, A.M., Christou, P., May 31, 2005. An alternative strategy for sustainable pest resistance in genetically enhanced crops. Proc. Natl. Acad. Sci. U. S. A. 102 (22), 7812–7816.

Mehta, S., Lal, S.K., Sahu, K.P., Venkatapuram, A.K., Kumar, M., Sheri, V., Varakumar, P., Vishwakarma, C., Yadav, R., Jameel, M.R., Ali, M., 2020. CRISPR/Cas9-edited rice: a new frontier for sustainable agriculture. In: New Frontiers in Stress Management for Durable Agriculture. Springer, Singapore, pp. 427–458.

Miles, P.W., 1989. Specific responscs and damage caused by Aphidoidea. In: Minks, A.K., Harrewijn, P. (Eds.), Aphids, their Biology, Natural Enemies and Control, vol. 100, p. 23141.

Morrison, W.P., Peairs, F.B., 1998. Response model concept and economic impact. In: Response Model for an Introduced Pest—The Russian Wheat Aphid. Entomological Society of America, Lanham, MD.

Okuzaki, A., Toriyama, K., February 2004. Chimeric RNA/DNA oligonucleotide-directed gene targeting in rice. Plant Cell Rep. 22 (7), 509–512.

Olson, S., October 1, 2015. An analysis of the biopesticide market now and where it is going. Outlooks Pest Manag. 26 (5), 203–206.

Pesticide Consumption in India. Government of India, Ministry of Agriculture and Farmers welfare: pqs, n.d. gov.in/statistical-database (Accesssed on August 6, 2020).

Pingali, P.L., July 31, 2012. Green revolution: impacts, limits, and the path ahead. Proc. Natl. Acad. Sci. U. S. A. 109 (31), 12302–12308.

Puinean, A.M., Foster, S.P., Oliphant, L., Denholm, I., Field, L.M., Millar, N.S., Williamson, M.S., Bass, C., June 24, 2010. Amplification of a cytochrome P450 gene is associated with resistance to neonicotinoid insecticides in the aphid *Myzus persicae*. PLoS Genet. 6 (6), e1000999.

Rahbe, Y., Deraison, C., Bonadé-Bottino, M., Girard, C., Nardon, C., Jouanin, L., April 1, 2003. Effects of the cysteine protease inhibitor oryzacystatin (OC-I) on different aphids and reduced performance of *Myzus persicae* on OC-I expressing transgenic oilseed rape. Plant Sci. 164 (4), 441–450.

Ran, Y., Patron, N., Kay, P., Wong, D., Buchanan, M., Cao, Y.Y., Sawbridge, T., Davies, J.P., Mason, J., Webb, S.R., Spangenberg, G., December 2018. Zinc finger nuclease-mediated precision genome editing of an endogenous gene in hexaploid bread wheat (*Triticum aestivum*) using a DNA repair template. Plant Biotechnol. J. 16 (12), 2088–2101.

Ribeiro, A.D., Pereira, E.J., Galvan, T.L., Picanco, M.C., Picoli, E.D., Da Silva, D.J., Fari, M.G., Otoni, W.C., March 2006. Effect of eggplant transformed with oryzacystatin gene on *Myzus persicae* and *Macrosiphum euphorbiae*. J. Appl. Entomol. 130 (2), 84–90.

Rossi, M., Goggin, F.L., Milligan, S.B., Kaloshian, I., Ullman, D.E., Williamson, V.M., August 18, 1998. The nematode resistance gene Mi of tomato confers resistance against the potato aphid. Proc. Natl. Acad. Sci. U. S. A. 95 (17), 9750–9754.

Sabater-Munoz, B., Legeai, F., Rispe, C., Bonhomme, J., Dearden, P., Dossat, C., Duclert, A., Gauthier, J.P., Ducray, D.G., Hunter, W., Dang, P., March 1, 2006. Large-scale gene discovery in the pea aphid *Acyrthosiphon pisum* (Hemiptera). Genome Biol. 7 (3), R21.

Scorza, R., Callahan, A., Dardick, C., Ravelonandro, M., Polak, J., Malinowski, T., Zagrai, I., Cambra, M., Kamenova, I., October 1, 2013. Genetic engineering of Plum pox virus resistance:'HoneySweet' plum—from concept to product. Plant Cell Tissue Organ Cult. 115 (1), 1–2.

Smigocki, A.C., Ivic-Haymes, S., Li, H., Savić, J., February 26, 2013. Pest protection conferred by a beta vulgaris serine proteinase inhibitor gene. PloS One 8 (2), e57303.

Svitashev, S., Schwartz, C., Lenderts, B., Young, J.K., Cigan, A.M., November 16, 2016. Genome editing in maize directed by CRISPR–Cas9 ribonucleoprotein complexes. Nat. Commun. 7 (1), 1–7.

Svitashev, S., Young, J.K., Schwartz, C., Gao, H., Falco, S.C., Cigan, A.M., October 1, 2015. Targeted mutagenesis, precise gene editing, and site-specific gene insertion in maize using Cas9 and guide RNA. Plant Physiol. 169 (2), 931–945.

Tabashnik, B.E., Brevault, T., Carriere, Y., June 2013a. Insect resistance to Bt crops: lessons from the first billion acres. Nat. Biotechnol. 31 (6), 510–521.

Tabashnik, B.E., Fabrick, J.A., Unnithan, G.C., Yelich, A.J., Masson, L., Zhang, J., Bravo, A., Sober6n, M., November 7, 2013b. Efficacy of genetically modified Bt toxins alone and in combinations against pink bollworm resistant to Cry1Ac and Cry2Ab. PloS One 8 (11), e80496.

Tabashnik, B.E., Huang, F., Ghimire, M.N., Leonard, B.R., Siegfried, B.D., Rangasamy, M., Yang, Y., Wu, Y., Gahan, L.J., Heckel, D.G., Bravo, A., December 2011. Efficacy of genetically modified Bt toxins against insects with different genetic mechanisms of resistance. Nat. Biotechnol. 29 (12), 1128–1131.

Tagu, D., Klingler, J.P., Moya, A., Simon, J.C., June 2008. Early progress in aphid genomics and consequences for plant–aphid interactions studies. Mol. Plant-Micr. Interac. 21 (6), 701–708.

Turner, C.T., Davy, M.W., MacDiarmid, R.M., Plummer, K.M., Birch, N.P., Newcomb, R.D., June 2006. RNA interference in the light brown apple moth, *Epiphyas postvittana* (Walker) induced by double-stranded RNA feeding. Insect Mol. Biol. 15 (3), 383–391.

Vachon, V., Laprade, R., Schwartz, J.L., September 15, 2012. Current models of the mode of action of *Bacillus thuringiensis* insecticidal crystal proteins: a critical review. J. Invertebr. Pathol. 111 (1), 1–2.

Van Eck, L., Schultz, T., Leach, J.E., Scofield, S.R., Peairs, F.B., Botha, A.M., Lapitan, N.L., December 2010. Virus-induced gene silencing of WRKY53 and an inducible phenylalanine ammonia-lyase in wheat reduces aphid resistance. Plant Biotechnol. J. 8 (9), 1023–1032.

Walters, F.S., deFontes, C.M., Hart, H., Warren, G.W., Chen, J.S., May 15, 2010. Lepidopteran-active variable-region sequence imparts coleopteran activity in eCry3. 1Ab, an engineered *Bacillus thuringiensis* hybrid insecticidal protein. Appl. Environ. Microbiol. 76 (10), 3082–3088.

Walters, F.S., Stacy, C.M., Lee, M.K., Palekar, N., Chen, J.S., January 15, 2008. An engineered chymotrypsin/cathepsin G site in domain I renders *Bacillus thuringiensis* Cry3A active against western corn rootworm larvae. Appl. Environ. Microbiol. 74 (2), 367–374.

Wang, E., Wang, R., DeParasis, J., Loughrin, J.H., Gan, S., Wagner, G.J., April 2001. Suppression of a P450 hydroxylase gene in plant trichome glands enhances natural-product-based aphid resistance. Nat. Biotechnol. 19 (4), 371–374.

Wang, Z., Zhang, K., Sun, X., Tang, K., Zhang, J., December 1, 2005. Enhancement of resistance to aphids by introducing the snowdrop lectin gene gna into maize plants. J. Biosci. 30 (5), 627–638.

Yu, X.D., Pickett, J., Ma, Y.Z., Bruce, T., Napier, J., Jones, H.D., Xia, L.Q., May 2012. Metabolic engineering of plant-derived (E)-β-farnesene synthase genes for a novel type of aphid-resistant genetically modified crop plants. J. Integr. Plant Biol. 54 (5), 282–299.

Zhang, R., Liu, J., Chai, Z., Chen, S., Bai, Y., Zong, Y., Chen, K., Li, J., Jiang, L., Gao, C., May 2019. Generation of herbicide tolerance traits and a new selectable marker in wheat using base editing. Nat. Plants 5 (5), 480–485.

Zhu, T., Mettenburg, K., Peterson, D.J., Tagliani, L., Baszczynski, C.L., May 2000. Engineering herbicide-resistant maize using chimeric RNA/DNA oligonucleotides. Nat. Biotechnol. 18 (5), 555–558.

Chapter 17

Medicinal plants associated microflora as an unexplored niche of biopesticide

Ved Prakash Giri[a,c], Shipra Pandey[a,b], Satyendra Pratap Singh[a,d], Bhanu Kumar[d], S.F.A. Zaidi[e] and Aradhana Mishra[a,b]

[a]*Division of Microbial Technology, CSIR-National Botanical Research Institute, Lucknow, Uttar Pradesh, India;* [b]*Academy of Scientific and Innovative Research (AcSIR), Ghaziabad, Uttar Pradesh, India;* [c]*Department of Botany, Lucknow University, Lucknow, Uttar Pradesh, India;* [d]*Pharmacognosy and Ethnopharmacology Division, CSIR-National Botanical Research Institute, Lucknow, Uttar Pradesh, India;* [e]*Department of Soil Science and Agriculture Chemistry, Acharya Narendra Deva University of Agriculture and Technology, Faizabad, Uttar Pradesh, India*

17.1 Introduction

17.1.1 Medicinal plant diversity in India

India is one of the most diverse countries in the world having a rich repository of high value, endemic and rare medicinal plants (Kamboj, 2000; Krishnan et al., 2011). India is one of the 12 mega diversity countries in the world having four biodiversity hotspots. Concerning plant diversity, India's ranks 10th in the world and fourth in Asia (Singh and Chaturvedi, 2017). The reason behind this vast diversity is the presence of different climatic conditions such as alpine in the Himalayas to arid zones in Rajasthan. There are tropical forests in the Western Ghats while plateaus, mountains and valleys in North-Eastern states (Ganie et al., 2020). Apart from varying topography, soil, rainfall, temperature, humidity conditions also differ from place to place which gives rise to huge phytodiversity. The microclimatic variations further leads to differences in the phenology, metabolism, physiology, chemical profile and even morphology of plants in addition to growth patterns across the geography (Ncube et al., 2012).

India is a repository of vast traditional knowledge and a deep-rooted system of indigenous medicine. According to a report from the Government of India, about 75% of the Indian population including the majority of tribal and ethnic communities are mostly dependent on the traditional knowledge and practices for primary health care needs (Kala et al., 2006; Dhakal et al., 2020). The age-old Indian traditional medicine system "Ayurveda" is very extensive in terms of the plants used, owing to the great phytodiversity of the country. In India, there are several traditional systems of the medicine being practiced in different regions.

According to an estimate, more than 45,000 plant species are commonly found in India out of which flowering plants constitute around 15,000−18,000; members of bryophytes are around 1800; algal species are 2500; 1600 lichens; 23,000 fungal species exist in India (Bharucha, 2006; Sharma et al., 2008). The surveys conducted by several workers have revealed that approximately 20,000 plant species are having one or the other medicinal properties (Mukherjee, 2008; Kumar et al., 2019). From Indian Himalayan Region (IHR) itself, 357 species of medicinal plants belonging to 237 genera and 98 families were recorded. Asteraceae, Lamiaceae, Rosaceae, and Ranunculaceae were the dominant families in the IHR region. The IHR alone supports about 8000 species of angiosperms (40% endemics), 44 species of gymnosperms (15.91% endemics), 600 species of pteridophytes (25% endemics), 1737 species of bryophytes (32.53% endemics), 1159 species of lichens (11.22% endemics) and 6900 species of fungi (27.39% endemics) (Sharma et al., 2014).

The worldwide consumption of herbal medicines has markedly increased. According to the Secretariat of the Convention on Biological Diversity, global sales of herbal products were estimated to be the US $60 billion in 2000. The sale of herbal medicines is expected to get higher at an average annual growth rate of 6.4% (Inamdar et al., 2008). In 2008, the global market for herbal remedies was about the US $83 billion with a steady growth rate ranging between 3% and 12% per year (Zhang et al., 2012a,b). The market for herbal drugs has seen a good tendency of growth at a fast rate worldwide.

Biopesticides. https://doi.org/10.1016/B978-0-12-823355-9.00014-6

There are several factors responsible for growth like increased general awareness in people to protect from the side effects of synthetic medicine (Zahra et al., 2020), more inclination of masses toward Ayurveda and herbal treatment; require upgradation in quality and evaluation of efficacy and safety of herbal medicines in minimal cost (Calixto, 2000; Krishna et al., 2020).

In India, the medicinal plant market is mostly unorganized at present. Most of the herbal drug manufacturers procure the raw material from the wild by overexploitation of available natural resources (Laladhas et al., 2015). Due to unavailability of sufficient quantity of raw material, adulteration of inferior quality raw material or similar-looking plant species to the genuine drug is common practice in many of the herbal drug industries (Dubey, 2004; Kunle et al., 2012; Shaheen et al., 2019a,b). The value of medicinal plants related trade in India is the US $5.5 billion, although its share in the global export market of herbal drugs is less than 0.5%. The export potential of China in medicinal plants is nearly INR 18,000–22,000 crores. India exports crude drugs mainly to developed countries like the USA, Germany, France, Switzerland, the UK and Japan. The Indian herbal drugs exported to foreign countries mainly include Aconite, Aloe, Belladonna, Acorus, Cinchona, Cassia tora, Dioscorea, Digitalis, Ephedra, Plantago and Senna, etc. (Joshi, 2019). About 165 herbal drugs and their extract are exported from India (Prajapati et al., 2003; Ali, 2009). Overall, it can be said that despite having huge biodiversity and endemic medicinal plants, whereas our herbal drug market has not yet grown to its full potential. We are lagging behind in terms of herbal drug manufacture and export in comparison to countries like China due to a lack of proper attention and governmental policy for the Indian herbal drug market potential. However, in recent years the Ministry of AYUSH and related departments are taking care of these issues.

17.1.2 Niche of microflora

Microorganisms are considered as pillars of the existence of life on earth and represent the finest repertoire in molecular, protein as well as chemical versatility in nature (Chatterjee, 2019). After the origin of life on earth, they are evolved in the basics of life such as ecological processes, biogeochemical cycles and food chains even maintaining critical relationships between themselves as well as with other organisms existing on earth (Dick, 2019; Matthews et al., 2020). As a result of all contributions, microbes are efficiently reconstructing the geographical conditions, ecosystems and consequently providing better conditions for the development and proliferation of multicellular organisms (Hunter-Cevera, 1998).

17.2 Plant-microbe association

Traditional medicinal plants have a great impact on pharmaceutical industries by contributing bio-active compounds as herbal supplements and medicine development for human health care along with a nontoxic and cost-effective manner. The World Health Organization (WHO) defined the medicinal plants as "the plant which one or more of their organs contains substances that can be used for therapeutic purposes as well as used as precursors for chemosynthesis of pharmaceutical drugs." Many countries; Asia, China, Egypt and Africa's primary health care is dependent on native medicinal plants as written in their historic background. Bioactive compounds of medicinal plants known as their primary and secondary metabolites *viz*: phenolics, alkaloids, steroids, flavonoids, tannins, terpenes, essential oils, saponins, and anthraquinones, etc. used for the treatment of various diseases and body ailments (Egamberdieva and de silva, 2015). The plant microbiome is an important factor for increasing the synthesis of bioactive compounds and the production of secondary metabolites. They commonly reside along with the rhizospheric, phyllospheric, and endospheric region of the plants.

17.2.1 Rhizospheric association of microbes

The relationship between medicinal plants and microbes plays a pivotal role in the biosynthesis of metabolites. Soil, a reservoir of bacterial, fungal and actinomycetes and their activities are the major driving factor for soil and plant health (Compant et al., 2010; Aislabie et al., 2013; Müller et al., 2016). The rhizosphere is the surrounding area of the soil which is intimately associated with the root system of the plant have great availability and activity of heterogeneous microorganisms due to presence of root exudates and other organic nutrients (Hartmann et al., 2008; Poole, 2017; Hu et al., 2018). Root exudates are partially translocate to the carbon and excretory substances that are fixed by photosynthesis and other metabolic pathways in plants (Bais et al., 2006). The rhizospheric microbiome is highly productive than other part of plant microflora associated with the rhizospheric region shows an array of interactions that influence the growth and metabolome of medicinal plants (Huang et al., 2018). Several beneficial microbe's interactions with plants and their functional features mentioned in Table 17.1.

TABLE 17.1 Beneficial microbe's interaction with plants.

Sr. No.	Microbes	Beneficial activity with plant	References
Bacteria			
	Agrobacterium sp.	Indol-3-acetic acid producing bacteria enhances plant growth and development.	Mohite (2013)
	Rhizobium leguminosarum	Indol-3-acetic acid production; promoting growth after inoculation on axenically grown rice seedlings.	Ruzzi and Aroca (2015)
	Enterobacter sp.	Fixed significantly higher amounts of atmospheric nitrogen and produced higher amounts of Indol 3 acetic acid.	Kumar et al. (2017)
	Azospirillum brasilense	Mutual exchange of resources involved in producing and releasing the phytohormone; production of IAA by the bacterium, using tryptophan and thiamine.	Palacios et al. (2016)
	Bacillus subtilis	Plant growth promotion by spermidine-production.	Xie et al. (2014)
	Paenibacillus polymyxa	Produce plant growth regulating substances such as cytokinin.	Poehlein et al. (2018)
	Methylobacterium	Induces the synthesis of cytokinin in soybean plants	Holland et al. (2002)
	Pseudomonas protegens	Assessing the influence of fatty acid on antibiotic and siderophore production.	Quecine et al. (2016)
	Rhizobium leguminosarum	Nodulation, nitrogen fixation and plasmid transfer.	Boyer and Wisniewski-Dye (2009)
	Staphylococcus arlettae	Reduction of Arsenic and availability of phosphorus.	Srivastava et al. (2013)
	Pseudomonas koreensis	Prevent Heavy metal toxicity like Zn, Cd, As, Pb.	Babu et al. (2015)
	Pseudomonas sp.	Phosphate solubilizing activity	Otieno et al. (2015)
	Gluconacetobacter diazotrophicus	Colonization in rice plant and showing plant growth promotion.	Santoyo et al. (2016)
	Pantoea agglomerans	Up-regulation of aquaporin genes and induction of salt tolerance in tropical corn.	Gond et al. (2015)
	Pseudomonas vancouverensis	Tolerance to cold/chillimg stress and reduction of ROS.	Subramanian et al. (2015)
	Frankia sp.	Induce the formation nodules on the roots of their dicotyledonous host plants.	Van Nguyen and Pawlowski (2017)
	Nocardia sp.	Root nodule formation in host plant and promoting seedling growth.	Ghodhbane-Gtari et al. (2018)
	Kitasatospora sp.	Indole-3-acetic acid production for soil applications.	Shrivastava et al. (2008)
Fungus			
	Piriformospora indica	Colonization in root and induces the plant innate immunity evaluated by determining the phytoalexin and camalexin concentration.	Peskan Peskan-Berghöfer et al. (2015)
	Trichoderma viride	Produce auxins, small peptides, volatile compounds and other active metabolites that promote root branching along with plant growth and development.	López-Bucio et al. (2015)
	Talaromyces wortmannii	Emitted several terpenoids including β-caryophyllene which inducing resistance of *Brassica campestris* L. var. *perviridis* along with growth of plants.	Yamagiwa et al. (2011)
	Aspergillus spp., *Fusarium* spp., *Penicillium* spp., *Piriformospora* spp., *Phoma* spp., and *Trichoderma* spp.	Well-known fungal genera for plant growth promotion activity.	Hossain et al. (2017)

Continued

TABLE 17.1 Beneficial microbe's interaction with plants.—cont'd

Sr. No.	Microbes	Beneficial activity with plant	References
	Piriformospora indica	Symbiotic interaction with *Arabidopsis thaliana* and induces the performance of plant and tolerance against stress.	Vahabi et al. (2015)
	Neotyphodium lolii	Superoxide dismutase (SOD) activity changed in host plants.	Tian et al. (2008)
	Westerdykella aurantiaca	Promotes protein and carotenoid production.	Srivastava et al. (2012)
	Trichoderma longibrachiatum	Increases salt tolerance of Wheat by improving the antioxidative defense system and gene expression	Zhang et al. (2016)
	Aspergillus niger	Promotes accumulation of phenolic, salicylic acid, and chlorophyll contents.	Anwer and Khan (2013)
	Fusarium equiseti	Inhibits proliferation of pathogen and disease resistance.	Kojima et al. (2013)
	Trichoderma asperellum	Biocontrol activity against phytopathogens.	Islam et al. (2016)
	Penicillium chrysogenum	Induces systemic acquired resistance (SAR), which enhances defenses in plants.	Chen et al. (2018)
	Trichoderma virens	Antagonize biocontrol agent against pathogens of crop plants.	Lamdan et al. (2015)
	Aureobasidium pullulans	Contribution in biological treatment slight increase contents of tocols, alkylresorcinols and sterols in grains.	Wachowska et al. (2016)
Actinomycetes			
	Streptomyces rochei	Promotes soil enzyme productivity.	Jog et al. (2012)
	Streptomyces thermolilacinus		
	Streptomyces toxytricini	Promotes the accumulation of phenolics and chlorophyll.	Patil et al. (2011)
	Streptomyces coelicolor *Streptomyces olivaceus*	Promotes the production of ammonia, siderophore, IAA and prevent water stress tolerance.	Yandigeri et al. (2012)
	Streptomyces spp.	Production of Siderophore, ammonia, phosphate solulization activity, nitrogen fixation.	Kaur et al. (2013)
	Thermomonaspora fusca	Production of siderophore.	Dimise et al. (2008)

Several microbes present in the rhizospheric area shows plant growth-promoting (PGPR) activity (Kloepper, 1978), they provide soil nutrients to plants and control the biotic and abiotic stresses. Mainly *Bacillus, Pseudomonas*, *Azospirillum, Burkholderia, Bacillus, Enterobacter, Rhizobium, Erwinia, Serratia, Alcaligenes, Arthrobacter, Acinetobacter* and *Flavobacterium* has the potential to be a competent rhizospheric bacteria and express the PGPR activity (Berg et al., 2011; Kushwaha et al., 2020). The PGPRs used as mainly bio-fertilizers have shown symbiotic behavior by root-nodulation and nitrogen-fixing property, whereas phosphate solubilizing microbial inoculant provides insoluble or bound phosphate into a soluble form (Bhat et al., 2015). Some species of *Bacillus* produce volatile organic compounds for plant growth promotion (Bitas et al., 2013; Köberl et al., 2013). Similarly, Phyto-stimulators produce Auxins which involves in root elongation and development. Several strains of *Azospirillum* enables plant growth promotion by producing the auxins, cytokinins and gibberellins that are essential for plant health and growth (Çakmakçõ et al., 2020). Even though, rhizospheric microbial load distinct in medicinal plants due to the secretion of specific bio-active secondary metabolites (Qi et al., 2012). PGPRs indirectly boots the plant's immune system by secretion of proteins and carbohydrate compounds which initiate signaling and plant system recognized between pathogenic and non-pathogenic microbes (Macho and Zipfel, 2014; Pusztahelyi, 2018). Rhizoremediators; plant microbiome association reveal as a promising tool for the removal of soil pollutants and contaminants. The rate of degradation of pollutants accelerates in the rhizospheric region due to the production of organic acid and biofilm formation (Kumar et al., 2020; Saravanan et al., 2020).

17.2.2 Phyllospheric association of microbes

Above ground portion of plants including stem, leaves, flowers and fruits are prominent compartments where the abundance of the microbial community can be made a direct effect with the host plant (Mechan Llontop, 2020). Phyllospheric microbiome performs several constitutive roles subjected to plant growth and development, in terms of N_2 fixation, 60 kg N/ha only fixed by tropical plant phyllosphere and biosynthesis of various phytohormones for the protection of associated plant against pathogenic invaders. Furthermore, they also have a lot of potentialities which can be useful for the development of new strategies in agriculture practices. The phyllosphere microbial communities containing bacteria, fungi, viruses, and algae their density can be reached up to 105–107 per cm^2 (Alam, 2014). Phyllospheric microbial communities are also beneficial to the survival of plants in harsh conditions such as, limited concentrations of organic substances, variable pH, O_2 concentration, temperature, UV, humidity, etc. (Verma et al., 2017). Because of the close attachment with several environmental factors, the microbial load at the phyllosphere drastically fluctuating in the same species of plants as well as at the same developmental stage (Bulgarelli et al., 2013). These significant alterations in microbial dynamics are also the possible reason that imprinted the great versatility in the nutritional depositions at the phyllospheric region. The appearance of leaf and other areal parts of the plant largely influenced by the microbial load on the plant. Therefore, the narrow leaf containing grasses and wax containing broad-leaf plants having less microbial load as compared to cucumber and beans plants (Sivakumar et al., 2020). Different microbial communities are associated with plants at specific sites presumably because of differences in light or UV intensity, air flow rate, humidity, etc. For instance, pigment-producing bacterial strains are mostly inhabiting at the epiphytic region whereas, mineral and humic acid utilizing bacterial communities are found at the rhizosphere (Rana et al., 2020). This evidence was further authenticated by other findings where common root colonizers such as *Rhizobium* and *Bradyrhizobiaum* are unable to colonize the epiphytic regions of the same plant (Martínez-Hidalgo and Hirsch, 2017).

17.2.3 Endophytic microbiome association with medicinal plants

Plant associated endophytic microbiome strongly affects the quality and synthesis of bioactive secondary metabolites by medicinal plants. Endophytes protect plants against abiotic and biotic stresses by producing secondary metabolites (El-Deeb et al., 2013; Egamberdieva et al., 2017). Recently, Mishra et al. (2018) have observed the effects of endophytic bacteria *B. amyloliquefaciens* (BA) and *Pseudomonas fluorescens* (PF) individual as well as in combination on *W. somnifera* during *A. alternata* (AA) infection. Significant reductions in disease incidence and biotic stress amelioration have been recorded after the treatment of endophytic inoculants, their visual observation represented in Fig. 17.1. Several reports are highlighted the increased secondary metabolites production by endophytes and plant associations. Secondary metabolites rich source of pharmaceutical and modern therapeutic products (Pan et al., 2013), because microbes can produce a diverse range of metabolites includes terpenoids, alkaloids, antibiotics, alkaloids, polypeptides, isocoumarins, quinones, phenylpropanoids, lignans and aromatic compounds (Zhang et al., 2006; Gao et al., 2010). Various novel metabolites have been synthesized to the production of novel products for the anticancer, immune-modulatory agent, anti-parasitic, insecticidal, pesticidal, antiviral, antimicrobial agents at the industrial level, some microbes known for increasing the production of medicinal plant metabolites mentioned in Table 17.2. Apart from this, novel metabolites opens-up an opportunity for the development of new drugs for antimicrobial resistance and *anti*-HIV. Due to the increasing demand for potent metabolites and less availability of medicinal plants, endophytes are grown at a commercial level to enhance the production at large amounts of metabolites. In addition, fungal endophytes are also an essential component of medicinal microflora. Their symbiotic relationship with the mediational plant can considerably influence the secondary metabolite production by participating in a mechanistic way of the metabolic pathway (Gupta and Chaturvedi, 2019).

FIG. 17.1 Effects of endophytic bacteria *B. amyloliquefaciens* (BA) and *P. fluorescens* (PF) singly as well as in combination on *W. somnifera* during *A. alternata* (AA) infection. *Image adopted from Mishra et al. (2018).*

TABLE 17.2 Microbial association with medicinal plants.

Sr. No.	Plant	Microbes	Function	References
1.	*Andrographis paniculata*	*Glomus mosseae* and *Trichoderma harzianum*	Improve Phosphorous uptake and alkaloid production	Arpana and Bagyaraj (2007)
2.	*Neptunia oleracea*	*Rhizobium undicola*	IAA production	Ghosh et al. (2015)
3.	*Ocimum sanctum, Coleus forskohlii, Catharanthus roseus, Aloe vera*	*Azospirillum* *Azotobacter* *Pseudomonas*	N_2 fixation	Karthikeyan et al. (2008)
4.	*Ocimum basilicum,*	*Bacillus lentus* and *Pseudomonas*	ACC-deaminase activity	Golpayegani and Tilebeni (2011)
5.	*Mentha arvensis*	*Bacillus pumilus, Halomonas desiderata* and *Exiguobacterium oxidotolerans*	ACC-deaminase activity	Bharti et al. (2014)
6.	*Origanum vulgare*	*Pseudomonas, Stenotrophomonas*	Antioxidant activity increases	Solaiman and Anawar (2015)
7.	*Mentha piperita*	*Pseudomonas fluorescens*	Essential oil contents (+) pulegone and (−) menthone enhance	Santoro et al. (2011)
8.	*Mucuna pruriens*	*Rhizobium meliloti*	Siderophore production	Arora et al. (2001)
9.	*Piper nigrum*	*Pseudomonas and Azospirillum* sp.	phosphate-solubilizing ability	Ramachandran et al. (2007)
10.	*Ocimum sanctum*	*Achromobacter xylosoxidants*	ACC-deaminase activity and lower ethylene level	Barnawal et al. (2012)
11.	*Bacopa monnieri*	*Glomus mosseae*	Enhance plant growth and salinity tolerance	Khaliel et al. (2011)
12.	*Sorghum bicolor*	*Glomus mosseae* or *Glomus intraradices*	Enhanced production of alcohols, alkenes, ethers and acids	Sun and Tang (2013)
13.	*Artemisia annua*	*Glomus mosseae* and *Bacillus subtilis*	Enhance yield of artemisinin	Awasthi et al. (2011)
14.	Musli	*Piriformospora indica* and *Pseudomonas* *Fluorescens*	Enhance survival rate	Gosal et al. (2010)
15.	*Sphaeranthus amaranthoides*	*Glomus walkeri*	Increases the production of phenols, ortho-dihydroxy phenols, flavonoids, alkaloids, and tannins	Sumithra and Selvaraj (2011)
16.	*Zingiber cassumunar*	*Arthrinium* sp.	Antioxidant and antimicrobial activity against human pathogens	Pansanit and Pripdeevech al. (2018)
17.	Basil	*Bacillus subtilis*	α-terpineol and eugenol	Banchio et al. (2009)
18.	*Teucrium polium*	*Bacillus* sp. *and Penicillium* sp.	IAA production and antimicrobial activity	Hassan (2017)
19.	*Azadirachta indica*	*Phomopsis* sp., *Xylaria* sp.	Ten-membered lactones, Sesquiterpenes	Wu et al. (2008), Huang et al. (2015)
20.	*Rauwolfia tetraphylla*	*Curvularia* sp. and *Aspergillus* sp.	Synthesis of antimicrobial metabolites	Alurappa and Chowdappa (2018)
21.	*Taxus brevifolia*	*Taxomyces andreanae*	Biosynthesis of anticancer; taxol component	Stierle et al. (1995)
22.	*Musa acuminata*	*Phomopsis* sp.	Synthesis of anticancerous compound; Oblongolide	Kharwar et al. (2011), Mishra et al. (2012)

TABLE 17.2 Microbial association with medicinal plants.—cont'd

Sr. No.	Plant	Microbes	Function	References
23.	*Cynara cardunculus*	*Glomus intraradices, G. mosseae*	Increased total phenolic content in leaves and flower heads of *Cynara cardunculus*	Ceccarelli et al. (2010)
24.	*Medicago sativa* L.	*Sinorhizobium meliloti*	Enhance flavonoids in roots of legume plants	Catford et al. (2006)
25.	*Trifolium repens,*	*Glomus intraradices,*	Increases flavonoid content	Ponce et al. (2004)
26.	*Forsythia suspensa*	*Colletotrichum gloeosporioides*	Antioxidant activity, phillyrin	Zhang et al. (2012a,b)
27.	*Mentha arvensis*	*G. fasciculatum*	Increase oil content	Gupta et al. (2002)
28.	*Glycyrrhiza uralensis*	Glomus mosseae and Glomus veriforme	Triterpenoid saponin, Glycyrrhizic acid	Liu et al. (2007)
29.	*Ociimum basilicum*	*G. mosseae*	Enhanced oil yield, Rosmarinic acids, and caffeic acids	Toussaint et al. (2008)
30.	*Pinellia ternata*	*Bacillus cereus, Aranicola proteolyticus, Serratia liquefaciens, Bacillus thuringiensis,* and *Bacillus licheniformis*	Alkaloid production, Guanosine and inosine	Liu et al. (2015)
31.	Opium poppy (*Pappaver sominiferum)*	*Marmoricola* sp.	Enhance alkaloid production, the baine and codeine	Pandey et al. (2016)
32.	*Catharanthus roseus*	*Staphylococcus sciuri and Micro-coccus* sp.	Vindoline, ajmalicine and serpentine production	Tiwari et al. (2013)
33.	*Cynodon dactylon*	*Rhizoctonia* sp.	*Anti-Helicobacter pylori* activity, Rhizoctonic acid	Ma et al. (2004)
34.	*Angelica archangelica*	*G. mosseae, G. intraradices*	Enhance monoterpenoids and coumarins	Zitterl-Eglseer et al. (2015)
35.	*Salvia officinalis*	*G. intraradices*	Enhance essential oil content, 1,8-cineole, bornyl acetate, camphor, α-thujone, and β-thujone	Geneva et al. (2010)

Biocontrol activity; many of the microbial inoculants have been recognized for antagonistic activity against phytopathogens. Recently, *Bacillus amyloliquefaciens* and *Pseudomonas fluorescens* have investigated for the biocontrol activity against *Alternaria alternata* causing leaf spot disease in *Withania sominifera* (Mishra et al., 2018). Scanning electron micrographs of biocontrol activity of bacterial endophytes *B. amyloliquefaciens* (BA) and *P. fluorescens* (PF) against *A. alternata* (AA) represented in (Fig. 17.2). Raptured mycelia of AA have shown after the treatment of bio-inoculant BA and PF while untreated control remained healthy mycelia.

Moreover, plant's root endophyte *Arbuscular mycorrhiza* (AM), colonization with the medicinal plant has shown activities in plant growth promotion. 80% of terrestrial plant's roots weaved with AM fungi (Manoharachary and Kunwar, 2015). AM fungi colonize in the root of plants and provide nutrition as well as enhance plant immune system by promoting abiotic and biotic stress amelioration efficacy (Ceccarelli et al., 2010; Hart and Forsythe, 2012). Mycorrhiza *Glomus* colonize with plants and enhance the metabolites *viz*: alcohol, ether, acids (Sun and Tang, 2013).

Some microbes have shown prime importance in pathogen suppression by antibiotic production, which has tremendous industrial importance as *Streptomyces* gram-positive and spore-forming filamentous *Actinobacteria,* used for the production of the largest family of antibiotic for controlling pathogenic microbes (Kemung et al., 2018). *Pseudomonas, Bacillus* and *Trichoderma* spp. are well known for antibiosis responses (Sansinenea and Ortiz, 2013; Contreras-Cornejo et al., 2016; Pandey et al., 2018). These microbes control phytopathogens by producing cell wall degrading enzymes, toxins, bio-surfactants, minerals, etc. (Berg, 2009).

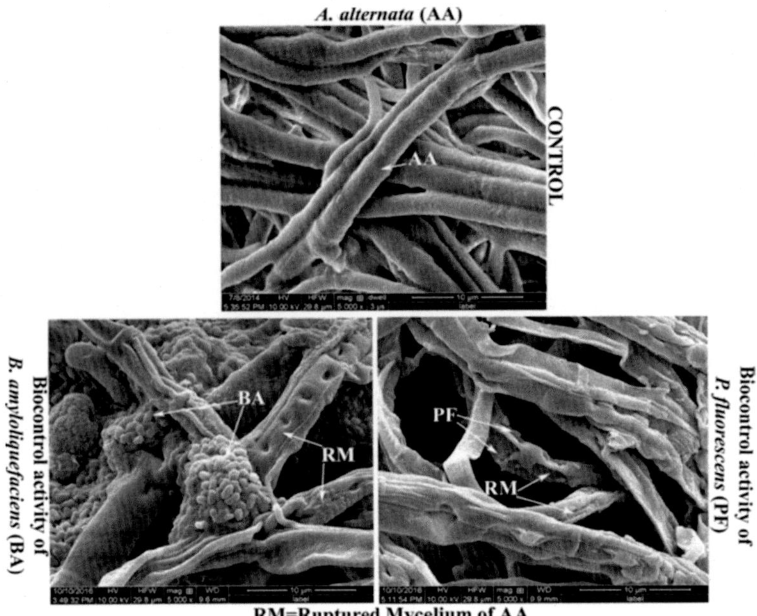

FIG. 17.2 Scanning electron micrographs of biocontrol activity of bacterial endophytes *Bacillus amyloliquefaciens* (BA) and *Pseudomonas fluorescens* (PF) against *Alternaria alternata* (AA). *Image adopted from Mishra et al. (2018).*

17.3 Relative factors between microflora and plants

Endophytes are the next important factor for microbial colonization at phyllospheric region of the plants. This could be also possible that a particular microbial community is found from the plant habitat but the spores are migrating through the flow of wind and colonize at the aerial part of the plant. Based on several studies has been found that air and erosols, water and soil are the most important sources of microbial cells that able to appointed the microbial dynamics at the phyllospheric region of the plants (Bulgarelli et al., 2013).

As similar, plant genotypic variation is also the significant driver of microbial diversity. Even though several plant species are found in the same habitat and environmental conditions but they have specific microbial communities due to diversity of genetic as well as metabolic variations. Geographical parameters also play a constitutive role in the designing of the microbial matrix that influences the quality of the end products manufactured by the host plant (Saad et al., 2020). However, it could be possible to analyze the distinct distribution of microbial matrix. These fluctuations are because of the variations in carbon substrates (i.e. amino acids, glucose, xylose) and nutrients present in the host plants. Despite all, the most common microbial colonizing communities are belongs to proteobacteria, actinobacteria, bacteroidetes and firmicutes (Bodenhausen et al., 2013). Therefore, the introduction of new techniques is should be needed to modify with other taxa of microbial communities associated with the diversified medicinal plants.

17.4 Conclusion and future perspectives

Biodiversity hotspots of India revealed as a rich repository of symbiotically beneficial microbes with endemic and rare medicinal plants. Diverse microflora of medicinal plants leads to exploring an evolutionary relationship with the host plant. Emphasis on novel applications of microbes for developing bio-based solutions that can avoid environmental damage and health effects for humans. Microbiome engineering required purposeful strategies for isolation and identification of indigenous communities for the dynamics of specific host and pathogen partners. A broad group of medicinal plants associated microflora summarized in this chapter that an unexplored biopesticide agent. There is increasing interest in the exploration of microbial inoculants for disease management as well as a mechanistic role in the biosynthesis of the bioactive compound of medicinal plants. Aim of this chapter, introduce novel insight into the microbiome of medicinal plants and their association with a specific host, a noticeable number of phytotherapeutic compounds produces due to the microbial interactions with medically important plant. Besides, it highlighted the possibilities for elevating plant protection along with plant growth and development and encouraging the commercial cultivation of medicinal plants to large scale production of bioactive phytochemicals.

References

Aislabie, J., Deslippe, J.R., Dymond, J., 2013. Soil microbes and their contribution to soil services. In: Ecosystem Services in New Zealand—conditions and Trends. Manaaki Whenua Press, Lincoln, New Zealand, pp. 143–161.

Alam, M., 2014. Microbial Status of Irrigation Water for Vegetables as Affected by Cultural Practices, vol. 2013. No. 97.

Ali, M., 2009. Present status of herbal medicines in India. J. Herb. Med. & Toxicol. 3 (2), 1–7.

Alurappa, R., Chowdappa, S., 2018. Antimicrobial activity and phytochemical analysis of endophytic fungal extracts isolated from ethno pharmaceutical plant *Rauwolfia tetraphylla* L. J. Pure Appl. Microbiol. 12 (1), 317–333.

Anwer, M.A., Khan, M.R., 2013. *Aspergillus niger* as tomato fruit (*Lycopersicum esculentum* Mill.) quality enhancer and plant health promoter. J. Postharvest Technol. 1 (1), 36–51.

Arora, N.K., Kang, S.C., Maheshwari, D.K., 2001. Isolation of siderophore-producing strains of *Rhizobium meliloti* and their biocontrol potential against *Macrophomina phaseolina* that causes charcoal rot of groundnut. Curr. Sci. 673–677.

Arpana, J., Bagyaraj, D.J., 2007. Response of kalmegh to an arbuscular mycorrhizal fungus and a plant growth promoting rhizomicroorganism at two levels of phosphorus fertilizer. Am-Euras. J. Agric. Environ. Sci. 2, 33–38.

Awasthi, A., Bharti, N., Nair, P., Singh, R., Shukla, A.K., Gupta, M.M., Darokar, M.P., Kalra, A., 2011. Synergistic effect of *Glomus mosseae* and nitrogen fixing *Bacillus subtilis* strain Daz26 on artemisinin content in *Artemisia annua* L. Appl. Soil Ecol. 49, 125–130.

Babu, A.G., Shea, P.J., Sudhakar, D., Jung, I.B., Oh, B.T., 2015. Potential use of *Pseudomonas koreensis* AGB-1 in association with *Miscanthus sinensis* to remediate heavy metal (loid)-contaminated mining site soil. J. Environ. Manag. 151, 160–166.

Bais, H.P., Weir, T.L., Perry, L.G., Gilroy, S., Vivanco, J.M., 2006. The role of root exudates in rhizosphere interactions with plants and other organisms. Annu. Rev. Plant Biol. 57, 233–266.

Banchio, E., Xie, X., Zhang, H., Pare, P.W., 2009. Soil bacteria elevate essential oil accumulation and emissions in sweet basil. J. Agric. Food Chem. 57 (2), 653–657.

Barnawal, D., Bharti, N., Maji, D., Chanotiya, C.S., Kalra, A., 2012. 1-Aminocyclopropane-1-carboxylic acid (ACC) deaminase-containing rhizobacteria protect *Ocimum sanctum* plants during waterlogging stress via reduced ethylene generation. Plant Physiol. Biochem. 58, 227–235.

Berg, G., 2009. Plant–microbe interactions promoting plant growth and health: perspectives for controlled use of microorganisms in agriculture. Appl. Microbiol. Biotechnol. 84 (1), 11–18.

Berg, G., Zachow, C., Cardinale, M., Müller, H., 2011. Ecology and human pathogenicity of plant-associated bacteria. In: Regulation of Biological Control Agents. Springer, Dordrecht, pp. 175–189.

Bharti, N., Barnawal, D., Awasthi, A., Yadav, A., Kalra, A., 2014. Plant growth promoting rhizobacteria alleviate salinity induced negative effects on growth, oil content and physiological status in *Mentha arvensis*. Acta Physiol. Plant. 36 (1), 45–60.

Bharucha, E., 2006. Textbook of Environmental Studies. Universities Press (India) Private Limited, Hyderabad (India).

Bhat, T.A., Ahmad, L., Ganai, M.A., Khan, O.A., 2015. Nitrogen fixing biofertilizers; mechanism and growth promotion: a review. J. Pure Appl. Microbiol. 9 (2), 1675–1690.

Bitas, V., Kim, H.S., Bennett, J.W., Kang, S., 2013. Sniffing on microbes: diverse roles of microbial volatile organic compounds in plant health. Mol. Plant-Micr. Interact. 26 (8), 835–843.

Bodenhausen, N., Horton, M.W., Bergelson, J., 2013. Bacterial communities associated with the leaves and the roots of *Arabidopsis thaliana*. PloS One 8 (2).

Boyer, M., Wisniewski-Dye, F., 2009. Cell–cell signalling in bacteria: not simply a matter of quorum. FEMS Microbiol. Ecol. 70 (1), 1–19.

Bulgarelli, D., Schlaeppi, K., Spaepen, S., Van Themaat, E.V.L., Schulze-Lefert, P., 2013. Structure and functions of the bacterial microbiota of plants. Annu. Rev. Plant Biol. 64, 807–838.

Çakmakçõ, R., Mosber, G., Milton, A.H., Alatürk, F., Ali, B., 2020. The effect of auxin and auxin-producing bacteria on the growth, essential oil yield, and composition in medicinal and aromatic plants. Curr. Microbiol. 1–14.

Calixto, J.B., 2000. Efficacy, safety, quality control, marketing and regulatory guidelines for herbal medicines (Phytotherapeutic Agents). Braz. J. Med. Biol. Res. 33, 179–189.

Catford, J.G., Staehelin, C., Larose, G., Piché, Y., Vierheilig, H., 2006. Systemically suppressed isoflavonoids and their stimulating effects on nodulation and mycorrhization in alfalfa split-root systems. Plant Soil 285 (1–2), 257–266.

Ceccarelli, N., Curadi, M., Martelloni, L., Sbrana, C., Picciarelli, P., Giovannetti, M., 2010. Mycorrhizal colonization impacts on phenolic content and antioxidant properties of artichoke leaves and flower heads two years after field transplant. Plant Soil 335 (1–2), 311–323.

Chatterjee, S., 2019. The Protein/RNA World and the Origin of Life.

Chen, Z., Wang, J., Li, Y., Zhong, Y., Liao, J., Lu, S., Wang, L., Wang, X., Chen, S., 2018. Dry mycelium of *Penicillium chrysogenum* activates defense via gene regulation of salicylic acid and jasmonic acid signaling in *Arabidopsis*. Physiol. Mol. Plant Pathol. 103, 54–61.

Compant, S., Clément, C., Sessitsch, A., 2010. Plant growth-promoting bacteria in the rhizo-and endosphere of plants: their role, colonization, mechanisms involved and prospects for utilization. Soil Biol. Biochem. 42 (5), 669–678.

Contreras-Cornejo, H.A., Macías-Rodríguez, L., del-Val, E., Larsen, J., 2016. Ecological functions of *Trichoderma* spp. and their secondary metabolites in the rhizosphere: interactions with plants. FEMS Microbiol. Ecol. 92 (4), fiw036.

Dhakal, P., Chettri, B., Lepcha, S., Acharya, B.K., 2020. Rich yet undocumented ethnozoological practices of socio-culturally diverse indigenous communities of Sikkim Himalaya, India. J. Ethnopharmacol. 249, 112386.

Dick, G.J., 2019. The microbiomes of deep-sea hydrothermal vents: distributed globally, shaped locally. Nat. Rev. Microbiol. 17 (5), 271–283.

Dimise, E.J., Widboom, P.F., Bruner, S.D., 2008. Structure elucidation and biosynthesis of fuscachelins, peptide siderophores from the moderate thermophile *Thermobifida fusca*. Proc. Natl. Acad. Sci. U. S. A. 105 (40), 15311–15316.

Dubey, N.K., 2004. Flora of BHU Campus. Banaras Hindu University. BHU Press, Varanasi, India.

Egamberdieva, D., da Silva, J.A.T., 2015. Medicinal plants and PGPR: a new frontier for phytochemicals. In: Plant-Growth-Promoting Rhizobacteria (PGPR) and Medicinal Plants. Springer, Cham, pp. 287–303.

Egamberdieva, D., Wirth, S., Behrendt, U., Ahmad, P., Berg, G., 2017. Antimicrobial activity of medicinal plants correlates with the proportion of antagonistic endophytes. Front. Microbiol. 8, 199.

El-Deeb, B., Fayez, K., Gherbawy, Y., 2013. Isolation and characterization of endophytic bacteria from *Plectranthus tenuiflorus* medicinal plant in Saudi Arabia desert and their antimicrobial activities. J. Plant Interact. 8 (1), 56–64.

Ganie, A.H., Tali, B.A., Nawchoo, I.A., Khuroo, A.A., Reshi, Z.A., Dar, G.H., 2020. Diversity in medicinal and aromatic flora of the Kashmir Himalaya. In: Biodiversity of the Himalaya: Jammu and Kashmir State. Springer, Singapore, pp. 545–563.

Gao, F.K., Dai, C.C., Liu, X.Z., 2010. Mechanisms of fungal endophytes in plant protection against pathogens. Afr. J. Microbiol. Res. 4 (13), 1346–1351.

Geneva, M.P., Stancheva, I.V., Boychinova, M.M., Mincheva, N.H., Yonova, P.A., 2010. Effects of foliar fertilization and arbuscular mycorrhizal colonization on *Salvia officinalis* L. growth, antioxidant capacity, and essential oil composition. J. Sci. Food Agric. 90 (4), 696–702.

Ghodhbane-Gtari, F., Nouioui, I., Hezbri, K., Lundstedt, E., D'Angelo, T., McNutt, Z., Laplaze, L., Gherbi, H., Vaissayre, V., Svistoonoff, S., ben Ahmed, H., 2018. The plant-growth-promoting actinobacteria of the genus Nocardia induces root nodule formation in *Casuarina glauca*. Antonie Van Leeuwenhoek 1–16.

Ghosh, P.K., Kumar De, T., Maiti, T.K., 2015. Production and metabolism of indole acetic acid in root nodules and symbiont (*Rhizobium undicola*) isolated from root nodule of aquatic medicinal legume *Neptunia oleracea* Lour. J. Bot. 2015.

Golpayegani, A., Tilebeni, H.G., 2011. Effect of biological fertilizers on biochemical and physiological parameters of basil (*Ociumum basilicm* L.) medicine plant. Am.-Eurasian J. Agric. Environ. Sci. 11, 411–416.

Gond, S.K., Torres, M.S., Bergen, M.S., Helsel, Z., White Jr., J.F., 2015. Induction of salt tolerance and up-regulation of aquaporin genes in tropical corn by rhizobacterium *P antoea agglomerans*. Lett. Appl. Microbiol. 60 (4), 392–399.

Gosal, S.K., Karlupia, A., Gosal, S.S., Chhibba, I.M., Varma, A., 2010. Biotization with *Piriformospora Indica* and *Pseudomonas Fluorescens* Improves Survival Rate, Nutrient Acquisition, Field Performance and Saponin Content of Micropropagated *Chlorophytum* Sp.

Gupta, M.L., Prasad, A., Ram, M., Kumar, S., 2002. Effect of the vesicular–arbuscular mycorrhizal (VAM) fungus *Glomus fasciculatum* on the essential oil yield related characters and nutrient acquisition in the crops of different cultivars of menthol mint (*Mentha arvensis*) under field conditions. Bioresour. Technol. 81 (1), 77–79.

Gupta, S., Chaturvedi, P., 2019. Enhancing Secondary Metabolite Production in Medicinal Plants Using Endophytic Elicitors: A Case Study of *Centella Asiatica* (Apiaceae) and Asiaticoside. Endophytes for a Growing World, pp. 310–323.

Hart, M.M., Forsythe, J.A., 2012. Using arbuscular mycorrhizal fungi to improve the nutrient quality of crops; nutritional benefits in addition to phosphorus. Sci. Hortic. 148, 206–214.

Hartmann, A., Rothballer, M., Schmid, M., 2008. Lorenz Hiltner, a pioneer in rhizosphere microbial ecology and soil bacteriology research. Plant Soil 312 (1–2), 7–14.

Hassan, S.E.D., 2017. Plant growth-promoting activities for bacterial and fungal endophytes isolated from medicinal plant of *Teucrium polium* L. J. Adv. Res. 8 (6), 687–695.

Holland, M.A., Long, R.L.G., Polacco, J.C., 2002. *Methylobacterium* spp.: phylloplane bacteria involved in cross-talk with the plant host? Phyllosphere Microbiol. 125–135.

Hossain, M.M., Sultana, F., Islam, S., 2017. Plant growth-promoting fungi (PGPF): phytostimulation and induced systemic resistance. In: Plant-Microbe Interactions in Agro-Ecological Perspectives. Springer, Singapore, pp. 135–191.

Hu, L., Robert, C.A., Cadot, S., Zhang, X., Ye, M., Li, B., Manzo, D., Chervet, N., Steinger, T., van der Heijden, M.G., Schlaeppi, K., 2018. Root exudate metabolites drive plant-soil feedbacks on growth and defense by shaping the rhizosphere microbiota. Nat. Commun. 9 (1), 2738.

Huang, R., Xie, X.S., Fang, X.W., Ma, K.X., Wu, S.H., 2015. Five new guaiane sesquiterpenes from the endophytic fungus *Xylaria* sp. YM 311647 of *Azadirachta indica*. Chem. Biodivers. 12 (8), 1281–1286.

Huang, W., Long, C., Lam, E., 2018. Roles of plant-associated microbiota in traditional herbal medicine. Trends Plant Sci. 23 (7), 559–562.

Hunter-Cevera, J.C., 1998. The value of microbial diversity. Current Opinion in Microbiology 1 (3), 278–285.

Inamdar, N., Edalat, S., Kotwal, V.B., Pawar, S., 2008. Herbal drugs in milieu of modern drugs. Int. J. Green Pharm. 2, 2–8.

Islam, M.M., Hossain, D.M., Rahman, M.M.E., Suzuki, K., Narisawa, T., Hossain, I., Meah, M.B., Nonaka, M., Harada, N., 2016. Native *Trichoderma* strains isolated from Bangladesh with broad spectrum antifungal action against fungal phytopathogens. Arch. Phytopathol. Plant Protect. 49 (1–4), 75–93.

Jog, R., Nareshkumar, G., Rajkumar, S., 2012. Plant growth promoting potential and soil enzyme production of the most abundant *Streptomyces spp.* from wheat rhizosphere. J. Appl. Microbiol. 113 (5), 1154–1164.

Joshi, M.C., 2019. Hand Book of Indian Medicinal Plants. Scientific Publishers.

Kala, C.P., Dhyani, P.P., Sajwan, B.S., 2006. Developing the medicinal plants sector in Northern India: challenges and opportunities. J. Ethnobiol. Ethnomed. 2 (1), 32.

Kamboj, V.P., 2000. Herbal medicine. Curr. Sci. 78 (1), 35–39.

Karthikeyan, B., Jaleel, C.A., Lakshmanan, G.A., Deiveekasundaram, M., 2008. Studies on rhizosphere microbial diversity of some commercially important medicinal plants. Colloids Surf. B Biointerfaces 62 (1), 143–145.

Kaur, T., Sharma, D., Kaur, A., Manhas, R.K., 2013. Antagonistic and plant growth promoting activities of endophytic and soil actinomycetes. Arch. Phytopathol. Plant Protect. 46 (14), 1756–1768.

Kemung, H.M., Tan, L.T.H., Khan, T.M., Chan, K.G., Pusparajah, P., Goh, B.H., Lee, L.H., 2018. *Streptomyces* as a prominent resource of future anti-MRSA drugs. Front. Microbiol. 9.

Khaliel, A.S., Shine, K., Vijayakumar, K., 2011. Salt tolerance and mycorrhization of *Bacopa monneiri* grown under sodium chloride saline conditions. Afr. J. Microbiol. Res. 5 (15), 2034–2040.

Kharwar, R.N., Mishra, A., Gond, S.K., Stierle, A., Stierle, D., 2011. Anticancer compounds derived from fungal endophytes: their importance and future challenges. Nat. Prod. Rep. 28 (7), 1208–1228.

Kloepper, J.W., 1978. Plant growth-promoting rhizobacteria on radishes. In: Proc. of the 4th Internet. Conf. on Plant Pathogenic Bacter, Station de Pathologie Vegetale et Phytobacteriologie, vol. 2. INRA, Angers, France, pp. 879–882.

Köberl, M., Schmidt, R., Ramadan, E.M., Bauer, R., Berg, G., 2013. The microbiome of medicinal plants: diversity and importance for plant growth, quality and health. Frontiers in Microbiology 4, 400.

Kojima, H., Hossain, M.M., Kubota, M., Hyakumachi, M., 2013. Involvement of the salicylic acid signaling pathway in the systemic resistance induced in *Arabidopsis* by plant growth-promoting fungus *Fusarium equiseti* GF19-1. J. Oleo Sci. 62 (6), 415–426.

Krishna, S., Dinesh, K.S., Nazeema, P.K., 2020. Globalizing ayurveda-opportunities and challenges. Int. J. Health Sci. Res. 10 (3), 55–68.

Krishnan, P.N., Decruse, S.W., Radha, R.K., 2011. Conservation of medicinal plants of Western Ghats, India and its sustainable utilization through in vitro technology. Vitro Cell Dev. Biol. Plant 47 (1), 110–122.

Kumar, A., Devi, S., Agrawal, H., Singh, S., Singh, J., 2020. Rhizoremediation: a unique plant microbiome association of biodegradation. In: Plant Microbe Symbiosis. Springer, Cham, pp. 203–220.

Kumar, A., Maurya, B.R., Raghuwanshi, R., Meena, V.S., Islam, M.T., 2017. Co-inoculation with *Enterobacter* and Rhizobacteria on yield and nutrient uptake by wheat (*Triticum aestivum* L.) in the alluvial soil under Indo-Gangetic plain of India. J. Plant Growth Regul. 36 (3), 608–617.

Kumar, K., Raj, A., Sivakumar, K., 2019. Etnobotanical studies on Solanum species from Nilgiri Biosphere Reserve of Western Ghats, Tamil Nadu, India. World Sci. News 115, 104–116.

Kunle, O.F., Egharevba, H.O., Ahmadu, P.O., 2012. Standardization of herbal medicines-A review. Int. J. Biodivers. Conserv. 4 (3), 101–112.

Kushwaha, R.K., Rodrigues, V., Kumar, V., Patel, H., Raina, M., Kumar, D., 2020. Soil microbes-medicinal plants interactions: ecological diversity and future prospect. In: Plant Microbe Symbiosis. Springer, Cham, pp. 263–286.

Laladhas, K.P., Preetha, N., Baijulal, B., Oommen, V., 2015. Conservation of medicinal plant resources through community born biodiversity management committee, Kerala, India. Biodivers. Conserv. Chall. Future 27.

Lamdan, N.L., Shalaby, S., Ziv, T., Kenerley, C.M., Horwitz, B.A., 2015. Secretome of the Biocontrol Fungus *Trichoderma Virens* Co-cultured with Maize Roots: Role in Induced Systemic Resistance. Molecular & Cellular Proteomics, pp. mcp–M114.

Liu, J., Wu, L., Wei, S., Xiao, X., Su, C., Jiang, P., Song, Z., Wang, T., Yu, Z., 2007. Effects of arbuscular mycorrhizal fungi on the growth, nutrient uptake and glycyrrhizin production of licorice (*Glycyrrhiza uralensis* Fisch). Plant Growth Regul. 52 (1), 29–39.

Liu, Y., Liu, W., Liang, Z., 2015. Endophytic bacteria from *Pinellia ternata*, a new source of purine alkaloids and bacterial manure. Pharmaceut. Biol. 53 (10), 1545–1548.

López-Bucio, J., Pelagio-Flores, R., Herrera-Estrella, A., 2015. *Trichoderma* as biostimulant: exploiting the multilevel properties of a plant beneficial fungus. Sci. Hortic. 196, 109–123.

Ma, Y.M., Li, Y., Liu, J.Y., Song, Y.C., Tan, R.X., 2004. Anti-*Helicobacter pylori* metabolites from *Rhizoctonia* sp. Cy064, an endophytic fungus in *Cynodon dactylon*. Fitoterapia 75 (5), 451–456.

Macho, A.P., Zipfel, C., 2014. Plant PRRs and the activation of innate immune signaling. Mol. Cell 54 (2), 263–272 (j).

Manoharachary, C., Kunwar, I.K., 2015. Arbuscular mycorrhizal fungi: the nature's gift for sustenance of plant wealth. In: Plant Biology and Biotechnology. Springer, New Delhi, pp. 217–230.

Martínez-Hidalgo, P., Hirsch, A.M., 2017. The nodule microbiome: N_2-fixing rhizobia do not live alone. Phytobiomes 1 (2), 70–82.

Matthews, N., Zhang, W., Bell, A.R., Treemore-Spears, L., 2020. Ecosystems and ecosystem services. In: The Food-Energy-Water Nexus. Springer, Cham, pp. 237–258.

Mechan Llontop, M.E., 2020. Identification, Characterization, and Use of Precipitation-Borne and Plant-Associated Bacteria. Doctoral Dissertation. Virginia Tech.

Mishra, A., Gond, S.K., Kumar, A., Sharma, V.K., Verma, S.K., Kharwar, R.N., 2012. Sourcing the fungal endophytes: a beneficial transaction of biodiversity, bioactive natural products, plant protection and nanotechnology. In: Microorganisms in Sustainable Agriculture and Biotechnology. Springer, Dordrecht, pp. 581–612.

Mishra, A., Singh, S.P., Mahfooz, S., Singh, S.P., Bhattacharya, A., Mishra, N., Nautiyal, C.S., 2018. Endophyte-mediated modulation of defense-related genes and systemic resistance in *Withania somnifera* (L.) Dunal under *Alternaria alternata* stress. Appl. Environ. Microbiol. 84 (8) e02845-17.

Mohite, B., 2013. Isolation and characterization of indole acetic acid (IAA) producing bacteria from rhizospheric soil and its effect on plant growth. J. Soil Sci. Plant Nutr. 13 (3), 638–649.

Mukherjee, P.K., 2008. Quality Control of Herbal Drugs, first ed. Business Horizones Pharmaceutical Publications, New Delhi.

Müller, D.B., Vogel, C., Bai, Y., Vorholt, J.A., 2016. The plant microbiota: systems-level insights and perspectives. Ann. Rev. Genet. 50, 211–234.

Ncube, B., Finnie, J.F., Van Staden, J., 2012. Quality from the field: the impact of environmental factors as quality determinants in medicinal plants. South Afr. J. Bot. 82, 11–20.

Otieno, N., Lally, R.D., Kiwanuka, S., Lloyd, A., Ryan, D., Germaine, K.J., Dowling, D.N., 2015. Plant growth promotion induced by phosphate solubilizing endophytic *Pseudomonas* isolates. Front. Microbiol. 6, 745.

Palacios, O.A., Gomez-Anduro, G., Bashan, Y., de-Bashan, L.E., 2016. Tryptophan, thiamine and indole-3-acetic acid exchange between *Chlorella sorokiniana* and the plant growth-promoting bacterium *Azospirillum brasilense*. FEMS Microbiol. Ecol. 92 (6), fiw077.

Pan, S.Y., Zhou, S.F., Gao, S.H., Yu, Z.L., Zhang, S.F., Tang, M.K., Sun, J.N., Ma, D.L., Han, Y.F., Fong, W.F., Ko, K.M., 2013. New perspectives on how to discover drugs from herbal medicines: CAM's outstanding contribution to modern therapeutics. Evid. Based Compl. Altern. Med. 2013, 627375.

Pandey, C., Dheeman, S., Negi, Y.K., Maheshwari, D.K., 2018. Differential response of native *Bacillus* spp. isolates from agricultural and forest soils in growth promotion of *Amaranthus hypochondriacus*. Biotechnol. Res. 4 (1), 54—61.

Pandey, S.S., Singh, S., Babu, C.V., Shanker, K., Srivastava, N.K., Kalra, A., 2016. Endophytes of opium poppy differentially modulate host plant productivity and genes for the biosynthetic pathway of benzylisoquinoline alkaloids. Planta 243 (5), 1097—1114.

Pansanit, A., Pripdeevech, P., 2018. Antibacterial secondary metabolites from an endophytic fungus, Arthrinium Sp. MFLUCC16-1053 isolated from *Zingiber* cassumunar. Mycology 1—9.

Patil, H.J., Srivastava, A.K., Singh, D.P., Chaudhari, B.L., Arora, D.K., 2011. Actinomycetes mediated biochemical responses in tomato (*Solanum lycopersicum*) enhances bioprotection against *Rhizoctonia solani*. Crop Protect. 30 (10), 1269—1273.

Peskan-Berghöfer, T., Vilches-Barro, A., Müller, T.M., Glawischnig, E., Reichelt, M., Gershenzon, J., Rausch, T., 2015. Sustained exposure to abscisic acid enhances the colonization potential of the mutualist fungus *Piriformospora indica* on *Arabidopsis thaliana* roots. New Phytol. 208 (3), 873—886.

Poehlein, A., Hollensteiner, J., Granzow, S., Wemheuer, B., Vidal, S., Wemheuer, F., 2018. First insights into the draft genome sequence of the endophyte *Paenibacillus amylolyticus* strain GM1FR, isolated from *Festuca rubra* L. Genome Announc. 6 (4), e01516—e01517.

Ponce, M.A., Scervino, J.M., Erra-Balsells, R., Ocampo, J.A., Godeas, A.M., 2004. Flavonoids from shoots and roots of *Trifolium repens* (white clover) grown in presence or absence of the arbuscular mycorrhizal fungus Glomus intraradices. Phytochemistry 65 (13), 1925—1930.

Poole, P., 2017. Shining a light on the dark world of plant root—microbe interactions. Proc. Natl. Acad. Sci. U. S. A. 114 (17), 4281—4283.

Prajapati, N.D., Purohit, S.S., Sharma, A.K., Kumar, T.A., 2003. A Handbook of Medicinal Plants. Agrobios (India), Jodhpur.

Pusztahelyi, T., 2018. Chitin and chitin-related compounds in plant—fungal interactions. Mycology 9 (3), 189—201.

Qi, X., Wang, E., Xing, M., Zhao, W., Chen, X., 2012. Rhizosphere and non-rhizosphere bacterial community composition of the wild medicinal plant *Rumex patientia*. World J. Microbiol. Biotechnol. 28 (5), 2257—2265.

Quecine, M.C., Kidarsa, T.A., Goebel, N.C., Shaffer, B.T., Henkels, M.D., Zabriskie, T.M., Loper, J.E., 2016. An interspecies signaling system mediated by fusaric acid has parallel effects on antifungal metabolite production by *Pseudomonas protegens* strain Pf-5 and antibiosis of *Fusarium* spp. Appl. Environ. Microbiol. 82 (5), 1372—1382.

Ramachandran, K., Srinivasan, V., Hamza, S., Anandaraj, M., 2007. Phosphate solubilizing bacteria isolated from the rhizosphere soil and its growth promotion on black pepper (*Piper nigrum* L.) cuttings. In: First International Meeting on Microbial Phosphate Solubilization. Springer, Dordrecht, pp. 325—331.

Rana, K.L., Kour, D., Yadav, A.N., Yadav, N., Saxena, A.K., 2020. Agriculturally important microbial biofilms: biodiversity, ecological significances, and biotechnological applications. In: New and Future Developments in Microbial Biotechnology and Bioengineering: Microbial Biofilms. Elsevier, pp. 221—265.

Ruzzi, M., Aroca, R., 2015. Plant growth-promoting rhizobacteria act as biostimulants in horticulture. Sci. Hortic. 196, 124—134.

Saad, M.M., Eida, A.A., Hirt, H., 2020. Tailoring plant-associated microbial inoculants in agriculture: a roadmap for successful application. J. Exp. Bot. 71 (13), 3878—3901.

Sansinenea, E., Ortiz, A., 2013. An antibiotic from *Bacillus thuringiensis* against Gram-negative bacteria. Biochem. Pharmacol. 2, e142.

Santoro, M.V., Zygadlo, J., Giordano, W., Banchio, E., 2011. Volatile organic compounds from rhizobacteria increase biosynthesis of essential oils and growth parameters in peppermint (*Mentha piperita*). Plant Physiol. Biochem. 49 (10), 1177—1182.

Santoyo, G., Moreno-Hagelsieb, G., del Carmen Orozco-Mosqueda, M., Glick, B.R., 2016. Plant growth-promoting bacterial endophytes. Microbiol. Res. 183, 92—99.

Saravanan, A., Jeevanantham, S., Narayanan, V.A., Kumar, P.S., Yaashikaa, P.R., Muthu, C.M., 2020. Rhizoremediation—A promising tool for the removal of soil contaminants: a review. J. Environ. Chem. Eng. 8 (2), 103543.

Shaheen, S., Ramzan, S., Khan, F., Ahmad, M., 2019a. Marketed herbal drugs: how adulteration affects. In: Adulteration in Herbal Drugs: A Burning Issue. Springer, Cham, pp. 51—55.

Shaheen, S., Ramzan, S., Khan, F., Ahmad, M., 2019b. Adulteration in Herbal Drugs: A Burning Issue. Springer International Publishing.

Sharma, A., Shanker, C., Tyagi, L.K., Singh, M., Rao, C.V., 2008. Herbal medicine for market potential in India: an overview. Acad. J. Plant Sci. 1, 26—36.

Sharma, J., Gairola, S., Sharma, Y.P., Gaur, R.D., 2014. Ethnomedicinal plants used to treat skin diseases by Tharu community of district Udham Singh Nagar, Uttarakhand, India. J. Ethnopharmacol. 158, 140—206.

Shrivastava, S., D'Souza, S.F., Desai, P.D., 2008. Production of indole-3-acetic acid by immobilized actinomycete (*Kitasatospora* sp.) for soil applications. Curr. Sci. 1595—1604.

Singh, J.S., Chaturvedi, R.K., 2017. Diversity of ecosystem types in India: a review. Proc. Ind. Natl. Sci. Acad.—INSA 83 (3), 569—594.

Sivakumar, N., Sathishkumar, R., Selvakumar, G., Shyamkumar, R., Arjunekumar, K., 2020. Phyllospheric microbiomes: diversity, ecological significance, and biotechnological applications. In: Plant Microbiomes for Sustainable Agriculture. Springer, Cham, pp. 113—172.

Solaiman, Z.M., Anawar, H.M., 2015. Rhizosphere microbes interactions in medicinal plants. In: Plant-Growth-Promoting Rhizobacteria (PGPR) and Medicinal Plants. Springer, Cham, pp. 19–41.

Srivastava, P.K., Shenoy, B.D., Gupta, M., Vaish, A., Mannan, S., Singh, N., Tewari, S.K., Tripathi, R.D., 2012. Stimulatory effects of arsenic-tolerant soil fungi on plant growth promotion and soil properties. Microbes and Environments ME11316.

Srivastava, S., Verma, P.C., Chaudhry, V., Singh, N., Abhilash, P.C., Kumar, K.V., Sharma, N., Singh, N., 2013. Influence of inoculation of arsenic-resistant *Staphylococcus arlettae* on growth and arsenic uptake in *Brassica juncea* (L.) Czern. Var. R-46. J. Hazard Mater. 262, 1039–1047.

Stierle, A., Strobel, G., Stierle, D., Grothaus, P., Bignami, G., 1995. The search for a taxol-producing microorganism among the endophytic fungi of the Pacific yew, *Taxus brevifolia*. J. Nat. Prod. 58 (9), 1315–1324.

Subramanian, P., Mageswari, A., Kim, K., Lee, Y., Sa, T., 2015. Psychrotolerant endophytic *Pseudomonas* sp. strains OB155 and OS261 induced chilling resistance in tomato plants (*Solanum lycopersicum* Mill.) by activation of their antioxidant capacity. Mol. Plant Microbe Interact. 28 (10), 1073–1081.

Sumithra, P., Selvaraj, T., 2011. Influence of Glomus walkeri Blaszk and Renker and plant growth promoting rhizomicroorganisms on growth, nutrition and content of secondary metabolites in *Sphaeranthes amaranthoides* (L.) Burm. Int. J. Agric. Technol. 7 (6), 1685–1692.

Sun, X.G., Tang, M., 2013. Effect of arbuscular mycorrhizal fungi inoculation on root traits and root volatile organic compound emissions of *Sorghum bicolor*. South Afr. J. Bot. 88, 373–379.

Tian, P., Nan, Z., Li, C., Spangenberg, G., 2008. Effect of the endophyte *Neotyphodium lolii* on susceptibility and host physiological response of perennial ryegrass to fungal pathogens. Eur. J. Plant Pathol. 122 (4), 593–602.

Tiwari, R., Awasthi, A., Mall, M., Shukla, A.K., Srinivas, K.S., Syamasundar, K.V., Kalra, A., 2013. Bacterial endophyte-mediated enhancement of in planta content of key terpenoid indole alkaloids and growth parameters of *Catharanthus roseus*. Ind. Crop. Prod. 43, 306–310.

Toussaint, J.P., Kraml, M., Nell, M., Smith, S.E., Smith, F.A., Steinkellner, S., Schmiderer, C., Vierheilig, H., Novak, J., 2008. Effect of *Glomus mosseae* on concentrations of rosmarinic and caffeic acids and essential oil compounds in basil inoculated with *Fusarium oxysporum* f. sp. *basilici*. Plant Pathol. 57 (6), 1109–1116.

Vahabi, K., Sherameti, I., Bakshi, M., Mrozinska, A., Ludwig, A., Reichelt, M., Oelmüller, R., 2015. The interaction of Arabidopsis with *Piriformospora indi*ca shifts from initial transient stress induced by fungus-released chemical mediators to a mutualistic interaction after physical contact of the two symbionts. BMC Plant Biol. 15 (1), 58.

Van Nguyen, T., Pawlowski, K., 2017. Frankia and actinorhizal plants: symbiotic nitrogen fixation. In: Rhizotrophs: Plant Growth Promotion to Bioremediation. Springer, Singapore, pp. 237–261.

Verma, P., Yadav, A.N., Kumar, V., Singh, D.P., Saxena, A.K., 2017. Beneficial plant-microbes interactions: biodiversity of microbes from diverse extreme environments and its impact for crop improvement. In: Plant-Microbe Interactions in Agro-Ecological Perspectives. Springer, Singapore, pp. 543–580.

Wachowska, U., Tańska, M., Konopka, I., 2016. Variations in grain lipophilic phytochemicals, proteins and resistance to *Fusarium spp.* growth during grain storage as affected by biological plant protection with *Aureobasidium pullulans* (de Bary). Int. J. Food Microbiol. 227, 34–40.

Wu, S.H., Chen, Y.W., Shao, S.C., Wang, L.D., Li, Z.Y., Yang, L.Y., Li, S.L., Huang, R., 2008. Ten-membered lactones from *Phomopsis* sp., an endophytic fungus of *Azadirachta indica*. J. Nat. Prod. 71 (4), 731–734.

Xie, S.S., Wu, H.J., Zang, H.Y., Wu, L.M., Zhu, Q.Q., Gao, X.W., 2014. Plant growth promotion by spermidine-producing *Bacillus subtilis* OKB105. Mol. Plant Microbe Interact. 27 (7), 655–663.

Yamagiwa, Y., Inagaki, Y., Ichinose, Y., Toyoda, K., Hyakumachi, M., Shiraishi, T., 2011. *Talaromyces wortmannii* FS2 emits β-caryphyllene, which promotes plant growth and induces resistance. J. Gen. Plant Pathol. 77 (6), 336–341.

Yandigeri, M.S., Meena, K.K., Singh, D., Malviya, N., Singh, D.P., Solanki, M.K., Yadav, A.K., Arora, D.K., 2012. Drought-tolerant endophytic actinobacteria promote growth of wheat (*Triticum aestivum*) under water stress conditions. Plant Growth Regul. 68 (3), 411–420.

Zahra, W., Rai, S.N., Birla, H., Singh, S.S., Rathore, A.S., Dilnashin, H., Keswani, C., Singh, S.P., 2020. Economic importance of medicinal plants in Asian countries. In: Bioeconomy for Sustainable Development. Springer, Singapore, pp. 359–377.

Zhang, J., Wider, B., Shang, H., Li, X., Ernst, E., 2012a. Quality of herbal medicines: challenges and solutions. Compl. Ther. Med. 20 (1–2), 100–106.

Zhang, H.W., Song, Y.C., Tan, R.X., 2006. Biology and chemistry of endophytes. Nat. Prod. Rep. 23 (5), 753–771.

Zhang, Q., Wei, X., Wang, J., 2012b. Phillyrin produced by *Colletotrichum gloeosporioides*, an endophytic fungus isolated from *Forsythia suspensa*. Fitoterapia 83 (8), 1500–1505.

Zhang, S., Gan, Y., Xu, B., 2016. Application of plant-growth-promoting fungi *Trichoderma longibrachiatum* T6 enhances tolerance of wheat to salt stress through improvement of antioxidative defense system and gene expression. Front. Plant Sci. 7, 1405.

Zitterl-Eglseer, K., Nell, M., Lamien-Meda, A., Steinkellner, S., Wawrosch, C., Kopp, B., Zitterl, W., Vierheilig, H., Novak, J., 2015. Effects of root colonization by symbiotic arbuscular mycorrhizal fungi on the yield of pharmacologically active compounds in *Angelica archangelica* L. Acta Physiol. Plant. 37 (2), 21.

Chapter 18

Trichoderma: a potential biopesticide for sustainable management of wilt disease of crops

Narasimhamurthy Konappa[b,1], Nirmaladevi Dhamodaran[c,1], Soumya Satyanand Shanbhag[c], Manjunatha Amitiganahalli Sampangi[c], Soumya Krishnamurthy[d], Udayashankar C. Arakere[a], Srinivas Chowdappa[b] and Sudisha Jogaiah[e]

[a]*Department of Studies in Biotechnology, University of Mysore, Mysore, Karnataka, India;* [b]*Department of Microbiology and Biotechnology, Bangalore University, Bengaluru, Karnataka, India;* [c]*Department of Microbiology, Ramaiah College of Arts, Science and Commerce, Bengaluru, Karnataka, India;* [d]*Department of Microbiology, Field Marshal K. M. Cariappa College, A Constituent College of Mangalore University, Madikeri, Karnataka, India;* [e]*Laboratory of Plant Healthcare and Diagnostics, PG Department of Biotechnology and Microbiology, Karnatak University, Dharwad, Karnataka, India*

18.1 Introduction

The population of world is expected to range 8.5 billion by 2030, 9.7 billion by 2050 and exceed 11 billion in 2100 which would require rising of overall food production by some 70% (UOECD environmental outlook to 2050). India is considered to be largest agricultural economy and has millions of people dependent on farming and its allied activities for a living. Two-third of present world people depends upon farming for their basic requirement, but now a days, yield and production of cultivated crops are receiving destructed (Elumalai and Rengasamy, 2012). Agriculture has been facing threats due to activities of pests including bacteria, fungi, virus, weeds, nematodes, and insects since long, which cause yield loss about 31%−42% (Agrios, 2005). Yield damages produced by biotic constraints are a main task to farming production that necessity to be controlled.

Several plant diseases are accountable for the loss of about one-third of the crop yields at worldwide level (Lugtenberg et al., 2002). Approximately 25% of the yield is vanished each year because of plant diseases caused by different pathogens all over the world (According to Lugtenberg, 2015). Yield losses due to different diseases can consequence in starvation and hunger particularly in developing nations and soil borne wilt causing pathogens are among the main causal agents decreasing crop productivity. Wilt infections are the utmost extensive and damaging soil borne plant diseases, which attack an enormous number of plant types all over the world. Vascular wilt diseases account for 2%−90% yield loss recorded in wide range of crops (Pataky et al., 2000). Vascular wilts diseases are devastating plant diseases that can disturb both annual yields in addition to woody perennials, therefore influence main food damages and destructive valuable natural environments (Yadeta and Thomm, 2013).

Fusarium oxysporum species complex, a ubiquitous soil borne filamentous fungus, includes devastating pathogenic forms that cause wilt of flowers, ornamentals, vegetables, and other significant crops. Numerous host plants are attacked by *Fusarium oxysporum* and above 150 host specific forms or forme speciales are reported (Baayen, 2000). The wilt caused by the bacterium *Ralstonia solanacearum* is significant since it extensively affects various plant crops. *Ralstonia solanacearum* is familiar to cause wilt diseases in humid and subtropical weathers with excessive rainfall and the wilt is documented in 200 plant species of economically significant florae (Lyons et al., 2001). The wilt pathogen enters via root, host tissues over injuries, usual openings or can be distributed into the xylem by pest vectors. Nematodes appear to ease the

1. Authors contributed equally.

Biopesticides. https://doi.org/10.1016/B978-0-12-823355-9.00003-1

infection by wilt pathogen in as a minimum certain of the wilts. Wilts occur as a consequence of the existence and actions of pathogens in the plant xylem vascular tissues. Afterward the wilt affecting pathogen infects the plant, they twitch to increase and transfer through the vessels of xylem. Wilt pathogens affect the water movement and nutrients which will cause wilting, drooping, and the death of the host plants (Agrios, 2005). Wilt pathogens typically continue to extent within through the xylem vessels till the death of the whole host plant (Agrios, 2005). Wilting might be the consequence of damage to the root structure upon the limited plugging of water conductivity of the vessels and toxic materials produced by the wilt pathogens (Deacon, 2004). The fungal chlamydospores and bacterial cells overwinter in host remains in soil, vegetative material, seeds, or in insect vectors as latent cells (Agrios, 2005).

Management of wilt causing pathogens can be difficult because there are not at all effective methods to control the diseased hosts, so usually pathogens should be detached from the diseased areas. Moreover, numerous of the wilt initiating pathogens are soil inhabitants that create persistent dormant structures that can persist for extensive times, which create management of these kind of diseases more tough (Yadeta and Thomm, 2013). Chemical treatments, soil solarization are unsuccessful in pathogen control. Further, high genetic variability of *F. oxysporum* poses a major challenge to development of effective disease management strategies (Nirmaladevi et al., 2016). Farmers everywhere in the world use organic pesticides to control plant infections in order to sustain the superiority and severance of agricultural yields (Junaid et al., 2013). The inappropriate and excessive use of chemical pesticides over the past decades is no longer sustainable due to health hazards to human, environmental pollution, toxicity to other living organisms, negative effects impacting soil fertility and soil biodiversity (Lade et al., 2017) and also pesticides can lead to improvement of some resistant pathogens, as well as adverse impacts on natural enemies makes these pesticides ecologically unacceptable (Naher et al., 2014; Griffin, 2014; Lamichhane et al., 2017; Carvalho, 2017). Chemical pesticides cause main side effects for example endocrine disturbance, cytotoxicity, reproductive toxicity and breast cancer, etc. (Nicolopoulou-Stamati et al., 2016).

It is very important to development of alternatives strategies for wilt disease control without disdaining the agricultural yield and cost-effectiveness. Finding environmentally friendly means of controlling vascular wilt pathogens is a necessity (Alsohiby et al., 2016; Carmona-Hernandez et al., 2019). The world recently attention resort to finding sustainable, nontoxic and environmental friendly replacements. Biological control agents (BCAs) mention to the use of certain living microbes to defeat the development of wilt pathogens. Farmers are shifting toward alternative environmentally safe and ecologically approach for the management of plant diseases, i.e., BCAs based formulations, referred to as biopesticides (Iqbal and Ashraf, 2017; Hashmi et al., 2018). Biopesticides are derived from microorganisms and are an ecofriendly alternative to the synthetic pesticides (Heimpel and Mills, 2017; Islam, 2018; Siyar et al., 2019; van Lenteren et al., 2018). Biopesticides are gaining interest because of its less risk and more environmental safety, target specificity, efficacy and biodegradability, no consequence on the produced food, fruits, vegetables, low production charge and effective solution for the eradication of phytopathogenic diseases (Kumar and Singh, 2015; Murthy and Srinivas, 2012).

Currently, various microorganisms have been recognized and are available as BCAs are *Trichoderma* spp., *Pseudomonas* spp., *Bacillus* spp., *Ampelomyces* spp., *Actinomycetes* spp., *Candida* spp., *Agrobacterium radiobacter*, nonpathogenic *Fusarium* spp., *Coniothyrium* spp. and atoxigenic *Aspergillus niger* (Singh et al., 2014; Naher et al., 2014; Keswani et al., 2015). Among the fungal BCAs, *Trichoderma* spp. have extended much interest due to their great reproductive ability, persistence under harsh environments, having biocontrol capacities against destructive pathogens, and productive producers of secondary metabolites (Srivastava et al., 2014; Contreras-Cornejo et al., 2016; Devi et al., 2017; Oladipo et al., 2018; Saravanakumar et al., 2018).

Trichoderma spp. are free existing filamentous molds which are widely distributed nature and can be isolated from soil, plant roots, decaying wood, bark and many other substrates (Manganiello et al., 2018; Macias-Rodríguez et al., 2018; Etschmann et al., 2015). More than 100 *Trichoderma* spp. has been recorded worldwide (Pandya et al., 2011), many of which are potential biopesticides (Benitez et al., 2004). Many researchers have demonstrated the potential of *Trichoderma* spp. in control of different plant diseases. The most effective *Trichoderma* spp. include *Trichoderma longibrachiatum*, *T. pseudokoningii*, *T. virens*, *T. viride*, *T. virens*, *T. koningii*, *T. polysporum*, *T. hamatum* and *T. harzianum*. (Asad et al., 2015; Lamichhane et al., 2017; Majeed et al., 2017; Abbas et al., 2017; Manganiello et al., 2018). Many commercial biopesticides with *Trichoderma* spp. as are available in developed countries (Junaid et al., 2013; Atanasova et al., 2013; Shaw et al., 2016; Kumar et al., 2017). *Trichoderma* spp. grow quickly in soil application, meanwhile *Trichoderma* spp. are obviously resistant to various lethal compounds, herbicides, fungicides, and pesticides (Seethapathy et al., 2017), bioremediation agents for heavy metal and xenobiotic contamination (Zhang et al., 2018).

Biocontrol mechanisms of *Trichoderma* spp. involve mycoparasitism (attack and killing of pathogen), antibiosis, competition (Spadaro and Droby, 2016), tolerance to abiotic stress, lytic enzymes production, secondary metabolites production, enhanced host resistance (Keswani et al., 2015; Pandey et al., 2016; de Medeiros et al., 2017; Oros and Naár, 2017; Mendoza-Mendoza et al., 2018) (Fig. 18.1) and increase of nutrients uptake such as phosphate solubilization, iron

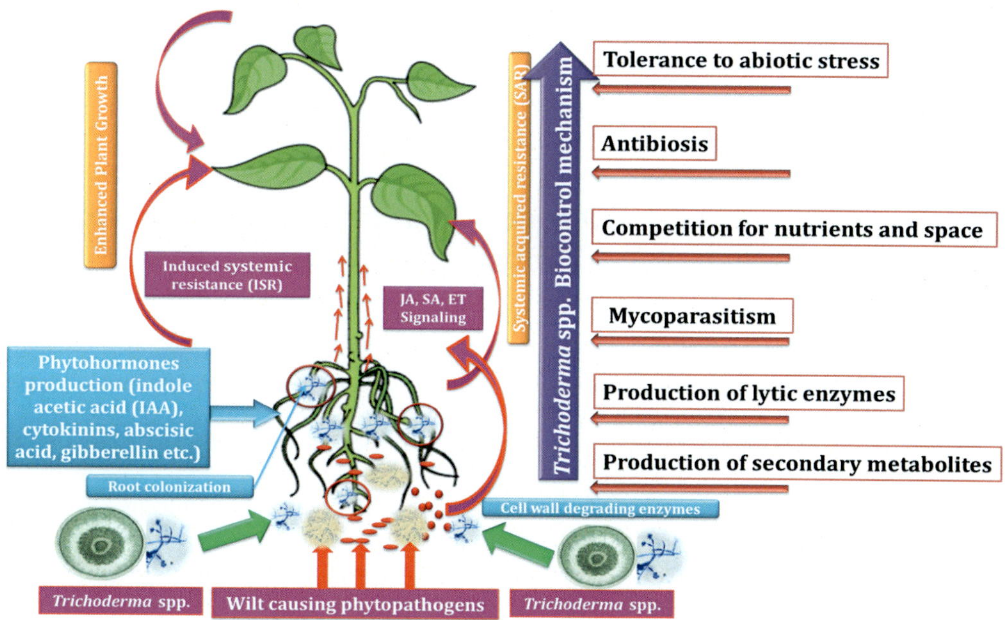

FIG. 18.1 Biocontrol mechanisms of *Trichoderma* spp. against wilt causing diseases.

sequestration, synthesis of phytohormones and non-volatile or volatile compounds (Bisen et al., 2016; Garnica-Vergara et al., 2016; Guzmán-Guzmán et al., 2017, 2019). The hydrolytic enzyme complexes produced by *Trichoderma* spp. are accomplished of decaying the cell wall of phytopathogens, thus allowing hyphae penetration, colonization and onset of mycoparasitism (Hermosa et al., 2012). Hyperparasites attack and kill spores, mycelium, and resting buildings of fungal plant pathogens and cells of bacterial plant pathogens (Ghorbanpour et al., 2018; Abbas et al., 2017; Zaidi and Singh, 2018; Iqbal and Ashraf, 2019).

In the past few years, investigators demonstrated their attention on plant disease resistance as induced systemic resistance (ISR) and systemic acquired resistance (SAR) induced by the *Trichoderma* root association (Pieterse et al., 2009; Hermosa et al., 2012). *Trichoderma* spp. are known by their ability to activate ISR responses based on jasmonic aid (JA), salicylic acid (SA) and ethylene dependent pathways (Hermosa et al., 2012; Tucci et al., 2011; Martínez-Medina et al., 2013). *Trichoderma* act systemically that contain signaling cascade and initiation and accumulation of defense related antimicrobial complexes which comprises defense enzymes as polyphenol oxidase, phenyl ammonia lyase, lipoxygenase, peroxidase, pathogenesis related proteins (PR proteins), phytoallexin as rishitin, terpenoid, lubimin, phytotuberol, sol-evetivone, resveratol, coumarin, and antioxidant as glutathione, ascorbic acid, etc. (Howell et al., 2000). In this chapter, we have presented the application of different *Trichoderma* spp. for the control of vascular wilt diseases in various crops. The major biocontrol mechanisms exerted by *Trichoderma* spp. against wilt pathogens and plant growth promotion has been highlighted.

18.2 *Trichoderma* in the control of wilt disease

Control of wilt diseases is a difficult task. Integrated disease management strategies are used to prevent and manage diseases in crops. Different prevention measures can be taken including rotation of the crop, use of disease resistant varieties, removal of infected parts of the plants etc., still the effectiveness of all these treatments fail to overcome the severe damage caused by the disease. Hence, the alternative method which can be effectively used is use of potential biopesticides. *Trichoderma* is a versatile biocontrol organism which can be effectively used for control of wilt diseases in different crop plants. Several investigators have explored the potential of various *Trichoderma* spp. to manage wilt disease in crop plants.

Trichoderma spp. have been evaluated for biocontrol efficacy against the destructive vascular wilt pathogens *Fusarium oxysporum* and *Ralstonia solanacearum* initially under laboratory conditions using culture based assays helping the identification of antagonistic and growth inhibitory activities. Potential strains are further used to check efficacy in pot trials, green house and field conditions. The biocontrol mechanisms are elucidated based on observations of root colonization, effect on seed germination, seedling growth, growth promotion, reduction of disease incidence, effect on disease severity, yield etc. The inhibitory or lethal activity of *Trichoderma* spp. on pathogen for example competition for nutrients and space, mycoparasitism, secretion of antimicrobial metabolites, secretion of cell wall degrading enzymes, plant growth promoters, ISR, gene expression studies in plants explain the biocontrol mechanism involved in wilt disease control and plant protection. The details are summarized in Table 18.1.

18.3 Mechanism of biocontrol by *Trichoderma* in the control of wilt pathogens

18.3.1 Competition

Among the various mechanisms recruited by biocontrol agents for the control of soil borne phytopathogens, competition for available space and nutrients is significant. Although competition is a natural interaction seen among microbial populations inhabiting in a common ecological niche and having common nutritional requirements, the phenomenon is important from the biocontrol point of view. Competition for vital nutrients contains nitrogen, carbon, and iron may result in the decrease germination of spores and inhibition of development of pathogens because of starvation (Alabouvette et al., 2009).

The potential biocontrol agent *Trichoderma* spp. competes for a nutrient which is evident by the display of fast outgrowth, utilization of the available nutrients more effectively compared to the phytopathogens, thus conferring disease suppression and plant protection. The production of various siderophores, effective iron chelating compounds, aids in solubilizing the available iron, making it accessible for host absorption and additionally inhibiting the development of pathogens due to iron deprivation (Benitez et al., 2004).

Competition for iron by siderophore production have been well documented in various *Trichoderma* spp., such as *T. viride*, *T. harzianum* and *T. lignorum*, which are higher in comparison to phytopathogenic Fusaria including *F. oxysporum*. Antagonistic interaction between *T. asperellum* and *F. oxysporum* result due to competition for iron, which is advantageous to the plant due to iron solubilizing activity (Segarra et al., 2010).

Competition for space and vital nutrients, nitrogen, carbon is a significant antagonistic attribute of *Trichoderma*. *Trichoderma* is usually deliberated as a destructive competitor against fungal pathogens that produces very fast toward pathogen and rapidly colonizes it (Cuervo-Parra et al., 2014). During competitive interaction, *Trichoderma* suppress the development of the pathogens in the rhizosphere and thus decrease the development of disease. Starvation is the most general cause leading to the destruction of soil borne microbes. Competition for nutrients has been considered as a crucial mechanism of biocontrol by *Trichoderma* spp. (Benitez et al., 2004; Harman, 2000). Establishment of nutrient limited situation in soils and root vicinity, places a stress on a microbe to compete for the accessible nutrients. Competition for space and nutrients, also depend on the *Trichoderma* spp. and the pathogen type (Infante et al., 2013).

The rhizosphere and root exudates are rich basis of nutrients that include sugars, vitamins, amino acids, iron, organic acids etc. Application of *T. harzianum* spores to soil inhibited infestations of *F. oxysporum* f. sp. *vasinfectum* and *F. oxysporum* f. sp. *melonis* (Sivan and Chet, 1989). Competition for carbon is active manner not only in *Trichoderma* but also certain other fungi like strains of *F. oxysporum* and *R. solani* (Alabouvette et al., 2009). Colonies of *T. harzianum* suppressed the development of *F. culmorum* in altered environmental situations and the macroscopic study showed that *T. harzianum* competed with *F. culmorum* for nutrients and space (Saravanakumar et al., 2008; Ferre and Santamarina, 2010).

There are many reports showing siderophore involvement the inhibition of soil borne plant pathogenic fungi (Vinale et al., 2013). Numerous *Trichoderma* spp. such as *T. viride*, *T. lignorum* and *T. harzianum* are potent siderophore producers than *F. solani* and *F. oxysporum*. So *Trichoderma* spp. accesses the little amounts of obtainable iron with high competence (Dutta et al., 2006). The iron competition has been defined as one of the main aspects in the antagonism of *T. asperellum* to *F. oxysporum* and iron solubilizing action of siderophores are also useful for plants (Segarra et al., 2010).

18.3.2 Mycoparasitism

Mycoparasitism, a complex mechanism, involves the direct attack by one fungus on another. The cascades of events include the recognition of host fungus followed by attack, hyphal diffusion and lysis. The interaction between the parasitic fungus and the host results in coiling, formation of haustoria, secretion of degradative enzymes that help in hyphal penetration, production of antimicrobial metabolites, killing and finally deriving nutrients from the plant (Holzlechner et al., 2016; Omann et al., 2012). Among the various biocontrol mechanisms exerted by *Trichoderma*, mycoparasitism is a major weapon contributing to the control of devastating plant pathogens.

TABLE 18.1 Efficacy of *Trichoderma* species against vascular wilt pathogens.

Crop	Pathogen	Biocontrol agent	Mode of application	Biocontrol mechanism	References
Chickpea (*Cicer arietinum* L.).	*Fusarium oxysporum* f. sp. *ciceris*	*Trichoderma viride*, *T. harziarum*, *T. virens* *T. cerinum*	Seed treatment Soil application	Increased seed germination Plant growth promotion and increased yield Decreased wilt incidence Production of volatile and non-volatile compounds Production of cell wall degrading enzymes (β-1, 3-glucnase, chitinase and protease), peptaibol antibiotics, indole acetic acid (IAA), siderophore and phosphate solubilization Induced systemic resistance (ISR) based increased activities of phenylalanine ammonia lyase (PAL), polyphenol oxidase (PPO)	Dubey et al. (2006) Dubey et al. (2011) Kaur and Mukhopadhyay (1992) Haidar et al. (2012) Shahid et al. (2015) Jayalakshmi et al. (2009) Khare et al. (2018)
Tomato (*Lycopersicon esculentum)*	*F. oxysporum* f. sp. *lycopersici*	*Trichoderma* spp. *T. asperellum* *T. harzianum* *T. virens* *T. atroviride* *T. longibrachiatum*	Seed treatment Soil application Root dip	Increased seed germination Reduced disease incidence Plant growth promotion and increased yield Production of non-volatile compounds Induction of defense enzyme such as peroxidase (PO), polyphenol oxidase (PPO) and phenylalanine ammonia lyase (PAL), ammonium lyase Increased total phenolic content Induction of jasmonic acid and salicylic acid signaling cascades Inhibited reactive oxygen species (ROS) production	Kumar et al. (2011) Patel and Saraf (2017) Srivastava et al. (2010) Sundaramoorthy and Balabaskar (2013) Ghazalibiglar et al. (2016) Sreenu and Zacharia (2017) Singh et al. (2015) Arenas et al. (2018) Christopher et al. (2010) Patel and Saraf (2017) Sallam et al. (2019) Jogaiah et al. (2018) Herrera-Téllez et al. (2019) Arenas et al. (2018) Ramaswamy and Sundaram (2020)
Piegonpea	*Fusarium udum*	*Trichoderma* consortium *T. viride* *T. harzianum*	Seed treatment Soil application	Reduced disease severity ISR based induction of enzymes PO, PPO, PAL, total phenolics.	Ravikumara et al. (2017) Prasad et al. (2002)
Carnation (*Dianthus caryophyllus L.*)	*F. oxysporum* f. sp. *dianthi*	*T. asperellum*	Compost amended with *T. asperellum*	Improved suppressive capacity of growth medium against Fusarium wilt	Sant et al. (2010)
Cucumber (*Cucumis sativus* L.)	*F. oxysporum* f. sp. *cucumerinum*	*T. asperellum* *T. harzianum* *T. pseudokoningii* *T. longibrachiatum*	Seedling treatment Soil application	Increase in the plant growth and yield Reduced disease incidence Mycoparasitism, hydrolytic enzymes, volatile organic compounds Increase in activity of defense enzymes PO, PPO, superoxide dismutase (SOD), catalase (CAT) and ascorbate oxidase (AAO) Improving plant tolerance to biotic stress Metabolism stimulation in cucumber	Li et al. (2019) Kareem et al. (2016) Saravanakumar et al. (2015) Zhang et al. (2014) Chen et al. (2019)

Continued

TABLE 18.1 Efficacy of *Trichoderma* species against vascular wilt pathogens.—cont'd

Crop	Pathogen	Biocontrol agent	Mode of application	Biocontrol mechanism	References
Chrysanthemum	*F. oxysporum* f. sp *chrysanthemi*	*T. harzianum*	Soil application	Antagonistic activity and disease control	Singh and Kumar (2011)
Eggplant *Solanum melongena*	*F. oxysporum*	*T. harzianum*	Soil application	Reduced disease incidence Increase in seed germination Plant growth promotion	Kareem and Al-Araji (2017)
Bean *Phaseolus vulgaris*	*F. solani* *F. oxysporum* f. sp. *phaseoli*	*T. harzianum,* *T. viride* *T. virens*	Combined seed and soil treatment Seed treatment	Reduced disease incidence Increase in seed germination Plant growth promotion and ISR based increase in production of PO, PPO, and chitinase.	Abd-El-Khair et al. (2019) Carvalho et al. (2014)
Banana (Panama disease)	*F. oxysporum* f. sp. *cubense*	*Trichoderma spp.* *T. harzianum* *T. viride* *T. koningii* *T. peudokoningii* *T. hamatum*	Root dip Soil application	Plant growth promotion Reduced disease incidence Increased yield ISR based induction of POs, PAL and increase in total phenolics	Bubici et al. (2019) Thangavelu et al. (2004) Thangavelu and Mustafa (2010) Thangavelu and Gopi (2015)
Chilli (*Capsicum annuum*)	*F. solani*	*T. harzianum* *T. viride* *Trichoderma* spp.	Seed treatment and seedling dip	Reduced disease incidence	Mohiddin et al. (2018) Naik et al. (2009)
Potato	*F. solani*	*T. brevicompactum* *T. longibrachiatum* *T. asperellum*	Soil application	Antagonistic activity and production of volatile and non-volatile inhibitors	Ommati and Zaker (2012)
Lentil (*Lens culinaris*)	*F. oxysporum* f. sp. *lentis*	*T. harzianum* *T. viride* *T. koningii*	Soil application	Antagonistic activity Reduced wilt incidence	Akter et al. (2018) Sharfuddin and Mohanka (2012)
Cotton	*F. oxysporum* f. sp. *vasinfectum*	*T. harzianum*	Seed treatment Soil application	Reduced disease incidence	Sivan and Chet (1986)
Melon	*F. oxysporum* f. sp. *melonis*	*T. harzianum*	Seed treatment Soil application	Reduced disease incidence	Sivan and Chet (1986)
Grape (*Vitis vinifere* L.)	*F. oxysporum f. sp herbemantis*	*T. harzianum*	Soil application	Reduced disease severity Improved yield	El-Mohamedy et al. (2010)
Wheat	*F. culmorum*	*T. harzianum*	Seed treatment Soil application	Reduced disease incidence	Sivan and Chet (1986)
Tobacco (*Nicotiana tabacum* L.)	*Ralstonia solanacearum*	*T. harzianum*	Soil application	Antagonistic effect Enhanced plant growth Reduced disease severity ISR based increase in POX, PAL, PPO activities	Yuan et al. (2016) Maketon et al. (2008)
Tomato (*Lycopersicon esculentum*)	*R. solanacearum*	*Trichoderma spp.* *T. asperellum*	Soil application Root dip	Antagonistic effect Increased seed germination Enhanced plant growth and improved yield Reduced disease severity ISR based increase in POX, PAL, PPO, β-1,3-glucanase and total phenol	Murthy et al. (2013) Murthy and Srinivas (2013) Yendyo et al. (2017) Konappa et al. (2018)

Electron microscopic observations of the mycoparasitic activity of *Trichoderma harzianum* antagonistic to plant pathogenic *Fusarium oxysporum* f. sp. *cucumerinum* were made by Zhang et al. (2014). In dual cultures, Trichoderma mycelium made physical contact with the pathogen hyphae on the third day of inoculation. This was followed by overgrowth of *Trichoderma*, twisting, coiling around the pathogen hyphae, parallel growth and parasitizing the host. The authors suggest the involvement of upregulated expression of hydrolytic enzymes and production of several volatile organic compounds aiding the mycoparasitic action of *Trichoderma*. Mycoparasitism utilized by *Trichoderma* occurs in response to the presence of the host, involving the sequential expression of an array of genes encoding cell wall degrading enzymes (CWDEs) and secondary metabolites during the time course of predation (Atanasova et al., 2013).

18.3.3 Cell wall degrading enzymes

Biological control of phytopathogens by Trichoderma is brought about in multi-step mechanism involving a range of cellular activities. Cell wall degradation with the aid of secreted hydrolytic enzymes is critical to its bioactivity. Synthesis and secretion of a battery of extracellular Cell wall degrading enzymes (CWDEs) is one of the prime activities in *Trichoderma* mediated biological control. Attachment of *Trichoderma* to the target fungal hyphae and coiling is followed by the secretion of CWDEs that facilitate cell wall degradation, penetration and mycoparasitism. Chitinases and glucanases are among the major hydrolytic enzymes secreted by Trichoderma involved in the cell wall degradation of pathogens. Further the cell wall degradation products such as β-1,3 glucan and chitin are utilized as nutrients for growth promotion of *Trichoderma* (Gajera et al., 2012; Viterbo et al., 2002).

The secretion of chitinase and glucanase has also been reported in *T. longibrachiatum* inhibitory to the growth of *F. oxysporum*. *Trichoderma harzianum* antagonistic the chickpea wilt pathogen *F. oxysporum* f. sp. *ciceri* produced the enzymes ß-glucanase, chitinase, protease and xylanase. Further, the activity of *F. oxysporum* protease was totally inhibited, conferring resistance to the root colonization by the wilt pathogen (Hernandez, 2016; Jayalakshmi et al., 2009). High chitinolytic activity has been reported by *T. viride* that potentially suppressed the development of *F. oxysporum* (Khatri et al., 2017). The upregulation of secreted enzymes for example β-1,3-glucanase, chitinase, protease, xylanase, and cellulase in was noticed in *Trichoderma harzianum* mycoparasitic to the cucumber wilt pathogen *Fusarium oxysporum* f. sp. *cucumerinum* (Zhang et al., 2014). Murthy et al. (2013) reported *Trichoderma asperellum* in the biological management of bacterial wilt of tomato. Seed treatment and green house studies revealed important reduction of disease frequency and severity along with development in seed germination, vigor index, root, shoot lengths. Evaluation of enzymatic activities of potent *T. asperellum* isolates antagonistic to the *R. solanacearum* showed the secretion of CWDEs chitinase, β-1,3-glucanase, protease along with amylase, lipase and cellulase (Murthy et al., 2013). The exo and *endo*-glucanases act by hydrolyzing the β-glucans present in the fungal cell wall of pathogens. The major endoglucanase being BGN 13.1 which is expressed on induction with the presence of fungal cell walls (De la Cruz et al., 1995). The secretion and characterization of characterization of chitinases from mycoparasitic *T. harzianum* has been reported by several researchers. The chitinolytic system has been well explored in *T. harzianum* and is comprised of two *N*-acetylglucosaminidases, four endochitinases and an exochitinase (De La Cruz et al., 1992; Haran et al., 1995). Seidl et al. (2005) reported the presence of 20 and 36 different chitinase encoding genes including *ech42*, *chi33*, *nag1*, *chi18-13* in *Trichoderma genome.* The secretion of an array of isozymes could offer better mycoparasitic activity of *Trichoderma* against phytopathogens (Seidl et al., 2005).

The genes intricate in the expression and synthesis of chitinases such as *chit33* and *chit42; N*-acetylglucosaminidase (*exc1 and exc2*), glucanase encoding *bgn13.1* gene and *prb1* gene involved in synthesis of protease have been reported in *Trichoderma* mediated biological control. Quantitative reverse transcription and polymerase chain reaction (qRT-PCR) based reports on the expression of these genes was correlated with the biocontrol action against melon wilt causing pathogen *F. oxysporum* f. sp. *melonis. Trichoderma harzianum* displays the ability to secrete an array of hydrolytic enzymes in response to the presence of phytopathogenic *F. oxysporum.* Further, the expression patterns of the genes and secretion of hydrolytic enzymes encompassing β-1,3-glucanases, chitinases, and proteases occur in synchrony enabling efficient biocontrol activity against the wilt pathogen (Mondejar et al., 2011). The proteases play a dual role by dissolving the protein component of the pathogenic cell wall as well as inactivating the pathogenicity related enzymes produced by the wilt causing pathogen which aid in controlling the development or progression of the disease (Markovich and Kononova, 2003).

18.3.4 Antibiosis by antimicrobial metabolites

Filamentous fungi including *Trichoderma* are known to produce a collection of secondary (SMs) metabolites with broad chemical and functional diversity. Among the several mechanisms in controlling plant pathogens, antibiosis is one of the critical antagonistic interactions, brought about by the production of diffusible antimicrobial secondary metabolites or antibiotics that damage pathogens (Singh et al., 2018). More than 373 molecules encompassing a variety of agriculturally important bioactive metabolites include peptaibiotics such as peptaibols, gliotoxin, gliovirin, viridin, viridol, koninginins, pyrones, Non-ribosomal peptides (NRPs), siderophores, nitrogen containing compounds, volatile and non-volatile terpenes are implicated in the biocontrol of phytopathogens by *Trichoderma* (Contreras-Cornejo et al., 2016). These secondary metabolites confer the mycoparasitic property to the biocontrol agent, involve in plant-microbe interactions, and contribute to host growth promotion in addition to elicit plant defense responses.

Several researchers have attempted the production and purification of these antibiotics from cultures of *Trichoderma* spp. along with the evaluation of the efficacy of these molecules in pathogen control and their phytotoxic effects. Ethyl acetate extracts of secondary metabolites of Trichoderma spp. grown on different growth media were evaluated for antibacterial activity against the wilt pathogen *Ralstonia solanacearum* by Khan et al. (2020). Highest growth inhibition was exhibited by strains of *T. pseudoharzianum* cultured on solid wheat medium followed by *T. asperelloides* and *T. viride*. Scanning electron microscopic observations of *R. solanacearum* cells treated with *T. pseudoharzianum* extracts displayed extensive morphological changes including cell wall and membrane disintegration as well as leakage of cytoplasmic contents (Khan et al., 2020). Leylaie and Zafari (2018) reported culture extracts *T. brevicompactum, T. asperellum, T. koningiopsis*, and *T. longibrachiatum* endophytic to *Vinca* spp. displayed bacteriostatic and bactericidal activities against plant pathogenic *R. solanacearum*. The GC—MS analyses of culture extracts of *T. koningiopsis* generated profiles corresponding to major classes of volatile metabolites including pyrones (lactones), acids, esters, monoterpene, alcohols, and furanes, lipids. The most profusely produced compound was 6-pentyl-alpha-pyrone (6-PP) which is implicated crucial to the biocontrol property of *T. koningiopsis, T. harzianum* and *T. koningii* (Hanson, 2005; Leylaie and Zafari, 2018). Pyrones extracted from cultures of *Trichoderma koningii* possess potential antifungal properties inhibiting the growth of several phytopathogens including soil borne *F. oxysporum* (Simon et al., 1988). Scarselletti and Faull (1994) suggested a correlation between the secretion of the pyrone 6-pentyl-α-pyrone [6-p-p] by *T. harzianum* and its fungicidal effects on *F. oxysporum*. The pyrone has significant growth inhibitory effects as well as completely blocked spore germination of the wilt pathogen. Koninginins, a major pyrane produced by *T. koningii* effectively suppressed the growth of many soil borne fungi including *F. oxysporum* (Dunlop et al., 1989).

In their attempt to unravel the mechanism involved in *Trichoderma* as a biofungicide in plant protection against *F. oxysporum* complex Li et al. (2018) analyzed the antimicrobial metabolites secreted by *Trichoderma virens* and *T. viride* in reaction to *F. oxysporum*. Volatile compounds generated by *T. virens* and *T. harzianum* such as acids, alcohols, ketones, esters, and sesquiterpenes. The most important ones were 3-octanone and 1-octen-3-ol which have fungistatic and fungicidal activities (Li et al., 2018). Categorization of SMs produced by *Trichoderma viride*, *T. virens* and *T. harzianum* which effectively reduced disease severity of chickpea wilt affected by *F. oxysporum* f. sp. *ciceris* was done by Dubey et al. (2011). Mass spectral analyses of culture extracts of these biocontrol *Trichoderma* displayed compounds, viz., koningin-A, 3-methyl-heptadecanol, 6-nonylene alcohol, massoilactone, 3-(propenone)-4-(hexa-2′-4′-dineyl)-2-(5H)-furanone, methyl-cyclopentane, methyl cyclohexane, N-methyl pyrollidine, 3-(2′-hydroxypropyl)-4-(hexa-2′-4-dineyl)-2-(5H)-furanone, dermadin, ketotriol, and 2-methyl heptadecanol, palmitic acid. The authors suggest that the observed fungicidal activity against the chickpea wilt pathogen could be attributed to the production of these diverse metabolites. In their study, Saravanakumar et al. (2015) reported the potential of *T. asperellum* in significantly reducing the disease severity and inhibitory effects on phytopathogenic *Fusarium oxysporum* f. sp. *cucumerinum* which causes yield losses in the important vegetable crop cucumber. Metabolite profiling of *T. asperellum* extract by GC—MS showed the presence of several metabolites. Interestingly, 1,6-diphenylhexane-1,3,4,6-tetrone was found potent in inhibiting the activity and down regulating the expression levels of the target protein Snt2 in *F. oxysporum* f. sp. *cucumerinum*.

The genus *Trichoderma* produces more than 190 peptaibols, which are a major class of antimicrobial peptides. Among the major peptaibols, trichokonin VI, VII and VII display antibiotic effects against wide range of microorganisms (Song et al., 2006). In a study on the mode of action of Trichokonin VI from *Trichoderma pseudokoningii*, Shi et al. (2012), used the vascular wilt fungus *Fusarium oxysporum* as an experimental system. Trichokonin VI exhibited significant antifungal activity trigerring morphological changes in the pathogen involving alteration of membrane permeability, loss of mitochondrial membrane potential, accumulation of intracellular ROS, disintegration of subcellular structures, DNA damage, cytoplasmic vacuolation, and so on leading to the metacaspase-independent apoptotic cell death.

18.3.5 Induced systemic resistance

Biocontrol microorganisms including *Trichoderma* find wide spread application in sustainable agriculture for management of crop diseases and plant growth promotion. Among the various biocontrol mechanisms of *Trichoderma* in combating pathogen attack is its ability to trigger plant defense response through ISR. *Trichoderma* mediated ISR in crop plants involve the activation of a network of different defense pathways involved in the expression and synthesis defense biochemicals. Trichoderma activates the build-up of signal molecules such as JA and SA and the expression of PR proteins for example chitinases and glucanases and mechanical strengthening of cell walls by lignifications. Other key defense enzymes in plants, such as phenylalanine-ammonialyase, peroxidase and polyphenol oxidase are crucial in the biosynthesis of phenolic antimicrobials (Patel and Saraf, 2017).

In an attempt to identify an efficient *Trichoderma* strain for the control of Fusarium wilt of banana, *Trichoderma* spp. counting *T. viride, T. harizianum, T. koningii, T. peudokoningii, T. hamatum* from rhizosphere soils of different banana cultivating regions of India by Thangavelu and Mustaffa (2010). These isolates reduced mycelial development and germination of *F. oxysporum* f. sp. *cubense* spore. Pot trials and field assessment of *T. viride* rice chaffy grain formulation, proved effective in reducing disease incidence and also inducing defense enzymes peroxidase, phenylalanine ammonia lyase in treated plants. Further, the increase in phenolic content in treated plants showed the ability of *T. viride* in ISR in plants and wilt protection (Thangavelu and Mustaffa, 2010). Seed treatment of the legume *Vigna mungo,* with spores of biocontrol agent *T. viride* resulted in increased accumulation and enhanced activities of defense related molecules POX, PPO and PAL and total phenolic contents in *Fusarium oxysporum* infected plants showed the potential contribution in developing disease resistance to wilt disease in legumes (Surekha et al., 2014).

Isolates of *Trichoderma* have also been explored as a tool to combat the destructive pathogen *Ralstonia solanacearum*, the bacterial wilt agent. Konappa et al. (2018) explained the mechanism of induced systemic resistance and its implication in the management *Ralstonia solanacearum* disease in tomato. Green house trials and field application *Trichoderma asperellum* was potential in reducing disease incidence and enhancing plant development and tomato yield. Seed and seedling treatments significantly increase activities of defense enzymes POX, PAL, PPO, β-1,3-glucanase and total phenol activities in *T. asperellum* pre-treated plants challenge inoculated with *R. Solanacearum* compared to control (Murthy et al., 2013; Konappa et al., 2018). Improved biochemical responses and activities of defense enzymes in *Trichoderma* treatments challenged with pathogen related to untreated plants, signifies the potential of *Trichoderma* in combating phytopathogens and resistance development against *Ralstonia solanacearum.*

The biosynthesis and expression levels of phytohormones and small signaling molecules such as JA, SA and ET show a significant role in host defense combating pathogen attack. The potential of these signal cascades in development of plant response and eliciting ISR in plants against plant pathogens is well established. Beneficial biocontrol microorganisms can trigger and enhance the expression of these signaling molecules thus contributing to the development of ISR (Pieterse et al., 2009).

Jogaiah et al. (2018) studied the interactions between Trichoderma-plant pathogen systems and elucidated the role of *T. virens* in plant development regulation and eliciting defense responses in tomato plants infected with *F. oxysporum* f. sp. *lycopersici*. Application of spores and culture filtrates of *T. virens* reduced the Fusarium wilt incidence in tomato. Further investigations on the expression of *PDF1* and *PR1a* genes involved in JA-responsive defensin gene and SA-inducible PR protein 1 acidic respectively, in wild type tomato plants showed a clear upregulation of these genes in *Trichoderma* treatments. However, in Fusarium wilt susceptible tomato varieties, JA-deficient mutant *def1* and SA-deficient mutant *NahG* plants the expression levels of JA and SA respectively, were very low compared to wild type plants on treatment with *T. virens*. The expression of defense enzymes functional in Trichoderma mediated ISR against Fusarium wilt of tomato was carried out by Christopher et al. (2010). Seed treatments with *T. virens* and talc based formulations for soil application induced the build up of defense enzymes PO, PPO and PAL whose activities found maximum between the 1−2 weeks of application. Patel and Saraf (2017) elucidated the mechanism of ISR by *T. asperellum* against *F. oxysporum* in tomato plant. Along with plant growth promotion and reduction of wilt incidence in treated plants, *T. asperellum* mediated ISR displayed elevated levels of total phenolics, PO, PPO and PAL activities enhancing plant defenses. Enhanced activities of enzymes such as chitinase and glucanase involved in fungal cell wall degradation clearly suggest the bioefficacy of *T. asperellum* in managing Fusarium wilt of crops. Sallam et al. (2019) isolated *Trichoderma* spp. in their attempt to find a potential agent for the control of Fusarium wilt in tomato. *Trichoderma atroviride* and *T. longibrachiatum* were effective in disease suppression and further their ability in inducing systemic resistance was studied by real time RT-PCR. The expression levels of defense genes encoding PR proteins including β-1,3-glucanase gene was found to be upregulated in roots of tomato plants application with *Trichoderma* challenged with the pathogen.

Li et al. (2019) studied the efficacy of *Trichoderma* in the management of Fusarium wilt in cucumber as well as the physiological modifications induced. Treatments with *T. pseudokoningii, T. asperellum* and *T. harzianum* revealed reduced wilt incidence, enhanced seedling growth, improved quality and yield. Further, *T. pseudokoningii* was most effective in increasing the activities of enzymes superoxide dismutase, POX, catalase, PPO, and ascorbate oxidase which have a role in stress responses in cucumber seedlings.

Reactive oxygen species (ROS) and Reactive Nitrogen species (RNS) are vital to plant growth and stress responses, which alter various signaling networks solely or in coordination (Zehra et al., 2017; Lindermayr, 2018). Among the various mechanisms of ISR based defense responses, *Trichoderma asperellum* potentially reduces the levels of ROS in tomato plants, protecting the plants during infection by *F. oxysporum* (Herrera-Tellez et al., 2019). In an investigation on the mechanism of ISR in cucumber roots by *Trichoderma harzianum* against Fusarium wilt, Chen et al. (2019), reported the potential of *Trichoderma* to reduce the oxidative and nitrostative stresses by decreasing the accumulation of reactive species including superoxide, hydrogenperoxide and nitric oxide. Further, *Trichoderma* improved the antioxidant potential of the plant which was evident by the enhanced expression levels of the enzymes crucial to the Ascorbate (AsA)-glutathione (GSH) pathway and Oxidative pentose phosphate pathway (OPPP). These metabolic pathways contribute significantly toward plant tolerance to biotic stresses and improve plant defense against *F. oxysporum* (Chen et al., 2019).

18.4 Conclusion

This chapter has shown that several research attempts have been successful with *Trichoderma* spp. as a promising biocontrol agent in crop protection from vascular wilt disease. Several biocontrol and plant growth promoting traits of *Trichoderma* spp. including its potent antagonistic activity against pathogens, induced systemic resistance in the host, enhancement of plant nutrient uptake and increase in abiotic stress tolerance and so on, make it an efficient and reliable biocontrol agent. Production of CWDEs and SMs are noteworthy in the mechanism of antibiosis by *Trichoderma*. The *Trichoderma* genome harbors several genes that work in coordination, controlling the expression of important biocontrol traits. This clearly shows the efficacy and advantages of biological control in managing devastating diseases such as wilt when other methods fail to control the disease. However, future efforts can focus on improving the field performance of existing biocontrol agents including *Trichoderma* spp., by developing integrated disease control strategies, designing better application methods, improving formulations for better efficacy under various environmental conditions. Genetic engineering approaches for enhanced biocontrol attributes needs a better understanding, which is vital for the development of strains with better market and field potentials.

References

Abbas, A., Jiang, D., Fu, Y., 2017. *Trichoderma* spp. as antagonist of *Rhizoctonia solani*. J. Plant Pathol. Microbiol. 8, 402. https://doi.org/10.4172/2157-7471.1000402.

Abd-El-Khair, H., Elshahawy, I.E., Haggag, H.E.K., 2019. Field application of *Trichoderma* spp. combined with thiophanate-methyl for controlling *Fusarium solani* and *Fusarium oxysporum* in dry bean. Bull. Natl. Res. Cent. 43 (1). https://doi.org/10.1186/s42269-019-0062-5.

Agrios, G.N., 2005. In: Burlington, M.A. (Ed.), Plant Pathology, figth ed. Elsevier Academic Press.

Akter, F., Ahmed, M.G.U., Alam, M.F., Alam, M.J., Begum, N., 2018. Bio-control of lentil wilt disease by *Trichoderma harzianum*. Int. J. Agric. Environ. Bio-res. 3 (6), 158–171.

Alabouvette, C., Olivain, C., Migheli, Q., Steinberg, C., 2009. Microbiological control of soil–borne phytopathogenic fungi with special emphasis on wilt–inducing *Fusarium oxysporum*. New Phytol. 184, 529–544.

Alsohiby, F.A.A., Yahya, S., Humaid, A.A., 2016. Screening of soil isolates of bacteria for antagonistic activity against plant pathogenic fungi. PSM Microbiol. 1 (1), 05–09.

Arenas, O.R., Olguín, J.F.L., Ramón, D.J., Jarquín, D.M.S., Lezama, C.P., Morales, P.S., Lara, M.H., 2018. Biological control of *Fusarium oxysporum* in tomato seedling production with Mexican strains of *Trichoderma*. In: Fusarium - Plant Diseases, Pathogen Diversity, Genetic Diversity, Resistance and Molecular Markers. https://doi.org/10.5772/intechopen.72878.

Asad, S.A., Tabassum, A., Hameed, A., Hassan, F., Afzal, A., Khan, S.A., et al., 2015. Determination of lytic enzyme activities of indigenous *Trichoderma* isolates from Pakistan. Braz. J. Microbiol. 46 (4), 1053–1064.

Atanasova, L., Le Crom, S., Gruber, S., Coulpier, F., Seidl-Seiboth, V., Kubicek, C.P., et al., 2013. Comparative transcriptomics reveals different strategies of *Trichoderma* mycoparasitism. BMC Genom. 14, 121. https://doi.org/10.1186/1471-2164-14-121.

Baayen, R.P., 2000. Diagnosis and detection of host-specific forms of *Fusarium oxysporum*. EPPO Bull. 30, 489–491.

Benitez, T., Rincon, A.M., Limon, M.C., Codon, A.C., 2004. Biocontrol mechanisms of *Trichoderma* strains. Int. Microbiol. 7, 249–260.

Bisen, K., Keswani, C., Patel, J.S., Sarma, B.K., Singh, H.B., 2016. *Trichoderma* spp.: efficient inducers of systemic resistance in plants. In: Chaudhary, D.K., Verma, A. (Eds.), Microbial–Mediated Induced Systemic Resistance in Plants. Springer, Singapore, pp. 185–195.

Bubici, G., Kaushal, M., Prigigallo, M.I., Cabanás, C.G.L., Mercado-Blanco, J., 2019. Biological control agents against Fusarium wilt of banana. Front. Microbiol. 10 (616), 1–33. https://doi.org/10.3389/fmicb.2019.00616.

Carmona-Hernandez, S., Reyes-Pérez, J.J., Chiquito-Contreras, R.G., Rincon- Enriquez, G., Cerdan-Cabrera, C.R., Hernandez-Montiel, L.G., 2019. Biocontrol of postharvest fruit fungal diseases by bacterial antagonists: a review. Agron 9 (3), 121.

Carvalho, F.P., 2017. Pesticides, environment, and food safety. Food. Energy. Secur. 6, 48–60.

Carvalho, D.C.D., Junior, M.L., Martins, I., Inglis, P.W., Mello, S.C.M., 2014. Biological control of *Fusarium oxysporum* f. sp. *phaseoli* by *Trichoderma harzianum* and its use for common bean seed treatment. Trop. Plant. Pathol. 39 (5), 384–391. https://doi.org/10.1590/S1982-56762014000500005.

Chen, S.C., Ren, J.J., ZhaoH, J., Wang, X.L., Wang, T.H., Jin, S.D., et al., 2019. *Trichoderma harzianum* improves defense against *Fusarium oxysporum* by regulating ROS and RNS metabolism, redox balance, and energy flow in cucumber roots. Phytopathology 109 (6), 972–982.

Christopher, D.J., Raj, T.S., Rani, S.U., Udhayakumar, R., 2010. Role of defense enzymes activity in tomato as induced by *Trichoderma virens* against Fusarium wilt caused by *Fusarium oxysporum* f. sp. *lycopersici*. J. Biopestic. 3 (1), 158–162.

Contreras-Cornejo, H.A., Macías-Rodríguez, L., del-Val, E., Larsen, J., 2016. Ecological functions of *Trichoderma* spp. and their secondary metabolites in the rhizosphere: interactions with plants. FEMS Microbiol. Ecol. 92 (4). https://doi.org/10.1093/femsec/fiw036.

Cuervo–Parra, J.A., Snchez–Lpez, V., Romero–Cortes, T., Ramrez–Lepe, M., 2014. *Hypocrea/Trichoderma viridescens* ITV43 with potential for biocontrol of *Moniliophthora roreri* cif par, *Phytophthora megasperma* and *Phytophthora capsici*. Afr. J. Microbiol. Res. 8, 1704–1712.

De La Cruz, J., Rey, M., Lorca, J.M., Hidalgo- Gallego, A., Dominguez, F., Pintor-Toro, J.A., et al., 1992. Isolation and characterization of three chitinases from *Trichoderma harzianum*. Eur. J. Biochem. 206, 859–867.

De la Cruz, J., Pintor-Toro, J.A., Benítez, T., Llobell, A., Roero, L.A., 1995. Novel endo-β- 1,3-glucanase, BGN13.1, involved in the mycoparasitism of *Trichoderma harzianum*. J. Bacteriol. Res. 177, 6937–6945.

de Medeiros, H.A., Filho, J.V.A., de Freitas, L.G., Castillo, P., Rubio, M.B., Hermosa, R., et al., 2017. Tomato progeny inherit resistance to the nematode *Meloidogyne javanica* linked to plant growth induced by the biocontrol fungus *Trichoderma atroviride*. Sci. Rep. 7, 40216.

Deacon, J., 2004. Fungal Biology, fourth edition. Blackwell Publishing Ltd., pp. 293–297

Devi, N.O., Singh, N.I., Devi, R.K.T., Chanu, W.T., 2017. In vitro evaluation of *Alternaria solani* (Ellis and Mart.) Jones and Grout causing fruit rot of tomato by plant extracts and bio-control agents. Int. J. Curr. Microbiol. Appl. Sc. 6 (11), 652–661. https://doi.org/10.20546/ijcmas.2017.611.078.

Dubey, S.C., Suresh, M., Singh, B., 2006. Evaluation of Trichoderma species against *Fusarium oxysporum* f. sp. *ciceris* for integrated management of chickpea wilt. Biol. Contr. 40 (1), 118–127. https://doi.org/10.1016/j.biocontrol.2006.06.006.

Dubey, S.C., Tripathi, A., Dureja, P., Grover, A., 2011. Characterization of secondary metabolites and enzymes produced by *Trichoderma* species and their efficacy against plant pathogenic fungi. Indian J. Agric. 81 (5), 455–461.

Dunlop, R.W., Simon, A., Sivasithamparam, K., Ghisalberti, E.L., 1989. An antibiotic from *Trichoderma koningii* active against soilborne plant pathogens. J. Nat. Prod. 52, 67–74.

Dutta, S., Kundu, A., Chakraborty, M., Ojha, S., Chakrabarti, J., Chatterejee, N., 2006. Production and optimization of Fe(III) specific ligand, the siderophore of soil inhabiting and wood rotting fungi as deterrent to plant pathogens. Acta Phytopathol. Entomol. Hung. 41, 237–248.

El-Mohamedy, R.S.R., Ziedan, E.H., Abdalla, A.M., 2010. Biological soil treatment with *Trichoderma harzianum* to control root rot disease of grapevine (*Vitis vinifera* L.) in newly reclaimed lands in Nobaria province. Arch. Phytopathol. Plant Protect. 43 (1), 73–87.

Elumalai, L.K., Rengasamy, R., 2012. Synergistic effect of seaweed manure and *Bacillus sp.* on growth and biochemical constituents of Vigna radiata L. J. Biofert. Biopestic. 3, 121–128.

Etschmann, M.M., Huth, I., Walisko, R., et al., 2015. Improving 2- phenylethanol and 6-pentyl-α-pyrone production with fungi by microparticle-enhanced cultivation (MPEC). Yeast 32, 145–157.

Ferre, F.S., Santamarina, M.P., 2010. Efficacy of *Trichoderma harzianum* in suppression of *Fusarium culmorum*. Annales. De. Microbiologie. 60, 335–340.

Gajera, H.P., Bambharolia, R.P., Patel, S.V., Khatrani, T.J., Goalkiya, B.A., 2012. Antagonism of *Trichoderma* spp. against *Macrophomina phaseolina*: evaluation of coiling and cell wall degrading enzymatic activities. J. Plant Pathol. Microbiol. 3 (7), 1–8.

Garnica–Vergara, A., Barrera–Ortiz, S., Munoz–Parra, E., Raya–Gonzalez, J., Mendez–Bravo, A., et al., 2016. The volatile 6–pentyl–2H–pyran–2–one from *Trichoderma atroviride* regulates *Arabidopsis thaliana* root morphogenesis via auxin signaling and ethylene insensitive 2 functioning. New Phytol. 209, 1496–1512.

Ghazalibiglar, H., Kandula, D.R.W., Hampton, J.G., 2016. Biological control of Fusarium wilt of tomato by *Trichoderma* isolates. N. Z. Plant Protect. 69, 57–63. https://doi.org/10.30843/nzpp.2016.69.5915.

Ghorbanpour, M., Omidvari, M., Abbaszadeh-Dahaji, P., Omidvar, R., Kariman, K., 2018. Mechanisms underlying the protective effects of beneficial fungi against plant diseases. Biol. Contr. 117, 147–157. https://doi.org/10.1016/j.biocontrol.2017.11.006.

Griffin, M.R., 2014. Biocontrol and bioremediation: two areas of endophytic research which hold great promise. In: Verma, V.C., Gange, A.C. (Eds.), Advances in Endophytic Research. Springer, India, pp. 257–282.

Guzmán-Guzmán, P., Alemán-Duarte, M.I., Delaye, L., Herrera-Estrella, A., Olmedo- Monfi, V., 2017. Identification of effector-like proteins in *Trichoderma spp.* and role of a hydrophobin in the plant-fungus interaction and mycoparasitism. BMC Genet. 18 (16).

Guzmán–Guzmán, P., Porras–Troncoso, M.D., Olmedo–Monfil, V., Herrera–Estrella, A., 2019. *Trichoderma* species: versatile plant symbionts. Phytopathology 109, 6–16.

Hanson, J.R., 2005. The chemistry of the bio-control agent, *Trichoderma harzianum*. Sci. Prog. 88, 237–248. https://doi.org/10.3184/003685005783 238372.

Haran, S., Schickler, H., Oppenheim, A., Chet, I., 1995. New components of chitinolytic system of *Trichoderma harzianum*. Mycol. Res. 99 (9), 441–446.

Harman, G.E., 2000. Myths and dogmas of biocontrol: changes in perceptions derived from research on *Trichoderma harzianum* T– 22. Plant Dis. 84, 377–393.

Hashmi, I.H., Aslam, A., Farooq, T.H., Zaynab, M., Munir, N., Tayyab, M., Abbasi, K.Y., 2018. Antifungal activity of biocontrol agents against corm rot of *Gladiolus grandiflorus* L. caused by *Fusarium oxysporum*. Int. J. Mol. Microbiol. 1 (1), 29–37.

Haidar, M., Bahramnejad, B., Amini, J., Siosemardeh, A., Haji-allahverdipoor, K., 2012. Suppression of chickpea ('Cicer arietinum' L.) 'Fusariums' wilt by 'Bacillus subtillis' and 'Trichoderma harzianum'. Plant Omics 5, 68–74.

Heimpel, G.E., Mills, N., 2017. Biological Control - Ecology and Applications. Cambridge University Press, Cambridge.

Hermosa, R., Viterbo, A., Chet, I., Monte, E., 2012. Plant-beneficial effects of *Trichoderma* and of its genes. Microbiology 158 (1), 17–25.

Hernández, E.O., Morales, J.H., Conde-Martínez, V., Michel-Aceves, A.C., Lopez-Santillan, J.A., Torres-Castillo, J.A., 2016. In vitro activities of *Trichoderma* species against *Phytophthora parasitica* and *Fusarium oxysporum*. Afr. J. Microbiol. Res. 10 (15), 521–527.

Herrera-Téllez, V.I., Cruz-Olmedo, A.K., Plasencia, J., Gavilanes-Ruíz, M., Arce-Cervantes, et al., 2019. The protective effect of *Trichoderma asperellum* on tomato plants against *Fusarium oxysporum* and *Botrytis cinerea* diseases involves inhibition of reactive oxygen species production. Int. J. Mol. Sci. 20 (8), 1–13. https://doi.org/10.3390/ijms20082007.

Holzlechner, M., Reitschmidt, S., Gruber, S., Zeilinger, S., Marchetti-Deschmann, M., 2016. Visualizing fungal metabolites during mycoparasitic interaction by MALDI mass spectrometry imaging. J. Proteomics. 16 (11), 1742–1746.

Howell, C.R., Hanson, L.E., Stipanovic, R.D., Puckhaber, L.S., 2000. Induction of terpenoid synthesis in cotton roots and control of *Rhizoctonia solani* by seed treatment with *Trichoderma virens*. Phytopathology 90, 248–252.

Infante, D., Martinez, B., Peteira, B., Reyes, Y., Herrera, A., 2013. Molecular identification of thirteen isolates of *Trichoderma* spp. and evaluation of their pathogenicity towards *Rhizoctonia solani* Kühn. Biotecnol. Apl. 30, 23–28.

Iqbal, M.N., Ashraf, A., 2017. Antagonism in Rhizobacteria: application for biocontrol of soil-borne plant pathogens. PSM Microbiol. 2 (3), 78–79.

Iqbal, M.N., Ashraf, A., 2019. *Trichoderma*: a potential biocontrol agent for soil borne fungal pathogens. Int. J. Mol. Microbiol. 2 (1), 22–24.

Islam, S., 2018. Microorganisms in the rhizosphere and their utilization in agriculture: a mini review. PSM Microbiol. 3 (3), 105–110.

Jayalakshmi, S.K., Raju, S., Usha Rani, S., Benagi, V.I., Sreeramulu, K., 2009. *Trichoderma harzianum* L1 as a potential source for lytic enzymes and elicitor of defense responses in chickpea (*Cicer arietinum* L.) against wilt disease caused by *Fusarium oxysporum* f. sp. *ciceri*. Aust. J. Crop. Sci. 3 (1), 44–52.

Jogaiah, S., Abdelrahman, M., Tran, L.S.P., Ito, S.I., 2018. Different mechanisms of *Trichoderma virens*-mediated resistance in tomato against Fusarium wilt involve the jasmonic and salicylic acid pathways. Mol. Plant Pathol. 19 (4), 870–882. https://doi.org/10.1111/mpp.12571.

Junaid, J., Dar, M.N.A., Bhat, T.A., Bhat, A.H., Bhat, M.A., 2013. Commercial biocontrol agents and mechanism of action in the management of plant pathogens. J. Mod. Plant. Anim. Sci. 1 (2), 39–57.

Kareem, H.J., Al-Araji, A.M., 2017. Evaluation of *Trichoderma harzianum* biological control against *Fusarium oxysporum* f. sp. *melongenae*. J. Sci. 58 (4B), 2051–2060.

Kareem, T., Ugoji, O., Aboaba, O., 2016. Biocontrol of Fusarium wilt of cucumber with *Trichoderma longibrachiatum* NGJ167 (Rifai). Br. Microbiol. Res. J. 16 (5), 1–11. https://doi.org/10.9734/bmrj/2016/28208.

Kaur, N.P., Mukhopadhyay, A.N., 1992. Integrated control of 'chickpea wilt complex' by *Trichoderma* and chemical methods in India. Trop. Pest Manag. 38 (4), 372–375. https://doi.org/10.1080/09670879209371730.

Keswani, C., Bisen, K., Singh, V., Sarma, B.K., Singh, H.B., 2015. Formulation technology of biocontrol agents: present status and future prospects. Bioform. Sustain. Agric. 35–52.

Khan, R.A.A., Najeeb, S., Mao, Z., Ling, J., Yang, Y., Li, Y., Xie, B., 2020. Bioactive secondary metabolites from *Trichoderma* spp. against phytopathogenic bacteria and root-knot nematode. Microorganisms 8 (3), 401. https://doi.org/10.3390/microorganisms8030401.

Khare, E., Kumar, S., Kim, K., 2018. Role of peptaibols and lytic enzymes of *Trichoderma cerinum* Gur1 in biocontrol of *Fusraium oxysporum* and chickpea wilt. Environ. Sustain. 1, 39–47.

Khatri, D.K., Tiwari, D.N., Bariya, H.S., 2017. Chitinolytic efficacy and secretion of cell wall-degrading enzymes from *Trichoderma* spp. in response to phytopathological fungi. J. Appl. Biol. Biotechnol. 5 (6), 1–8.

Konappa, N., Krishnamurthy, S., Siddaiah, C.N., Ramachandrappa, N.S., Chowdappa, S., 2018. Evaluation of biological efficacy of *Trichoderma asperellum* against tomato bacterial wilt caused by *Ralstonia solanacearum*. Egypt. J. Biol. Pest. Co. 28 (1), 1–11. https://doi.org/10.1186/s41938-018-0069-5.

Kumar, S., Singh, A., 2015. Biopesticides: present status and the future prospects. J. Fertil. Pestic. 6 (2), 100–129.

Kumar, D.P., Thenmozhi, R., Anupama, P.D., Nagasathya, A., Thajuddin, N., Paneerselv, A., 2011. Selection of potential antagonistic *Bacillus* and *Trichoderma* isolates from tomato rhizospheric soil against *Fusarium oxysporum* f.sp. *lycoperscisi*. Res. J. Biol. Sci. 6 (10), 523–531. https://doi.org/10.3923/rjbsci.2011.523.531.

Kumar, G., Maharshi, A., Patel, J., Mukherjee, A., Singh, H.B., Sharma, B.K., 2017. *Trichoderma*: a potential fungal antagonist to control plant diseases. SATSA Mukhapatra Ann. Tech. Iss. 21, 206–218.

Lade, B.D., Gogle, D.P., Nandeshwar, S.B., 2017. Nano bio pesticide to constraint plant destructive pests. J. Nano Res. 6 (3), 00158. https://doi.org/10.15406/jnmr.2017.06.00158.

Lamichhane, J.R., Durr, C., Schwanck, A.A., Robin, M.H., Sarthou, J.P., Cellier, V., Messean, A., Aubertot, J.N., 2017. Integrated management of damping- off diseases. A review. Agron. Sustain. Devel. 37 (2), 25. Springer Verlag/EDP Sciences/INRA.

Leylaie, S., Zafari, D., 2018. Antiproliferative and antimicrobial activities of secondary metabolites and phylogenetic study of endophytic *Trichoderma* species from *Vinca* plants. Front. Microbiol. 9, 1484. https://doi.org/10.3389/fmicb.2018.01484.

Li, M., Ma, G., Hua, L., Su, X., Tian, Y., Huang, W.K., et al., 2019. The effects of *Trichoderma* on preventing cucumber Fusarium wilt and regulating cucumber physiology. J. Integr. Agric. 18 (3), 607–617. https://doi.org/10.1016/S2095-3119(18)62057-X.

Li, N., Wang, W., Bitas, V., Subbarao, K., Liu, X., Kang, S., 2018. Volatile compounds emitted by diverse *Verticillium* species enhance plant growth by manipulating auxin signaling. Mol. Plant Microbe Interact. 31 (10), 1021–1031. https://doi.org/10.1094/MPMI-11-17-0263-R.

Lindermayr, C., 2018. Crosstalk between reactive oxygen species and nitric oxide in plants: key role of S-nitrosoglutathione reductase. Free Radic. Biol. Med. 122, 110–115. https://doi.org/10.1016/j.freeradbiomed.2017.11.027.

Lugtenberg, B., 2015. Life of microbes in the rhizosphere. In: Lugtenberg, B. (Ed.), Principles of Plant-Microbe Interactions. Springer International Publishing Switzerland, Heidelberg, pp. 7–15.

Lugtenberg, B., Chin-A-Woeng, T., Bloemberg, G., 2002. Microbe-plant interactions: principles and mechanisms. Antonie Leeuwenhoek 81, 373–383.

Lyons, N., Cruz, L., Santos, M.S., 2001. Rapid field detection of *Ralstonia solanacearum* in infected tomato and potato plants using the *Staphylococcus aureus* slide agglutination test. Bull. OEPP 31, 91–93.

Macías–Rodríguez, L., Guzmán–Gómez, A., García–Juárez, P., Contreras– Cornejo, H.A., 2018. *Trichoderma atroviride* promotes tomato development and alters the root exudation of carbohydrates, which stimulates fungal growth and the biocontrol of the phytopathogen *Phytophtora cinnamomic* in a tripartite interaction system. FEMS Microbiol. Ecol. 94.

Majeed, M., Hassan, M.G., Hassan, M., Mohuiddin, F.A., Paswal, S., Farooq, S., 2017. Damping off in chilli and its biological management-A review. J. Curr. Microbiol. App. Sci. 7 (4), 2175–2185.

Maketon, M., Apisitsantikul, J., Siriraweekul, C., 2008. Greenhouse evaluation of *Bacillus subtilis* AP-01 and *Trichoderma harzianum* AP-001 in controlling tobacco diseases. Braz. J. Microbiol. 39 (2), 296–300. https://doi.org/10.1590/S1517-83822008000200018.

Manganiello, G., Sacco, A., Ercolano, M.R., Vinale, F., Lanzuise, S., Pascale, A., Napolitano, M., Lombardi, N., Lorito, M., Woo, S.L., 2018. Modulation of tomato response to *Rhizoctonia solani* by *Trichoderma harzianum* and its secondary metabolite harzianic acid. Front. Microbiol. 9, 1966. https://doi.org/10.3389/fmicb.2018.01966.

Markovich, N.A., Kononova, G.L., 2003. Lytic enzymes of *Trichoderma* and their role in plant defense from fungal diseases: a review. J. Appl. Biochem. Microbiol. 39, 389–400.

Martínez-Medina, A., Fernández, I., Sánchez-Guzmán, M.J., Jung, S.C., Pascual, J.A., Pozo, M.J., 2013. Deciphering the hormonal signalling network behind the systemic resistance induced by *Trichoderma harzianum* in tomato. Front. Plant Sci. 4 (206).

Mendoza-Mendoza, A., et al., 2018. Molecular dialogues between *Trichoderma* and roots: role of the fungal secretome. Fungal Biol. Rev. 32, 62–85.

Mohiddin, F.A., Bhat, F.A., Bhat, K.A., Bhat, Z.A., Bhat, M.A., Hamid, B., 2018. Development of *Trichoderma* based bio-formulations for the management of chilli wilt. J. Pharmacogn. Phytochem. 7 (1), 2118–2122.

Mondejar, R.L., Ros, M., Pascual, J.A., 2011. Mycoparasitism-related genes expression of Trichoderma harzianum isolates to evaluate their efficacy as biological control agent. Biol. Control. 56, 59–66. https://doi.org/10.1016/j.biocontrol.2010.10.003.

Murthy, K.N., Srinivas, C., 2012. In vitro screening of bioantagonistics agents and plant extracts to control bacterial wilt (*Ralstonia solanacearum*) of tomato (*Lycopersicon esculentum*). Int. J. Agri. Technol. 8 (3), 999–1015.

Murthy, K.N., Srinivas, C., 2013. Efficacy of *Trichoderma asperellum* against *Ralstonia solanacearum* under greenhouse conditions. Ann. Plant Sci. 2, 342–350.

Murthy, K.N., Uzma, F., Srinivas, C., 2013. Induction of systemic resistance by *Trichoderma asperellum* against bacterial wilt of tomato caused by *Ralstonia solanacearum*. Int. J. Adv. Res. 1, 181–194.

Naher, L., Yusuf, U., Ismail, A., Hossain, K., 2014. *Trichoderma* spp.: a biocontrol agent for sustainable management of plant diseases. Pakistan J. Bot. 46 (4), 1489–1493.

Naik, M.K., Madhukar, H.M., Devika, Rani, G.S., 2009. Evaluation of biocontrol efficacy of trichoderma isolates and methods of its application against wilt of chilli (capsicum annuum l.) Caused by Fusarium solani (Mart) Sacc. J. Biol. Control. 23, 31–36.

Nicolopoulou-Stamati, P., Maipas, S., Kotampasi, C., Stamatis, P., Hens, L., 2016. Chemical pesticides and human health: the urgent need for a new concept in agriculture. Front. Pub. Health 4 (148). https://doi.org/10.3389/fpubh.2016.00148.

Nirmaladevi, D., Venkataramana, M., Srivastava, R.K., Uppalapati, S.R., Gupta, V.K., Yli-Mattila, T., et al., 2016. Molecular phylogeny, pathogenicity and toxigenicity of *Fusarium oxysporum* f. sp. *Lycopersici*. Sci. Rep. 6, 21367. https://doi.org/10.1038/srep21367.

Omann, M.R., Lehner, S., Escobar, R.C., Brunner, K., Zeilinger, S., 2012. The seven-transmembrane receptor Gpr1 governs processes relevant for the antagonistic interaction of *Trichoderma atroviride* with its host. J. Microbiol. 158, 107–118.

Ommati, F., Zaker, M., 2012. In vitro and greenhouse evaluations of *Trichoderma* isolates for biological control of potato wilt disease (*Fusarium solani*). Arch. Phytopathol. Plant Protect. 45 (14), 1715–1723. https://doi.org/10.1080/03235408.2012.702467.

Oladipo, O.G., Ezeokoli, O.T., Maboeta, M.S., Bezuidenhout, J.J., Tied, t L.R., Jordaan, A., Bezuidenhout, C.C., 2018. Tolerance and growth kinetics of bacteria isolated from gold and gemstone mining sites in response to heavy metal concentrations. J. Envi. Manag. 212, 357–366.

Oros, G., Naár, Z., 2017. Mycofungicide: *Trichoderma* based preparation for foliar applications. Am. J. Plant Sci. 8, 113–125.

Pandey, V., Ansari, M.W., Tula, S., Yadav, S., Sahoo, R.K., Shukla, N., et al., 2016. Dose-dependent response of *Trichoderma harzianum* in improving drought tolerance in rice genotypes. Planta 243, 1251–1264.

Pandya, J.R., Sabalpara, A.N., Chawda, S.K., 2011. *Trichoderma*: a particular weapon for biological control of phytopathogens. J. Agric. Technol. 7, 1187–1191.

Pataky, J.K., Michener, P.M., Freeman, N.D., Weinzierl, R.A., Teyker, R.H., 2000. Control of Stewart's wilt in sweet corn with seed treatment insecticides. Plant Dis. 84, 1104–1108.

Patel, S., Saraf, M., 2017. Biocontrol efficacy of *Trichoderma asperellum* MSST against tomato wilting by *Fusarium oxysporum* f. sp. Lycopersici. Arch. Phytopathol. Plant Protect. 50 (5–6), 228–238. https://doi.org/10.1080/03235408.2017.1287236.

Pieterse, C.M.J., Leon-Reyes, A., Van der Ent, S., Van Wees, S.C.M., 2009. Networking by small-molecule hormones in plant immunity. Nat. Chem. Biol. 5, 308–316.

Prasad, R.D., Rangeshwaran, R., Hegde, S.V., Anuroop, C.P., 2002. Effect of soil and seed application of *Trichoderma harzianum* on pigeonpea wilt caused by *Fusarium udum* under field condition. Crop Protect. 21 (4), 293–297. https://doi.org/10.1016/S0261-2194(01)00100-4.

Ramasamy, P., Sundaram, L., 2020. Biocontrol potential of *Trichoderma* spp. against *Fusarium oxysporum* in *Solanum lycopersicum* L. Asian J. Pharmaceut. Clin. Res. 13 (5), 156–161.

Ravikumara, B.M., Naik, M.K., Sharma, M., Sunkad, G., et al., 2017. Induced systematic resistance and evaluation of bio-control agents for management of Pigeonpea wilt caused by *Fusarium udum*. J. Pure Appl. Microbiol. 11 (1), 291–305.

Sallam, N.M.A., Eraky, A.M.I., Sallam, A., 2019. Effect of *Trichoderma* spp. on Fusarium wilt disease of tomato. Mol. Biol. Rep. 46, 4463–4470.

Sant, D., Casanova, E., Segarra, G., Avilés, M., Reis, M., Trillas, M.I., 2010. Effect of *Trichoderma asperellum* strain T34 on Fusarium wilt and water usage in carnation grown on compost-based growth medium. Biol. Contr. 53 (3), 291–296. https://doi.org/10.1016/j.biocontrol.2010.01.012.

Saravanakumar, D., Ciavorella, A., Spadaro, D., Garibaldi, A., Gullino, L.M., 2008. Metschnikowia pulcherrima strain MACH1 out competes Botrytis cinerea, Alternaria alternata and Penicillium expansum in apples through iron depletion. Postharvest Biol. Technol. 49, 121–128.

Saravanakumar, K., Yu, C., Dou, K., Wang, M., Li, Y., Chen, J., 2015. Synergistic effect of Trichoderma-derived antifungal metabolites and cell wall degrading enzymes on enhanced biocontrol of Fusarium oxysporum F. Sp. Cucumerinum. Biolog. Contr. https://doi.org/10.1016/j.biocontrol.2015.12.001.

Saravanakumar, K., Wang, S., Dou, K., Lu, Z., Chen, J., 2018. Yest two–hybrid and label–free proteomics–based screening of maize root receptor to cellulase of *Trichoderma harzianum*. Physiol. Mol. Plant Pathol. 104, 86–94.

Scarselletti, R., Faull, J.L., 1994. In vitro activity of 6-pentyl-U-pyrone, a metabolite of *Trichoderma harzianum*, in the inhibition of *Rhizoctonia solani* and *Fusarium oxysporum* f. sp. lycopersici. Mycol. Res. 98 (10), 1207–1209.

Seethapathy, P., Kurusamy, R., Kuppusamy, P., 2017. Soil borne diseases of major pulses and their biological management. An Int. J. Agri. 2 (1), 1–11.

Segarra, G., Casanova, E., Aviles, M., Trillas, I., 2010. *Trichoderma asperellum* strain T34 controls Fusarium wilt disease in tomato plants in soilless culture through competition for iron. Microb. Ecol. 59, 141–149.

Seidl, V., Huemer, B., Seiboth, B., Kubicek, C.P., 2005. A complete survey of *Trichoderma* chitinases reveals three distinct subgroups of family 18 chitinases. FEBS J. 272, 5923–5939.

Shahid, M., Srivastava, M., Pandey, S., Kumar, V., Singh, A., Trivedi, S., et al., 2015. Management of Fusarium wilt using mycolytic enzymes produced by *Trichoderma harzianum* (Th. Azad). Afr. J. Biotechnol. 14 (38), 2748–2754.

Shaw, S., Le Cocq, K., Paszkiewicz, K., Moore, K., Winsbury, R., de Torres Zabala, M., et al., 2016. Transcriptional reprogramming underpins enhanced plant growth promotion by the biocontrol fungus *Trichoderma hamatum* GD12 during antagonistic interactions with *Sclerotinia sclerotiorum* in soil. Mol. Plant Pathol. 17, 1425–1441. https://doi.org/10.1111/mpp.12429.

Sharfuddin, C., Mohanka, R., 2012. In vitro antagonism of indigenous Trichoderma isolates against phytopathogen causing wilt of lentil. Int. J. Life Sci. Pharm. Res. 2, 195–202.

Shi, M., Chen, L., Wang, X.W., Zhang, P., Zhao, B., Song, X.Y., et al., 2012. Antimicrobial peptaibols from *Trichoderma pseudokoningii* induce programmed cell death in plant fungal pathogens. Microbiology 158, 166.

Simon, A., Dunlop, R.W., Ghissalberti, E.L., Sivasithamparam, 1988. *Trichoderma koningii* produces a pyrone compound with antibiotic properties. Soil Biol. Biochem. 20, 263–264.

Singh, P.K., Kumar, V., 2011. Biological control of Fusarium wilt of Chrysanthemum with *Trichoderma* and botanicals. Int. J. Agr. Technol. 7 (6), 1603–1613. http://www.ijat-aatsea.com/pdf/v7_n6_11_November/13_IJAT_2011_7_6__Pawan_Kumar_Singh_FX_confirmed.pdf.

Singh, H.B., Singh, A., Sarma, B.K., Upadhyay, D.N., 2014. *Trichoderma viride* 2% WP (Strain No. BHU–2953) formulation suppresses tomato wilt caused by *Fusarium oxysporum* f. sp. *lycopersici* and chilli damping–off caused by *Pythium aphanidermatum* effectively under different agroclimatic conditions. Int. J. Agric. Environ. Biotechnol. 7, 313–320.

Singh, R., Biswas, S.K., Nagar, D., Singh, J., Singh, M., Mishra, Y.K., 2015. Sustainable integrated approach for management of Fusarium wilt of tomato caused by *Fusarium oxysporum* f. sp. lycopersici (sacc.) synder and hansen. Sustain. Agric. Res. 4 (1), 138. https://doi.org/10.5539/sar.v4n1p138.

Singh, A., Shukla, N., Kabadwal, B.C., Tewari, A.K., Kumar, J., 2018. Review on plant *Trichoderma*-pathogen interaction. Int. J. Curr. Microbiol. App. Sci. 7 (02), 2382–2397. https://doi.org/10.20546/ijcmas.2018.702.291.

Sivan, A., Chet, I., 1986. Biological control of *Fusarium* spp. in cotton, wheat and muskmelon by *Trichoderma harzianum*. J. Phytopathol. 116 (1), 39–47. https://doi.org/10.1111/j.1439-0434.1986.tb00892.x.

Sivan, A., Chet, I., 1989. The possible role of competition between *Trichoderma harzianum* and *Fusarium oxysporum* on rhizosphere colonization. Phytopathology 79, 198–203.

Siyar, S., Inayat, N., Hussain, F., 2019. Plant growth promoting Rhizobacteria and plants' improvement: a mini-review. PSM Biol. Res. 4 (1), 1–5.

Song, X.Y., Shen, Q.T., Xie, S.T., Chen, X.L., Sun, C.Y., Zhang, Y.Z., 2006. Broad-spectrum antimicrobial activity and high stability of trichokonins from *Trichoderma koningii* SMF2 against plant pathogens. FEMS Microbiol. Lett. 260, 119–125.

Spadaro, D., Droby, S., 2016. Development of biocontrol products for postharvest diseases of fruit: the importance of elucidating the mechanisms of action of yeast antagonists. Trends Food Sci. Technol. 47, 39–49. https://doi.org/10.1016/j.tifs.2015.11.003.

Sreenu, B., Zacharia, S., 2017. In vitro screening of plant extracts, *Trichoderma harzianum* and carbendazim against *Fusarium oxysporium* f. sp. lycopersici on tomato. Int. J. Curr. Microbiol. Appl. Sci. 6 (8), 818–823. https://doi.org/10.20546/ijcmas.2017.608.103.

Srivastava, R., Khalid, A., Singh, U.S., Sharma, A.K., 2010. Evaluation of arbuscular mycorrhizal fungus, fluorescent Pseudomonas and *Trichoderma harzianum* formulation against *Fusarium oxysporum* f. sp. lycopersici for the management of tomato wilt. Biol. Contr. 53 (1), 24–31. https://doi.org/10.1016/j.biocontrol.2009.11.012.

Srivastava, M., Shahid, M., Pandey, S., Singh, A., Kumar, V., Gupta, S., et al., 2014. *Trichoderma* genome to genomics: a review. J. Data Min. Genom. Proteonomics 5, 1000172.

Sundaramoorthy, S., Balabaskar, P., 2013. Biocontrol efficacy of *Trichoderma* spp. against wilt of tomato caused by *Fusarium oxysporum* f. sp. lycopersici. J. Appl. Biol. Biotechnol. 1 (03), 36–040. https://doi.org/10.7324/JABB.2013.1306.

Surekha, C.H., Neelapu, N.R.R., Siva Prasad, B., Sankar Ganesh, P., 2014. Induction of defense enzymes and phenolic content by *Trichoderma viride* in *Vigna mungo* infested with *Fusarium oxysporum* and *Alternaria alternata*. Indian J. Agric. 4 (4), 31–40.

Thangavelu, R., Gopi, M., 2015. Field suppression of Fusarium wilt disease in banana by the combined application of native endophytic and rhizospheric bacterial isolates possessing multiple functions. Phytopathol. Mediterr. 54 (2), 241–252.

Thangavelu, R., Mustaffa, M., 2010. A potential isolate of *Trichoderma viride* NRCB1 and its mass production for the effective management of Fusarium wilt disease in banana. Tree For. Sci. Biotechnol. 4, 76–84.

Thangavelu, R., Palaniswami, A., Velazhahan, R., 2004. Mass production of *Trichoderma harzianum* for managing Fusarium wilt of banana. Agric. Ecosyst. Environ. 103 (1), 259–263. https://doi.org/10.1016/j.agee.2003.09.026.

Tucci, M., Ruocco, M., De Masi, L., De Palma, M., Lorito, M., 2011. The beneficial effect of *Trichoderma* spp. on tomato is modulated by the plant genotype. Mol. Plant Pathol. 12, 341–354.

van Lenteren, J.C., Bolckmans, K., Köhl, J., Ravensberg, W.J., Urbaneja, A., 2018. Biological control using invertebrates and microorganisms: plenty of new opportunities. BioControl 63, 39–59. https://doi.org/10.1007/s10526-017-9801-4.

Vinale, F., Nigro, M., Sivasithamparam, K., Flematti, G., Ghisalberti, E.L., Ruocco, M., et al., 2013. Harzianic acid: a novel siderophore from *Trichoderma harzianum*. FEMS Microbiol. Lett. 347, 123–129.

Viterbo, A., Montero, M., Ramot, O., Friesem, D., Monte, E., Llobell, A., et al., 2002. Expression regulation of the endochitinase chit36 from *Trichoderma asperellum* (*T. harzianum* T-203). Curr. Genet. 42, 114–122.

Yadeta, K.A., Thomm, B.P., 2013. The xylem as battleground for plant hosts and vascular wilt pathogens. Front. Plant Sci. 4, 1–12.

Yendyo, S., Ramesh, G.C., Pandey, B.R., 2017. Evaluation of *Trichoderma* spp., *Pseudomonas fluorescence* and *Bacillus subtilis* for biological control of Ralstonia wilt of tomato. F1000Research 6 (0), 2028. https://doi.org/10.12688/f1000research.12448.1.

Yuan, S., Li, M., Fang, Z., Liu, Y., Shi, W., Pan, B., et al., 2016. Biological control of tobacco bacterial wilt using *Trichoderma harzianum* amended bioorganic fertilizer and the arbuscular mycorrhizal fungi *Glomus mosseae*. Biol. Contr. 92, 164–171. https://doi.org/10.1016/j.biocontrol.2015.10.013.

Zaidi, N.W., Singh, U.S., 2018. *Trichoderma* an impeccable plant health booster. In: MA, A. (Ed.), Biopesticides and Bioagents: Novel Tools for Pest Management. Apple Academic Press, USA, pp. 17–42.

Zehra, A., Meena, M., Dubey, M.K., Aamir, M., Upadhyay, R.S., 2017. Synergistic effects of plant defense elicitors and *Trichoderma harzianum* on enhanced induction of antioxidant defence system in tomato against Fusarium wilt disease. Bot. Stud. 58 (44).

Zhang, F., Yang, X., Ran, W., Shen, Q., 2014. *Fusarium oxysporum* induces the production of proteins and volatile organic compounds by *Trichoderma harzianum* T-E5. FEMS Microbiol. Lett. 359, 116–123.

Zhang, J., Wu, C., Wang, W., Wang, W., Wei, D., 2018. A versatile *Trichoderma reesei* expression system for the production of heterologous proteins. Biotechnol. Lett. 40 (6), 965–972.

Chapter 19

Biological inoculants and biopesticides in small fruit and vegetable production in California

Surendra K. Dara
University of California Cooperative Extension, San Luis Obispo, CA, United States

Biopesticides based on botanical extracts, entomopathogens, or other beneficial microorganisms have been available for several decades for controlling arthropod pests or diseases around the world. Botanical and microbial inoculants and soil amendments have also been used in multiple cropping systems for promoting plant growth, improving water or nutrient uptake, or resisting abiotic stressors. The use of biostimulants and biopesticides has seen a steady growth in the recent years due to the increased need for sustainable crop production practices and increased scientific interest in exploring their efficacy and role in crop production and protection. Biostimulants and biopesticides belong to different categories and are regulated differently. While biostimulants are meant to stimulate natural processes for improving plant growth and tolerating abiotic stresses (Du Jardin, 2015), biopesticides are meant to manage biotic stressors such as arthropod pests, pathogens, or weeds (EPA, 2016). However, several biostimulants induce resistance in plants to abiotic as well as biotic stressors and directly or indirectly contribute to crop protection beyond their agronomic benefit. For example, in addition to promoting plant growth and yields, the extract of *Ascophyllum nodosum* in tomato and sweet pepper reduced foliar diseases caused by *Xanthomonas campestris* pv. *vesicatoria* and *Alternaria solani* (Ali et al., 2019). The extract of *Inula viscosa* (a bushy flowering plant of Asteraceae) not only promoted the growth and flowering of succulents *Oscularia deltoides* and *Corpuscolaria lehmanii*, but also controlled the oleander aphid, *Aphis nerii* (Domenico, 2019). In addition to its biostimulation properties, seaweed extract stimulates plant defenses against pests and diseases (Mukherjee and Patel, 2020). Arbuscular fungi such as *Glomus* spp. and plant growth-promoting fungi such as *Trichoderma* spp. are marketed both as biostimulants and biofungicides (Szczałba et al., 2019). Plant growth promoting rhizobacteria *Bacillus subtilis* and *Bacillus amyloliquefaciens* are also used for disease management (Yu et al., 2011; Yuan et al., 2013). Similarly, some biopesticide active ingredients have biostimulation properties and could be benefiting plants in multiple ways. *Bacillus thuringiensis*, a popular entomopathogenic bacterium, also has plant growth-promoting properties (Azizoglu, 2019). Similarly, entomopathogenic fungi can promote plant growth and antagonize pathogens in addition to suppressing arthropod pests (Dara, 2019a). With this new knowledge of multifaceted interactions of biostimulants and biopesticides with plants, several of these products could be used as holistic tools for improving crop growth, health, and yield. Several recent studies in California explored the efficacy of various biologicals in small fruits and vegetables for managing pests and diseases or improving plant growth and yields. This chapter will summarize those studies and suggest strategies that can be applicable to multiple crops to promote sustainable food production.

19.1 Bioinoculants in strawberry

Multiple field studies evaluated various products based on bacteria, fungi, worm extract, seaweed extract, botanical extracts, and other materials which were applied as transplant treatments or periodical soil or foliar applications. The objective of these studies was to evaluate the impact of these materials on improving crop growth, health, and yield. Application rates, frequency, timing of the products, soil and environmental conditions, soil microbial diversity, agronomic and pest management practices including fumigation, and other factors influence soil-plant-microbial interactions.

Biopesticides. https://doi.org/10.1016/B978-0-12-823355-9.00001-8

While specific interactions were not investigated, depending on the scope of the study, various parameters, including the canopy growth, chlorophyll content, fruit yield and quality, disease occurrence in the field and postharvest storage, were measured in these studies.

In 2013, a commercial formulation of the entomopathogenic fungus *Beauveria bassiana* (strain GHA) was compared against a microbial inoculant of unknown composition in strawberries grown in raised beds (Dara, 2013a). While fruit yield could not be measured, *B. bassiana*-treated plants had better health than untreated or those treated with the other microbial product during two months of observation. Based on these results, another study was conducted in a commercial strawberry field during the 2013–2014 production season where periodic soil application of *B. bassiana* (GHA) showed no impact on the plant growth or fruit yield (Dara, 2016a,b). At the same commercial field, another study was conducted during the 2014–2015 production season where commercial formulations of three entomopathogenic fungi, *B. bassiana* (GHA), *Cordyceps fumosorosea* (strain FE9901), and *Metarhizium brunneum* (strain F52), along with other products based on *Streptomyces lydicus* (strain WYEC 108); a consortium of *B. amyloliquefaciens, Bacillus licheniformis, Bacillus magaterium, Bacillus pumilus, B. subtilis, Trichoderma harzianum*, and *Trichoderma reesei*; and another consortium of *Azotobacter chroococcum, Azospirillum lipoferum, Cellulomonas cellulans, Lactobacillus acidophilus, Pseudomonas fluorescens*, and *Aspergillus niger* were compared with untreated control and the grower standard practice of using a humic acid-based product (Dara and Peck, 2016). A product containing hydrogen dioxide and peroxyacetic acid was also used alone and in combination with the product containing *Bacillus* spp. and *Trichoderma* spp. Using hydrogen dioxide and peroxyacetic acid with beneficial microbes is to suppress pathogenic organisms with the former and repopulate the soil with beneficial microbes which are present in the latter. There were no significant differences ($P > .05$) in any of the measured parameters, but marketable yield was numerically higher from many microbial treatments compared to the grower standard. However, fruit yield in untreated control was also numerically higher than the grower standard and some other treatments. This field has been fumigated with chemical fumigants every year, and untreated control did not receive any fertilizers or bioinoculants after transplanting except for pre-plant fertilizers during field preparation.

In a follow up study in summer-planted strawberries in 2016 at the same field, another set of microbial products was compared with untreated control and the grower standard of humic acid-based product and transplant dip in a fungicide containing cyprodinil and fludioxonil (Dara and Peck, 2017). Treatments included a product containing *Glomus aggregatum, Glomus etunicatum, Glomus intraradices*, and *Glomus mosseae* applied as a transplant dip and/or through drip application at different rates and frequencies, *S. lydicus* (WYEC 108), and another product containing *B. subtilis* and *Saccharomyces cerevisiae*. Some of these treatments resulted in improved canopy growth and reduced powdery mildew (*Podosphaera aphanis*) ($P < .03$), but there was no impact on postharvest occurrence of *Botrytis cinerea* or the marketable fruit yield ($P > .05$). However, the grower standard had significantly ($P = .04$) higher number of dead or dying plants from an unknown cause on a single observation date five months after transplanting. Marketable yield was numerically higher in untreated control and some of the treatments compared to the grower standard or the remaining treatments. A final study conducted at this field during the 2017–2018 production season included the following treatments: (i) untreated control, (ii) grower standard fertilizer program with transplant dip in cyprodinil + fludioxonil, (iii) a product containing a blend of polyhydroxy carboxylic acids and a product containing carboxylic acids with calcium and boron, (iv) a microbe-rich worm extract, (v) multiple products based on one or more of marine algae, seaweed extract, silicon, phosphorus, and potassium, (vi) a silica fertilizer, botanical extract from undisclosed wild plants, and a consortium of *Azotobacter* spp., *Bacillus* spp., *Paenibacillus* spp., *Pseudomonas* sp., *Trichoderma* spp. and *Streptomyces* spp., (vii) and (viii) lower and higher rates of a product containing *Citrobacter freundii, Comamonas testosterone, Enterobacter cloacae* and *Pseudomonas putida*, and alfalfa extract, respectively (ix) a biodegradable fertilizer additive containing thermal polyaspartate, and (x) – (xiii) an experimental blend of beneficial microbes at different rates alone, and with a consortium of *B. amyloliquefaciens, B. lichenoformis, B. pumilus*, and *B. subtilis* (Dara and Peck, 2018). These treatments were applied as transplant dips, through drip irrigation, or as foliar sprays as required by the manufacturer. While various measured parameters did not show statistically significant differences, numerical differences were seen in marketable fruit yields from some treatments. Untreated control continued to perform better than the grower standard as in the previous studies. When untreated control was excluded from the analysis, some treatments had significantly ($P = .03$) higher yield than the grower standard. While the fertilizer additive, which is not a biological product, resulted in a 15.1% increase, yield improvement varied anywhere from 2.7% in the treatment with the worm extract to 16.2% in the treatment with marine algae and seaweed extract, compared to the grower standard.

During the 2017–2018 production season, a field study was conducted in strawberry at a research station comparing a product containing glycine betaine with the grower standard and two fertilizer programs (Dara, 2019b). Only one of the fertilizer programs had numerical increase in the marketable fruit yield and glycine betaine had no impact on yield improvement. During the following production season of 2018–2019, another study was conducted at this research station, where the grower standard program was supplemented with botanical, microbial, and organic acid products in the following treatments: (i) cold-pressed neem, (ii) and (iii) low and high rates of *B. amyloliquefaciens*, *T. harzianum* and humic acid, (iv) multiple products with *B. bassiana*, botanical extracts, or humic acids, and (v) products with fertilizers or organic acids (Dara, 2019c). There were no statistically significant ($P > .05$) differences among the measured parameters, but there was a numerical increase in marketable fruit yield varying from 9.8% to 35.7% among various treatments compared to the grower standard alone. Another unpublished study conducted during the same time next to this study showed a numerical increase in the marketable fruit yield from two rates of 24-epibrassinolide and a reduction from the application of a chitin product.

19.2 Bioinoculants in tomato

Two field studies were conducted at a research station during the summer of 2017 (Dara and Lewis, 2019) and late spring to fall of 2018 (Dara, 2019d) evaluating various nutrient and biostimulant materials in processing tomatoes. Treatments in the 2017 study were applied as supplements to the grower standard and included (i) a product containing potassium silicate, (ii) a product containing a consortium of *C. freundii*, *C. testosterone*, *E. cloacae*, and *P. putida*, and alfalfa extract, (iii) multiple products containing polyhydroxy carboxylic acids, carboxylic acids, calcium and boron, and a consortium of *A. chroococcum*, *Bacillus megaterium*, *Bacillus mycoides*, *B. subtilis* and *T. harzianum*, and (iv) multiple products providing macro and micronutrients. Statistically significant ($P > .05$) differences in seasonal yields were also not seen, but there was a numerical increase of 27.1% from potassium silicate, 29.9% from the multi-product treatment containing carboxylic acids and microbes, and 31.7% from the microbial consortium along with alfalfa extract. However, the treatment that received multiple macro and micronutrients had a 7.4% decrease in tomato yields compared to the grower standard. The study conducted in 2018 contained (i) grower standard program, (ii) grower standard supplemented with a product containing diammonium phosphate along with *B. amyloliquefaciens*, *B. licheniformis*, *B. pumilus*, and *B. subtilis*, (iii) grower standard program at 85% supplemented with the diammonium phosphate and microbial product, and (iv) grower standard program supplemented with a product containing tree extracts and potassium hydroxide. Statistically significant ($P = .04$) improvement was seen in the seasonal tomato yields where the diammonium phosphate with microbial blend had an 8% increase at the full rate of the grower standard program and 13.2% at the 85% rate of the grower standard program. The product with tree extracts resulted in a 26.6% increase compared to the grower standard alone. There were no noticeable disease issues during the studies to measure the impact of the treatments. Whitefly infestations occurred during both years, but they were neither monitored nor pesticide applications were made.

These strawberry and tomato studies suggest that biostimulants and other biological agricultural inputs can have a positive impact on yield, but they did not result in statistically significant improvements in a consistent manner. Variations in the efficacy of products in these studies could be due to the plant species, varieties, soil and environmental conditions, and other complex interactions among numerous biotic and abiotic factors. It is also important to note that when the farming practices generally good or the crop is not under any biotic or abiotic stress, impact of the biological treatments can be insignificant or less pronounced. Klokić et al. (2020) reported that biostimulants based on amino acids and humic acids improved tomato yields under a reduced fertilizer regimen helped prevent yield loss for only one of the two varieties. Among various impacts on different kinds of crops, biostimulants maintained yield and quality of vegetables under reduced fertilizer regimens (Bulgari et al., 2014). Dong et al. (2020) compared various biostimulants in strawberry and tomato where they noticed better responses in tomato than in strawberry. Under controlled conditions, strawberry plants showed different responses to the plant growth promoting bacteria, *Enterobacter rooggenkampii* and *Klebsiella pneumoniae*, for the yield and sugar content of the fruit (Pérez-Fragero et al., 2020).

19.3 Biopesticides in strawberry and grapes

From 2012 to 2015 a study was conducted each year in commercial strawberry fields with various chemical, botanical, and microbial (based on entomopathogenic fungi) pesticides in different combinations and rotations to develop effective management strategies for the western tarnished plant bug, *Lygus hesperus*, and other pests (Dara, 2016b; Dara et al., 2018a,b). Variation in the efficacy or lack of significant differences was seen with both chemical and non-chemical alternatives from year to year, but certain combinations of microbial and botanical insecticides appeared to be good substitutes for certain chemical pesticides either for reducing pest numbers or preventing their increase. These studies also indicate that field efficacy is quite variable and influenced by multiple factors but helped California strawberry growers understand the potential of microbial control and develop their own integrated pest management (IPM) strategies with one or more options evaluated in these studies.

A laboratory assay conducted in 2018 against a re-emerging grape pest, the western grapeleaf skeletonizer, *Harrisina metallica*, with biopesticides provided more promising results (Dara et al., 2019a). *Harrisina metallica*, first discovered in 1941, had been under control from the natural infections of *Harrisina brillians* granulovirus (Stern and Federici, 1990). However, it started re-emerging in some areas and is known to be a problem, especially in organic vineyards. The laboratory assay evaluated commercial formulations of azadirachtin, spinosad, *Bacillus thuringiensis* subsp. *aizawai* (Bta), and *B. thuringiensis* subsp. *kurstaki* (Btk), and unformulated California isolates of *B. bassiana* (ARSEF 8318) and *Metarhizium anisopliae* s.l. (ARSEF 8319) which resulted in 78.3%, 100%, 73.3%, 45%, 81.3%, and 100% mortality, respectively, in fourth-fifth instar larvae.

19.4 Biopesticides in vegetables

Multiple field studies were conducted for controlling insect pests in broccoli, celery, lettuce, and zucchini evaluating chemical and biological pesticides (Dara, 2013b, 2015; Dara et al., 2018a,b). While the zucchini study was conducted a research station, the remaining were conducted in commercial fields. In the 2011 study, *B. bassiana* and multiple chemical pesticides (acetamiprid, spinetoram, tolfenpyrad, tolfenpyrad + flubendiamide, and tolfenpyrad + methomyl) were evaluated against the western flower thrips, *Frankliniella occidentalis*, an important pest in lettuce (Dara, 2013b). Thrips populations occurred at low numbers, but control from *B. bassiana* was comparable to some of the chemical treatments. In 2012, a similar study was conducted against the cabbage aphid, *Brevicoryne brassicae*, and the green peach aphid, *Myzus persicae*, in broccoli with *B. bassiana* and chemical pesticides (acetamiprid, pyrifluquinazon, sulfoxaflor, tolfenpyrad, and tolfenypard + flubendiamide). Aphid populations were not controlled by *B. bassiana* after the first spray application, but they were reduced by the end of the second application similar to many chemical pesticide treatments. During late 2014 and early 2015, a field study was conducted in organic celery where the rice root aphid, *Rhopalosiphum rufiabdominale*, and the honeysuckle aphid, *Hyadaphis foeniculi*, emerged as new pests, stunting and killing plants (Dara, 2015). When organically approved biopesticides based on azadirachtin, a rosemary and peppermint oil blend, *B. bassiana*, *Burkholderia rinojensis*, and *Chromobacterium subtsugae* were evaluated for their efficacy, only *B. rinojensis*, *C. subtsugae*, and the combination of *B. bassiana* and azadirachtin suppressed aphid numbers with 24.3%, 29.4%, and 61.6% reduction, respectively. Where *B. bassiana* was used alone, aphid numbers increased by 128.9%. Another study was conducted in zucchini during the summer of 2017 where *B. rinojensis*, *I. fumosorosea*, different rates of an experimental botanical pesticide, and GS-omega/kappa-Hxtx-Hv1a (spider venom peptide) were compared with chemical pesticides flupyradifurone and sulfoxaflor (Dara et al., 2018a,b). Arthropod pests monitored in the study included the pacific spider mite (*Tetranychus pacificus*), the silver leaf whitefly (*Bemisia tabaci*), *F. occidentalis*, and an unidentified species of aphids. Except for *B. tabaci*, other pest populations gradually declined by the end of the study. There was a significant ($P < .03$) difference in controlling both egg and nymphal stages of *B. tabaci* from some treatments during two spray applications. Flupyradifurone and sulfoxaflor reduced both egg and nymphal stages, while *B. rinojensis* and the botanical pesticide at two higher rates reduced the nymphal stages by the end of the second spray application. The spider venom peptide product reduced nymphs after the first spray and limited their increase after the second spray. While the two chemical pesticides provided good control of all the pests, biopesticides appeared to provide moderate to fair control, indicating their importance in IPM strategies.

Both sets of studies with bioinoculants and biopesticides evaluated several experimental or commercial products on multiple crops under field conditions over several years. What can be learned from these studies is that (i) bioinoculants and biopesticides can play an important role in improving yields or crop health, (ii) variability is common for both chemical and biological inputs, especially under field conditions, and can be influenced by multiple factors, (iii) efficacy of some biopesticides can be enhanced in combinations or rotations with other pesticides, and (iv) continuous experimentation is necessary to understand the efficacy of biologicals and to develop strategies for their effective use.

19.5 Non-entomopathogenic roles of hypocrealean entomopathogenic fungi

Several entomophthoralean fungi play a critical role in suppressing arthropod populations through naturally occurring epizootics (Dara and Humber, 2020), while mass production of hypocrealean fungi has led to their use as biopesticides around the world (Lacey, 2017). Research in the past two decades investigated additional roles of hypocrealean entomopathogenic fungi as endophytes directly or indirectly impacting herbivores, as mycorrhizae improving nutrient and water absorption and promoting plant growth, or as pathogen antagonizers helping in disease control (Dara, 2019a; Mantzoukas and Elipoulos, 2020). Hypocrealean entomopathogens have a strong association with plants and promote their growth and health in multiple ways. In multiple California studies, they promoted growth and improved nutrient use in cabbage under water stress (Dara et al., 2017a), antagonized *Fusarium oxysporum* f. sp. *vasinfectum* Race 4 in cotton (Dara et al., 2017b), and antagonized *Macrophomina phaseolina* (Dara et al., 2019b) and *Botrytis cinerea* (Dara, 2019e) in strawberry. Several other studies also support such findings and demonstrated unconventional roles of entomopathogenic fungi. For example, Litwin et al. (2020) showed that entomopathogenic fungi can be used to remove environmental pollutants such as endocrine disrupting compounds, xenoestrogens, synthetic estrogens, triazines, dibutyltin, hydrocarbons, industrial dyes, and heavy metals, as well as for the biosynthesis of nanoparticles and biotransformation of flavonoids and steroids. Root inoculation of entomopathogenic fungi increased the biomass, reduced twospotted spider mite (*Tetranychus urticae*) numbers, or lessened the symptoms caused by the foliar pathogens *Mycosphaerella fragariae* and *Pestalotia longisetula* in strawberry (Canassa et al., 2019, 2020).

Studies on alternative uses of hypocrealean entomopathogenic fungi have generated valuable information on the applied aspects of microbial control in the recent years. As formulations of entomopathogenic fungi are generally more expensive, their additional roles in fighting diseases or improving plant growth would be beneficial to the growers and potentially reduce the need for using fungicides or bioinoculants.

19.6 Strategies and implications for sustainable food production

Synthetic agricultural inputs transformed modern agriculture by contributing to increased productivity. However, excessive use of certain inputs and their negative impact on environmental and human health has prompted an emphasis on natural solutions and led to the popularity of organic agriculture and to the eventual growth of the biological inputs industry. Research in the last few years has improved the understanding of biological inputs and their potential in both organic and conventional agriculture to promote sustainable food production. Building soil structure and microbial diversity through cover crops, organic fertilizers, biochar, seaweed or algal products, beneficial microbes, and other inputs is one of the areas of interest for growing healthy crops in recent years. At the same time, several small and large pesticide companies developed microbial, botanical, and other biological pesticides that further contributed to sustainable pest management options. Promoting and maintaining crop health through agronomic or cultural practices is the first step in IPM before pesticide or other control options are implemented (Dara, 2019f). While bioinoculants promote plant growth and health with potential reduction in fertilizer inputs, biopesticides further improve crop and environmental health through reduced reliance on chemical pesticides. The following are some suggested strategies for promoting bioinoculants and biopesticides and the responsibility of agricultural input industry, researchers and extension educators, and the farming community:

- **(i)** The agricultural input industry should invest in developing high quality products at affordable rates that build consumer conference and encourage their use.
- **(ii)** Researchers from both the private industry and public institutions should continue to develop strategies that improve the efficacy of biological inputs and optimize their costs.
- **(iii)** Since biological inputs have complex interactions in a crop environment, additional training on their modes of action, multifaced interactions, and use strategies will be useful for the farming community. Understanding the short- and long-term benefits of biologicals, combining and rotating biologicals with other agricultural inputs to improve efficacy, and developing strategies that maximize their potential and optimize costs would be some of the areas to focus outreach on.
- **(iv)** Despite the increased market for biological inputs, there is still skepticism among some researchers and farmers about their efficacy. Continued science-based research and effective outreach will help promote their use.

19.7 Conclusions

Although the knowledge of bioinoculants and biopesticides is continuously increasing, several scientific publications are based on laboratory or small-scale experiments using potted plants. There is a need to increase field experimentation, especially on commercial fields, to develop practical solutions in addition to producing scientific publications. Several studies presented in this chapter point out inherent variability in agricultural input efficacy and emphasize the need for continued research. A collective effort from the agricultural input industry, researchers, extension educators, and the farming community is necessary for the continued promotion of biological inputs for sustainable food production.

References

Ali, O., Ramsubhag, A., Jayaraman, S., 2019. Biostimulatory activities of *Ascophyllum nodosum* extract in tomato and sweet pepper crops in a tropical environment. PloS One 14 (5), e0216710. https://doi.org/10.1371/journal.pone.0216710.

Azizoglu, U., 2019. *Bacillus thuringiensis* as a biofertilizer and biostimulator: a mini-review of the little-known plant growth-promoting properties of Bt. Curr. Microbiol. 76, 1379−1385. https://doi.org/10.1007/s00284-019-01705-9.

Bulgari, R., Cocetta, G., Trivellini, A., Vernieri, P., Ferrante, A., 2014. Biostimulants and crop responses: a review. Biol. Agric. Hortic. (BAH) 31 (1), 1−17. https://doi.org/10.1080/01448765.2014.964649.

Canassa, F., D'Alessandro, C.P., Sousa, S.B., Demétrio, C.G.B., Meyling, N.V., Klingen, I., Delalibera Jr., I., 2019. Fungal isolate and crop cultivar influence the beneficial effects of root inoculation with entomopathogenic fungi in strawberry. Pest Manag. Sci. 76 (4), 1472−1482. https://doi.org/10.1002/ps.5662.

Canassa, F., Esteca, F.C.N., Moral, R.A., Meyling, N.V., Klingen, I., Delalibera, I., 2020. Root inoculation of strawberry with the entomopathogenic fungi *Metarhizium robertsii* and *Beauveria bassiana* reduces incidence of the twospotted spider mite and selected insect pests and plant diseases in the field. J. Pest. Sci. 93 (1), 261−274. https://doi.org/10.1007/s10340-019-01147-z.

Dara, S.K., 2013a. Entomopathogenic fungus *Beauveria bassiana* promotes strawberry plant growth and health. UCANR eJounal Entomol. Biol. https://ucanr.edu/blogs/blogcore/postdetail.cfm?postnum=11624 (Accessed on 19 May 2020).

Dara, S., 2013b. Field trials for managing aphids on broccoli and western flower thrips on lettuce. CAPCA Advis. 16 (20), 29−32.

Dara, S.K., 2015. Root aphids and their management in organic celery. CAPCA Advis. 18 (5), 65−70.

Dara, S.K., 2016a. First field study evaluating the impact of the entomopathogenic fungus *Beauveria bassiana* on strawberry plant growth and yield. UCANR eJournal Entomol. Biol. https://ucanr.edu/blogs/blogcore/postdetail.cfm?postnum=22546 (Accessed on 19 May 2020).

Dara, S.K., 2016b. Managing strawberry pests with chemical pesticides and non-chemical alternatives. Int. J. Fruit Sci. 16 (Suppl. 1), 129−141. https://doi.org/10.1080/15538362.2016.1195311.

Dara, S.K., 2019a. Non-entomopathogenic roles of entomopathogenic fungi in promoting plant health and growth. Insects 10 (9), 277. https://doi.org/10.3390/insects10090277.

Dara, S.K., 2019b. Evaluating the efficacy of anti-stress supplements on strawberry yield and quality. UCANR eJournal Entomol. Biol. https://ucanr.edu/blogs/blogcore/postdetail.cfm?postnum=31044 (Accessed on 20 May 2020).

Dara, S.K., 2019c. Improving strawberry yields with biostimulants: a 2018−2019 study. UCANR eJournal Entomol. Biol. https://ucanr.edu/blogs/blogcore/postdetail.cfm?postnum=31096 (Accessed on 19 May 2020).

Dara, S.K., 2019d. Effect of microbial and botanical biostimulants with nutrients on tomato yield. CAPCA Advis. 22 (5), 40−44.

Dara, S.K., 2019e. Five shades of gray mold control in strawberry: evaluating chemical, organic oil, botanical, bacterial, and fungal active ingredients. UCANR eJournal Entomol. Biol. https://ucanr.edu/blogs/blogcore/postdetail.cfm?postnum=30729 (Accessed on 23 May 2020).

Dara, S.K., 2019f. The new integrated pest management paradigm for the modern age. JIPM 10 (1), 12. https://doi.org/10.1093/jipm/pmz010.

Dara, S.K., Dara, S.S.R., Dara, S.S., 2017a. Impact of entomopathogenic fungi on the growth, development, and health of cabbage growing under water stress. Am. J. Plant Sci. 8 (6), 1224. https://doi.org/10.4236/ajps.2017.86081.

Dara, S.K., Dara, S.S.R., Dara, S.S., Anderson, T., 2017b. Fighting plant pathogenic fungi with entomopathogenic fungi and other biologicals. CAPCA Advis. 20 (1), 40−44.

Dara, S.K., Dara, S.S.R., Dara, S.S., Lewis, E., 2018a. Managing arthropod pests in zucchini with chemical, botanical, and microbial pesticides. CAPCA Advis. 21 (2), 40−46.

Dara, S.K., Dara, S.S., Jaronski, S., 2019a. Controlling the western grapeleaf skeletonizer with biorational products and California isolates of entomopathogenic fungi. CAPCA Advis 22 (2), 46−48.

Dara, S.S., Dara, S.S.R., Dara, S.K., Jaronski, S.T., 2019b. Entomopathogenic fungi antagonizing *Macrophomina phaseolina* in strawberries. Progr. Crop Consult. 4 (6), 20−24.

Dara, S.K., Humber, R.A., 2020. Entomophthoran. In: Amaresan, N., Senthil Kumar, M., Annapurna, K., Kumar, K., Sankaranarayanan, A. (Eds.), Beneficial Microbes in Agro-Ecology: Bacteria and Fungi. Academic Press, San Diego, pp. 757−776.

Dara, S.K., Lewis, E., 2019. Evaluating biostimulant and nutrient inputs to improve tomato yields and crop health. Progr. Crop Consult. 4 (5), 38−42.

Dara, S.K., Peck, D., 2016. Impact of entomopathogenic fungi and beneficial microbes on strawberry growth, health, and yield. UCANR eJournal Entomol. Biol. https://ucanr.edu/blogs/blogcore/postdetail.cfm?postnum=22709 (Accessed on 19 May 2020).

Dara, S.K., Peck, D., 2017. Evaluating beneficial microbe-based products for their impact on strawberry plant growth, health, and fruit yield. UCANR eJournal Entomol. Biol. https://ucanr.edu/blogs/blogcore/postdetail.cfm?postnum=25122 (Accessed on 19 May 2020).

Dara, S.K., Peck, D., 2018. Evaluation of additive, soil amendment, and biostimulant products in Santa Maria strawberry. CAPCA Advis 21 (5), 44−50.

Dara, S.K., Peck, D., Murray, D., 2018b. Chemical and non-chemical options for managing twospotted spider mite, western tarnished plant bug and other arthropod pests in strawberries. Insects 9 (4), 156. https://doi.org/10.3390/insects9040156.

Domenico, P., 2019. Possible use of Inula viscosa (*Dittrichia viscosa* L.) for biostimulation of *Oscularia deltoides* and *Corpuscolaria lehmanii* plants and protection agains *Aphis nerii*. GSC Biol. Pharm. Sci. 9 (3), 69−75. https://doi.org/10.30574/gscbps.2019.9.3.0231.

Dong, C., Wang, G., Du, M., Niu, C., Zhang, C., Zhang, P., Zhang, X., Ma, D., Ma, F., Bao, Z., 2020. Biostimulants promote plant vigor of tomato and strawberry after transplanting. Sci. Hortic. (Amst.) 267, 109355. https://doi.org/10.1016/j.scienta.2020.109355.

Du Jardin, P., 2015. Plant biostimulants: definition, concept, main categories and regulation. Sci. Hortic. (Amst.) 196, 3−14. https://doi.org/10.1016/j.scienta.2015.09.021.

EPA (Environmental Protection Agency), 2016. What Are Biopesticides? https://www.epa.gov/ingredients-used-pesticide-products/what-are-biopesticides (Accessed on 18 May 2020).

Klokić, I., Koleška, I., Hasanagić, D., Murtić, S., Bosančić, B., Todorović, V., 2020. Biostimulants' influence on tomato fruit characteristics at conventional and low-input NPK regime. Acta Agric. Scand., Sec. B-Soil & Plant Sci. 233–240. https://doi.org/10.1080/09064710.2019.1711156.

Lacey, L.A., 2017. Microbial Control of Insect and Mite Pests: From Theory to Practice. Academic Press, San Diego. https://doi.org/10.1016/C2015-0-00092-2.

Litwin, A., Nowak, M., Różalska, S., 2020. Entomopathogenic fungi: unconventional applications. Rev. Env. Sci. Bio/Tech. 19, 23–42. https://doi.org/10.1007/s11157-020-09525-1.

Mantzoukas, S., Eliopoulos, P.A., 2020. Endophytic entomopathogenic fungi: a valuable biological control tool against plant pests. Appl. Sci. 10 (1), 360. https://doi.org/10.3390/app10010360.

Mukherjee, A., Patel, J.S., 2020. Seaweed extract: biostimulator of plant defense and plant productivity. Int. J. Env. Sci. Tech. 17, 553–558. https://doi.org/10.1007/s13762-019-02442-z.

Pérez-Fragero, I.C., Camacho, M., Ridríguez, S.C., 2020. Characterization of plant growth promoting bacteria isolated from red fruits. Studies on growth promotion and fruit quality in strawberry plants. In: Biosaia: Revista de los másteres de Biotecnología Sanitaria y Biotecnología Ambiental, Industrial y Alimentaria, vol. 9. https://www.upo.es/revistas/index.php/biosaia/article/view/4760 (Accessed on 20 May 2020).

Stern, V.M., Federici, B.A., 1990. Granulosis virus: biological control for western grapeleaf skeletonizer. Calif. Agric. 44 (3), 21–22.

Szczałba, M., Kopta, T., Gąstoł, M., Sekara, A., 2019. Comprehensive insight into arbuscular mycorrhizal fungi, *Trichoderma* spp. and plant multilevel interactions with emphasis on biostimulation of horticultural crops. J. Appl. Microbiol. 127 (3), 630–647. https://doi.org/10.1111/jam.14247.

Yu, X., Ai, C., Xin, L., Zhou, G., 2011. The siderophore-producing bacterium, *Bacillus subtilis* CAS15, has a biocontrol effect on Fusarium wilt and promotes the growth of pepper. Eur. J. Soil Biol. 47 (2), 138–145. https://doi.org/10.1016/j.ejsobi.2010.11.001.

Yuan, J., Ruan, Y., Wang, B., Zhang, J., Waseem, R., Huang, Q., Shen, Q., 2013. Plant growth-promoting rhizobacteria strain Bacillus amyloliquefaciens NJN-6-enriched bio-organic fertilizer suppressed Fusarium wilt and promoted the growth of banana plants. J. Agric. Food Chem. 61 (16), 3774–3780. https://doi.org/10.1021/jf400038z.

Chapter 20

Development and regulation of microbial pesticides in the post-genomic era

Anirban Bhar[a,b], Akansha Jain[a] and Sampa Das[a]

[a]*Department of Botany, Ramakrishna Mission Vivekananda Centenary College, Kolkata, West Bengal, India;* [b]*Division of Plant Biology, Bose Institute, Kolkata, West Bengal, India*

20.1 Introduction

To ensure food securities for increasing populations on a global scale necessitates reorganizations in agro-economic policies and incorporations of novel agricultural techniques. The technological improvements have focused on qualities of the food grains as well as increased productivity (Sala et al., 2017). Quality improvements are sophisticated science that is ever-increasing for a particular type of crop whereas quantity improvement is overall increment of productivity which is utmost necessary for meeting the demand of increasing population. Increased productivity maintaining standard quality is further a challenge for scientists. Different abiotic and biotic stress factors further encompass a huge agricultural loss annually (Pandey et al., 2017). Insect pests are major yield limiting factors and cause a 30% loss of agricultural commodities (Sellamuthu et al., 2018). "Green revolution" was envisioned by Norman Borlaug and initiated in India by M.S. Swaminathan in mid 60s to meet the need of growing demand of high yielding rice (Nelson et al., 2019). After 70 years of the first green revolution, another revolution is needed to face future demand. Along with the use of high yielding varieties use of chemical fertilizers and chemical pesticides has been gaining priorities for the last couple of decades. Chemical pesticides have improved for their optimal utilization but the high risk to the environment makes environmentalists think twice for the excessive use of chemical pesticides. The residual effect of chemicals used as pesticides has a wide deleterious effect on environmental pollution (Arcury et al., 2002). The excess/residual chemicals in the soil not only change the soil chemistry but they also change the vital parameters and soil ecology of the agricultural field. These have long term defective impacts on agricultural productivity too. Agricultural effluents are also known to contaminate water bodies, rivers, etc. by rainwater disposal, and these chemical pollutants easily enter into the aquatic food chain (Zalidis et al., 2002). Bioaccumulation readily destroys each trophic level of the entire ecological pyramid.

USING Biopesticides are the deserving alternative and gaining continuous impact due to the above reasons. Biopesticides were classified into three distinct categories (i) biochemical pesticides, (ii) plant-incorporated protectants (PIPs) and (iii) microbial biopesticides by the United States Environmental Protection Agency (EPA) (Sinha, 2012). Microbial biopesticides embrace all beneficial microbial communities including bacteria, viruses, fungi, nematodes, and protozoans which were employed to get rid of insect pests. Microbial biopesticides were deployed directly or used as formulations alone or in cooperation with other chemical protectants (Chandler et al., 2008). Although, the association of microbes with plants had studied been for a long time but comprehensive knowledge and systematic application are further necessary to develop sustainability in the agricultural field. Restricted host range, slow activity and legislations to use of living materials are further challenges in proper utilization of microbial biopesticides. Sometimes the failure of a particular microbial biopesticides leads to loss of interest in industrial production of the same and long term propensity of that material has stunted (Persley and Gambley, 2013). In a nutshell although enormous experience has gathered for many years but considerable improvements in the efficiency of the microbial biopesticides, their formulations, industrial policies, and reprogramming of legislations are utmost needed for the future of these promising microorganisms. The present chapter describes different microbial biopesticides, their developments, and improvements that were made for quality enhancements. It also describes different laws and legislative regulations in shaping their proper agricultural utilizations. Further,

Biopesticides. https://doi.org/10.1016/B978-0-12-823355-9.00018-3

recent advancements of microbial biopesticides in the light of post-genomic concepts are also discussed. Overall, this book chapter will provide comprehensive knowledge in the development and use of microbial biopesticides along with their regulatory measures that would surely enrich the knowledge of the readers and also help researchers in shaping their future research in agricultural biotechnology.

20.2 Development of the microbial biopesticide

The history of agriculture is among the oldest practices of civilization ever acquired. Most of the crops of the present day have developed as a result of domestication along with migration. The microbes along with different pathogens and pests have also been co-evolved with plants inhabiting within rhizospheric as well as phyllospheric regions (Philippot et al., 2013). Friendly microorganisms are always associated with plants and help them to avert pathogenic pests through varied mechanisms (Arora et al., 2013). These dynamic soil ecology comprising plants, pests, and friendly microbial populations constitute the basis for the development of microbial biopesticides. Presently in modern agriculture, the use of biopesticides gains enormous acceleration due to their pollution-free environment-friendly nature (Thakore, 2006). The microorganisms used for pestiferous agents are also called biological control agents (BCAs) for pests. Although, BCA were used in agriculture for many years but non-comprehensive knowledge of the use and relatively less consistent research restricts their wide acceptance. Recently, BCA research particularly studies on microbial biopesticides reclaimed their obvious need in modern agriculture. Long term use of chemical fertilizers not only hampers the environment but also affects soil ecology brutally (Seiber et al., 2014). The dramatic loss of microbial population in the soil brings forth different new diseases, reduces soil quality, and hampers crop productivity too (Daughtrey and Benson, 2005). Development and application of microbial biopesticides not only counteract different non-vertebrate pests but they also help to revive soil ecology for sustainable agriculture. Interestingly, very recently it has reported that along with classical microorganisms, microalgae also have huge potentiality in the biopesticide industry (Costa et al., 2019).

Microbial pesticides include bacteria, viruses, fungi, nematodes, and protozoa which are associated with pathocidal activity. The development of any microbial biopesticides firstly relies on successful isolation and the formation of a pure culture of the individual potential candidates. Then these microbes are characterized by their morphological nature and identified by molecular characterization. Biocontrol capabilities were then determined thoroughly by checking antagonism against particular target pathogen and finally, bioassay has to be performed to confirm their specific roles. Epidemiological knowledge has also impacted a lot in determining the infectivity of a particular microbial biopesticide. Before, registration, legal activity, and release of biopesticides their mode of application also needs to be determined for their proper use (Fig. 20.1).

20.2.1 Plant growth regulators play crucial role in development of biopesticides

Many microorganisms have a healthy relationship with plants and they influence plant growth, development, and crop production, hence called plant growth promoters (PGPs) (Jain et al., 2020). These PGPs not only retard pathogenic organisms but also improves water and nutrient availability to the plants to make them healthy. Among many growth-promoting characteristic features 1-aminocyclopropane-1-carboxylic acid (ACC) deaminase activity, nitrogen fixation, solubilization and bioavailability of inorganic phosphorus (P), catalase activity, chitinase activity are important (Ahmad et al., 2011, 2013; Khan et al., 2009; Nadeem et al., 2014; Xiao et al., 2017; Zhang et al., 2011). US Environmental Protection Agency (USEPA) has its guidelines and regulatory compliance for the use of plant growth regulators (PGRs) as biopesticides (Lake et al., 2002). Some insect growth regulators (IGRs) have also proved to be potential biopesticides. The role of emamectin and spinetoram as biopesticide whereas, the activity of hexaflumuron and teflubenzuron as IGR against the growth of fourth instar larvae of *Spodoptera littoralis* had been checked (Assar et al., 2016). In a separate study it has been reported that azadiarchtin which has long been known for its antimicrobial activity can also act as an effective biopesticide. Some help plants to thrive well by producing phytohormones e.g., auxins, gibberellic acid, cytokinin, abscisic acid, ethylene, etc. (Ahmed et al., 2019). Different halo-tolerant fungi were tested for their prospective performance in the production of phytohormones. Indole acetic acid (IAA) and gibberellic acid (GA) were evident and most promising in *Aspergillus niger* (Samah et al., 2019). Application of the novel strain *Bacillus tequilensis* (SSB07) was found to enhance heat tolerance in Chinese cabbage seedlings by producing GA (GA1, GA3, GA5, GA8, GA19, GA24, and GA53) and auxin (IAA) as well. This bacterial strain was also found to induce stress hormone salicylic acid (SA), jasmonic acid (JA) and abscisic acid (ABA) in plant upon application (Kang et al., 2019). The role of microbe derived phytohormones in physiological responses, basal growth, and development of plants with the exploration of novel symbiotic associations were reviewed in detail by Frankenberger and Arshad (2020).

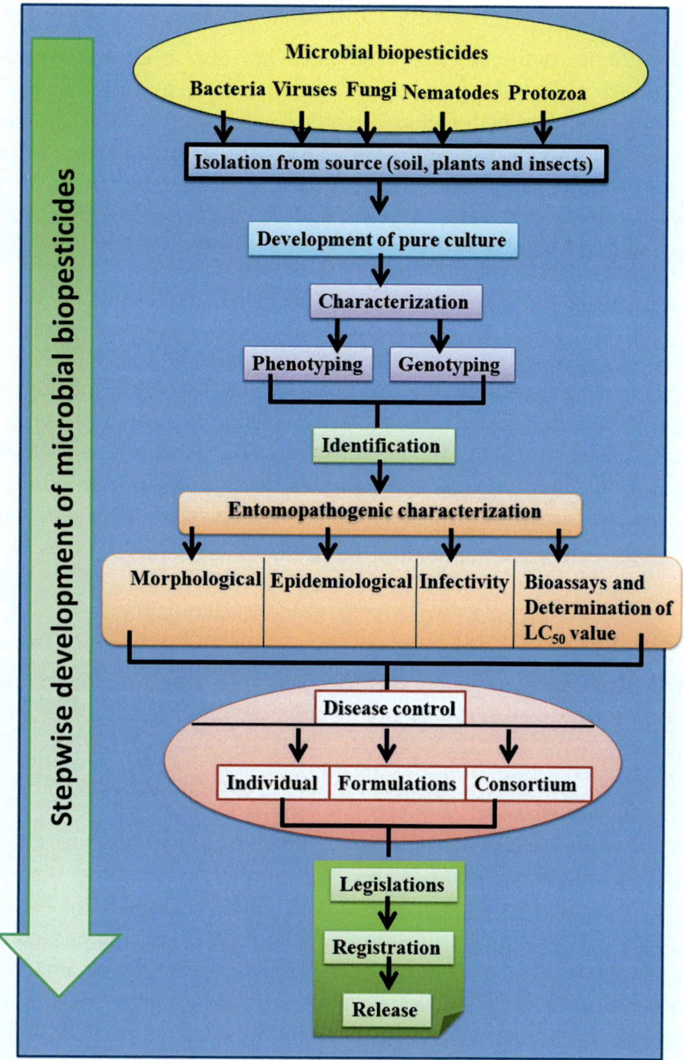

FIG. 20.1 Development of microbial pesticide from screening to commercialization.

20.2.2 Siderophores causes iron limiting conditions for many pathogenic pests

Siderophores are iron-chelating chemicals secreted by many chemotrophic bacteria. These siderophore help to chelate iron (Fe) from the soil and make them available to the plants by ligand exchange reactions (Miethke and Marahiel, 2007). Increased availability of Fe to plants due to siderophore causes iron limiting conditions for many pathogenic pests and inhibits their growth (Kloepper et al., 1980; Glick, 2012). Siderophore producing saprophytic fungi *Aureobasidium pullulansis* was reported to be suppressive toward different phytopathogenic fungi due to their fungal cell wall degrading capacity (Sun et al., 2019). Recently, it was demonstrated that siderophore is the key factor behind the biocontrol activity of endophyte *Burkholderia pyrrocinia* JK-SH007 in controlling canker disease (Min et al., 2019). In a separate study, it was reported that *Pseudomonas protegens* MP12 has broad spectrum antifungal activity evident by production of antimicrobial metabolites and siderophore. Genes phlD, pltB, and prnC were identified from *P. protegens* for the production of antifungal components e.g. 2,4-diacetylphloroglucinol (2,4-DAPG), pyoluteorin, and pyrrolnitrin, respectively (Andreolli et al., 2019). Actinobacteria were very well known for their bioprospecting activity. Different strains isolated from plant waste were exhibited siderophore production along with the production of some common antimicrobial metabolites e.g salicylic acid, cyanide components, and chitinase (Jurado et al., 2019). *Pseudomonas stutzeri* isolated from rhizospheric soil found to produce IAA along with siderophore and hydrogen cyanide (HCN). This bacterial strain was successfully used against phytopathogen *Stemphylium botryosum* (Mokrani et al., 2019).

20.2.3 Antibiosis, an important criterion for development of the microbial biopesticides

Antibiosis and production of different antimicrobial metabolites are also a common mode of action of this PGRs. Antibiosis is a mechanism by which chemical substances produced mainly by different microorganisms or secreted by plants, able to retard growth or kill other pathogenic microorganisms and pests. Antimicrobial substances are also chemicals produced by the metabolic activity of plants and those can act against a range of pathogenic organisms. These chemicals are largely grouped under bacteriocides (kills bacterial pathogen), fungicides (kills fungal pathogens), insecticides (kills insect pests mainly invertebrate pests), acaricides (kills insect members of Arachnida, subclass Acari/ Acarina) and herbicides (Copping and Menn, 2000).

20.3 Microbial pesticides: brief description

Pests are largely defined as any pathogenic organisms to crops i.e. any microbial pathogenic pests (virus, bacteria, or fungi), insect pests (mostly invertebrate insects) or herbs or weeds causing damage to the crop plants. Microbial pesticides are the tiny living warriors against these varied notorious organisms. These helpful microorganisms are employed directly to the soil or chemicals isolated from these organisms are specifically formulated for agronomic applications. Microbial biopesticides can further be classified according to their type's e.g. bacterial biopesticides, fungal biopesticides, viral biopesticides, and nematodal biopesticides.

20.3.1 Bacteria as biopesticides

Many bacterial species showed their anti insecticidal activity for a very long time but among all of them, members of Bacillaceae are famous. *Bacillus thuringiensis* (Bt) has long been known for their entomopathogenic role (Castagnola and Stock, 2014; Ruiu, 2015). *B. thuringiensis* is a Gram-positive, sporulating soil bacteria that are capable of producing Cry and Cyt proteins that are highly insecticidal mainly against lepidopteran, homopteran, and coleopteran insects (Palma et al., 2014). Mainly two varieties, *B. thuringiensis* var. *kurstaki* (DORBt-1, DORBt-5, PDBCBT1, NBAIIBTG4, NCIM2514) and *B. thuringiensis* subsp. *israelensis* (VCRCB-17,164) are widely used as biopesticides. *B. thuringiensis* var. *kurstaki* is known to be used against lepidopteran insect pests, whereas, *B. thuringiensis* subsp. *israelensis* also have pestiferous activity against mosquito larvae, black flies, and fungal gnat larvae (Kumar et al., 2018). These bacterial strains are known to produce crystal protein (Cry and Cyt) during sporulation that when enters into the mid gut of the targeted insects binds with specific receptors and produces pores on to the endothelial layers causing insect septicemia and death of the insects. These pore-forming toxins consists of three domains, (i) domain I is a α-helical (responsible for pore formation) and (ii) and (iii) are majorly β sheets in nature involved in receptor binding (Parker and Feil, 2005; Bravo et al., 2007; Sarkar et al., 2009). In separate studies the functional significance of alanine residue in Cry1 Ac domain III in binding with HaALP receptors of *Helicopverpa armigera* were established (Sarkar et al., 2009; Sengupta et al., 2013). Recently, some other *B. thuringiensis* strains were also commercially used as biopesticides. *Bacillus thuringiensis aizawai* has been used against armyworms and diamondback moth (commercial name: Able-WG, Agree-WP, Florbac, XenTari); *Bacillus thuringiensis tenebrionis* was employed against Colorado potato beetles (commercial names: Trident, Novodor); *Bacillus thuringiensis sphaericus* had worked against mosquitoes (commercial names: VectoMax, VectoLex) (Ruiu, 2018). *Bacillus firmusd* and *Lysinibacillus sphaericus* were utilized to combat boll worms, diamondback moth, and other Lepidoptera. They were also employed against plant pathogenic nematodes (Kumar et al., 2018).

Among other bacterial species *Burkholderia sp.* were widely used as biopesticide against chewing and sucking insects, mites, and nematodal pests. *Burkholderia cepacia* and *Burkholderia ambifaria* are the two strains gaining enormous scientific attention for their potentiality as biopesticides, herbicides as well as human health modulators (LiPuma and Mahenthiralingam, 1999; Holmes et al., 1998; Jones et al., 2001). They had been sold in commercial names Majestene, Venerate, etc. *Chromobacterium* were also used against chewing insects in commercial name Grandevo (Ruiu, 2018). Different species of *Pseudomonas* were also employed as biopesticides. For that reason, these bacteria are called plant probiotics (Anderson and Kim, 2018). *P. fluorescens* were implicated for their role as antifungal agents (Anbuselvi et al., 2010). These bacteria were also reported to be function against bacterial wilt of brinjal (Chakravarty and Kalita, 2011). Recently it was reported that *P. fluorescens* CL145A showed the pestiferous effect on zebra and quagga mussels (Molloy et al., 2013). Separately, a strain of *P. aeruginosa* PA 1201 was isolated and documented to produce a high amount of biopesticides Shenquinmycin and phenazine-1-carboxamide (Zhou et al., 2015). *P. syringae* ESC-10 and ESC-11 were showed their role against green and blue molds of *Citrus sp.* (Bull et al., 1997).

20.3.2 Viruses as biopesticides

Like bacteria, many viruses also act as biopesticide agents. Among many viruses, baculoviruses are most effective against invertebrate pests. Baculoviruses belong to Baculoviridae family and they spread through occlusion bodies. Among many morphological variations, nucleopolyhedroviruses (NPV) and granulovirus (GVs) were found mainly effective against lepidopteran insects and fall armyworm (Valicente, 2019). NPV were found to be effective against many insect pests. NPV isolated from *Anagrapha falcifera* and *Autographa californica* were formulated differently to apply on cotton plants (McGuire et al., 2001). Many plants on the other hand harbors entomopathogenic viruses to get rid of different pests. *Brassica oleracea* has badly targeted by aphids *Myzus persicae*. After the attack it was found that many *M. persicae* aphids were found to be infected with parvovirus, *M. persicae* densovirus (Van Munster et al., 2005). Many viruses like particles were found to be associated with many entomopathogenic fungi e.g. dsRNA of 910–3100 bp size was identified within fungus *Metarhizium anisopliae* (Bogo et al., 1996). The presence of Bassiana RNA Virus 1 (BbRV1), of the *totiviridae* family, was noticed within wide ranges of entomopathogenic fungus *Beauveria bassiana* (Herrero et al., 2012). This co-infection cases increased virulence of these phytopathogenic fungal species against their insect pests. *Anticarsia gemmatalisi* nucleopolyhedrovirus (AgMNPVs) was effective against major soybean caterpillar pest *Anticarsia gemmatalisi* (Moscardi et al., 2011). Another NPV called SfMNPV were extensively studied for their entomopathogenic role against *Spodoptera frugiperda* (Wolff et al., 2008). NPV were also effectively employed to counteract *Helicoverpa armigera* and application of baculoviruses was also examined against soybean looper, *Chrysodeixis* sp. (Valicente, 2019).

20.3.3 Fungi as biopesticides

Many soil fungi cause disease occurrence to the plants but many among them some members also have beneficial roles in combating phytopathogens and protect crop plants from insect attack. Most of the entomopathogenic fungi are classified largely under the class Entomophthorales of Zygomycota or Hyphomycetes under Deuteromycota (Shah and Pell, 2003). *Beauveria bassiana* and *B. brongniartii* were very common entomopathogenic fungi utilized to control a wide range of insect pests (Zimmermann, 2008). Corn borer larvae *Ostrinia nubilalis* (Hübner) was effectively controlled by the application of *Beauveria bassiana* on whole corn plants or as a foliar spray (Bing and Lewis, 1991). *B. brongniartii* strains were successfully utilized to restrict European cockchafer larvae in Switzerland (Enkerli et al., 2004). Another entomopathogenic fungus *Metarhizium anisopliae* also used as an effective biocontrol agent (Zimmermann, 1993; Kumar et al., 2018). *Lecanicillium sp.* previuosly classified as *Verticillium sp.* were also potentially important biocontrol agent against insects and nematodes to control plant diseases (Goettel et al., 2008). These group of fungus (*Lecanicillium longisporum* and *L. attenuatum*) were found to be antifungal (against cucumber powdery mildew disease caused by *Sphaeroheca fuliginea*) along with their role against aphids (Kim et al., 2008). Insecticide associated application of *L. muscarium* were also effective against sweet potato whitefly, *Bemisia tabaci* (Cuthbertson and Walters, 2005). Consortium activity of three entomopathogenic fungi, *Metarhizium anisopliae, Lecancillium psalliotae,* and *Beauveria bassiana* against *Rhipicephalus (Boophilus) annulatus* were examined (Pirali-Kheirabadi et al., 2007) infestation of aphids and root knot nematode *Meloidogyne incognita* in tomato plants (Lopez et al., 2014; Lopez and Sword, 2015; Singh et al., 2013). It was also evidenced that *P. lilacinum* can negatively control tick, *Rhipicephalus microplus* under laboratory situation (Angelo et al., 2012). Another warrior against insect pests is entomopathogen, *Isaria fumosorosea* which can act as broad-spectrum pesticide in *Citrus sp.* These fungi were reported to be effective against hemipteran insect, *Diaphorina citri* and diamondback moth, *Plutella xylostella* (Ali et al., 2010; Avery et al., 2011). Members of *Hirsutella thompsonii* were also employed as wonderful biological control agent against parasitic pests (Kanga et al., 2002).

Among biocontrol fungi, *Trichoderma* spp. are most popular. Priming plants with *Trichoderma* sp. induces a systemic response in plants against pathogens (Jain et al., 2015), while they may also stimulate plant growth and development (Singh and Nautiyal, 2012), provide resistance against abiotic stresses, and can enhance the population of other beneficial microorganisms (Jain et al., 2013; Singh et al., 2014). *Trichoderma* strains are reported as promising biocontrol agents for the *Sclerotium rolfsii* (Singh et al., 2013) and *Fusarium* (Javanshir Javid et al., 2016). Various other mycopathogenic fungi viz., *Ampelomyces quisqualis,* non-toxigenic *Aspergillus flavus, Coniothyrium minitans* fungi are posing a new trend in the markets worldwide (de la Cruz Quiroz et al., 2019).

20.3.4 Nematodes as biopesticides

Nematodes are also widely deployed as biocontrol agents to combat pests. *Steinernema carpocapsae* and *S. glaseri* were reported to infect engorged adult black-legged ticks *Ixodes scapularis* (Zhioua et al., 1995). The members of Dictyoptera e.g. cockroaches were infected by *Steinernema carpocapsae* (Koehler et al., 1992). In a separate study it was observed that larvae of *Galleria mellonella* were killed by *S. carpocapsae* but remained unaltered by *Heterorhabditis bacteriophora* (Han and Ehlers, 2000). Interestingly, it was demonstrated that the soil qualities play a pivotal role in entomopathogenicity of *S. carpocapsae* and *H. megidis* (Kruitbos et al., 2010). Corn earworm was efficiently controlled by *S. carpocapsae* and *S. riobravis* (Cabanillas and Raulston, 1996). Potato cyst nematode, *Globodera rostochiensis* was treated with entomopathogens, *S. feltiae* and *S. carpocapsae* (Perry et al., 1998). Field experimentation was performed to evaluate the function of heat killed and live *S. carpocapsae* against boxwood, *Buxus sp.* (Jagdale et al., 2002). In India, *S. thermophilum*, *H. indica*, *H. bacteriophora* were used against root grubs, weevils, cutworm, wide variety of lepidopteran insects, and termites (Kumar et al., 2018).

20.3.5 Protozoan as biopesticides

Some protozoans, particularly from the genus *Nosema,* are well known for the entomopathogenic activities. The spores of *Nosema sp.* which is also known as microsporidia were found to kill diamondback moth, *Plutella xylostella* by damaging midgut epithelial cells and vacuolization of cytosol. This infection was found to temperature-dependent and optimal infection took place at 35°C (Kermani et al., 2013). Protozoans are known to infect mainly insects belongs to family Acrididae. Besides Microsporidia, members from Amoebida, Neogregarinida, Eugregarinida, Ciliophora were also known to have entomopathogenicity (Johnson, 1997). *Nosema locustae* are very well known among all other *Nosema sp.* and used successfully in controlling locusts and grasshoppers across the world (Lange, 2005). Recently, it was observed that the combined effect of *N. locustae* and *Beauveria bassiana* on *Locusta migratoria* was pronounced (Tan et al., 2020). In laboratory condition another microsporidian, *Paranosema locustae* on grasshopper *Tropidacris collaris* was analyzed (Lange et al., 2017). *Nosema adaliae* was known to infect its natural host two-spotted lady beetle (*Adalia bipunctata*) but it was observed that this microsporidian horizontally transmitted also to Chinese praying mantis (*Tenodera sinensis*) (Preti, 2018). Another protozoan *Leptomonas wallacei* was found to infect *Oncopeltus fasciatus* and it was also observed that female insects were more infected than that of male ones (Vasconcellos et al., 2019). Hence, protozoans members could be an important option for the biological control of agricultural crops.

20.4 Genetic improvements of microbial pesticides

Biopesticides are an effective alternative to chemical pesticides but its narrow host range and unequal effectiveness on different insects are the major drawbacks. Genetic improvements of microbial biopesticides are mainly focused toward the increase in host range and greater virulence improvements. Impressive developments in genomics technologies and subsequent development in recombinant DNA technology (rDT) made some remarkable betterment of the existing microbial biopesticides. Among bacterial biopesticides, *Bacillus thuringiensis* has been studied for a long time. Bt toxins have wonderful insecticidal activity and many transgenic approaches had taken to introduce resistance against insect pests (Romeis et al., 2004; Ghosh et al., 2017). The *Cry1Ac7* gene from *B. thuringiensis* was cloned under the control of the *tac* promoter in broad host-specific vector and introduced into sugarcane associated strain of *Pseudomonas fluorescens*. At the same time chitinase gene, *ChiA* from *Serratia marcescens* was cloned similarly in the same bacteria. The effectiveness of both insecticides was observed against sugarcane borer *Eldana saccharina* and documented that the combination of *Cry1Ac* and *ChiA* was most effective (Downing et al., 2000). Non-toxic nature of *Bt* toxin in vertebrates and easily amendable nature made this toxin wide use, but at the same time, many insects were started developing resistance against endotoxin produced by this bacteria (Bauer, 1995). Additive and synergistic activities of delta endotoxins were influenced by transfer of *Bt genes* between non-homologous strains. Self transmissible plasmid bearing *Cry1A* was transferred from *aizawai* strain to *kurstaki* strain to improve target specificity, named Condor (EG2348). On the other hand, rDNA modified variant (ECX9399) was found to be effective against armyworm *Spodoptera frugiperda* (All et al., 1994). Black flies and mosquitoes are very common threats to the world and are very difficult to control. Scientists have developed a method to control these two notorious insects by *Bt* toxins from *Bacillus thuringiensis israelensis* and *B. sphaericus* (De Barjac and Sutherland, 2012). In a separate study *Cry1A (c)* was isolated from *B. thuringiensis* HD-1 and transferred to a broad host range vector pSUP204 to develop pSUP89A transferable plasmid. The transformation was done in *Pseudomonas cepacia* 526 bacteria and the effectiveness of this bacteria was tested against tobacco hornworm, *Manduca sexta* (Stock et al., 1990). Recently a wide conjugative self transferable plasmid pXO16 was reported from *B. thuringiensis israelensis* which is capable to mobilize or retro mobilize in extended host spectrum (Hinnekens et al., 2019). Comparative genomics study among *B. thuringiensis* and *B. anthracis* revealed the unique insecticidal clade with many toxin encoding genes were found in plasmid mini-replicons of *orf156* and *orf157* of *B. thuringiensis*. This information could be helpful in designing future biopesticides with multi-host specificities (Zheng et al., 2017). Many mutant *Cry* proteins were found to be enhanced insecticidal effects (Je et al., 2016). Hence, multiple site-directed mutagenesis were imposed to produce novel *Cry1Ac* with enhanced insecticidal efficacy (Kim et al., 2016).

Recently advancement of genome sequencing platforms revealed the complete genomic construction of many important biocontrol agents. Whole-genome sequence of mosquitocidal *Bacillus thuringiensis* LLP29 was performed that would provide huge beneficial information in shaping future biopesticides (Ma et al., 2020). Complete genome sequence of another important commercial biocontrol strain *Paenibacillus polymyxa* HY96-2 was performed and compared with other *P. polymyxa* strains to find out probable future biological control agents (Luo et al., 2018). Similarly, *Bacillus velezensis* 9D-6 was studied extensively for their potential role as biocontrol agent. LC MS/MS-based metabolomic screening revealed some surfactants like substances potentially function as antimicrobial materials. Genome analysis demonstrated a single circular chromosome consisting of 13 gene clusters which showed antimicrobial functions (Grady et al., 2019). Entomopathogenic fungi have also come under the genomic scanner for a better understanding of their mechanism of action and improvements of virulence. RNA Seq analyses were performed using entomopathogen *Beauveria bassiana* with bean bug insect pathosystem before and after infection (Lee et al., 2018). The evolution of acaropathogenic fungus *Hirsutella thompsonii* was demonstrated through mitochondrial genomic analysis (Wang et al., 2018). The mitochondrial genome of *Cordyceps tenuipes* was also analyzed recently (Li et al., 2019). Genomic analyses of Entomopathogenic fungi *Beauveria bassiana*, *Metarhizium sp.*, *Cordyceps militaris* and *Ophiocordyceps sinensis* unveiled that these fungi were once endophytes or plant pathogens (Wang et al., 2016). The development of virulence toward insects probably evolved through prolonged co-evolution. In a separate study with mutant *Metarhizium robertsii* it was demonstrated that DNA methyltransferases (DNMTases) play a crucial role in virulence and entomopathogenicity toward insect pests (Wang et al., 2017). Reactive oxygen species (ROSs) production is inevitable in any host-pathogen interaction. During infection, entomopathogens also cause oxidative stress in the insect bodies. Many scavenging enzymes play important roles in virulence of these pathogens. A very interesting study was conducted using biocontrol agent *Cordyceps pruinosa*, where the genomic structure of superoxide dismutase (SOD1) gene was analyzed. It was observed that *C. pruinosa* SOD1 is orthologous to other SOD1 from Ascomycetous fungi and composed of copper (Cu), zinc (Zn), and β barrel fold sites (Park et al., 2019). Fungal chitinases play a crucial role in pathogenesis. Glycoside hydrolase 18 (GH18) gene families were analyzed in *Isaria cicadae* providing its enormous possibilities in future application to develop biopesticides with enhanced virulence (Peng et al., 2020). These chitinases are effective in degrading the insect cuticles and enhances their virulence. An effort was made to generate cuticle degrading protease 1 (Pr1) overexpressing lines in *Metarhizium anisopliae* which significantly enhanced their insect infectivity (St Leger et al., 1996). These Pr1 hydrolase proteins were thought to involve only in the degradation of insect cuticle i.e. infection initiation. Recent detail evolutionary study with Pr1 gene family in *Beauveria bassiana* and *Metarhizium anisopliae* revealed that this gene family function more than expected. These hydrolases contributed to 19%–29% virulence of the entomopathogens (Gao et al., 2020). Recently a new class of M35 metalloproteases was also identified from *Metarhizium robertsii* which have a tremendous role in virulence by deactivating insect prophenoloxidases (Huang et al., 2020). In *Metarhizium acridium* a homeobox gene MaH1 was found to be inevitable in conidial development and virulence (Gao et al., 2019). Similarly, *Metarhizium acridium* PacC (MaPacC) was also found to be associated with conidiation and virulence (Zhang et al., 2019). In a separate study mycotoxin, IF8 produced by *Isaria fumosorosea* was analyzed and found to be effective against psyllid *Diaphorina citri* (Keppanan et al., 2019).

Among viral biopesticides, baculoviruses took the principal position because of their high sensitivity toward target insects and safe nature for the environment as well as human health. Baculoviruses are rod-shaped bacteria mainly differentiated into three subgroups, subgroup A are called NPV which colonizes into hard infectious core called occlusion body. The occlusion bodies of NPV made up of protein polyhedron. Subgroup B belongs to granulo viruses (GVs). GVs also produce occlusion bodies but a single bacterium present within each occlusion body. The occlusion bodies of GVs made up of another protein granulin. These occlusion bodies when comes in contact with the alkaline gut fluid of insects, virus release, and cause infection. Subgroup C contains free virus particles or non-occluded viruses. Subgroup A and B were gained the most attention for their insecticidal nature (Starnes et al., 1993). Granuloviridae *Cryptophlebia leucotreta* granulovirus (CrleGV-SA) was employed against an indigenous pest *Thaumatotibia leucotreta* which attacks *Citrus sp.* in Africa. The genomic stability of CrleGV-SA was analyzed by whole-genome sequencing of these virus particles (Van der Merwe et al., 2017).

The genetic improvements in protozoa are mainly focused on human disease-causing pathogens. Recently, the genome structural organization of *Nosema ceranae* was analyzed and expression study was performed for some genes

(Xiong et al., 2019). Genome sequence of cabbage butterfly parasite *Nosema sp.* isolate YNPr was analyzed and the structural organization was published (Xu et al., 2016). The mitochondrial genome of *Leptomonas seymouri* was analyzed to study the initiation of transcription (Vasil'eva et al., 2004). Along with genome organization, mitochondrial genes are also playing pivotal roles to determine the taxonomic categorization of entomopathogenic protozoa. An integrated morphological and molecular association study based on ultrastructure, microscopic data, and gene organization of glyceraldehydes phosphate dehydrogenase demonstrated polyphyly among *Leptomonas sp.* (Yurchenko et al., 2006).

20.5 Regulation and commercialization of microbial pesticides

After a microbial pesticide is developed and is ready to be marketed, its legal rights as the biotechnological invention have to be patented. Although a product of nature, biopesticides can be patented in synthesized form. Processes for developing biopesticides on large scale can also be patented (Montesinos, 2003). A patent allows the owner with a 20 years entitlement for industrial or commercial profiteering, as per the series of claims. Filing patents for microbial biopesticides is controlled by a legal course of national and international treaties. International treaties consist of the Treaty of the Union of Paris of 1883- signed up by 140 countries; the European Patent Agreement of 1973- is a multilateral treaty applicable in 18 European countries; Patent Cooperation Treaty (PCT) of 1970- provide patent protection simultaneously in the signing countries and the other 102 countries. Adopted in 1977, the Budapest Treaty on the international recognition of the deposit of microorganisms (culture collection) for the patent procedure is agreed upon by all countries of the World Intellectual Property Organization. All parties to the Treaty are compelled to identify the pure culture of microorganisms deposited for the patent procedure, irrespective of whether the depository authority location is outside the state. This Treaty increases the insurance of the depositor as it complies with a consistent system of deposit, recognition, and provision of microorganisms based pesticide.

Regardless of a large number of patents for microbial pesticides, merely a few of them are registered for agricultural use. The registration is guided by the definitive rules within each country. The registration in the United States is directed by the Biopesticides and Pollution Prevention Division established in the Office of Pesticide Programs in the Environmental Protection Agency (EPA). The EPA works based on the laws within the Federal Food, Drug and Cosmetic Act (FFDCA, 1938) and the Federal Insecticide, Fungicide and Rodenticide Act (FIFRA, 1947). The US EPA also classes few transgenes as biopesticides. The European Union (EU) assessment procedure was first laid down by the Directorate of the Consumer Health Protection under Directive 91/414/EEC, which has specific reformation for biopesticides by Directive 2001/36/EC. In EU microbial biopesticides are registered in a two-step process, where the active substance is assessed first under Regulation No. 283/2013 for data specifications. The second-step involves addition in the list of approved active substances using Regulation No. 1107/2009 (Regulation, 2009; Hauschild et al., 2011). Beginning from January 2000, 47 microbial biocontrol agents have been registered in the EU and 73 in the USA; 13 of them are included both in the EU and the USA (Frederiksa and Wesselerb 2019). The registration process in the EU is complex and time taking as compared with that in the USA (Frederiksa and Wesselerb, 2019). Registration of pesticides in Australia is authorized by the Agricultural and Veterinary Chemicals Code Act 1994 and administered by the Australian Pesticides and Veterinary Medicines Authority (APVMA). Genetically engineered microbes require approval from the Office of the Gene Technology Regulator (OGTR). In India, biopesticides regulation is governed by the Insecticide Act (1968), which requires registration of microbial pesticides with the Central Insecticides Board (CIB) of the Ministry of Agriculture.

A huge collaborative input is required starting from scientists to intermediate regulatory bodies and companies and final participants viz., farmers in the development and commercialization of biopesticides. The evaluation of efficacy and risk of the introduced biopesticide is assessed on the basis of scientific data and proof, and a different set of rules are designed as compared to chemical pesticides. Presently, result validation and regulatory guidance are being accurately required for biopesticides (Isman, 2015). Deposition criteria both at EU and Member State stages are tedious which the biggest obstacle in biopesticide development and release. It is time taking process for developed products to reach to consumers instantaneously, as strict procedures and time bounds are implied. Also, a comparatively higher price provoked due to the registration of the product is another drawback hindering the commercialization of microbial pesticides (Pavela, 2014). Thus for easy commercialization in this market, the registration process should be eased and simplified. The regulatory framework should thus guarantee to hasten registration of microbial pesticide-based products, advocating the application of secure technical and proper regulatory guideline from beginning itself. Additionally, the regulatory bodies should permit small enterprises to develop biopesticides, so that they can give farmers decent, cost-effective product, as per their requirement (Damalas and Koutroubas, 2018).

20.6 Microbial pesticides in the post-genomic era

Nearly 90% of the microbial biopesticides presently available in the market are Bt or *Bacillus thuringiensis* based (Kumar and Singh, 2015). Biopesticides account for nearly 5% of the total pesticide market worldwide, with a value of around $3 billion (Olson, 2015). With the availability of huge genomic databases, researchers all over the world are now focusing on the identification of genes favored by host plants, genes functioning in pest resistance, or immune response. The onset of the post-genomics era commits toward the improvement of plant-health by assisting in the identification of new targets for biopesticides and enhancing our knowledge of functional genes and mechanisms involved in the pathogenicity of microbes and pest resistance. Data mining and application of bioinformatics play crucial roles in the analysis and interpretation of the results obtained from large-scale sequencing platforms (Ngai and McDowell, 2017). The advent of genome sequencing and comparative genomics permits the subsidization of minute differences at the species level, providing propulsive microbial pesticide development with high specificity for an individual host-pathogen system with less or no toxic side effects toward for non-target organisms.

Comparative genomics has also assisted in unraveling the mechanisms involved in pesticide resistance. The incoming draft genomes of pathogens with gene annotations have helped in the excavating of protein targets for biopesticides. The application of Next generation sequencing provides new tools for improving our understanding of the complex cellular processes, allowing us to modulate plant pathogens and make the plants to respond faster to upcoming pathogens (Sarrocco et al., 2020). A profound knowledge into the microbial genes at the whole genome level can help us in understanding the inter-species or intra-species genetic diversity, so the most efficient biocontrol strain can be selected. A whole-genome sequencing of *Beauveria bassiana* (*Bb*) JEF-007 was done and genes involved in antagonism were characterized and were compared with other *Bb* isolates. The genetic diversity in *B. bassiana* can be exploited for the development of potential biopesticides with greater efficacy (Lee et al., 2018). The whole-genome sequencing of *Bacillus thuringiensis* DAR 81934 (a Bt strain with molluscicidal activity) unveiled six Bt toxin genes in both plasmid and chromosome sequences (Wang et al., 2013). An alternative strategy using attenuated or avirulent pathogens to compete for space and nutrients and/or reprogramming of plant defenses against plant pathogens can be advocated (Ghorbanpour et al., 2018). Employment of CRISPR-Cas9 has revolutionized genome editing technology and has allowed researchers to customize genomic sequences in a more decisive way (Knott and Doudna, 2018) of host plants or pathogens, as well as of beneficial microorganisms. CRISPR/Cas9 can be applied for the editing of well-known biocontrol agents with enhanced disease control ability (Vicente Muñoz et al., 2017). The silencing of the *ace1* gene in *Trichoderma atroviride* resulted in the up-regulation of four polyketide biosynthetic gene clusters, causing an increase in the antibiotics and secondary metabolites production, thereby, improving biocontrol potential against *Fusarium oxysporum* and *Rhizoctonia solani* (Fang and Chen, 2018). In a study, non-virulent CRISPR-mutant strains of *F. graminearum* and *F. culmorum* were released in the field to manage Fusarium Head Blight, as a strategy to compete with the pathogenic strain (Urban et al., 2003). CRISPR-Cas9 mediated fungal genome editing was used to introduce specific point mutations as well as gene deletions in *Aspergillus nidulans* (Nødvig et al., 2018). So in future, we can expect engineered or genome-edited microbes showing avirulent or less virulence (vaccine) as a pesticide in markets. A metagenomic approach using *Fusarium*-specific primers with universal primers for bacteria and fungi was applied to study the microbial communities and to unravel correlations between *Fusarium* spp. and other microbes (Cobo-Díaz et al., 2019). Similar studies can be used for screening novel antagonistic microbes against other plant pathogens for development of the efficient biopesticide formulations.

20.7 Future prospects

Biopesticides have protracted the global spotlight as a safer strategy compared to chemical pesticides, with a conceivably lower threat to humans and the surrounding. Despite the relatively huge number of patents for microbial pesticides, the commercialization process has not experienced impressive results. Search for organisms and/or consortium of microbes, research on formulation development and delivery systems can promote commercialization of microbial pesticides. Controlling the cost of these formulations would encourage farmers to switch to use the product. Although new products could be potential for use as biocontrol, more field data is needed to confirm the efficiency of microbial pesticide for specific crop plants. Environmental biosafety and regulatory concerns are also restricting the application of these new microbial pesticides. Harm to non-target organisms including beneficial rhizospheric microbes, mycorrhiza, as well as allergenicity, toxicity, or pathogenicity to other organisms including plants, animals, or humans by the microbial pesticide itself or due to contaminants direct for stringent screening and regulation before release or sell. The government should encourage the application of microbial pesticides by funding research and training programmes for their research and development. This could, in turn, promote academic-industrial collaborations for the development of new specific biopesticides. The agencies should also fund the toxicological studies to promote the registration of microbe based pesticides. Government labs at universities and institutes should distribute biopesticides to farmers for encouraging its use and building confidence before commercial production.

Looming food-borne diseases and diseases caused by microbes develops a socially challenging environment for microbial pesticides. Revolution in genomics has greatly added to our understanding of molecular mechanisms underlying pathogenesis, resistance, and the mechanism adopted by beneficial biocontrol agents in controlling the disease. Whole-genome sequencing at a lost cost has created interest for genome-based researches for applying the data to answer complex queries linked genes involved in cellular defense response, genes responsible for providing immunity, and for cheating host defense responses. The feasibility to edit microbial genomes described using CRISPR/Cas9 technology provides promising means of managing phytopathogens by designing new compelling microbial genotypes to work efficiently in field skipping the need of introducing the transgenes in nature. The future of microbial pesticides in the post-genomic era can be looked up for generating new active biomolecules from these beneficial microorganisms, and in generating engineered biocontrol agents and/or crop plants with genes from these microorganisms.

Acknowledgments

AJ acknowledges Department of Science and Technology women scientist scheme for financial support. SD acknowledges Indian National Science Academy for her senior scientist fellowship.

References

Ahmad, M., Zahir, Z.A., Asghar, H.N., Asghar, M., 2011. Inducing salt tolerance in mung bean through coinoculation with rhizobia and plant-growth-promoting rhizobacteria containing 1-aminocyclopropane-1- carboxylate deaminase. Can. J. Microbiol. 57, 578–589. https://doi.org/10.1139/w11-044.

Ahmad, M., Zahir, Z.A., Nadeem, S.M., Nazli, F., Jamil, M., Khalid, M., 2013. Field evaluation of *Rhizobium* and *Pseudomonas* strains to improve growth, nodulation and yield of mung bean under salt-affected conditions. Soil Environ. 32, 158–166.

Ahmed, T., Shahid, M., Noman, M., Hussain, S., Khan, M.A., Zubair, M., Ismail, M., Manzoor, N., Shahzad, T., Mahmood, F., 2019. Plant growth-promoting rhizobacteria as biological tools for nutrient management and soil sustainability. In: Plant Growth Promoting Rhizobacteria for Agricultural Sustainability. Springer, Singapore, pp. 95–110.

Ali, S., Huang, Z., Ren, S., 2010. Production of cuticle degrading enzymes by *Isaria fumosorosea* and their evaluation as a biocontrol agent against diamondback moth. J. Pest. Sci. 83 (4), 361–370.

All, J.N., Stancil, J.D., Johnson, T.B., Gouger, R., 1994. A genetically-modified *Bacillus thuringiensis* product effective for control of the fall armyworm (Lepidoptera: Noctuidae) on corn. Fla. Entomol. 437–440.

Anbuselvi, S., Rebecca, J., CM, K., 2010. Antifungal activity of *Pseudomonas fluorescens* and its biopesticide effect on plant pathogens. Natl. J. Chembiosis 1 (1).

Anderson, A.J., Kim, Y.C., 2018. Biopesticides produced by plant-probiotic *Pseudomonas chlororaphis* isolates. Crop Protect. 105, 62–69.

Andreolli, M., Zapparoli, G., Angelini, E., Lucchetta, G., Lampis, S., Vallini, G., 2019. Pseudomonas protegens MP12: a plant growth-promoting endophytic bacterium with broad-spectrum antifungal activity against grapevine phytopathogens. Microbiol. Res. 219, 123–131.

Angelo, I.C., Fernandes, É.K., Bahiense, T.C., Perinotto, W.M., Golo, P.S., Moraes, A.P.R., Bittencourt, V.R., 2012. Virulence of *Isaria* sp. and *Purpureocillium lilacinum* to *Rhipicephalus microplus* tick under laboratory conditions. Parasitol. Res. 111 (4), 1473–1480.

Arcury, T.A., Quandt, S.A., Russell, G.B., 2002. Pesticide safety among farmworkers: perceived risk and perceived control as factors reflecting environmental justice. Environ. Health Perspect. 110 (Suppl. 2), 233–240.

Arora, N.K., Tewari, S., Singh, R., 2013. Multifaceted plant-associated microbes and their mechanisms diminish the concept of direct and indirect PGPRs. In: Plant Microbe Symbiosis: Fundamentals and Advances. Springer, New Delhi, pp. 411–449.

Assar, A.A., El-Mahasen, M.A., Dahi, H.F., Amin, H.S., 2016. Biochemical effects of some insect growth regulators and bioinsecticides against cotton leafworm, *Spodoptera littoralis* (Boisd.)(Lepidoptera: Noctuidae). J. Biosci. & Appl. Res. 8, 582–589.

Avery, P.B., Wekesa, V.W., Hunter, W.B., Hall, D.G., McKenzie, C.L., Osborne, L.S., Powell, C.A., Rogers, M.E., 2011. Effects of the fungus *Isaria fumosorosea* (Hypocreales: Cordycipitaceae) on reduced feeding and mortality of the Asian citrus psyllid, *Diaphorina citri* (Hemiptera: Psyllidae). Biocontrol Sci. Technol. 21 (9), 1065–1078.

Bauer, L.S., 1995. Resistance: a threat to the insecticidal crystal proteins of *Bacillus thuringiensis*. Fla. Entomol. 78, 414–443.

Bing, L.A., Lewis, L.C., 1991. Suppression of *Ostrinia nubilalis* (Hübner)(Lepidoptera: Pyralidae) by endophytic *Beauveria bassiana* (Balsamo) Vuillemin. Environ. Entomol. 20 (4), 1207–1211.

Bogo, M.R., Queiroz, M.V., Silva, D.M., Giménez, M.P., Azevedo, J.L., Schrank, A., 1996. Double-stranded RNA and isometric virus-like particles in the entomopathogenic fungus *Metarhizium anisopliae*. Mycol. Res. 100 (12), 1468–1472.

Bravo, A., Gill, S.S., Soberon, M., 2007. Mode of action of *Bacillus thuringiensis* Cry and Cyt toxins and their potential for insect control. Toxicon 49 (4), 423–435.

Bull, C.T., Stack, J.P., Smilanick, J.L., 1997. *Pseudomonas syringae* strains ESC-10 and ESC-11 survive in wounds on citrus and control green and blue molds of citrus. Biol. Contr. 8 (1), 81–88.

Cabanillas, H.E., Raulston, J.R., 1996. Evaluation of *Steinernema riobravis, S. carpocapsae*, and irrigation timing for the control of corn earworm, *Helicoverpa zea*. J. Nematol. 28 (1), 75.

Castagnola, A., Stock, S.P., 2014. Common virulence factors and tissue targets of entomopathogenic bacteria for biological control of lepidopteran pests. Insects 5 (1), 139–166.

Chakravarty, G., Kalita, M.C., 2011. Comparative evaluation of organic formulations of *Pseudomonas fluorescens* based biopesticides and their application in the management of bacterial wilt of brinjal (*Solanum melongena* L.). Afr. J. Biotechnol. 10 (37), 7174–7182.

Chandler, D., Davidson, G., Grant, W.P., Greaves, J., Tatchell, G.M., 2008. Microbial biopesticides for integrated crop management: an assessment of environmental and regulatory sustainability. Trends Food Sci. Technol. 19 (5), 275–283.

Cobo-Díaz, J.F., Baroncelli, R., Floch, G.L., Picot, A., 2019. Combined metabarcoding and co-occurrence network analysis to profile the bacterial, fungal and *Fusarium* communities and their interactions in maize stalks. Front. Microbiol. https://doi.org/10.3389/fmicb.2019.00261.

Copping, L.G., Menn, J.J., 2000. Biopesticides: a review of their action, applications and efficacy. Pest Manag. Sci. Formerly Pestic. Sci. 56 (8), 651–676.

Costa, J.A.V., Freitas, B.C.B., Cruz, C.G., Silveira, J., Morais, M.G., 2019. Potential of microalgae as biopesticides to contribute to sustainable agriculture and environmental development. J. Environ. Sci. & Health Part B 54 (5), 366–375.

Cuthbertson, A.G., Walters, K.F., 2005. Pathogenicity of the entomopathogenic fungus, *Lecanicillium muscarium*, against the sweetpotato whitefly *Bemisia tabaci* under laboratory and glasshouse conditions. Mycopathologia 160 (4), 315–319.

Damalas, C.A., Koutroubas, S.D., 2018. Current status and recent developments in biopesticide use. Agriculture 8, 13. https://doi.org/10.3390/agriculture8010013.

Daughtrey, M.L., Benson, D.M., 2005. Principles of plant health management for ornamental plants. Annu. Rev. Phytopathol. 43, 141–169.

De Barjac, H., Sutherland, D.J. (Eds.), 2012. Bacterial Control of Mosquitoes and Black Flies: Biochemistry, Genetics & Applications of *Bacillus Thuringiensis Israelensis* and *Bacillus Sphaericus*. Springer.

de la Cruz Quiroz, R., Cruz Maldonado, J., Rostro Alanis, M., Torres, J.A., Saldívar, R.A., 2019. Fungi-based biopesticides: shelf-life preservation technologies used in commercial products. J. Pest. Sci. 92, 1003–1015. https://doi.org/10.1007/s10340-019-01117-5.

Downing, K.J., Leslie, G., Thomson, J.A., 2000. Biocontrol of the sugarcane borer eldana saccharina by expression of the *Bacillus thuringiensis* cry1Ac7 and *Serratia marcescens* chiA genes in sugarcane-associated bacteria. Appl. Environ. Microbiol. 66 (7), 2804–2810.

Enkerli, J., Widmer, F., Keller, S., 2004. Long-term field persistence of *Beauveria brongniartii* strains applied as biocontrol agents against European cockchafer larvae in Switzerland. Biol. Contr. 29 (1), 115–123.

Fang, C., Chen, X., 2018. Potential biocontrol efficacy of *Trichoderma atroviride* with cellulase expression regulator ace1 gene knock-out. Biotechnology 8, 302. https://doi.org/10.1007/s13205-018-1314-z.

FFDCA, 1938. https://ballotpedia.org/Federal_Food,_Drug,_and_Cosmetic_Act_of_1938. (Accessed 12 February 2021).

FIFRA, 1947. https://www.epa.gov/laws-regulations/summary-federal-insecticide-fungicide-and-rodenticide-act. (Accessed 25 March 2021).

Frankenberger Jr., W.T., Arshad, M., 2020. Phytohormones in Soils Microbial Production and Function. CRC Press.

Frederiksa, C., Wesselerb, J.H.H., 2019. A comparison of the EU and US regulatory frameworks for the active substance registration of microbial biological control agents. Pest Manag. Sci. 75, 87–103.

Gao, B.J., Mou, Y.N., Tong, S.M., Ying, S.H., Feng, M.G., 2020. Subtilisin-like Pr1 proteases marking the evolution of pathogenicity in a wide-spectrum insect-pathogenic fungus. Virulence 11 (1), 365–380.

Gao, P., Li, M., Jin, K., Xia, Y., 2019. The homeobox gene MaH1 governs microcycle conidiation for increased conidial yield by mediating transcription of conidiation pattern shift-related genes in *Metarhizium acridum*. Appl. Microbiol. Biotechnol. 103 (5), 2251–2262.

Ghorbanpour, M., Omidvari, M., Abbaszadeh-Dahaji, P., Omidvar, R., Kariman, K., 2018. Mechanisms underlying the protective effects of beneficial fungi against plant diseases. Biol. Contr. 117, 147–157. https://doi.org/10.1016/j.biocontrol.2017.11.006.

Ghosh, G., Ganguly, S., Purohit, A., Chaudhuri, R.K., Das, S., Chakraborti, D., 2017. Transgenic pigeonpea events expressing Cry1Ac and Cry2Aa exhibit resistance to *Helicoverpa armigera*. Plant Cell Rep. 36 (7), 1037–1051.

Glick, B.R., 2012. Plant Growth-Promoting Bacteria: Mechanisms and Applications. Scientifica. https://doi.org/10.6064/2012/963401. Article ID 963401.

Goettel, M.S., Koike, M., Kim, J.J., Aiuchi, D., Shinya, R., Brodeur, J., 2008. Potential of *Lecanicillium* spp. for management of insects, nematodes and plant diseases. J. Invertebr. Pathol. 98 (3), 256–261.

Grady, E.N., MacDonald, J., Ho, M.T., Weselowski, B., McDowell, T., Solomon, O., Renaud, J., Yuan, Z.C., 2019. Characterization and complete genome analysis of the surfactin-producing, plant-protecting bacterium *Bacillus velezensis* 9D-6. BMC Microbiol. 19 (1), 1–14.

Han, R., Ehlers, R.U., 2000. Pathogenicity, development, and reproduction of *Heterorhabditis bacteriophora* and *Steinernema carpocapsae* under axenic in vivo conditions. J. Invertebr. Pathol. 75 (1), 55–58.

Hauschild, R., Speiser, B., Tamm, L., 2011. Regulation according to EU directive 91/414: data requirements and procedure compared with regulation practice in other OECD countries. In: Ehlers, R.-U. (Ed.), Regulation of Biological Control Agents. Springer, Dordrecht, pp. 25–77.

Herrero, N., Dueñas, E., Quesada-Moraga, E., Zabalgogeazcoa, I., 2012. Prevalence and diversity of viruses in the entomopathogenic fungus *Beauveria bassiana*. Appl. Environ. Microbiol. 78 (24), 8523–8530.

Hinnekens, P., Koné, K.M., Fayad, N., Leprince, A., Mahillon, J., 2019. pXO16, the large conjugative plasmid from *Bacillus thuringiensis* serovar *israelensis* displays an extended host spectrum. Plasmid 102, 46–50.

Holmes, A., Govan, J., Goldstein, R., 1998. Agricultural use of *Burkholderia* (*Pseudomonas*) *cepacia:* a threat to human health? Emerg. Infect. Dis. 4 (2), 221. www.epa.gov/pesticides/biopesticides. https://www.wipo.int/treaties/en/registration/budapest/summary_budapest.html.

Huang, A., Lu, M., Ling, E., Li, P., Wang, C., 2020. A M35 family metalloprotease is required for fungal virulence against insects by inactivating host prophenoloxidases and beyond. Virulence 11 (1), 222–237.

Isman, M.B., 2015. A renaissance for botanical insecticides? Pest Manag. Sci. 71, 1587–1590.

Jagdale, G.B., Somasekhar, N., Grewal, P.S., Klein, M.G., 2002. Suppression of plant-parasitic nematodes by application of live and dead infective juveniles of an entomopathogenic nematode, *Steinernema carpocapsae*, on boxwood (*Buxus* spp.). Biol. Contr. 24 (1), 42–49.

Jain, A., Chakraborty, J., Das, S., 2020. Underlying mechanism of plant–microbe crosstalk in shaping microbial ecology of the rhizosphere. Acta Physiol. Plant. 42 (1), 1–13.

Jain, A., Singh, A., Singh, B.N., Singh, S., Upadhyay, R.S., Sarma, B.K., Singh, H.B., 2013. Biotic stress management in agricultural crops using microbial consortium. In: Maheshwari, D.K. (Ed.), Bacteria in Agrobiology. Springer-Verlag, pp. 427–448.

Jain, A., Singh, A., Singh, S., Singh, V., Singh, H.B., 2015. Phenols enhancement effect of microbial consortium in pea plants restrains *Sclerotinia sclerotiorum*. Biol. Contr. 89, 23–32.

Javanshir Javid, K., Mahdian, S., Behboudi, K., Alizadeh, H., 2016. Biological control of *Fusarium oxysporum* f. sp. radicis-cucumerinum by some *Trichoderma harzianum* isolates. Arch. Phytopathol. Plant Protect. 49, 471–484.

Je, Y.H., Choi, J.Y., Kim, S.E., Kim, J.S., 2016. U.S. Patent. Mutant *Bacillus thuringiensis* Proteins and Genes Encoding the Same with Improved Insecticidal Activity and Use There of, vol. 9. Seoul National University R&DB Foundation, p. 187 (512).

Johnson, D.L., 1997. Nosematidae and other Protozoa as agents for control of grasshoppers and locusts: current status and prospects. Mem. Entomol. Soc. Can. 129 (S171), 375–389.

Jones, A.M., Dodd, M.E., Webb, A.K., 2001. *Burkholderia cepacia*: current clinical issues, environmental controversies and ethical dilemmas. Eur. Respir. J. 17 (2), 295–301.

Jurado, M.M., Suárez-Estrella, F., López, M.J., López-González, J.A., Moreno, J., 2019. Bioprospecting from plant waste composting: *Actinobacteria* against phytopathogens producing damping-off. Biotechnol. Rep. 23, e00354.

Kang, S.M., Khan, A.L., Waqas, M., Asaf, S., Lee, K.E., Park, Y.G., Kim, A.Y., Khan, M.A., You, Y.H., Lee, I.J., 2019. Integrated phytohormone production by the plant growth-promoting rhizobacterium *Bacillus tequilensis* SSB07 induced thermotolerance in soybean. J. Plant Interact. 14 (1), 416–423.

Kanga, L.H.B., James, R.R., Boucias, D.G., 2002. *Hirsutella thompsonii* and *Metarhizium anisopliae* as potential microbial control agents of Varroa destructor, a honey bee parasite. J. Invertebr. Pathol. 81 (3), 175–184.

Keppanan, R., Krutmuang, P., Sivaperumal, S., Hussain, M., Bamisile, B.S., Aguila, L.C.R., Dash, C.K., Wang, L., 2019. Synthesis of mycotoxin protein IF8 by the entomopathogenic fungus *Isaria fumosorosea* and its toxic effect against adult *Diaphorina citri*. Int. J. Biol. Macromol. 125, 1203–1211.

Kermani, N., Abu-Hassan, Z.A., Dieng, H., Ismail, N.F., Attia, M., Ghani, I.A., 2013. Pathogenicity of *Nosema* sp.(microsporidia) in the diamondback moth, *Plutella xylostella* (Lepidoptera: Plutellidae). PloS One 8 (5).

Khan, M.S., Zaidi, A., Wani, P.A., Oves, M., 2009. Role of plant growth promoting rhizobacteria in the remediation of metal contaminated soils. Environ. Chem. Lett. 7, 1–19. https://doi.org/10.1007/s10311-008-0155-0.

Kim, J.J., Goettel, M.S., Gillespie, D.R., 2008. Evaluation of Lecanicillium longisporum, Vertalec® for simultaneous suppression of cotton aphid, Aphis gossypii, and cucumber powdery mildew, Sphaerotheca fuliginea, on potted cucumbers. Biol. Cont. 45, 404–409.

Kim, J.H., Kim, S.E., Choi, J.Y., Liu, Q., Lee, S.H., Fang, Y., Ha, K.B., Park, D.H., Kim, W.J., Je, Y.H., 2016. Improved insecticidal activities of novel *Bacillus thuringiensis* Cry1-type genes. J. Asia Pac. Entomol. 19 (1), 145–151.

Kloepper, J.W., Leong, J., Teintze, M., Schroth, M.N., 1980. Enhanced plant growth by siderophores produced by plant growth-promoting rhizobacteria. Nature 286 (5776), 885–886.

Knott, G.J., Doudna, J.A., 2018. CRISPR-Cas guides the future of genetic engineering. Science 361, 866–869. https://doi.org/10.1126/science.aat5011.

Koehler, P.G., Patterson, R.S., Martin, W.R., 1992. Susceptibility of cockroaches (Dictyoptera: Blattellidae, Blattidae) to infection by *Steinernema carpocapsae*. J. Econ. Entomol. 85 (4), 1184–1187.

Kruitbos, L.M., Heritage, S., Hapca, S., Wilson, M.J., 2010. The influence of habitat quality on the foraging strategies of the entomopathogenic nematodes *Steinernema carpocapsae* and *Heterorhabditis megidis*. Parasitology 137 (2), 303–309.

Kumar, K.K., Sridhar, J., Murali-Baskaran, R.K., Senthil-Nathan, S., Kaushal, P., Dara, S.K., Arthurs, S., 2018. Microbial biopesticides for insect pest management in India: current status and future prospects. J. Invertebr. Pathol. 165, 74–81.

Kumar, S., Singh, A., 2015. Biopesticides: present status and the future prospects. J. Fertil. & Pestic. 6, e129.

Lake, L.K., Shafer, W.E., Reilly, S.K., Jones, R.S., 2002. Regulation of biochemical plant growth regulators at the US Environmental Protection Agency. Hort Technol. 12 (1), 55–58.

Lange, C.E., 2005. The host and geographical range of the grasshopper pathogen *Paranosema* (Nosema) locustae revisited. J. Orthoptera Res. 14 (2), 137–141.

Lange, C.E., Bardi, C., Plischuk, S., 2017. Infectivity of *Paranosema locustae* (microsporidia) to the 'quebrachera' grasshopper, *Tropidacris collaris* (Orthoptera: Romaleidae), in the laboratory. Rev. Soc. Entomol. Argent. 67 (3–4).

Lee, S.J., Lee, M.R., Kim, S., Kim, J.C., Park, S.E., Li, D., Shin, T.Y., Nai, Y.S., Kim, J.S., 2018. Genomic analysis of the insect-killing fungus *Beauveria bassiana* JEF-007 as a biopesticide. Sci. Rep. 8 (1), 1–12.

Li, D., Zhang, G., Huang, L., Wang, Y., Yu, H., 2019. Complete mitochondrial genome of the important entomopathogenic fungus *Cordyceps tenuipes* (Hypocreales, Cordycipitaceae). Mitochondrial DNA Part B 4 (1), 1329–1331.

LiPuma, J.J., Mahenthiralingam, E., 1999. Commercial use of *Burkholderia cepacia*. Emerg. Infect. Dis. 5 (2), 305.

Lopez, D.C., Sword, G.A., 2015. The endophytic fungal entomopathogens *Beauveria bassiana* and *Purpureocillium lilacinum* enhance the growth of cultivated cotton (*Gossypium hirsutum*) and negatively affect survival of the cotton bollworm (*Helicoverpa zea*). Biol. Contr. 89, 53–60.

Lopez, D.C., Zhu-Salzman, K., Ek-Ramos, M.J., Sword, G.A., 2014. The entomopathogenic fungal endophytes *Purpureocillium lilacinum* (formerly *Paecilomyces lilacinus*) and *Beauveria bassiana* negatively affect cotton aphid reproduction under both greenhouse and field conditions. PloS One 9 (8).

Luo, Y., Cheng, Y., Yi, J., Zhang, Z., Luo, Q., Zhang, D., Li, Y., 2018. Complete genome sequence of industrial biocontrol strain *Paenibacillus polymyxa* HY96-2 and further analysis of its biocontrol mechanism. Front. Microbiol. 9, 1520.

Ma, W., Chen, H., Jiang, X., Wang, J., Gelbič, I., Guan, X., Zhang, L., 2020. Whole genome sequence analysis of the mosquitocidal *Bacillus thuringiensis* LLP29. Arch. Microbiol. 1–8.

McGuire, M.R., Tamez-Guerra, P., Behle, R.W., Streett, D.A., 2001. Comparative field stability of selected entomopathogenic virus formulations. J. Econ. Entomol. 94 (5), 1037–1044.

Miethke, M., Marahiel, M.A., 2007. Siderophore-based iron acquisition and pathogen control. Microbiol. Mol. Biol. Rev. 71 (3), 413–451.

Min, L.J., Wu, X.Q., Li, D.W., Chen, K., Guo, L., Ye, J.R., 2019. *Burkholderia pyrrocinia* JK-SH007 enhanced seed germination, cucumber seedling growth and tomato fruit via catecholate-siderophore-mediation. Int. J. Agric. Biol. 22 (4), 779–786.

Mokrani, S., Bejaoui, B., Belabid, L., Nabti, E., 2019. Potential of *Pseudomonas stutzeri* strains isolated from rhizospheric soil endowed with antifungal activities against phytopathogenic fungus *Stemphylium botryosum*. Asian J. Agric. 3 (02).

Molloy, D.P., Mayer, D.A., Giamberini, L., Gaylo, M.J., 2013. Mode of action of *Pseudomonas fluorescens* strain CL145A, a lethal control agent of dreissenid mussels (Bivalvia: Dreissenidae). J. Invertebr. Pathol. 113 (1), 115–121.

Montesinos, E., 2003. Development, registration and commercialization of microbial pesticides for plant protection. Int. Microbiol. 6, 245–252.

Moscardi, F., de Souza, M.L., de Castro, M.E.B., Moscardi, M.L., Szewczyk, B., 2011. Baculovirus pesticides: present state and future perspectives. In: Microbes and Microbial Technology. Springer, New York, pp. 415–445.

Nadeem, S.M., Ahmad, M., Zahir, Z.A., Javaid, A., Ashraf, M., 2014. The role of mycorrhizae and plant growth promoting rhizobacteria (PGPR) in improving crop productivity under stressful environments. Biotechnol. Adv. 32, 429–448. https://doi.org/10.1016/j.biotechadv.2013.12.005.

Nelson, A.R.L.E., Ravichandran, K., Antony, U., 2019. The impact of the green revolution on indigenous crops of India. J. Ethn. Foods 6 (1), 8.

Ngai, M., McDowell, M.A., 2017. The search for novel insecticide targets in the post-genomics era, with a specific focus on G-protein coupled receptors. The Memórias do Instituto Oswaldo Cruz 112 (1). https://doi.org/10.1590/0074-02760160345.

Nødvig, C.S., Hoof, J.B., Kogle, M.E., Jarczynska, Z.D., Lehmbeck, J., Klitgaard, D.K., Mortensen, U.H., 2018. Efficient oligo nucleotide mediated CRISPR-Cas9 gene editing in *Aspergilli*. Fungal Genet. Biol. 115, 78–89. https://doi.org/10.1016/j.fgb.2018.01.004.

Olson, S., 2015. An analysis of the biopesticide market now and where is going. Outlooks Pest Manag. 26, 203–206.

Palma, L., Muñoz, D., Berry, C., Murillo, J., Caballero, P., 2014. *Bacillus thuringiensis* toxins: an overview of their biocidal activity. Toxins 6 (12), 3296–3325.

Pandey, P., Irulappan, V., Bagavathiannan, M.V., Senthil-Kumar, M., 2017. Impact of combined abiotic and biotic stresses on plant growth and avenues for crop improvement by exploiting physio-morphological traits. Front. Plant Sci. 8, 537.

Park, N.S., Jin, B.R., Lee, S.M., 2019. Genomic structure of the Cu/Zn superoxide dismutase (SOD1) gene from the entomopathogenic fungus, *Cordyceps pruinosa*. Int. J. Ind. Entomol. 39 (2), 67–73.

Parker, M.W., Feil, S.C., 2005. Pore-forming protein toxins: from structure to function. Prog. Biophys. Mol. Biol. 88 (1), 91–142.

Pavela, R., 2014. Limitation of plant biopesticides. In: Singh, D. (Ed.), Advances in Plant Biopesticides. Springer, New Delhi, India, pp. 347–359.

Peng, Y., Wang, L., Gao, Y., Ye, L., Xu, H., Li, S., Jiang, J., Li, G., Dang, X., 2020. Identification and characterization of the glycoside hydrolase family 18 genes from the entomopathogenic fungus *Isaria cicadae* genome. Can. J. Microbiol. 66 (4), 274–287.

Perry, R.N., Hominick, W.M., Beane, J., Briscoe, B., 1998. Effect of the entomopathogenic nematodes, *Steinernema feltiae* and *S. carpocapsae*, on the potato cyst nematode, *Globodera rostochiensis* in pot trials. Biocontr. Sci. Technol. 8 (1), 175–180.

Persley, D., Gambley, C., 2013. Epidemiology and Management of Whitefly-Transmitted Viruses—A Cross-Industry Workshop.

Philippot, L., Raaijmakers, J.M., Lemanceau, P., Van Der Putten, W.H., 2013. Going back to the roots: the microbial ecology of the rhizosphere. Nat. Rev. Microbiol. 11 (11), 789–799.

Pirali-Kheirabadi, K., Haddadzadeh, H., Razzaghi-Abyaneh, M., Bokaie, S., Zare, R., Ghazavi, M., Shams-Ghahfarokhi, M., 2007. Biological control of *Rhipicephalus* (*Boophilus*) *annulatus* by different strains of *Metarhizium anisopliae, Beauveria bassiana* and *Lecanicillium psalliotae* fungi. Parasitol. Res. 100 (6), 1297–1302.

Preti, F., 2018. Effects of a Microsporidium Pathogen, *Nosema adaliae*, on the General Predator Chinese Praying Mantis. Tenodera sinensis.

Regulation, 2009. Regulation No. 1107/2009 of the European parliament and of the council of 21 October 2009 concerning the placing of plant protection products on the market and repealing council directives 79/117/EEC and 91/414/EEC. OJL 309, 1–50.

Romeis, J., Sharma, H.C., Sharma, K.K., Das, S., Sarmah, B.K., 2004. The potential of transgenic chickpeas for pest control and possible effects on non-target arthropods. Crop Protect. 23 (10), 923–938.

Ruiu, L., 2015. Insect pathogenic bacteria in integrated pest management. Insects 6 (2), 352–367.

Ruiu, L., 2018. Microbial biopesticides in agroecosystems. Agronomy 8 (11), 235.

Sala, S., McLaren, S.J., Notarnicola, B., Saouter, E., Sonesson, U., 2017. In quest of reducing the environmental impacts of food production and consumption. J. Clean. Prod. 140, 387–398.

Samah, N., Abdel-Monem, M.O., Abou-Taleb, K.A., Osman, H.S., El-Sharkawy, R.M., 2019. Production of plant growth regulators by some fungi isolated under salt stress. South Asian J. Res. Microbiol. 1–10.

Sarkar, A., Hess, D., Mondal, H.A., Banerjee, S., Sharma, H.C., Das, S., 2009. Homodimeric alkaline phosphatase located at *Helicoverpa armigera* midgut, a putative receptor of Cry1Ac contains α-GalNAc in terminal glycan structure as interactive epitope. J. Proteome Res. 8, 1838–1848. https://doi.org/10.1021/pr8006528.

Sarrocco, S., Herrera-Estrella, A., Collinge, D.B., 2020. Editorial: plant disease management in the post-genomic era: from functional genomics to genome editing. Front. Microbiol. 1, 107. https://doi.org/10.3389/fmicb.2020.00107.

Seiber, J.N., Coats, J., Duke, S.O., Gross, A.D., 2014. Biopesticides: state of the art and future opportunities. J. Agric. Food Chem. 62 (48), 11613–11619.

Sellamuthu, G., Narayanasamy, P., Padaria, J.C., 2018. *Bacillus thuringiensis:* genetic engineering for insect pest management. Micr. Clim. Resilient Agric. 255–278.

Sengupta, A., Sarkar, A., Priya, P., Dastidar, S.G., Das, S., 2013. New insight to structure-function relationship of GalNAc mediated primary interaction between insecticidal Cry1Ac toxin and HaALP receptor of *Helicoverpa armigera.* PloS One 8 (10). https://doi.org/10.1371/journal.pone.0078249.

Shah, P.A., Pell, J.K., 2003. Entomopathogenic fungi as biological control agents. Appl. Microbiol. Biotechnol. 61 (5–6), 413–423.

Singh, A., Jain, A., Sarma, B.K., Upadhyay, R.S., Singh, H.B., 2014. Rhizosphere competent microbial consortium mediates rapid changes in phenolic profiles in chickpea during *Sclerotium rolfsii* infection. Microbiol. Res. 169, 353–360.

Singh, P.C., Nautiyal, C.S., 2012. A novel method to prepare concentrated conidial biomass formulation of *Trichoderma harzianum* for seed application. J. Appl. Microbiol. 113, 1442–1450.

Singh, S., Pandey, R.K., Goswami, B.K., 2013. Bio-control activity of *Purpureocillium lilacinum* strains in managing root-knot disease of tomato caused by *Meloidogyne incognita.* Biocontrol Sci. Technol. 23 (12), 1469–1489.

Sinha, B., 2012. Global biopesticide research trends: a bibliometric assessment. Indian J. Agric. Sci. 82 (2), 95–101.

St Leger, R.J., Joshi, L., Bidochka, M.J., Roberts, D.W., 1996. Construction of an improved mycoinsecticide overexpressing a toxic protease. Proc. Natl. Acad. Sci. U. S. A. 93 (13), 6349–6354.

Starnes, R.L., Liu, C.L., Marrone, P.G., 1993. History, use, and future of microbial insecticides. Am. Entomol. 39 (2), 83–91.

Stock, C.A., McLoughlin, T.J., Klein, J.A., Adang, M.J., 1990. Expression of a *Bacillus thuringiensis* crystal protein gene in *Pseudomonas cepacia* 526. Can. J. Microbiol. 36 (12), 879–884.

Sun, P.F., Chien, I.A., Xiao, H.S., Fang, W.T., Hsu, C.H., Chou, J.Y., 2019. Intra specific variation in plant growth-promoting traits of *Aureobasidium pullulans.* Chiang Mai J. Sci. 46 (1), 15–31.

Tan, S.Q., Yin, Y., Cao, K.L., Zhao, X.X., Wang, X.Y., Zhang, Y.X., Shi, W.P., 2020. Effects of a combined infection with *Paranosema locustae* and *Beauveria bassiana* on *Locusta migratoria* and its gut microflora. Insect Sci. https://doi.org/10.1111/1744-7917.12776.

Thakore, Y., 2006. The biopesticide market for global agricultural use. Ind. Biotechnol. 2 (3), 194–208.

Urban, M., Mott, E., Farley, T., Hammond-Kosack, K., 2003. The *Fusarium graminearum* MAP1 gene is essential for pathogenicity and development of perithecia. Mol. Plant Pathol. 4, 347–359. https://doi.org/10.1046/j.1364-3703.2003.00183.x.

Valicente, F.H., 2019. Entomopathogenic viruses. In: Natural Enemies of Insect Pests in Neotropical Agroecosystems. Springer, Cham, pp. 137–150.

Van der Merwe, M., Jukes, M.D., Rabalski, L., Knox, C., Opoku-Debrah, J.K., Moore, S.D., Krejmer-Rabalska, M., Szewczyk, B., Hill, M.P., 2017. Genome analysis and genetic stability of the *Cryptophlebia leucotreta* Granulovirus (CrleGV-SA) after 15 years of commercial use as a biopesticide. Int. J. Mol. Sci. 18 (11), 2327.

Van Munster, M., Janssen, A., Clérivet, A., Van den Heuvel, J., 2005. Can plants use an entomopathogenic virus as a defense against herbivores? Oecologia 143 (3), 396–401.

Vasconcellos, L.R.C., Carvalho, L.M.F., Silveira, F.A., Gonçalves, I.C., Coelho, F.S., Talyuli, O.A., e Silva, T.L.A., Bastos, L.S., Sorgine, M.H., Reis, L.A., Dias, F.A., 2019. Natural infection by the protozoan *Leptomonas wallacei* impacts the morphology, physiology, reproduction, and lifespan of the insect *Oncopeltus fasciatus.* Sci. Rep. 9 (1), 1–13.

Vasil'eva, M.A., Bessolitsina, E.A., Merzlyak, E.M., Kolesnikov, A.A., 2004. Identification of the 12S rRNA gene promoter in *Leptomonas seymouri* mitochondrial DNA. Mol. Biol. 38 (6), 839–843.

Vicente Muñoz, I., Sarrocco, S., Vannacci, G., 2017. CRISPR-CAS for the genome editing of two *Trichoderma* spp. beneficial isolates. J. Plant Pathol. 99, S63. https://doi.org/10.1038/srep45763.

Wang, A., Pattemore, J., Ash, G., Williams, A., Hane, J., 2013. Draft genome sequence of *Bacillus thuringiensis* strain DAR 81934, which exhibits molluscicidal activity. Genome Announc. 1 (2), e0017512. https://doi.org/10.1128/genomeA.00175-12.

Wang, J.B., Leger, R.S., Wang, C., 2016. Advances in genomics of entomopathogenic fungi. Adv. Genet. 94, 67–105.

Wang, L., Zhang, S., Li, J.H., Zhang, Y.J., 2018. Mitochondrial genome, comparative analysis and evolutionary insights into the entomopathogenic fungus *Hirsutella thompsonii.* Environ. Microbiol. 20 (9), 3393–3405.

Wang, Y., Wang, T., Qiao, L., Zhu, J., Fan, J., Zhang, T., Wang, Z.X., Li, W., Chen, A., Huang, B., 2017. DNA methyltransferases contribute to the fungal development, stress tolerance and virulence of the entomopathogenic fungus *Metarhizium robertsii.* Appl. Microbiol. Biotechnol. 101 (10), 4215–4226.

Wolff, J.L.C., Valicente, F.H., Martins, R., de Castro Oliveira, J.V., de Andrade Zanotto, P.M., 2008. Analysis of the genome of *Spodoptera frugiperda* nucleopolyhedrovirus (SfMNPV-19) and of the high genomic heterogeneity in group II nucleopolyhedroviruses. J. Gen. Virol. 89 (5), 1202–1211.

Xiao, Y., Wang, X., Chen, W., Huang, Q., 2017. Isolation and identification of three potassium-solubilizing bacteria from rape rhizospheric soil and their effects on ryegrass. Geomicrobiol. J. 34, 873–880. https://doi.org/10.1080/01490451.2017.1286416.

Xiong, C., Tong, X., Chen, H., Geng, S., Zhuang, T., Zheng, Y., Fu, Z., Chen, D., Zhao, H., Guo, R., 2019. Optimization of gene structure and identification of novel genes in *Nosema ceranae.* J. Environ. Entomol. 41 (2), 373–379.

Xu, J., He, Q., Ma, Z., Li, T., Zhang, X., Debrunner-Vossbrinck, B.A., Zhou, Z., Vossbrinck, C.R., 2016. The genome of *Nosema* sp. Isolate YNPr: a comparative analysis of genome evolution within the *Nosema/Vairimorpha* clade. PloS One 11 (9).

Yurchenko, V., Lukeš, J., Xu, X., Maslov, D.A., 2006. An integrated morphological and molecular approach to a new species description in the Trypanosomatidae: the case of *Leptomonas podlipaevi* n. sp., a parasite of *Boisea rubrolineata* (Hemiptera: Rhopalidae). J. Eukaryot. Microbiol. 53 (2), 103–111.

Zalidis, G., Stamatiadis, S., Takavakoglou, V., Eskridge, K., Misopolinos, N., 2002. Impacts of agricultural practices on soil and water quality in the Mediterranean region and proposed assessment methodology. Agric. Ecosyst. Environ. 88 (2), 137–146.

Zhang, H.S., Wu, X.H., Li, G., Qin, P., 2011. Interactions between arbuscular mycorrhizal fungi and phosphate-solubilizing fungus (*Mortierella* sp.) and their effects on *Kostelelzkya virginica* growth and soil enzyme activities of rhizosphere and bulk soils at different salinities. Biol. Fertil. Soils 47, 543–554. https://doi.org/10.1007/s00374-011-0563-3.

Zhang, M., Wei, Q., Xia, Y., Jin, K., 2019. MaPacC, a pH-responsive transcription factor, negatively regulates thermotolerance and contributes to conidiation and virulence in *Metarhizium acridum*. Curr. Genet. 1–12.

Zheng, J., Gao, Q., Liu, L., Liu, H., Wang, Y., Peng, D., Ruan, L., Raymond, B., Sun, M., 2017. Comparative genomics of *Bacillus thuringiensis* reveals a path to specialized exploitation of multiple invertebrate hosts. mBio 8 (4) e00822-17.

Zhioua, E., Lebrun, R.A., Ginsberg, H.S., Aeschlimann, A., 1995. Pathogenicity of *Steinernema carpocapsae* and *S. glaseri* (Nematoda: Steinernematidae) to Ixodes scapularis (Acari: Ixodidae). J. Med. Entomol. 32 (6), 900–905.

Zhou, L., Jiang, H., Jin, K., Sun, S., Zhang, W., Zhang, X., He, Y.W., 2015. Isolation, identification and characterization of rice rhizobacterium *Pseudomonas aeruginosa* PA1201 producing high level of biopesticide "Shenqinmycin" and phenazine-1-carboxamide. Acta Microbiol. Sin. 55 (4), 401–411.

Zimmermann, G., 1993. The entomopathogenic fungus *Metarhizium anisopliae* and its potential as a biocontrol agent. Pestic. Sci. 37 (4), 375–379.

Zimmermann, G., 2008. The entomopathogenic fungi *Isaria farinosa* (formerly *Paecilomyces farinosus*) and the *Isaria fumosorosea* species complex (formerly *Paecilomyces fumosoroseus*): biology, ecology and use in biological control. Biocontrol Sci. Technol. 18 (9), 865–901.

Chapter 21

Microbial biopesticides for sustainable agricultural practices

Indu Kumari[a], Razak Hussain[b,1], Shikha Sharma[c], Geetika[c] and Mushtaq Ahmed[d]

[a]National Institute of Pathology, New Delhi, Delhi, India; [b]Department of Botany, Aligarh Muslim University, Aligarh, Uttar Pradesh, India; [c]Department of Environmental Sciences, School of Earth and Environmental Sciences, Central University of Himachal Pradesh, Kangra, Himachal Pradesh, India; [d]Centre for Molecular Biology, Central University of Jammu, Jammu, Jammu & Kashmir, India

21.1 Introduction

Ever since human started farming, pests are the most imperative factor in reducing agricultural production till date (Abdulkhair and Alghuthaymi, 2016). These pests can be any entity like viruses, bacteria, fungi, nematode, insects, birds, weeds and animals (Yadav et al., 2015). It is estimated that these pests are accountable for eradicating almost 40% of agricultural production directly (Tijjani et al., 2016). Hence, these pests are foremost threat to sustainable agricultural practices (Sharma et al., 2014). This is evident that human population is on rise and for humans survival on earth, food security is a key concern (Singh et al., 2018). Increasing population require more food to rise however, the land used for agriculture is limited and is decreasing day by day (Vitousek et al., 1997). In such circumstances, the main alternative to produce more food from less land is to promote crops yield and reduce the crop loss from the pathogenic microbes and the pests (Singh et al., 2018). But the pests contribute a jeopardized role in reducing the agronomics as whatever the effect of these pests have on the crop, it is irrevocable as well as they also spoil the quality of the foodstuff, due to which the health of the consumers also get negatively affected (Ferrigo et al., 2016; Singh et al., 2018). It is believed that about 50,000 fungal species, 10,000 insect's species, 15,000 nematode species and 1800 weed species are directly responsible for abolition of the crop yield (Koul, 2011). Now the primary solution that comes to avoid these pests is the use of pesticides (Singh et al., 2018). That's why in order to keep crop yields elevated, massive amount of chemical pesticides are used worldwide (Yadav et al., 2015). Hence, the pesticides play an essential role in crop protection. A pesticide is a synthetic chemical material that is used to diminish the effect of the pests on the crop yield (Yadav et al., 2015). Generally, organochlorines (e.g. endosulfan and DDT), organophosphates (e.g. glyphosate) carbamates (e.g. maneb) and synthetic pyrethroids (e.g. cypermethrin) are regularly used in the form of fungicides, herbicides and insecticides to avoid crop destruction from different kind of pests (Yadav and Devi, 2017). But these chemical pesticides are actually associated with awful impacts on the environment as they are highly toxic, persistent in nature and are biomagnified in the food chain (Yadav and Devi, 2017). Moreover, these conventional pesticides not only kill/affect the targeted pests but also affect non-targeted life forms such as earthworms, pollinators (like beetles, fruit flies, bees and birds) and natural predators that are beneficial to the crop plants (Ware, 1980; Yadav and Devi, 2017). These synthetic pesticides also encourage the evolution of pesticide resistant among pest organisms especially which make these pesticides ineffective that are useless in the future (Benítez et al., 2004). It is documented that about 200 weed species are found to be resistant to herbicides and about 500 arthropod species are already immune to insecticides (Hajek, 2004; Heap, 2010). The synthetic pesticides not only contaminate/pollute air, water, soil and food but also impose physical deformations, genetic disorders and various diseases in plants, birds, insects, mammals and humans (Sharma et al., 2018). The increased communal consciousness toward the ill consequences of the man-made pesticides on the environment and the consumers, the pesticide treated food products are being avoided, other safe alternatives are being searched and are gradually used to replace the conventional pesticide

1. Present Address: Department of Botany, Central University of Jammu, Samba, Jammu & Kashmir, India.

Biopesticides. https://doi.org/10.1016/B978-0-12-823355-9.00024-9

practices (Tijjani et al., 2016). Hence a safe environment friendly way to reduce the effect of the pests on crop yield is the use of biopesticides. Biopesticides are biological in origin (Lugtenberg et al., 2002). Biopesticides are the pesticides which are received from a natural substance for instance plants, animals, microbes and some natural resources (Sudakin, 2003). These biopesticides are further categorized as - (1) Microbial biopesticides (2) Plant-Incorporated-Protectants (PIPs) and (3) Biochemical pesticides. Microbial biopesticides include viruses, bacteria, fungi, protozoan and nematodes as active component (Sudakin, 2003). PIPs generally include proteins and genes that are introduced in to the host (called Genetically Modified Organisms [GMOs]) by using genetic engineering practice (Sharma et al., 2018). This amendment in the genetic constitution of the host made them competent of protecting themselves against the pests (Sharma et al., 2018). Biochemical pesticides encompass the biochemical that are gained from natural entities for example-black pepper, garlic oil, corn gluten, citronella oil, neem oil, capsaicin, fatty acids, canola oil and insects pheromones (that hinder the molting, mating and foraging behavior of insects) (Sharma and Malik, 2012; Sharma et al., 2018). Biopesticides are naturally less toxic, effective in small quantity, target specific, cost effective and frequently decompose rapidly in the environment in comparison to conventional non-natural pesticides (Sharma et al., 2018). Biopesticides mechanism of control is preventive rather than curative and their effect on plant physiology is negligible (Tijjani et al., 2016). Comparatively, biopesticides are different in structure (source) and mechanism (mode of action) than artificial pesticides (Sharma et al., 2018). The general mechanism for a biopesticide to eliminate the pests includes competition, antibiosis, hyperparasitism and synergism (Tijjani et al., 2016). When these biopesticides are used as a unit of Integrated Pest Management (IPM) strategy they habitually decline the use of man-made pesticides whereas agricultural production remained significant (Sharma and Malik, 2012). Since the dawn of biopesticides, plenty of ecofriendly commodities/formulations have been registered and launched (Sharma et al., 2018). These biopesticide formulations are easily available in dry (as dustable powders, granules, wettable powders and water dispersible granules) and liquid (as emulsions, suspension concentrates, suspo-emulsions, oil dispersions capsule suspension and ultra low volume liquids) formulations and are commonly used is seed dressing, foliar application and seedling dipping to lessen the effect of pests on crop product (Tijjani et al., 2016). In the majority of the cases active constituent of biopesticides are developed in the equivalent method as conventional artificial pesticides and most suitable for cultivators to employ the similar tool for its application (Slavica and Brankica, 2013). Globally, biopesticides contributes 4.5% of the whole pesticide manufacture. This manufacturing rate is 6% in USA while 3% in India as compared to the chemical pesticide production (Sharma et al., 2018). Presently, in India 12 types of formulations are registered as biopesticides that mostly comprise neem, *Trichoderma*, *Bacillus* and *Pseudomonas* based formulations that are used to control various pests, but in future production of these biological formulations will hike because of their virtue of environmental friendly behavior (Sharma et al., 2018). Hence, in nutshell, we can say that biopesticides are boon to sustainable agricultural practices as they are safe and effective in nature and hold a bright scope in the future in term of their precious role in the IPM approaches.

21.2 Microbial biopesticides

Among all biopesticides about 90% biopesticides are microbial based formulations (Koul, 2011). These microbial biopesticides are safe and highly specific. About 1500 naturally found microorganism are insecticidal in nature (Koul, 2011). In recent years the usage of microbial biopesticides have been hiked due to their low cost, lower risk of pest resistance and production of non-hazardous residues in nature as compared to the conventional chemical pesticides (Sudakin, 2003). These microbial biopesticides even have potential to control vectors of crop diseases (Koul, 2011). Microbial biopesticides used as Biological Control Agents are the largest group of broad-spectrum biopesticides with microorganism (bacteria, fungi, viral particles, nematodes or protozoans groups) as their active ingredients which occur in natural or genetically engineered form (Kachhawa, 2017; Sharma et al., 2018). The toxins produced by the microbial pathogens are generally peptides and are responsible for their pathogenic effects on different target organisms in the species specific manner (Burges, 1981; Pathak et al., 2017). These toxins vary significantly in their structure, specificity, mode of action and toxicity. Toxic metabolites cause disease in specific host and prevent establishment of other microorganisms by various modes of action like competition for space and nutrition (Kachhawa, 2017; Pathak et al., 2017). Microbial biopesticides are of following type depending upon the type of active ingredient present (Fig. 21.1):

21.2.1 Bacterial biopesticides

Bacterial biopesticides involve bacteria based commodities as active ingredient. Since 20th century bacterial populations have been used as biopesticides in control of insects/pests. Bacterial pesticides are widely used cheapest source of pest biocontrol. These bacterial biopesticides can be further divided into four types-crystalliferous spore forming (e.g. *Bacillus*

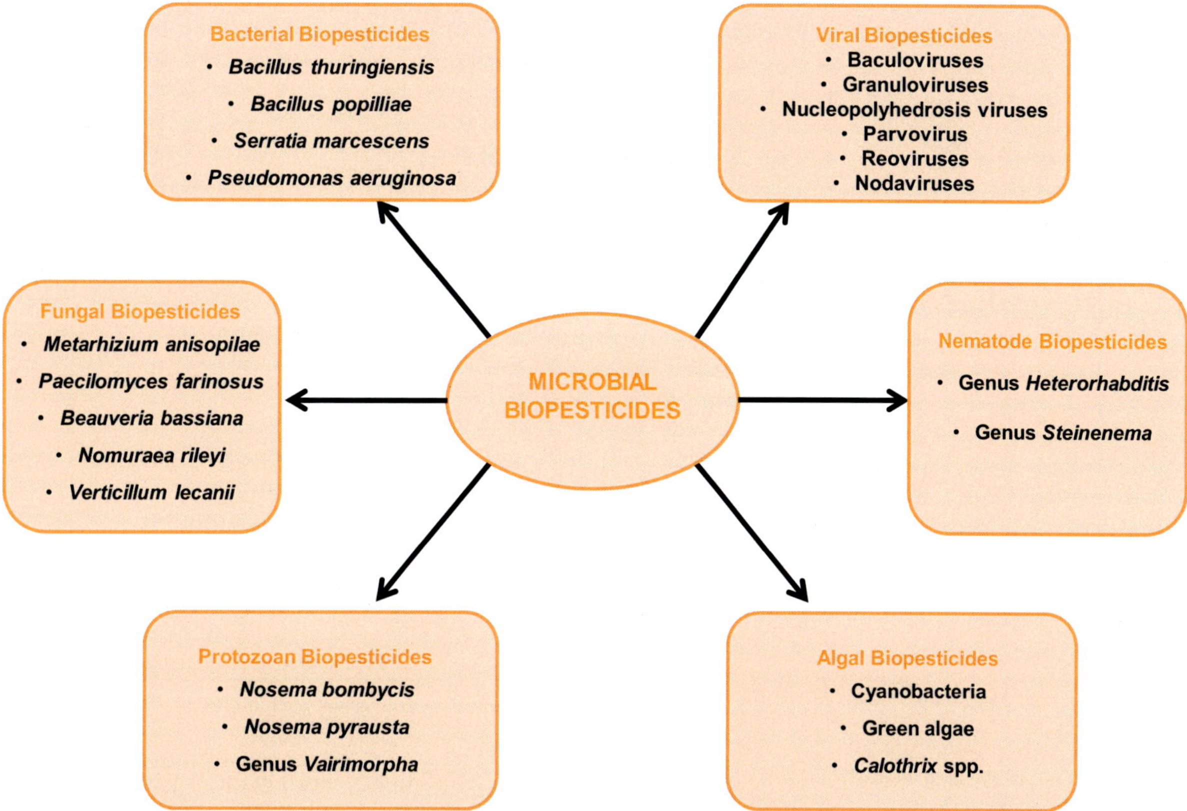

FIG. 21.1 A diagrammatic representation of six main types of microbial biopesticides with their examples. The genus/species from the respective class which are widely used and accepted as biopesticides are also shown.

thuringiensis), obligate pathogens (e.g. *Bacillus popilliae*), facultative pathogens (e.g. *Pseudomonas aeruginosa*) and potential pathogens (e.g. *Serratia marcesens*) (Koul, 2011). Out of these above mentioned categories, the spore forming bacteria are mostly preferred commercially for their effectiveness and safe nature (Koul, 2011). Many bacterial species have capability to infect different insects but member of Bacillaceae family, particularly *Bacillus* spp. are adopted as pesticides and received maximum attention (Koul, 2011; Kachhawa, 2017). The classic examples of the spore forming bacterium species are *Bacillus thuringiensis. B. thuringiensis* is a cosmopolitan, rod shaped, gram positive, motile and endospore producing bacterium that is most extensively exploited and investigated insecticidal bacterium (Sudakin, 2003) and second most studied and commonly used bacteria is *Bacillus sphaericus* (Koul, 2011; Ruiu, 2018). These bacterial species are specific to individual species of butterflies, moths, beetles, flies or mosquitoes and to be effectively come in contact with the target and ingests it (Kachhawa, 2017). *B. thuringiensis* produces a toxin (δ-endotoxin) a crystal (cry) protein encoded by plasmid DNA and has insecticidal effect (Fig. 21.2). Cry proteins have specialized insecticidal properties that contribute to ~25% of dry weight of bacterium (Kachhawa, 2017). Once insect ingests the crystals, they are solubilized as they enter midgut due to its alkaline conditions and through proteolytic activity thus fragmented into toxic core (Jisha et al., 2013). This proteolysis cleaves the N and C terminus peptide parts of full protein. This toxin binds to receptors of the membrane present in the apical microvillus of the insect's epithelial midgut cells. The binding of the toxin leads to conformational changes that paves way for its entry to the cell membrane. Finally, the oligomerization of the toxin lead to the formation of pore/ion channel within the receptor. Further, this ion channel disrupts the membrane transport that causes cell lysis and finally leads to the insect death (Khetan, 2001a,b; Jisha et al., 2013; Schünemann et al., 2014). Therefore, the larva dies due to the undernourishment and destruction of the gut epithelial cells (Senthil-Nathan, 2015). The popularity of Bt toxin is because of its non-toxicity toward animals and humans as it has been observed that the alteration of crystal protein into active toxin require an alkaline environment like insect's gut which is absent in case of most mammalians and binding site of the toxin is also specific (Khetan, 2001a,b; Kachhawa, 2017). *B. thuringiensis* based bio insecticides developed till date is used against different classes of lepidopteran, dipteran and coleopteran larvae. *Bacillus subtilis* found in various ecological niches is also proven as potential biocontrol agent against phytopathogenic

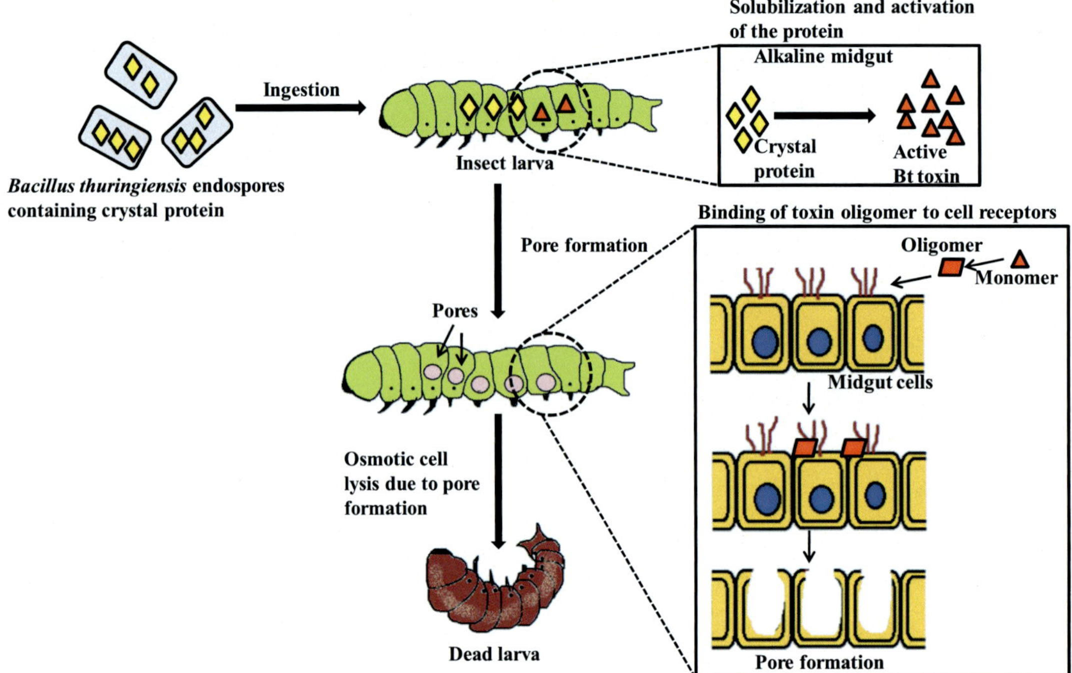

FIG. 21.2 Mode of parasitism of Bt toxin of *Bacillus thruningiensis* on lepidopteran insects.

fungi, bacteria, nematodes and mosquitoes (Ongena et al., 2010). SPBI is a lipopeptide based biosurfactant recently derived from *B. subtilis* is used against olive moth *Prays oleae* and have an LC_{50} value of 142 μg/mL (Ghribi et al., 2011; Mnif and Ghribi, 2015). Another species of *Bacillus, B. sphaericus* produces parasporal crystal inclusions having binary toxins Cry48Aa-Cry49Aa, BinA and BinB and Mtx1 and Mtx2 toxins. The toxin is specific against mosquito larva and causes its death (Mnif and Ghribi, 2015). Betaproteobacteria group also show significant potential to control biocontrol species. *Burkholderia rinojensis*, an insecticidal strain which is developed into a product whose ingestion causes death of insects and mites. Another strain *Chromobacterium subtsugae* showed broad spectrum insecticidal activity by production of secondary metabolites against species of Hemiptera, Coleoptera, Lepidoptera and Diptera (Ruiu, 2018). Other bacterial species are also used as biopesticides against pathogenic organisms such as *Agrobacterium radiobactor* (an aggressive inhibitor of pathogenic *Agrobacterium tumefaciens*), *Bacillus popilliae*, *Bacillus licheniformis* (produces anti-fungal enzymes), *Bacillus subtilis* (a competitive inhibitor of pathogenic fungi), *Saccharopolysporaspinosa*, *Pseudomonas cepacia*, *Pseudomonas solanacearum*, *Pseudomonas fluorescens*, *Pseudomonas chlororaphis* (a competitive inhibitor of different pathogenic fungi), *Pseudomonas syringae* (competitively inhibits the growth of bacteria that promote ice formation and prevents frost damage) and *Streptomyces griseoviridis* (a competitive inhibitor of pathogenic fungi and produces various anti-fungal metabolites) (Sudakin, 2003; Koul, 2011; Ruiu, 2018).

21.2.2 Viral biopesticides

About 700 species of viruses which have insect-infecting tendencies have been segregated from Hymenoptera (100), Lepidoptera (560), Orthoptera, Diptera and Coleoptera (40) (Khachatourians, 2009; Koul, 2011; Erayya et al., 2013). Insect specific viruses are submicroscopic, intracellular, obligate and pathogenic entities (Erayya et al., 2013; Kachhawa, 2017). Among these viral species DNA-containing baculoviruses (BVs), RNA-containing reoviruses, granuloviruses (GVs), nucleopolyhedrosis viruses (NPVs), iridoviruses, poxviruses, cytoplasmic polyhedrosis viruses, nodaviruse, acoviruses, picrona-like viruses, paraviruses, tetraviruses and polydna-viruses are commonly used to control pests of field crops, vegetables and plant chewing insects such as moths, worms and caterpillars (Lacey et al., 2008; Koul, 2011; Erayya et al., 2013). Baculoviruses act through oral ingestion and first infection takes place once contaminated food is ingested by target pathogen (Ruiu, 2018). The mode of action of these insecticidal viruses involves proliferation of the viruses in the cytoplasm or nuclei of the target cell (Fig. 21.3). Viral protein expression take places in three phases i.e. the early phase within few hours after infection (0–6 h), the late phase (6–24 h) and the very late phase (24–72 h) after the infection.

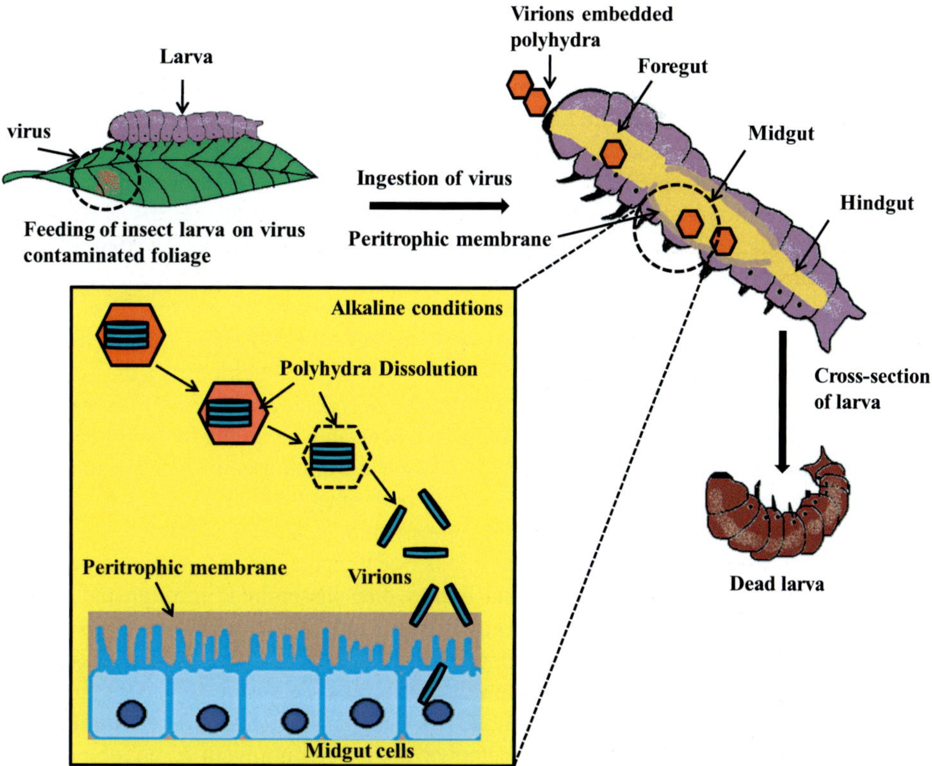

FIG. 21.3 Mode of parasitism of baculoviruses againstlepidopteran insects.

Ingested occlusion bodies once entered the host midgut; alkaline environment there triggers release of specific virions into midgut lumen through dissolution of occlusion bodies.

These specific virions interact with microvillar epithelial cells membrane with the help of their envelope proteins and are known as occlusion-derived viruses (ODVs) (Kachhawa, 2017; Ruiu, 2018). The synthesis of ODVs occurs in the late phase of viral infection. Then virions enter nucleus of the midgut cells and virus replicate itself within the nuclei. The susceptibility varies as NPVs can infect all tissue types whereas GVs are specific to particular tissue type (fat bodies cell only) (Koul, 2011; Kachhawa, 2017) (Fig. 21.3). NPVs are occluded within each occlusion body and form polyhedral structures whereas GVs are occluded in a single occlusion body and generate granular structures. The infection is then spread to other tissues like nerve cells, fat bodies and hemocytes in the hemolymph. A plethora of granules and polyhedra are produced within the infected nuclei/cytoplasm. They generate enzootics and eventually reduce the pest population (Koul, 2011). The majorities (60%) of known insect viruses are from family Baculoviridae and these baculoviruses are used for the development of most viral biopesticides against pests of food/fiber crops (Erayya et al., 2013). These are rod-shaped (40−70 nm × 250−400 nm in size) comprised of a capsidmade up of protein having DNA-protein core surrounded by a lipoprotein envelope. The viral DNA is circular, supercoiled and double stranded varying between a range of 80−180 kbp (Erayya et al., 2013; Kachhawa, 2017). Baculoviruses are considered safe for vertebrates as there is no case of pathogenicity reported till date. The virus causes its infection through production of infectious particles in the form of crystalline occlusion bodies (Kachhawa, 2017; Ruiu, 2018). The viruses are divided into two main groups depending upon the morphology of inclusion bodies: first type is NPVsand another is the granuloviruses (GVs). NPVs are those in which occlusion bodies are polyhedron-shaped and in GVs these are granular-shaped (Ruiu, 2018). Depending upon environmental conditions the death time may vary but generally it takes 3−7 days under optimal conditions and 3−4 weeks in non-ideal conditions. The infection is also able to cause some behavioral changes in target pest which affects their gene expression mechanisms (Kachhawa, 2017; Ruiu, 2018). Viral pesticides are among the safest biopesticides used and show almost no effects on non-target organisms like plants, vertebrates and beneficial insects (Haase et al., 2015).

21.2.3 Fungal biopesticides/mycopesticides

The fungi that represent fungal biopesticides may be commensal, symbiont, facultative or obligate parasite of insects (Barbara and Clewes, 2003; Pineda et al., 2007). The fungi grow in different habitats like aquatic or terrestrial and mainly associated with group known as entomopathogenic fungi (Koul, 2011). Entomopathogenic fungi are used to control insect population such as thrips, whiteflies, scale insects, aphids, mites and mosquitoes (Barbara and Clewes, 2003; Pineda et al., 2007). More than 700 fungal species from around 90 genera have insecticidal properties, four major groups are Pyrenomycetes, Hyphomycetes, Laboulbeniales and Zygomycete (Koul, 2011; Kachhawa, 2017). Most widely used fungal species are *Metarhizium anisopilae*, *Paecilomyces farinosus*, *Beauveria bassiana*, *Nomuraea rileyi* and *Verticillium lecanii* (Koul, 2011; Ruiu, 2018). Many commercial products are formulated by different fungal species (usually less than 10) and are available globally. Bassi in 1835 first time show the use of white muscardine fungus on silkworm and formulated the germ theory. Then in 2010, Gilbert and Gill described the use of insect infecting fungi for biocontrol of different pests (Shah and Goettel, 1999; Gilbert and Gill, 2010; Kachhawa, 2017). Entomopathogenic fungi showed varying mechanisms for pathogenesis in host and it chiefly depends upon contact with the insect (Fig. 21.4.) (Pekrul and Grula, 1979). The fungi provoke infection when spores are hold on the integument surface and establish conidia in the joints. As soon as formation of the germination tube starts, fungi start excreting enzymes such as chitinase, protease, lipoxygenase, upases and quitobiases. Fungus penetrates deep inside the host body through both enzymatic and mechanical action. Once enzymes degrade host's cuticle the mechanical pressure is applied and appressorium is developed internally in the germination tubes (Kachhawa, 2017; Ruiu, 2018). Specific environmental conditions like temperature and relative humidity are also required for spore germination to trigger the infection. The hyphal bodies then disseminate through the hemocoel to different organs/body parts like muscle tissues, malpighian tubes, mitochondria, fatty bodies and hemocytes and invade there. This will result into the insect death within 3−14 days of fungal infection. After death fungi invades all the organs of insect and hyphae penetrate the cuticle layer to emerge from interior to exterior surface. To ensure spread the fungi then initiates spore formation outside the infected host (Kachhawa, 2017; Ruiu, 2018). Fungus also releases a variety of metabolites during its vegetative growth and these metabolites act as virulence factors or toxins. Actinomycin A, novobiocin and cycloheximide

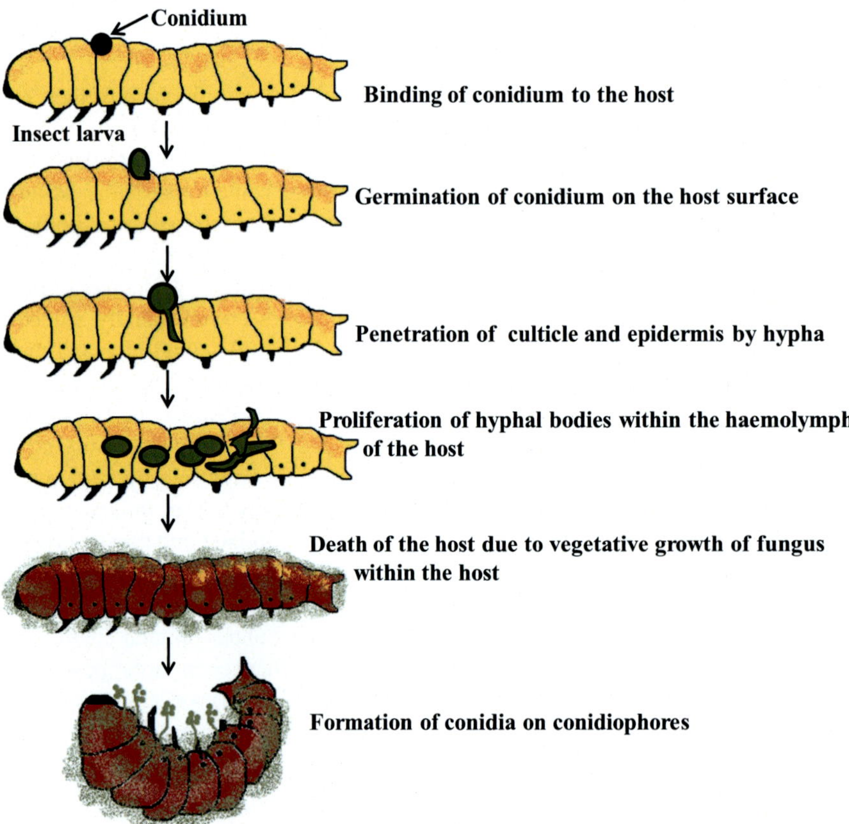

FIG. 21.4 Mode of parasitism of entomopathogenic fungi against lepidopteran insects.

are the most active toxins which are synthesized by these fungal species (Koul, 2011; Ruiu, 2018). *Saccharopolyspora spinosa*, an actinomycete also used for manufacturing commercially available biopesticides compound named as Spinosyns. Spinosyns are active against hymenopterans, thysanopterans, siphonaterans and dipterans (Koul, 2011). Entomopathogenic fungi have numerous modes of action for their pathogenesis and hence are very potential microbial biopesticides (Koul, 2011). One of the most used fungal biopesticides is *Beuveria bassiana*, which was also the first insect used for microbial control till the end of 19th century. *B. bassiana* and *B. brongniartii*, another species of same genus also showed virulence against diverse host range and used in diverse formulations of biopesticides as active substances (Zimmermann, 2007). Another well explored entomopathogenic fungus used as biopesticides is *Metarhizium anisopliae* having wide range of insect host species. *M. anisopliae*, natural predator of some insects such as grasshoppers and locusts (Lomer et al., 2001). This organism has no ill effect on environment and mammalians so far as this organism is unable to thrive inside mammals (Sudakin, 2003). The fungus has diverse strains active against different hosts and capable of producing insecticidal toxins and virulence factors (Zimmermann, 2007; Ruiu, 2018). Other fungal species that are used as microbial biopesticides frequently are- *Aspergillus flavus* strain AF36 (a competitive inhibitor of aflatoxin producing *Aspergillus flavus*), *Gliocladium catenulatum* (produces certain anti-fungal metabolites), *Lagenidium giganteum* (have larvicidal property against mosquitoes), *Myrothecium verrucaria* (have nematocidal effect) and *Trichoderma* spp. (competetive and antagonistic inhibitor of a number of fungal and bacterial species) (Sudakin, 2003; Kumari et al., 2020).

21.2.4 Nematode biopesticides

Another mesmerizing group of microorganisms that demolish the pests such as gnats, white grubs weevils and many species of Sesiidae family are entomopathogenic nematodes (Klein, 1990; Koul, 2011; Atwa, 2014). Entomopathogenic nematodes having significant insecticidal potential are used as biopesticides for biological management of insects/pests. The nematodes are nonsegmented, soft bodied roundworms and are obligate or facultative to some insects (Atwa, 2014; Kachhawa, 2017). They need to recycle in their host if they are obligate parasites, to maintain their presence in the environment. These microorganisms are capable of controlling the pests in cryptic environments such as stem borers and soil borne pests (Klein, 1990). Nematodes that are frequently used in pest management include the genera *Heterorhabditis* and *Steinernema* such as *H. megidis, H. bacteriophora, S. carpocapsae, S. scapterisci, S. glaseri* and *S. riobrave* (Ehlers et al., 1998; Shapiro-llan et al., 2006). These nematodes show symbiotic association with insect pathogenic bacteria (*Xenorhabdus* spp. and *Photorhabdus* spp.) having great insecticidal potential (Atwa, 2014; Ruiu, 2018). These nematodes can be easily produced in both lab conditions (in vivo and in vitro) in the form of mass culture (Grewal and Georgis, 1999). These nematodes invade the host as infective juveniles (Fig. 21.5). These juveniles are free living organisms that pierce the

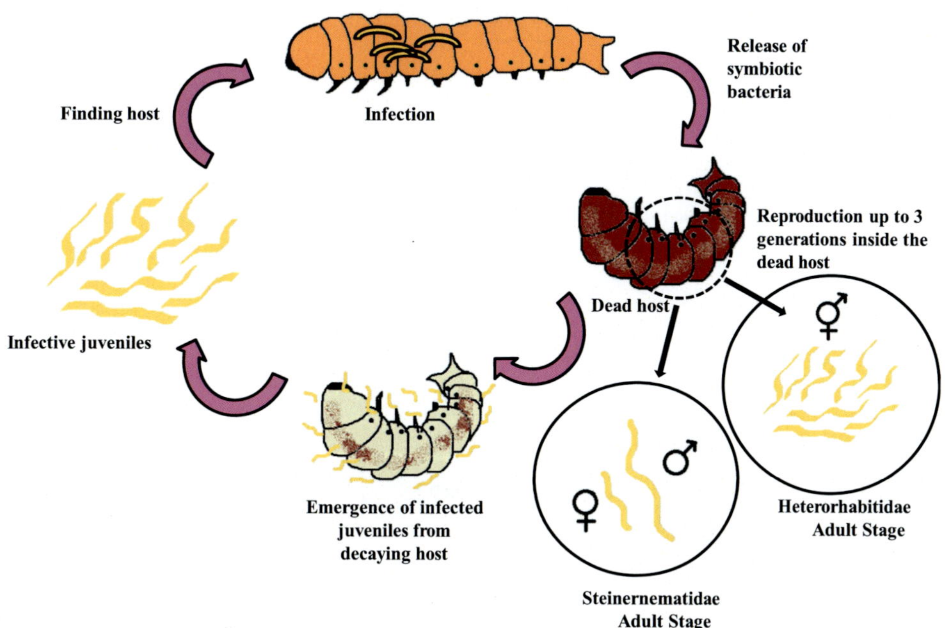

FIG. 21.5 Mode of parasitism of entomopathogenic nematodes against lepidopteran insects.

host through cuticle, mouth, spiracles or anus. After penetrating the host they discharge symbiont bacteria into the host's hemocoel and within 24–48 h after infection they kill the host (Dowds and Peters, 2002; Koul, 2011; Ruiu, 2018). These bacterial symbionts then proliferate (multiply in the insect hemolymph) with the release of virulence factors and toxins which help in the weakening of the host. The metabolites are

also produced that favor the generation of an environment which is suitable for nematode reproduction (Ruiu, 2018). Through four juvenile stages the progeny nematodes are developed into adult. These nematodes live in the dead host up to three generations and then the infective juveniles locate new host and leave the corpse (Dowds and Peters, 2002; Atwa, 2014). Depending upon the environmental conditions they will continue to infect new hosts, interfere their life cycle and kill them (Kachhawa, 2017).

21.2.5 Protozoan biopesticides

Entomopathogenic protozoans are microbial group which are extremely diverse and having around 1000 different species. They have capable of attacking invertebrates mainly pest species and are commonly named as microsporidians (Koul, 2011; Kachhawa, 2017). They naturally affect a wide range of pest's population by inducing debilitating and chronic effects (Maddox, 1987; Brooks, 1988). Among various protozoans, Microsporan protozoans have been scrutinized comprehensively to control pests as they are ubiquitous and intracellular obligatory parasites of several pests and are extensively important components of pest management programmes (Solter and Becnel, 2000). *Vairimorpha* and *Nosema* are two genera that have potential to attack Orthopteran and Lepidopteran insects (Lewis, 2002; Koul, 2011). These two species are commonly observed, slow acting, host specific and produces chronic infections. They can persist and recycle within the host and show debilitating effects on overall fitness including reproduction of target insects. The biological activities of these protozoans are complex and can grow inside living reported host or either required some intermediate host (Senthil-Nathan, 2015). First reported microsporidium, *N. bombycis*is a pathogen of pebrine disease (silkworm disease) which persisted in mid-nineteenth century in North America, Europe and Asia. During investigation of *Nosema pyrausta* parasitism on European corn borer (*Ostrinia nubilalis*), it is found that the protozoan is able to infect the host through horizontal and vertical transmission (Fig. 21.6) (Koul, 2011).

In horizontal transmission protozoan spore is engulfed by the host. The spore proliferates and produces more spores in the midgut of the host. These spores live in the host and released with feces. The host finally dies because of the infection but these spores stay viable and consumed by other larvae of the host and hence the infection cycle continues in this manner (Koul, 2011). If the spore is engulfed by female larva then this protozoan infection is passing down to next

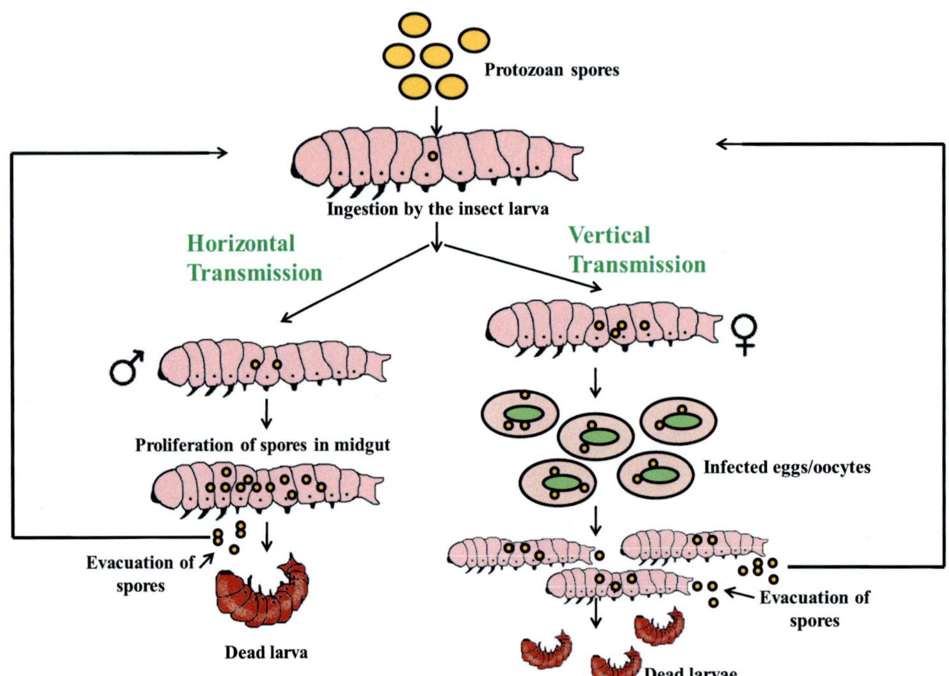

FIG. 21.6 Mode of parasitism of entomopathogenic protozon against lepidopteran insects.

generation through vertical transmission (Senthil-Nathan, 2015). In vertical transmission an ovarian tissue of the adult female get infected and produces infected oocytes. These infected oocytes ultimately generate infected larvae that sooner or later die with *Nosema* infection by leaving viable spores in the vicinity to initiate other *Nosema* infection cycle (Senthil-Nathan, 2015). The only registered protozoan biopesticides is *N. locustae*, which control grasshopper population. They effectively infect the grasshoppers in their nymphal stages and can kill them within 6 weeks (Bidochka and Khachatourians, 1991). *N. pyrausta*is is another effective microsporidian which causes death of the larvae and reduces longevity and fecundity of adult of European corn borer (Siegel et al., 1986).

21.2.6 Algal biopesticides

Microalgae are microbial entities having simple nutritional requirements for their growth and biomass production. Microalgae biomass can be used in sustainable agricultural practices as biopesticides, improve soil characteristics (bioavailability of macro and micronutrients), soil health, reduces soil erosion and also act as biofertilizer (Costa et al., 2019). Microalgae act as potent biopesticides due to its ability to produce different biocompounds like growth regulators, terpenes and phenolic compounds. Marine macroalgae/sea weeds are also prominent source of active secondary metabolites having antimicrobial properties (Rajesh et al., 2011; Costa et al., 2019). The compounds are used for the management of insects/pest or plant pathogens. These bioactive compounds also have biocontrol ability due to their antiviral, antibacterial and antifungal properties (Rajesh et al., 2011; Hamed et al., 2018). First biopesticidal activity from algae was reported from green alga *Chlorella*, the bioactive compound isolated was chlorellin which acts by inhibiting the bacterial species (Pratt et al., 1945). Cyanobacteria are one of the main biocontrol agent used to control pathogenic microbes, soil-borne diseases in plants and nematodes (El-Mougy and Abdel-Kader, 2013). The compounds like polyphenols, tocopherols, proteins, oils, carbohydrates, allelochemicals, saponins, sesquiterpenes and nitrogen-rich compounds are synthesized by the cyanobacteria. These compounds are responsible for the modifications in cytoplasmic membrane (structural and functional) required for its disruption, inhibition of protein synthesis and enzymes inactivation in the targeted microbe (Costa et al., 2019). The toxin produced from some stains of cyanobacteria can also behave as anti-feeding compound of larvae and therefore, decreases mosquito population when used as biofertilizer (Costa et al., 2019). Microalgae show inhibitory effect on early larval stages of predators and effects like inhibition of pupation processes and fecundity of adult predators (Matamoros and Rodriguez, 2016). The compounds which are mainly responsible for insecticidal activity of genus *Fischerella* are 12-epi-hapalindole C isonitrile and hapalindole L (Hernández-Carlos and Gamboa-Angulo, 2011). These compounds are reported to eliminate all larvae (around 100%) in 48 h from *Chironomus riparius*, a pest of rice cultivars (Becher et al., 2007). A sesquiterpene compound eremophilone, isolated from *Calothrix* spp. also act as effective biopesticides for pest control in rice plantations (Hernández-Carlos and Gamboa-Angulo, 2011).

21.3 Microbial products in biopesticides

Practicability of any biological product can be possibly affected by enormous factors such as impact of the bio-product on the pest, range of the bio-product to affect other pests/organisms, market size, field performance, production cost, technological constrains including formulation and delivery system (Ruiu, 2018). Major challenges that are confronted by microbial biopesticide formulations are complexity in creation, lack of correct formulation, obligation of high humid surroundings, sensitivity toward desiccation and ultra violet (UV) radiations (Koul, 2011). But in recent time due to incessant research and discoveries in microbial product production global marketplace of microbial biopesticides is cultivating (Ruiu, 2018). Nowadays, the microbial biopesticide formulations are formulated in such as way that these formulations are stable, simple to handle, apply and store and also have ability to withstand different environmental conditions (Ruiu, 2018). The ultimate product of a microbial biopesticide is sustained by addition of various adjvuants and carriers with active microbial ingredient such as addition of stabilizers, spreaders, synergists, stickers, surfactants, anti-freezing agents, coloring agents, additional nutrients, UV protectants, melting agents, dispersants, anticaking agents, adhesive materials, dispersing agents, polymers, resins and inert fillers to augment their adsorption and to shield the biopesticide from harsh environmental conditions (Tijjani et al., 2016). Some major microbial biopesticide formulations are shown in Table.21.1.

21.4 Current status of biopesticides in India

With limited land resources and growing human population, crop yields reduced due to pest-associated factors and it is becoming critical with time (Pandey and Seto, 2015; Kumar et al., 2019). The losses from pests on farm yield has been

TABLE 21.1 Microbial biopesticides formulation.

Category	Active ingredient	Target	Commercial/trade name	References
Bacteria	*Bacillus thuringiensis aizawai*	Armyworms, diamondback moth	Able-WG, Agree-WP, Florbac, XenTari	Ruiu (2018)
	Bacillus thuringiensis kurstaki	Lepidoptera	Biobit, Cordalene, Costar-WG, Crymax-WDG, Deliver, Dipel, Foray, Javelin-WG, Lepinox Plus, Lipel, Rapax	Ruiu (2018)
	Bacillus firmus	Nematodes	Bionemagon	Ruiu (2018)
	Bacillus subtilis	*Corticium invisum* and *C. theae*	Biotok	Rikita and Utpal (2014)
	Saccharopolyspora spinosa	Insects	Grandevo	Ruiu (2018)
Fungi	*T. viridae, Tricho-derma viridae and T. harzianum*	Pathogens of vegetables, pulse and cereals. *Rhizoctonia solani* and *Macrophomina phaseolina*	Bioderma, Antagon	Rikita and Utpal (2014)
	Beauveria bassiana	Wide range of insects and mites	Bio-Power, Biorin/Kargar, Botanigard, Daman, Naturalis, Nagestra, Beauvitech-WP, Bb-Protec, Racer, Mycotrol	Ruiu (2018)
	Metarhizium anisopliae	beetles and caterpillar pests; grasshoppers, termites	Biomet/Ankush, Bio-Magic, Devastra, Kalichakra, Novacrid, Met52/BIO1020 granular, Pacer	Ruiu (2018)
	Paecilomyces fumosoroseus	Insects, mites, nematodes, thrips	Bioact WG, No-Fly-WP, Paecilomite	Ruiu (2018)
	Heterorhabditis downesi	Black vine weevil *Otiorhynchus sulcatus*	NemaTrident-CT	Ruiu (2018)
	Phasmarhabditis hermaphrodita	Molluscs	Slugtech-SP	Ruiu (2018)
Nematodes	*Steinernema kraussei*	Vine weevil larvae	Kraussei-system	Ruiu (2018)
	Steinernema carpocapsae	Borer beetles, caterpillars, cranefly, moth larvae, *Rhynchophorus ferrugineus, Tipulidae.*	Capsanem, Carpocapsae-system, Exhibitline SC, Optinem-C, NemaGard, Nemastar, NemaTrident-T, NemaRed, Nemasys C, Palma-life	Ruiu (2018)
	Heterorhabditis bacteriophora	*Otiorhynchus* spp., chestnut moths, black vine weevil And soil-dwelling beetle Larvae, *Melolontha melolontha*, caterpillars, Cutworms, leafminers	Larvanem, Nemaplant, NemaShield-HB, Nematop, Nematech-H NemaTrident-H, NemaTrident-C, Nema-green, Optinem-H	Ruiu (2018)
	Nosema locustae	European cornborer caterpillars, grasshoppers and mormon crickets	NOLO Bait®, grasshopper Attack®	Usta (2013)
	Cydia pomonella (CpGV)	Codling moth, *C. pomonella*	Cyd-X, Cyd-X HP (Certis USA) Virosoft CP4 (BioTEPP) Carpovirusine (Arysta)	Arthurs and Dara (2019)
	Zucchini yellow mosaic virus, weak strain.	Zucchini yellow mosaic virus	Curbit	Chandler et al. (2011)
Protozoa	*Helicoverpa* spp.	Corn earworm (*H. zea*), cotton bollworm (*H. armigera*),	Gemstar LC (Certis), Heligen (AgBiTech)	Arthurs and Dara (2019)

TABLE 21.1 Microbial biopesticides formulation.—cont'd

Category	Active ingredient	Target	Commercial/trade name	References
Viruses	*Spodoptera litura* nucleopolyhedrovirus	*Spodoptera litura*	Biovirus–S, Somstar-SL	Ruiu (2018)
	Spodoptera littoralis nucleopolyhedrovirus (SpliNPV)	African cotton leaf worm (*Spodoptera littoralis*)	Littovir	Ruiu (2018)

estimated 10%–30% and the supplementary post harvest damages can be estimated more significantly, a country-wide estimated survey indicated that loss in cash crops and fruits is as follows cereals and pulses up to 6%, oil seeds 10%, fruits 18% and vegetables 13%. These losses are due to harvesting, manual/machine handling and storage and the use of some traditional chemicals have been quieted due to environment and health issues (Damalas and Eleftherohorinos, 2011; Thind, 2015; Jha et al., 2015). Till 2015, manufacturing of 32 pesticides have been stopped, import and use. 8 pesticides ave been withdrawn from Indian market and 13 more were limited for use (CIBRC, 2017). The process of insecticide resistance is also a hindrance for using some major field pesticides in India alarming the progress and adoption of strategies of non-chemical pest management (Kranthi et al., 2002; Mishra et al., 2015). The microbial biopesticides in India originated from bacteria, fungi, viruses and nematodes and their bioactive compounds may be used widely for many arthropod pests. The limited environmental toxicity, target meticulosity and due to safety reasons to non-target organisms microbial pesticides are increasingly becoming popular (Kumar and Singh, 2015; Senthil-Nathan, 2015; Kumar et al., 2019). The estimated biopesticides market is increasing globally and its growth is anticipated to be >$4.5 billion by 2023 (Olson, 2015). The increase in global agricultural trade and modifying regulatory environment measures and growing consumer choices resulted in a rapid change in biopesticides industry in India. The current status of biopesticides in Indian pesticide market is represented by nearly 5%, this ratio includes 15 microbial species and 970 formulations obtained from microbial origin have been registered by bureau of Central Insecticides Board and Registration Committee (CIBRC) (Kumar et al., 2019). The Indian biopesticides marketing suffers from lack off adoption, restricted resources for production at large scale, products remain unregistered and other challenges which are associated with the implementation, regulation, promotion and commercialization at various levels (Singh et al., 2016).

21.4.1 Registration norms and regulation of microbial biopesticides

In India, the regulation of biopesticides is governed by the CIBRC by the Insecticides Act of 1968 and Insecticides Rules of 1971. The CIBRC advisory give suggestions to the governments of center and state for the manufacturing, sale and distribution process and ensure safe use of all the insecticides including biopesticides. The issuance of the licenses is governed by registration committee of CIBRC to all the public funded and private firms for industrial scale production, distribution, promotion and sale of biopesticides (Kumar et al., 2019). In India the identification and development of entomopathogenic organisms as microbial biopesticides is carried out by both public and private institutions. According to a survey of 2017, 361 public sector and private laboratories working in the field of biocontrol were associated biopesticides production in India (DPPQS, 2017). Many extension agencies supply biopesticides to farmers free of cost (Mishra et al., 2015).

21.4.2 Evolution of microbial biopesticides for the management of insect pest in India

The microbial biopesticides production at commercial scale in India has been popularized recently. Till 2017, the total number of insecticidal products based on microbes are as follows: 188 fungal, 39 obtained from nematode, 51 from bacteria and 27 from NPV and have been registered with CIBRC. Moreover, 9 products of entomopathogenic nematodes (EPNs) exempted from CIBRC registration are sold in India (Table. 21.2). Many of the biofungicides registered, marketed and sold in India as wettable powders, comprising of *Trichoderma viride* (300 registrations), *T. harzianum* (294 registrations), 87 products of *B. bassiana*, 30 products of *M. anisopliae s.l.* and the bacteria *Pseudomonas fluorescens* (202 registrations) and *Bacillus subtilis* (5 registrations) and one product (Bio Jodi) which is a combination of both bacteria. These pathogens are known for colonizing the rhizophere/phylloplane and these products are marketed as soil inoculants for their ability against infection of fungal pathogens (Woo et al., 2014; Kumar et al., 2019).

TABLE 21.2 Microbial bioinsecticides and bionematicides sold/registered in India (CIBRC, 2017; Kumar et al., 2019).

	Microorganism	Strains	Registration
Fungi	*Beauveria bassiana*	AAI, BB-IARI-RJP, BB-ICAR-RJP, HaBa, IIHR	87
	B. brongniartii		1
	Metarhizium anisopliae s.1.	AAI, CPB/PSP-T26, KSCL/Ma-59, UMAIM	33
	Lecanicillium lecanii	AAI, AS-MEGH-VL, ICAR, NCIM 1312	62
	L.lecanii & Hirsutella thompsonii		1
	H. thompsonii		1
	Isaria fumosorosea		3
	Purpureocillium lilacinum	IIHR-PL-2	35
	Pochonia Chlamydosporia	IIHR VC-3	4
Bacteria	*Bacillus thuringiensis*	DOR Bt-1, DOR Bt-5, PDBCBT1, NBAIIBTG4, NCIM 2514	35
	B.thuringiensis	VCRC B-17,164	12
	Lysinibacillus sphaericus	B-101	3
	B. firmus	NCIB 2637	1
Viruses	*NPV of Helicoverpa armigera*	HaNPV	22
	NPV of Spodoptera litura	S1NPV	5
Nematodes	*Steinernema carpocapsae*		NA
	S. thermofilum		NA
	Heterorhabditis indica		NA
	H. bacteriophora		NA

The microbial community comprises a smaller part of the pesticide market and minute fractions of indigenous entomopathogenic isolates have been registered. Up to 2014, the ICAR-National Bureau of Agriculturally Important Microorganisms (NBAIMs) maintained 5375 microbial cultures under the National Agriculturally Important Microbial Culture Collection (NBAIM, 2017). A repository of potential microbial biopesticides is maintained by NBAIR including 210 strains of entomopathogenic fungi, 284 strains of *Bt*, six NPVs and 119 strains of EPN (NBAIR, 2017).

Despite many challenges, the bright future of Indian microbial pesticide market is expected. The subcontinental biopesticides research is evolving in a rapid way with an increasing focus on identification of effective indigenous isolates, with improved formulation and manufacturing technologies at a low costs and increased productivity. Due to little agricultural education of rural farmers in India, there is a need for working with farmers and other stake holders by universities, federal and state agencies for improving their knowledge and effective plans for accepting biopesticides. Microbial pesticides can prove beneficial for improving environment and human health through a reduced effect of toxic chemical pesticides.

21.5 Current advancement in the microbial biopesticides in the field of genomics, transcriptomics and proteomics

There are studies that have explored the microbe-pest interactions or potential of microbial biopesticides by genomic, transcriptomic and proteomic analysis. Whole genome analysis of the microbes has revealed the molecular systems which are involved in the entomapathogenic acitivity of the microbes. The proteins involved in the insecticidal acitivity have been described by the genomic analysis and these are namely adhesins, detoxification and toxin biosynthesis, cuticle degradation genes, hemolysins, proteases and lipases, insecticidal toxins, proteases, putative hemolysins, lethal chemical compound (HCN), novel secondary metabolites to suppress the growth, infect and kill the insects and contains a wide array of antibiotic synthesizing genes. The pan-genomic study of the entomopthogenic fungi is being carried out that will help to understand the speciation and evolution of these fungi which is not well illustrated. Till date the genome of 21 entomopathogenic fungal genera has been sequenced to understand their adaptive evolution in relation to the pest. These fungal genera are as follows: *Beauveria brongniartii* RCEF 3172, *Cordyceps bassiana, Trichoderma* spp., *Akanthomyces lecanii* RCEF 1005, *Metarhizium* spp. (*Metarhizium rileyi* RCEF 4871, *Metarhizium album, Metarhizium guizhouense, Metarhizium brunneum, Metarhizium majus, Metarhizium frigidum, Metarhizium pinghaense cultivar, Metarhizium robertsii*), *Metacordyceps brittlebankisoides, Moelleriella libera, Isaria farinosa, Isaria javanica, Aschersonia badia, Lecanicillium fungicola, Lecanicillium saksenae, Sporothrix insectorum, Sporothrix schenckii, Sporothrix pallida, Sporothrix globosa, Moelleriella libera* RCEF 2490, *Cordyceps fumosorosea* ARSEF 2679, *Hirsutella vermicola* AS3.7877, *Hirsutella rhossiliensis* OWVT, *Hirsutella thompsonii, Hirsutella minnesotensis, Tolypocladium inflatum, Tolypocladium cylindrosporum, Tolypocladium paradoxum, Ophiocordyceps sinensis, Basidiobolus meristosporus, Ascosphaera apis, entomophthora muscae* s. s. *and E. muscae* s. l. There are studies which have focused on the genomic analysis of the bacterial biopesticides. Some of the sequenced bacteria are as follows *Pasteuria, Pseudomonas entomophila* (Vodovar et al., 2006), *Photorhabdus luminescens* (Duchaud et al., 2003), *Serratia marcescens* (Wang et al., 2011), *Bacillus thuringiensis* (Cheng et al., 2015), *Burkholderia ambifaria* (Mullins et al., 2019). Entomopathogenic nematodes have shown potential for the biological control of the pests. Therefore, the genome based analysis has been carried out. The entomopathogenic nematodes for which whole genome sequence is available are namely *Heterorhabditis bacteriophora* (Sandhu et al., 2006), *Steinernema, Steinernema feltiae, Steinernema monticolum, Steinernema scapterisci, Steinernema glaseri, Steinernema borjomiensis n. sp.*(Lu et al., 2016). *Anticarsia gemmatalis* (de Castro Oliveira et al., 2006), the mostly exploited virus pesticide in the world has been sequenced. In addition to this, *Choristoneura rosaceana entomopoxvirus* "L" and *Choristoneura biennis entomopoxvirus* "L", *Heliothis zea* nudivirus, *Helicoverpa armigera* SNPV are also sequenced. The number of species of the insect pathogenic microbes which are whole genome sequenced are increasing and there is a need to analyze this big data generated from genomics. However, the big data obtained from the genomics will be of limited use if it is not curated. Further, studies that involve mechanistic understanding of the genes and their structural counterparts that are important for biological control should be explored. Therefore, the integrated approach involving transcriptomics and proteomics will help for better understanding of the mechanism of action of microbial biopesticides on the pests. Transcriptomics approach will help to unravel the expression of different gene invovled at the interface of the host response to the entomopathogenic microbe infection, mode of action of entomopathogenic microbe toxin-anti-toxin system and gene clusters invovled in the secondary metabolites biosynthesis and secretion and receptors of the entomopathogenic microbe invovled host recognition and lead to cellular signaling which mediate pathogenesis (Harith Fadzilah et al., 2019). The integrated study invovling transcriptomics and proteomics of *Bacillus thuringiensis* has unraveled the sporulation and parasporal crystal formation. It was observed that the metabolic regulation mechanism at the level of proteases and the amino acid metabolism, poly-β-hydroxybutyrate and acetoin, the pentose phosphate shunt, tricarboxylic acid cycle, increase in the concentration of enzymes and cytochromes and most F_0F_1-ATPase subunits was altered which regulate the sporulation (Wang et al., 2013). The immune response of the migratory locust, *Locusta migratoria* was documented against the mycoinsecticide *Metarhizium acridum* by comparative transcriptomic analysis. It was observed that diverse strategies have been adopted by the fat body cells of locust and the hemocytes for its survival against *M. acridum* infection (Zhang et al., 2015). The whole body changes in the insects have been reported at the transcriptional levels against *Beauveria bassiana* (Xia et al., 2013). The comparative genomic and transcriptomic study of *Metarhizium anisopliae* and *M. acridum* has syntenic genome. *M. anisopliae* have diverse gene families that belong to proteases, cell wall proteins like chitinases, cytochrome P450s, proteins required for cuticle lysis like polyketide synthases, and nonribosomal peptide synthetases, detoxification and toxin biosynthesis (Gao et al., 2011). Mullins et al. (2019) has reported a plant-protective metabolite cepacin from the genomic analysis of *Burkholderia ambifaria.* Vongsangnak et al. (2017) has reported the genome scale network which provide details about the genes of *Cordyceps militaris* (*i*WV1170) that interacts with its counterparts. It was unraveled that *C. militaris* has various genes encoding secreted enzymes involved in degradation of biomolecules (lipids, carbohydrates and proteins) (Vongsangnak et al., 2017). Such studies that are focused on the metabolic network may help to unfold the events occurring at the interface of insect-host which lead to pathogenicity and the secondary metabolite production. Therefore, the integration of genomics, transcriptomics, proteomics, metabolomics and bioinformatics approaches are the advanced and need of the time which will excel and will help the researchers to explore and devise novel strategies to combat the pests with more efficient Biopesticides products. RNA interference is a promising biopesticide tool that can be employed to control the devastating pests which affect the production of the crops (Christiaens et al., 2016).

21.6 Conclusion and future directions

Microbial bioinsecticides are promising alternative to the harmful and hazardous chemical pesticides. These are natural, target specific, environment friendly and pests can't develop resistance against them. These are better alternative to chemical insecticides and their pesticidal action is either from organism itself or through the production of products/byproducts (e.g. toxins) by these microbes (Kachhawa, 2017). The toxins produced by the microbial pathogens are generally peptides and are responsible for their pathogenic effect on different target organisms in species specific manner (Burges, 1981; Pathak et al., 2017). These toxins vary significantly in their structure, specificity and toxicity. Toxic metabolites cause disease in specific host and prevent establishment of other microrganisms by various modes of action like competition (Kachhawa, 2017; Pathak et al., 2017). These microbial biopesticides are safe for humans as well as beneficial organisms and less/no residue left in the soil/food. They have no threat on natural enemies thus preserve biodiversity and are ecologically safe. In last three decades microbial biopesticides are developed as biological control agents for crop protection due to target specificity against the pests (Pathak et al., 2017). These microbial biopesticides fit nicely in IPM as they are precise to their target pests and harmless to humans (Kachhawa, 2017). They show detrimental effects against pathogenic microbes, plant parasitic nematodes like foliar nematodes, ring nematodes, potato cyst nematodes, root-knot nematodes, sting nematodes, stubby nematodes, root-lesion nematodes and stunt nematodes (Atwa, 2014). For the efficient, effective and reasonable microbial biopesticidal product, the biopesticide preparations should be appropriate, stable, easily available and target specific. The microbial biopesticides market in India has many shortcomings which need to be addressed by proper implementation of the policies made by the government. There is a need to provide awareness about the present microbial formulations to the farmers through Krishi Vikas Kendras (KVKs) and interactive sessions and discussion between the scientists and the field workers. The molecular events which can increase the efficiency of the microbial pesticides need to be investigated by interdisciplinary studies. These studies include the researchers from basic sciences, agriculture sciences, biostatistics, bioinformaticians and molecular biologists. The integrated approaches comprising of in vitro, in vivo and in silico may help to develop the novel interventions to combat the pathogenic microbes and pests. These approaches reveal the evolutionary features which play role during coevolving of the host-pathogen interactions and adaptation of the pest to its ambience. The factors which lead to adaptations in the host/pest in their microenvironment can be deciphered by such analysis. The comparative studies of genome, transcriptome, proteome and metabolome will provide a larger picture about the events occurring at the molecular levels which may give some novel biomolecules that can have application in agriculture as well as bio-medical field.

References

Abdulkhair, W.M., Alghuthaymi, M.A., 2016. In: Plant Pathogen. IntechOpen, pp. 49–61.

Arthurs S., Dara, S.K., 2019. Microbial biopesticides for invertebrate pests and their markets in the United States. J. Invertebr. Pathol. 165, 13–21.

Atwa, A.A., 2014. Entomopathogenic nematodes as biopesticides. In: Basic and Applied Aspects of Biopesticides. Springer, New Delhi, pp. 69–98.

Barbara, D., Clewes, E., 2003. Plant pathogenic *Verticillium* species: how many of them are there? Mol. Plant Pathol. 4, 297–305.

Becher, P.G., Keller, S., Jung, G., Süssmuth, R.D., Jüttner, F., 2007. Insecticidal activity of 12-epi-hapalindole J isonitrile. Phytochemistry (Oxf.) 68 (19), 2493–2497.

Benítez, T., Rincón, A.M., Limón, M.C., Codón, A.C., 2004. Biocontrol mechanisms of *Trichoderma* strains. Int. Microbiol. 7, 249–260.

Bidochka, M.J., Khachatourians, G.G., 1991. Microbial and protozoan pathogens of grasshoppers and locusts as potential biocontrol agents. Biocontrol Sci. Technol. 1 (4), 243–259.

Brooks, F.M., 1988. Entomogenous protozoa. In: Ignoffo, C.M., Mandava, M.B. (Eds.), Handbook of Natural Pesticides, Vol V, Microbial Insecticides, Part A, Entomogenous Protozoa and Fungi. CRC Press Inc, Boca Raton, pp. 1–149.

Burges, H.D., 1981. Microbial Control of Pests and Plant Diseases 1970–1980. Academic Press.

Chandler, D., Bailey, A.S., Tatchell, G.M., Davidson, G., Greaves, J., Grant, W.P., 2011. The development, regulation and use of biopesticides for integrated pest management. Philos. Trans. R. Soc. Lond., B, Biol. Sci. 366 (1573), 1987–1998.

Christiaens, O., Prentice, K., Pertry, I., Ghislain, M., Bailey, A., Niblett, C., Gheysen, G., Smagghe, G., December 12, 2016. RNA interference: a promising biopesticide strategy against the African Sweetpotato Weevil Cylas brunneus. Sci. Rep. 6, 38836.

CIBRC, 2017. Central Insecticides Board and Registration Committee. Ministry of Agriculture and Farmers Welfare, Government of India (Accessed 25 June 2017). http://cibrc.nic.in.

Cheng, F., Wang, J., Song, Z., Cheng, J.E., Zhang, D., Liu, Y., 2015. Complete genome sequence of *Bacillus thuringiensis* YC-10, a novel active strain against plant-parasitic nematodes. J. Biotechnol. 210, 17–18.

Costa, J.A., Freitas, B.C., Cruz, C.G., Silveira, J., Morais, M.G., 2019. Potential of microalgae as biopesticides to contribute to sustainable agriculture and environmental development. J. Environ. Sci. Heal. Part B 54 (5), 366–375.

Damalas, C.A., Eleftherohorinos, I.G., 2011. Pesticide exposure, safety issues, and risk assessment indicators. Int. J. Environ. Res. Publ. Health 8, 1402–1419. https://doi.org/10.3390/ijerph8051402.

de Castro Oliveira, J.V., Wolff, J.L.C., Garcia-Maruniak, A., Ribeiro, B.M., de Castro, M.E.B., de Souza, M.L., et al., 2006. Genome of the most widely used viral biopesticide: *Anticarsia gemmatalis* multiple nucleopolyhedrovirus. J. Gen. Virol. 87 (11), 3233–3250.

Dowds, B.C.A., Peters, A., 2002. Virulence mechanisms. In: Gaugler, R. (Ed.), Entomopathogenic Nematology. CAB International, New York, pp. 79–98.

DPPQS, 2017. Directorate of Plant Protection Quarantine and Storage. Ministry of Agriculture and Farmers Welfare, Government of India. http://ppqs.gov.in/divisions/integrated-pest-managment/bio-control-labs. (Accessed 30 June 2017).

Duchaud, E., Rusniok, C., Frangeul, L., Buchrieser, C., Givaudan, A., Taourit, S., et al., 2003. The genome sequence of the entomopathogenic bacterium *Photorhabdus luminescens*. Nat. Biotechnol. 21 (11), 1307–1313.

Ehlers, R.-U., Lunau, S., Krasomi-Osterfeld, K., Osterfeld, K.H., 1998. Liquid culture of the entomopathogenic nematode-bacterium complex *Heterorhabditis megidisl Photorhabdus luminescens*. BioControl 43, 77–86.

El-Mougy, N.S., Abdel-Kader, M.M., 2013. Effect of commercial cyanobacteria products on the growth and antagonistic ability of some bioagents under laboratory conditions. J. Pathogens 2013. https://doi.org/10.1155/2013/838329.

Erayya, J.J., Sajeesh, P.K., Vinod, U., 2013. Nuclear Polyhedrosis Virus (NPV), a potential biopesticide: a review. Res. J. Agric. For. Sci. 1 (8), 30–33.

Ferrigo, D., Raiola, A., Causin, R., 2016. Fusarium toxins in cereals: occurrence, legislation, factors promoting the appearance and their management. Molecules 21, 627.

Gao, Q., Jin, K., Ying, S.H., Zhang, Y., Xiao, G., Shang, Y., et al., 2011. Genome sequencing and comparative transcriptomics of the model entomopathogenic fungi *Metarhizium anisopliae* and *M. acridum*. PLoS Genet. 7 (1), e1001264.

Ghribi, D., Mnif, I., Boukedi, H., Kammoun, R., Ellouze-Chaabouni, S., 2011. Statistical optimization of low-cost medium for economical production of *Bacillus subtilis* biosurfactant, a biocontrol agent for the olive moth Prays oleae. Afr. J. Microbiol. Res. 5 (27), 4927–4936.

Gilbert, L.I., Gill, S.S. (Eds.), 2010. Insect Control: Biological and Synthetic Agents. Academic Press.

Grewal, P.S., Georgis, R., 1999. Entomopathogenic nematodes. In: Hall, F.R., Menn, J.J. (Eds.), Biopesticides: Use and Delivery. Humana Press, Totowa: NJ, pp. 271–299.

Haase, S., Sciocco-Cap, A., Romanowski, V., 2015. Baculovirus insecticides in Latin America: historical overview, current status and future perspectives. Viruses 7 (5), 2230–2267.

Hajek, A., 2004. Natural Enemies: An Introduction to Biological Control. Cambridge University Press, Cambridge United Kingdom.

Hamed, S.M., Abd El-Rhman, A.A., Abdel-Raouf, N., Ibraheem, I.B., 2018. Role of marine macroalgae in plant protection & improvement for sustainable agriculture technology. Beni-Suef. Univ. J. Basic Appl. Sci. 7 (1), 104–110.

Harith Fadzilah, N., Abdul-Ghani, I., Hassan, M., 2019. Proteomics as a tool for tapping potential of entomopathogens as microbial insecticides. Arch. Insect Biochem. Physiol. 100 (1), e21520.

Heap, I., 2010. The International Survey of Herbicide Resistant Weeds. See. www.weedscience.org/in.asp. http://ec.europa.eu/environment/integration/research/newsalert/pdf/134na5.pdf(2008).

Hernández-Carlos, B., Gamboa-Angulo, M.M., 2011. Metabolites from freshwater aquatic microalgae and fungi as potential natural pesticides. Phytochem. Rev. 10 (2), 261–286.

Jha, S.N., Vishwakarma, R.K., Ahmad, T., Rai, A., Dixit, A.K., 2015. Report on Assessment of Quantitative Harvest and Post-harvest Losses of Major Crops and Commodities in India. All India Coordinated Research Project on Post-Harvest Technology. ICAR-CIPHET.

Jisha, V.N., Smitha, R.B., Benjamin, S., 2013. An overview on the crystal toxins from *Bacillus thuringiensis*. Adv. Microbiol. 3 (5), 462.

Kachhawa, D., 2017. Microorganisms as a biopesticides. J. Entomol. & Zool. Stud. 5 (3), 468–473.

Khachatourians, G.G., 2009. Insecticides, microbials. Appl. Microbiol. Agro/Food 95–109.

Khetan, S.K., 2001a. Commercialization of biopesticides. In: Microbial Pest Control. Marcel Dekker, New York, pp. 245–264.

Bacterial insecticide: *Bacillus thuringiensis*. In: Khetan, S.K. (Ed.), 2001b. Microbial Pest Control. Marcel Dekker, New York, pp. 3–42.

Koul, O., 2011. Microbial biopesticides: opportunities and challenges. CAB Rev. 17 (6), 1–26.

Klein, M.G., 1990. Efficacy against soil-inhabiting insect pests. In: Gaugler, R., Kaya, H.K. (Eds.), Entomopathogenic Nematodes in Biological Control. CRC Press, Boca Raton, FL, pp. 195–214.

Kranthi, K.R., Jadhav, D.R., Kranthi, S., Wanjari, R.R., Ali, S., Russell, D., 2002. Insecticide resistance in five major insect pests of cotton in India. Crop Protect. 21, 449–460.

Kumar, K.K., Sridhar, J., Murali-Baskaran, R.K., Senthil-Nathan, S., Kaushal, P., Dara, S.K., Arthurs, S., 2019. Microbial biopesticides for insect pest management in India: current status and future prospects. J. Invertebr. Pathol. 165, 74–81.

Kumar, S., Singh, A., 2015. Biopesticides: present status and the future prospects. J. Fertil. Pestic. 6. https://doi.org/10.4172/2471-2728.1000e129.

Kumari, I., Sharma, S., Ahmed, M., 2020. Tripartite interactions between plants, *Trichoderma* and the pathogenic fungi. In: Molecular Aspects of Plant Beneficial Microbes in Agriculture. Academic Press, pp. 391–401.

Lacey, L.A., Headrick, H.L., Arthurs, S.P., 2008. Effect of temperature on long-term storage of codling moth granulovirus formulations. J. Econ. Entomol. 101, 288–294.

Lewis, L.C., 2002. Protozoan control of pests. In: Pimental, D. (Ed.), Encyclopedia of Pest Management. Taylor & Francis, UK, pp. 673–676.

Lomer, C.J., Bateman, R.P., Johnson, D.L., et al., 2001. Biological control of locusts and grasshoppers. Annu. Rev. Entomol. 46, 667–702.

Lu, D., Baiocchi, T., Dillman, A.R., 2016. Genomics of entomopathogenic nematodes and implications for pest control. Trends Parasitol. 32 (8), 588–598.

Lugtenberg, B.J.J., Chin-A-Woeng, T.F.C., Bloemberg, G.V., 2002. Microbe–plant interactions: principles and mechanisms. Antonie Leeuwenhoek 81, 373–383.

Maddox, J.V., 1987. Protozoan diseases. In: Fuxa, J.R., Tanada, Y. (Eds.), Epizootiology of Insect Diseases. Wiley, New York, pp. 417–452.

Matamoros, V., Rodriguez, Y., 2016. Batch vs continuous-feeding operational mode for the removal of pesticides from agricultural run-off by microalgae systems: a laboratory scale study. J. Hazard Mater. 309, 126–132.

Mishra, J., Tewari, S., Singh, S., Arora, N.K., 2015. Biopesticides: where we stand? In: Arora, N.K. (Ed.), Plant Microbes Symbiosis: Applied Facets. Springer, New Delhi, pp. 37–75.

Mnif, I., Ghribi, D., 2015. Potential of bacterial derived biopesticides in pest management. Crop Protect. 77, 52–64.

Mullins, A.J., Murray, J.A., Bull, M.J., Jenner, M., Jones, C., Webster, G., et al., 2019. Genome mining identifies cepacin as a plant-protective metabolite of the biopesticidal bacterium *Burkholderia ambifaria*. Nat. Microbiol. 4 (6), 996–1005.

NBAIM, 2017. ICAR-national Bureau of Agriculturally Important Microorganisms, Current Status of Microbial Holdings in NAIMCC. http://nbaim.org.in/uploads/. (Accessed 30 June 2017).

NBAIR, September 2017. ICAR-national Bureau of Agricultural Insect Resources, Newsletter, vol. 9. Bengaluru, India, 4pp.

Olson, S., 2015. An analysis of the biopesticide market now and where it is going. Outlooks Pest Manag. 26, 203–206. https://doi.org/10.1564/v26_oct_04.

Ongena, M., Henry, G., Thonart, P., 2010. The roles of cyclic lipopeptides in the biocontrol activity of *Bacillus subtilis*. In: Recent Developments in Management of Plant Diseases. Springer, Dordrecht, pp. 59–69.

Pandey, B., Seto, K.C., 2015. Urbanization and agricultural land loss in India: comparing satellite estimates with census data. J. Environ. Manag. 148, 53–66.

Pathak, D.V., Yadav, R., Kumar, M., 2017. Microbial pesticides: development, prospects and popularization in India. In: Plant-Microbe Interactions in Agro-Ecological Perspectives. Springer, Singapore, pp. 455–471.

Pekrul, S., Grula, E.A., 1979. Mode of infection of the corn earworm (*Heliothis zea*) by *Beauveria bassiana* as revealed by scanning electron microscopy. J. Invertebr. Pathol. 34, 238–247.

Pineda, S., Alatorre, R., Schneider, M., Martinez, A., 2007. Pathogenicity of two entomopathogenic fungi on *Trialeurodes* vaporariorum and field evaluation of a *Paecilomyces fumosoroseus* isolate. Southwest. Entomol. 32, 43–52.

Pratt, R., Oneto, J.F., Pratt, J., 1945. Studies on *Chlorella vulgaris*. X. Influence of the age of the culture on the accumulation of chlorellin. Am. J. Bot. 1, 405–408.

Rajesh, S., Asha, A., Kombiah, P., Sahayaraj, K., 2011. Biocidal activity of algal seaweed on insect pest and fungal plant pathogen. In: Proceedings of the National Seminar on Harmful/beneficial Insects of Agricultural Importance with Special Reference to the Nuisance Pest Luprops Tristis in Rubber Plantations, pp. 86–91.

Rikita, B., Utpal, D., 2014. An overview of fungal and bacterial biopesticides to control plant pathogens/diseases. Afr. J. Microbiol. Res. 8, 1749–1762.

Ruiu, L., 2018. Microbial biopesticides in agroecosystems. Agronomy 8 (11). https://doi.org/10.3390/agronomy8110235. In this issue.

Sandhu, S.K., Jagdale, G.B., Hogenhout, S.A., Grewal, P.S., 2006. Comparative analysis of the expressed genome of the infective juvenile entomopathogenic nematode, *Heterorhabditis bacteriophora*. Mol. Biochem. Parasitol. 145 (2), 239–244.

Schünemann, R., Knaak, N., Fiuza, L.M., 2014. Mode of Action and Specificity of *Bacillus thuringiensis* Toxins in the Control of Caterpillars and Stink Bugs in Soybean Culture. ISRN.

Senthil-Nathan, S., 2015. A review of biopesticides and their mode of action against insect pests. In: Environmental Sustainability. Springer, New Delhi, pp. 49–63.

Shah, P.A., Goettel, M.S., 1999. Directory of Microbial Control Products and Services. Microbial Control Division, Society for Invertebrate Pathology, Gainesville, FL, p. 31.

Shapiro-Ilan, D.I., Gouge, D.H., Piggott, S.J., Patterson Fife, J., 2006. Application technology and environmental considerations for use of entomopathogenic nematodes in biological control. Biol. Contr. 38, 124–133.

Sharma, K.R., Raju, S.V., Jaiswal, D.K., Thakur, S., 2018. Biopesticides: an effective tool for insect pest management and current scenario in India. Indian. J. Agric. Allied Sci. 4 (2), 59–62.

Sharma, S., Malik, P., 2012. Biopesticides: types and applications. Int. J. Adva Pharm. Biol. Chem. (IJAPBC) 1 (4), 2277–4688.

Sharma, P., Sharma, M., Raja, M., Shanmugam, V., 2014. Status of *Trichoderma* research in India: a review. Ind. Phytopathol. 67 (1), 1–19.

Siegel, J.P., Maddox, J.V., Ruesink, W.G., 1986. Lethal and sublethal effects of *Nosema pyrausta* on the European corn borer (*Ostrinia nubilalis*) in central Illinois. J. Invertebr. Pathol. 48 (2), 167–173.

Singh, A., Shukla, N., Kabadwal, B.C., Tewari, A.K., Kumar, J., 2018. Review on plant-trichoderma- pathogen interaction. Int. J. Curr. Microbiol. App. Sci. 7 (2), 2382–2397.

Singh, H.B., Keswani, C., Bisen, K., Sarma, B.K., Kumar, P., 2016. Development and application of agriculturally important microorganisms in India. In: Singh, H.B., Sarma, B.K., Keswani, C. (Eds.), Agriculturally Important Microorganisms: Commercialization and Regulatory Requirements in Asia, pp. 167–182. https://doi.org/10.1007/978-981-10-2576-1-10.

Slavica, G., Brankica, T., 2013. Biopesticide formulations, possibility of application and future trends. Pestic. Fitomedicina 28 (2), 97–102.

Solter, L.F., Becnel, J.J., 2000. Entomopathogenic microsporida. In: Lacey, L.A., Kaya, H.K. (Eds.), Field Manual of Techniques in Invertebrate Pathology: Application and Evaluation of Pathogens for Control of Insects and Other Invertebrate Pests. Kluwer Academic, Dordrecht, pp. 231–254.

Sudakin, D.L., 2003. Biopesticides. Toxicolog. Rev. 22 (2), 83–90.

Szewczyk, B., Hoyos-Carvajal, L., Paluszek, M., Skrzecz, I., De Souza, M.L., 2006. Baculoviruses—re-emerging biopesticides. Biotechnol. Adv. 24 (2), 143–160.

Thind, T.S., 2015. Perspectives on crop protection in India. Outlooks Pest Manag. 26, 121–127.

Tijjani, A., Bashir, K.A., Mohammed, I., Muhammad, A., Gambo, A., Habu, M., 2016. Biopesticides for pests control: a review. J Biopest Agric 3 (1), 6–13.

Usta, C., 2013. Microorganisms in biological pest control—a review (bacterial toxin application and effect of environmental factors). Curr. Progress Biol. Res. 287–317.

Vitousek, P.M., Mooney, H.A., Lubchenco, J., Melillo, J.M., 1997. Human domination of Earth's ecosystems. Science 277 (5325), 494–499.

Vodovar, N., Vallenet, D., Cruveiller, S., Rouy, Z., Barbe, V., Acosta, C., et al., 2006. Complete genome sequence of the entomopathogenic and metabolically versatile soil bacterium *Pseudomonas entomophila*. Nat. Biotechnol. 24 (6), 673–679.

Vongsangnak, W., Raethong, N., Mujchariyakul, W., Nguyen, N.N., Leong, H.W., Laoteng, K., 2017. Genome-scale metabolic network of *Cordyceps militaris* useful for comparative analysis of entomopathogenic fungi. Gene 626, 132–139.

Wang, J., Mei, H., Zheng, C., Qian, H., Cui, C., Fu, Y., et al., 2013. The metabolic regulation of sporulation and parasporal crystal formation in *Bacillus thuringiensis* revealed by transcriptomics and proteomics. Mol. Cell. Proteomics 12 (5), 1363–1376.

Wang, Y., Fang, X., An, F., Wang, G., Zhang, X., 2011. Improvement of antibiotic activity of *Xenorhabdus bovienii* by medium optimization using response surface methodology. Microb. Cell Factories 10 (1), 98.

Ware, G.W., 1980. Effect of pesticides on non-target organisms. Residue Rev. 76, 173–201.

Woo, S.L., Ruocco, M., Vinale, F., Nigro, M., Marra, R., Lombardi, N., Pascale, A., Lanzuise, S., Manganiello, G., Lorito, M., 2014. *Trichoderma*-based products and their widespread use in agriculture. Open Mycol. J. 8 (1), 71–126.

Xia, J., Zhang, C.R., Zhang, S., Li, F.F., Feng, M.G., Wang, X.W., Liu, S.S., 2013. Analysis of whitefly transcriptional responses to *Beauveria bassiana* infection reveals new insights into insect-fungus interactions. PloS One 8 (7), e68185.

Yadav, I.C., Devi, N.L., 2017. Pesticides classification and its impact on human and environment. Environ. Sci. Engg. Toxicol. 6, 140–158.

Yadav, I.C., Devi, N.L., Syed, J.H., Cheng, Z., Li, J., Zhang, G., Jones, K.C., 2015. Current status of persistent organic pesticides residues in air, water, and soil, and their possible effect on neighboring countries: a comprehensive review of India. Sci. Total Environ. 511, 123–137.

Zimmermann, G., 2007. Review on safety of the entomopathogenic fungi *Beauveria bassiana* and *Beauveria brongniartii*. Biocontrol Sci. Technol. 17 (6), 553–596.

Zhang, W., Chen, J., Keyhani, N.O., Zhang, Z., Li, S., Xia, Y., 2015. Comparative transcriptomic analysis of immune responses of the migratory locust, *Locusta migratoria*, to challenge by the fungal insect pathogen, *Metarhizium acridum*. BMC Genom. 16 (1), 1–21.

Chapter 22

Use of microbial consortia for broad spectrum protection of plant pathogens: regulatory hurdles, present status and future prospects

Ratul Moni Ram[a], Ashim Debnath[b], Shivangi Negi[c] and H.B. Singh[d]

[a]*Department of Plant Pathology, A. N. D. University of Agriculture and Technology, Ayodhya, Uttar Pradesh, India;* [b]*Department of Genetics and Plant Breeding, A.N.D. University of Agriculture and Technology, Ayodhya, Uttar Pradesh, India;* [c]*Department of Seed Technology, A. N. D. University of Agriculture and Technology, Ayodhya, Uttar Pradesh, India;* [d]*Department of Biotechnology, GLA University, Mathura, Uttar Pradesh, India*

22.1 Introduction

Soil microorganisms have gained prominence in sustainable agriculture in last couple of decades. These microorganisms have paved the path for advanced biotechnological researches and actually cleared the way for manageable advancement driven studies (Timmis et al., 2017). They are getting popular among various researchers, analysts, specialists, strategy makers and other international organizations, as lot of research is being carried out to unravel more advantages of soil microorganisms in agriculture, bioremediation (Zabbey et al., 2017; Eze et al., 2018), modern applications, pharmaceuticals and some more (Odoh, 2017). Since the quest for food security to the overgrowing heightened and a constant pressure has arisen to minimize crop losses (Glick, 2012), there always have been an un-extinguishing want for expanded yield per unit area. Thus the major challenge confronting us is to enhance food production from ever-reducing per capita land availability. Microorganisms establish one of the most varying natural networks in the soil environment. The interaction plants with beneficial soil microbes not only facilitates a hike in food production moreover it provides ecological stability (Tringe et al., 2005; Hansel et al., 2008).

Microbial consortium (MC) comprises the symbiotic association of two microbial groups and even more (Clark et al., 2009), for improved yield development. The prime objective of any consortia is fulfill the gap created by one microorganism by the other, and is assumed if one fails to bestow its features, then the next one will demonstrate its action to its full potential. In several researches carried worldwide, it has been discovered that application of different blends of BCAs imparted a more significant level of defense contrasted with their single application (Dunne et al., 1998). Association of diverse microbial groups increases uptake of soil organic matters and regulates nutrient mobilization which further facilitates plant growth promotion (Nuti and Giovannetti, 2015). Owing to these advantages, the acceptability and applicability of MC have increased by practitioner. MC often responds better to ecological stress compared to single organism as they have likelihood to adapt and act in the plant vicinity compared to any single strain (Nuti and Giovannetti, 2015; Jain et al., 2013a,b). So, the focus on the use of multiple microorganisms in biological control is gaining momentum as MC play a key role in disease suppression and plant growth promotion which ultimately strengthens our food security (de Boer et al., 2003).

The monetary losses incurred as a result of pests and diseases in India have been accounted to be around 42.66 million dollars (Subash et al., 2017). Not many synthetic pesticides are accessible in the Indian pesticide market which could provide good results along with safeguarding the environment. So there is an urgent need to search for some better alternatives and technologies dependent on natural procedures for pest management (Kumar, 2012; Arora et al., 2016). Biopesticides seems to be an appropriate contender in this regard; however in developing countries in India and others face

Biopesticides. https://doi.org/10.1016/B978-0-12-823355-9.00017-1

various obstacles in their commercialization (Singh et al., 2002, 2004; Dutta, 2015). The major constraint is the moderate activity and lower shelf life, higher production input and lack of awareness among the growers. Besides this, the inadequate knowledge regarding benefits offered by biopesticides, absence of technical skill in the later phases of improvement, trouble in quality control testing time of these bioactive products before enlistment and commercialization are few hinderances in their successful registration and commercialization (Keswani et al., 2016).

This chapter comprehensively discusses about biological control and how microbial consortium fits into current biopesticide scenario along with various regulatory and commercialization issues associated with them. Moreover the authors have tried to highlight their present status of biopesticides along with the possible future prospects and general recommendations that could enhance their promotion and acceptance for boosting agriculture in India.

22.2 Biological control

Biological control is regarded as an environment friendly and holistic approach for plant disease management along with quality crop production endeavor. According to "Biological control of plant disease may be precisely defined as any condition or practice where by survival or activity of a pathogen is reduced through the agency of any living organism with the result that is reduction in incidence of the disease caused by the pathogen." However, Cook (1988) defined biological control as "use of natural or modified organisms, genes, or gene products to reduce the effects of pests and diseases." Wilson (1997) again defined biological control as "the control of a plant disease with a natural biological process or the product of a natural biological process." In biological control a wide range of microbes have been commercially released as biocontrol agents and this provides an alternate strategy to chemical disease management. Biocontrol strategy mainly depends on the synergistic relationship among the different genera of microbial community acting together to confront various biotic and abiotic stresses in plants (Saxena et al., 2013; Ram et al., 2018).

The antagonistic microorganisms use following mechanisms to facilitate disease control.

1. Antibiosis

   It is usually the process of secretion of anti-microbial compounds by BCAs to suppress or kill pathogens in the vicinity of their growth area. In antibiosis, microbial toxins, secondary metabolites, enzymes and other toxic substances are produced which at low concentration inhibit or kill the target microorganisms. The resulting action leads to disruption of cell membranes, inhibition of metabolic machinery and induction of plant defense system. Usually antagonistic bacteria and fungi produces numerous metabolites such as Bacillomycin D, harzianic acid, viridin, gliovirin, tricholin, Iturin A, peptaibols, alamethicins, massoilactone antibiotics, 6-pentyl pyrone, glisoprenins and heptelidic acid etc. playing key role in imparting plant defense.

2. Mycoparasitism

   Mycoparasitism is regarded as one of the most typical mechanism exhibited by antagonistic fungal species for management of phytopathogenic fungi. This process involves tropical growth of the antagonist on the pathogen followed by attachment and coiling around their target. In this phenomenon, one fungus is being parasitic on another fungus. The parasitizing fungus is termed hyperparasite and the parasitized fungus as hypoparasite. The lysis of hyphal cell walls of pathogens is carried out by various enzymes such as chitinases, β-1,3-glucanases and proteases (Keswani et al., 2014). β- 1,3-glucanases as the ability to degrade the cell wall, inhibit growth of host mycelium and spore germination.

3. Competition

   Iron is a vital element required by majority of microorganism to run various cellular and metabolic and cellular processes. Mostly stable complexes are made by oxidation of Fe^{2+} to Fe^{3+}. But in low iron concentration, few organisms form iron-binding ligands called siderophores having high affinity to quench iron from the micro-environment for proper growth of the microorganisms. Siderophore production under Fe deficient condition has been demonstrated by few strains of *P. aeruginosa* and *Trichoderma* spp. against *Pythium* sp., the causal agent of damping-off and root rot of many crops (Charest et al., 2005; Singh et al., 2012a). Competition is regarded as an indirect interaction between the BCA and the pathogen; and the pathogens gets eliminated by the exhaustion of food source and niche exclusion (Jain et al., 2013a).

4. Hydrogen cyanide

   Many PGPR strains produces hydrogen cyanide which directly or indirectly attack on the pathogen and in turn decrease disease incidence and increases the yield. The best example is florescent pseudomonas which produces HCN and shows ability to confront the pathogen (Sharma et al., 2013).

5. Lytic Enzymes

In simpler terms, lysis is usually complete or partial destruction of cells by enzymes. BCAs secrete various metabolites which interfere or restrict pathogen growth and activities. For example *Serratia plymuthica* C48 produces chitinase which inhibit spore germination and germ-tube elongation in *Botrytis cinerea* (Frankowski et al., 2001). Similarly, bacterial chitinases and b-glucanases are antifactor for several fungal plant pathogens and wood deteriorating fungi (Arora et al., 2007). In simpler words it can be stated that lysis leads to cell wall collapse of target which frequently results in cell mortality (Jain et al., 2012; Singh et al., 2017).

6. Induced Systemic Resistance

Induction of resistance in host plants by the action of PGPR and PGPF is one of the possible mechanisms in imparting plant defense. It emerged as a vital tool by which selected BCAs impart defense against a broad range of phytopathogens (Mishra et al., 2015a; Ram et al., 2018). Plants usually generate induced resistance as a result of interaction by pest/pathogen on roots inhabited by BCAs or even after treatment with specific chemical (Singh et al., 2012a). Production of plant defense compounds such as phytoalexins, phenolics, flavonoids, essential oil, and enhancement of enzyme activities such as chitinase, phenylalanine ammonia lyase, peroxidase, polyphenol oxidase, suoeroxise dismutase, ascorbate peroxidase etc. takes place as a part of induced systemic resistance (Jain et al., 2012, 2013b).

7. Plant growth promotion

BCAs also produce growth promoting hormones viz., Auxins, Gibberellins, Cytokinins etc. which suppress the pathogens and facilitate plant growth promotion and in turn increase yield. The studies conducted during past resolve that PGPR enhances plant growth either by producing plant growth regulators or volatile compounds and phytohormones, enhanced nutrient uptake capacity, lowering disease incidence, encouragement of other beneficial symbiosis, reduction of the ethylene level in plant and stimulation of disease-resistance mechanisms etc. (Jain et al., 2012).

22.3 Microbial consortium

Microbial consortium is usually referred as group of diverse microorganisms that have the ability to act together in a community. Microbial communities are ubiquitous in their natural environment and key players in global carbon and nitrogen cycles (Stolyar et al., 2007). The Plant growth promoting rhizobacteria (PGPR) strains live in together with the non PGPR in soil or rhizosphere in various combinations (Goswami et al., 2016). The fungi dominant in the rthizospheric region are termed as PGPF. The current biocontrol strategy employs the blending of various BCAs of different microbial species having plant growth-promoting attributed in order to achieve desired outcomes. A reliable biocontrol needs compatibility of all the co-inoculated microorganisms, their establishment in the rhizosphere, and the lack of competition among them. Evaluation is possibly the most important phase during development of microbial consortium because it provides an understanding of its contribution in decreasing disease intensity and increasing plant growth. Attempts are being made to develop microbial consortium for disease suppression and plant growth promotion. In consortium, two or more microbial groups living symbiotically and these consortiums are efficient, robust, modular and reliable in nature (Ram and Singh, 2017). According to Koornneef and Pieterse (2008) plant microbe cross talk supports them to work as a community which ultimately brings excellent symbiosis. The central idea behind using consortium is that, a single microorganism may not necessarily give protection against a wide range of pathogen so using a group of micro-organism will surely provide protection against multitargeted pathogens. Thus the use of microbial consortium not only accelerates disease suppression but also give a positive impact on plant growth promotion (Stockwell et al., 2011; Bhatia et al., 2018). The benefits bestow by microbial consortia along with the possible defense mechanisms is depicted in Fig. 22.1.

Before the development of a microbial consortium, an initial footstep is needed which includes, the compatibility of microorganisms to be used for the concerned host plant and the co-inoculation of these organisms should effect the host directly or indirectly. Combined inoculation of beneficial microorganism showed increased growth and yield characters as well as germination, nutrient uptake, plant height, number of branches, nodulation, yield, and total biomass of crops. Microbial biopesticides symbolize the foremost group of broad range biopesticides, which are pest specific and useful in disease suppression among all the biopesticides using in recent time (Khachatourians, 2009). Consortium application improves efficiency, consistency and reliability of microbes under different soil conditions. Combinations of bio control agent in consortium are expected to result in a higher level of protection and having potential to suppress multiple plant diseases.

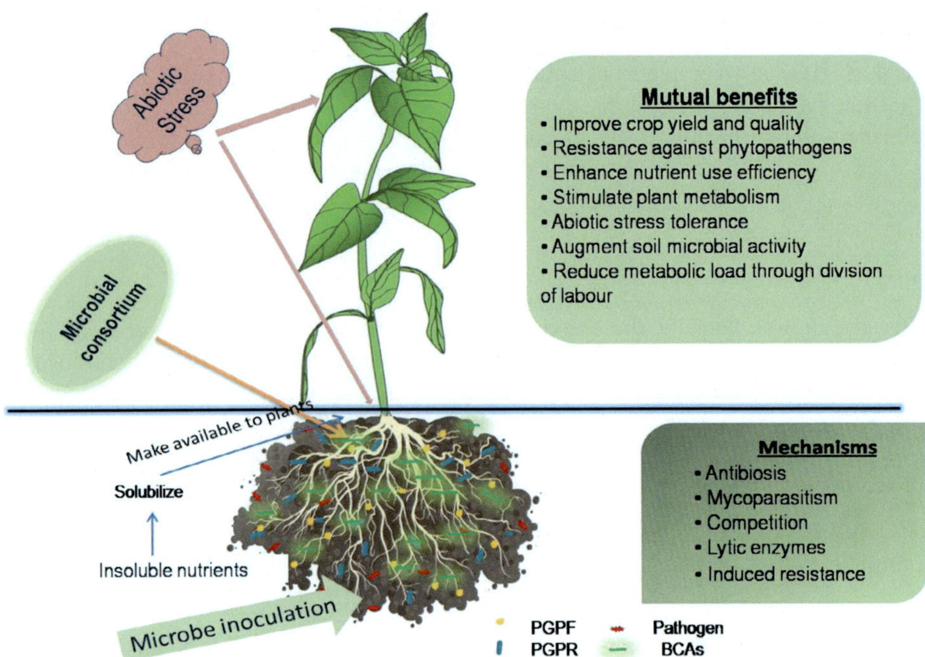

FIG. 22.1 Effect of application of microbial consortium based products on overall plant development.

22.4 Characteristics of microbial consortium

The development of a suitable consortium demands few basic features of microbes which needed to be taken into account. The selected microbes need to be compatible with each other and also give resistance to environmental calamities (Singh et al., 2012a). In addition to it they should be non pathogenic, have longer shelf life, synergistically active, fast acting, easy to handle and have high efficacy. The overall development process should be cheap and economical. Most of the reported biological agents have lead to provide high levels of disease suppression compared with the individual antagonist (Jain et al., 2013a,b).

22.5 Microbial consortium mediated plant defense mechanism in biological control

When the plants are bioprimed with beneficial rhizospheric microbes the level of activation of the defense responses are much higher as compare with the natural beneficial microorganisms (Shoresh et al., 2010). Some successful and effective microbial consortia against soil-borne pathogens have been developed such as *Sclerotinia sclerotiorum* and *Sclerotium rolfsii.* Some of the host-mediated defense responses involved in suppression of plant diseases by microbial consortia consisting of potential bio control agents viz., *Trichoderma harzianum, B. subtilis* and *P. aeruginosa* against *S. sclerotiorum* in pea plants (Jain et al., 2012) and *P. aeruginosa, T. harzianum* and *Mesorhizobium* sp. against *S. rolfsii* in chickpea plants (Singh et al., 2012a). Microbial strain *S. sclerotiorum* when treated individually showed enhancement of the defense parameters of the pea plants up to only 1.1−1.7 fold increment but when treated with triple microbial consortium consisting of compatible strains of *P. aeruginosa* PJHU15, *T. harzianum* TNHU27 and *B. subtilis* BHHU100 enhanced the defense parameters of the same plants up to 1.4−2.3 folds (Jain et al., 2012).

22.6 Different types of microbial consortium

22.6.1 Fungal and fungal

Co-inoculation of fungal microorganisms has been extensively used for preparation of consortium. *Trichoderma* one of the major fungal antagonist which is very effective to plant disease management and different species *i. e T. viride, T. asperellum* and *T. harzianum* were used to suppress chickpea wilt (Yadav et al., 2017). In sugarcane, corn and soybean the colonization of arbuscular mycorrhizal and plant growth were studied by using five plant growth promoting fungi: *T. asperella*, *Pochonia chlamydosporia*, *Purpureocillium lilacinum*, *Beauveria bassiana* and *Metarhizium anisopliae.*

The results of the experiment revealed that, the root growth in corn and soybean was mediate by fungi consortium and the colonization of arbuscular mycorrhizal in sugarcane and soybean. The fungal antagonist also produced phytohormones like JA, SA and ABA to regulation of mycorrhizal colonization and auxins for root growth (Christyan et al., 2018). For biological management of disease causing pathogens through different modes of action the microbial consortium is applied to crops. Theses microorganisms act together and induced resistance in plants for the targeted phytopathogen (Kohli et al., 2019). In maize, different *Trichoderma* spp. were used against *Fusarium verticillioides* (Sobowale et al., 2007). Begum and their co-workers in 2010 also evaluated the efficacy of *T. harzianum* along with *T. virens* for the control of damping-off of soybean caused by *Colletotrichum truncatum* and also observed reduction in pre or post-emergence damping off. The beneficial fungal microbes enhanced plant growth and health, induced the disease reduction, ecological enrichment with nutrient availability and also increased stress tolerance in the rhizosphere (Lugtenberg et al., 2002; Morrissey et al., 2004).

22.6.2 Bacterial and bacterial

Different spp. of beneficial bacteria were used from the past decades for the biological control of pathogens. Combine application of these bacterial strains are very effective against to disease suppression, biotic or abiotic conditions and multiplication of phytopathogens. Use of co-inoculated bacterial antagonist shows better disease control as compared to the single antagonist (Srivastava et al., 2010). The bacterial consortium activates the phenylpropanoid pathway and enzyme activities which accumulate of total PR proteins, proline and phenolics in order to pathogen inhibition. Bacterial antagonist mainly worked on antibiosis, competition and induced systemic resistance mechanisms. Combining application of endophytic starins *P. fluorescens* EBC6 and *P. fluorescens* EBC5 has been recorded for the management chilli damping off caused by *Pythium aphanidermatum* (Muthukumar et al., 2010). Two strains of *P. putida viz* WCS358 and RE8 have different mechanisms *i.e* p seudobactin mediated competition and ISR. Combining these mechanisms by applying a mixture of the biocontrol strains leads to more effective biological control of *Fusarium* wilt of radish (de Boer et al., 2003).

Consortium of *Pseudomonas* spp. was used to suppress the take-all disease of wheat (Duffy and Weller, 1995). The co-inoculation of *Rhizobia* sp., *B. cereus* strain BS03, *P. aeruginosa* RRLJ04 was found to induced the PAL, PO, and PPO activities in order to control the Fusarial wilt in pigeon pea (Dutta et al., 2008). Similarly, *Bacillus amyloliquefaciens* IN937a and *B. pumilus* IN937b were used together and showed broad spectrum protection against *Ralstonia solanacearum, C. gloeosporioides, Sclerotium rolfsii* in different crops (Jetiyanon et al., 2003). *Pseudomonas* and *Rhizobium* were co-inoculated and decreased the multiplication of nematodes as compared to when they were inoculated singly. They also enhanced the nitrogen and phosphorus availability in soil and these nutrients had unpleasant effect on nematode (Pant et al., 1983). Six *Rhizobium* strains with three biofertilizers were used and *Rhizobium* strains (BINAR P6 and BINAR P36) and BINA biofertilizer resulted maximum reduction of seed rot, foot and root rot of Bush bean. Moreover, these microbial consortiums also increased germination, plant stand, vigor index, plant height, number of green pods per plant, weight of green pods per plant and weight of seeds per plant (Khalequzzaman and Hossain, 2008).

Bacterial consortium of *Bacillus* sp. MML2551, *B. licheniformis* MML2501, *Streptomyces fradiae* MML1042, *P. aeruginosa* MML2212 were used in biological control of sunflower necrosis virus disease and reported that theses strains are effective adjacent to SNVD and induced the plant growth promotion (Srinivasan and Mathivanan, 2009). Three different strains of *P. fluorescens viz.* CHA0, CoT1, and CPO1 were used to treat the tomato seedlings, which increased shoot length and suppress the tomato spot wilt virus (Kandan et al., 2005). Siddiqui and Akhtar (2009) observed that consortium of *Rhizobium* spp. and *Paecilomyces lilacinus* KIA were very effective in reducing the root-knot nematode caused by *Meloidogyne javanica* in chickpea roots and strain *Paecilomyces lilacinus* KIA also minimize the capability to parasitize females and eggs of nematodes while *Rhizobium* strain produced phytoalexins and antibiotics. Combined effects of the microbial consortium lead to inhibit the nematodes growth and increased plant growth.

22.6.3 Fungal and bacterial

Combine application of Fungal and bacterial antagonists were being used as microbial consortium for the disease management and yield enhancement. They showed promising way to improve efficacy of biocontrol strains. When the plant pathogenic microorganisms comes in contact with the host plant, than these co-inoculated antagonist works together and triggered defense responses through production of PR proteins, secondary metabolites, plant growth promotion enzymes or produced mechanisms to tolerate biotic and abiotic stress (Shoresh et al., 2010). *Trichoderma* and *Pseudomans* spp. are the commonly used antagonists, which showed compatibility with each other. Thakkar and Saraf (2010) isolate thirty isolates of bacteria and six isolates of fungus from the soil and evaluated for their antagonistic activity against phytopathogens like *Macrophomina phaseolina* and *Sclerotinia sclerotiorum,* under in vitro conditions. They reported *Pseudomonas aeruginosa* (MBAA1), *Bacillus cereus* (MBAA2) and *Bacillus amyloliquefaciens* (MBAA3) and *Trichoderma citrinoviride* (MBAAT) were screened as the most effective strains and developed as plant growth promoting microbial consortium to reduce the disease incidence in *Glycine* max both under in vitro and in vivo conditions. Similarly *P. fluorescens* and *T. viride* reduced incidence of sheath blight disease as compared to control (Mathivanan et al., 2005). In an experiment conducted by Singh et al., 2012b, combined application of *Trichoderma harzianum* and *Bacillus lentimorbus* has shown significant plant growth promotion in aromatic plant *Artemisia annua* (Fig. 22.2). Likewise, Rudresh et al. (2005) reported that co-inoculation of *Rhizobium*, *B. megaterium* sub sp., *Phospaticum* strain-PB, with *Trichoderma* spp. enhanced nutrient uptake, plant height, germination, noduation, pod yield on chick pea as compared to the untreated control.

Activation of phenylpropanoid pathway and antioxidant mechanism was stimulated by consortium of *P. aeruginosa* PJHU15, *T. harzianum* TNHU27 and *B. subtilis* BHHU100, when it used against *Sclerotinia sclerotiorum* in Pea (Jain et al., 2012). *Trichoderma harzianum* Tr6 and *Pseudomonas spp.* Ps14 were isolated from cucumber rhizosphere to test their combined effect against *Fusarium oxysporum* f. sp. *radicis cucumerinum* (Alizadeh et al., 2013). Sugarcane plant inoculated with consortia of *Trichoderma harzianum* and *Bacillus lentimorbus* demonstrated disease reduction and enhanced plant growth and vigor (Fig. 22.3). In the same way, combine application of *T. harzianum, Pseudomonas* and *G. intraradices* against *Fusarium* wilt, which suppressed disease severity and also improved plant growth and nutrient uptake (Srivastava et al., 2010). Siddiqui and Singh (2005) used combined application of *Rhizobium* with *P. straita* tested against *Meloidogyne incognita* in *Pisum sativum* and revealed that consortium of theses microorganisms decline the multiplication of nematodes and increased the growth characters. Effect of *Aphanomyces euteiches* f. sp. *pisi* was reduced by *T. harzianum* and *P. fluorescens* strain 2-79RN10 which produced antibiotic phenazine to control the disease (Dandurand and Knudsen, 1993). Plant growth promoting rhizobacteria provide resistance against a broad spectrum of pathogens. Consortium of PGPR strains *B. cereus* AR156, *B. subtilis* SM21, and Serratia sp. XY21 were showed promising results against phytophthora blight and enhanced fruit quality of sweet pepper (Zang et al., 2019).

FIG. 22.2 Effect of plant growth promoting microorganisms on growth of *Artemisia annua* (Left to Right) C=Control, 1. Inoculated with *Trichoderma harzianum* 2. *Bacillus lentimorbus* 3. *Pseudomonas aeruginosa* 4. Consortia of *Trichoderma harzianum* and *B. lentimorbus*.

FIG. 22.3 Sugarcane inoculated with consortia of *Trichoderma harzianum* and *Bacillus lentimorbus*. (a) Control (b) treated with consortia.

22.6.4 Algae and bacteria

The applicability of algal based consortia has not so much tested in case of plant diseases but is quite popular in bioremediation process. Algae cultivation in gaining popularity in agricultural, industrial and municipal wastewater as they play a crucial role in treatment of wastewater and bio manufacturing (Safonova et al., 2004). The wastewater effluents holds a bulk of nutrients (especially nitrogen and phosphorous) for growth of eukaryotic green algal species. The consortia of bacteria and cyanobacteria/microalgae have proved to be efficient in detoxification of organic and inorganic pollutants, and depletion of nutrients from wastewater assemblance, compared to a single microbe (Perera et al., 2019) Both the species enjoys a symbiotic relationship as the cyanobacterial/algal photosynthesis releases oxygen, acting as main electron acceptor to the heterotrophic bacteria. In turn, bacteria facilitate photoautotrophic growth of the partners through carbon dioxide and other stimulatory release (Subhaschandrabose et al., 2011). Apart from wastewater treatment the consortia act as a tool for biohydrogen production (Shetty et al., 2019). The benefits of utilizing them as the center on which to assemble the consortia lies on the way that they are broadly accessible, to deliver a variety of items with noteworthy significance in the welfare of human and ecology.

22.7 Need for development of biopesticides containing microbial consortium

The indiscriminate application of chemicals since last few decades has resulted in degradation of the quality of the ecosystem. The rising concern toward detoriation of ecosystem has forced the scientists/researchers to design for some better alternatives. Biological control sustained as an efficient management tool for some time until the target pathogen starts develop resistant against the particular BCA due to rapid evolution. In nature there is a constant battle occurs between the pathogen and antagonists where one tries to overpower the other. So, these obstacles have motivated the scientists to go for blend of antagonists which could perform diverse function and act synergistically. This concept paved the path for development of microbial consortium. Likewise human, microbial consortium too works on the concept of division of labor (DoL) where the metabolic tasks are being divided into various organisms and overall performance is synergistic (Roell et al., 2019). Recent years have witnessed more emphasis on combined use of BCAs with diverse disease control mechanisms so as to facilitate disease management process efficiently. The sparkling results of consortia application have mesmerized scientists to develop consortium involving microbes from different genera *i.e.,* bacteria and fungi, bacteria and actinomycetes, bacteria and mycorrhiza etc. (Singh et al., 2012a). A list of microbial consortium exhibiting significant results against multiple pathogens is depicted (Table 22.1). Currently in biopesticide industry majority of the biopesticides include only one BCA, which could act upon only few pests/pathogens. For an efficient management strategy, biopesticides of microbial consortia need to be developed which could act upon multiple sites of pathogen.

22.7.1 Biopesticide

Organic farming is considered to be the main driving force behind biopesticide production. In exceptionally broad terms, as per the US Natural Insurance Organization (USEPA), biopesticides are products derived from natural elements, for example, animals, plants, microorganisms, minerals etc (Mishra et al., 2015b). Biopesticides usually comprises living organisms that control pest and pathogens. The EPA classified biopesticides into three significant categories depending upon the type of active ingredient being used i.e. biochemical, plant incorporated protectants and microbial pesticides. Biochemical pesticides are products either derived from natural sources or possess similar structure and capacity as the naturally occurring substances. Biochemical pesticides vary from conventional pesticides in their structure and action mechanism (O'Brien et al., 2009). Biochemical pesticides usually involve attractants, insect sex pheromones; moult hormones, juvenile hormones, natural plant regulators semiochemicals etc. (Rajapakse et al., 2016). The plant incorporated protectants are usually the plant derived products which have antagonistic property against the target organism. They are macromolecular in nature and usually get produced in genetically modified (GM) crops *viz.*, plants incorporated with Bt gene, kinase inhibitor gene etc. (Parker and Sander, 2017).

22.7.2 Microbial pesticides

Microbial pesticides are commonly termed as BCAs. They are highly selective with moderate or even no toxicity in contrast with chemicals (MacGregor, 2006). The primary constituent of a microbial pesticide is the antagonistic microorganism which may be fungi, bacteria, virus, nematode, protozoa or algae. The most regularly utilized microbial biopesticides are live microbes or their spores, which are pathogenic for the targeted pest/pathogen. These comprise biofungicides (*Trichoderma*, *Pseudomonas*, *Bacillus*, *Burkholderia* etc.), bioherbicides (*Phytophthora*), and bioinsecticides (*Bt*) (Gupta and Dikshit, 2010).

TABLE 22.1 Microbial consortia effective against various diseases/pathogens along with their action mechanism.

S·N.	Microbial consortium	Fighting against	Crop	Growth promoting activities	References
	P. fluorescens strain (2-79RN10) and *T. harzianum*	*Aphanomyces euteiches* f. sp. *pisi*	Pea	- Triggered induced systemic resistance - Produced siderophore - Inhibit hyphal growth	Dandurand and Knudsen (1993)
	Rhizobium tropici (UMR, 1899) and *B. subtilis* (MBI600)	*Fusarium solani* f. sp. *phaseoli*	French bean	- Possible siderophore production - Enhanced plant biomass	Estevez de Jensen et al. (2002)
	Bacillus sp. strain mixture (IN937b + SE49 + T4 + INRN)	*Colletotrichum gloeosporioides, Sclerotium rolfsii,* cucumber mosaic virus	Cucumber	- Induced systemic resistance	Jetiyanon et al. (2003)
	Trichoderma harzianum (DB11) and *Gliocladium catenulatum* Gliomix	*Phytophthora cactorum, P. fragariae*	Strawberry	- Plant growth promotion - Antimicrobial activity	
	T. viride (NCC 34) and *P. fluorescens* (MTCC 1749)	*Rhizoctonia solani*	Rice	- Enhanced germination, plant height, plant biomass - Induced systemic resistance	Mathivanan et al. (2005)
	B. cereus strain BS03, *Rhizobia spp.,* and *P. aeruginosa* RRLJ04	*Fusarial* wilt	Pigeon pea	- Induced PAL (phenylalanine ammonia lyase) and PO (peroxidase) activity - Incresed germination and plant growth - Enhanced yield	Dutta et al. (2008)
	Rhizobium spp., *P. straita* and *Glomus intraradices*	*Macrophomina phaseolina* and *Meloidogyne incognita*	Chickpea	- Increased plant growth, number of pods, chlorophyll, nitrogen, phosphorus and potassium content	Akhtar and Siddiqui (2008)
	B. subtilis isolate 16 (Bs16) and *P. fluorescens* isolate (Pf1, Py15)	*Macrophomina phaseolina.*	Mulberry	- Induced PO,PPL, PPO, β-1, 3-glucanase, chitinase and phenol activity - Inhibit mycelial growth of *M. phaseolina* - Promote growth characters	Ganeshmoorthi et al. (2008)
	B. subtilis (7612), *Gigaspora margarita* (AA), *Glomus intraradices* (KA), *Burkholderia cepacia* (4684), *Penicillium chrysogenum* (CA1), and *A. niger* (CA)	*Meloidogyne incognita*	Tomato	- Increased the length and shoot dry mass - Increase in shoot dry mass	Siddiqui and Akhtar (2009)
	P. fluorescens (WCS365) and *P. chlororaphis* (PCL1391)	*Colletotrichum lindemuthianum*	French bean	- Reduced sporulation - Decreased pathogen growth - Increased plant height, growth and yield characters	Bardas et al. (2009)
	B. amyloliquefaciens, B. macerans, B. atrophaeus, B. subtilis, B. pumils, Flavobacter balastinium, P. putida and Burkholderia cepacia	*Fusarium sambucinum, Fusarium culmorum* and *Fusarium oxysporum*	Potato	- Suppress potato dry rot - Incresed growth	Recep et al. (2009)
	G. intraradices, Pseudomonas, and *T. harzianum*	*Fusarium* wilt	Tomato	- Decresed disease incidence - Enhanced plant growth and nutrient uptake	Srivastava et al. (2010)

	B. pumilus, A. niger, A. awamori, T. harzianum and *P. putida*	*Fusarium* root-rot	Pea	- Increase in growth, chlorophyll, catalase and peroxidase activities - Reduced disease severity	Akhtar and Azam (2013)
	T. harzianum TNHU27, *P. aeruginosa* (PJHU15) and *B. subtilis* (BHHU100)	*Sclerotinia sclerotiorum*	Pea	- Activation of phenylpropanoid pathway and antioxidant mechanism - ROS and antioxidant activity	Jain et al. (2012), Jain et al. (2013a,b)
	Pseudomonas sp. (Ps14) and *T. harzianum* (Tr6)	*Fusarium oxysporum* f. sp. *radicis cucumerinum*	Cucumber	- Triggered defense related genes - Induced systemic resistance - Suppress disease severity	Alizadeh et al. (2013)
	Trichoderma harzianum (THU0816), *Mesorhizobium* sp. (RL091) and *P. aeruginosa* (PHU094)	*Sclerotium rolfsii*	Chickpea	- Activation of PP pathway - Increased lignin deposition - Increased germination and growth yield	
	Trichoderma harzianum (THU0816) and *Pseudomonas aeruginosa* (PJHU15)	*Sclerotinia sclerotiorum*	French bean	- Enhanced activity of plant defense enzymes - Suppress disease severity	Ram et al. (2015)
	Serratia sp. XY21, *B. cereus* AR156 and *B. subtilis* SM21,	Phytophthora blight	Sweet pepper	- Suppress disease severity - Improved fruit quality - Improved soil property	Zang et al. (2019)
	Trichoderma harzianum (TNHU27) and *Pseudomonas aeruginosa* (PJHU15)	*Sclerotinia sclerotiorum*	Cauliflower	- Activation of phenylpropanoid pathway and antioxidant mechanism	Ram et al. (2019)
	Trichoderma viride, PGPR-1 and *Rhizobium* strain-B1	*Rhizoctonia* root rot and Angular leaf spot	French bean	- Improved plant growth, pod yield, seed yield, seed quality and seed vigor - Suppress disease severity	Negi et al. (2019)

Microbial pesticides are the products of naturally occurring or genetically modified microbes such as bacteria, fungi, viruses, algae or protozoans. They suppress pests/pathogens either by releasing toxic metabolites specific to the targeted pest thus preventing establishment of pathogens through mycoparasitism, competition, or other mechanisms (Clemson, 2007; Bisen et al., 2016). These biopesticides can be applied in various forms such as live organisms, dead organisms, and spores, and moreover in different formulations of microbe-based pesticides that are commercially available.

Since its initiation, the worldwide production and distribution of biopesticides has witnessed a tremendous rise and is further expected to touch a compound annual growth rate of 20% (Market and Market, 2013). The global biopesticide market is projected to achieve US$ 6.6 billion by end of 2020 with a hike of CAGR of 18.8% from 2015 to 2020. In India, biopesticide market is forecasted to show high growth with projected compounded annual rate of 19% over the 2015–20 periods (Keswani et al., 2016). Out of the total global biopesticide present in market for all crop types, the contribution of bacterial biopesticides is 74%, fungal biopesticides 10%, viral biopesticides 5%, predator biopesticides 8% while rest biopesticides accounts for 3% (Mishra et al., 2015b).

22.8 Current status of Indian biopesticide sector

In India, various government and private organizations are engaged in identification and development of as microbial biopesticide at commercial scale. Since 2017, 361 administrative and private biocontrol research centers are involved biopesticide production business. A few legislative offices legitimately or in a roundabout way support biopesticide production and distribution. The Ministry of Agriculture and Farmers Welfare, and the Department of Biotechnology are the major funding body for biopesticide research and commercialization. Moreover, a few Indian Council of Agricultural Research (ICAR) institutes, mainly National Research Center for Integrated Pest Management and National Bureau of Agricultural Insect Resources (NBAIRs), with some State Agricultural Universities (SAUs) facilitate research and perform quality check of biopesticides (Rabindra, 2005). Few of these organizations also impart trainings to officials of State Agriculture Departments concerning quality control of biopesticides (Singh et al., 2016). Numerous biopesticides are provided free of cost to farmers through government extension offices (Mishra et al., 2015b). The National Policy for actualized by the federal Department of Agriculture and Cooperation, has prescribed large scale utilization of biopesticides to farmers for improving and maintaining health of the overall ecosystem.

According to annual report of Directorate of Plant Protection, Quarantine and Storage (DPPQS), in India, a total of 361 biocontrol laboratories and units are working while only few of them are involved in production sector (Fig. 22.4). At global scale, biopesticides comprises just 4% of the total pesticide consumption while organochloride pesticides top the chart having a share of 40% (Fig. 22.5). Till date, 970 biopesticide products have been registered by Central Insecticides Board and Registration Committee (CIBRC) which is the major governing body related to registration and regulation of

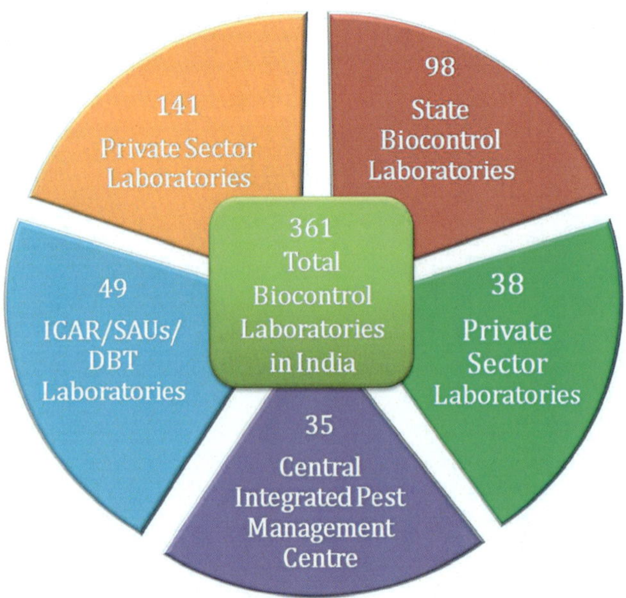

FIG. 22.4 Current structure of biocontrol laboratories and units working in India. *Modified from Mishra et al. (2020).*

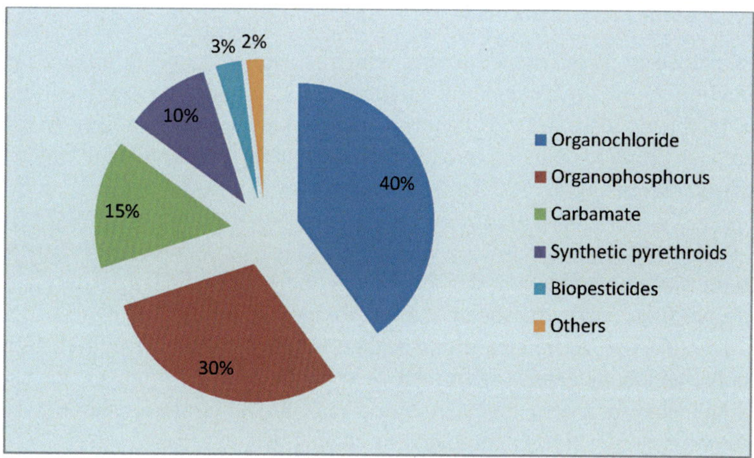

FIG. 22.5 Consumption of synthetic pesticides and biopesticides in India. *Source: Mordor intelligence.*

biopesticides in our country. According to Mishra et al. (2020), among the total biopesticide market for all type of crops, fungal biopesticides shares the major portion i.e. 66%; bacterial biopesticides 29%; viral biopesticides 4%; while the "rest" biopesticides accounts for 1% (Fig. 22.6). Among the bacterial biopesticides, *P. fluorescens* holds the major share of 71% followed by Bacillus spp. 28%; while the fungal biopesticide market is ruled by *Trichoderma* spp. which accounts for 56% of the total share.

22.9 Hurdles in commercialization of microbial based products in India

Commercialization of biopesticides in our country always face certain obstacles time to time, and the duty to tackle such issues lies with the private sector yet in addition with other important components of the society *viz.*, central and state government agencies, public and private firms, academicians, marketing professionals, etc. So it is the prime duty of all these organizations to evaluate the issues separately/altogether and bring it into light of general public. However, the different factors influencing the commercialization of biopesticides are as follows-

❖ Lack of knowledge and perception

The lack of information, knowledge and faith of farmers for such products is one of the key issues behind the slacking performances in the global market. The biopesticide are not quick in action compared to chemical pesticides which makes them less appealing to farmers who crave for immediate results. In few cases, farmers misinterpret the mentioned guidelines and safety measures during application. Moreover, maximum farmers lack the fundamental skill needed for proper application and storage of biopesticides. So in order to minimize such issues, training programs on handling of biopesticides should be conducted for farmers.

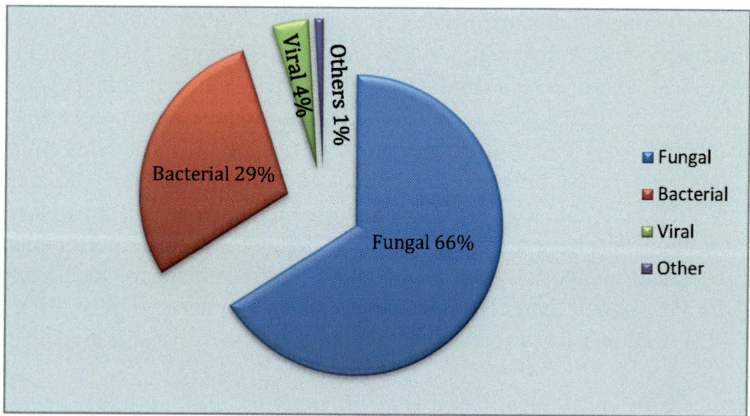

FIG. 22.6 Share of different categories of biopesticides. *Source: DPPQS.*

❖ Low Reliability and Inconsistent Field Performance

The biopesticides are subjected to inconsistent stability which is a major concern among farmers. Moreover, the organisms involved in production highly dependent on environmental parameters such as temperature, pH, moisture, exposure to radiation along with various other soil factors which influence their viability (Arora et al., 2010). Besides, production, biopesticides are also prone to contamination, which ultimately reduces the tally of active microbes and in turn leads to their underperformance (Evans et al., 1993).

❖ **Low Quality and Shelf Life**

Maintenance of cell count and contamination free product is the major challenges faced by the manufactures and farmers (Alam, 2000). These products need utmost care during storage and transit, failing to which leads to contamination. Farmers find it difficult to maintain the sterile conditions in their home for a longer period. Contamination brings about death of viable microbial cells, which adversely affect their field performance. Moreover, the shelf life of biopesticide is limited which render them applicable for a particular season or a year. So the farmers sometimes feel hesitant in purchasing the product as its expiry period is quite short as compared to chemicals.

❖ **More investment and fluctuating profit**

The biopesticides require utmost care during each stage of development till transit and storage and a huge amount of money is incurred in every process. Moreover, registration and licensing also adds to the overall cost. Study of the market is also necessary in order to popularize the product among local shopkeepers and farmers. So overall a large capital is invested by the manufactures in development of biopesticides. All these cost can be recovered only if the product gets excelled in the market and accepted by the farming community. So a huge risk is associated with biopesticide production and these products will be developed only when the associated firms could obtain a long-term profit. For example, few biopesticides based on baculovirus are highly selective in nature, so these products carry low profit potential and reduced market size (Tripathi et al., 2020).

❖ **Health and Ecological Risks**

Biopesticides may evoke a threat to human health if not used according to the instruction manual of the commercial product. However, the risk associated with them is quite low compared to chemical pesticides. In few cases it has been found that occupational exposure to biopesticides lead to health hazards *viz.*, microbial product where Bt as active ingredient, have not reported to create adverse effect on human, but exceptionally few cases have been found (Doekes et al., 2004). Several reports have also mentioned that spores of entomopathogenic fungi such as *B. bassiana* and *M. anisopliae* are allergens to immune-compromized patients (Keswani et al., 2014). Moreover, few strains of *Trichoderma* sp. is also reported to be detrimental to few crustaceans and marine organisms as well as humans (Kredics et al., 2011).

22.9.1 Regulatory framework and challenges for biopesticides in India

In India, the Central Insecticides Board and Registration Committee (CIBRC) is the apex body which regulates biopesticides under the Insecticides Act (1968) and Insecticides Rules (1971). This panel provides technical guidance to central and state government agencies as well as private firms on key issues related to manufacture, sale, transit and use of biopesticides (Tripathi et al., 2020; Singh et al., 2016). The CIB maintains a strong panel of eminent and dynamic scientists from all concerned disciplines who work collaborately for its proper functioning. The registration committee of CIBRC holds the right to grant license to all agencies for manufacture, distribution and sale of biopesticides to various firms. The novel biopesticide formulation has to undergo several quality check protocols, with proper analyzation of its associated potential threat to human health and ecosystem. A definite protocol is maintained during this entire process i.e. procuring a sample till licensing of the product. There are two sections under which manufacturers can apply for new products *i.e.*, section 9(3B) (provisional registration for a new active ingredient used in India) or 9(3) (regular registration) section of the Insecticides Act. However, the applicants have to submit a dossier mentioning critical information regarding their product *i.e.*, specifications (source, chemical composition etc.), toxicology, target organism, bioefficacy, weight and handling instructions while applying for registration under each category. Till date under section 9 (3B), 791 products have been registered whereas section 9(3) accounts for a total of 68 products. Further classification of these products has been depicted in Fig. 22.7.

The enclosure of biopesticides in government based IPM programmes was the initial venturing step headed for the ascent of microbe based products in our country. The primary government authorities responsible for advancement and

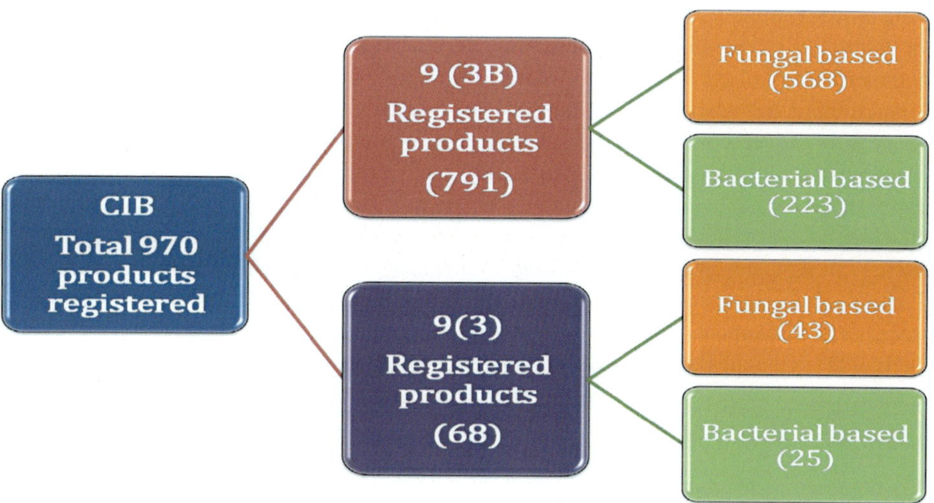

FIG. 22.7 Commercially registered biopesticides with CIBRC under 9 (3B) and 9(3). *Source: www.cibrc.nic.in/G_biopesticides.doc.*

creating awareness among farmers regarding biopesticides in India are the Ministry of Agriculture and Farmers Welfare and the Department of Biotechnology (DBT). The Directorate of Biological Control, Central IPM Center (Faridabad), and National Center for IPM (NCIPM) under ICAR, are likewise among the key players advancing dispersion of biopesticides (Alam, 1994). The DBT provides financial assistance being carried out on biopesticides and furthermore helps to generate toxicological information for product registration. Likelihood, few agencies such as National Agricultural Research System (NARS) and the National Board of Accreditation (NBA) whose essential obligation lies in directing different quality control trials of these microbes based products. Moreover they impart training to government officials of agriculture department in testing quality control. The state government's duty in executing IPM programs is imperative as it involves in procurement and regulation of biopesticides to farmers at moderate cost. Thus it creats a business opportunity and furthermore empowers the commercialization of microbial pesticides (Rabindra and Grzywacz, 2010).

Registration of biopesticide is the primary obstacle in the development, and in most cases registration is much more expensive than the production. The Indian registration committee is quite robust in nature. The registration process is not only costly plus time consuming too (Ehlers, 2006). The key issue is that biopesticides include active microbe cells (live organisms), and sometimes these live cells are treated like pathogens by the government authorities which render delay in their clearance (Keswani et al., 2016). Quality control testing of products is still a key issue to their dissemination in agricultural fields. Moreover, the registration committee involves experts from different fields, and few cases revealed that sometimes they have less experience, resulting in pending decisions which delays the registration process. The administration strategy additionally demands that adequacy of a biopesticide product should be guaranteed and quantified. The approved biopesticides can be promoted lawfully for crop security. Despite the fact that microbial consortia based products act multifunctionally, limited attention is paid for product development and registration (Jain et al., 2013a,b). Stated that these microbe-based products demand fastidious alignment as far as social strategies and their microbial composition in the overall production process. Moreover the registration of biopesticide containing two or more than two organisms requires extensive research regarding quality control issue. In few cases the bioagents act as both biopesticide and biofertilizers so the committee might face an awkward situation that in which category the licensing should be provided.

22.9.2 Future prospects

Biopesticides have been emerged as a safer alternative than chemical pest control measures, with conceivably less hazard to human and the ecosystem. To this end, collaboration of public and private sectors is needed to facilitate the development, manufacturing and trade of this ecofriendly alternative. The Indian biopesticide market is struggling with various challenges, the primary being the robust registration laws, poor quality control measures, lack of innovation and product diversification, less popularization of microbial products among peasants, long waiting span for certification of new product etc. (Keswani et al., 2016). Moreover, the development pace of biopesticide industry in India is demand based as it shares only 3% of the Indian pesticide market. In order to enhance manufacture and commercialization of biopesticides at a global scale, the regulatory barrier needs to be lessened. CIBRC should facilitate easy and speedy registration of products while maintaining the efficacy, quality, and consistency (Singh et al., 2016). The underlying procedure for registration and certification should be in public domain so that the manufacturers get an advanced idea regarding the overall procedure. Moreover, other commercialization barriers such as unawareness, poor shelf life, and field inconsistency can be removed by further research and training activities. Availability of superior quality products to farmers at a minimal price is the main objective. The detection of novel constituents and further research on formulation and delivery would surely boost commercialization and utilization of biopesticides. Recent advances in omics based technologies can lead to the production of advanced biopesticides holding longer shelf life and consistency (Keswani et al., 2016). In addition, on field evaluation of currently available biopesticides should get accomplished. Recombinant DNA technology can prove as an excellent tool to improve the efficacy of products in different farming systems, which ultimately lessen the thrust for new active ingredients. The advancement in nanotechnology science facilitated in development of new formulations such as nanoemulsion, nanosuspension and nanoencapsulation which could aid in their efficient delivery at target site with minimum residual losses (Glick, 2012; Damalas and Koutroubas, 2018).

22.10 Conclusion

Biological control is regarded as an ecofriendly substitute to chemical pest management. Previously biological control is usually attained by targeting one or few pests/pathogens by using a single BCA. However, the rapid evolutions of the pathogen under higher selection pressure lead to development of new strains carrying resistance against the former antagonist. Moreover, a single BCA usually have only action target site. This scenario has laid the foundation for further research with blending of two or more antagonists commonly known as microbial consortium. Till date microbial consortium is successfully used in many crops to ward off pathogens. However, the compatibility of all the antagonists is the key criteria in development of any consortium. The current target of sustainable agriculture demands use of microbial pesticides which conserves ecological harmony through reduced dependence on chemicals. A wide range of biopesticides are commercially available in different formulations. Despite being so much effective an economical, still the microbial pesticides accounts for only 3% of the global biopesticide production. Since their inception, the scenario of biopesticides remains in quandary. The main factor responsible for their poor performance is product commercialization and certain regulatory hurdles. The availability of biocontrol products is lagging the demand due to stringent registration policies or sometimes the product does not end up meeting quality standards. Moreover, the farmers often get perplexed and hesistant in picking biopesticides over the chemicals. Therefore, it is very important for the policy makers to gear up registration process but become tedious also to stop the regulation of unscrupulous products in the market. Overall a combined effort of different research institutes, SAUs, government organizations as well as NGOs is needed to boost the stature of biopesticides. A strong fortitude at global scenario is needed to promote these green bioproducts and minimize the hazardous chemicals in order to maintain harmony of ecosystem.

References

Akhtar, M.S., Siddiqui, Z.A., 2008. Biocontrol of a root-rot disease complex of chickpea by Glomus intraradices, *Rhizobium* sp. and *Pseudomonas straita*. Crop Protect. 27 (3–5), 410–417.

Akhtar, M.S., Azam, T., 2013. Effects of PGPR and antagonistic fungi on the growth, enzyme activity and Fusarium root-rot of pea. Arch. Phytopathol. Plant Protect. 47 (2), 138–148.

Alam, G., 1994. Biotechnology and Sustainable Agriculture: Lessons from India. Technical paper no. 103. OECD Development Centre, Paris.

Alam, G., 2000. A Study of Biopesticides and Biofertilizers in Haryana, India Gatekeeper Series No. 93. IIED, UK.

Alizadeh, H., Behboudi, K., Ahmadzadeh, M., Javan-Nikkhah, M., Zamioudis, C., Pieterse, C.M.J., Bakker, P.A.H.M., 2013. Induced systemic resistance in cucumber and *Arabidopsis thaliana* by the combination of *Trichoderma harzianum* Tr6 and *Pseudomonas* sp. Ps14. Biol. Contr. 65, 14–23.

Arora, N.K., Khare, E., Maheshwari, D.K., 2010. Plant growth promoting rhizobacteria: constraints in bioformulation, commercialization, and future strategies. In: Maheshwari, D.K. (Ed.), Plant Growth and Health Promoting Bacteria. Springer, Berlin, pp. 97–116.

Arora, N.K., Kim, M.J., Kang, S.C., Maheshwari, D.K., 2007. Role of chitinase and b-1,3-glucanase activity produced by a fluorescent pseudomonad and in vitro inhibition of *Phytophthora capsici* and *Rhizoctonia solani*. Can. J. Microbiol. 53, 207–212.

Arora, N.K., Verma, M., Prakash, J., Mishra, J., 2016. Regulation of biopesticides: global concerns and policies. In: Bioformulations: For Sustainable Agriculture. Springer, New Delhi, pp. 283–299.

Bardas, G.A., Lagopodi, A.L., Kadoglidou, K., Tzavella-Klonari, K., 2009. Biological control of three *Colletotrichum lindemuthianum* races using *Pseudomonas chlororaphis* PCL1391 and *Pseudomonas fluorescens* WCS365. Biol. Contr. 49, 139–145.

Bhatia, S.K., Bhatia, R.K., Choi, Y.K., Kan, E., Kim, Y.G., Yang, Y.H., 2018. Biotechnological potential of microbial consortia and future perspectives. Crit. Rev. Biotechnol. 38 (8), 1209–1229.

Bisen, K., Keswani, C., Patel, J.S., Sarma, B.K., Singh, H.B., 2016. *Trichoderma* spp.: efficient inducers of systemic resistance in plants. In: Chaudhary, D.K., Verma, A. (Eds.), Microbial-mediated Induced Systemic Resistance in Plants. Springer, Singapore, pp. 185–195.

Charest, M.H., Beauchamp, C.J., Antoun, H., 2005. Effects of the humic substances of de-inking paper sludge on the antagonism between two compost bacteria and *Pythium ultimum*. FEMS Microbiol. Ecol. 52, 219–227.

Christyan, P.F., De Carvalho, R., Resende, F.M.L., Azevedo, L.C.B., 2018. Consortium of five fungal isolates conditioning root growth and arbuscular mycorrhiza in soybean, corn, and sugarcane. An. Braz. Acad. Sci. 90 (4), 3649–3660.

Clark, D.P., Dunlap, P.V., Madigan, M.T., Martinko, J.M., 2009. Brock Biology of Microorganisms. Pearson, San Francisco, p. 485.

Clemson, H.G.I.C., 2007. Organic Pesticides and Biopesticides, Clemson Extension, Home and Garden Information Center. Clemson University, Clemson.

Cook, R.J., 1988. Biological control and holistic plant- health care in agriculture. Am. J. Alternative Agric. 3 (2), 51–62.

Damalas, C.A., Koutroubas, S.D., 2018. Current status and recent developments in biopesticide use. Agriculture 8, 1–6.

Dandurand, L.M., Knudsen, G.R., 1993. Influence of *Pseudomonas fluorescens* on hyphal growth and biocontrol activity of *Trichoderma harzianumin* in the spermosphere and rhizosphere of pea. Phytopathology 83, 265–270.

de Boer, M., Bom, P., Kindt, F., Keurentjes, J.J.B., van der Sluis, I., van Loon, L.C., Bakker, P.A.H.M., 2003. Control of Fusarium wilt of radish by combining *Pseudomonas putida* strains that have different disease-suppressive mechanisms. Biol. Contr. 93, 626–632.

Doekes, G., Larsen, P., Sigsgaard, T., Baelum, J., 2004. IgE sensitization to bacterial and fungal biopesticides in a cohort of Danish greenhouse workers: the BIOGART study. Am. J. Ind. Med. 46, 404–407.

Duffy, B.K., Weller, D.M., 1995. Use of *Gaeumannomyces graminis* var. graminis alone and in combination with fluorescent *Pseudomonas* spp. to suppress take-all of wheat. Plant Dis. 79, 907–911.

Dunne, C., Loccoza, Y.M., McCarthya, J., Higginsa, P., Powellb, J., Dowlinga, N., O'Gara, F., 1998. Combining proteolytic and phloroglucinol-producing bacteria for improved biocontrol of Pythium-mediated damping-off of sugar beet. Plant Pathol. 47, 299–307.

Dutta, S., 2015. Biopesticides: an eco-friendly approach for pest control. World J. Pharm. Pharmaceut. Sci. 4 (6), 250–265.

Dutta, S., Mishra, A.K., Kumar, B.S.D., 2008. Induction of systemic resistance against Fusarial wilt in pigeon pea through interaction of plant growth promoting rhizobacteria and rhizobia. Soil Biol. Biochem. 40, 452–461.

Ehlers, R.U., 2006. Einsatz der Biotechnologie im biologischen Pflanzenschuz. Schnreihe dtsch. Phytomed. Ges. 8, 17–31.

Estevez de Jensen, C., Percich, J.A., Graham, P.H., 2002. Integrated management strategies of bean root rot with *Bacillus subtilis* and *Rhizobium* in Minnesota. Field Crop. Res. 74, 107–115.

Evans, J., Wallace, C., Dobrowolski, N., 1993. Interaction of soil type and temperature on the survival of *Rhizobium leguminosarum* bv Viciae. Soil Biol. Biochem. 25, 1153–1160.

Eze, C.N., Odoh, C.K., Eze, E.A., Enemuor, S.C., Orjiakor, I.P., Okobo, U.J., 2018. Chromium (III) and its effects on soil microbial activities and phytoremediation potentials of *Arachis hypogea* and *Vigna unguiculata*. Afr. J. Biotechnol. 17 (38), 1207–1214.

Frankowski, J., Lorito, M., Scala, F., Schmidt, R., Berg, G., Bahl, H., 2001. Purification and properties of two chitinolytic enzymes of *Serratia plymuthica* HRO-C48. Arch. Microbiol. 176, 421–426.

Ganeshamoorthi, P., Anand, T., Prakasam, V., Bharani, M., Ragupathi, N., Samiyappan, R., 2008. Plant growth promoting rhizobacterial (PGPR) bio-consortia mediates induction of defense-related proteins against infection of root rot pathogen in mulberry plants. J. Plant Intract. 3 (4), 233–244.

Glick, B.R., 2012. Plant Growth promoting bacteria: mechanisms and applications. Science p15. Article ID 963401.

Goswami, D., Thakker, J.N., Dhandhukia, P.C., 2016. Portraying mechanics of plant growth promoting rhizobacteria (PGPR): a review. Cogent. Food Agric. 2 (1), 1127500.

Gupta, S., Dikshit, A.K., 2010. Biopesticides: an ecofriendly approach for pest control. J. Biopestic. 3, 186–188.

Hansel, C.M., Fendorf, S., Jardine, P.M., Francis, C.A., 2008. Changes in bacterial and archaeal community structure and functional diversity along a geochemically variable soil profile. Appl. Environ. Microbiol. 74 (5), 1620–1633.

Jain, A., Singh, S., Sarma, B.K., Singh, H.B., 2012. Microbial consortium-mediated reprogramming of defence network in pea to enhance tolerance against *Sclerotinia sclerotiorum*. J. Appl. Microbiol. 112, 537–550.

Jain, A., Singh, A., Singh, B.N., Singh, S., Upadhyay, R.S., Sarma, B.K., Singh, H.B., 2013a. Biotic stress management in agricultural crops using microbial consortium. In: Maheshwari, D.K. (Ed.), Bacteria in Agrobiology: Disease Management, vol. 5. Springer-Verlag, Berlin/Heidelberg, pp. 427–448.

Jain, A., Singh, A., Singh, S., Singh, H.B., 2013b. Microbial consortium induced changes in oxidative stress markers in pea plants challenged with *Sclerotinia sclerotiorum*. J. Plant Growth Regul. 32, 388–398.

Jetiyanon, K., Fowler, W.D., Kloepper, J.W., 2003. Broad spectrum protection against several pathogens by PGPR mixtures under field conditions in Thailand. Plant Dis. 87, 1390–1394.

Kandan, A., Ramaiah, M., Vasanthi, V.J., Radjacommare, R., Nandakumar, R., Ramanathan, A., Samiyappan, R., 2005. Use of *Pseudomonas fluorescens* based formulations for management of tomato spot wilt virus (TSWV) and enhanced yield in tomato. Biocontrol Sci. Technol. 15, 553–569.

Keswani, C., Mishra, S., Sarma, B.K., Singh, S.P., Singh, H.B., 2014. Unraveling the efficient application of secondary metabolites of various *Trichoderma*. Appl. Microbiol. Biotechnol. 98, 533–544.

Keswani, C., Sarma, B.K., Singh, H.B., 2016. Synthesis of policy support, quality control, and regulatory management of biopesticides in sustainable agriculture. In: Singh, H.B., Sarma, B.K., Keswani, C. (Eds.), Agriculturally Important Microorganisms: Commercial and Regulatory Requirement in Asia. Springer, Singapore, pp. 167–181.

Khachatourians, G.G., 2009. Insecticides, microbials. Appl. Microbiol. Agro/Food 95–109.

Khalequzzaman, K.M., Hossain, I., 2008. Effect of rhizobium strains and biofertilizers on foot, root rot and yield of bush bean in *Sclerotinia sclerotiorum* infested soil. J. Biol. Sci. 16, 73–78.

Kohli, J., Kolnaar, R., Ravensberg, W.J., 2019. Mode of action of microbial biological control agents against plant diseases: relevance beyond efficacy. Front. Plant Sci. 10, 845.

Koornneef, A., Pieterse, C.M., 2008. Cross talk in defense signaling. Plant Physiol. 146 (3), 839–844.

Kredics, L., Hatvani, L., Manczinger, L., Vagvolgyi, C., Antal, Z., 2011. Trichoderma. In: Liu, D. (Ed.), Molecular Detection of Human Fungal Pathogens. Taylor & Francis Group, London, UK, pp. 509–526.

Kumar, S., 2012. Biopesticides: a need for food and environmental safety. J. Biofert. Biopestic. 3 (4), 1–3.

Lugtenberg, B.J.J., Chin-A-Woeng, T.F.C., Bloemberg, G.V., 2002. Microbe plant interactions: principles and mechanisms. Anton Leeuw Int. J. 81, 373–383.

MacGregor, J.T., 2006. Genetic toxicity assessment of microbial pesticides: needs and recommended approaches. Intern. Assoc. Environ. Mutagen. Soc. 1–17.

Market and Market, 2013. Report Code: CH 1266 Global Biopesticides Market – Trends and Forecasts (2012–2017). India.

Mathivanan, N., Prabavathy, V.R., Vijayanandraj, V.R., 2005. Application of talc formulations of *Pseudomonas fluorescens* Migula and *Trichoderma viride* Pers. Ex S.F. gray decrease the sheath blight disease and enhance the plant growth and yield in rice. J. Phytopathol. 153, 697–701.

Mishra, S., Singh, A., Keswani, C., Saxena, A., Sarma, B.K., Singh, H.B., 2015a. Harnessing plant-microbe interactions for enhanced protection against phytopathogens. In: Arora, N.K. (Ed.), Plant Microbe Symbiosis– Applied Facets. Springer, New Delhi, pp. 111–125.

Mishra, J., Tewari, S., Singh, S., Arora, N.K., 2015b. Biopesticides: where we stand?. In: Plant Microbes Symbiosis: Applied Facets. Springer, New Delhi, pp. 37–75.

Mishra, J., Dutta, V., Arora, N.K., 2020. Biopesticides in India: technology and sustainability linkages. 3 Biotech 10, pp1–12.

Morrissey, J.P., Dow, J.M., Mark, L., O'Gara, F., 2004. Are microbes at the root of a solution to world food production? EMBO Rep. 5, 922–926.

Muthukumar, A., Bhaskaran, R., Sanjeev, K., 2010. Efficacy of endophytic *Pseudomonas fluorescens* (Trevisan) migula against chilli damping-off. J. Biopestic. 3, 105–109.

Negi, S., Bharat, N.K., Kumar, M., 2019. Effect of seed biopriming with indigenous PGPR, *Rhizobia* and *Trichoderma* sp. on growth, seed yield and incidence of diseases in French bean (*Phaseolus vulgaris* L.). Legume Res. https://doi.org/10.18805/LR-4135.

Nuti, M., Giovannetti, G., 2015. Borderline products between bio-fertilizers/bio-effectors and plant protectants: the role of microbial consortia. J. Agric. Sci. Technol. 5, 305–315.

O'Brien, K.P., Franjevic, S., Jones, J., 2009. Green chemistry and sustainable agriculture: the role of biopesticides, advancing green chemistry. Ecology 90, 2223–2232.

Odoh, C.K., 2017. Plant growth promoting rhizobacteria (PGPR): a bioprotectant bioinoculant for sustainable agrobiology. A review. Int. J. Adv. Res. Biol. Sci. 4 (5), 123–142.

Pant, V., Hakim, S., Saxena, S.K., 1983. Effect of different levels of N, P, K on the growth of tomato Marglobe and on the morphometrics of root knot nematode *Meloidogyne incognita*. Indian J. Nematol. 13, 110–113.

Parker, K.M., Sander, M., 2017, April. Environmental fate of double-stranded RNA (dsRNA) biopesticides from RNA interference (RNAi)-based crop protection. In: In 253rd ACS National Meeting: Advanced Materials, Technologies, Systems & Processes (ACS Spring 2017).

Perera, I.A., Abinandan, S., Subashchandrabose, S.R., Venkateswarlu, K., Naidu, R., Megharaj, M., 2019. Advances in the technologies for studying consortia of bacteria and cyanobacteria/microalgae in wastewaters. Crit. Rev. Biotechnol. 39 (5), 709–731.

Rabindra, R.J., 2005. Current status of production and use of microbial pesticides in India and the way forward. In: Rabindra, R.J., Hussaini, S.S., Ramanujam, B. (Eds.), Microbial Biopesticide Formulations and Applications: Technical Document No. 55. Project Directorate of Biological Control, Bangalore, pp. 1–12.

Rabindra, R.J., Grzywacz, D., 2010. India. In: Kabaluk, J.T., Svircev, A.M., Goettel, M.S., Woo, S.G. (Eds.), The Use and Regulation of Microbial Pesticides in Representative Jurisdictions Worldwide. IOBC Global, p. 99.

Rajapakse, R.H.S., Ratnasekera, D., Abeysinghe, S., 2016. Biopesticides research: current status and future trends in Sri Lanka. In: Agriculturally Important Microorganisms. Springer, Singapore, pp. 219–234.

Ram, R.M., Jain, A., Singh, A., B Singh, H., 2015. Biological management of Sclerotinia rot of bean through enhanced host defense responses triggered by *Pseudomonas* and *Trichoderma* species. J. Pure App. Microbial. 9 (1), 523–532.

Ram, R.M., Singh, H.B., 2017. Microbial consortium in biological control: an explicit example of teamwork below ground. J. Ecofriend. Agric. 13, 1–12.

Ram, R.M., Keswani, C., Bisen, K., Tripathi, R., Singh, S.P., Singh, H.B., 2018. Biocontrol technology: eco-friendly approaches for sustainable agriculture. In: Brah, D., Azevedo, V. (Eds.), Omics Technologies and Bio-Engineering: Towards Improving Quality of Life Volume II Microbial, Plant, Environmental and Industrial Technologies. Academic, London, pp. 177–190.

Ram, R.M., Tripathi, R., Birla, H., Dilnashin, H., Singh, S.P., Keswani, C., 2019. Mixed PGPR consortium: an effective modulator of antioxidant network for management of collar rot in cauliflower. Arch. Phytopathol. Plant Protect. 52 (7–8), 844–862.

Recep, K., Fikrettin, S., Erkol, D., Cafer, E., 2009. Biological control of the potato dry rot caused by *Fusarium* species using PGPR strains. Biol. Contr. 50 (2), 194–198.

Roell, G.W., Zha, J., Carr, R.R., Koffas, M.A., Fong, S.S., Tang, Y.J., 2019. Engineering microbial consortia by division of labor. Microb. Cell Factories 18 (1), 1–11.

Rudresh, D.L., Shivaprakash, M.K., Prasad, R.D., 2005. Effect of combined application of Rhizobium, phosphate solubilizing bacterium and *Trichoderma* spp. on growth, nutrient uptake and yield of chickpea (*Cicer aritenium* L.). Appl. Soil Ecol. 28, 139–146.

Safonova, E., Kvitko, K.V., Iankevitch, M.I., Surgko, L.F., Afti, I.A., Reisser, W., 2004. Biotreatment of industrial wastewater by selected algal-bacterial consortia. Eng. Life Sci. 4 (4), 347–353.

Saxena, A., Mishra, S., Raghuwanshi, R., Singh, H.B., 2013. Biocontrol agents: basics to biotechnological applications in sustainable agriculture. In: Tiwari, S.P., Sharma, R., Gaur, R. (Eds.), Recent Advances in Microbiology, vol. 2. Nova Publishers, USA, pp. 141–164.

Sharma, A., Diwevidi, V.D., Singh, S., Pawar, K.K., Jerman, M., Singh, L.B., Singh, S., Srivastawav, D., 2013. Biological control and its important in agriculture. Int. J. Biotechnol. Bioeng. Res. 4 (3), 175–180.

Shetty, P., Boboescu, I.Z., Pap, B., Wirth, R., Kovacs, K.L., Biro, T., Futo, Z., White III, R.A., Maroti, G., 2019. Exploitation of algal-bacterial consortia in combined biohydrogen generation and wastewater treatment. Front. Energy Res. 7, 52.

Shoresh, M., Harman, G.E., Mastouri, F., 2010. Induced systemic resistance and plant responses to fungal biocontrol agents. Annu. Rev. Phytopathol. 48, 1–23.

Siddiqui, Z.A., Akhtar, M.S., 2009. Effects of antagonistic fungi, plant growth-promoting rhizobacteria, and arbuscular mycorrhizal fungi alone and in combination on the reproduction of *Meloidogyne incognita* and growth of tomato. J. Gen. Plant Pathol. 75, 144.

Siddiqui, Z.A., Singh, L.P., 2005. Effect of fly ash, *Pseudomonas striata* and Rhizobium on the reproduction of nematode *Meloidogyne incognita* and on the growth and transpiration of pea. J. Environ. Biol. 26, 117–122.

Singh, A., Sarma, B.K., Upadhyay, R.S., Singh, H.B., 2012a. Compatible rhizosphere microbes mediated alleviation of biotic stress in chickpea through enhanced antioxidant and phenylpropanoid activities. Microbiol. Res. 168, 33–40.

Singh, H.B., Singh, A., Nautiyal, C.S., 2002. Commercialization of biocontrol agents: problems and prospects. In: Rao, G.P. (Ed.), Frontiers of Fungal Diversity in Indian Subcontinent. International Book Distributing Company, Lucknow, India, pp. 847–861.

Singh, H.B., Singh, A., Singh, S.P., Nautiyal, C.S., 2004. Commercialization of biocontrol agents: the necessity and its impact on agriculture. In: Singh, S.P., Singh, H.B. (Eds.), Ecoagriculture with Bioaugmentation: An Emerging Concept. Rohitashwa Printers, Lucknow, pp. 1–20.

Singh, H.B., Singh, B.N., Singh, S.P., Sarma, B.K., 2012b. Exploring different avenues of *Trichoderma* as a potent bio-fungicidal and plant growth promoting candidate-an overview. Rev. Plant Pathol. 5, 315–426.

Singh, H.B., Keswani, C., Bisen, K., Sarma, B.K., Chakrabarty, P.K., 2016. Development and application of agriculturally important microorganisms in India. In: Agriculturally Important Microorganisms. Springer, Singapore, pp. 167–182.

Singh, H.B., Sarma, B.K., Keswani, C., 2017. Advances in PGPR Research. CABI, Wallingford, ISBN 9781786390325, p. 408.

Sobowale, A.A., Cardwell, K.F., Odebode, A.C., Bandyopadhyay, R., Jonathan, S.G., 2007. Persistence of *Trichoderma* species within maize stem against *Fusarium verticillioides*. Arch. Phytopathol. Plant Protect. 40 (3), 215–231.

Srinivasan, K., Mathivanan, N., 2009. Biological control of sunflower necrosis virus disease with powder and liquid formulations of plant growth promoting microbial consortia under field conditions. Biol. Contr. 51, 395–402.

Srivastava, R., Khalid, A., Singh, U.S., Sharma, A.K., 2010. Evaluation of arbuscular mycorrhizal fungus, fluorescent *Pseudomonas* and *Trichoderma harzianum* formulation against *Fusarium oxysporum* f. sp. lycopersici for the management of tomato wilt. Biol. Contr. 53, 24–31.

Stockwell, V.O., Johnson, K.B., Sugar, D., Loper, J.E., 2011. Mechanistically compatible mixtures of bacterial antagonists improve biological control of fire blight of pear. Phytopathology 101, 113–123.

Stolyar, S., Van Dien, S., Hillesland, K.L., Pinel, N., Lie, T.J., Leigh, J.A., Stahl, D.A., 2007. Metabolic modeling of a mutualistic microbial community. Mol. Syst. Biol. 3, 92.

Subash, S.P., Chand, P., Pavithra, S., Balaji, S.J., Pal, S., 2017. Pesticide use in Indian agriculture: trends, market structure and policy issues. In: Policy Brief. ICAR-National Centre for Agricultural Economics and Policy Research, New Delhi, India, p. 43.

Subashchandrabose, S.R., Ramakrishnan, B., Megharaj, M., Venkateswarlu, K., Naidu, R., 2011. Consortia of cyanobacteria/microalgae and bacteria: biotechnological potential. Biotechnol. Adv. 29 (6), 896–907.

Thakkar, A., Saraf, M., 2010. Development of microbial consortia as a biocontrol agent for effective management of fungal diseases in Glycine max L. Arch. Phytopathol. Plant Protect. 48 (6), 459–474.

Timmis, K., de Vos, W.M., Ramos, J.L., Vlaeminck, S.E., Prieto, A., Danchin, A., Verstraete, W., de Lorenzo, V., Lee, S.Y., Brüssow, H., Timmis, J.K., Singh, B.K., 2017. The contribution of microbial biotechnology to sustainable development goals. Microb. Biotechnol. 10 (5), 984–987.

Tringe, S.G., von Mering, C., Kobayashi, A., Salamov, A.A., Chen, K., Chang, H.W., Podar, M., Short, J.M., Mathur, E.J., Detter, J.C., Bork, P., Hugenholtz, P., Rubin, E.M., 2005. Comparative metagenomics of microbial communities. Science 308, 554–557.

Tripathi, Y.N., Divyanshu, K., Kumar, S., Jaiswal, L.K., Khan, A., Birla, H., Gupta, A., Singh, S.P., Upadhyay, R.S., 2020. Biopesticides: current status and future prospects in India. In: Bioeconomy for Sustainable Development. Springer, Singapore, pp. 79–109.

Wilson, C.L., 1997. Biological control and plant diseases- a new paradigm. J. Ind. Microbiol. Biotechnol. 19, 158–159.

Yadav, M.L., Simon, S., Lal, A.A., 2017. Efficacy of *Trichoderma* spp. *P. Fluorescence* and Neem cake against the chick pea wilt. Ann. Plant Protect. Sci. 25 (2), 347–350.

Zabbey, N., Sam, K., Onyebuchi, A.T., 2017. Remediation of contaminated lands in the Niger Delta, Nigeria: prospects and challenges. Sci. Total Environ. 586, 952–965.

Zang, L.-N., Wang, D.-C., Hu, Q., Dai, X.-D., Xie, Y.-S., Li, Q., Liu, H.-M., Guo, J.-H., 2019. Consortium of plant growth-promoting rhizobacteria strains suppresses sweet pepper disease by altering the rhizosphere microbiota. Front. Microbiol. 10, 1668.

Chapter 23

Biocides through pyrolytic degradation of biomass: potential, recent advancements and future prospects

Avedananda Ray[a], Sabuj Ganguly[b] and Ardith Sankar[c]

[a]*Department of Agricultural & Environmental Sciences, Tennessee State University, Nashville, TN, United States;* [b]*Department of Entomology and Agricultural Zoology, Institute of Agricultural Sciences, Banaras Hindu University, Varanasi, Uttar Pradesh, India;* [c]*Department of Agronomy, Institute of Agricultural Sciences, Banaras Hindu University, Varanasi, Uttar Pradesh, India*

23.1 Introduction

Changing global pesticide policies and Integrated Pest Management (IPM) initiatives are aimed at reducing environmental and human health threats to pesticides. This was explicitly indicated that the risks to health and the atmosphere from the use of pesticides must be decreased and there must be reduced reliance on chemical controls. Currently, extensive pesticides had been applied to control various crop pests. Production & Consumption of Synthetic pesticides had significantly increased due to many advantages like reliable controlling method, economically feasible, easy application strategy. The Synthetic pesticides definitely contributed for the increased agricultural production but ecological damage is the unprecedented side effect of extensive application of these pesticides. Synthetic pesticides leach down to ground water and contaminate the water reservoir. Moreover, workers of pesticide industry have faced serious health hazard during manufacturing, formulating and application of synthetic pesticides. By now it's clear that dependence of the synthetic chemicals must be reduced and considerable adaptations of bio-based formulations are required. For overcoming these challenges of chemically formulated synthetic pesticides, one of the best measures is the use of eco-friendly technology which can use plant biomass-originated products.

Numerous policies focus on "*alternative plant protection techniques*" to replace the chemically formulated pesticides. Compared to Europe, the adaptation of natural pesticides in Asia and the USA had been significantly increased for plant protection purposes. In Asia and the USA, there is already an established green chemical market. Concomitantly, due to increased environmental concern in Europe, the Knowledge Based Bio-Economy (KBBE) concept has been developed to highlight the conversion products of plant biomass. The popularity of the efficient use of plant biomass products have increased now-a-days due to their biodegradability and least persistence. Moreover, these products are less toxic to non-target organisms, economically viable and easy available worldwide. Numerous researches had been done on plant essential oils about their bio-cide property such as the use of essential oils as an insecticide, fungicide, nematicide and bioherbicide. The major drawback lies in the essential bio-oil is that the process employs extensive mechanical repertoire and very less quantity is obtained.

From an ample production of biomass (146 billion metric tons/year), maximum proportions of this are agricultural waste or residues such as husk, cobs, and straws (Demirbaş, 2002). These waste materials either be inefficiently discarded or burnt ineffectively, which leads to formation of aerosols, semi-volatile organic carbons, polycyclic aromatic hydrocarbons (PAHs), polychlorinated dibenzo-p-dioxins, large particulates, residual ash and gasses like CO, N_2O, NO_x, CO_2, CH_4 which can contaminate the environmental surrounding (Gadde et al., 2009; Ferek et al., 1998). This ineffective burning must be reduced by proper waste conversion technologies to avert the environmental and toxicological threats (Lemieux et al., 2004). Pyrolysis is one of the fast-growing bio-based conversion technologies. Biomass Pyrolysis is the thermal degradation of biomass in the oxygen deficient or oxygen deprived environment (Balat and Balat, 2009). During pyrolysis, in high temperature the plant

Biopesticides. https://doi.org/10.1016/B978-0-12-823355-9.00025-0

biomass are dehydrolysed, decomposed, decarboxylated to produce volatiles, water vapor, tar which can be condensed by cold environment trap (Mansur, 2013). The conventional pyrolysis products are generally bio-fuel and bio-char, but recently many other byproducts have been recognized. The crude reddish brown aqueous liquid fraction is known as pyroligneous acid (PA) or wood vinegar which is rich in oxygenated compounds (Mathew et al., 2014). The term "pyroligneous" includes "pyrolysis" (process of thermo conversion) and "lignin," (the components of the plant biomass).

Pyrolytically produced wood vinegar has been potentially used as a bio-cide in less synthetic chemical consuming countries. Wood vinegar is also used in countries where number of marginal farmers is more. The production and application of wood vinegar is already reported in countries like china, Japan, Korea (commonly known as "Mokusaku-eki" in South Korea and Japan), Thailand, Taiwan, Indonesia, France, Sweden, Norway, Brazil and the USA (Yatagai et al., 2002; Imamura and Watanabe, 2007; Lee et al., 2012; Wang et al., 2012; Guillen and Manzanos, 2005). Substantial research had been focused on the bio-based economy surrounded by the pyrolytic products and their potential as emerging bio-cide. Although ample research had been done on the potential of pyrolytic fluids as biocide, very few number of review articles had been found. The objective of this review chapter was to delineate the potential of pyrolytic liquids in pesticide applications. The first section describes various pyrolysis from biomass, liquids and wood tar in the past. The second section focuses on the chemical efficacy of pyrolysis products as biocide.

23.1.1 Bio-pesticides: a green alternative to synthetic pesticide

Biopesticides are prepared from sources that are living, such as plants, animals and microorganisms. Some of the bio-pesticides include chitin, fulvic acid and plant extracts derived from moringa, soursop, tannic acid, wood vinegar, and spirulina. Plant extracts from spices such as onions, garlic, red pepper, black pepper, poppy seed and cinnamon also contain potential antibacterial compounds (Nabavi et al., 2015). Chitin promotes expression of unique early responsive and defense-related plant genes (Minami et al., 1998; Nishizawa et al., 1999). Fulvic acid is one of the humic substances with size <3500 Da comprised of carbon, hydrogen, oxygen, nitrogen and sulfur (HS), made up from lignites, sapropels, and peat coupled with non-living organic matter of the soil and water ecosystems (Thurman, 1985; Orlov, 1990; Swift, 1993). Fulvicacid is a novel antimicrobial molecule that has sometimes been shown to have antibacterial and antifungal properties (Van Rensburg et al., 2000). Bio-fungicides are fungicides that contain a microorganism (usually a bacterium or fungus) as the active ingredient (Swain et al., 2014). They have a compounded mode of action; for example, they can control pathogens through competition for critical nutrients, and/or physical space; and are applied prior to disease onset (Swain et al., 2014). Additionally their mode of action could be through antibiosis; where the microorganism produces an antibiotic or a toxin that interferes with the pathogen. These bioactive compounds may include peptides, glycosides, alkaloids, saponins, terpenoids and flavonoids.

Recently, advancement in the use of wood vinegar has earned much attention, especially in the agriculture industry (Tiilikkala et al., 2010) There are minimal negative effects associated with its usage to the environment as compared to the synthetic pesticides which can run off or leach to the ground water (Tiilikkala et al., 2010). On plants, wood vinegar has been used as a foliar spray and as a replacement for synthetic fertilizers and pesticides (Yatagai et al., 2000) Wood vinegar is a brown, flavorful liquid that is a result of pyrolysis. Pyrolysis is the process by which wood is heated in a closed vessel or airtight container leading to carbonization. During this process smoke is given off which is cooled, and a liquid is collected (Yang et al., 2016b) When this liquid settles, three distinct layers form: an oily liquid occupies the top layer, the middle layer consists of a transparent, yellowish−brown liquid called raw wood vinegar; and the thick wood tar settles at the bottom. Burning different kinds of wood, for example eucalyptus (Amen-Chen et al., 1997), oak (Guillén and Manzanos, 2002) bamboo (Mu and Høy, 2004) coconut shell (Wititsiri, 2011) and apple trees in an airtight vessel can produce various kinds of wood vinegar. Many wood-vinegar sources are identified as safe, natural inhibitors with numerous bioactive compounds that make them suitable for use in antifungal, termiticidal and repellent applications (Mu and Høy, 2004) Wood vinegar has antibacterial activities as well (Mu and Høy, 2004) (Table 23.1).

23.2 Pyrolysis-an efficient technology

The lignocellulosic biomass can be thermo-chemically converted into biofuels by means of pyrolysis, gasification and liquefaction. The pyrolysis, an old technology, is apparently gain more attention among energy utilization technique. In pyrolysis, various process parameters (temperature, pressure, charring time, rate of heating), chemical nature of feedstock biomass decides the pyrolytic yield. Basically, Biomass is converted into biochar, bio-condensate and syn-gas.

TABLE 23.1 Plant sources and their anti-microbial properties.

Plant source	Anti-microbial activity	References
Bio-active molecules from *Spirulinaplatensis* (Spiriluna),	Plant-pathogenic bacteria, due their antimicrobial property	Usharani and Balu (2015)
Moringaoleifera	Antibacterial compounds including; 4-(4′−O-acetyl-α-L-rhamnopyranosyloxy) benzyl Isothiocy-anate, 4-(α-L-rhamnopyranosyloxy) benzyl isothiocy-anate,niazimicin, Pterygospermin, benzyl isothiocyanate and 4-(α-L-rhamnopyranosyloxy) benzyl Glucosinolate	Jed and Fahey (2005)
Soursop (*Annonamuricata)*	Antibacterial agent	Sundarrao et al. (1993)
Sonata containing *Bacillus pumilus*	Control bacterial spot, downy mildew and powdery mildew, white mold, fire blight, scab, Early and late blight	Wilbur Ellis Company Inc. (n.d.)
Regalia contains extract of knotweed (*Reynoutriasachalinensis*)	Preventative Biofungicide	Evergreen Growers (n.d.)

23.3 Pyrolytic feedstock

Biomass can be defined as vegetation as well as various organic wastes which can fulfill human need. In present scenario, once again its utilization for generation of pyrolytic oil is raised due to GHG emissions. The potential of various kinds of biomass availability exists in India. Dry and wet biomass of Crops that have been used for energy include: sugarcane, corn, grains, pulses, rubber sugar beets etc. Prominent biomasses used currently as energy plantation are kadam, babul, bamboo, Julie flora, Meliadubia There are several factors, proportions of fixed carbon and volatiles, ash/residue content, alkali metal content, and cellulose/lignin ratio. Typically, agricultural lignocellulosic biomass comprises 40%−50% cellulose, 20%−30% hemicellulose, and 10%−25% lignin. As per Ministry of New and Renewable 200 million tones of agro processing and domestic wastes are generated annually in India and disposed in a distributed manner, because these areas are managed by poor farmers and the unorganized sector, rural worker and the low income small agro based industry sector. A large amount of agriculture residues are produced in agriculture based country like India. Agricultural residues can be field based residues (Primary residue) which are generated in the field at the time of harvest (e.g. rice straw, sugar cane tops), or processing based residues (secondary residues) those co-produced during processing (e.g. rice husk and bagasse) (Fig. 23.1).

Discharge from the industries such as black liquor from paper and pulp industry, milk processing units, breweries, vegetable packaging industry, and animal manure can be used. The dissolved organic matters in the wastewater can anaerobically digested to produce biogas and fermented to produce ethanol. Food wastes like Vegetable peels, waste foods from household and industry levels, discarded food, pulp and fiber material waste, confectionary industry wastes, vegetable scraping and peeling wastes, coffee grounds, blanching vegetables, meat industry wastes are generally dumped into open environments, could be potentially use as pyrolytic feedstock. Major compositions of Municipal solid waste (MSW) are paper in India which are 80% of total MSW. The organic fractional of Municipal solid waste can be anaerobic digested or directly combusted.

Animal manure is principally composed of organic material, moisture and ash. Decomposition of animal manure can occur either in an aerobic or anaerobic environment. Under aerobic conditions, CO_2 and stabilized organic materials (SOMs) are produced, while extra CH_4 is also produced under anaerobic conditions. Potential of CH_4 production is notable in India due to the more production of animal manure, which enables the huge energy potential.

23.4 Products of pyrolysis

During pyrolysis, various biochemical framework viz., cellulose, hemicelluloses and lignin undergoes different biochemical degradation pathway and produces different biochemical products; Cellulose and hemicelluloses

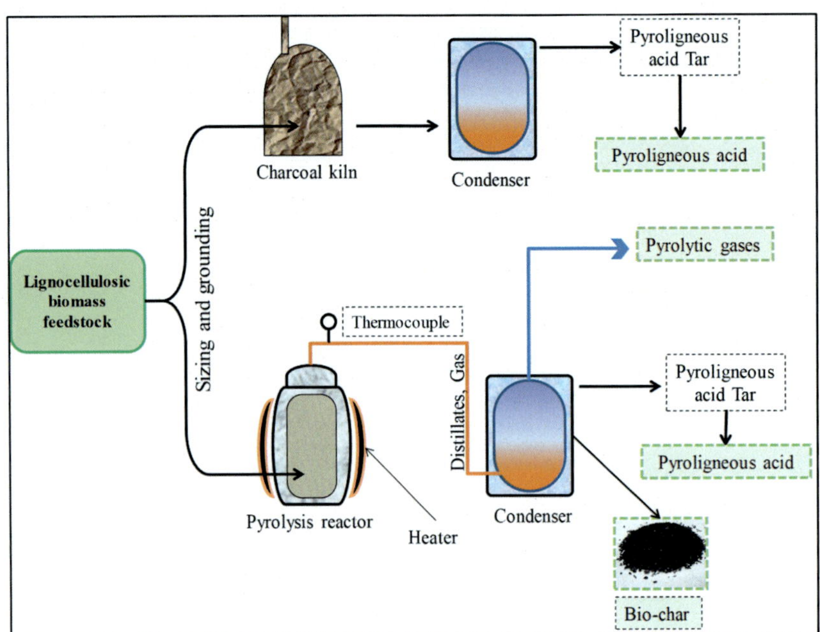

FIG. 23.1 Diagrammatic representation of pyrolytic degradation of biomass for production of bio-pesticides.

(holocellulose) primarily decompose into furan compounds, anhydro cellulose and levoglucosan whereas hemicullulose degradation produces acetic acid furan derivatives. Lignin depolymerizes into phenols and methoxyphenols, such as guaiacols and syringols. Because of this, it does not exhibit thermal stability and cannot be effectively fractionated by conventional techniques.

As known, the water in liquid product is originated from both a dehydration reaction of organic compounds (especially cellulose and hemicellulose) and free water (moisture) in the biomass. A fraction of the formed water can exist in a separate aqueous phase whereas the rest can be dispersed into oil. Water in bio-oils results from the original moisture in the feedstock and as a product of the dehydration reactions occurring during pyrolysis. Therefore, the water content varies over a wide range (15%–30%) depending on the feedstock and process conditions. At this concentration, water is usually miscible with the oligomeric lignin-derived components because of the solubilizing effect of other polar hydrophilic compounds (low-molecular-weight acids, alcohols, hydroxyaldehydes, and ketones) mostly originating from the decomposition of carbohydrates. Under increased heating rate, the oil composition would not change but amount of water-soluble compounds and water in oil was proportionally decreased. The amount of the water-soluble compounds again be increased when temperature reached above 600 C, may be due to the degradation of pyrolytic lignin to water-soluble compound at higher temperatures.

Slow pyrolysis (reduced heating rate with longer residence times) at low temperature mainly produce bio-char whereas, fast pyrolysis (increased heating rate with lower residence times) at high temperature mainly produce higher yield of bio-fuel (Hagner et al., 2015). Pyrolytic bio-fuels are usually dark brown, free-flowing liquids having a distinctive smoky odor. Pyrolytic end product quality depends upon the crop variety, climatic condition, production practices. The pines produced more terpenes and less methoxylated phenols than the spruces. Methoxylated phenols are produced from decomposition of lignin. The lignin composition is known to be specific for a certain species. The degree and type of substition of functional groups in these phenols will depend on the biomass fuel. The changes in production of simple terpenes between spruce and pine species can be due to individual variations within the same tree species (Janson, 1993). Even in the same species of woody biomass, the chemical compositions of sapwood (SW) and heartwood (HW) zones are significantly different. The lignin contents in teak HW and SW are 37.3 and 35.4 wt%, respectively. In addition, the lignin compositions, such as H-lignin (p-hydroxyphenyl subunits), G-lignin (guaiacyl subunits) and S-lignin (syringyl subunits) vary in different wood zones, e.g. the content of S-lignin is higher in HW than SW of teak wood. The extractive content of HW is usually higher than that of SW, e.g. the total extractives in the HW of *Acacia melanoxylon* was about twice of that in the SW. The product distributions of organic compounds from the different tree species studied show common features. For example, pyrolysis produced a wide array of aldehydic and ketonic oxygenates as well as terpenes. Substantial amounts of phenols and methoxylated phenols were also formed constituting about half of the chemicals produced Table 23.3.

Due to their varied chemical composition, bio-oils exhibit different boiling temperature. In addition to water and volatile organic components, biomass pyrolysis oils contain substantial amounts of nonvolatile materials such as sugars and oligomericphenolics. The oil is a complicated organic compound that mainly consists of water, acids, phenols, aldehydes and heterocyclic compounds. But no clear trend can be identified due to the interaction between the two pyrolysis parameters and the complexity of the chemical composition. The complete chemical profiling of pyrolytic bio-oil is impossible. 25%–40% of volatile oil compounds can be identified by GC–MS methods (Mullen and Huang, 2009) HPLC can be used to quantify some water-soluble, non-volatile species (Mullen and Huang, 2009). Water soluble fractions comprised of low-molecular-mass carboxylic acids and alcohols, aldehydes, ketones, sugars and lignin monomers besides water (Oasmaa et al., 2003) whereas water-insoluble fractions consists of lignin-depolymerized compounds (Mungkunkamchao et al., 2013). Water-insoluble fractions again categorized into low molecular weight lignin (LMWL) and high molecular weight lignin (HMWL) compounds (Oasmaa et al., 2003). Lignin degradation increases with the rising temperature, so the quantity of water insoluble pyrolytic lignin in bio-oil was increased by elevating temperature. An increase in pyrolysis severity reduces the organic liquid yield due to cracking of the vapors and formation of gases but leaves the organic liquid with less oxygen. The other major groups of compounds identified are hydroxyaldehydes, hydroxyketones, sugars, carboxylic acids, and phenolics. Most of the phenolic compounds are present as oligomers having a molecular weight ranging from 900 to 2500. The presence of oxygen in many oil components is the primary reason for differences in the properties and behavior seen between hydrocarbon fuels and biomass pyrolysis oils.

23.5 Acetic acid as potential product

The oil obtained from biomass can be used as an energy source and as a feedstock for chemical production. Pyrolysis is one of the alternative method to produce acetic acid. Fast pyrolysis of lignocellulosic biomass with a short residence time produces acetic acid rich bio-oil. Diebold showed that concentrations of acetic acid in bio-oils obtained from the fast pyrolysis of wood biomass ranged from 0.5 wt% to 12 wt% depending upon the biomass. Chemicals produced are hydroxyketones, sugars, carboxylic acids and phenolics, hydroxyaldehydes. An acetic acid concentration of 27.3% from corn stalk, 30 wt%–39 wt%, from palm kernel shell bio-oil obtained from pyrolysis shell at ~300–400°C. Some researcher reported that bio-oils dominated by small molecules, such as acetic acid, furfural and simple phenols, can be produced by slow pyrolysis. Acetic acid (21.04% ± 1.08%) was obtained from slow pyrolysis of heartwood at 500°C (Yang et al., 2016). It is expected that the aqueous phase also contains considerable amount of cellulose-derived sugars and anhydro sugars, such as levoglucosan (Mohan et al., 2007) which are valuable chemicals. The aqueous phase can be considered as feedstock for acetic acid. Some studies using hydrothermal process showed acetic acid was a stable intermediate in hydrothermal oxidation of organics, which implied that it was possible to obtain a high yield of acetic acid. In order to enhance the acetic acid yield, Jin et al. performed a two-step continuous-flow hydrothermal conversion of vegetable wastes using both a hydrothermal reaction without oxygen and a hydrothermal oxidation reaction and reported a maximum acetic acid yield of 16% from vegetable mixture.

23.5.1 Chemical composition of wood vinegar

The characterization of liquid products from pyrolysis has been continued a long time. The products contain many organic components and the composition is very complicated. During the last twenty years the interest has mainly been focused on the liquid product from fast pyrolysis. According to literature, the main organic components of wood vinegar are methanol and acetic acid. Other components are acetone, methyl acetone, acetaldehyde, allyl alcohol, furan and furfural, and formic, propionic and butyric acids.

23.5.2 Eco-toxicology of pyrolytic products

An extensive review of the different biomass sources and particulars of production and application of pyrolytic bio-pesticides prepared from those sources have been provided in Tables 23.2.

23.6 Acetic acid eco-toxicology

Hahn and Brown (1967) demonstrated that when *E. coli* is treated with organic acids, such as acetate help to increase the intracellular ratio of homocysteine to methionine; Inhibition of growth will be happened as the consequences of declining of $tRNA_{Met}$. As homocysteine competitively inhibit methionyl-tRNA synthetase, adding methionine could restore $tRNA_{Met}$ pools by decrease the intracellular ratio of homocysteine to methionine. Levine & Fellers showed that sub-lethal concentrations of acetic acid reduce the thermal death points of bacteria.

TABLE 23.2 Particulars of chemical derivatives of biomass pyrolysis (attached).

Chemical name	Type	Chemical formula	Toxicity	Toxicity references
3-Methylbutanal	Aldehydes	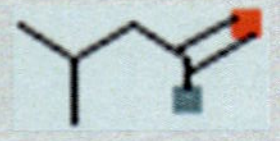	LD50 value is 5600 mg/kg	Golubev and Lyublina (1962)
Acetaldehyde	Aldehydes	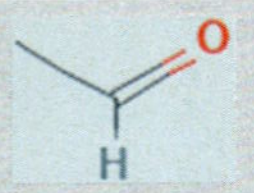	Cells were unable to repair DNA single- and double-strand breaks caused by acetaldehyde	Singh and Khan (1995)
2-Furancarboxaldehyde (furfural)	Furan and pyran derivatives	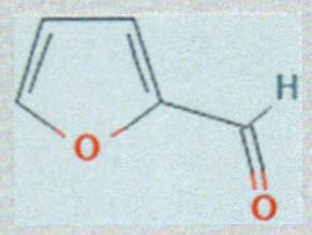	A3; confirmed animal carcinogen with unknown relevance to humans.	Andersen et al. (1998)
2(5H)-Furanone (γ-crotonolactone)	Furan and pyran derivatives	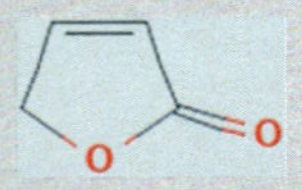	–	–
1-Hydroxy-2-propanone	Ketones	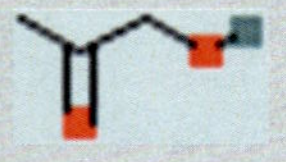	LD50-2200 mg/kg (2200 mg/kg)	SmythJr and Carpenter (1948)
Acetone	Ketones	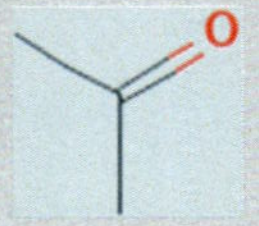	3500 mg/kg-day by oral	ATSDR (1994)
1-Acetoxy-propan-2-one	Ketones	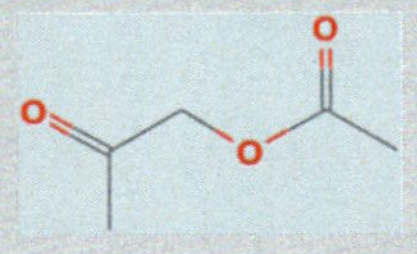	–	–
3-Methyl-1,2-cyclopentanedione (cyclotene)	Diketones	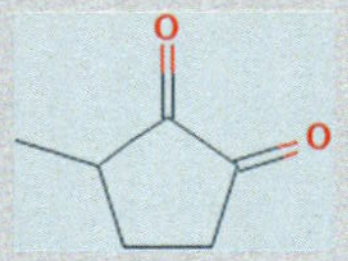	–	–
Ethyl butyrate	Alcohols, esters, and acids		Rat LD50 13 gm/kg	Jenner et al. (1964)

Acetic acid	Alcohols, esters, and acids		Cytogenetic analysis 40 mmol/L	–
Pentanoic acid	Alcohols, esters, and acids		140 mg/m^3	–
Methanol	Alcohols, esters, and acids	H H-C-H OH	56 mg/L	–
1-Methyl-2,5-pyrrolidinedione	Nitrogenated compounds		LD50 >6 gm/kg (6000 mg/kg)	French demande patent document., #2509610
Phenol	Phenol and derivatives		EC_{50}···244	–
4-Methylphenol	Phenol and derivatives		30000 μg/L/20H (-enzymatic activation step)	NIOSH toxicity data
3-Ethylphenol	Phenol and derivatives		–	–

Continued

TABLE 23.2 Particulars of chemical derivatives of biomass pyrolysis (attached).—cont'd

Chemical name	Type	Chemical formula	Toxicity	Toxicity references
Vanillin	Phenol and derivatives	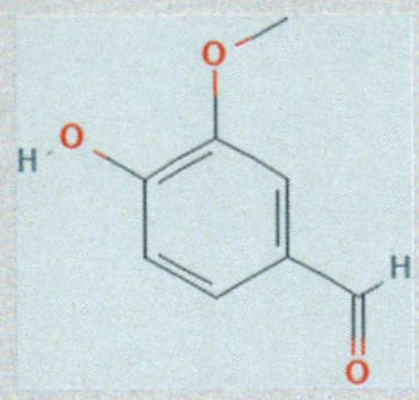	Cytogenetic analysis - 4 mmol/L	NIOSH toxicity data March 2019
2,6-Dimethoxyphenol (syringol)	Syringol and derivatives	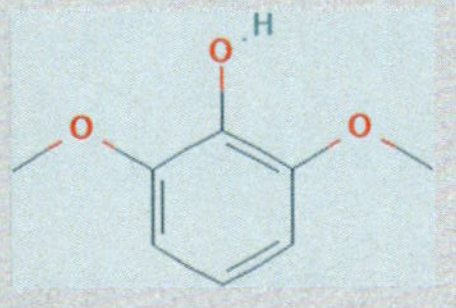	LD50 550 mg/kg	Orłowski and Boruszak (1991)
4-Methyl-2,6-dimethoxyphenol (4-methylsyringol)	Syringol and derivatives	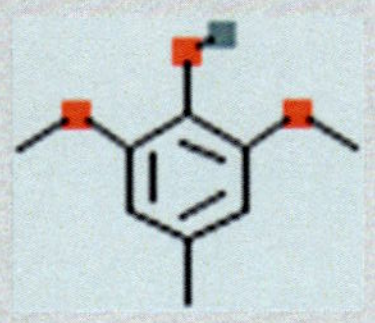	–	–
2-Methoxyphenol (guaiacol)	Guaiacol and derivatives	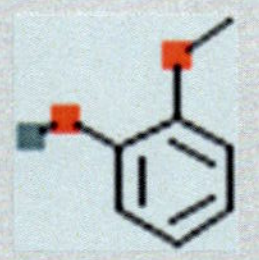	LD50 621 mg/kg	Drugs of the future., 5 (539), 1980
1,2-Benzenediol (pyrocatechol)	Pyrocatechol and derivatives	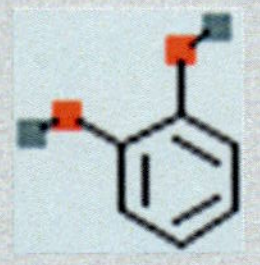	Cytogenetic analysis 3 μmol/L	NIOSH toxicity data December 2018

TABLE 23.3 Prevailing bio-cide models from various biomass sources.

Biomass/model biomass	Pyrolysis temp	Major chemical area basis	Major possible biocide chemical	Test for organism	Result for bio-cide activity	References
Tomato plant waste	500°C	Neophytadiene	Neophytadiene + phytol	Colorado potato beetle (leptinotarsa decemlineata),	45% mortality rate	Cáceres et al. (2015)
Palm kernel shells	500°C	Acetic acid (CAS) ethylic acid	Phenol and acetic acid compounds	*Phytophthora palmivora fungus*	4% liquid smoke of pyrolytic oil decrease the spot diameter of culture fungus and increase the incubation time	Faisal et al. (2018)
Durian (*Durio zibethinus*), peel	300	Acetic acid (CAS) ethylic acid, Piperidine, 1-methyl- (CAS) n-methylpiperidine	Phenol and acetic acid	–	–	Faisal et al. (2018)
	340	Acetic acid (CAS) ethylic acid, Carbamic acid, phenyl ester (CAS) phenyl carbamate				
	380	Acetic acid (CAS) athylic acid Phenol (CAS) izal				
Tobacco leaves	450	Phenol derivatives and indole	Phenol and D-limonene	*Pythium ultimum*	Phenol and D-limonene significantly decrease the spot diameter of 3cultures	Booker et al. (2010)
				Clavibacter michiganensis		
				Streptomyces scabies		
Rice husk	450–500	Phenol; 2,4,-Dimethyl-3-(methoxycarbonyl)-5- ethylfuran;	–	Soybean pests, armyworm (*Spodoptera* sp.), pod borer (*Etiella* sp.), pod sucking (*Nezara viridulla, Riptortus linearis,* and *Piezodorus hybneri*)	Decrease pest attack significantly	Risfaheri et al. (2018)
	330	2-Methoxy-4-methylphenol (4-methylguaiacol); 2-Ethylphenol;				
	160	Acetic acid (CAS) ethylic acid; Phenol, 2-methoxy-(CAS) guaiacol;				

Continued

TABLE 23.3 Prevailing bio-cide models from various biomass sources.—cont'd

Biomass/model biomass	Pyrolysis temp	Major chemical area basis	Major possible biocide chemical	Test for organism	Result for bio-cide activity	References
Wet coffee grounds	500	Phenol, 3-methyl	Possibly phenolic compounds but not detectable in some other fraction.	Colorado potato beetles	Approx 100% mortality rate in specific fraction of bio-oil	Bedmutha et al. (2011)
Finely ground tobacco leaves	350–550	Phenolic compounds, nicotine	Not detectable	*Streptomyces scabies, ClaVibacter michiganensis,* and *Pythium ultimum*	Significantly decrease the spot diameter of 3 cultures	Booker et al. (2009)
Waste husks from the *C. megalocarpus* trees	300, 400, 450	Acetic acid, methanol	–	–	–	Browning et al. (2019)
Kraft lignin, organosolv lignin, xylan, -cellulose	550	–	–	CPB 1st instar larvae	Mixture of the three bio-oil components was neither synergistic Nor antagonistic, but rather additive. Hemicellulose bio-oil alone has less LD50 values	Hossain et al. (2013)
Small wood materials Or bark-free birch (*Betula pendula)* material	450	Furfural + acetic acid compound	Acetic acid mainly with synergistic effect	The snails (*A. arbustorum*)	100% repellent effect	–
Eucalyptus and pine sawdust	400	*acetic acid, acetaldehyde*	*acetic acid*	–	–	Faluku (2016)
Ground tobacco leaves, tomato plants	400 –565°C	Nicotine	–	Colorado potato beetle, cabbage looper, spider mite *Clavibacter michiganensis, Xanthomonas campestris*	Two bio-oils pyrolyzed at 300 –400 0 C shows inhibition to *Clavibacter michiganensis, Xanthomonas campestris and significantly effect mortality rate of insects*	Hossain (2016)
Canola *B. napus* and mustard *B. juncea* and *B. carinata*	300	1-Hexadecanol	Hexadecanoic acid or the octadecanoic methyl ester	Colorado potato beetle (CPB), *Leptinotarsa decemlineata* say (Coleoptera: Chrysomelidae)	ESP solvent soluble Phase of crude oil has highest mortality rate after 72 h	Suqi et al. (2014)
	500	Glycerol				
Apple tree Pruning residues	300	1-(3,4-Dimethoxyphenyl)-ethanone 3-Methoxy-5-methylphenol	Phenolic components	*Aspergillus niger* *Saccharomyces cerevisiae*	Significantly decrease the spot diameter	García et al. (2017)

Slab cuts from rubberwood logs	400	Acetic acid	–	–	–	Ratanapisit et al. (2009)
Litchi chinensis	600°C	2,6-Dimethoxyphenol (syringol)	Syringol, catechol	*Staphylococcus aureus, Acinetobacter baumannii, Pseudomonas aeruginosa, Staphylococcus aureus*	Significantly decrease the spot diameter	–
Red meranti (*Shorea* sp.) chips	500	Organic acids such As acetic acid	OH groups in the alkyl chain or phenols	*T. versicolor* and *T. palustris*	Significantly decrease the spot diameter	Nakai et al. (2007)
Oil palm trunk wood meal	450	Acetic acid	Total phenol and total acid (Table 23.2) Together with the characteristic existence of 3-hydroxy-2-butanone And propanoic acid	*T. versicolor, F. palustris.*	Significantly decrease the spot diameter	Oramahi and Yoshimura (2013)
LABAN wood (*Vitex pubescens* vahl) wood meal	450	Phenolic compounds (analyzed)	Phenolic compounds	*T. versicolor, F. palustris.*	Significantly decrease the spot diameter	Oramahi and Yoshimura (2013)

23.7 Quinone eco-toxicology

Wang et al. (2006), have demonstrated that endoplasmic reticulum stress have induced along with arylating quinone toxicity. Arylating quinone electrophiles disrupt the disulfide bond formation. Free thiols on both cysteinyl proteins and oxido reductases should be available to react with arylating quinones during disulfide shuffling (Lamé et al., 2003). Arylating Quinones basically involves in nucleophilic 1, 4-addition to thiols (involved in folding process of secretory proteins) on proteins or catalytic reduction of oxygen with ETC electron involved mitochondrial metabolism. Disruption of disulfide bond causes ER stress due to buildup of misfolded proteins.

23.8 Catechol eco-toxicology

Catechol induces breaking of DNA strands (Melikian et al., 2008; Navasumrit et al., 2008), increasing rate of DNA recombination, Formation of DNA adducts (Oikawa et al., 2001), translocations and chromosomes damage in DNA-level. Catechol also increase the free radical production (Shen et al., 2009) by hampering the respiratory chain (Yang et al., 2016a) catechol derivatives also interfere with energy transducing membrane and redox cycling.

23.9 Phenol eco-toxicology

Phenolic compounds have long been used for antiseptic properties. In cell membrane level, Phenol induced potential damage that can leak cellular constituents. From the demonstration of Hugo and Bloomfield (1971) it is cleared that chlorinated bis-phenol fenticlor had a bactericidal activity by leaking of cellular 260-nm-absorbing material. Same leaking mechanism is also responsible for anti-fungal activity. From the demonstration of Pulvertaft and Lumb (1948), it is well established that even low concentrations of phenols (320 mg/mL) can cause rapid lysis in growth cultures of staphylococci, streptococci and *E. coli* with the help of autolytic enzymes. Srivastava et al. (2008) showed that phenol sensitivity is higher in young bacterial cells than older cell as phenolic compounds act at the separation point of daughter cells. *S. aureus* and *E. coli* metabolic activities are hampered by Fenticlor, Chlorocresol due to induced increment of proton permeability with destruction of proton motive force (PMF); so, oxidative phosphorylation is mostly affected due to uncoupling. At higher concentration exposure of phenol, an irreversible cellular damage can happen due to cytoplasmic constituent coagulation (Hugo and Longworth, 1966).

23.10 Other alcohol

Aliphatic alcohol had also potential to damage cellular membrane. Aiphaticalhols are basically lipid soluble though hydrophobic interaction. Organic groups of the silanols have similar hydrophobic interaction with lipid of cellular membranes. This hydrophobic interaction can cause losses of membrane permeability and ultimately bacteria will die. Antimicrobial effect may depend upon the hydrophilic to hydrophobic portions of the molecule as long alkyl chains in the anti-microbial molecule shows reduced bio-activity. The hydroxyl functional group of alcohols helps to localize the near-membrane cellular fatty acid constituents through hydrogen bonding with ester linkages. Strong hydrogen bonding or depends upon the H-bond acidity as silanols show better hydrogen bonding compared to other analogous alcohols due to H-bond acidity. The eco-toxicology of fast pyrolysis liquids have been summarized and described in Fig. 23.2.

23.10.1 Disadvantage

In spite of many ostensible benefits, the feasibility of the biomass energy conversion especially in terms of GHG emissions is still debatable. Another important factor is process to make the fuel. In the case of ethanol production, the use of cogeneration with a gas-fired turbine over conventional production schemes has been claimed to reduce GHG emissions by 45%. However, using coal would wipe out the GHG savings as compared to gasoline. By-product usage is another important factor. It makes a difference if the by-products of the distillation process used as animal feed or for fuel. The lowest overall GHG emissions (up to 80% reduction over fossil fuels) are attainable when bio-diesel is made from waste cooking oil or methane from liquid manure. In view of the above the implementation of bio-fuels in any country, the environmental impact is the major factor to be considered before we consider the exploitation of bio-energy. As mentioned in the last paragraph, the land-use change is the major contributor to the GHG emissions. These considerations are necessary at the time of making life cycle assessment of any bio-energy generation process.

While considering the utilization of agro-fuels, the social impacts are also important and cannot be overlooked due to many reasons. In the process of implementing wood vinegar, particularly agro-fuels, we need to ensure that the land as well as resources, we are allocating to the harvesting of these crops do not interfere the food crops sector. Secondly, in this

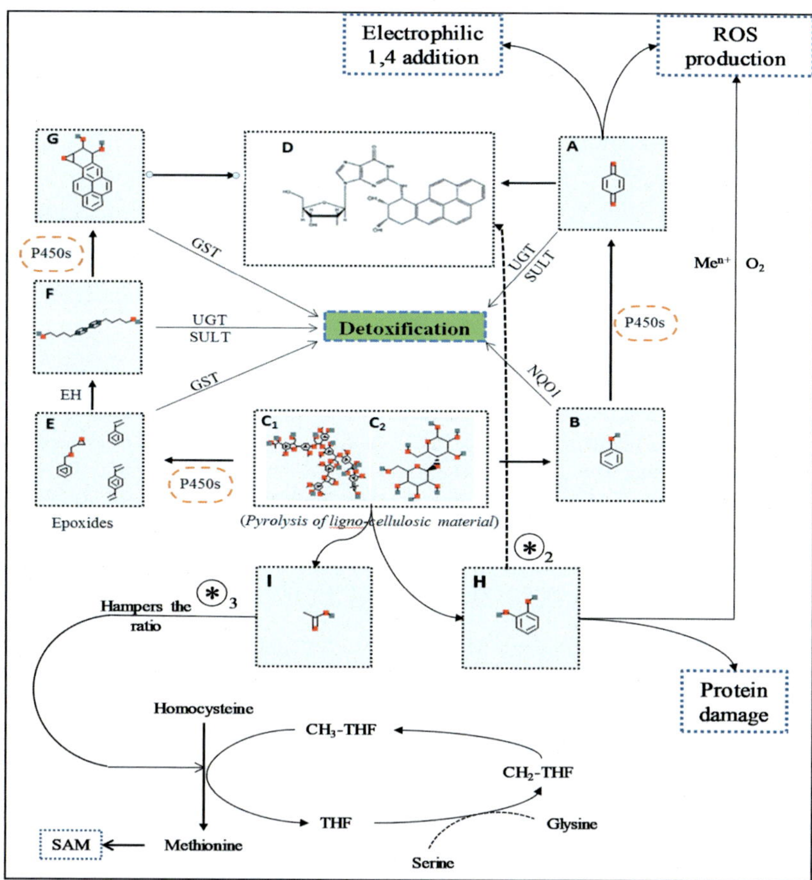

FIG. 23.2 Schematic diagram of ecotoxicology of pyrolytic products. A: Quinone compound; B: Phenolic compounds; C: Lignin/Cellulose; D: DNA abducts; E: Epoxide compounds; F: Diolcompounds; G: Diol Epoxides compounds; H: Catechol compounds; I: Acetic acid compounds; Hahn and Brown, 1967.

process lest we make some of the farmers jobless. In a country like India agriculture is one of the most important sectors of its economy. It is the means of livelihood of almost two-thirds of the work force in the country and according to the economic data for the financial year 2006–07, agriculture accounts for 18% of India's GDP. About 43% of India's geographical area is used for agricultural activity. Though the share of Indian agriculture in the GDP has steadily declined, it is still the single largest contributor to the GDP and plays a vital role in the overall socio-economic development of India. The poor farmers, not only earn their livelihood, but the food for their family is also coming from their crops. In Europe, the commercialization of biological control agents including wood vinegar is practically impossible for small to medium enterprises (SMEs).

23.11 Future prospects

As mentioned in potential challenges, syn-gas produced from pyrolytic plants is a major environmental concern. The inorganic catalyst can be used in the pyrolysis for e.g., Calcium magnesium acetate is used for controlling the SO_x and NO_x emissions. Therefore, further substantiate research should be initiated on the effect of inherent and added inorganic compounds on the pyrolysis process.

In 2000, it was estimated that world pesticide market shares around 32 billion US$ whereas in the developing country shares approximately 3 billion US$. Bio-based products had accounted for only 2%–3% of annual turnover for products allied with plant protection. This creates a massive opportunity to replace the existing plant protection products with bio-pesticides. There has been a growing market of wood vinegar replacing the existing fossil fuel industry. Integration of bio-energy and sustainable chemical production are the promising technologies. Carbonization can be a rising element for integrating chemical sustainability. In japan, the potential market of biochar can concomitantly heighten the wood vinegar production. Due to large population of India, a major part of GDP shared is based from agriculture. Therefore, significant amount of agriculture biomass residue as well as livestock dung is produced. Rural energy consumption is dominated (approx. 80% of total energy consumed) by biomass-fuels. Existing bio-fuels in rural India are fuel wood (accounting for 54% of bio-fuels), plantation crop residue and dung. Paucity and higher prices of fuel wood, people are tend to shift toward dung and various crop residues. Thus, improving the biomass conversion technology could improve the lifestyle of rural India. This will create a potential opportunity for bio-pesticide for side upshot.

References

Amen-Chen, C., Pakdel, H., Roy, C., 1997. Separation of phenols from *Eucalyptus* wood tar. Biomass Bioenergy 13 (1–2), 25–37.

Andersen, M.E., MacNaughton, M.G., Clewell, H.J., March 1, 1998. Exposure indices for 1992–1993. Cincinnati, OH.* ACGIH. 1998. Threshold limit values for chemical substances and physical agents. Biological exposure Indices. 1998 TLVs and BEIs. In: American Conference of Governmental Industrial Hygienists, 10, pp. 541–550. Health.

ATSDR, 1994. Toxicological Profile for Acetone.

Balat, M., Balat, H., 2009. Recent Trends in Global Production and Utilization of Bio-Ethanol Fuel Applied Energy, vol. 86.

Bedmutha, R., Booker, C. J., Ferrante, L., Briens, C., Berruti, F., Yeung, K. K. C.,et al.). Insecticidal and bactericidal characteristics of the bio-oil from the fast pyrolysis of coffee grounds. J. Anal. Appl. Pyrol., 90(2), 224–231.

Booker, C. J., Yeung, K. C., Bedmutha, R., Vogel, T., Berruti, F., Briens, C.,et al. Characterization of the Pesticide Properties of Tobacco Bio-Oil.

Booker, C. J., Bedmutha, R., Vogel, T., Gloor, A., Xu, R., Ferrante, L.,et al. Experimental investigations into the insecticidal, fungicidal, and bactericidal properties of pyrolysis bio-oil from tobacco leaves using a fluidized bed pilot plant. Ind. Eng. Chem. Res., 49(20), 10074–10079.

Browning, S., Lawrence, T., Joshi, C., Seay, J., 2019. Analysis of green pesticide production by valorization of husks from *Croton megalocarpus* tree nuts. Environ. Prog. Sustain. Energy e13312.

Cáceres, L.A., McGarvey, B.D., Briens, C., Berruti, F., Yeung, K.K.C., Scott, I.M., 2015. Insecticidal properties of pyrolysis bio-oil from greenhouse tomato residue biomass. J. Anal. Appl. Pyrol. 112, 333–340.

Demirbaş, A., 2002. Biodiesel from vegetable oils via transesterification in supercritical methanol. Energy Conv. & Manag. 43 (17), 2349–2356.

Evergreen Growers, n.d. https://www.evergreengrowers.com/regalia-biofungicide.

Faisal, M., Chamzurni, T., Daimon, H., 2018. A study on the effectiveness of liquid smoke produced from palm kernel shells in inhibiting black pod disease in cacao fruit in vitro. Int. J. Geomate 14, 36–41.

Faluku, N., 2016. Pyrolysis of Eucalyptus and Pine Sawdust for Vinegar Pesticide Production. Doctoral Dissertation, Makerere University.

Ferek, R.J., Reid, J.S., Hobbs, P.V., Blake, D.R., Liousse, C., 1998. Emission factors of hydrocarbons, halocarbons, trace gases and particles from biomass burning in Brazil. J. Geophys. Res.: Atmos. 103 (D24), 32107–32118.

Gadde, B., Bonnet, S., Menke, C., Garivait, S., 2009. Air pollutant emissions from rice straw open field burning in India, Thailand and the Philippines. Environ. Pollut. 157 (5), 1554–1558.

García, A., Spigno, G., Labidi, J., 2017. Antioxidant and biocide behaviour of lignin fractions from apple tree pruning residues. Ind. Crop. Prod. 104, 242–252.

Golubev, A.A., Lyublina, E.I., 1962. Gigienatruda i professional'nyezabolevaniya. Occup. Hyg. & Occup. Dis. 4, 26–32.

Guillén, M.D., Manzanos, M.J., 2002. Study of the volatile composition of an aqueous oak smoke preparation. Food Chem. 79 (3), 283–292.

Guillen, M.D., Manzanos, M.J., 2005. Characteristics of smoke flavorings obtained from mixtures of oak (Quercus) wood and aromatic plants (*Thymus vulgaris* and *Salvia lavandulifolia* Vahl). Flavour Fragrance J. 20, 676–685.

Hagner, M., Kuoppala, E., Fagernäs, L., Tiilikkala, K., Setälä, H., 2015. Using the copse snail Ariantaarbustorum (Linnaeus) to detect repellent compounds and the Quality of wood Vinegar. Int. J. Environ. Res. 9 (1), 53–60.

Hahn, G.A., Brown, J.W., 1967. Properties of a methionyl-tRNAsysthetase from Sarcinalutea. Biochimica et Biophys. Acta (BBA) Enzymol. 146 (1), 264–271.

Hossain, M.M., Scott, I.M., McGarvey, B.D., Conn, K., Ferrante, L., Berruti, F., Briens, C., 2013. Toxicity of lignin, cellulose and hemicellulose-pyrolyzed bio-oil combinations: estimating pesticide resources. J. Anal. Appl. Pyrol. 99, 211–216.

Hossain, M.M., 2016. Recovery of Valuable Chemicals from Agricultural Waste through Pyrolysis.

Hugo, W.B., Bloomfield, S.F., 1971. Studies on the mode of action of the phenolic antibacterial agent fentichlor against *Staphylococcus aureus* and *Escherichia coli* II. The effects of fentichlor on the bacterial membrane and the cytoplasmic constituents of the cell. J. Appl. Bacteriol. 34 (3), 579–591.

Hugo, W.B., Longworth, A.R., 1966. The effect of chlorhexidine on the electrophoretic mobility, cytoplasmic constituents, dehydrogenase activity and cell walls of *Escherichia coli* and *Staphylococcus aureus*. J. Pharm. Pharmacol. 18 (9), 569–578.

Imamura, E., Watanabe, Y., 2007. U.S. Patent No. 7214393. U.S. Patent and Trademark Office, Washington, DC.

Janson, R.W., 1993. Monoterpene emissions from Scots pine and Norwegian spruce. J. Geophys. Res.: Atmos. 98 (D2), 2839–2850.

Jed, W.F., Fahey, S.D., 2005. Moringaoleifera: a review of the medical evidence for its nutritional, therapeutic, and prophylactic properties. Trees Life J. 1 (5).

Jenner, P.M., Hagan, E.C., Taylor, J.M., Cook, E.L., Fitzhugh, O.G., 1964. Food flavourings and compounds of related structure I. Acute oral toxicity. Food Chem. Toxicol. 2, 327–343.

Lamé, M.W., Jones, A.D., Wilson, D.W., Segall, H.J., 2003. Protein targets of 1, 4-benzoquinone and 1, 4-naphthoquinone in human bronchial epithelial cells. Proteomics 3 (4), 479–495.

Lee, D., Kim, H.S., Kim, J.K., 2012. The role of self-construal in consumers' electronic word of mouth (eWOM) in social networking sites: a social cognitive approach. Comput. Hum. Behav. 28 (3), 1054–1062.

Lemieux, P.M., Lutes, C.C., Santoianni, D.A., 2004. Emissions of organic air toxics from open burning: a comprehensive review. Prog. Energy Combust. Sci. 30 (1), 1–32.

Mansur, E.T., 2013. Prices versus quantities: environmental regulation and imperfect competition. J. Regul. Econ. 44 (1), 80–102.

Mathew, S., Yahayu, M.A., Mahmud, K.N., Zakaria, Z.A., 2014. Pyroligneous acid from plant biomass and its applications. In: Biotechnology Development in Agriculture, Industry and Health. Advanced Conversion Technologies for Lignocellulosic Biomass, vol. 3. UTM Press, Awaiting final publication.

Melikian, A.A., Chen, K.M., Li, H., Sodum, R., Fiala, E., El-Bayoumy, K., 2008. The role of nitric oxide on DNA damage induced by benzene metabolites. Oncol. Rep. 19 (5), 1331–1337.

Minami, S., Suzuki, H., Okamoto, Y., Fujinaga, T., Shigemasa, Y., 1998. Chitin and chitosan activate complement via the alternative pathway. Carbohydr. Polym. 36 (2–3), 151–155.

Mohan, S.V., Babu, V.L., Bhaskar, Y.V., Sarma, P.N., 2007. Influence of recirculation on the performance of anaerobic sequencing batch biofilm reactor (AnSBBR) treating hypersaline composite chemical wastewater. Bioresour. Technol. 98 (7), 1373–1379.

Mu, H., Høy, C.E., 2004. The digestion of dietary triacylglycerols. Prog. Lipid Res. 43 (2), 105–133.

Mullen, T.R., Huang, Q., 2009. U.S. Patent Application No. 11/947908.

Mungkunkamchao, T., Kesmala, T., Pimratch, S., Toomsan, B., Jothityangkoon, D., 2013. Wood vinegar and fermented bioextracts: natural products to enhance growth and yield of tomato (*Solanumlycopersicum* L.). Sci. Hortic. 154, 66–72.

Nabavi, S.M., Marchese, A., Izadi, M., Curti, V., Daglia, M., Nabavi, S.F., 2015. Plants belonging to the genus Thymus as antibacterial agents: from farm to pharmacy. Food Chem. 173, 339–347.

Nakai, T., Kartal, S.N., Hata, T., Imamura, Y., 2007. Chemical characterization of pyrolysis liquids of wood-based composites and evaluation of their bio-efficiency. Build. Environ. 42 (3), 1236–1241.

Navasumrit, P., Arayasiri, M., Hiang, O.M.T., Leechawengwongs, M., Promvijit, J., Choonvisase, S., et al., 2008. Potential health effects of exposure to carcinogenic compounds in incense smoke in temple workers. Chem. Biol. Interact. 173 (1), 19–31.

Nishizawa, S., Kato, Y., Teramae, N., 1999. Fluorescence sensing of anions via intramolecularexcimer formation in a pyrophosphate-induced self-assembly of a pyrene-functionalized guanidinium receptor. J. Am. Chem. Soc. 121 (40), 9463–9464.

Oasmaa, A., Kuoppala, E., Solantausta, Y., 2003. Fast pyrolysis of forestry residue. 2. Physicochemical composition of product liquid. Energy Fuels 17 (2), 433–443.

Oikawa, S., Hirosawa, I., Hirakawa, K., Kawanishi, S., 2001. Site specificity and mechanism of oxidative DNA damage induced by carcinogenic catechol. Carcinogenesis 22 (8), 1239–1245.

Oramahi, H.A., Yoshimura, T., 2013. Antifungal and antitermitic activities of wood vinegar from *Vitexpubescens* Vahl. J. Wood Sci. 59 (4), 344–350.

Orlov, D.S., 1990. Humic Acids of Soils and General Theory of Humification, vol. 1. Moscow State University, p. 325.

Orłowski, J., Boruszak, D., 1991. Toxicologic investigations of selected phenolic compounds. I. Acute and subacute toxicity of guaiacol, methyl-guaiacol and syringol. Folia Med. Cracov. 32 (3–4), 309–317.

Pulvertaft, R.J.V., Lumb, G.D., 1948. Bacterial lysis and antiseptics. Epidemiol. Infect. 46 (1), 62–64.

Ratanapisit, J., Apiraksakul, S., Rerngnarong, A., Chungsiriporn, J., Bunyakarn, C., 2009. Preliminary evaluation of production and characterization of wood vinegar from rubberwood. Songklanakarin J. Sci. Technol. 31 (3).

Risfaheri, R., Hoerudin, H., Syakir, M., 2018. Utilization of rice husk for production of multifunctional liquid smoke. J. Adv. Agric. Technol. 5 (3).

Shen, F.T., Lin, J.L., Huang, C.C., Ho, Y.N., Arun, A.B., Young, L.S., Young, C.C., 2009. Molecular detection and phylogenetic analysis of the catechol 1, 2-dioxygenase gene from *Gordonia* spp. Syst. Appl. Microbiol. 32 (5), 291–300.

Singh, N.P., Khan, A., 1995. Acetaldehyde: genotoxicity and cytotoxicity in human lymphocytes. Mutat. Res. DNA Repair 337 (1), 9–17.

Smyth Jr., H.F., Carpenter, C.P., 1948. Further experience with the range finding test in the industrial toxicology laboratory. J. Ind. Hyg. Toxicol. 30 (1), 63.

Srivastava, C., Nikles, D.E., Thompson, G.B., 2008. Tailoring nucleation and growth conditions for narrow compositional distributions in colloidal synthesized FePt nanoparticles. J. Appl. Phys. 104 (10), 104314.

Sundarrao, K., Burrows, I., Kuduk, M., Yi, Y.D., Chung, M.H., Suh, N.J., Chang, I.M., 1993. Preliminary screening of antibacterial and antitumor activities of Papua New Guinean native medicinal plants. Int. J. Pharmacogn. 31 (1), 3–6.

Suqi, L., Caceres, L., Schieck, K., Booker, C.J., McGarvey, B.M., Yeung, K.K.C., et al., 2014. Insecticidal activity of bio-oil from the pyrolysis of straw from *Brassica* spp. J. Agric. Food Chem. 62 (16), 3610–3618.

Swain, M.R., Anandharaj, M., Ray, R.C., Rani, R.P., 2014. Fermented fruits and vegetables of Asia: a potential source of probiotics. Biotechnol. Res. Int. 2014.

Swift, G., 1993. Directions for environmentally biodegradable polymer research. Accounts Chem. Res. 26 (3), 105–110.

Thurman, E.M., 1985. Amount of organic carbon in natural waters. In: Organic Geochemistry of Natural Waters. Springer, Dordrecht, pp. 7–65.

Tiilikkala, K., Fagernäs, L., Tiilikkala, J., 2010. History and Use of Wood Pyrolysis Liquids as Biocide and Plant Protection Product.

Usharani, K., Balu, A.R., 2015. Structural, optical, and electrical properties of Zn-doped CdO thin films fabricated by a simplified spray pyrolysis technique. Acta Metall. Sinica (Eng. Lett.) 28 (1), 64–71.

Van Rensburg, S.J., Daniels, W.M.U., Van Zyl, J.M., Taljaard, J.J.F., 2000. A comparative study of the effects of cholesterol, beta-sitosterol, beta-sitosterolglucoside, dehydro-epiandrosteronesulphate and melatonin on in vitro lipid peroxidation. Metab. Brain Dis. 15 (4), 257–265.

Wang, X., Thomas, B., Sachdeva, R., Arterburn, L., Frye, L., Hatcher, P.G., et al., 2006. Mechanism of arylatingquinone toxicity involving Michael adduct formation and induction of endoplasmic reticulum stress. Proc. Natl. Acad. Sci. U. S. A. 103 (10), 3604–3609.

Wang, M., Han, J., Dunn, J.B., Cai, H., Elgowainy, A., 2012. Well-to-wheels energy use and greenhouse gas emissions of ethanol from corn, sugarcane and cellulosic biomass for US use. Environ. Res. Lett. 7 (4), 045905.

Wilbur Ellis Company Inc., n.d. www.ag.wilburellis.com.

Wititsiri, S., 2011. Production of wood vinegars from coconut shells and additional materials for control of termite workers, *Odontotermes* sp. and striped mealy bugs, Ferrisiavirgata. Songklanakarin J. Sci. Technol. 33 (3).

Yang, B., Wang, Y., Qian, P.Y., 2016. Sensitivity and correlation of hypervariable regions in 16S rRNA genes in phylogenetic analysis. BMC Bioinf. 17 (1), 1–8.

Yang, J.F., Yang, C.H., Liang, M.T., Gao, Z.J., Wu, Y.W., Chuang, L.Y., 2016. Chemical composition, antioxidant, and antibacterial activity of wood vinegar from Litchi chinensis. Molecules 21 (9), 1150.

Yatagai, F., Saito, T., Takahashi, A., Fujie, A., Nagaoka, S., 2000. rpsL mutation induction after space flight on MIR. Mutat. Res. 453 (1), 1–4.

Yatagai, M., Nishimoto, M., Hori, K., Ohira, T., Shibata, A., 2002. Termiticidal activity of wood vinegar, its components and their homologues. J. Wood Sci. 48 (4), 338–342.

Chapter 24

Trichoderma: agricultural applications and beyond

R.N. Pandey[a], Pratik Jaisani[a] and H.B. Singh[b,c]

[a]*Department of Plant Pathology, B. A. College of Agriculture, Anand Agricultural University, Anand, Gujarat, India;* [b]*Department of Plant Pathology, Institute of Agricultural Sciences, Banaras Hindu University, Varanasi, Uttar Pradesh, India;* [c]*Department of Biotechnology, GLA University, Mathura, Uttar Pradesh, India*

24.1 Introduction

Agriculture is the prime source of livelihood of peoples all over the world, as plants are major source of raw materials used by human-being, animals and the industries. Hundreds of crops belonging to cereals, pulses, oilseeds, vegetables, floriculture, horticultural, plantation, fodder, fiber, etc. are cultivated by the farmers for their livelihood and to fullfill the need of the growing population. The estimated population of 9.8 billion people worldwide till 2050 would require the raising of nutritional food production in surplus, against the challenges of prevailing and emerging new pests in the wake of climate change and enhancing genetic productivity of crops by some 70% (FAO, 2017). Among different production constraints of obtaining potential yield of these crops, biotic stresses viz. pests i.e. diseases incited by fungi, bacteria, viruses, nematodes, insects, weeds, etc., play important role in reduction of quantitative and qualitative yield losses. In general the pests causes crop yield losses of about 31%–42% (Agrios, 2005). One of the ways of ensuring higher agricultural production is to reduce the losses due to these stresses at pre and post harvest stages of the production system.

24.2 Achieving UN sustainable development goals (SDGs)

The prime goals of UN which it focusses on include poverty (1); lowering hunger (2); Good state of health (3); Climate action (13) and partnership for their accomplishment, are directly or indirectly related to plant health and agriculture, for which efforts are needed to save immense crop yield losses through eco-friendly and sustainable management of biotic and abiotic stresses. Therefore, United Nations has decided to commemorate the year 2020 as "International Year of Plant Health." FAO focussed on ensuring sound plant health to produce sufficient food for growing population and their economic growth to end poverty. Prevailing and emerging pests viz. plant pathogens, nematodes, insect pests, weeds, and abiotic stresses viz. extreme weathers due to climate change, deteriorating soil health due to salinity, alkalinity, acidity, soil pollutants, droughts, flooding, etc. are the potential threat to different countries for robust plant health and crop production.

24.3 Pesticides consumption in the management of pests

World over the consumption of pesticides ranges from 17 kg/ha in developed countries like Taiwan, China, Japan, etc. to 0.6 kg/ha in India. But in India and other developing countries a large number of farmers use the pesticides injudiciously due to ignorance of proper use of pesticides for managing diseases and insect pests. The world pesticides consumption includes insecticides 44%, fungicides 21%, herbicides 30% and other 5%. The injudicious and over use of chemical pesticides has detrimental effects on human health, ecosystems, groundwater, the pollinators, parasitoids, predators and wild animals, resistance development in insect-pest, plant pathogens, pests resurgence, environmental hazards, food safety hazards, etc.

Biopesticides. https://doi.org/10.1016/B978-0-12-823355-9.00013-4

24.4 Benefits of microbes in rhizosphere

The earth's top strata carrying fertile soil with organic matter supporting the plant nutrition is the home to $>10^9$ (1 billion) microbes/g consisting of beneficial and harmful bacteria, fungi, viruses, archaea, etc. Plant Rhizosphere, are the playground for soil-borne plant pathogens and beneficial microorganisms (PGPMs), influencing biology, chemistry, function and structure of the soil by utilizing compounds exuded by the root. The root exudates stimulate biochemical and phenological interactions between roots and soil organisms for root and plant growth and survival of the microbes. The beneficial microbes known as bioagents, PGPR, PGPM, PGPF, symbionts, VAM, endophytes, etc. play vital role in suppression of abiotic and biotic stresses, etc. by their activities of biological control, modifying the biochemical and physical properties of the rhizosphere, degrading organic matters, solubilzation of phosphates and minerals, facilitating uptake of micronutrients, thereby creating a favorable environment for the plants growth, by altering the levels of peroxidases, superoxide dismutases, L-proline, Polyphenoloxydase production of ACC deaminase.

24.5 Soil borne diseases and plant pathogens

The fungi viz. species of *Fusarium, Rhizoctonia, Sclerotinia, Sclerotium, Pythium,* etc. surviving in the soil in the form of hyphae, sclerotia, conidia, chlamydospores, fruiting structures, initiate infection in the hosts and cause wilt, root rot, stem & collar rot, damping-off, etc. Often these diseases are difficult to manage due to limitations of chemical control measures, availability of resistant host cultivars, breakdown of host resistance, occurrence of fungicide resistance in pathogens, etc. Escalating public concern over possible health hazards of chemical pesticides and also sharp rise in cost of production and shrinking income of farmers, has led to the search of cost effective and eco-friendly pest management strategies viz. IPM/IDM, with the aims at restraining the pest species by integrating pests management technologies with leading roles of bio-agents and minimum emphasis on the use of pesticides. Under the changing agriculture scenario, the more use of bioagents seems to be promising for managing biotic and abiotic stresses without disturbing the equilibrium of environment and ecosystem.

24.5.1 *Trichoderma*—a fungus of unique characteristics

24.5.1.1 Description

The filamentous fungal genus: *Trichoderma* (teleomorph: Hypocrea), belongs to Kingdom: Fungi (obtain nutrients from organic matter); Phylum: Ascomycota (possess the ascus, for reproduction); Sub-phylum: Pezizomycotina (includes majority of filamentous, ascoma-producing species); Class: Eu-ascomycetes or Sordariomycetes (tendency to form lichen with other organisms); Order: Hypocreales (structures producing spores are brightly colored); Family: Hypocreaceae (perithecial ascomata that are brightly red, yellow, etc.); contains several species. *Trichoderma* spp. are extensively studied fungi for biotic and abiotic stress management, crop growth enhaners, enzymes and antibiotics production/industrial uses, molecular biology, transgenic crop, metabolizers of xenobiotics, and commercial biofungicides etc. They possess multiple characteristics which qualify them as wonderful bioagents.

Growth

The growth of *Trichoderma* spp. are readily seen on Potato dextrose agar (PDA), compared to other nutrient media. On rear side of culture, it appears as pale, tan, or yellowish (Sutton et al., 1998). The wooly, compact colonies on PDA grow rapidly and get matured within 4–5 days at 25°C. Initially the colonies are white, which turn to blue-green or yellow-green patches, usually in the form of concentric rings as the conidia are produced on the mycelium. The growth of *Trichoderma* on culture media can be easily identified, as it forms huge amount of greenish to whitish conidia attached on phialides located on plentifully as well as scantily branched conidiophores (Samuels, 1996).

Morphological description

Mycelium The mycelia are floccose to arachnoid and mostly appeare whitish in color.

Conidiophores These are highly branched, loosely compactly tufted. Phialides are produced on the chief axis and located next to the tip. The branches often re-branch and produce secondary branches and the longest branches are usually found located nearest to the main axis.

Phialides These are characteristically distended in the middle often forming a cylindrical or nearly sub-globose shape. Phialides may be usually penicillate (*Gliocladium*-like), or compactly grouped on wide main axis or arises solitary.

Conidia They are ellipsoidal, often 3−5 × 2−4 μm in diameter, with smooth surface, tuberculate to warted in some species. Formation of synanamorphs is observed in some species, thereby producing typical pustules. Such structures are mostly recognized by their lonely branched conidiophores, which have conidia in a green liquidy drop observed on the end of individual phialide.

Chlamydospores These are typically unicellular or multicellular (*T. stromaticum*), sub-globose, terminal at short hyphae or formed inside hyphal cells.

Teleomorphs *Trichoderma,* are the anamorph specieses of the ascomycete genus *Hypocrea* Fr., which form light to dark brown, yellow to orange fleshy, discoidal to pulvinate or effused stromata. Perithecia are completely immersed. Ascospores are hyaline or green, usually spinulose, and bicelled.

24.5.1.2 Ecology and biodiversity

Habitat

Trichoderma spp. are fast growing, asexually reproducing, and sporulating, opportunistic invader; possess powerful hydrolytic enzymes and antibiotic to compete with the other microbes for space, nutrients, etc. (Herrera-Estrella and Chet, 2004; Schuster and Schmoll, 2010; Montero-Barrientos et al., 2011; Zachow et al., 2009). They are cosmopolitan, ecologically very dominant and ubiquitous in all types of agricultural soils, desert soils, pastures, salt marshy land, lakes, salt affected marine soils in various ecosystems (Harman et al., 2004; Sadfi-Zouaoui et al., 2009). They are known for playing an important role in the breaking down or degradation of plant polysaccharides of decaying organic matter, decaying barks of trees; and can be found in rhizosphere and rhizoplane of woody and herbacious plants root ecosystems of different climatic zones (Chet et al., 1997; Chaverri et al., 2011; De Bellis et al., 2007); and sclerotia and propagules of other fungi (Montero-Barrientos et al., 2011; Mukherjee et al., 2013; Gal-Hemed et al., 2011).

Distribution

Trichoderma spp. i.e. *T. viride, T. spirale,* etc. have been isolated from the different ecosystems of Tunisia, Tenerife-Canary Island; having strong bioclimatic and edaphic variability (Christian et al., 2009; Sadfi-Zouaoui et al., 2009); South East Asia i.e. Taiwan and Indonesia (Kubicek et al., 2003); New species *T. eijii* and *T. pseudolacteum*, from Japan (Kim et al., 2013); *T. harzianum,* the anamorph of *Hypocrea orientalis* from nine geographic locations, representing nineteen different habitats of northern half of the Nile valley in Egypt (Gherbawy et al., 2004).

The effect of abiotic contributors on the species allotment of *Trichoderma* among the disturbed and undisturbed environments of forest, viz. shrub, grazed grass steppes of the land mass of Sardinia, Italy, the Mediterranean hot spot of biodiversity revealed the presence of the large population of pan-European and/or pan-global *Hypocrea/Trichoderma* species of the sections *Trichoderma* and *Pachybasium*. A significant decrement in local *Hypocrea/Trichoderma* diversity was observed which were found to be suceeded by invasive species from Eurasia, Africa and the Pacific Basin was observed (Migheli et al., 2009). The geographical distribution of *T. hypocrea* has been documented (Samuels et al., 2012b; Chaverri and Samuels, 2003; Jaklitsch, 2009). *H. lixii/T. harzianum* was found to be the most copious species (57%) showing its presence in 10, of the 15 sites of soils, followed by *T. spirale* and *T. gamsii,* as revealed in the Mediterranean (Sardinia, Tyrrhenian Islands) investigation (Migheli et al., 2009).

However, *T. aggressivum, T. pleurotophilum*, and *T. fulvidum*, have been reported to cause green mold and met with considerable crop yield losses in *Agaricus bisporus* and *Pleurotus ostreatus*, mushroom industry (Hatvani et al., 2007; Samuels et al., 2012a). Similarly, *T. longibrachiatum*, has been reported to infect immune suppressed humans, and is listed as emerging fungal human pathogens (Kredics et al., 2003, 2004).

Trichoderma as endophytes

Many novel species of endophytic or endosymbiotic *Trichoderma* with antagonistic potential against pathogens have been found in stem, leaves and roots of annual and woody plants (Petrini, 1991; Gazis and Chaverri, 2010; Patel et al., 2017). Phylogenetic analysis revealed that endophytism is recently featured trait in *Trichoderma* spp. (Samuels et al., 2006; Samuels and Ismaiel, 2009; Chaverri et al., 2011; Druzhinina et al., 2011). They also influence community biodiversity and microbial interactions (Zhang et al., 2005).

Endophytic strains during interaction with plants roots, alter the gene activation pattern in the shoots, which alter the plant physiology i.e. production of phytohormones which improve uptake of nitrogen fertilizers; production of bioactive secondary toxic metabolites/alkaloids that deter herbivores and pathogens, Improve resistance to insect pests and herbivore and helps the plants in Drought acclimatization, Increased competitiveness, Enhanced tolerance to heavy metal, low pH, high salinity, photosynthetic efficiency for higher yield of the plants (Harman et al., 2012; Chaverri and Samuels, 2013; El_Komy et al., 2015). Endophytic *Trichoderma* spp. have been found to decrease the necrosis caused by *Gremmeniella abietina* in *Pinus halepensis* seedlings (Romeralo et al., 2015). *T. asperellum* and *T. asperelloides* turns endophytic after introduction through seed and soil treatment in rice variety Pusa Basmati-1 at ICAR- IARI, New Delhi (Leon et al., 2017). *T. chlorosporum* in *Dendrobium nobile* in China (Yuan et al., 2008); *T. martiale, T. hamatum* and *T. asperellum* from *Theobroma cacao* respectively, from Brazil (Hanada et al., 2008); USA (Bae et al., 2009), and Indonesia (Rosmana et al., 2015); *T. amazonicum* from *Hevea* spp. in Amazon basin (Chaverri et al., 2011); *T. flagellatum, Trichoderma* sp. in Coffee from Ethiopia (Mulaw et al., 2013) have been identified as endophytes. Similarly, *T. gamsii* from *Lens esculenta* (Rinu et al., 2014); *T. gamsii* and *T. gamsii* YIM PH30019 isolated from Indian Himalayan mountain ecosystem have been recognized as promising biocontrol agents of root-rot fungal pathogens i.e. *Phoma herbarum, Fusarium flocciferum, E. nigrum* of *Panax notoginseng* (Chen et al., 2016) due to their properties of secretions of ammonia and salicylic acid, phosphate solubilization, chitinase activity, mycoparasitism, suppressive action of induced volatile compounds.

24.5.1.3 *Identification of* Trichoderma *spp. and their strains*

Morphological

The name of genus *Trichoderma* was first proposed by Persoon (1794) in relation to macroscopic resemblance. The first real generic description of *Trichoderma* and systematic identification of the species was proposed by Rifai (1969), based on the characteristics growth rate of colony and microscopically observed characters (Biolog system) i.e. conidiophore branching pattern and morphology of conidium. On the basis of these characters, the available isolates of *Trichoderma* were sub-divided into nine species. Since then a number of workers (Bissett, 1984; Bissett, 1991a, 1991b; Samuels, 1996; Samuels et al., 1998; Manczinger et al., 2012) have discussed the methods of identification of *Trichoderma*. However, there are certain limitations in identifying the species correctly. Earlier Rifai's classification based on morphologica traitsl of *Trichoderma* comprised of 9 species aggregates (Rifai, 1969); however the taxonomic position of this genus has been changing from time to time (Bissett, 1991).

Molecular

Molecular data studies obtained from DNA and key enzymes has been most oftenly used to characterize strains of *Trichoderma* precisely and classify them phylogenetically at species level than the common conventional methods (Lieckfeldt et al., 1998; Hermosa et al., 2000; Kubicek et al., 2003; Jaisani and Pandey, 2017). Random amplified polymorphic DNA (RAPD) profiles have also been used to record genetic diversity (Muthumeenaksi et al., 1994; Zimand et al., 1994 Arisan-Atac et al., 1995; Ospina-Giraldo et al., 1998; Chen et al., 1999; Kullnig et al., 2000; Hermosa et al., 2001) along with recording the information about carbon source utilization patterns, which has shown its importance in studies pertaining to assessment of *Trichoderma* biodiversity (Manczinger and Polner, 1987). Additionaly the isoenzyme data analysis of hydrolytic enzymes established intraspecific groupings within *T. harzianum*, on the basis of morphological and biochemical characters. First of all Zamir and Chet (1985) tried successfully to characterize *Trichoderma* using isoenzyme analysis. Likewise, for phylogenetic analysis of *T. longibrachiatum* rDNA translation elongation factor (tef1) gene was used (Kuhls et al., 1997), *T. harzianum* (Gams and Meyer, 1998; Lee and Hseu, 2002; Chaverri et al., 2003), *T. viride* (Lieckfeldt et al., 1999), biocontrol strains (Hermosa et al., 2000; Sanz et al., 2004), biotypes causing the green mold of mushrooms (Muthumeenaksi et al., 1994; Ospina-Giraldo et al., 1998; Samuels et al., 2002), evolution studies and the disruption of new haplotypes in the genus (Kullnig-Gradinger et al., 2002; Kubicek et al., 2003).

Internal transcribed spacers (ITSs) and multiple gene sequences have been used for developing online interactive algorithms for the identification of *Trichoderma* species based on (Druzhinina et al., 2005). Different isolates of *Trichoderma* till date have been identified at the species level by the exclusive oligonucleotide BarCode for *Hypocrea/Trichoderma* (*TrichO*KEY), sequence similarity analysis (*Tricho*BLAST) and phylogenetic determinations. The online search tool (TricoBLAST) and Oligonucleotide barcode (TrichOKEY) methods with newer versions of accurate identification of the new species are now available online at http://www.isth.info/biodiversity/index. php (Druzhinina et al., 2005; Kopchinskiy et al., 2005). A *tef1*-gene based DNA bar code algorithm, *Tricho*CHIT, has been developed for rapid identification of the strains of *Hypocrea lixii/Trichoderma harzianum* (Nagy et al., 2007). Since the availability of the molecular tools, a number of novel species viz. *T. aggressivum, T. amazonicum, T. strigosellum* sp. nov, etc. have been described and identified (Lopez-Quintero et al., 2013), besides, the existing ones.

However, Advanced method as Intact-cell MALDI-TOF MS (ICMSs) has been found suitable for fast and to the point identification of hydrophobin class II group of compounds which might provide a platform for identification of

Trichoderma and other fungal species and their strains from even minute quantity of biomass (Neuhof et al., 2007; Nakari-Seta et al., 2007). The "MALDI-TOF MS" technique is based on the analysis of peptides which embody a draft equivalent to sequencing for determination of indepth information on species limits (Respinis et al., 2010).

However, phenotype microarray which is based on analysis of carbon utilization can be used (Bochner et al., 2001). Similarly, habitat preference and nutrition mode can be used for analysis of endophytic species (Chaverri and Samuels, 2013). The technique of Isozyme profiles based on cellulose-acetate electrophoresis has been used for biochemical diversity determination of intra and inters species of the genus (Manczinger et al., 2012). Subsequently, more than 200 phylogenetically classified species of *Trichoderma* were discovered (Atanasova et al., 2013). The name *Trichoderma* is now used for the known green forms of the fungi characterized by the initial *T. viride* species described by Persoon (1794), Rifai (1969), and Samuels (2006). Because of its unique characteristics and beneficial uses, it has been extensively studied by different workers. Now volume of literature Google scholar database 4, numbers of Patents (Google patent database), etc. are available on this wonderful fungus.

24.5.2 *Trichoderma* spp. in agricultural application

Trichoderma spp. are the fungi of multifarious activities viz,. acting as biocontrol agent bio-fertilizer, bioprotectant, biostimulant, phosphate solubelizer, decomposer of organic waste, bioremediation, and sourse of various useful genes etc. (Fig. 24.2). *Trichoderma* spp. are very effective in managing various soil borne fungal plant pathogens (Papavizas, 1985), due to presence of strong cell wall hydrolytic enzymes and antibiotics (Lorito et al., 1998; Keszler et al., 2000); competence in utilizing nutrients (Samuels, 1996; Howell, 2003), rapidly colonizing substrates (Grondona et al., 1997); capacity to alter the Rhizosphere and to grow under non favorable conditions; stimulants of defense response, ability to induce systemic acquired resistance (SAR) in plants and enhancer of crop growth (Enkerly et al., 1999; Inbar et al., 1994); acting as Bio-fertilizers, biodecomposers of organic matter, bioremediation, increased uptake of nutrients; enzymes production for industrial uses, etc. (Harman et al., 2004; Woo et al., 2006; Schuster and Schmoll, 2010; Dagurere et al., 2014). The significant expression and variety of *Trichoderma* enzymes (Lorito et al., 1998) has led to their tremendous use in biodegradation, organic matter composting, textiles industry etc. *Trichoderma* spp. studied extensively in relation to its application in agriculture and beyond has been summarized as under:

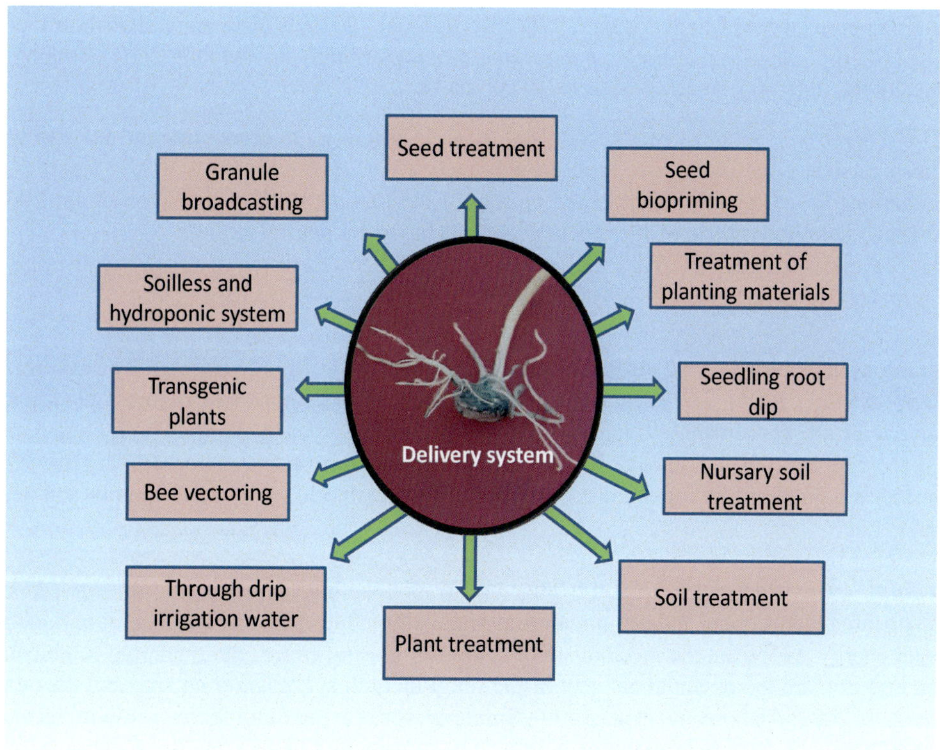

FIG. 24.1 Different delivery methods of *Trichoderma* formulations.

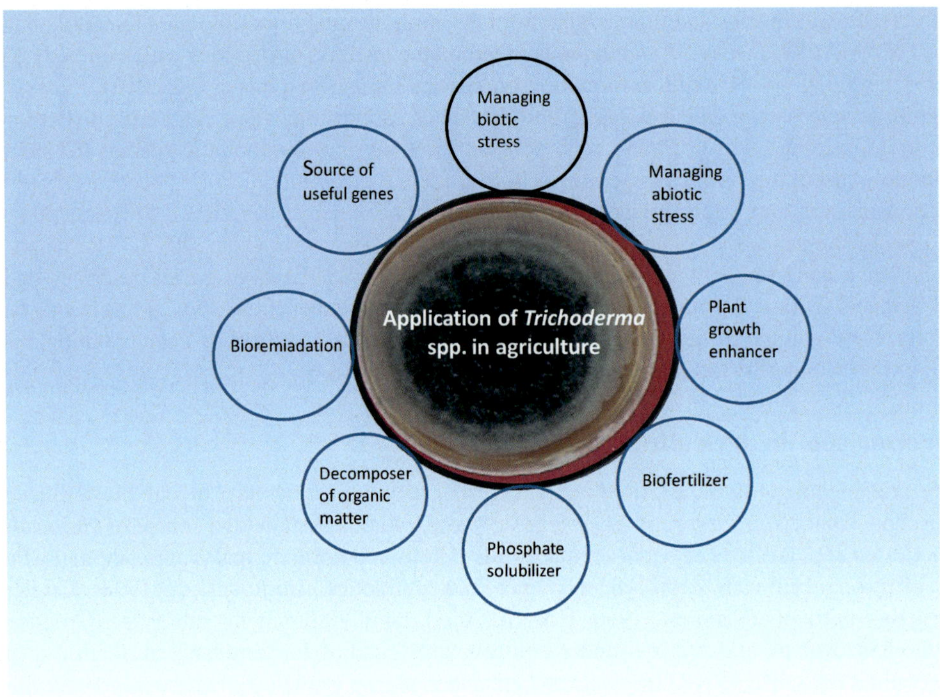

FIG. 24.2 Various method of application of *Trichoderma* spp. in agriculture.

24.5.2.1 Trichoderma *spp. as biocontrol agents for management of diseases in crops*

Many specieses of *Trichoderma* have been known since 1930 as plant disease suppressive agents (Weindling, 1932). They are now used widely extensively world -wide as most promising, economical, eco-friendly, biocontrol agents, biofertilizers and biostimulants (Elad et al., 1984; Liu and Baker, 1986; Lorito et al., 2010; Vinale et al., 2006), in the management of fungal plant diseases of crop plants viz. *Pythium, Sclerotium, Macrophomina*, etc. (Monte, 2001; Domingues et al., 2000; Anand and Reddy, 2009).

- More than 200 *Trichoderma* spp. have been recognized for their efficiency as biocontrol and as organisms for inducing resistance in hosts against plant pathogens, crop growth, etc.
- Plenty of formulations based on *Trichoderma* are present in the market world wide (Woo et al., 2006; Vernner and Bauer, 2007). They share about 60% of the available fungal based biocontrol agents.
- In addition, the biocontrol efficacy of *Trichoderma* spp. has also been explored against several foliar diseases.

Mode of action

Trichoderma spp. are well equipped with different strategies which make them successful biocontrol agents, growth promoters, etc. (Benítez et al., 2004). *Trichoderma* spp. applies various mechanisms to suppress plant pathogens, control disease and improve over all plant health (Vinale et al., 2006; Pal and Gardener, 2006). The biological mechanisms i.e. Competition for space and nutition; Antibiosis; Mycoparasitism; Induction of host defense; etc. (Howell, 2003) help to manage biotic stresses as a good option for chemicals (Chet, 1987) and abiotic stresses due to poor soil health and climate change (Table 24.1).

Competition for food, space and nutrients Like other fungal phytopathogens viz. *Fusarium* spp., *Rhizoctonia solani, etc., Trichoderma* spp. are heterotrophic and acquire nutrients i.e. carbon, nitrogen, phosphorus, iron, minerals, etc. from its vicinity. They obtain energy ATP from the metabolism of cellulose, glucan, chitin, etc. rendering to glucose. The nutrition particularly corbon and iron are the essential element of the fungi including *Trichoderma* spp. and they exert competition for them (Sarrocco et al., 2009; Alabouvette et al., 2009). Similarly, for iron the fungi secretes low-molecular-weight ferric-iron specific chelators known as Siderophore to obtain iron from surrounding and arising the situation of starvation of iron for others, affecting the growth of adjoining fungi (Chet and Inbar, 1994).

TABLE 24.1 Mechanisms of biological control of phytopathogens by *Trichoderma* spp.

Type	Mechanism	Function
Indirect antagonism	Competition for space, food, nutrition	Colonizes the roots very rapidly and utilizes nutrients i.e. Exudates/leachates of plant, Siderophore scavenging, Physical niche occupation.
	Induction of host resistance	The association with plant involves molecular recognition through a signaling network mediated by the plant hormones salicylic acid (SA), jasmonic acid (JA) and ethylene (ET). It initiates a series of biochemical alterations of the plant defense response -mediated through production of related metabolites and enzymes viz. phenylalanine ammonia lyase (PAL), chitinase, glucanase, etc.
Direct antagonism	Hyperparasitism/Predation	Cell wall degrading enzymes viz. chitinase, β -1, 3 glucanase, glycoside hydrolases, proteases, etc. causes lysis and death of cell wall of fungal plant pathogens.
Mixed-path antagonism	Antibiosis/Lysis	Antagonistic interaction between bioagents and the plant pathogens mediated by low molecular weight diffusible secondary metabolites or antibiotics of the bioagents that are detrimental for the growth of pathogens, eg. gliovirin, gliotoxin, pyrones and peptaibols, etc.

- Competition occurs between micro-organisms dwelling in soils when space or nutrients are the limiting factors. In general, soil-borne plant pathogens, viz. species of *Pythium, Fusarium, Rhizoctonia, Sclerotium,* etc. infecting the host plant through mycelial interaction are more prone to receiving tough competition from other microbes (Tari and Anderson, 1988).
- *Trichoderma* spp. are efficient competitors for space and nutritions in their vicinity as they produce metabolites that suppress the growth and development of pathogens. They are voracious saprophyte by virtue of possessing different enzymes, which help in digestion of cellulose, glucan, chitin and others, in organic matters for its nutrition, growth and increasing its population by utilizing energy through glucose (Chet et al., 1997).
- They are better rhizosphere colonizers than the plant pathogens, therefore, grow faster and occupy the space and utilize the available organic and inorganic matters for food, increases in the population, thereby taking over the pathogens.

Antibiosis and lysis Antibiosis is the phenomenon of inhibition of the growth of one organism by a compounds or antibiotics secreted by another microbe. Most *Trichoderma* strains release harmful metabolites viz. gliovirin, a-pyrone, peptaibols, trichotoxins, asperelines, trichovirins, trichorzins, trichorzianins, trichopolyns, harzianic acid, alamethicins, tricholin,6-penthyl-a- pyrone, massoilactone, viridian, glisoprenins, heptelidic acid, etc., (Röhrich et al., 2014); that are antagonistic or act as signaling molecules in all the processes of interaction with other living organisms and affecting their growth (Harman et al., 2004; Benítez et al., 2004), impede colonization by antagonizing microbes and inhibit the pathogens viz. *R. solani, B. cinerea,* etc.

Antibiotics are toxins of microbial origin which even at lower concentration have the potential of killing other microbes. Most of the microbes secrete multiple compounds having antibiotic activity (Thomashow et al., 2002). Usually in most of the cases, antibiotics secreted by microorganisms have been found effective at suppressing plant pathogens/ diseases. Strains of *T. virens* produce Gliotoxin, which is effective against *R. solani* (Abbas et al., 2017). The hydrolytic enzymes and antibiotics secreted together by a particular microbe have shown the enhanced level of antagonism (Howell, 1998). A higher level of synergistic antagonism were observed between the enzyme endochitinase from *T. harzianum* and gliotoxin; hydrolytic enzymes and peptaibols; extracellular enzymes and of α-pyrone in case of *B. cinerea* and *R. solani.*

- **Lysis** is the process of breakdown of biological material including plants and animals. *T. harzianum* is capable of secreting cell wall lysis enzyme that helps in dissolving a huge number of polymeric compounds (Bull et al., 2002).

Hyperparasitism/mycoparasitism and predation Some *Trichoderma* spp. due to their efficiency in production of many extra cellular enzymes i.e. cellulases, chitinases, glucanases i.e. β-1,3-glucanase and β-1,6-glucanase, glycoside hydrolases, proteases, etc., cause mycoparasitism on fungal phytopathogens (Harman et al., 2004; Zeilinger et al., 1999). These enzymes are tempted against cell walls of fungi (Kiss, 2003) by which they attack the mycelia and other propagules of fungi acting as plant pathogens of diseases in crops.

- The mechanism involve are chemotrophic growth, recognition of the host fungus, coiling around them, produce appressorium containing osmotic solute viz. glycerol, penetration and lysis of the fungus and propagules through sequential production of pathogenesis related peptides i.e.
- peptaibols and peptaibiotics and cell wall degrading hydrolytic enzymes viz. chitinases, β-1,3-glucanases and N-acetyl-β-D-glucosaminidases, thereby facilitating the entry of the mycelium and digestion of the cell contents (Howell, 2003; Sanz et al., 2004; Benhamou and Chet, 1997) for nutrition and growth, which cause reduction of population of the pathogens. Pearl millet transformed by gene β-1,3-glucanase (gluc78) from *T. atroviride,* showed 58% reduction in the incidence of downy mildew caused by *S. graminicola* (O'Kennedy et al., 2011).

Induction and exploitation of induced systemic resistance against different stresses The review on interactions of *Trichoderma* spp. with plants (Harman et al., 2004), has highlighted that as per the recent discoveries, *Trichoderma* spp. have been found as opportunistic, avirulent plant symbionts. Efficient strains of root-colonizing *Trichoderma* spp. have been identified which colonize the root, penetrate the cells and tissues and bring about a series of alterations in the plant, characteristic of the systemic defense response in plants (De Meyer et al., 1998; He et al., 2004). The endophytic plants symbionts, having the potential to flourish and grow in soil, also infect, penetrate and establish long lasting colonization into the epidermis and a few adjoining cells of plant roots and establish communication with the plants through chemical dialog which continue and offer life long profit to plants. It produces certain compounds that encourage on site or systemic resistance response in host plants. Root colonization by *Trichoderma* helps in enhancing the growth of plant, increasing crop productivity, resistance to multiple diseases along with helping in uptake and use of the nutrients.

- Studies conducted by different groups (Alfano et al., 2007; Bae et al., 2011; Djonovic et al., 2007; Marra et al., 2006; Shoresh and Harman, 2008; Shoresh et al., 2010) revealed that root colonization and biochemical signaling by *Trichoderma* strains sturdily alter the physiological state of plant due to change in plant gene expression.
- Chemical elicitors of innate defensive system of SAR and ISR including salicylic acid, siderophore, 2,3-butanediol, lipopolysaccharides and other volatile substances are plentiful and ranging from hydrophobin-like proteins, peptides, and smaller molecules may be produced by the plant growth promoting strains of *Trichoderma* spp. in compatible host (Table 24.2).
- Peptaibols: A class of linear peptides having strong antimicrobial activity against fungi, act in synergy with enzymes responsible for degrading cell wall and stop the development of fungal pathogens and also invoke resistance in plants against pathogens (Wiest et al., 2002). Peptaibole alamethicin from *T. viride* have been found to induce long distance electrical signals in plants (Maischak et al., 2010; Heiko et al., 2010). The triggering of the systemic resistance in host was found to be due to defense mechanisms involving the production of enzymes and metabolites (Djonovic et al., 2006; Seidi et al., 2006; Mukherjee et al., 2012) which included pathogenesis related (PR) proteins and enzymes engaged in the reply to oxidative stresses (Gajera et al., 2013).

The knowledge about the beneficial traits of *Trichoderma* spp., facilitated their use as bioagents using propagules, protein based formulations and transgenic plants modified with useful genes (Kubicek et al., 2001; Monte, 2001). The strains of *T. asperellum, T. harzianum, T. atroviride,* etc. have been found to induce systemic resistance by stimulating metabolic changes that respond to inhance higher tolerance (Saksirirat et al., 2009; Sriram et al., 2009; De Meyer et al., 1998; Contreras-Cornejo et al., 2011; Salas-Marina et al., 2011; Yoshioka et al., 2012; Harman et al., 2012) against plant pathogens (Table 24.2). Strain T-39 of *T. harzianum* as soil and seed application was able to induce the resistance in plants, against soil-borne and foliar pathogens viz. *C. lindemuthianum, B. cinerea, P. viticola* in different crops (Bigirimana et al., 1997).

- Some strains of *Trichoderma* spp. have been reported to enhance resistance to various biotic as well as abiotic stresses viz. water deficits i.e. drought, salt, atmospheric temperature (Bae et al., 2009; Shoresh et al., 2010). An abiotic stress tolerance is imparted by enhancing the levels of enzymes involved in scavenging the dreadful reactive oxygen species (ROS) (Shoresh et al., 2010). Enhanced production of enzymes in the pathways such as the Glutathione-ascorbate cycle, help in recycling the antioxidants more rapidly, which help in reducing the effects of stresses. Besides, the other benefits imparted by these endophytic plant symbionts is increase in nitrogen use efficiency (NUE) in plants, which could possibly help in reducing N application rates by significant number with no significant reduction in yield (Harman, 2011; Shoresh et al., 2010).

24.5.2.2 Management of biotic and abiotic stresses in crop plants

- Plant pathogens/disease management due to strategies like fight for space, nutrition; production of iron chelators; mycoparasitism; antibiosis; Multi-hydrolytic enzymes i.e. Cellulase, Esterase, Proteases, Chitinases, 1,6-ß-Glucanases, 1,3-ß-Glucanases, Chitobiosidases; Associated antimicrobials i.e. production of alkyl pyrones; pentyl analogs; dermadin; isonitriles; poly-ketides; peptaibols; steroids di-keto piperazines; etc. for ISR, SAR in plants are associated with *Trichoderma* spp. and interaction with plant pathogens and host plants.

TABLE 24.2 Induced Systemic Resistance (ISR) in plants triggered by *Trichoderma* spp.

Trichoderma spp., crop and pathogen	Induced systemic resistance	References
T. harzianum (T-39): • **Bean:** *C. lindemuthianum, B. cinerea*	No disease on leaves	Bigirimana et al. (1997)
• **Tomato, lettuce, pepper, tobacco, bean:** *B. cinerea*	No disease on leaves	De Meyer et al. (1998)
• **Tomato:** *B. cinerea*	Decline in disease severity	Meller et al. (2013)
• **Grapes:** *Plasmopara viticola*	Triggering of mechanisms related to defense	Levy et al. (2015)
• **Cucumber, strawberry, bean, tomato:** • *B.* cinerea; *Podosphaera xanthii*	Protection from diseases caused in foliage parts	Levy et al. (2015)
***T. harzianum* T-22; *T. atroviride* P1** • **Bean:** *B. cinerea, X. campestris* pv. *phaseoli*	Triggering of pathways associated with antifungal compounds in leaves	Harman et al. (2004)
***T. harzianum* T-1 & T-22; *T. virens* T3** • **Cucumber:** *Green-mottle, mosaic virus*	Root inoculation of strains protected the leaves from the disease	Lo et al. (2000)
***T. harzianum* T-22** • **Tomato:** *Alternaria solani*	Root inoculation of the strain protected the disease in leaves	Seaman (2003)
***Trichoderma* GT3-2** • **Cucumber:** *C. orbiculare, P. syringae* pv. *lachrymans*	Activation of defense related genes related to lignifications	Koike et al. (2001)
T. harzianum • **Pepper:** *Phytophthora capsici*	Enhanced the production of the phytoalexins capsidiol, toxic to pathogen	Ahmed et al. (2000)
***T. asperellum* (T-203)** • **Cucumber:** *Pseudomonas syringae* pv. *lachrymans*	Modulated expression of proteins related to jasmonic acid/ethylene signaling	Shoresh et al. (2005)
T. asperellum* SKT-1** • ***A. thaliana: *P. syringae* pv. *tomato* DC3000 • ***A. thaliana:****Cucumber mosaic virus*	ISR to colonization by SKT-1 and its cell-free culture filtrate Improved defense mechanism against infection of CMV	Yoshioka et al. (2012) Elsharkawy et al. (2013)
***T. harzianum* Tr6, & *Pseudomonas* sp. Ps14** • **Cucumber:** *F. oxysporum* f. sp. *radicis*	Both the strains activated the set of defense-related genes	Alizadeh et al. (2013)
T. virens & T. atroviride • **Tomato:** *Alternaria solani, B. cinerea,* and *P. syringae* pv. *tomato* (Pst DC3000)	Secreted proteins- Sm1 and Epl, related to induction of SAR	Salas-Marina et al. (2015)
***T. virens* G-6, G-6-5 and G-11** • **Cotton:** *Rhizoctonia solani*	Protected plant by triggering fungitoxic terpenoid phytoalexins	Howell et al. (2000)

- *Trichoderma* species possess natural enzymatic proteins that may help in the plant survival against various biotic and abiotic stressors. Many proteins/enzymes, viz. mitogen-activated protein kinase were found to confer resistance against soil-borne and foliage pathogens (Howell et al., 2000; Perazzoli et al., 2012; Viterbo et al., 2005).
- *T. hamatum* GD12 which encourages plant growth of *A. thaliana,* biocontrol of *Sclerotinia sclerotiorum,* and ISR to *Magnaporthe oryzae* in rice contained unique genomic regions of 47-kbp with the potential to programme novel bioactive metabolites (Studholme et al., 2013).
- Cloning and functional illucidation of *hsp*70 gene from *T. harzianum* T34 was done and the expressed protein in *Arabidopsis* was found to provide higher level of forbearance to heat, salt, osmotic and oxidative stresses (Montero-Barrientos et al., 2008).
- Similarly, Putative kelch-repeat protein coding gene, Thkel 1 from *T. harzianum* known for governing the glucosidase activity was cloned and characterized and expressed protein was able to enhance tolerance level against salt and osmotic stresses in *A. thaliana* (Hermosa et al., 2011). Aspartyl proteases produced *in planta* with the attachment of *T. asperellum,* in cucumber plants play a role as mycoparasite and as plant opportunistic symbiont (Viterbo et al., 2004).

Management of biotic stresses i.e. diseases of crops

- *Trichoderma* spp. are effective biocontrol agent against fungal plant pathogens/diseases viz. damping-off (*Pythium* spp.), stem canker/rot (*Phytophthora* spp.) Wilt (*Fusarium* spp.), Root rot, sheath blight (*Rhizoctonia* sp., *Macrophomina* sp., *Sclerotium sp., Sclerotinia* sp., *Botrytis* sp.), collar rot (*Aspergillus* sp.), Leaf spots (*Alternaria* spp.), and many others, occurring in the crops viz. Cereals: maize, wheat, rice, sorghum; Pulses: mung bean, chickpea, pigeon pea, cowpea; Oilseed crops: Groundnut, sunflower, mustard; Fruits and Vegetables: tomato, brinjal, potato, Commercial crops: Cotton, Sugarcane; and other crops of economic importance (Table 24.3).

TABLE 24.3 ***Trichoderma*** **spp. in the management of diseases in crops.**

Crop, disease and pathogens	Biocontrol agent	Treatment	References
Rice (*Oryza sativa*): **Blast** (*Magnaporthe oryzae*)	*T. harzianum*	Seed, Foliar	Chou et al. (2019)
Sheath blight (*Rhizoctonia solani*)	*T. harzianum,T. Viride,* *T. virens*	Seed, Seedlings	Bhat et al. (2009); Pal et al. (2015)
Brown spot (*Drechslera oryzae*)	*T. viride*	Seed	Gomathinayagam et al. (2010)
False smut (*Ustilaginoidea virens*)	*T.harzianum, T. viride,* *T. virens, T. reesei*	Seed	Kannahi et al. (2016)
Wheat (*Triticum aestivum*): **Karnal bunt** (*Neovossia indica*)	*T. viride, T. harzianum*	Seed	Aggarwal et al. (1996)
Loose smut (*Ustilago segatum tritici*)	*T. viride, T. harzianum,* *T. lignorum*	Seed &Soil	Aggarwal et al. (1991)
Maize (*Zea mays*) *Fusarium* stalk rot (*F. graminearum*) Post-flowering stalk rot complex	*T. harzianum* *T. harzianum* (Native IARI)	Seed &Soil Seed	Kandasamy et al. (2017) Meena et al. (2010)
Sorghum (*Sorghum bicolor* L) **Zonate leaf spot** (*G. sorghi*)	*T. harzianum*	Foliar	Purohit et al. (2013)
Anthracnose (*C. sublineolum*)	*T. harzianum* WKY1	Seed, Soil	Wesam et al. (2017)
Chickpea (*Cicer arietinum* L) **Wilt** (*F. oxysporum* f. sp. *ciceris*)	*T. harzianum, T. viride,* *Trichoderma* spp. *T. viride*	Seed and Soil as bio-pillets Seed biopriming Soil, enriched FYM	Dubey et al. (2013) Pandey et al. (2017), Jaisani et al. (2016)
Seed rot, Wet root rot (*R. solani*) **Dry root rot** (*R. bataticola*) **Collar rot** (*S. rolfsii*)	*T. virens* *T. harzianum* *T. viride* *T. viride*	Seed, soil-biopillets Seed and soil as biopillets, Seed- biopriming Soil -enriched FYM Seed, Soil Biopellet	Dubey et al. (2012) Pandey et al. (2017) Mandal et al. (2015), Mukherjee (1997)
Pigeonpea (*Cajanus cajan*) Wilt (*Fusarium udum*)	*T. viride, T. harzianum* *T. harzianum, T. koningii,* *T. hamatum*	Seed, soil in FYM Soil in FYM	Kumar et al. (2009) Ram and Pandey (2011) Mahesh et al. (2010)
Cowpea (*Vigna sinensis*): **Damping off/Root rot** (*Macro-phomina. phaseolina)*	*T. harzianum,* *T. koningii*	Seed and Soil Seed application	Timothy et al. (2001)
Mung bean (*Vigna radiata*): Root rot (*M. phaseolina)* **Web blight** (*R. solani*)	*Trichoderma* spp. *T. viride, T. harzianum* *T. viride* *T. viride* *Trichoderma virens* *T. viride*	Seed, Soil biopillets Seed- biopriming Soil, enriched FYM Seed pelleting Seed, Soil in pulse bran and saw dust	Dubey et al. (2009) Meena et al. (2016) Pandey et al. (2017) Raghuchander et al. (1997), Dubey and Patel (2002).
Soybean (*Glycine* max L.): **Root rot** (*M. phaseolina)* Myrothecium Leaf Spot, Anthrac-nose and Rhizoctonia Arial Blight	*T.harzianum, T.viride* *T.viride*	Seed biopriming Soil -enriched FYM Seed through polymer coating	Pandey and Gohel, (2017) Falah Kuchlan et al. (2018).

TABLE 24.3 ***Trichoderma*** **spp. in the management of diseases in crops.—cont'd**

Crop, disease and pathogens	Biocontrol agent	Treatment	References
Groundnut (*Arachis hypogaea L.*): **Stem and Pod rot (*S. rolfsii*)** **Root rot** (*M. phaseolina*) **Collar rot** (*Aspergillus niger*) **Leaf spot** (*Cercospora arachidicola; Cercosporidium personatum*)	• *T. harzianum* • *T. viride* • *T. viride* • *T. harzianum*	• Seed • Seed or soil • Seed and soil • Seed and foliar spray	Raman and Korikanthimath (2006), Rakholiya et al. (2010), Patel (2009), Hossain and Hossain (2014)
Mustard (*Brassica juncea* (L.): **Alternaria blight** (*A.brassicae*) **Damping off** (*P.aphanidermatum*)	*T. harzianum* *T. viride*	Foliar spray Soil	Kumar et al. (2019) Khare et al. (2010)
Sunflower (*Helianthus annuus*): **Root rot** (*R. solani*) & **Collar rot** (*S.rolfsii*) **Blight** (*Sclerotinia minor & S. sclerotiorum*)	*T.viride* *T. harzianum, T. virens*	Seed and Soil application Seed, soil with compost	Mathivanan et al. (2000) Abdollahzadeh et al. (2006)
Cotton (*Gossypium hursutum*): **Wilt** (*F.oxy. f. sp. vasinfectum*) & **Root knot** (*M. incognita*) **Root rot**(*R. solani & M. phasiolina*)	*T. virens* (*G. virens*) *T. viride*	Seed Seed and Soil application	Zhang et al. (1996) Mathivanan et al. (2000)
Potato Black-scurf (*R. solani*):	*T. virens, T. atroviride* and *T. barbatum*	Soil treatment	Hicks et al. (2014)
Brinjal (*Solanum melongena L.*): **Damping off** (*P.ahanidermatum*) **Wilt** (*F. solani*)&Collar rot (*S. sclerotiorum*),	*T. harzianum* *T.viride*	Seed & Soil Seed and Soil application	Nirmalkar et al. (2018) Mathivanan et al. (2000)
Tomato (*Solanum lycopersicum*): Damping off (*Pythium indicum*) Root knot nematode (*M. javanica*) Wilt (*F. oxy* f.sp. *lycopersici*)	*T. harzianum* *T. harzianum* *T. harzianum*	Seed treatment Soil pretreatment Soil treatment	Jayaraj et al. (2006) Nirmalkar et al. (2018) Sharon et al. (2001) Singh et al. (2004)
Banana (*Musa* sp.): **Panama disease** (*F. oxysporum* f. sp.*cubense*)	*T. viride, T.harzianum, T. asperellum*	Root and rhizome dipping, Soil ammendments.	Bubici et al. (2019)

Management of abiotic stresses

Protection of crops against abiotic stresses viz. drought, salinity, alkalinity, allelopathic, oxidative stresses, heavy metal accumulation, etc.; plant and root growth, increased nutritional uptake, and inducing protection against oxidative stress, similar to rhizobacteria, was also obtained with *Trichoderma* spp. (Hidangmayum and Dwivedi, 2018; Yasmeen and Siddiqui, 2017).

Drought stress

- Drought affects rice crops, when soil moisture drop down below saturation. Drought or even a short period of water deficit is highly detrimental to rice productivity. Colonization of *T. harzianum* T-22 around roots was found to augment the level of plant enzymes namely peroxidases, chitinases; lipoxygenase-pathway and compounds like phytoalexins and phenolic acids; which imparted robust resistance in the plant against stresses. Similarly, the enzymes viz. Superoxide dismutases (SOD), Ascorbate peroxidases and catalases (CAT) have been reported to act as main enzymatic scavengers or removal of harmful ROS, presumably superoxide (O_2—) and H_2O_2 and also help in redox maintenance. This ability of plants enables it to resists to water deficit i.e. drought, methyl viologen (MV) exposure, and other abiotic stresses (Mastouri et al., 2012).

- Seed bioprimming of drought tolerant *T. harzianum* strains @ 10 g/Kg seeds in drought susceptible rice variety PB 1121, has shown the characteristics of increased root and shoot length, leaf area index, membrane stability index and relative water content; enhanced catalase and peroxidase activity; and cause a reduction of leaf rolling and free proline content. This enabled the plant to greater scavenging activities for the free radicals and harmful compounds, generated by the drought stressed plants and thereby delayed the drought response in rice.
- Similar findings have been reported for management of drought in tomato with *Trichoderma* spp. (Shoresh et al., 2010; Mastouri et al., 2012); in Cacao with *T. hamatum* DIS2196 (Bae et al., 2009); and in maize plant treated with *T. hazianum* (Harman, 2000).

Salinity stress

- Salt stress have been reported to cause a significant decline in germination percentage, plant growth parameters, photosynthetic pigments, stability of membranes, content of water in tissues, and total content of lipids in *Ochradenus baccatus.* It encouraged the synthesis and increased level of diacylglycerol, phenols, sterol esters, non-esterified fatty acids; enzymatic antioxidants; lipid peroxidation; Na + content in shoot and a significant decline in other divalent cations. Application of *T. hamatum* has helped to alleviate the above metabolic processes along with reducing the negative impact of salt stress on plant growth (Hashem et al., 2014). Similarly, characterization of "1-aminocyclopropane-1-carboxylate (ACC) deaminase (ACCD)" from *T. asperellum* was done (Viterbo et al., 2010). *T. asperellum* ACCC30536 like other PGPM, producing gene TaACCD was found to enhance tolerance of *Populus* sp. to salt (Zhang et al., 2016). Similar findings were reported for management of salinity with *T. hazianum* in tomato (Mastouri et al., 2012); Mustard (Ahmad et al., 2015); Maize and Rice (Yasmeen and Siddiqui, 2017) and other crops (Hidangmayum and Dwivedi, 2018).

Stress of heavy metals in soils

- *T. asperellum* caused reduction in Cu accumulation and translocation to onion leaves (Téllez-Vargas et al., 2017); *T. atroviride* improved photoextraction efficiency and alleviated cellular toxicity of cadmium and nickel in mustard (Cao et al., 2008) and *T, harzianum* enhanced tolerance to heavy metals, salts, and resistance against *P. syringae* and *R. solani* (Dana et al., 2006).

Stress of extreme temperatures i.e. cold and heat waves

- Plants are unable to scavenge the accumulation ROS under severe stress condition and leads to damages of cellular components (Mitler, 2002). In such cases, *Trichoderma* inoculated plants were able to protect the stresses by increasing the ROS scavenging abilities of the plants. In a study, *T. harzianum* AK 20G strain decresed the lipid peroxidation rate and electrolyte leakage and helped in overall increasing the leaf water content and proline build up, thereby, mitigated the negative effects of chilling stress (Ghorbanpour et al., 2018). Similarly, *T. harzianum* inoculated *Arabidopsis* plants were able to tolerate heat stress by producing heat shock proteins (Montero-Barrientos et al., 2008, 2010).

24.5.2.3 Trichoderma *spp. as crops growth enhancer*

- *Trichoderma* spp. are multifunctional plant symbionts and act as novel agricultural inputs known for enhancing the plant growth, and productivity; besides acting as efficient biocontrol agent for managing various plant diseases (Harman, 2011). *Trichoderma* treated plants have shown enhanced growth and yield in many crops viz., rice, maize, sugarcane (Fig. 24.3), chickpea (Fig. 24.4), Gladiolus (Fig. 24.5) etc. due to production of phytohormons (Chowdappa et al., 2013 Singh et al., 2012), cytokinin like molecules (Zeatin) and gibberellins related molecules (GA3 or GA4) (Tucci et al., 2011; Idowu et al., 2016; Kashyap et al., 2017); higher photosynthetic rate (Doni et al., 2013), increased nutrient uptake, and tolerance to abiotic and biotic stress, etc. which could be helpful in enhancing growth parameters and crop growth. Besides, it also acts as biofertilizer.

Biofertilizer

Trichoderma spp. also serves the plants as bio-fertilizers due to their phosphorus solubilizing and organic matter decomposition activities, resulting in increased supply of micronutrients to the plants (Vinale et al., 2006).

Phosphate solubilization *Trichoderma* spp. has been found to increase the plant growth parameters and biomass of chickpea (Rudresh et al., 2005; Altomare et al., 1999) and showed the soil remediation activity (Baker, 1988; Esposito and da-Silva, 1998). They solubilize tricalcium phosphate (TCP) by organic acid and other mechanisms. *T. viride, T. virens* and *T. harzianum* were found to be efficient in solubilizing TCP, as these at 9–10 μg/mL successfully solubilized 70% TCP in comparison to *Bacillus megaterium* which showed solubilization at 12.43 μg/mL and significantly increased the uptake of P in chickpea.

FIG. 24.3 Effect of *Trichoderma harzianum* (Strain NBRI-1055) on growth of sugarcane (a)-upper plate inoculated and (b)-lower uninoculated control). (c). Profuse rooting in *Trichoderma* inoculated plants of sugarcane can be seen (left) as compared to uniniculated control.

FIG. 24.4 Effect of *Trichoderma harzianum* (strain −NBRI1055) on growth of chickpea. Upper plate uninoculated control and lower inoculated with *Trichoderma harzianum* (strain −NBRI1055).

FIG. 24.5 Effect of *Trichoderma harzianum* (strain-NBRI 1055) inolulation on growth and flowering behavior of Gladiolus. Note: Ten days advanced flowering was observed in inoculated plants (lower plate).

Increase in minerals and yield in crops Application of 100% *Trichoderma* biofortified biofertilizer as BioF/compost in tomato, resulted in increased mineral contents, leaf greenness, enhanced plant growth, and ascorbic acid, β-carotine, and lycopene in fruits and higher yield of 12.9% over recommended doses of treatments. The increased yield and yield parameters in tomatoes was attributed to increased soil fertility and population of microbes in the Rhizosphere of tomato, and suggested to use it in place of chemical fertilizers (Khan et al., 2016).

Decomposer of organic matter

Trichoderma spp. have been found to be great decomposers of organic matter by converting them to nutrient-enriched biocompost, from which micronutrients, NPK, etc. are released and made available to the plants. The decomposition of Teak and Bamboo leaf litter with maximum nutritional compositions and humic acid was obtained with the cowdung slurry @10% and decomposing culture of *Trichoderma* spp. @ 1kg/ton (Wagh and Gangurde, 2015). Similarly, crop residues i.e. trashes of cane, paddy, wheat; press mud cake (PMC) and on farm organic matters were successfully recycled into good quality biofortified compost by utilizing *Trichoderma* spp (Sharma et al., 2012; Saju et al., 2002).

Bioremediation

Trichoderma spp. were found useful in degrading polycylic aromatic hydrocarbons (Azcbn-Aguilar and Barea, 1997); synthetic dyes, pentachlorophenols, endosulphan, DDT (Katayama and Matsumura, 1993) by the enzymes i.e. hydrolyzes, peroxidase, lactases and other lytic enzymes to improve health of soil and plant and production of crop yield. It has also the natural ability to resist the pesticides viz. fungicides, insecticides, herbicides; and phenolic compounds, etc. (Chet et al., 1997). The mechanism of ABC transporter protein systems in *Trichoderma* strains seems to play the role in imparting resistance against toxic compounds, therefore, such strains can be explored and used in the remediation of soil contaminated with pesticides viz. organochlorines, organophosphates, etc. (Harman et al., 2004).

24.5.2.4 Delivery systems for management of stresses and plant growth

Delivery of *Trichoderma* spp. in appropriate and suitable ways to deliver proper colony forming units (CFUs)/gram of the product i.e. carrier materials viz. talc, at the cite of action with convenient delivered methods to users is important consideration to develop the delivery system for the use of bioagents. Several delivery systems have been recommended for the use of *Trichoderma* spp. (Fig. 24.1)

Seed treatment

Seeds of cereals, pulses, oilseeds, vegetables, etc. are treated with commercial formulations of products carrying 2×10^6 cfu/g @ of 5–10 g/kg seed at the time of sowing.

Seed biopriming

Seed biopriming is the recent and beneficial seed treatment with bioagents for the management of the diseases in crop plants (Pandey, 2017). Seed biopriming is the treating of seeds with the bioagents viz. *Trichoderma* sp., followed by incubating the treated seeds under the warm and humid conditions, until just prior to germination of seeds i.e. radical emergence (Ramanujam et al., 2010). In one case the seeds of chickpea were soaked in *T. asperellum* T42 spore (cell) suspensions mixed with 1% Carboxy Methyl Cellulose (CMC) for the duration of 2 h; followed by air drying at room temperature for another 2 h; and then incubated in moist chamber for 24 h (Yadav et al., 2013). In another case the seeds were surface sterilized, dried and soaked in spore suspensions of *Trichoderma* sp. The spore suspension was prepared by collecting the spores in sterilized saline (NaCl 0.85%) water, and filtered with sterilized muslin cloth. The OD value of 1.026 of the suspension which contained 2.26×10^7 spores/mL was ascertained at 600 nm. The pellets obtained were resuspended in same volume of autoclaved 1.5% CMC (Carboxy-methyl-cellulose). Drying of seeds was done aseptically under air in laminar air flow for 2 h. The air dried seeds were kept in the moist chamber at 98% relative humidity and 28–30°C, for 24 h (Singh et al., 2016). However, some other methods of seed bioprimings have also been suggested (Entesari et al., 2013; Pandey, 2017).

Treatment of planting materials

Certain vegetable, crops viz. potato, onion, garlic, ginger, yam, elephant foot, colocasia; floriculture crops, gladiolus, tulips, etc.; medicinal & aromatic plants, fruit crops, sugarcane, etc. are raised by using planting materials viz. tubers, bulbs, corm, cuttings of stem, roots, etc. These seeding materials can be treated by *Trichoderma* formulations alone @ 1 or 1.5 kg per 10 lit. of water, or mixed with 100 g of well decomposed FYM/liter of water by dipping the planting materials for 15 min, then shade dried and used for planting.

Seedlings root dip

The seedlings of rice, tobacco, chilli, at the time of transplanting can be treated by talc based *Trichoderma* formulations @ 1 or 1.5 kg per 10 lit. of water and dipping the roots of seedlings for 15 min, at the time of transplanting.

Nursery soil treatment

Seedlings of many crops stated above are raised under nursary to avoid the risk of loosing them due to abiotic and biotic stresses. Therefore, it is suggested to apply *Trichoderma* enriched FYM (1 kg *Trichoderma* in 100 kg of FYM) in the nursery soils @ 100 g per m^2 at the time of sowing or drenching *Trichoderma* @ 10 g mixed with 100 g of well decomposed FYM per liter of water at the initiation of the disease.

Soil treatment

Apply talc based formulation of *Trichoderma* (1%) @ 5 Kg per hectare in the soil after turning-up of sun hemp or dhaincha grown for green manuring. Or use *Trichoderma* enriched FYM, prepared by thoroughly mixing of 1 kg *Trichoderma* formulation (1%) in 50 kg of well decomposed moist (15%–20%) farmyard manure, followed by covering the heap with polythene and keep for 12–15 days. However, to facilitate good growth of *Trichoderma,* the heap should be turned at 6–8 days and the moisture is maintained by sprinkling the heap with water intermittently.

Plant treatment

Soil drenching of plants near stem or collar region with water suspension of *Trichoderma* 10 g/L of water; or with sand @ 5 Kg/ha can help to manage the stem or collar rot of crops.

Through drip irrigation water

T. harzianum and *T. viride* were found good to manage white mold (*S. sclerotiorum*) of tomato, with three applications @ 1.0 L/ha of formulation having 1×10^9 CFU/mL starting from 30 days after transplanting and at 10 day intervals, through drip irrigation (Aguiar et al., 2014). The studies conducted at some other places have shown encouraging results.

Bee vectoring

Bees have been found useful for pollinations of crops. Similarly, the Honey Bees and Bumble Bees were tested for their efficiency of transmission of the *Trichoderma* and found useful for propagation of *Trichoderma harzianum* in strawberries for the effective management of *Botrytis* (Kovach et al., 2000).

Transgenic plants

Tobacco, pearl millet and potato plants transformed with endochitinase gene from *Trichoderma* has shown the enhanced resistance against the infection of fungal pathogens. Tansgenic lines having such genes were found highly tolerant to foliar pathogens viz. *Alternaria alternata, A. solani, Botrytis cirerea, etc.*; as well as soil-borne pathogen viz. *Rhizoctonia* spp. (O'Kennedy et al., 2011; Kashyap et al., 2011). The sound knowledge of genetic transformation and genetic engineering technique has now made possible to clone and insert desirable genes into the crop plants having good agronomic traights to impart resistance against different biotic and abiotic stresses (Sanghera et al., 2011; Nicolás et al., 2014).

Soilless and hydroponic systems

Trichoderma spp. have been found to perform good in the management of the diseases of crops grown in soil less medium ie. Peat, saw dust, rockwood, sand, etc. *Trichoderma* spp. have effectively managed Wilt of tomato (*Fusarium* sp.) and damping-off of cucumber (*Pythium* sp.); and *T. virens* managed damping-off in crops caused by *Pythium* sp. and *Rhizoctonia* sp. (Paulitz, 1997).

24.5.3 *Trichoderma* spp. in sustainable environment

- Different species of *Trichoderma* has a wide adoptability to various habitats due to their diversified metabolic pathways led by important enzymes viz. cellulases, 1–3 beta glucanases, hydrolyzes, peroxidases, lactases, and secondary metabolites (Harman et al., 2004; Sandhya et al., 2004; Ahamed and Vermette, 2008; Keswani et al., 2014). Because of the good characteristics, these metabolites have been found to play a role in the synthesis of silver nanoparticles (Maliszewska et al., 2009; Vahabi et al., 2011).

24.5.3.1 Mitigating nitrous oxide emissions from cultivated crop fields

Increased nitrous oxide (N_2O) emissions, was recorded in tea field, which was partly due to excessive use of nitrogenous fertilizer. Bioaugmentation of tea field with *T. viride* has helped in mitigation of N_2O from the field (Xu et al., 2014).

24.5.3.2 Wood preservation

Since *Trichoderma* spp. are antagonistic to many fungi, therefore, their utility for wood preservation may be tested. *T. viride* was found efficient to inhibit the growth of decay fungi i.e. *Gloeophyllum* sp. and *G. sepiarium* under field conditions in tropical environment (Ejechi, 1997).

24.5.3.3 Industrial bioreactors

Biofuel are now important source of energy to deal with the global warming effect of environment and ill effect of petroleum on human health. In this area *T. reesei* has been found most efficient in the production of cellulase and act as a study model for fundamental studies on secretion system of proteins (Ahamed and Vermette, 2009; Li et al., 2013). The recent advancement in molecular studies of the mechanism of the cellulose degrading pathway *vis-a-vis* genome sequence of *T. reesei* has paved the way for exploration of novel way of metabolite engineering (Kubicek et al., 2009). *T. reesei* posses the least number of genes accountable for production of enzymes viz, cellulase, responsible for degradation of plant cell wall chiefly made up of cellulose (Martinez et al., 2008). The knowledge can be used to develope a refined and improved process of manufacturing of biofuels like methanol-an alternative of fossil fuels, by using agricultural wastes, and cellulases and hemicellulases produced by *T. reesei* or other important isolates, followed by additional fermentation by other microbes like yeast (Schuster and Schmoll, 2010).

24.5.4 Commercialization of *Trichoderma* spp.

Procedures for commercialization includes: Isolation of native efficient strains from the natural ecosystem; Identification of the *Trichoderma* spp.; Testing the bioefficacy of the bioagents against the fungal pathogens under lab. i.e. antagonism, disease management under pot culture and field experiments; Sensitivity against agro-chemicals; Standardization of solid and liquid state Mass production; Develop Formulations for better shelf-life; Standardization of different Delivery systems for better efficiency; Multilocation testing for bioefficacy of disease management and Toxicological Data generation for registration; Registration of the bioagents with complete dossier with the respective Environment Protection Agencies (EPAs) and to get the license from the respective agencies for the sale to the farmers.

24.5.4.1 Isolation

Trichoderma spp. can be isolated from root surface i.e. rhizosphere and rhizoplane soils, decomposed organic matter, decaying tree barks; and propagules like sclerotia of other fungi. *Trichoderma* spp. can make use of an extensive array of compounds as sole carbon and nitrogen sources.

24.5.4.2 Sensitivity against agro-chemicals

- The pesticides viz. fungicides, insecticides, herbicides, etc. have been found to affect the growth and efficiency of *Trichoderma,* when used concurrently during crop production. *Trichoderma* strains were tested for the sensitivity and tolerance to the pesticides by different workers (Sawant and Mukhopadhyay, 1990; Pandey and Upadhyay, 1998; Sharma et al., 2001; Nallathambi et al., 2001; Sushir and Pandey, 2001; Bhatt and Sabalpara, 2001; Patibanda et al., 2002; Lal and Maharshi, 2007; Madhusudan et al., 2010). In general *Trichoderma* has shown greater tolerance to the insecticides, nematicides, herbicides, in comparison to some fungicides.
- However, it has the ability to overcome the toxicity rapidly (Mercy, 2004) and quickly colonizes the pesticides treated soils (Oros et al., 2011).

24.5.4.3 Standardization of mass production of Trichoderma spp. in solid and liquid state media

In mass multiplication of *Trichoderma* spp., on readily available inexpensive agricultural produce like grains, stem, husks, cakes, etc., the protocols needs to be developed as refined technology of higher spores and mycelia production. The level of carbon, nitrogen and minerals can also be artificially balanced to give maximum yield of spores, superior shelf life and high percentage of viable spores. Conidia and chlamydospores are used as spore biomass in making *Trichoderma* based products (Harman et al., 1991; Jin et al., 1991; Eyal et al., 1997).

Solid state fermentation

A broad array of substrates like cereals grains viz. rice (broken), millets, ragi, sorghum, etc. (Jeyarajan, 2006; Upadhyay and Mukhopadhyay, 1986); Agricultural waste i.e. bagasse of sugarcane; trashes of sugarcane and groundnut; straw of rice; husks of rice, wheat, bajra; wheat bran, etc. (Jagadeesh and Geeta, 1994; Tewari and Bhanu, 2003; Tewari and Bhanu, 2004), Banana pseudostem; dry weeds, banana and mango leaves, Chickpea husk, etc.(Balasubramanian et al., 2008); Organic matter i.e. Well decomposed FYM; Seasoned press mud; Poultry manure (Tewari and Bhanu, 2004; Kousalya and Jeyarajan, 1990), Cakes of Neem, Mustard, Groundnut and Caster (Rini and Sulochana, 2007); Fresh cow dung, Vermicompost (Pramod and Palakshappa, 2009); etc. have been found useful for mass production of *Trichoderma* spp. The detail protocols of mass production of the bioagents can be obtained from these workers.

Liquid state fermentation

A wide range of substrates like Molasses and Brewers yeast, Czapeck's Dox Broth and V8 Broth (Harman et al., 1991); Potato Dextrose Broth, V8 juice and molasses yeast medium (Prasad et al., 2002); etc. have been found to support good production of *Trichoderma* spp.

Substrate/carrier materials for solid formulationns

Talc, Peat, Multani soil, Kaolin (Aliminum silicate), Bentonite (Montmorillonite), etc. can be used as carrier of the bioagents.

24.5.4.4 Formulations of Trichoderma spp.

Different formulations i.e. Dusts (Woods, 2003); Wettable powders (Tadros, 2005; Brar et al., 2006; Knowles, 2008); Granules (Tadros, 2005) (Lyn et al., 2010); Water dispersible granules (Knowles, 2008); Capsule suspension (Brar et al., 2006) nanoemulsion, nanosuspension, nano capsule suspension, (Glare et al., 2012; Slavica and Tanović, 2013); Oil-based emulsifiable suspension (Brar et al., 2006); Whole culture; Aqueous suspension; Oil-flowable; Shellac latex formulation, etc. have been developed.

Carriers based formulations of *Trichoderma* spp. i.e

Talc based formulation (Jeyarajan et al., 1994); Lignite based formulations; Vermiculite-wheat bran based formulation (Lewis, 1991); Pesta granules based formulation (Connick et al., 1991); Alginate prills based formulation (Fravel et al., 1999); Press mud based formulation; Coffee husk based formulation (Sawant and Sawant, 1996); Oil-based formulations (Batta, 2005) have been prepared for use of the farmers.

Out of these, Talc based formulation being produced in large quantities in India for supply to the farmers. *Trichoderma* grown on potato dextrose broth or any other media is harvested with in short period of time (6–7 days). Spores and mycelial biomass along with broth @ 150 g/1000 mL is homogenized with the help of Homogenizer or Mixer and mixed with sterilized talc powder in the ratio of 1: 10; i.e. 100 mL homogenized spore and mycelial biomass in broth is mixed in 1000 g talc powder; and 10%–12% moisture is maintained. The 1.5% formulation of *Trichoderma* with CFUs ranging from $10^6 - 10^{10}$ is prepared and packed in polythene bags of 250 g, 500 g, 1000 g, capacity (Woo et al., 2014). It has shelf life of 5–6 months.

Liquid formulations

Liquid formulations viz. Water-based, Oil-based, Polymer-based, or based on these combinations, have been developed. Water-based formulations i.e. Suspension concentrates, Suspo-emulsions, Capsule suspension, etc., are adding with inert ingredients, such as stabilizers, stickers, surfactants, coloring agents, anti-freeze compounds, and additional nutrients.

- Globally, about 1400 biopesticide products are being sold by different agencies. Among the countries USA consumes about 40% of biopesticides of the global production, followed by Europe and Oceanic Countries, consuming 20% each. A number of commercial companies, IPM centers and some of the SAU's produce biocontrol (Rabindra, 2005; Singh et al., 2012).
- Bio-formulations must carry the required number of colony forming units (cfu), broad spectrum of action, and effective in controlling the pathogens; should possess longer shelf-life; tolerance to wide range of temperature; UV radiations, desiccation, oxidizing agents, (Jeyarajan and Nakkeeran, 2000); competent to colonize rhizosphere; abled saprophytic ability and potentials to promote and enhanced plant growth; Safe for plant, animal and human being; economical to develop and cost effective while application; easy to apply; etc.

24.5.5 Advantages, challenges, constaints in sustainability of *Trichoderma* based disease management technology and future course of action

As stated above, there are many advantages of using *Trichoderma* spp. for plant disease management and achieving plant health. It is easy to adopt and apply, cost effective, help in reduction of input cost of fertilizers and pesticides and can help in achieving food security all over the world (Harman, 2011; Shoresh et al., 2010).

However, despite the promising impacts of this technology in agriculture, the biopesticide industry is somehow not growing at the desired momentum. Therefore, some of the challenges which need to be addressed are given below:

- Poor quality of bioagents i.e. lack of prolonged shelf-life and desired CFU count of bioactive ingradient (10^8/g) in commercially available formulations, higher costs of formulations, short spore viability due to poor transportation and storage facilities.
- Maintaining a "Gold Standard" of the formulations to explore full potential of the bioagents.
- Issuing instruction to production units for compliance of standard quality production in handling, processing and packaging to ensure to national standards and guidelines and adhering to Good Laboratory Practices/Good Manufacturing Practices (GL/MPs).
- Quality standard testing (pH, moisture content, particle size, carrier material, shelf life packaging material and weight, purity, labeling of products and seed adhesion of the finished product) of commercial formulated products.
- There should be stringent technical inspection by EPA's before granting registration for production of bioagents and random inspection of manufacturing sites.
- Strict vigilance on Quality Assurance/Quality Control by sampling of the bioagents from different production units at quarterly intervals. Criteria for checking the entry of spurious products in the market.
- Molecular diagnosis of the purity of strains used in formulation by using ITS and specific primers for accurate identification of the strain.
- *Trichoderma* colonize host plants without showing any visible symptoms, suggesting their involvement in multipartite interactions with hosts, which needs to be investigated to understand the metabolic pathways and signaling functions. Studies on interactions between the endophytes, host plant and pathogen should be done for exploring better management practices for plant diseases.

- There is a need for novel endophytic *Trichoderma* in India, with competence in production of novel antimicrobial compounds to deal with the pathogenic and abiotic stresses. India and other developing country must enter into this new area of research for economical, sustainable, and ecofriendly control of crop diseases caused by various plant pathogens, soil and environmental factors.
- Studies on effect of biopriming of seeds with microbial consortia for strengthening the resistance in plants against biotic/abiotic stresses and enhancing soil supressiveness needs to be done.
- The factors playing key roles on performance of the bioagents under different environmental conditions and its rhizospheric competence and establishment in soil and acclimatize to environment (Viterbo et al., 2002; Bénitez et al., 2004; Harman et al., 2004) needs to be studied.
- The endemic gene pool of *Trichoderma* should be identified as it may contain equally competent or superior strain as to exotic strains due to specific influence on plant physiological and biochemical processes (Cho and Tiedji, 2000; Nawrocka and Małolepsza, 2013).
- Compatibility of the bioagents with pesticides and other soil microbes should be studied.
- Consortium is a formulation comprising of a group of compatible microbes capable of suppressing the growth of pathogens and promoting plant growth by mineralization of nutrients for sustainable crop health management. Therefore, useful consortia formulations comprising of 1–4 microbes should be developed as in case of Indonesian Institute of Sciences (LIPI), Indonesia, who has developed a microbial cocktail using six microbes viz. *Bacillus* sp., *Pseudomonas* sp., *Burkholderia* sp., *Bervundimonas* sp., *Brevibacillus* sp., *Ochrobactrum* sp. in one formulation.
- Seed industries need to be sensitized about potential use of *Trichoderma* for seed treatment.
- High initial inputs cost drive a burden to adopt biopesticides manufacturing, which needs to be subsidized by the agencies to the entreprneurs and farmers.
- Transparency in procurement policy of *Trichoderma* for farmers needs to be ascreated.
- Lack of skilled personnels for establishing production unit is an important issue for developing entrepreneurship for commercial production of biocontrol agents, which needs to be addressed.
- Non-availability of GLP laboratories for quality certification of bio-products is a matter of concern, which should be created.
- Lack of awareness about the benefits of *Trichoderma* and practical experience for its use among the farmers and extension personnel are the bottle neck for adoption of the technology, which needs to be addressed.
- Registration guidelines for microbial formulations with multiple biocontrol agents/consortium having multi-trait properties are not framed, which needs to be taken up.

Presently process of registration of efficient native strains of *Trichoderma* is cumbersome and costly, which needs to be simplified and financial help need to be provided by the government or any other suitable agencies.

References

Abbas, A., Jiang, D., Fu, Y., 2017. *Trichoderma* spp. as antagonist of *Rhizoctonia solani*. J. Plant Pathol. Microbiol. 8 (3), 402–409.

Abdollahzadeh, J., Mohammadi, E., Goltapeh, Rouhani, H., 2006. Biological control of Sclerotinia stem rot (*S. minor*) of sunflower using *Trichoderma* species. Plant Pathol. J. 5, 228–232.

Aggarwal, R., Srivastava, K.D., Singh, D.V., Bahadur, P., Nagarajan, S., 1991. Possible biocontrol of loose smut of wheat. J. Bio Contr. 6, 114–115.

Aggarwal, R., Singh, D.V., Srivastava, K.D., Bahadur, P., 1996. The potential of antagonists for biocontrol of *Neovossia indica* causing Karnal bunt of wheat. Indian J. Biol. Contr. 9, 69–70.

Agrios, G.N., 2005. Plant Pathology, fifth ed. Elsevier Academic Press, London, UK, p. 922.

Aguiar, R.A., CunhaM, G., Murillo Jr., L., 2014. Management of white mold in processing tomatoes by *Trichoderma* spp. and chemical fungicides applied by drip irrigation. Biol. Contr. 74, 1–5.

Ahamed, A., Vermette, P., 2008. Culture-based strategies to enhance cellulase enzyme production from *Trichoderma reesei* RUT-C30 in bioreactor culture conditions. Biochem. Eng. J. 40, 399–407.

Ahamed, A., Vermette, P., 2009. Effect of culture medium composition on *Trichoderma reesei's* morphology and cellulase production. Bioresour. Technol. 100, 5979–5987.

Ahmad, P., Abeer, H., Elsayed, F.A.A., Alqarawi, A.A., Riffat, J., Dilfuza, E., et al., 2015. Role of *Trichoderma harzianum* in mitigating NaCl stress in Indian mustard (*Brassica juncea* L.) through antioxidative defense system. Front. Plant Sci. 6, 868.

Ahmed, A.S., Sanchez, C.P., Candela, M.E., 2000. Evaluation of induction of systemic resistance in pepper plants (*Capsicum annuum*) to *Phytopthora capsici* using *Trichoderma harzianum* and its relation with capsidiol accumulation. Eur. J. Plant Pathol. 106, 817–824.

Alabouvette, C., Olivain, C., Migheli, Q., Steinberg, C., 2009. Microbiological control of soil-borne phytopathogenic fungi with special emphasis on wilt-inducing *Fusarium oxysporum*. New Phytol. 184, 529–544.

Alfano, G., Ivey, M.L., Cakir, C., Bos, J.I., Miller, S.A., Madden, L.V., Kamoun, S., Hoitink, H.A.J., 2007. Systemic modulation of gene expression in tomato by *Trichoderma hamatum* 382. Phytopathology 97, 429–437.

Alizadeh, H., Behboudi, K., Ahmadzadeh, M., Javan-Nikkhah, M., Zamioudis, C., Pieterse, C.M., Bakker, P.A., 2013. Induced systemic resistance in cucumber and *Arabidopsis thaliana* by the combination of *Trichoderma harzianum* Tr6 and *Pseudomonas* sp. Ps14. Biol. Contr. 65 (1), 14–23.

Altomare, C., Norvell, W.A., Bjorkman, T., Harman, G.E., 1999. Solubilization of phosphates and micronutrients by the plant growth promoting and biocontrol fungus *Trichoderma harzianum* Rifai. Appl. Environ. Microbiol. 65, 2926–2933.

Anand, S., Reddy, J., 2009. Biocontrol potential of *Trichoderma* sp. against plant pathogens. Int. J. Agric. Sci. 1, 30–39.

Arisan-Atac, I., Heidenreich, E., Kubicek, C.P., 1995. Randomly amplified polymorphic DNA fingerprinting identifies subgroups of *Trichoderma viride* and other *Trichoderma* sp. capable of chestnut blight biocontrol. FEMS Microbiol. Lett. 126, 249–256.

Atanasova, L., Druzhinina, I.S., Jaklitsch, W.M., 2013. Two hundred *Trichoderma* species recognized on the basis of molecular phylogeny. In: *Trichoderma*: Biology and Applications. CABI, Wallingford, pp. 10–42.

Azcbn-Aguilar, C., Barea, J.M., 1997. Applying mycorrhiza biotechnology to horticulture: significance and potentials. Sci. Hortic. 68, 1–24.

Bae, H., Sicher, R.C., Kim, M.S., Kim, S.H., Strem, M.D., Melnice, R.L., Bailey, B.A., 2009. The beneficial endophyte *Trichoderma hamatum* isolate DIS 219b promotes growth and delays the onset of drought response in *Theobrama cacao*. J. Exp. Bot. 60, 3279–3295.

Bae, H., Roberts, D.P., Lim, H.S., Strem, M., Park, S.C., Ryu, C.M., 2011. Endophytic *Trichoderma* isolates from tropical environments delay disease and induce resistance against *Phytophthora capsici* in hot pepper using multiple mechanisms. Mol. Plant Microbe Interact. 24, 336–351.

Baker, R., 1988. *Trichoderma* spp. as plant-growth stimulants. CRC Crit. Rev. Biotechnol. 7 (2), 97–106.

Balasubramanian, C., Udaysoorian, P., Prabhu, C., Kumar, G.S., 2008. Enriched compost for yield and quality enhancement in sugarcane. J. Ecobiol. 22, 173–176.

Batta, Y.A., 2005. Postharvest biological control of apple grey mold by *Trichoderma harzianum* Rifai formulated in an invert emulsion. Crop Protect. 23, 19–26.

Benhamou, N., Chet, I., 1997. Cellular and molecular mechanisms involved in the intersection between *Trichoderma harzianum* and *Pythium ultimum*. Appl. Environ. Microbiol. 63, 2095–2099.

Benítez, T., Rincón, A.M., Limón, M.C., Codón, A.C., 2004. Biocontrol mechanisms of *Trichoderma* strains. Int. Microbiol. 7, 249–260.

Bhat, K.A., Ali, A., Wani, A.H., 2009. Evaluation of biocontrol agents against *Rhizoctonia solani* Kuhn and sheath blight disease of rice under temperate ecology. Plant Dis. Res. 24 (1), 15–18.

Bhatt, T.K., Sabalpara, A.N., 2001. Sensitivity of some bio-inoculants to pesticides. J. Mycol. Plant Pathol. 31, 114–115.

Bigirimana, J., Meyer, G de, Poppe, J., Elad, Y., Hofte, M., 1997. Induction of systemic resistance on bean (*Phaseolus vulgaris*) by *Trichoderma harzianum*. Meded. Fac. Landbouwwet. Univ. Gent 62, 1001–1007.

Bissett, J., 1984. A revision of the genus *Trichoderma*. I. Sect. *Longibrachiatum* sect. nov. Can. J. Bot. 62, 924–931.

Bissett, J., 1991. A revision of the genus *Trichoderma*. II. Infrageneric classification. Can. J. Bot. 69, 2357–2372.

Bissett, J., 1991a. A revision of the genus *Trichoderma*. III. Sect. *Pachybasium*. Can. J. Bot. 69, 2373–2417.

Bissett, J., 1991b. A revision of the genus *Trichoderma*. IV. Additional notes on section *Longibrachiatum*. Can. J. Bot. 69, 2418–2420.

Bochner, B.R., Gadzinski, P., Panomitros, E., 2001. Phenotype microarrays for high-throughput phenotypic testing and assay of gene function. Genome Res. 11, 1246–1255.

Brar, S.K., Verma, M., Tyagi, R.D., Valero, J.R., 2006. Recent advances in downstream processing and formulations of *Bacillus thuringiensis* based biopesticides. Process Biochem. 41 (2), 323–342.

Bubici, G., Kaushal, M., Prigigallo, M.I., Gómez-Lama Cabanás, C., Mercado-Blanco, J., 2019. Biological control agents against *Fusarium* wilt of banana. Front. Microbiol. 10, 616.

Bull, C.T., Shetty, K.G., Subbarao, K.V., 2002. Interactions between Myxobacteria, plant pathogenic fungi, and biocontrol agents. Plant Dis. 86, 889–896.

Cao, L., Jiang, M., Zeng, Z., Du, A., Tan, H., Liu, Y., 2008. *Trichoderma atroviride* F6 improves phytoextraction efficiency of mustard (*Brassica juncea* (L.) Coss.var. *foliosa Bailey*) in Cd, Ni contaminated soils. Chemosphere 71 (9), 1769–1773.

Chaverri, P., Samuels, G.J., 2003. *Hypocrea/trichoderma* (Ascomycota, Hypocreales, Hypocreaceae): species with green ascospores. Stud. Mycol. 48, 1–116.

Chaverri, P., Samuels, G.J., 2013. Evolution of habitat preference and nutrition mode in a cosmopolitan fungal genus with evidence of inter kingdom host jumps and major shifts in ecology. Evolution 67, 2823–2837.

Chaverri, P., Catlebury, L.A., Samuels, G.J., Geiser, M.D., 2003. Multilocus phylogenetic structure within the *Trichoderma harzianum/Hypocrea lixii* complex. Mol. Phylogenet. Evol. **27**, 302–313.

Chaverri, P., Gazis, R., Samuels, G.J., 2011. *Trichoderma amazonicum*, a new endophytic species on *Hevea brasiliensis* and *guianensis* from the Amazon basin. Mycologia 103, 139–151.

Chen, X., Romaine, C.P., Tan, Q., Schlagnhaufer, B., Ospina-Giraldo, M.D., Royse, D.J., Huff, D.R., 1999. PCR-based genotyping of epidemic and pre-epidemic *Trichoderma* isolates associated chestnut blight biocontrol. FEMS Microbiol. Lett. 126, 249–256.

Chen, J.-L., Sun, S.-Z., Miao, C.-P., Wu, K., Chen, Y.-W., Xu, L.-H., Guan, H.-L., Zhao, L.-X., October 2016. Endophytic *Trichoderma gamsii* YIM PH30019: a promising biocontrol agent with hyperosmolar, mycoparasitism, and antagonistic activities of induced volatile organic compounds on root-rot pathogenic fungi of *Panax notoginseng*. J. Ginseng Res. 40 (4), 315–324.

Chet, I., 1987. *Trichoderma*-application, mode of action, and potential as a biocontrol agent of soil-born pathogenetic fungi. In: Chet, I. (Ed.), Innovative Approaches to Plant Disease Control. John Wiley and Sons, pp. 137–160.

Chet, I., Inbar, J., 1994. Biological control of fungal pathogens. Appl. Biochem. Biotechnol. 48, 37–43.

Chet, I., Inbar, J., Hadar, I., 1997. Fungal antagonists and mycoparasites. In: Wicklow, D.T., Söderström, B. (Eds.), The Mycota IV: Environmental and Microbial Relationships. Springer-Verlag, Berlin, pp. 165–184.

Cho, J.-C., Tiedji, J.M., 2000. Biogeography and degree of endemicity of *Fluorescent pseudomonas* strains in soil. Appl. Environ. Microbial. 66, 5446–5448.

Chou, C., NancyCastilla, B.H., Tanaka, T., Chib, S., Sato, I., 2019. Rice blast management in Cambodian rice fields using *Trichoderma harzianum* and a resistant variety. Crop Protect. 104864. Available Online 28 June 2019.

Chowdappa, P., Mohan Kumar, S.P., Jyothi Lakshmi, M., Upreti, K.K., 2013. Growth stimulation and induction of systemic resistance in tomato against early and late blight by *Bacillus subtilis* OTPB1 or *Trichoderma harzianum* OTPB3. Biol. Contr. 65 (1), 109–117.

Christian, R., Röhrich, W.M., Jaklitsch, H., Voglmayr, A., Iversen, C.Z., Christian, B., Henry, M., Meincke1, R., Komon-Zelazowska, M., Druzhinina, I., Christian, S., Kubicek, P., Berg, G., 2009. Fungal diversity in the Rhizosphere of endemic plant species of Tenerife (Canary Islands): relationship to vegetation zones and environmental factors. ISME J. 3, 79–92.

Connick, W., Daigle, D., Quimby, P., 1991. An improved invert emulsion with high water retention for mycoherbicide delivery. Weed Technol. 5, 442–444.

Contreras-Cornejo, H.A., Macias-Rodriguez, L., Beltran-Pena, E., Herrera-Estrella, A., Lopez-Bucio, J., 2011. *Trichoderma*-induced plant immunity likely involves both hormonal and camalexin dependent mechanisms in *Arabidopsis thaliana* and confers resistance against necrotrophic fungi *Botrytis cinerea*. Plant Signal. Behav. 6, 1554–1563.

Dagurere, Y., Siegel, K., Edel-Hermann, V., Steinberg, C., 2014. Fungal proteins and genes associated with biocontrol mechanism of soil borne pathogens: a review. Fungal Biol. Rev. 28, 97–125.

Dana, M.M., Pintor-Toro, J.A., Cubero, B., 2006. Transgenic tobacco plants overexpressing chitinases of fungal origin show enhanced resistance to biotic and abiotic stress agents. Plant Physiol. 142, 722–730.

De Bellis, T., Kernaghan, G., Widden, P., 2007. Plant community influences on soil microfungal assemblages in boreal mixed-wood forests. Mycologia 99 (3), 356–367.

De Meyer, G., Bigirimana, J., Elad, Y., Hofte, M., 1998. Induced systemic resistance in *Trichoderma harzianum* T39 biocontrol of *Botrytis cinerea*. Eur. J. Plant Pathol. 104, 279–286.

Djonovic, S., Pozo, M.J., Dangott, L.J., Howell, C.R., Kenerley, C.M., 2006. Sm1, a proteinaceous elicitor secreted by the biocontrol fungus *Trichoderma virens* induces plant defense responses and systemic resistance. Mol. Plant Microbe Interact. 19, 838–853.

Djonovic, S., Vargas, W.A., Kolomiets, M.V., Horndeski, M., Weist, A., Kenerley, C.M., 2007. A proteinaceous elicitor Sm1 from the beneficial fungus *Trichoderma virens* is required for systemic resistance in maize. Plant Physiol. 145, 875–889.

Domingues, F.C., Queiroz, J.A., Cobral, J.M.C., Fonceca, L.P., 2000. The influence of culture conditions on mycelial structure and cellulose production by *Trichoderma reesei* Rut C-30. Enzyme Microb. Technol. 26, 394–401.

Doni, F., Al-Shorgani, N.K.N., Tibin, E.M.M., Abuelhassan, N.N., Anizan, I., Che-Radziah, C.M.Z., 2013. Microbial involvement in growth of paddy. Curr. Res. J. Biol. Sci. 5 (6), 285–290.

Druzhinina, I.S., Kopchinskiy, A.G., Komoń, M., Bissett, J., Szakacs, G., Kubicek, C.P., 2005. An oligonucleotide barcode for species identification in *Trichoderma* and *Hypocrea*. Fungal Genet. Biol. 42, 813–928.

Druzhinina, I.S., Seidl-Seiboth, V., Herrera-Estrella, A., Horwitz, B.A., Kenerley, C.M., Monte, E., Mukherjee, P.K., Zeilinger, S., Grigoriev, I.V., Kubicek, C.P., 2011. *Trichoderma*: the genomics of opportunistic success. Nat. Rev. Microbiol. 16, 749–759.

Dubey, S.C., Patel, B., 2002. Mass multiplication of antagonists and standardization of effective dose for management of web blight of urd and mung bean. Indian Phytopathol. 55, 338–341.

Dubey, S.C., Bhavani, R., Singh, B., 2009. Development of Pusa 5SD for seed dressing and Pusa Biopellet 10G for soil application formulations of *Trichoderam harzianum* and their evaluation for integrated management of dry root rot of mungbean (*Vigna radiata*). Biol. Contr. 50, 231–242.

Dubey, S.C., Tripathi, A., Singh, B., 2012. Combination of soil application and seed treatment formulations of *Trichoderma* species for integrated management of wet root rot caused by *Rhizoctonia solani* in chickpea (*Cicer arietinum*). Indian J. Agric. Sci. 82 (4), 357–364.

Dubey, S.C., Tripathi, A., Singh, B., 2013. Integrated management of *Fusarium* wilt by combined soil application and seed dressing formulations of *Trichoderma* species to increase grain yield of chickpea. Int. J. Pest Manag. 59 (1), 47–54.

Ejechi, B.O., 1997. Biological control of wood decay in an open tropical environment with *Penicillium* spp. and *Trichoderma viride*. Int. Biodeterior. Biodegrad. 39, 295–299.

El_Komy, M.H., Saleh, A.A., Eranthodi, A., Molan, Y.Y., 2015. Characterization of novel *Trichoderma asperellum* isolates to select effective biocontrol agents against tomato *Fusarium* wilt. Plant Pathol. J. 31 (1), 50–60.

Elad, Y., Barak, R., Chet, I., 1984. Parasitism of sclerotia of *Sclerotium rolfsii* by *Trichoderma harzianum*. Soil Biol. Biochem. 16, 381–386.

Elsharkawy, M.M., Shimizu, M., Hideki, T., Kouichi, O., Mitsuro, H., 2013. Induction of systemic resistance against cucumber mosaic virus in *Arabidopsis thaliana* by *Trichoderma asperellum* SKT-1. Plant Pathol. J. 29, 193–200.

Enkerly, J., Felix, G., Boller, T., 1999. The enzymatic activity of fungal xylanase is not necessary for its elicitor activity. Plant Physiol. 121, 391–398.

Entesari, M., Sharifzadeh, F., Ahmadzadeh, M., Farhangfar, M., 2013. Seed biopriming with *Trichoderma* species and *Pseudomonas* fluorescent on growth parameters, enzymes activity and nutritional status of soybean. Int. J. Agron. Plant Prod. 4 (4), 610–619.

Esposito, E., da-Silva, M., 1998. Systematics and environmental application of the genus *Trichoderma*. Crit. Rev. Microbiol. 24, 89–98.

Eyal, J.C.P., Baker, J.D., Reeder, W., Devane, E., Lumsden, R.D., 1997. Large scale production of chlamydospores of *Gliocladium virens* strain GL 21 in submerge culture. J. Ind. Microbiol. Biotechnol. 19, 163–168.

Falah Kuchlan, P., Kuchlan, M.K., Ansari, M.M., 2018. Efficient application of *Trichoderma viride* on Soybean [Glycine Max (L.) Merrill] seed using thin layer polymer coating. Legume Res. https://doi.org/10.18805/LR-3834.

FAO, 2017. The Future of Food and Agriculture – Trends and Challenges, p. 163. Rome.

Fravel, D.R., Rhodes, D.'J., Larkin, R.P., 1999. Production and commercialization of biocontrol products. In: Albajes, R., Lodovica Gullino, M., Van Lenteren, J.C., Elad, Y. (Eds.), Integrated Pest and Disease Management in Greenhouse Crops. Kluwer Academic Publishers, Boston, pp. 365–376.

Gajera, H., Domadiya, R., Patel, S., Kapopara, M., Golakiya, B., 2013. Molecular mechanism of *Trichoderma* as bio-control agents against phytopathogen system – a review. Curr. Res. Microbiol. Biotechnol. 1, 133–142.

Gal-Hemed, I., Atanasova, L., Komon-Zelazowska, M., Druzhinina, I.S., Viterbo, A., Yarden, O., 2011. Marine isolates of *Trichoderma* spp. as potential halotolerant agents of biological control for arid-zone agriculture. Appl. Environ. Microbiol. 77, 5100–5109.

Gams, W., Meyer, W., 1998. What exactly is *Trichoderma harzianum* Rifai? Mycologia 90, 904–915.

Gazis, R., Chaverri, P., 2010. Diversity of fungal endophytes in leaves and stems of rubber trees (*Hevea brasiliensis*) in Tambopata, Peru. Fungal Ecol. 4, 94–102.

Gherbawy, Y., Druzhinina, I., Shaban, G.M., Wuczkowsky, M., Yaser, M., El- Naghy, M.A., Prillinger, H.J., Kubicek, C.P., 2004. *Trichoderma* populations from alkaline agricultural soil in the Nile valley, Egypt, consist of only two species. Mycol. Prog. 3, 211–218.

Ghorbanpour, A., Salimi, A., Ghanbary, M.A.T., Pirdashti, H., Dehestani, A., 2018. The effect of *Trichoderma harzianum* in mitigating low temperature stress in tomato (*Solanum lycopersicum* L.) plants. Sci. Hortic. 230, 134–141.

Glare, T., Caradus, J., Gelernter, W., Jackson, T., Keyhani, N., Köhl, J., Stewart, A., 2012. Have biopesticides come of age. Trends Biotechnol. 30 (5), 250–258.

Gomathinayagam, S., Rekha, M., Murugan, S.S., Jagessar, J.C., 2010. The biological control of paddy disease brown spot (*Bipolaris oryzae*) by using *Trichoderma viride* in vitro condition. J. Biopestic. 3 (1), 93–95.

Grondona, I., Hermosa, M.R., Tejada, M., Gomis, M.D., Mateos, P.F., Bridge, P.D., Monte, E., Garcõa-Acha, I., 1997. Physiological and biochemical characterization of *Trichoderma harzianum*, a biological control agent against soilborne fungal plant pathogens. Appl. Environ. Microbiol. 63, 3189–3198.

Hanada, R.E., de Jorge Souza, T., Pomella, A.W., Hebbar, K.P., Pereira, J.O., Ismaiel, A., Samuels, G.J., 2008. *Trichoderma martiale* sp. nov., a new endophyte from sapwood of *Theobroma cacao* with a potential for biological control. Mycol. Res. 112 (Pt 11), 1335–1343, 10.1016.

Harman, G.E., 2000. Myths and dogmas of biocontrol. Changes in perceptions derived from research on *Trichoderma harzianum* T-22. Plant Dis. 84, 377–393.

Harman, G.E., 2011. Multifunctional fungal plant symbionts: new tools to enhance plant growth and productivity. New Phytol. 189 (3), 647–649.

Harman, G.E., 2011. *Trichoderma*- not just for biocontrol anymore. Phytoparasitica 39, 103–108.

Harman, G.E., Jin, X., Stasz, T.E., Peruzzotti, G., Leopold, A.C., Taylor, A.G., 1991. Production of conidial biomass of *T. harzianum* for biological control. Biol. Contr. 1, 23–28.

Harman, G.E., Howell, C.R., Viterbo, A., Chet, I., Lorito, M., 2004. *Trichoderma* species – opportunistic, avirulent plant symbionts. Nat. Rev. Microbiol. 2, 43–56.

Harman, G.E., Herrera-Estrella, A.H., Benjamin, A., Matteo, L., 2012. Special issue: *Trichoderma* – from basic biology to biotechnology. Microbiology 58, 1–2.

Hashem, A., Abd_Allah, E.F., Alqarawi, A.A., Al Huqail, A.A., Egamberdieva, D., 2014. Alleviation of abiotic salt stress in *Ochradenusbaccatus* (Del.) by *Trichodermahamatum* (bonord.) bainier,. J. Plant Interact. 9 (1), 857–868.

Hatvani, L., Antal, Z., Manczinger, L., Szekeres, A., Druzhinina, I.S., Kubicek, C.P., Nagy, A., Nagy, E., Vagvolgyi, C., Kredics, L., 2007. Green mold diseases of *Agaricus* and *Pleurotus* spp. are caused by related but phylogenetically different *Trichoderma* species. Phytopathology 97, 532–537.

He, P., Chintamanani, S., Chen, Z., Zhu, L., Kunkel, B.N., Alfano, J.R., Tang, X., Zhou, J.M., 2004. Activation of a COI1-dependent athway in *Arabidopsis* by *Pseudomonas syringae* type III effectors and coronatine. Plant J. 37, 589–602.

Heiko, M., Zimmermann, M.R., Felle, H.H., Boland, W., Mithöfer, A., 2010. Alamethicin-induced electrical long distance signaling in plants. Plant Signal. Behav. 5 (8), 988–990.

Hermosa, M.R., Grondona, I., Iturriaga, E.A., Dõaz-Mõnguez, J.M., Castro, C., Monte, E., Garcõa-Acha, I., 2000. Molecular characterization and identification of biocontrol isolates of *Trichoderma* spp. Appl. Environ. Microbiol. 66, 1890–1898.

Hermosa, M.R., Grondona, I., Dõaz-Mõnguez, J.M., Iturriaga, E.A., Monte, E., 2001. Development of a strain-specific SCAR marker for the detection of *Trichoderma atroviride* 11, a biological control agent against soilborne fungal plant pathogens. Curr. Genet. 38, 343–350.

Hermosa, R., Botella, L., Keck, E., Jiménez, J.A., Montero-Barrientos, M., Arbona, V., Gómez-Cadenas, A., Monte, E., Nicolás, C., 2011. The overexpression in *Arabidopsis thaliana* of a *Trichoderma harzianum* gene that modulates glucosidase activity, and enhances tolerance to salt and osmotic stresses. J. Plant Physiol. 168, 1295–1302.

Herrera-Estrella, A., Chet, I., 2004. The biological control agent *Trichoderma*: from fundamentals to applications. In: Arora, D. (Ed.), Handbook of Fungal Biotechnology, vol. 2. Dekker, New York, pp. 147–156.

Hicks, E., Damian, B., Braithwaite, M., Mclean, K., Falloon, R., Stewart, A., 2014. *Trichoderma* strains suppress Rhizoctonia diseases and promote growth of potato. Phytopathol. Mediterr. 53 (3), 502–514.

Hidangmayum, A., Dwivedi, P., 2018. Plant esponses to *Trichoderma* spp. and their tolerance to abiotic stresses: a review. J. Pharmacogn. Phytochem. 7 (1), 758–766.

Hossain, M.H., Hossain, I., 2014. Evaluation of three botanicals, bavistin and BAU-biofungicide for controlling Leaf spot of groundnut caused by *Cercospora arachidicola* and *Cercosporidium personatum*. The Agriculturists 12 (1), 41–49.

Howell, C.R., 1998. The role of antibiosis in biocontrol. In: Harman, G.E., Kubicek, C.P. (Eds.), Trichoderma & Gliocladium, vol. 2. Taylor & Francis, Padstow, pp. 173–184.

Howell, C.R., 2003. Mechanisms employed by *Trichoderma* species in the biological control of plant diseases: the history and evolution of current concepts. Plant Dis. **87**, 4–10.

Howell, C.R., Hanson, L.E., Stipanovic, R.D., Puckhaber, L.S., 2000. Induction of terpenoid synthesis in cotton roots and control of *Rhizoctonia solani* by seed treatment with *Trichoderma virens*. Phytopathology 90, 248–252.

Idowu, O.O., Oni, A.C., Salami, A.O., 2016. The interactive effects of three *Trichoderma* species and damping-off causative pathogen *Pythium aphanidermatum* on emergence indices, infection incidence and growth performance of sweet pepper. Int. J. Recent Sci. Res. 7, 10339–10347.

Inbar, J., Abramski, M., Coen, D., Chet, I., 1994. Plant growth enhancement and disease control by *Trichoderma harzianum*in vegetable seedlings grown under commercial conditions. Eur. J. Plant Pathol. 100, 337–346.

Jagadeesh, K.S., Geeta, G.S., 1994. Effect of *Trichoderma harzianum* grown on different food bases on thebiological control of *Sclerotium rolfsii* Sacc. in groundnut. Environ. Ecol. 12, 471–473.

Jaisani, P., Pandey, R.N., 2017. Morphological and molecular characterization for identification of isolates of *Trichoderma* spp. from rhizospheric soils of crops in middle Gujarat. Indian Phytopathol. 70 (2), 238–245.

Jaisani, P., Prajapati, H.N., Yadav, D.L., Pandey, R.N., 2016. Seed Biopriming and *Trichoderma* enriched FYM based soil application in management of chickpea (*Cicer arietinum* L.) wilt complex. J. Pure Appl. Microbiol. 10 (3), 2453–2460.

Jaklitsch, W.M., 2009. European species of *Hypocrea* part I. The green-spored species. Stud. Mycol. 63, 1–91.

Jayaraj, J.N., Radhakrishnan, R., Velazhahan, 2006. Development of formulations of *Trichoderma harzianum* strain M1 for control of damping-off of tomato caused by *Pythium aphanidermatum*. Arch. Phytopathol. Plant Protect. 39 (1), 1–8.

Jeyarajan, R., 2006. Prospects of Indigeneous Mass Production and Formulation of *Trichoderma*, Incurrent Status of Biological Control of Plant Disease Using Antagonistic Organism in India. Project Directorate of Biological Control, Banglore, pp. 74–80, 445.

Jeyarajan, R., Nakkeeran, S., 2000. Exploitation of microorganisms and viruses as biocontrol agents for crop disease mangement. In: Upadhyay, et al. (Eds.), Biocontrol Potential and Their Exploitation in Sustainable Agriculture. Kluwer Academic/Plenum Publishers, USA, pp. 95–116.

Jeyarajan, R., Ramakrishnan, G., Dinakaran, D., Sridar, R., 1994. Development of products of *Trichoderma viride* and *Bacillus subtilis* for biocontrol of root rot diseases. In: Dwivedi, B.K. (Ed.), Biotechnology in India. Bioved Research Society, Allahabad, pp. 25–36.

Jin, X., Harman, G.E., Taylor, A.G., 1991. Conidial biomass and desiccation tolerance of *Trichoderma harzianum* produced at different medium water potentials. Biol. Contr. 7, 243–267.

Kandasamy, S., Li, Y., Yu, C., Wang, Q., Wang, M., Sun, J., Gao, J.-X., Chen, J., May 11, 2017. Effect of *Trichoderma harzianum* on maize rhizosphere microbiome and biocontrol of *Fusarium* Stalk rot. Sci. Rep. 7, 1771 doi: 10.1038/s41598 -017-01680-w.

Kannahi, M., Dhivya, S., Senthilkumar, R., 2016. Biological control on rice false smut disease using *Trichoderma* Species. Int. J. Pure App. Biosci 4 (2), 311–316.

Kashyap, P.L., Gulzar, S.S., Wani, S.H., Shafi, W., Kumar, S., Srivastava, A.,K., Haribhushan, A., Arora, D.K., 2011. Genes of microorganisms: paving way to Tailor next generation fungal disease resistant crop plants. Not. Sci. Biol. 3 (4), 147–157.

Kashyap, P.L., Kumar, S., Srivastava, A.K., 2017. Nanodiagnostics for plant pathogens. Environ. Chem. Lett. 15, 7–13.

Katayama, A., Matsumura, F., 1993. Degradation of organochlorine pesticides, particularly endosulfan, by *Trichoderma harzianum*. Environ. Toxicol. Chem. 12, 1059–1065.

Keswani, C., Mishra, S., Sarma, B., Singh, S., Singh, H., 2014. Unraveling the efficient applications of secondary metabolites of various *Trichoderma* spp. Appl. Microbiol. Biotechnol. 98, 533–544.

Keszler, A., Forgacs, E., Kotali, L., Vizcaõno, J.A., Monte, E., Garcõa- Acha, I., 2000. Separation and identification of volatile components in the fermentation broth of *Trichoderma atroviride* by solid-phase extraction and gas chromatography–mass spectroscopy. J. Chromatogr. Sci. 38, 421–424.

Khan, M.Y., Haque, M.M., Molla, A.H., Rahman, M.M., Alam, M.Z., 2016. Antioxidant compounds and minerals in tomatoes by *Trichoderma* enriched biofertilizer and their relationship with the soil environments. J. Integ. Agric. 15, 60345–60347.

Khare, A., Singh, B.K., Upadhyay, R.S., 2010. Biological control of *Pythium aphanidermatum*causingdamping -off of mustard by mutants of *Trichoderma viride* 1433. J. Agric. Technol. 6 (2), 231–243.

Kim, C.S., Shirouzu, T., Nakagiri, A., Sotome, K., Maekawa, N., 2013. *Trichoderma eijii* and *T. pseudolacteum*, two new species from Japan. Mycol. Prog. 15. https://doi.org/10.1007/s11557-012-0886-y (Online version).

Kiss, L., 2003. A review of fungal antagonists of powdery mildews and their potential as bio agents. Pest Manag. Sci. 59, 475–483.

Knowles, A., 2008. Recent developments of safer formulations of agrochemicals. Environmentalist 28 (1), 35–44. https://doi.org/10.1007/s10669-007-9045-4.

Koike, N., Hyakumachi, M., Kageyama, K., Tsuyumu, S., Doke, N., 2001. Induction of systemic resistance in cucumber against several diseases by plant growth-promoting fungi: lignification and superoxide generation. Eur. J. Plant Pathol. 107, 523–533.

Kopchinskiy, A., Komoń, M., Kubicek, C.P., Druzhinina, I.S., 2005. TrichoBLAST: a multilocus database for *Trichoderma* and *Hypocrea* identifications. Mycol. Res. 109, 658–660.

Kousalya, G., Jeyarajan, R., 1990. Mass multiplication of *Trichoderma* spp. J. Biol. Contr. 4, 70–71.

Kovach, J., Petzoldt, R., HarmanG, E., 2000. Use ofhoney bees and bumble bees to disseminate T*richoderma harzianum* 1295-22 to strawberries for botrytis control. Biol. Contr. 18 (3), 235–242.

Kredics, L., Antal, Z., Doczi, I., Manczinger, L., Kevei, F., Nagy, E., 2003. Clinical importance of the genus *Trichoderma*. A review. Acta Microbiol. Immunol. Hung. 50, 105–117.

Kredics, L., Antal, Z., Szekeres, A., Manczinger, L., Doczi, I., Kevei, F., Nagy, E., 2004. Production of extracellular proteases by human pathogenic *Trichoderma longibrachiatum* strains. Acta Microbiol. Immunol. Hung. 51, 283–295.

Kubicek, C.P., Mach, R.L., Peterbauer, C.K., Lorito, M., 2001. *Trichoderma*: from genes to biocontrol. J. Plant Pathol. 83, 11–23.

Kubicek, C.P., Bissett, J., Druzhinina, I., Kullnig-Gradinger, C., Szakacs, G., 2003. Genetic and metabolic diversity of *Trichoderma*: a case study on South-East Asian isolates. Fungal Genet. Biol. 38, 310–319.

Kubicek, C.P., Mikus, M., Schuster, A., Schmoll, M., Seiboth, B., 2009. Metabolic engineering strategies for the improvement of cellulose production by *Hypocrea jecorina*. Biotechnol. Biofuels 2, 19.

Kuhls, K., Lieckfeldt, E., Samuels, G.J., Meyer, W., Kubicek, C.P., Borner, T., 1997. Revision of *Trichoderma* section *Longibrachitaum* including related teleomorphs based on an analysis of ribosomal DNA internal transcribed spacer sequences. Mycologia 89, 442–460.

Kullnig, C., Szakacs, G., Kubicek, C.P., 2000. Molecular identification of *Trichoderma* species from Russia, Siberia and Himalaya. Mycol. Res. **104**, 1117–1125.

Kullnig-Gradinger, C., Szakacs, G., Kubicek, C.P., 2002. Phylogeny and evolution of the genus *Trichoderma*: a multigene approach. Mycol. Res. 106, 757–767.

Kumar, S., Upadhyay, J.P., Rani, A., 2009. Evaluation of *Trichoderma* species against *Fusarium udum* Butler causing wilt of pigeonpea. J. Biol. Contr. 23 (3), 329–332.

Kumar, M., Zacharia, S., Lal, A.A., 2019. Management of alternarial blight of mustard (*Brassica juncea* L.) by botanicals, *Trichoderma harzianum* and fungicides. Plant Archiv. 19 (Suppl. 1), 1108–1113.

Lal, B., Maharshi, R.P., 2007. Compatibility of biocontrol agents *Trichoderma* spp. with pesticides. J. Mycol. Plant Pathol. 37, 295–300.

Lee, C.F., Hseu, T.H., 2002. Genetic relatedness of *Trichoderma* sect. *Pachybasium* species based on molecular approaches. Can. J. Microbiol. 48, 831–840.

Leon, V.C., Raja, M., Pandian, R.T.P., Kumar, A., Sharma, P., 2017. Studies on opportunistic endophytism of *Trichoderma* species in rice (Pusa Basmati-1 (PB1)). Indian J. Exper. Biol. 56, 121–128.

Levy, N.O., Meller, H.Y., Haile, Z.M., Elad, Y., David, E., Jurkevitch, E., Katan, J., 2015. Induced resistance to foliar diseases by soil solarization and *Trichoderma harzianum*. Plant Pathol. 64, 365–374.

Lewis, J.A., 1991. Formulation and delivery system of biocontrol agents with emphasis on fungi *Beltsville symposia*. In: Keister, D.L., Cregan, P.B. (Eds.), The Rhizosphere and Plant Growth, vol. 14. Agric. Res., pp. 279–287

Li, C., Yang, Z., Zhang, R., Zhang, D., Chen, S., 2013. Effect of pH on cellulase production and morphology of *Trichoderma reesei* and the application in cellulosic material hydrolysis. J. Biotechnol. 168, 470–477.

Lieckfeldt, E., Kuhls, K., Muthumeenakshi, M., 1998. Molecular taxonomy of *Trichoderma* and *Gliocladium* and their teleomorphs. In: Kubicek, C.P., Harman, G.E. (Eds.), *Trichoderma* and *Gliocladium*, Volume 1. Basic Biology, Taxonomy and Genetics. Taylor & Francis, London, pp. 35–74.

Lieckfeldt, E., Samuels, G.J., Nirenberg, H.I., Petrini, O., 1999. A morphological and molecular perspective of *Trichoderma viride*: is it one or two species. Appl. Environ. Microbiol. 65, 2418–2428.

Liu, S., Baker, R., 1986. Mechanism of biological control in soil suppressive to *Rhizoctonia solani*. Phytopathology 70, 404–412.

Lo, C.T., Liao, T.F., Deng, T.C., 2000. Induction of systemic resistance of cucumber to cucumber green mosaic virus by the root-colonizing *Trichoderma* spp. Phytopathology 90, S47.

Lopez-Quintero, C.A., Lea, A., Esperanza, F.-M.A., Walter, G., Monika, K.-Z., Bart, T., Müller Wally, H., Teun, B., Irina, D., 2013. DNA barcoding survey of *Trichoderma* diversity in soil and litter of the Colombian lowland Amazonian rainforest reveals *Trichoderma strigosellum* sp. nov. and other species. Antonie van Leeuwenhoek J. Microbiol. 104 (2), 657–674.

Lorito, M., Woo, S.L., Garcia Fernandez, I., Colucci, G., Harman, G.E., Pintor-Toro, J.A., Filippone, E., Mucciflora, S., Lawrence, C.B., Zoina, A., Tuzun, S., Scala, F., 1998. Genes from mycoparasitic fungi as a source for improving plant resistance to fungal pathogens. Proc. Natl. Acad. Sci. U.S.A. 95, 7860–7865.

Lorito, M., Woo, S.L., Harman, G.E., Monte, E., 2010. Translational research on *Trichoderma*: from omics to the field. Annu. Rev. Phytopathol. 48, 395–417.

Lyn, M.E., Burnett, D., Garcia, A.R., Gray, R., 2010. Interaction of water with three granular biopesticide formulations. J. Agric. Food Chem. 58 (1), 1804–1814.

Madhusudan, P., Gopal, K., Haritha, V., Sangale, U.R., Rao, S.V.R.K., 2010. Compatability of *Trichoderma viride* with fungicides and efficiency against *Fusarium solani*. J. Plant Dis. Sci. 5, 23–26.

Mahesh, M., Muhammad, S., Sreenivasa, S., Shashidhar, K.R., 2010. Integrated management of pigeonpea wilt caused by *Fusarium udum* butler. EJBS 2 (1), 1–7.

Maischak, H., Zimmermann, M.R., Felle, H.H., Boland, W., et al., 2010. Alamethicin-induced electrical long distance signaling in plants. Plant Signal. Behav. 5, 988–990.

Maliszewska, I., Aniszkiewicz, L., Sadowski, Z., 2009. Biological synthesis of gold nanostructures using the extract of *Trichoderma koningii*. Acta Phys. Pol., A 116, 163–165.

Manczinger, L., Polner, G., 1987. Cluster analysis of carbon source utilization patterns of *Trichoderma* isolates. Syst. Appl. Microbiol. 9, 214–217.

Manczinger, L., Rákhely, G., Vágvölgyi, C., Szekeres, A., 2012. Genetic and biochemical diversity among *Trichoderma* isolates in soil samples from winter wheat fields of the Pannonian Plain. Acta Biol. Szeged. 56, 141–149.

Mandal, A.K., Dubey, S.C., Tripathi, A., 2015. Bio-agent based integrated management strategy against stem rot of chickpea. Indian Phytopathol. 68 (4), 402–409.

Marra, R., Ambrosino, P., Carbone, V., Vinale, F., Woo, S.L., Ruocco, M., Ciliento, R., Lanzuise, S., Ferraioli, S., Soriente, I., Gigante, S., Turra, D., Fogliano, V., Scala, F., Lorito, M., 2006. Study of the three-way interaction between *Trichoderma atroviride*, plant and fungal pathogens by using a proteomic approach. Curr. Genet. 50, 307–321.

Martinez, D., Larrondo, L.F., Putnam, N., Sollewijn-Gelpke, M.D., Huang, K., Chapman, J., Helfenbein, K.G., Ramaiya, P., Detter, J.C., Larimer, F., Coutinho, P.M., Henrissat, B., Berka, R., Cullen, D., Rokhsar, D., 2008. Genome sequence of the lignocellulose degrading fungus *Phanerochaete chrysosporium* strain RP78. Nat. Biotechnol. 22, 695–700.

Mastouri, F., Thomas, B., Harman, G.E., 2012. *Trichoderma harzianum* enhances antioxidant defense of tomato seedlings and resistance to water deficit. Mol. Plant Microbe Interact. 25 (9), 1264–1271.

Mathivanan, N., Srinivasan, K., Chelliah, S., 2000. Biological control of soil-borne diseases of cotton, eggplant, okra and sunflower by *Trichoderma viride*. J. Plant Dis. Prot. 107 (3), 235–244.

Meena, S., Dutta, R., Kumar, S., 2010. Potential bio-control agent for management of post-flowering stalk rot complex and maize plant health. Arch. Phytopathol. Plant Protect. 43, 1392–1395.

Meena, B.N., Pandey, R.N., Ram, D., 2016. Seed biopriming for management of root rot and blight of mungbean incited by *Macrophomina phaseolina* (Tassi) Goid. and *Rhizoctonia solani* Kuhn. J. Pure Appl. Microbiol. 10 (2), 0973–7510.

Meller, H.Y., Haile, M.Z., David, D., Borenstein, M., Shulchani, R., Elad, Y., 2013. Induced systemic resistance against grey mould in tomato (*Solanum lycopersicum*) by benzothiadiazole and *Trichoderma harzianum* T39. Phytopathology 104, 150–157.

Mercy, M., 2004. Protein Profiling of Isolates of *Trichoderma harzianum* Rifai Tolerant to Pesticides. M. Sc. (Agri.) thesis submitted to AAU, Anand, p. 91.

Migheli, Q., Balmas, V., Komoñ-Zelazowska, M., Scherm, B., Fiori, S., Caria, R., Alexey, G., Kopchinskiy, A., Kubicek, C.P., Druzhinina, I.S., 2009. Soils of a Mediterranean hot spot of biodiversity and endemism (Sardinia, Tyrrhenian Islands) are inhabited by pan-European, invasive species of *Hypocrea/Trichoderma*. Environ. Microbiol. 11 (1), 35–46.

Mitler, R., 2002. Oxidative stress, antioxidants and stress tolerance. Trends Plant Sci. 7, 405–410.

Monte, E., 2001. Understanding *Trichoderma*: between agricultural biotechnology and microbial ecology. Int. Microbiol. 4, 1–41.

Montero-Barrientos, M., Hermosa, R., Cardoza, R.E., Gutierrez, S., Nicolás, C., Monte, E., 2010. Transgenic expression of the *Trichoderma harzianum* HSP70 gene increases *Arabidopsis* resistance to heat and other abiotic stresses. J. Plant Physiol. 167, 659–665.

Montero-Barrientos, M., Hermosa, R., Cardoza, R.E., Gutiérrez, S., Monte, E., 2011. Functional analysis of the *Trichoderma harzianum nox1* gene, encoding an NADPH oxidase, relates production of reactive oxygen species to specific biocontrol activity against *Pythium ultimum*. Appl. Environ. Microbiol. 77, 3009–3016.

Montero-Barrientos, M., Hermosa, R., Nicolas, C., Cardoza, R.E., Gutierrez, S., Monte, E., 2008. Over expression of a *Trichoderma hsp70* gene increases fungal resistance to heat and other abiotic stresses. Fungal Genet. Biol. 45, 1506–1513.

Mukherjee, P.K., 1997. *Trichoderma* sp. as a microbial suppressive agent of *Sclerotium rolfsii* on vegetables. World J. Microbiol. Biotechnol. 13, 497–499.

Mukherjee, P.K., Buensanteai, N., Moran-Diez, M.E., Druzhinina, I.S., Kenerley, C.M., 2012. Functional analysis of non-ribosomal peptide synthetases (NRPSs) in *Trichoderma virens* reveals a polyketide synthase (PKS)/NRPS hybrid enzyme involved in induced systemic resistance response in maize. Microbiology 158, 155–165.

Mukherjee, P.K., Horwitz, B.A., Singh, U.S., Mukherjee, M., Schmoll, M., 2013. *Trichoderma* in agriculture, industry and medicine: an overview. In: Mukherjee, P.K., Horwitz, B.A., Singh, U.S., Mukherjee, M., Schmoll, M. (Eds.), Trichoderma: Biology and Applications. CABI, Nosworthy, Way, Wallingford, Oxon, UK, pp. 1–9.

Mulaw, T.B., Druzhinina, I.S., Kubicek, C.P., Atanasova, L., 2013. Novel endophytic *Trichoderma* spp. isolated from healthy *Coffea arabica* roots are capable of controlling coffee tracheomycosis. Diversity 5, 750–766.

Muthumeenaksi, S., Mills, P.R., Brown, A.E., Seaby, D.A., 1994. Intra- specic molecular variation among *Trichoderma harzianum* isolates colonizing mushroom compost in the British Isles. Microbiology 140, 769–777.

Nagy, V., Seidl, V., Szakacs, G., Komoń-Zelazowska, M., Kubicek, C.P., Druzhinina, I.S., 2007. Application of DNA Bar codes for screening of industrially important fungi: the haplotype of *Trichoderma harzianum* Sensu Stricto indicates superior chitinase formation. Appl. Environ. Microbiol. 73 (21), 7048–7058.

Nakari-Seta, T., Penttila, M., von Dö hren, H., 2007. Direct identification of hydrophobins and their processing in *Trichoderma* using intact-cell MALDI-TOF MS. FEBS J. 274, 841–852.

Nallathambi, P., Padmanaban, P., Mohanraj, D., 2001. Fungicide resistance in sugarcane associated *Trichoderma* isolates. J. Mycol. Plant Pathol. 31, 125.

Nawrocka, J., Małolepsza, U., 2013. Diversity in plant systemic resistance induced by *Trichoderma*. Biol. Contr. 67, 149–156.

Neuhof, T., Dieckmann, R., Druzhinina, I.S., Kubicek, C.P., von Do-hren, H., 2007. Intact-cell MALDI-TOF mass spectrometry analysis of peptaibol formation by the genus *Trichoderma/Hypocrea*: can molecular phylogeny of species predict peptaibol structures? Microbiology 153, 3417–3437.

Nicolása, C., Hermosab, R., Rubiob, B., Mukherjeec, P.K., Monteb, E., 2014. *Trichoderma* genes in plants for stress tolerance- status and prospects. Plant Sci. 228, 71–78.

Nirmalkar, V.K., Tiwari, R.K.S., Singh, S., 2018. Efficacy of bio-agents against damping off in solanaceous crops under nursery conditions. Int. J. Plant Protect. 11 (1), 1–9.

O'Kennedy, M.M., Crampton, B.G., Lorito, M., Chakauya, E., Breese, W.A., Burger, J.T., Botha, F.C., 2011. Expression of a β-1,3-glucanase from a biocontrol fungus in transgenic pearl millet. South Afr. J. Bot. 77, 335–345.

Oros, G., Naar, Z., Cserhati, T., 2011. Growth response of *Trichoderma* species to organic solvents. Mol. Inf. 30, 276–285.

Ospina-Giraldo, M.D., Royse, D.J., Chen, X., Romaine, C.P., 1998. Molecular phylogenetic analyses of biological control strains of *Trichoderma harzainum* and other biotypes of *Trichoderma* spp. associated with mushroom green mould. Phytopathology 89, 308–313.

Pal, K.K., Gardener, B.M.S., 2006. Biological Control of Plant Pathogens. The Plant Health Instructor, pp. 1–25.

Pal, R., Biswas, M.K., Mandal, D., Naik, B.S., 2015. Management of sheath blight disease of rice through bio control agents in west central table land zone of Odisha. Int. J. Adv. Res. 3 (11), 747–753.

Pandey, R.N., 2017. Seed bio-priming in the management of seed- and soil-borne diseases. Indian Phytopathol. 70 (2), 164–168.

Pandey, R.N., Gohel, N.M., 2017. Evaluation of bioagents for management of charcoal rot [*Macrophomina phaseolina* (Tassi) Goid] in soybean [*Glycine max* (L.) Merrill] through seed treatment and soil application. Trends Biosci. 10 (15), 2667–2670.

Pandey, K.K., Upadhyay, J.P., 1998. Sensitivity of different fungicides to *Fusarium udum, Trichoderma harzianum* and *Trichoderma viride* for integrated approach of disease management. Veg. Sci. 2, 89–92.

Pandey, R.N., Gohel, N.M., Pratik, J., 2017. Management of wilt and root rot of chickpea caused by *Fusarium oxysporum* f. sp. *ciceri* and *Macrophomina phaseolina* through seed biopriming and soil application of bio-Agents. Int. J. Curr. Microbiol. App. Sci. 6 (5), 2516–2522.

Papavizas, G.C., 1985. *Trichoderma* and *Gliocladium* : biology, ecology and potential for bio control. Annu. Rev. Phytopathol. 23, 23–24.

Patel, P.R., 2009. *Trichoderma harzianum* Rifai in the Management of Collar Rot of Groundnut by *Aspergillns niger* Van Tieghem. M. Sc. (Agri.) thesis submitted to AAU, Anand, pp. 1–79.

Patel, J.S., Kharwar, R.N., Singh, H.B., Upadhyay, R.S., Sarma, B.K., 2017. *Trichoderma asperellum* (T42) and *Pseudomonas fluorescens* (OKC)-enhances resistance of pea against *Erysiphe pisi* through enhanced ROS generation and lignifications. Front. Microbiol. 8, 306. https://doi.org/10.3389/fmicb.2017.00306.

Patibanda, A.K., Upadhyay, J.P., Mukhopadhyay, A.N., 2002. Efficacy of *Trichoderma harzianum* Rifai alone or in combination with fungicides against *Sclerotium* wilt of groundnut. J. Biol. Contr. 16, 57–63.

Paulitz, I.C., 1997. Biological control of root pathogens in soilless and hydroponic systems. Hortscience 32, 193–196.

Perazzoli, M., Moretto, M., Fontana, P., Ferrarini, A., Velasco, R., Moser, C., Delledonne, M., Pertot, I., 2012. Downy mildew resistance induced by *Trichoderma harzianum* T39 in susceptible grapevines partially mimics transcriptional changes of resistant genotypes. BMC Genom. 13, 660.

Persoon, C.H., 1794. Neuer Versuch einer systematischen Einteilung der Schwämme. Racodium Römer's Neues Magazin der Botanik 1, 123.

Petrini, O., 1991. Fungal endophytes of tree leaves. In: Andrews, J.H., Hirano, S.S. (Eds.), Microbial Ecology of Leaves. Springer Verlag, New York, pp. 179–197.

Pramod, K.T., Palakshappa, M.G., 2009. Evaluation of suitable substrates for on-farm production of antagonist *Trichoderma harzianum*. Karnataka J. Agric. Sci. 22, 115–117.

Prasad, R.D., Rangeshwaran, R., Anuroop, C.P., Phanikumar, P.R., 2002. Bioefficacy and shelf -life of conidial and chlamydospore formulations of *Trichoderma harzianum* Rifai. J. Biol. Contr. 16, 145–148.

Purohit, J., Singh, Y., Bisht, S., Srinivasaraghvan, A., 2013. Evaluation of antagonistic potential of *Trichoderma harzianum* and *Pseudomonas fluorescens* isolates against *Gloeocercospora sorghi* causing zonate leaf spot of sorghum. Bioscan 8 (4), 1327–1330.

Rabindra, R.J., 2005. Current status of production and use of microbial pesticides in India and the way forward.p1-12. Technical Document No.55. Project Directorate of Biological Control. In: Rabindra, R.J., Hussaini, S.S., Ramanujam, B. (Eds.), Microbial Biopesticde Formulations and Application.

Raghuchander, T., Jayashree, K., Samiyaplan, R., 1997. Management of *Fusarium* wilt of banana using antagonistic micro organisms. Biol. Contr. 11, 101–105.

Rakholiya, K.B., Jadeja, K.B., 2010. Effect of seed treatment of biocontrol agents and chemicals for the management of stem and pod rot of groundnut. Int. J. Plant Protect. 3 (2), 276–278.

Ram, H., Pandey, R.N., 2011. Efficacy of bio-control agents and fungicides in the management of wilt of pigeon pea. Indian Phytopathol. 64 (3), 269–271.

Raman, R., Korikanthimath, V.S., 2006. Management of groundnut root rot by *Trichoderma viride* and *Pseudomonas fluorescens* under Rainfed conditions. Indian J. Plant Protect. 34 (2), 239–241.

Ramanujam, B., Prasad, R.D., Sriram, S., Rangeswaran, R., 2010. Mass production, formulation, quality control and delivery of *Trichoderma* for plant disease management. J. Plant Protect. Sci. 2 (2), 1–8.

Respinis, S.D., Vogel, G., Benagli, C., Tonolla, M., Petrini, O., Samuels, G.J., 2010. MALDI-TOF MS of *Trichoderma*: a model system for the identification of microfungi. Mycol. Prog. 9, 79–100.

Rifai, M.A., 1969. A revision of the genus *Trichoderma*. Mycol. Pap. 116, 1–56.

Rini, C.R., Sulochana, K.K., 2007. Substrate evaluation for multiplication of *Trichoderma* spp. J. Trop. Agric. 45, 58–60.

Rinu, K., Priyanka, S., Pandey, A., 2014. *Trichoderma gamsii* (NFCCI 2177): a newly isolated endophytic, psychrotolerant, plant growth promoting, and antagonistic fungal strain. J. Basic Microbiol. 54 (5), 408–417.

Romeralo, C., Santamaría, O., Pando, V., Diez, J.J., 2015. Fungal endophytes reduce necrosis length produced by *Gremmeniella abietina* in *Pinus halepensis* seedlings. Biol. Contr. 80, 30–90.

Rosmana, A., Samuels, G.J., Ismaiel, A., Ibrahim, E.S., Chaverri, P., Herawati, Y., Asman, A., 2015. *Trichoderma asperellum*: a dominant endophyte species in cacao grown in Sulawesi with potential for controlling vascular streak dieback disease. Trop. Plant Pathol. 40, 19–25.

Rudresh, D.L., Shivaprakash, M.K., Prasad, R.D., 2005. Tricalcium phosphate solubilizing abilities of *Trichoderma* spp. in relation to P uptake & growth yield parameters of chickpea (*Cicer arietinum* L.). Can. J. Microbiol. 51, 217–226.

Röhrich, C.R., Jaklitsch, W.M., Voglmayr, H., Iversen, A., Vilcinskas, A., Nielsen, K.F., Thrane, U., von Döhren, H., Brückner, H., Degenkolb, T., 2014. Front line defenders of the ecological niche! screening the structural diversity of peptaibiotics from saprotrophic and fungicolous *Trichoderma/Hypocrea* species. Fungal Divers. 69, 117–146.

Sadfi-Zouaoui, N., Hannachi, I., Rouaissi, M, Hajlaoui, M.R., Rubio, M.B., Monte, E., Boudabous, A., Hermosa, M.R., et al., 2009. Biodiversity of *Trichoderma* strains in Tunisia. Can. J. Microbiol. 55, 154–162.

Sadfi-Zouaoui, N., Hannachi, I., Rouaissi, M., Hajlaoui, M.R., Rubio, M.B., Monte, E., Saksirirat, W., Chareerak, P., Bunyatrachata, W., 2009. Induced systemic resistance of biocontrol fungus, *Trichoderma* spp. against bacterial and gray leaf spot in tomatoes. Asian J. Food Agro-Ind. 2, S99–S104.

Saju, K.A., Anandraj, M., Sarma, Y.R., 2002. On-farm production of *Trichoderma harzianum* using organic matter. Indian Phytopathol. 55 (3), 277–281.

Saksirirat, W., Chareerak, P., Bunyatrachata, W., 2009. Induced systemic resistance of biocontrol fungus, *Trichoderma* spp. against bacterial and gray leaf spot in tomatoes. Asian J. Food Agro-Ind. 2, S99–S104.

Salas-Marina, M.A., Silva-Flores, M.A., Uresti-Rivera, E.E., Castro- Longoria, E., Herrera-Estrella, A., Casas-Flores, S., 2011. Colonization of *Arabidopsis* roots by *Trichoderma atroviride* promotes growth and enhances systemic disease resistance through jasmonic acid/ethylene and salicylic acid pathways. Eur. J. Plant Pathol. 131, 15–26.

Salas-Marina, M.A., Isordia-Jasso, M., Islas-Osuna, M.A., Delgado-Sánchez, P., Jiménez-Bremont, J.F., Rodríguez-Kessler, M., Rosales-Saavedra, M.T., Herrera-Estrella, A., Casas-Flores, S., 2015. The Epl1 and Sm1 proteins from *Trichoderma atroviride* and *Trichoderma virens* differentially modulate systemic disease resistance against different life style pathogens in *Solanum lycopersicum*. Front. Plant Sci. 23, 77.

Samuels, G.J., 1996. *Trichoderma*: a review of biology and systematic of the genus. Mycol. Res. 100, 923–935.

Samuels, G.J., 2006. *Trichoderma:* systematic, the sexual state, and ecology. Phytopathology 96, 195–206.

Samuels, G.J., Ismaiel, A., 2009. *Trichoderma evansii* and *T. lieckfeldtiae:* two new *T. hamatum*-like species. Mycologia 101, 142–152.

Samuels, G.J., Petrini, O., Kuhls, K., Lieckfeldt, E., Kubicek, C.P., 1998. The *Hypocrea schweinitzii* complex and *Trichoderma* sect. Longibrachiatum. Stud. Mycol. 41, 1–54.

Samuels, G.J., Dodd, S.L., Gams, W., Castlebury, L.A., Petrini, O., 2002. *Trichoderma* species associated with the green mould epidemic of commercially grown *Agaricus bisporus.* Mycologia 94, 146–170.

Samuels, G.J., Dodd, S., Lu, B.S., Petrini, O., Schroers, H.J., Druzhinina, I.S., 2006. The *Trichoderma koningii* aggregate species. Stud. Mycol. 56, 67–133.

Samuels, G.J., Dodd, S.L., Gams, W., Castlebury, L.A., Petrini, O., 2012a. *Trichoderma* species associated with the green mold epidemic of commercially grown *Agaricus bisporus.* Mycologia 94, 146–170.

Samuels, G.J., Ismaiel, A., Mulaw, T.B., Szakacs, G., Druzhinina, I.S., Kubicek, C.P., Jaklitsch, W.M., 2012b. The Longibrachiatum clade of *Trichoderma*: a revision with new species. Fungal Divers. https://doi.org/10.1007/s13225-012-0152-2.

Sandhya, C., Adapa, L.K.K., Nampoothri, M., Binod, P., Szakacs, G., Pandey, A., 2004. Extracellular chitinase production by *Trichoderma harzianum* in submerged fermentation. J. Basic Microbiol. 44, 49–58.

Sanghera, G.S., Wani, S.H., Singh, G., Kashyap, P.L., Singh, N.B., 2011. Designing crop plants for biotic stresses using transgenic approach. Vegetos 24 (1), 1–25.

Sanz, L., Montero, M., Grondona, I., Vizcaíno, J.A., Llobell, A., Hermosa, R., Monte, E., 2004. Cell wall-degrading isoenzyme profiles of *Trichoderma* biocontrol strains show correlation with rDNA taxonomic species. Curr. Genet. 46, 277–286.

Sarrocco, S., Guidi, L., Fambrini, S., DesI'Innocenti, E., Vannacci, G., 2009. Competition for cellulose exploitation between *Rhizoctonia solani* and two *Trichoderma* isolated in the decomposition of wheat straw. J. Plant Pathol. 91, 331–338.

Sawant, I.S., Mukhopadhyay, A.N., 1990. Integration of metalaxyl MZ with *Trichoderma harzianum* for the control of *Pythium* damping-off in sugarbeet. Indian Phytopathol. 43, 535–541.

Sawant, I.S., Sawant, S.D., 1996. A simple method for achieving high cfu of *Trichoderma harzianum* on organic wastes for field applications. Indian Phytopathol. 9, 185–187.

Schuster, A., Schmoll, M., 2010. Biology and biotechnology of *Trichoderma.* Appl. Microbiol. Biotechnol. 87, 787–799.

Seaman, A., 2003. Efficacy of OMRI-Approved Products for Tomato Foliar Disease Control, vol. 129. Integrated Pest Management Program publication, New York State, pp. 164–167.

Seidi, V., Marchetti, M., Schandl, R., Allmaier, G., Kubicek, C.P., 2006. EPL1, the major secreted protein of *Hypocrea atroviridis* on glucose, is a member of a strongly conserved protein family comprising plant defense response elicitors. FEBS J. 273, 4346–4359.

Sharma, S.D., Mishra, A., Pandey, R.N., Patel, S.J., 2001. Sensitivity of *Trichoderma harzianum* to fuingicides. J. Mycol. Plant Pathol. 31, 251–253.

Sharma, B.L., Singh, S.P., Sharma, M.L., 2012. Bio-degradation of crop residues by *Trichoderma* species vis-à vis nutrient quality of the prepared compost. Sugar Tech. 14 (2), 174–180.

Sharon, E., Bar-Eyal, M., Chet, I., Herrera-Estrella, A., Kleifeld, O., Spiegel, Y., 2001. Biological control of the root-knot nematode *Meloidogyne javanica* by *Trichoderma harzianum.* Phytopathology 91, 687–693.

Shoresh, M., Harman, G.E., 2008. The molecular basis of shoot responses of maize seedlings to*Trichoderma harzianum* T22 inoculation of the root: a proteomic approach. Plant Physiol. 147, 2147–2163.

Shoresh, M., Yedidia, I., Chet, I., 2005. Involvement of jasmonic acid/ethylene signaling pathway in the systemic resistance induced in cucumber by *Trichoderma asperellum* T203. Phytopathology 95, 76–84.

Shoresh, M., Mastouri, F., Harman, G.H., 2010. Induced systemic resistance and plant responses to fungal biocontrol agents. Annu. Rev. Phytopathol. 48, 21–43.

Singh, F., Hooda, I., Sindhan, G.S., 2004. Biological control of tomato wilt caused by *Fusarium oxysporum* f. sp. *lycopersici.* J. Mycol. Plant Pathol. 34 (2), 568–570.

Singh, H.B., Singh, B.N., Singh, S.P., Sarma, B.K., 2012. Exploring different avenues of *Trichoderma* as a potent bio-fungicidal and plant growth promoting candidate-an overview. Rev. Plant Pathol. 5, 315–426.

Singh, V., Upadhyay, R.S., Sarma, B.K., Singh, H.B., 2016. Seed bio-priming with *Trichoderma asperellum* effectively modulate plant growth promotion in pea. IJAEB 9 (3), 361–365.

Slavica, G., Tanović, B., 2013. Biopesticide formulations, possibility of application and future trends. Pestic.Phytomed. (Belgrade) 28 (2), 97–102.

Sriram, S., Manasa, S.B., Savitha, M.J., 2009. Potential use of elicitors from *Trichoderma* in induced systemic resistance for the management of *Phytophthora capsici* in red pepper. J. Biol. Contr. 23, 449–456.

Studholme, D.J., Harris, B., Kate, L.C., Winsbury, R., Perera, V., Ryder, L., Ward, J.L., Beale, H.M., Thornton, C.R., Grant, M., 2013. Investigating the beneficia ltraits of *Trichoderma hamatum* GD12 for sustainable agriculture-insights from genomics. Front. Plant Sci. Plant-Micr. Interact. 4, 1–13.

Sushir, M.A., Pandey, R.N., 2001. Tolerance of *Trichoderma harzianum* Rifai to insecticides and weedicides. J. Mycol. Plant Pathol. 31, 102.

Sutton, D.A., Fothergill, A.W., Rinaldi, M.G., 1998. Guide to Clinically Significant Fungi, first ed. Williams & Wilkins, Baltimore.

Tadros, F., 2005. Applied Surfactants, Principles and Applications. Wiley-VCH Verlag GmbH and Co.KGaA, pp. 187–256.

Tari, P.H., Anderson, A.J., 1988. *Fusarium* wilt suppression and agglutinability of *Pseudomonas putida*. Appl. Environ. Microbiol. 54, 2037–2041.

Téllez-Vargas, J., Rodríguez-Monroy, M., López-Meyer, M., Montes-Belmont, R., Sepúlveda-Jiménez, G., 2017. *Trichoderma asperellum* ameliorates phytotoxic effects of copper in onion (*Allium cepa* L.). Environ. Exp. Bot. 136, 85–93.

Tewari, L., Bhanu, C., 2003. Screening of various substrates for sporulation and mass multiplication of bio-control agent *Trichoderma harzianum* through solid state fermentation. Indian Phytopathol. 56 (4), 476–478.

Tewari, L., Bhanu, C., 2004. Evaluation of agro-industrial wastes for conidia based inoculum production of bio-control agent: *Trichoderma harzianum*. J. Sci. Ind. Res. 63, 807–812.

Thomashow, L.S., Bonsall, R.F., Weller, D.M., 2002. Antibiotic Production by Soil and Rhizosphere Microbes in Situ, second ed. ASM Press, Washington DC, pp. 638–647.

Timothy, A., Cardwell, K., Allyn Florini, D., Ikotun, T.A., 2001. Seed treatment with *Trichoderma s*pecies for control of damping-off of cowpea caused by *Macrophomina phaseolina*. Biocontrol Sci. Technol. 11 (4), 449–457.

Tucci, M., Ruocco, M., Masi, L.D., Palma, M.D., Lorito, M., 2011. The beneficial effect of *Trichoderma* spp. on tomato is modulated by the plant genotype. Mol. Plant Pathol. 12, 341–354.

Upadhyay, J.P., Mukhopadhyay, A.N., 1986. Biological bioefficacy of *Trichoderma harzianum* control of *Sclerotiumrolfsii* by *Trichoderma harzianum* in sugarbeet. Trop. Pest Dis. Manag. 32, 215–220.

Vahabi, K., Mansoori, G.A., Karimi, S., 2011. Biosynthesis of silver nanoparticles by fungus *Trichoderma reesei*: a route for large scale production of AgNPs. Insciences J. 1, 65–79.

Vernner, R., Bauer, P., 2007. Q-TEO, a formulation concept that overcomes the incompability between water and oil. Pfalzenschutz-Nachr. Bayer 60 (1), 7–26.

Vinale, F., Marra, R., Scale, F., Ghisalberti, E.L., Lorito, M., Sivasithamparam, K., 2006. Major secondary metabolites produced by two commercial *Trichoderma* strains active different phytopathogens. Lett. Appl. Microbiol. 43, 143–148.

Viterbo, A., Harel, M., Chet, I., 2004. Isolation of two aspartyl proteases from *Trichoderma asperellum* expressed during colonization of cucumber roots. FEMS Microbiol. Lett. 238, 151–158.

Viterbo, M., Harel, B., Horwitz, A., Chet, I., Mukherjee, P.K., 2005. *Trichoderma* mitogen-activated protein kinase signaling is involved in induction of plant systemic resistance. Appl. Environ. Microbiol. 71, 6241–6246.

Viterbo, A., Landau, U., Kim, S., Chernin, L., Chet, I., 2010. Characterization of ACC deaminase from the biocontrol and plant growth-promoting agent *Trichoderma asperellum* T203. FEMS Microbiol. Lett. 305, 42–48.

Viterbo, A., Montero, M., Ramot, O., Friesem, D., Monte, E., Llobell, A., Chet, I., 2002. Expression regulation of the endochitinase chit36 from *Trichoderma asperellum (T. harzianum* T-203). Curr. Genet. 42, 114–122.

Wagh, S.P., Gangurde, S.V., 2015. Effect of Cow-dung slurry and *Trichoderma*spp.on quality and decomposition of Teak and Bamboo leaf compost. Res. J. Agric. For. Sci. 3 (2), 1–4.

Weindling, R., 1932. *Trichoderma lignorum* as a parasite of other soil fungi. Phytopathology 22, 837–845.

Wesam, I.A.S., Ghoneem, K.M., Rashad, Y.M., Al-Askar, A.A., 2017. *Trichoderma harzianum* WKY1: an indole acetic acid producer for growth improvement and anthracnose disease control in sorghum. Biocontrol Sci. Technol. 27 (5), 654–676.

Wiest, A., Grzegorski, D., Xu, B., Goulard, C., Rebuffat, S., Ebbole, D.J., Bodo, B., Kenerley, C., 2002. Identification of peptaibols from *Trichoderma virens* and cloning of a peptaibol synthetase. J. Biol. Chem. 277, 20862–20868.

Woo, S.L., Scala, F., Ruocco, M., Lorito, M., 2006. The molecular biology of the interactions between *Trichoderma* spp., pathogenic fungi, and plants. Phytopathology 96, 181–185.

Woo, S.L., Ruocco, M., Vinale, F., Nigro, M., Marra, R., Lombardi, N., Pascale, A., Lanzuise, S., Manganiello, G., Lorito, M., 2014. *Trichoderma*-based products and their widespread use in agriculture. Open Mycol. J. 8 (Suppl. 1, M 4), 71–126.

Woods, T.S., 2003. Pesticide formulations. In: AGR 185 in Encyclopedia of Agrochemicals. Wiley & Sons, New York, pp. 1–11.

Xu, S.X.F., Ma, S., Bai, Z., Xiao, R., Li, Y., Zhuang, G., 2014. Mitigating nitrous oxide emissions from tea field soil using bioaugmentation with a *Trichoderma viride* biofertilizer. Hindawi Publishing Corporation Sci.World J. 9. https://doi.org/10.1155/2014/793752. Volume, Article ID 793752.

Yadav, S.K., Dave, A., Sarkar, A., Singh, H.B., Sarma, B.K., 2013. Co-inoculated biopriming with *Trichoderma, Pseudomonas* and *Rhizobium* improves crop growth in *Cicer arietinum* and *Phaseolus vulgaris*. IJAEB 6 (2), 255–259.

Yasmeen, R., Siddiqui, Z.S., 2017. Physiological responses of crop plants against *Trichoderma harzianum* in saline environment. Acta Bot. Croat. 76 (2), 154–162.

Yoshioka, Y., Ichikawa, H., Naznin, H.A., Kogure, A., Hyakumachi, M., 2012. Systemic resistance induced in *Arabidopsis thaliana* by *Trichoderma asperellum* SKT-1, a microbial pesticide of seed-borne diseases of rice. Pest Manag. Sci. 68, 60–66.

Yuan, Z.L., Chen, Y.C., Zhang, C.L., Lin, F.C., Chen, L.Q., 2008. *Trichoderma chlorosporum*, a new record of endophytic fungi from *Dendrobium nobile* in China (in Chinese). Mycosystema 27, 608–610.

Zachow, C., Berg, C., Müller, H., Meincke, R., Komon-Zelazowska, M., Druzhinina, I.S., Kubicek, C.P., Berg, G., 2009. Fungal diversity in the Rhizosphere of endemic plant species of Tenerife (Canary Islands): relationship to vegetation zones and environmental factors. ISME J. 3, 79–92. https://doi.org/10.1038/ismej.2008.87.

Zamir, D., Chet, I., 1985. Application of enzyme electrophoresis for the identification of isolates in *Trichoderma harzianum*. Can. J. Bot. 31, 578–580.

Zeilinger, S., Galhaup, C., Payer, K., Woo, S.L., Mach, R.L., Fekete, C., Lorito, M., Kubicek, C.P., 1999. Chitinase gene expression during mycoparasitic interaction of *Trichoderma harzianum* with its host. Fungal Genet. Biol. 26, 131–140.

Zhang, J.C., Howell, R., Starr, J.L., 1996. Suppression of *Fusarium* colonization of cotton roots and *Fusarium* wilt by sted treatments with *Gliocladium virens* and *Bacillus subtilis*. Biocontrol Sci. Technol. 6 (2), 175–188.

Zhang, C.L., Druzhinina, I.S., Kubicek, C.P., Xu, T., 2005. *Trichoderma* biodiversity in China: evidence for a north to south distribution of species in East Asia. FEMS Microbiol. Lett. 251, 251–257.

Zhang, F., Zhihua, L., Mijiti, G., Yucheng, W., Fan, H., Wang, Z., 2016. Functional analysis of the 1-aminocyclopropane-1- carboxylate deaminase gene of the biocontrol fungus *Trichoderma asperellum* ACCC30536. Can. J. Plant Sci. 96, 265–275.

Zimand, G., Valinsky, L., Elad, Y., Chet, I., Manulis, S., 1994. Use of the RAPD procedure for the identification of *Trichoderma* strains. Mycol. Res. 98, 531–534.

Chapter 25

Exploring the potential role of *Trichoderma* as friends of plants foes for bacterial plant pathogens

Narasimhamurthy Konappa[a], Udayashankar C. Arakere[b], Soumya Krishnamurthy[c], Srinivas Chowdappa[a] and Sudisha Jogaiah[d]

[a]*Department of Microbiology and Biotechnology, Bangalore University, Bengaluru, Karnataka, India;* [b]*Department of Studies in Biotechnology, University of Mysore, Mysore, Karnataka, India;* [c]*Department of Microbiology, Field Marshal K M Cariappa College, A Constituent College of Mangalore University, Madikeri, Karnataka, India;* [d]*Laboratory of Plant Healthcare and Diagnostics, PG Department of Biotechnology and Microbiology, Karnatak University, Dharwad, Karnataka, India*

25.1 Introduction

Agriculture is still considered to be backbone of Indian economy, even after seven decades after independence, with more than 60% of the people dependent directly for their living. Farming sector in India contributes significantly to the national economy. Plant pathogens are a serious danger to worldwide food safety, about 37% of crop damage is because of pests; of which 12% crop damage because of pathogens (Sharma, 2012) and causing in cost of billions of dollars of farm crop. Biotic stress of plant diseases includes fungi, viruses, viroids, bacteria, phytoplasmas, protozoa and few nematodes. Bacterial phytopathogens are reported to cause maximum damaging diseases and enforce maximum damages to both normal and production methods.

Most plant disease control methods use agrochemicals and it usage has reached to a peek. There are numerous agriculture chemicals including fungicides, bactericides, insecticides fertilizers and many more (Sudisha et al., 2005; Jogaiah et al., 2007). Currently used agrochemicals share a wide range among them, with total pesticide usage has also increasing day by day to about 45.39 thousand tons are used worldwide. Use of chemical pesticides entails an exorbitant financial burden to the population and developing nation like India (Satapute et al., 2019a,b).

A decrease or removal of organic pesticides in agronomy is greatly essential. In order to overwhelmed such harmful management methods, investigators, researchers around the world paid more care to improvement of alternate approaches like, by definition, harmless in surroundings, non lethal to humans, animals and are ecofriendly. One such approach is usage of biological control agents (BCAs) to manage plant pathogens (Harman et al., 2004; Jogaiah et al., 2013). Biocontrol is the inhibition of pathogens by the treatment of a beneficial organisms or BCAs generally a mold, bacterium or virus, or mixture of these to plant or soil (Junaid et al., 2013). The cost of the worldwide microbial based pesticide market by 2019 is predictable to attain about $4556.37 Million, the yearly progress rate of 15.30% from 2014 to 2019. There are up to 175 reported biopesticide active agents and they have been used in control of pathogens, weeds, insects and nematodes (Singh et al., 2014). Presently, the number of beneficial microorganisms have been revealed and are existing as marketable products against plant diseases; several biocontrol agents viz., *Ampelomyces* sp., *Bacillus* sp., *Streptomyces* sp., *Candida* sp., *Coniothyrium* sp., *Cryptococuss* sp., *Fusarium* sp., *Gliocladium* sp., *Penicillium* sp., *Phythium* sp., *Phlebiopsis* sp., *Pseudomonas* sp., *Agrobacterinum* and *Trichoderma* sp. (Ramamoorthy et al., 2002; Jeyaseelan et al., 2012; Murali et al., 2013; Jogaiah et al., 2013, 2016; Naher et al., 2014).

Among the advantageous fungi, *Trichoderma* sp. have increased much attention because of their great generative capability, survival in adverse circumstances, productive producers of secondary metabolites and capacity to defense against different phytopathogens with diverse mode of action (Contreras-Cornejo et al., 2016; Jogaiah et al., 2018).

Biopesticides. https://doi.org/10.1016/B978-0-12-823355-9.00002-X

Hence, *Trichoderma* sp. is currently known as useful BCAs global. *Trichoderma* sp. is successful antagonists having biocontrol abilities against economically important plant parasitic soil-borne pathogens and present abundantly in almost all type of soils (Shahid et al., 2014). Globally, members of genus *Trichoderma* occupy the upper slot in the section among fungal BCAs (Choudhary et al., 2013). It is one of the well known filamentous fungi, usually broadly disseminated and ubiquitous in almost all types of soils (Woo et al., 2014; Singh et al., 2014, 2016), rotting vegetation, plant material, wood (Brotman et al., 2013).

Numerous *Trichoderma* sp. can be categorized as avirulent plant symbionts (Harman et al., 2004; Hafez et al., 2013). Around year 2013, several new *Trichoderma* species were discovered and involves more than 200 phylogenetically divergent species on the basis of rpb2 sequences (Atanasova et al., 2013). In India around 250 *Trichoderma* products are obtainable for field treatments (Woo et al., 2014) but the % of share of biofungicides is only a minor share of the fungicides marketplace and led by artificial chemicals. The various *Trichoderma* sp. has long existed recognized not only controlled pathogens, but additionally for their capability to improve plant growth (biofertilizer) and stimulate plant defense mechanisms (Hermosa et al., 2012; Keswani et al., 2014; Chen et al., 2015). Different biocontrol mechanisms in management of severity of crop diseases comprises mycoparasitism, induced systemic resistance, antibiosis, stimulates the stress tolerance, production of antibiotics, metabolites and hydrolytic enzymes, phytohormones, solubilization nutrients and antagonism for nutrients and space (Fig. 25.1) (Singh, 2014; Bisen et al., 2015; Keswani et al., 2016). Numerous *Trichoderma* sp. are able to produce several plant defense eliciting microbe associated molecular patterns (MAMPs) for example swollenins, peptaibol, xylanases, and ceratoplatanins (Druzhinina et al., 2011). Very recently, the spores and culture filtrate of *Trichoderma virens* differentially induce resistance and elicitation of resistance against *Fusarium oxysporum* was through Jasmonic and salicylic acid signaling cascades (Jogaiah et al., 2018). The current chapter emphases on studies of mode of action of *Trichoderma* sp. against phytopathogenic bacteria and highlights the importance of Trichogenic-nanoparticles and its application in plant disease control.

25.2 Mechanisms

25.2.1 Competition with pathogens for space and nutrients

Starvation is the utmost reason for the decease of soil microbes (Benitez et al., 2004), therefore that competition for blocking nutrients consequences in biocontrol of plant pathogens. Competition is a phenomenon in which *Trichoderma* sp. and pathogen compete for limited nutrient and space availability (Harman, 2000). Both the BCAs and the phyto pathogens compete with one another for nutrients to grow in the surroundings. Among all the mechanisms, nutrient competition are the most significant that prevent infection from the pathogen (Verma et al., 2007). *Trichoderma* sp. has

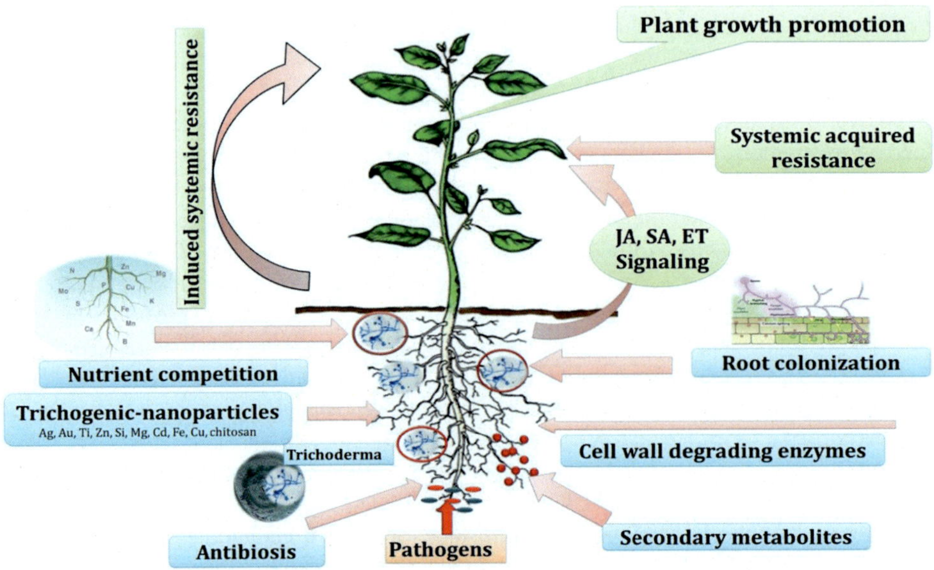

FIG. 25.1 Mechanism of action by *Trichoderma* species in plants against plantpathogens.

more ability to mobilize soil nutrients and their take up as compared to other microbes. Members of genus *Trichoderma* are fine rhizosphere proficient and efficient soil colonizers, which make the main proficient competitor alongside other soil microflora. They inhabit same places as the pathogen use the same nutrients and can inhabit access points to the plant tissues that would be used by the pathogen thus averting disease by the pathogen. *Trichoderma* sp. is fast developing fungi and hence effective contenders for space and nutrients and commonly considered as an aggressive competitor against soil borne pathogens and rapidly colonizes substrates to exclude pathogens (Cuervo-Parra et al., 2014). Significant information on nutrition of *Trichoderma* sp. are available in works but very slight is known about specific carbon and nitrogen nutrients on mass production of *Trichoderma* antagonists (Rajput et al., 2014). Competition for micro nutrients occurs for BCAs have more effective consuming acceptance system for the elements than the pathogens (Nelson, 1990).

The main mechanism of pathogen control through competition for nutrients comprises the production of composites like siderophores that competently sequester iron and eliminate the phytopathogen from this essential component (Raaijmakers et al., 2002). In most of the filamentous fungi, iron application is necessary for feasibility and most molds produce small molecular ferric iron definite chelators known as siderophores under iron starvation that activate ecological iron (Eisendle et al., 2004). In certain conditions, siderophore secretion and modest attainment in obtaining Fe^{3+} is the mechanism by which BCAsmanage plant infections (Santoyo et al., 2012). In *Trichoderma* sp., intracellular siderophores are produced by three non-ribosomal protein synthases (NRPs), which are present as a cluster in the genome (Mukherjee et al., 2012a; Perazzoli et al., 2012; Zeilinger et al., 2016). Lehner et al. (2013) assessed eight different strains of Trichoderma, together with *T. asperellum*, *T. reesei T. atroviride*, *T. virens*, *T. polysporum*, *T. gamsii*, *T. hamatum*, and *T. harzianum* and observed that on typical *Trichoderma* sp. formed 12 to 14 siderophores, with six communal to altogether species.

It is reported that *T. harzianum* CECT 2413 have a gene which codes a great affinity glucose transporter (Gtt1). This strain is present in environments extremely deprived in nutrients and it depends on extracellular enzymes for existence. The important exudates of glucose absorption comprise absorption permeases and enzymes, simultaneously by proteins intricate in membrane with cell wall alterations. Whereas the part of the glucose transport structure leftovers to be exposed, its competence may be vital in antagonism (Delgado-Jarana et al., 2003). Fascinatingly, Gtt1 is said at extremely low glucose attentions alike to the circumstances of competence among microbes (Benõtez et al., 2004).

25.2.2 Antibiosis

The antibiosis may be defined as antagonistic communication connecting low molecular weight diffusible secondary metabolites (SMs) or non-specific metabolite of microorganism source by lytic enzyme, volatile compounds or additional toxic material produced by *Trichoderma* strains that have a straight effect on the development of plant pathogen (Howell, 2003; Limon et al., 2004; Viterbo et al., 2007). Secondary metabolites and antibiotics formed by *Trichoderma* sp. show a vital role in antagonistic activity (Ajitha and Lakshmedevi, 2010). *Trichoderma* species cause decay of phytopathogenic microbes without any physical contact between microorganisms by producing the antimicrobial compounds. This process generally called as "antibiosis" and term secondary metabolites is a group of heterogeneous chemically divergent natural complexes may show significant roles in the defense reaction, symbiosis, stimulating or inhibiting spore formation and germination, metal transport, differentiation and competition against organisms etc. for the producing organism (Vinale et al., 2012; Keswani et al., 2014; Bisen et al., 2015). A multiple chemical contact is recognized among *Trichoderma* sp. and their hosts involving volatile and diffusible SMs, small peptides, and/or antibiotics, which effect root development, branching and absorptive ability (Lopez-Bucio et al., 2015).

Antibiotic production is one of the most vital biocontrol abilities. *Trichoderma* sp. produces several secondary metabolites with antibiotic activities and their production is species/strain dependent (Zeilinger et al., 2016). The SMs produced by *Trichoderma* sp. are of 3 categories: (i) volatile metabolites, namely 6-pentyl-a-pyrone (6 PP) and maximum of the isocyanide bases; (ii) water soluble metabolites, i.e. koningic acid or heptelidic acid and (iii) peptaibols, which are lined oligopeptides of 12–22 amino acids rich in α-aminoisobutyric acid, N-acetylated at the N-terminus and comprising an amino alcohol (Pheol or Trpol) at the C-terminus (Rebuffat et al., 1989). The analytical reports portray that 373 different compounds or antibiotics produced from *Trichoderma* sp. against their phytopathogens for example volatile and nonvolatile toxic metabolites for example alamethicins, tricholin, harzianic acid, peptaibols, 6 PP, formic aldehyde, glisoprenins, acetaldehydes gliotoxin, viridian, Terpenoids, heptelidic acid, harzianopyridone, harziandione, gliovirin, peptaibols, massoilactone, viridin, trichodermin, epipolythio dioxopi perazines (ETPs), (Gajera et al., 2013; Hermosa et al., 2014; Strakowska et al., 2014; Bae et al., 2016; Contreras-Cornejo et al., 2016). *Trichoderma* sp. also produces siderophores and a great number of peptaibiotics known as peptaiboles, which comprise with great incidence non standard amino acids (Degenkolb et al., 2006). *Trichoderma* can yield a mass of composites with antagonistic effects comprising cell wall degrading enzymes (CWDEs) viz., lipase, xylanase, amylase, pectinase, glucanase, cellulase, arabinase, and protease (Hermosa et al., 2014; Strakowska et al., 2014).

Gliotoxin and glyoviridin from *T. virens* (Wilhite et al., 2001), viridin, alkylpyrones, isonitriles, polyketides, peptaibols, diketopiperazines and sesquiterpenes isolated from *Trichoderma* sp. (Reino et al., 2008), trichodermin (Tijerino et al., 2011) and 6 PP (Garnica–Vergara et al., 2016) are some of the examples which are found effective against target phytopathogen in vitro and/or in situ conditions. For instance, gliotoxin is produced by Q strains of *T. virens* whereas gliovirin is synthesized by the P strains; together have tough antagonistic actions (Mukherjee et al., 2012b, 2013; Ruano-Rosa et al., 2014; Scharf et al., 2016). Marfori et al. (2002) reported about dual culture of *T. harzianum* with *Catharanthus roseus* callus formed an antagonistic complex trichosetin with an amazing action against Gram positive bacteria *B. subtilis* and *S. aureus*. *T. harzianum* NF-9 acts antagonistic against *Xanthomonas oryzae* in rice plants (Harman et al., 2004). Sadykova et al. (2015) conducted the antagonistic activity of 42 *Trichoderma* sp. amongst 8 strains viz., *T. asperellum*, *T. viride*, *T. citrinoviride*, *T. hamatum*, *T. koningii*, *T. atroviride*, *T. harzianum* and *T. longibrachiatum*.

The collection of CWDEs and metabolites exhibited in an advanced level of antagonistic activity than that achieved by whichever mechanism only (Monte, 2001). Dubey (2002) described that *Trichoderma viride* and *T. harzianum* were greatest antagonistic activity for inhibition of growth of numerous soil and seed borne phytopathogens. Mishra et al. (2011) observed that in clear zone of inhibition was showing antibiosis among pathogen and *Trichoderma* in dual culture method. Vinale et al. (2014) displayed that the 6 PP filtered from the culture filtrate of several *Trichoderma* sp. such as *Trichoderma viride*, *Trichoderma koningii*, *Trichoderma atroviride* and *Trichoderma harzianum* has revealed equally laboratory and in plant antimicrobial actions to numerous plant pathogens.

Plant growth promoting properties of *Trichoderma* are also reliant on auxin buildup in tomato plants (Martinez-Medina et al., 2014). Earlier studies shown that *T. atroviride* formed 6 PP a volatile compound which stimulated plant development and controlled root construction, preventing primary root development and stimulating lateral root development (Garnica-Vergara et al., 2016). *A. thaliana* seedling when exposed to volatile compound produced by *Trichoderma* improved root dividing and biomass creation and enhanced flowering (Contreras-Cornejo et al., 2014). The siderophores binds essential metals ions such as iron (Vinale et al., 2013) and they promote nutrients uptake and growth of plant (Vinale et al., 2013). Siderophores produced by beneficially microorganism is deliberated vital for host iron uptakes, mostly in calcareous soil (Vinale et al., 2013).

25.2.3 Cell wall degrading enzymes

Several species of *Trichoderma* are "rhizosphere competent" along with are capable to break carbohydrates and chlorophenolic compounds, the xenobiotic insecticides used in agronomy (Harman et al., 2004). Some biological control agents secrete CWDEs capable to degrade proteins, chitin, and cellulose, hemicellulose, thus showing to direct inhibition of various phytopathogens. Secretion and release of CWDEs is the main mechanism usage of BCAs to manage soil borne pathogens (Kumar et al., 2015). *Trichoderma* sp. produces CWDEs that degrade cell wall of pathogens (Howell, 2003) and particular of these CWDEs are of commercial significance (Srivastava et al., 2014). These enzymes act the mechanical integrity of cell wall of the pathogen (Budi et al., 2000). The well known *Trichoderma* sp. are straight antagonize a variety of phytopathogens as they are capable of identifying others in the soil and its shattered other phytopathogens over appearance of CWDEs, usually glucanases chitinases with proteases (Harman et al., 2004). The CWDEs, are not only intricate in parasitism for nutritious drives, but are also significant for cell wall recycling through mature and autolysis as well as cell wall restoration through vigorous growth (hyphal dividing) (Gruber and Seidl Seiboth, 2012). Exudation of a variability of carbohydeate degrading enzymes comprising glucanases, chitinases, proteases and cellulases is a communal feature of fungal and bacterial BCAs (Jan et al., 2011). Certain CWDEs can be intricate in both antagonistic and saprophytic progressions provided that an evolutionary benefit to species with both degrading and incompatible possible, for the effective establishment of diverse biological niches in soil.

An important portion of the *Trichoderma* capability contains in the synthesis and exudation of an excessive variability of extracellular CWDEs, such as chitin-1,4-β-chitobiosidases, amylases, phospholipases, endochitinases, xylanases, N– acetyl-β-D-glucosaminidase, β-1,6-glucanases, proteases, β-1,3- glucanases, mananases, lipases, pectinases, DNAses, RNAses, etc. (Sandhya et al., 2004; Vahabi et al., 2011; Lopes et al., 2012; Geraldine et al., 2013; Vos et al., 2015). The oligocarbohydrares formed from breakdown of pathogen cell walls act as indicating particles to induce the host resistance mechanisms. Certain have been filtered, categorized and their coding genes cloned (Viterbo et al., 2002; Montero et al., 2007). Once filtered, several enzymes from genus *Trichoderma* have revealed to have great antimicrobial action against an extensive range of plant pathogens.

Genome examination allowed valuation of the entire numbers of CWDEs encrypted in the *Trichoderma* sp. genomes and unraveled even additional multifaceted enzymic deprivation mechanism for cell walls of fungi than earlier predicted (Kubicek et al., 2011). Earlier evidences proposed the involvement of chit42, chit3, bgn13.1, Bgn2 Bgn3 and prb1 are involved in chitinases, glucanases and proteases in biocontrol (Mondejar et al., 2011). Concurrently genes coding degrading enzymes such as glucanases, and chitinases and those for SMs like nonribosomal peptide synthetases (NRPSs) path be stated to destroy the phytopathogens (Kubicek et al., 2011; Perazzoli et al., 2012). Genetic indication has been providing for the association of 11 & 14 modules peptaibols via an only NRPS Tex2 from *Trichoderma virens* species (Mukherjee et al., 2011).

Due to higher production and action of reactive oxygen species (ROSs) scavenging enzymes for example catalase ascorbate, peroxidase, superoxide dismutase and are induced in tomato plants colonization of *T. harzianum* T22 in water shortage condition consequences in improved plant growth (Mastouri et al., 2012). In the previous few years investigation of signaling pathways essential genus Manczinger et al. (2002) partitioned 18 *Trichoderma* sp. for their capability to damage bacterial cells by *Trichoderma harzianum* T19, which damage cells of *Bacillus subtilis* through exudation as a minimum 3 trypsin like proteases, 6-chymotrypsin like proteases with 4 NAGases, which are chief enzymes associated by degradation of bacterial cells. Other lytic enzymes and peroxidase, lactases and formed by different *Trichoderma* sp. are possible influences helping in degradation of aromatic hydrocarbons (PAHs). Deprivation underlying of rhizosphere competent *Trichoderma* sp. alongside numerous artificial dyes, pentachlorophenol, endosulfan, and dichlorodiphenyl trichloroethane (DDT) were confirmed earlier (Katayama and Matsumura, 1993). *Trichoderma* sp. may show an important role in the biodegradation of soil polluted with insecticides and takes the capacity to degrade an extensive variety of pesticides (Harman et al., 2004).

25.2.4 Plant growth promotion

Several BCAs, for example both fungal and bacterial, are not only able to control the plant pathogens, but are too able to stimulate plant development and growth (Hermosa et al., 2012; Chen et al., 2015). Numerous saprotrophic fungi, particularly certain isolates of *Trichoderma* sp. are described to escalation the fertility of soils plus expressively developed the plant growth outside control of crop diseases (Oliveira et al., 2012; Shukla et al., 2012). The several independent research groups have observed the increased growth of various crops with the use of antagonist *Trichoderma* sp. (Hermosa et al., 2013; Stewart and Hill, 2014). Advancement of plant growth is one of the advantageous characters of genus *Trichoderma* (Contreras-Cornejo et al., 2016). *Trichoderma* sp. inhabits roots and affords signaling to the plants to elicit the formation of plant growth regulators and to influence systemic resistance against phytopathogens. *Trichoderma* sp. has been recommended as plant growth promoting fungi (PGPF) because of its capability to secrete siderophores, phosphates solubilizing enzymes with plant hormones (auxin, cytokinin or gibberellins) or growth factors (Doni et al., 2013; Contreras-Cornejo et al., 2014; Martínez Medina et al., 2014).

The promotion of plant growth influence of *Trichoderma* sp. treatments has been qualified to numerous mechanisms intricate on plants, such as the liberation of materials with auxin activity, lesser peptides in addition to VOCs, which increase root architecture and solubilization macro and micronutrients, thus increasing plant development and crop yield, blockage of enzymatic pathways of pathogens, defense to different phytopathogens (Lorito and Woo, 2015; Diánez Martínez et al., 2016; Rouphael et al., 2017). The plant growth improvement stimulated by *Trichoderma* sp. can be clarified by an up regulation of photosynthesis associated proteins and a greater photosynthetic effectiveness (Shoresh et al., 2010). Properties of the *Trichoderma* sp. in stimulating plant establishment stimulate plant growth of several forms of plants and elicitation of host defense reaction to defend them against phytopathogens (Singh et al., 2016). Increased shoot and roots length, weight, of pod weight and nodules numbers per plant was indicated in peanut with *T. harzianum* (Kamaruzzaman et al., 2016). Plant development dignified as lengths of root and shoot were expressively greater in *T. harzianum* treated chilli seeds (Joshi et al., 2010).

Zhang et al. (2014) revealed that *T. longibrachiatum* was found fairly effective for stimulating plant development and nematode control in wheat. Growth stimulating capacity of *Trichoderma harzianum* Th3 has been demonstrated in wheat cultivar Raj 3765 at agriculturalist's field in Rajasthan's Jaipur and Kota district. Its application enhances rootlets, tillers, grains weight and grain yield 36.25–46.73Q/ha (29% in Jaipur) and 36.88 to 50.12Q/ha (36% in Kota) significantly (Sharma et al., 2012). Brotman et al. (2010) observed *Trichoderma* sp. can stimulate growth in tomato and tobacco plants, improvement of up to 300%. Treatments of *Trichoderma viride* and *Trichoderma harzianum* only plus mixed with biofertilizers stimulate plant development successfully. It was detected that 50% *Trichoderma* sp. only or in mixture with biofertilizers was expressively improved the shoot length, root weight, shoot weight, root length, nodulation and grain crop observed expressively (Hermosa et al., 2001).

Bezuidenhout et al. (2012) described *T. harzianum* can synthesize a compound gliotoxin which can simulant the phytohormone gibberellic acid which is concerned in seed germination. The seed or soil applications of *Trichoderma* sp. increases percent seed germination, triggering enzymes, phytohormones and modifying soil microorganisms and nutrient obtainability in soil. Among 89 *Trichoderma* sp., *T. harzianum* which commonly multiplies at the plants rhizosphere have several mechanisms to control fungal pathogens thus management numerous plant infections, and stimulate host health, development and seedling vigor (Hajieghrari, 2010; Sultana et al., 2012).

In maize, *Trichoderma* sp. increased growth, improved root biomass production with improved root hair growth (Harman et al., 2004). Seeds treated with *T. viride* enhanced seedling fresh with dry weight of shoot, root plus broad beans nodules (Woo et al., 2006). Christopher et al. (2010) worked with several *Trichoderma virens* isolates under in vitro plus greenhouse condition for promoting plant growth in tomato. Among the various isolates tested, *T. virens* (Tv1) increased the plant growth. In maize seedlings, *Trichoderma* sp. inoculation moves root system structure comprising improved biomass of root and improved root hair growth (Harman et al., 2004). Sundaramoorthy and Balabaskar (2013) evaluated that efficacy of *Trichoderma* sp. to stimulate the plant development and yield factors of tomato under in vitro and in vivo conditions.

Secretion of phytohormones by *Trichoderma* sp. treated plant stands the key for plant growth improvement (Chowdappa et al., 2013). *Trichoderma* sp. are reported to produce cytokinin like molecules and gibberellins related molecules (GA3 or GA4) which could be beneficial for developments of plant potency (Idowu et al., 2016; Kashyap et al., 2017). *Trichoderma* sp. synthesizes IAA, the major auxin in florae (Yue et al., 2014; Enders and Strader, 2015), acting as a plant growth promoter. Contreras-Cornejo et al. (2009) have explained the signaling appliance by *Trichoderma virens* stimulates plant growth with improvement in *A. thaliana*. *Trichoderma virens* synthesizes indolic compounds, viz. IAA, IAAld, tryptophol and ICAld (Contreras-Cornejo et al., 2011). Saxena et al. (2015) demonstrated the differential ability of *Trichoderma* sp. in the IAA producing isolate and plant development aspect in which effective results were obtained in *T. asperellum* and *T. longibrachiatum* isolates in both laboratory along with in vivo experiments. Brotman et al. (2012) revealed that *A. thaliana* transcriptional and metabolic profiling including significant changes of amino acids involved in the biosynthesis of plant hormones and plant defense metabolites. The promotion of plant growth may in part reflect an increased energy source required for the activation of host defense and growth advancement effects mediated by *Trichoderma* sp. (Brotman et al., 2012). Martínez-Medina et al. (2014) described that auxin production was induced and a reduction in cytokinins and abscisic acid content by *Trichoderma* sp. that stimulated the plant growth.

Kotasthane et al. (2015) reported that possible IAA producing strain *T. viride* T14 was stimulated maximum plant growth, and chlorophyll contents, when seed treatment to bottle gourd, cucumber, and bitter gourd. *Trichoderma harzianum* pointedly improved the capacity in rice seedlings to stand drought pressure with progress rice water holding ability (Shukla et al., 2012). Rice plant treated with *Trichoderma* sp. has better uptake of nutrient (Saba et al., 2012). Doni et al. (2014) described that numbers of leaf and tillers has been establish expressively greater in *Trichoderma* sp. applied rice plants and increase in net photosynthetic rate in rice seedlings. Rice plant treated with *T. harzianum* considerably improved the capacity to stand drought and water deficit condition contributing to the better uptake of nutrient and plant development (Doni et al., 2014).

Borges et al. (2015) confirmed the maximum rice biomass with greatest phosphate solubilization effectiveness in rice that had been treated with *T. asperelloides* and *T. harzianum*. The biological activities of *Trichoderma* sp. in soil help in bioavailability of different nutrients and minerals either through chelation or solubilization, and thus make them available at the root surface of the plants which outcomes in improved growth (Jiang et al., 2011; Menezes-Blackburn et al., 2014). Yedidia et al. (2001) recommended the role of *T. harzianum* in nutrient uptake by the host at an actual initial phase of fungal plant interaction. Inhibition of ethylene represents the finest studied mechanism of plant development induced by microorganisms (Nascimento et al., 2014).

Trichoderma asperelloides T203 possesses aα-1-aminocyclopropane-1-carboxylic acid (ACC) deaminase gene acc1 which codes an enzyme that slices ACC, the instantsign of the phytohormone ethylene, to form α-ketobutyrate with ammonia (Kubicek et al., 2011). Inoculation of *Trichoderma* sp. that produces ACC deaminase boosts plant growth elevation by decrease of ethylene (Viterbo et al., 2010). In addition, it has been recommended that equally to ACC deaminase generating bacteria, *Trichoderma* can enhance plant development in abiotic stress situations, by depressing the levels of ethylene in addition to stimulating an increment in anti oxidant action and by modulation polyamine content (Brotman et al., 2013; Salazar-Badillo et al., 2015). Brotman et al. (2010) described that the plant development elevation seems to be interceded by the production of IAA by *Trichoderma* sp. and the enzyme activity of ACC deaminase, which reduces the high levels of ethylene accumulated in plants through diverse types of stresses.

The 6 PP separated from culture filtrate of diverse species for example *Trichoderma viride*, *Trichoderma harzianum*, *Trichoderma atroviride*, *Trichoderma koningii* with 6 PP is intricate in plant development actions and stimulation of disease resistance (Vinale et al., 2008b). In *A. thaliana* root response to 6 PP involves constituents of auxin passage, signaling with ethylene EIN2 response modulator (Garnica-Vergara et al., 2016). The 6 PP foliar application on tomato plant, 6 PP at 0.166 mg/L concentration exhibited moredevelopment and abroad root organization (Vinale et al., 2009; Kotasthane et al., 2015). Harzianic acid is a secreted siderophore molecule synthesized by *T. harzianum* and displays plant development stimulation action and was defined as a new siderophore because of its capacity to bind with decent attraction vital metallic ions (Vinale et al., 2013). Cai et al. (2013) described that harzianolide formed by *Trichoderma* sp. which develops the initial phase of plant growth over improvement of root length. Aplethora of chemically diverse SMs produced by *Trichoderma* sp. have been described as equally plant development agents along with antagonistic activity (Vinale et al., 2009).

25.2.5 Induced systemic resistance (ISR)

Another mechanism usually connected with defense of hosts by BCAs is stimulation of the host protection pathways. Induced systemic resistance is supposed to be among significant mechanisms of biocontrol properties of *Trichoderma* (Harman, 2006). There are 2 different classes of systemic resistance can be discussed to hosts as induced systemic resistance and systemic acquired resistance (SAR) referred as similar phenotypically (Mathys et al., 2012), however biochemical pathways involved differentiates them (Birkenbihl et al., 2017). Any distant contact or permeation in plant roots trigger their immune scheme, though *Trichoderma* sp. modify the host resistant system and documented as nonpathogenic (Samules and Hebbar, 2015). Different defense pathways happen in florae that can stop the antagonistic influence of phytopathogens, which might be incited subsequent contact to pathogens, advantageous microorganisms or some chemical substances. It is characterized by extensive range resistance to phytopathogens of numerous varieties in addition to abiotic stresses are also induced (Pieterse et al., 2014). *Trichoderma* sp. is significant for variable several genes intricate in host defense to stresses or for improving the host basal metabolism (Domínguez et al., 2016).

Jasmonic acid (JA), salicylic acid (SA), ethylene (ET) and abscisic acid (ABA) responses to different stresses by inducing several signaling paths that have multifaceted systems of antagonistic relations (Verma et al., 2016). The systemic acquired resistance is reported to be salicylic acid dependent pathway, whereas induced systemic resistance is independent of salicylic acid pathway. Numerous known *Trichoderma* sp. are capable to stimulate JA and ET production which are intricate in ISR increase (Nawrocka and Malolepsza, 2013). It is usually activated by indigenous infection, associating with the association of SA with stimulation of pathogenesis related proteins (PRs) genes (Kumar et al., 2015). It can offer extended period resistance through the plant to following different pathogen attack. Following reactions include SA and JA/ET signaling, which results in complete host obtaining variable steps of resistance to phytopathogen attack (Shoresh et al., 2010). This happens as an importance of the production of elicitors (antibiotics, proteins, and volatiles) by the BCAs that activate the expression of the genes of the SA or the JA/ET pathways (Pieterse et al., 2014; Birkenbihl et al., 2017). *Trichoderma* colonization on root and leaf tissues improved initiation of JA related defense reactions important to greater resistance to phytopathogens (Lorito et al., 2010; Tucci et al., 2011; Hermosa et al., 2012).

Saksirirat et al. (2009) described *T. harzianum* T9 induces resistance in tomato against *X. campestris* pv. vesicatoria, 14 days post inoculation with growing actions of β-1,3 glucanase and chitinase. Narasimhamurthy et al. (2018) observed *T. asperellum* could protect tomato plants against *R. solanacearum* (Fig. 25.2). The two strains of *Trichoderma asperellum* isolate documented effective prevention against *R. solanacearum* (Narasimhamurthy et al., 2013a,b). Outcome of different *Trichoderma* sp. to bacterial phytopathogens was listed in Table 25.1.

FIG. 25.2 Effect of *T. asperellum* seed treatment in tomato on resistance against bacterial wilt pathogen *R. solanacearum* under greenhouse conditions. (a) Control, (b) Tomato plants grown from seeds treated with *T. asperellum*.

TABLE 25.1 Effect of different *Trichoderma* sp. against bacterial phytopathogens.

Trichoderma sp.	Crop	Pathogen	References
T. asperelloides T203	Tomato	*P. syringae* pv. *tomato*	Brotman et al. (2012)
T. asperelloides	Cucumber	*P. syringae*	Brotman et al. (2008)
T. asperellum	Cucumber, Maize	*P. syringae*	Viterbo and Chet (2006)
T. asperellum	Cucumber	*P. syringae* pv. *lachrymans*	Shoresh et al. (2005)
T. asperellum	Tomato	*R. solanacearum*	Narasimhamurthy et al. (2013a,b, 2018)
T. asperellum SKT-1	Arabidopsis	*P. syringae* pv. *tomato* DC3000	Yoshioka et al. (2012)
T. asperellum T-203	Cucumber	*P. syringae* pv. *lachrymans*	Yedida et al. (2003)
T. asperellum T 34	Arabidopsis	*P. syringae* pv. *tomato*	Segarra et al. (2009)
T. atroviride & *T. virens*	Tomato	*P. syringae* pv. *tomato*	Salas-Marina et al. (2015)
T. harzianum	Tomato	*Clavibacter michiganensis* subsp. *michiganensis*	Utkhede and Koch (2004)
T. harzianum	Tomato	*Xhantomonas campestris* pv. *vesicatoria*	Suarez-Estrella et al. (2013)
T. harzianum	Tomato	*Ralstonia Solanacearum*	Liza and Bora (2009)
T. harzianum	Cotton	*Xhantomonas campestris* pv. *malvacearum*	Raghavendra et al. (2013)
T. harzianum	Tobacco	*R. solanacearum*	Yuan et al. (2016)
T. harzianum	Tomato	*Xanthomonas campestris* *Clavibacter michiganensis*	Jawad and Zafar (2017)
T. harzianum	Brinjal	*R. solanacearum*	Barua and Bora (2008)
T. harzianum	Potato, Tobacco, Tomato	*R. solanacearum*	Maketon et al. (2008)
T. harzianum	Bean	*P. syringae* pv. *phaseolicola*	Gailte et al. (2005)
T. harzianum	Tobacco	*Pseudomonas syringae*	Dana et al. (2006)
T. harzianum & *T. asperellum*	Tomato	*Xantomonas campestris* pv. *vesicatoria*	Saksirirat et al. (2009)
T. harzianum & *T. viride*	Potato	*Erwinia carotovora*	Sandipan et al. (2015)
T. harzianum & *T. viride*	Banana	*Ralstonia solanacearum* race 2	Ceballos et al. (2014)
T. harzianum T22, *T. atroviride* P1	Bean	*X. campestris* pv. *phaseoli*	Harman et al. (2004)
T. harzianum T23	In vitro	*Clavibacter michiganensis* and *Erwinia amylovora*	El–Hasan et al. (2009)
T. harzianum, *T. viride*, *T. koningii* and *T. longibrachiatum*	Ginger	*R. solanacearum*	Roop and Jagtap (2017)
T. harzianum, *T. virens* and *T. asperellum*	Tomato	*Ralstonia* sp.	Yendyo and Pandey (2018)
T. Koningii SMF2	In vitro	*Ralstonia solanacearum, Erwinia carotovora* pv. *catotovora* and *Clavibacter michiganensis* ssp. *michiganensis*	Xiao–Yan et al. (2006)
T. reesei	Arabidopsis and Tomato	*Clavibacter michiganensis*	Saloheimo et al. (2002)

TABLE 25.1 Effect of different *Trichoderma* sp. against bacterial phytopathogens.—cont'd

Trichoderma sp.	Crop	Pathogen	References
T. virens	Cucumber	*Pseudomonas syringae*	Viterbo et al. (2007)
T. virens PS1–7	Rice	*Burkholderia plantarii*	Wang et al. (2013)
T. harzianum	Rice	*X. oryzae* pv. *oryzae*	Harman et al. (2004), Hastuti et al. (2012)
Trichoderma GT3–2	Cucumber	*P. syringae* pv. *lachrymans*	Koike et al. (2001)
Trichoderma hamatum 382	Radish and tomato	*X. campestris* pv. *armoraciae* and *Xanthomonas campestris* pv. *vesicatoria*	Aldahmani et al. (2005)
Trichoderma harzianum	Rice	*Xantomonas oryzae* pv. *oryzae*	Gokil-Prasad and Sinha (2012)
T. harzianum, *T. viride*, *T. koningii*, *T. virens*	Ginger	*R. solanacearum*	Raghu et al. (2013)
Trichoderma sp.	Cabbage	*Xanthomonas campestris*	Gan et al. (2010)
Trichoderma sp.	Tomato	*Xhantomona euvesicatoria*	Fontenelle et al. (2011)
Trichoderma sp.	Tomato	*R. solanacearum*	Sudisha et al. (2013)
Trichoderma sp. *T. harzianum*	Tomato	*Xanthomonas euvesicatoria*	Fontenelle et al. (2011)
Trichoderma viride, *T. harzinum*	Chilli	*Xanthomonas axonopodis* pv. *vesicatoria*	Sharma (2018)

Jayalaksmi et al. (2009) described increased PPO activity in chickpea by the treatment with *T. harzianum* strain L1, implicating it in induced defense responses against root rot in chickpea. Certain *Trichoderma* SMs may performance as elicitors of host resistance mechanisms tophytopathogens. Different strain of *Trichoderma* produces small SMs and shown to induce pathogenesis-related (PR) protein and reduces diseases symptoms systemically (Vinale et al., 2008a; Maischak et al., 2010). Peptaibols are linear peptide antibiotics of 5–20 amino acid remains produced by NRPs activity. Alamethicin, a 20 mer peptaibol from *Trichoderma viride*, stimulates JA and SA biosynthesis (Engelberth et al., 2001), whereas 18 mer from *T. virens* stimulate ISR in cucumber against *P. syringae* pv. Lachrymans (Viterbo et al., 2007; Luo et al., 2010). Treatment of alamethicin of *Trichoderma viride* stimulated SA and JA mediated resistance responses in *Phaseolus lunatus* (Engelberth et al., 2001). Harzianolides, pyrones (esters), harzianopyridone and other oxygen heterocyclic compounds groups, were defined as latentelicitors of plant resistance appliances and plant development (Mukherjee et al., 2012b).

25.3 Trichogenic-nanoparticles and its application in crop protection

Alongside natural and biological control of plant pathogens, in current years, planned nanoparticles (NPs) have attained specific care as a potential competitor for increase crop yield, resistance and disease control tools (Elamawi and Al-Harbi, 2014; Elamawi and El-Shafey, 2013; Abdelmalek and Salaheldin, 2016). Role of NPs in farming is known as nano-agriculture; novel skill is frequently useful to increase the crop yields (Duhan et al., 2017). Nanoscale science and nanotechnologies are intended to have the possible to transform farming and food structures (Norman and Hongda, 2013) and has given birth to the new era of agro-nanotechnology. Nanopesticides are one of a new approaches being used to address the problems of pesticides. Nanopesticides cover varied types of products, some of which are already in the market. The term nanotechnology, byword of current day science owes its origin from the Greek word "nano" literally meaning dwarf. When it is stated in terms of measurement one nanometer is one billionth of one meter ($1 \text{ nm} = 10^{-9} \text{ m}$) (Morones et al., 2005).

Nanoparticles produced from various natural sources can be useful in farming (Kaur et al., 2018). In green nanotechnology, for the production of NPs from plants and microorganisms were used. *Trichoderma* sp. succeeds under variable ecological conditions since high flexibility to development regulate, sporulation and production of degrading enzymes (Harman et al., 2012). The usage of *Trichoderma* sp. in making metallic NPs has established important attention as they proposal definite benefits over the use of bacteria for the production of NPs. Most *Trichoderma* sp. has a great tolerance to metals and high wall binding competences, in addition to essential modest nutrient, as well as intracellular metal uptake abilities. Initial studies showed the potential of NPs in increase seed germination and development, pathogen detection, plant protection, and pesticide/herbicide remainder detection (Bhainsa and D'Souza, 2006).

Biosynthesis of selenium, gold, silver, platinum, gold, silica, zirconium, zinc oxide, copper, titanium, chitosan and magnetite NPs by fungi, bacteria, viruses, and yeast have been reported (Narayanan and Sakthivel, 2010; Kaur et al., 2018). Treatment of silver nanoparticles (AgNPs) in soil and as seed/seedling coatings may not only control the plant pathogen, but also improve plant development by numerous known and unidentified mechanisms. Silver is also an excellent plant growth stimulator (Sharon et al., 2010).

Different fungal species are a capable claimant for the creation of nanoparticles in both intra as well as extracellular. Some *Trichoderma* sp. such as *Trichoderma harzianum*, *Trichoderma asperellum*, *Trichoderma pseudokoningii*, *Trichoderma longibrachiatum* and *Trichoderma virens* being used for production of AgNPs (Devi et al., 2013). Gaikwad et al. (2013) described that AgNPs were synthesized from *T. asperellum*. Potential anti-microbial ability of AgNPs have been studied by some researchers (Prabahu et al., 2010; Roy et al., 2013; Abbas et al., 2015; Zakharova et al., 2017). The NPs have latent scenarios of use in plant disease control in different methods (Khan and Rizvi, 2014; Huang et al., 2015; Mishra et al., 2016). Influence of AgNPs on crops such as wheat (Jhanzab et al., 2015), barley (Gruyer, 2014), Brassica (Pallavi et al., 2016), radish (Zuverza-Mena et al., 2016) has been described and no harmful influence of NPs was found on several plants. Gan et al. (2010) studied antibacterial activity of nanosilver against *X. campestris* toward control of black rot in cabbage. Supernatant of seed culture was used for the biosynthesis of AgNPs from *T. koningii* and assessment of their antibacterial action (Tripathi et al., 2013). Kamran et al. (2011) described that the AgNPs and TiO_2NPs with a decent potential may be used for eliminating the bacterial pathogens in the tobacco. The selenium nanoparticles were formed from different *Trichoderma* sp. for example *T. asperllum*, *T. atrovirie*, *T. harzianum*, *T. virens*, *T. longibrachiatum* and *T. brevicompactrum* found hexagonal, near spherical and irregular shape and achieved helpful outcomes (Yadav et al., 2018).

25.4 Conclusions

The increasing concern for serious environmental and health problems because of chemical pesticides application has resulted in improvement of alternate methods for management of plant pathohens. *Trichoderma* sp. possess numerous potentials and they have abundant potential use in farming contains greater plant growth, crop yield, improving physiological response to stresses. The members of *Trichoderma* sp. are extensively used in farming and production areas because of its production of significant different CWDEs such as glucanases, chitinases, and proteases and these enzymes have an important role in biocontrol activity. Nanoparticles with unique properties can be easily produced from biological sources and can be useful in farming. Nanotechnology uses in basic farming, value addition, protection of crops and food can therefore carry a change in the agriculture scenario of India. Biosynthesized NPs could be used successfully against phytopathogens to defend the numerous crop plants and their yields, all the NPs have no harmful. Mycosynthesis in specific is costeffective and ecofriendly; it is now a significant branch of Bionanotechnology and is denoted to as myconanotechnology. Nanotechnology can be applied in all features of the food chain, both for successful food security and quality control, and as new food constituents or flavors and NPs have promising upcoming for the current farming practices like precision delivery of nutrients and fertilizers and disease identification at an initial stage.

References

Abbas, A., Naz, S.S., Syed, S.A., 2015. Antimicrobial activity of silver nanoparticles (AgNPs) against *Erwinia carotovora* pv. *carotovora* and *Alternaria solani*. Int. J. Biosci. 6, 9–14.

Abdelmalek, G.A.M., Salaheldin, T.A., 2016. Silver nanoparticles as a potent fungicide for citrus phytopathogenic fungi. J. Nano Res. 3, 15406.

Ajitha, P.S., Lakshmedevi, N., 2010. Effect of volatile and von–volatile compounds from *Trichoderma* spp. against *Colletotrichum capsici* incitant of anthracnose on Bell peppers. Nat. Sci. 8, 265–296.

Aldahmani, J.H., Abbasi, P.A., Sahin, F., Hoitink, H.A.J., Miller, S.A., 2005. Reduction of bacterial leaf spot severity on radish, lettuce, and tomato plants grown in compost-amended potting mixes. Can. J. Pl. Pathol. 27, 186–193.

Atanasova, L., Knox, B.P., Kubicek, C.P., Druzhinina, I.S., Baker, S.E., 2013. The polyketide synthase gene pks4 of *Trichoderma reesei* provides pigmentation and stress resistance. Eukaryot. Cell 12, 1499–1508.

Bae, S.J., Mohanta, T.K., Chung, J.Y., Ryua, M., Park, G., Shim, S., Hong, S.B., Seo, H., Bae, D.W., Bae, I., Kima, J.J., Bae, H., 2016. *Trichoderma* metabolites as biological control agents against *Phytophthora* pathogens. Biol. Contr. 92, 128–138.

Barua, L., Bora, B.C., 2008. Comparative efficacy of *Trichoderma harzianum* and *Pseudomonas fluroscens* against *Meloidogyne incognita* and *Ralstonia solanacearum* complex in brinjal. Indian J. Nematol. 38, 86–89.

Benitez, T., Rincon, A.M., Limon, M.C., Codon, A.C., 2004. Biocontrol mechanisms of *Trichoderma* strains. Int. Microbiol. 7, 249–260.

Bezuidenhout, C.N., Van Antwerpen, R., Berry, S.D., 2012. An application of principal component analyses and correlation graphs to assess multivariate soil health properties. Soil Sci. 177, 498–505.

Bhainsa, K.C., D'Souza, S.F., 2006. Extracellular biosynthesis of silver nanoparticles using the fungus *Aspergillus fumigates*. Colloids Surf., B 47, 160–164.

Birkenbihl, R.P., Liu, S., Somssich, I.E., 2017. Transcriptional events defining plant immune responses. Curr. Opin. Plant Biol. 38, 1–9.

Bisen, K., Keswani, C., Mishra, S., Saxena, A., Rakshit, A., Singh, H.B., 2015. Unrealized potential of seed biopriming for versatile agriculture. In: Rakshit, A., Singh, H.B., Sen, A. (Eds.), Nutrient Use Efficiency, from Basics to Advances. Springer, New Delhi, pp. 193–206.

Borges, C., Chagas Junior, L.F., Rodrigues, de, Carvalho, A.F., de Oliveira Miller, M.L., Orozco Colonia, B.S., 2015. Evaluation of the phosphate solubilization potential of *Trichoderma* strains and effects on rice biomass. J. Soil Sci. Plant Nutr. 15, 794–804.

Brotman, Y., Briff, E., Viterbo, A., Chet, I., 2008. Role of swollenin, an expansin–like protein from *Trichoderma* in plant root colonization. Plant Physiol. 147, 779–789.

Brotman, Y., Gupta, J.K., Viterbo, A., 2010. Trichoderma. Curr. Biol. 20, 390–391.

Brotman, Y., Landau, U., Pninic, S., Lisec, J., Balazadeh, S., et al., 2012. The LysM receptor–like kinase LysMRLK1 is required to activate defense and abiotic–stress responses induced by overexpression of fungal chitinases in *Arabidopsis* plants. Mol. Plant 5, 1113–1124.

Brotman, Y., Landau, U., Cuadros–Inostroza, Á., Takayuki, T., et al., 2013. *Trichoderma*–plant root colonization, escaping early plant defense responses and activation of the antioxidant machinery for saline stress tolerance. PLoS Pathog. 9, e1003221.

Budi, S.W., Van, T.D., Arnould, C., Dumas–Gaudot, E., Gianinazzi–Pearson, V., Gianinazzi, S., 2000. Hydrolytic enzyme activity of *Paenibacillus* sp. strain B2 and effects of the antagonistic bacterium on cell integrity of two soil borne pathogenic bacteria. Appl. Soil Ecol. 15, 191–199.

Cai, F., Yu, G., Wang, P., Wei, Z., Fu, L., et al., 2013. Harzianolide, a novel plant growth regulator and systemic resistance elicitor from *Trichoderma harzianum*. Plant Physiol. Biochem. 73, 106–113.

Ceballos, G., Álvarez, E., Bolaños, M.A., 2014. Reduction in populations of *Ralstonia solanacearum* race 2 in plantain (Musa AAB Simmonds) with extracts from Trichoderma sp. and antagonistic bacteria. Acta Agron. 63 (1), 2014.

Chen, J.L., Sun, S.Z., Miao, C.P., Wu, K., Chen, Y.W., Xu, L.H., Guan, H.L., Zhao, L.X., 2015. Endophytic *Trichoderma gamsii* YIM PH30019, a promising biocontrol agent with hyperosmolar, mycoparasitism, and antagonistic activities of induced volatile organic compounds on root–rot pathogenic fungi of *Panax notoginseng*. J. Ginseng Res. 315–324.

Choudhary, C.S., Jain, S.C., Kumar, R., Jaipal Singh, C., 2013. Efficacy of different fungicides, biocides and botanical extract seed treatment for controlling seed–borne *Colletotrichum* sp. in chilli (*Capsicum annuum* L.). Bioscan 8, 123–126.

Chowdappa, P., Kumar, S.P.M., Lakshmi, M.J., Upreti, K.K., 2013. Growth stimulation and induction of systemic resistance in tomato against early and late blight by *Bacillus subtilis* OTPB1 or *Trichoderma harzianum* OTPB3. Biol. Contr. 65, 109–117.

Christopher, D.J., Raj, T.S., Shanmugapackiam, S., Udhayakumar, R., Usharani, S., 2010. Ecofriendly management of Fusarial wilt disease in Tomato. Ann. Plant Protect. Sci. 18, 447–450.

Contreras–Cornejo, H.A., Macias–Rodriguez, L., Cortes–Penagos, C., López–Bucio, J., 2009. *Trichoderma virens*, a plant beneficial fungus, enhances biomass production and promotes lateral root growth through an auxin–dependent mechanism in *Arabidopsis*. Plant Physiol. 149, 1579–1592.

Contreras–Cornejo, H.A., Macías–Rodríguez, L., Beltrán–Peña, E., Herrera–Estrella, A., López–Bucio, J., 2011. *Trichoderma*–induced plant immunity likely involves both hormonal and camalexin–dependent mechanisms in *A. thaliana* and confers resistance against necrotrophic fungus Botrytis cinerea. Plant Signal. Behav. 6, 1554–1563.

Contreras–Cornejo, H.A., Macías–Rodríguez, L., Herrera–Estrella, A., López–Bucio, J., 2014. The 4– phosphopantetheinyl transferase of *Trichoderma virens* plays a role in plant protection against *Botrytis cinerea* through volatile organic compound emission. Plant Soil 379, 261–274.

Contreras–Cornejo, H.A., Macías–Rodríguez, L., del–Val, E., Larsen, J., 2016. Ecological functions of *Trichoderma* spp. and their secondary metabolites in the rhizosphere, interactions with plants. FEMS Microbiol. Ecol. 92, fiw03. https://doi.org/10.1093/femsec/fiw03.

Cuervo–Parra, J.A., Snchez–Lpez, V., Romero–Cortes, T., Ramrez–Lepe, M., 2014. *Hypocrea/Trichoderma* viridescens ITV43 with potential for biocontrol of *Moniliophthora roreri* Cif Par, *Phytophthora megasperma* and *Phytophthora capsici*. Afr. J. Microbiol. Res. 8, 1704–1712.

Dana, M.M., Pintor-Toro, J.A., Cubero, B., 2006. Transgenic tobacco plants overexpressing chitinases of fungal origin show enhanced resistance to biotic and abiotic stress agents. Plant Physiol. 142, 722–730.

Degenkolb, T., Gräfenhan, T., Berg, A., Nirenberg, H.I., Gams, W., Brückner, H., 2006. Peptaibiomics: Screening for polypeptide antibiotics (peptaibiotics) from plant-protective Trichoderma species. Chem. Biodivers. 3, 593–610.

Delgado–Jarana, J., Moreno–Mateos, M.A., Benítez, T., 2003. Glucose uptake in *Trichoderma harzianum*, role of gtt1. Eukaryot. Cell 2, 708–717.

Devi, T.P., Kulanthaivel, S., Kamil, D., Borah, J.L., Prabhakaran, N., Srinivasa, N., 2013. Biosynthesis of silver nanoparticles from *Trichoderma* species. Indian J. Exp. Biol. 51, 543–547.

Diánez Martínez, F., Santos, M., Carretero, F., Marín, F., 2016. *Trichoderma saturnisporum*, a new biological control agent. J. Sci. Food Agric. 96, 1934–1944.

Domínguez, S., Rubio, M.B., Cardoza, R.E., Gutiérrez, S., Nicolás, C., et al., 2016. Nitrogen metabolism and growth enhancement in tomato plants challenged with *Trichoderma harzianum* expressing the *Aspergillus nidulans* Acetamidase amd S gene. Front. Microbiol. 7, 1182.

Doni, F., Al−Shorgani, N.K.N., Tibin, E.M.M., Abuelhassan, N.N., Anizan, I., Che−Radziah, C.M.Z., et al., 2013. Microbial involvement in growth of paddy. Curr. Res. J. Biol. Sci. 5 (6), 285−290.

Doni, F., Isahak, A., Zain, C.R.C.M., Ariffin, S.M., Mohamad, W.N.W., Yusoff, W.M.W., 2014. Formulation of *Trichoderma* sp. SL2 inoculants using different carriers for soil treatment in rice seedling growth. Springer Plus 3, 532.

Druzhinina, I.S., Seidl-Seiboth, V., Herrera-Estrella, A., Horwitz, B.A., Kenerley, C.M., Monte, E., Mukherjee, P.K., Zeilinger, S., Grigoriev, I.V., Kubicek, C.P., 2011. Trichoderma: the genomics of opportunistic success. Nature Rev. 9, 749−759.

Dubey, S.C., 2002. Bio−agent based integrated management of collar rot of French bean. Indian Phytopathol. 55, 230−231.

Duhan, J.S., Kumar, J., Kumar, N., Kaur, P., Nehra, K., Duhan, S., 2017. Nanotechnology, the new perspective in precision agriculture. Biotechnol. Rep. 15, 11−23.

Eisendle, M., Oberegger, H., Buttinger, R., Illmer, P., Haas, H., 2004. Biosynthesis and uptake of siderophores is controlled by the PacC−mediated ambient−pH regulatory system in *Aspergillus nidulans*. Eukaryot. Cell 3, 561−563.

Elamawi, R.M., Al−Harbi, R.E., 2014. Effect of biosynthesized silver nanoparticles on *Fusarium oxysporum* fungus the cause of seed rot disease of faba bean, tomato and barley. J. Plant Protect. Pathol. Mansoura Univ. 5, 225−237.

Elamawi, R.M.A., El−Shafey, R.A.S., 2013. Inhibition effects of silver nanoparticles against rice blast disease caused by *Magnaporthe grisea*. Egypt. J. Agric. Res. 91, 1271−1283.

Enders, T.A., Strader, L.C., 2015. Auxin activity, past, present and future. Am. J. Bot. 102, 180−196.

Engelberth, J., Koch, T., Schüler, G., Bachmann, N., Rechtenbach, J., Boland, W., 2001. Ion channel−forming alamethicin is a potent elicitor of volatile biosynthesis and tendril coiling. Cross talk between jasmonate and salicylate signaling in lima bean. Plant Physiol. 125, 369−377.

Fontenelle, A.D.B., Guzzo, S.D., Lucon, C.M.M., Harakava, R., 2011. Growth promotion and induction of resistance in tomato plant against *Xanthomonas euvesicatoria* and *Alternaria solani* by Trichoderma spp. Crop Protect 30, 1492−1500. https://doi.org/10.1016/j.cropro.2011.07.019.

Gaikwad, S., Birla, S.S., Ingle, A.P., Gade, A.K., Marcato, P.D., Rai, M.K., Duran, D., 2013. Screening of different Fusarium species to select potential species for the synthesis of silver nanoparticles. J. Braz. Chem. Soc. 24, 1974−1982.

Gailīte, A., Samsone, I., Ievinsh, G., 2005. Ethylene is involved in Trichoderma-induced resistance of bean plants against *Pseudomonas syringae*. Acta Universitatis Latviensis 691, 59−70.

Gajera, H., Domadiya, R., Patel, S., Kapopar, M., Golakiya, B., 2013. Molecular mechanism of Trichoderma as bio-control agents against phytopathogen system − a review. Curr. Res. Microbiol. Biotechnol. 1, 133−142.

Gan, N., Yang, X., Xie, D., Wu, Y., Wen, W.A., 2010. Disposable organophosphorus pesticides enzyme biosensor based on magnetic composite nanoparticles modified screen printed carbon electrode. Sensors 10, 625−638.

Garnica−Vergara, A., Barrera−Ortiz, S., Munoz−Parra, E., Raya−Gonzalez, J., Mendez−Bravo, A., et al., 2016. The volatile 6−pentyl−2H−pyran−2−one from *Trichoderma atroviride* regulates *Arabidopsis thaliana* root morphogenesis via auxin signaling and ETHYLENE INSENSITIVE 2 functioning. New Phytol. 209, 1496−1512.

Geraldine, A.M., Cardoso Lopes, F.A., Costa Carvalho, D.D., Barbosa, E.T., Rodrigues, A.R., Brandão, R.S., Ulhoa, C.J., Junior, M.L., 2013. Cell wall−degrading enzymes and parasitism of sclerotia are key factors on field biocontrol of white mold by *Trichoderma* spp. Biol. Contr. 67, 308−316.

Gokil-Prasad, G., Sinha, A.P., 2012. Compartive antagonistic potential of fungal and bacterial bioagents against isolates of Xhantomonas oryzae pv oryzae. Ann. Plant Prot. Sci. 20 (1), 154−159.

Gruber, S., Seidl−Seiboth, V., 2012. Self−versus non−self, fungal cell wall degradation in *Trichoderma*. Microbiology 158, 26−34.

Gruyer, N., 2014. Interaction between silver nanoparticles and plant growth. Acta Hortic. 1037, 795−800.

Hafez, E.E., Meghad, A., Elsalam, H.A.A., Ahmed, S.A., 2013. *Trichoderma viride*−Plant pathogenic fungi interactions. World Appl. Sci. J. 21, 1821−1828.

Hajieghrari, B., 2010. Effects of some Iranian *Trichoderma* isolates on maize seed germination and seedling vigor. Afr. J. Biotechnol. 9, 4342−4347.

Harman, G.E., 2000. Myths and dogmas of biocontrol − changes in perceptions derived from research on *Trichoderma harzianum* T-22. Plant Dis. 84, 377−393.

Harman, G.E., 2006. Overview of mechanisms and uses of *Trichoderma* spp. Phytopathology 96, 190−194.

Harman, G.E., Howell, C.R., Viterbo, A., Chet, I., Lorito, M., 2004. *Trichoderma* species− opportunistic, avirulent plant symbionts. Nat. Rev. Microbiol. 2, 43−56.

Harman, G.E., Herrera−Estrella, A.H., Horwitz, B.A., Lorito, M., 2012. Special issue, *Trichoderma* from basic biology to biotechnology. Microbiology 158, 1−2.

Hermosa, M.R., Grondona, I., Díaz−Mínguez, J.M., Iturriaga, E.A., Monte, E., 2001. Development of a strain−specific SCAR marker for the detection of the *Trichoderma atroviride* 11, a biological control agent against soil borne fungal plant pathogens. Curr. Genet. 38, 343−350.

Hermosa, R., Viterbo, A., Chet, I., Monte, E., 2012. Plant−beneficial effects of *Trichoderma* and of its genes. Microbiology 158, 17−25.

Hermosa, R., Rubio, M.E., Cardoza, M.B., Nicolás, E., Monte, E., Gutiérrez, S., 2013. The contribution of *Trichoderma* to balancing the costs of plant growth and defense. Int. Microbiol. 16, 69−80.

Hastuti, R.D., Lestari, Y., Suwanto, A., Saraswati, R., 2012. Endophytic Streptomyces spp. as biocontrol agents of rice bacterial leaf blight pathogen (*Xanthomonas oryzae* pv. oryzae). HAYATI J. Biosci. 194, 155−162.

Hermosa, R., Cardoza, M.B., Rubio, M.E., Gutiérrez, S., Monte, E., 2014. Secondary metabolism and antimicrobial metabolites of *Trichoderma*. In: Gupta, V.K., Schmoll, M., Herrera− Estrella, A., Upadhyay, R.S., Druzhinina, I., Tuohy, M. (Eds.), Biotechnology and Biology of *Trichoderma*. Elsevier, Netherlands, pp. 125−137.

Howell, C.R., 2003. Mechanisms employed by *Trichoderma* species in the biological control of plant diseases, the history and evolution of current concepts. Plant Dis. 87, 4–10.

Huang, S., Wang, L., Liu, L., Hou, Y., Li, L., 2015. Nanotechnology in agriculture, livestock, and aquaculture in China. A review. Agri. Sus. Dev. 35, 369–400.

Idowu, O.O., Oni, A.C., Salami, A.O., 2016. The interactive effects of three *Trichoderma* species and damping–off causative pathogen *Pythium aphanidermatum* on emergence indices, infection incidence and growth performance of sweet pepper. Int. J. Recent Sci. Res. 7, 10339–10347.

Jan, A.T., Azam, M., Ali, A., Haq, Q.M.R., 2011. Novel approaches of beneficial Pseudomonas in mitigation of plant diseases an appraisal. J. Plant Interact. 6, 195–205.

Jawad, A., Zafar, I., 2017. Effect of growth conditions on antibacterial activity of *Trichoderma harzianum* against selected pathogenic bacteria. Sarhad J. Agric. 33 (4), 501–510.

Jayalaksmi, S.K., Raju, S., Usha Rani, S., Bengai, V.I., Sreeramulu, K., 2009. *Trichoderma harzianum* L1 as a potential source for lytic enzymes and elicitor of defense responses in chickpea against wilt disease caused by *Fusarium oxysporum* f. sp. ciceri. Aust. J. Crop. Sci. 3, 44–52.

Jeyaseelan, C.E., Tharmila, S., Niranjan, K., 2012. Antagonistic activity of *Trichoderma* spp. and *Bacillus* spp. against *Pythium aphanidermatum* isolated from tomato damping off. Arch. Appl. Sci. Res. 4, 1623–1627.

Jhanzab, H.M., Razzaq, A., Jilani, G., Rehman, A., Hafeez, A., Yasmeen, F., 2015. Silver nano– particles enhance the growth, yield and nutrient use efficiency of wheat. Int. J. Agro. Agri. Res. 7, 15–22.

Jiang, X., Geng, A., He, N., Li, Q., 2011. New Isolate of *Trichoderma viride* strain for enhanced cellulolytic enzyme complex production. J. Biosci. Bioeng. 111, 121–127.

Jogaiah, S., Mitani, S., Amruthesh, K.N., Shekar Shetty, H., 2007. Activity of cyazofamid against Sclerospora graminicola., a downy mildew disease of pearl millet. Pest Manag. Sci. 63, 722–727.

Jogaiah, S., Mostafa, A., Tran, L.S.P., Shin–ichi, I., 2013. Characterization of rhizosphere fungi that mediate resistance in tomato against bacterial wilt disease. J. Exp. Bot. 64, 3829–3842.

Jogaiah, S., Mahantesh, K., Sharathchnadra, R.G., Shetty, H.S., Vedamurthy, A.B., Tran, L.–S.P., 2016. Isolation and evaluation of proteolytic actinomycete isolates as novel inducers of pearl millet downy mildew disease protection. Sci. Rep. 6, 30789. https://doi.org/10.1038/srep30789.

Jogaiah, S., Abdelrahman, M., Tran, L.–S.P., Ito, S.–I., 2018. Different mechanisms of *Trichoderma virens*–mediated resistance in tomato against *Fusarium* wilt involve the jasmonic and salicylic acid pathways. Mol. Plant Pathol. 19, 870–882.

Joshi, B.B., Bhatt, R.P., Bahukhandi, D., 2010. Antagonistic and plant growth activity of *Trichoderma* isolates of Western Himalayas. J. Environ. Biol. 31, 921–928.

Junaid, J.M., Dar, N.A., Bhat, T.A., Bhat, A.H., Bhat, M.A., 2013. Commercial biocontrol agents and their mechanism of action in the management of plant pathogens. Int. J. Mod. Plant Anim. Sci. 1, 39–57.

Kamaruzzaman, M., Rahman, M.M., Ahmad, M.U., 2016. Efficacy of four selective *Trichoderma* isolates as plant growth promoters in two peanut varieties. Int. J. Biol. Res. 4, 152–156.

Kamran, S., Forogh, M., Mahtab, E., Javad Asgar, M., 2011. In vitro antibacterial activity of nanomaterials for using in Tobacco plants tissue culture. World Acad. Sci. Eng. & Tech. vol. 79, 372–373.

Kashyap, P.L., Kumar, S., Srivastava, A.K., 2017. Nanodiagnostics for plant pathogens. Environ. Chem. Lett. 15, 7–13.

Katayama, A., Matsumura, F., 1993. Degradation of organochlorine pesticides, particularly endosulfan, by *Trichoderma harzianum*. Environ. Toxicol. Chem. 12, 1059–1065.

Kaur, P., Thakur, R., Duhan, J.S., Chaudhury, A., 2018. Management of wilt disease of chickpea in vivo by silver nanoparticles; biosynthesized by rhizospheric microflora of chickpea (*Cicer arietinum*). J. Chem. Technol. Biotechnol. https://doi.org/10.1002/jctb.5680.

Keswani, C., Mishra, S., Sarma, B., Singh, S., Singh, H., 2014. Unraveling the efficient applications of secondary metabolites of various *Trichoderma* spp. Appl. Microbiol. Biotechnol. 98, 533–544.

Keswani, C., Bisen, K., Singh, V., Sarma, B.K., Singh, H.B., 2016. Formulation technology of biocontrol agents, present status and future prospects. In: Arora, N.K., Mehnaz, S., Balestrini, R. (Eds.), Bioformulations, for Sustainable Agriculture. Springer, India, pp. 35–52.

Khan, M.R., Rizvi, T.F., 2014. Nanotechnology, scope and application in plant disease management. Plant Pathol. J. 13, 214–231.

Koike, N., Kyakumachi, M., Kageyama, K., Tsuyumu, S., Doke, N., 2001. Induction of systemic resistance in cucumber against several diseases by plant growth-promoting fungi: Lignification and superoxide generation. Eur. J. Plant Pathol. 107, 523–533.

Kotasthane, A., Agrawal, T., Kushwah, R., Rahatkar, O.V., 2015. In vitro antagonism of *Trichoderma* spp. against *Sclerotium rolfsii* and *Rhizoctonia solani* and their response towards growth of cucumber, bottle gourd and bitter gourd. Eur. J. Plant Pathol. 141, 523–543.

Kubicek, C.P., Herrera–Estrella, A., Seidl–Seiboth, V., Martinez, D.A., Druzhinina, I.S., Thon, M., et al., 2011. Comparative genome sequence analysis underscores mycoparasitism as the ancestral life style of *Trichoderma*. Genome Biol. 12, R40.

Kumar, A., Vandana Yadav, A., Giri, D.D., Singh, P.K., Pandey, K.D., 2015. Rhizosphere and their role in plant–microbe interaction. In: Chaudhary, K.K., Dhar, D.W. (Eds.), Microbes in Soil and Their Agricultural Prospects. Nova Sci. Publisher, Inc., Hauppauge, pp. 83–97.

Lehner, S.M., Atanasova, L., Neumann, N.K., Krska, R., Lemmens, M., et al., 2013. Isotope–assisted screening for iron–containing metabolites reveals a high degree of diversity among known and unknown siderophores produced by *Trichoderma* spp. Appl. Environ. Microbiol. 79, 18–31.

Limon, M.C., Chacón, M.R., Mejías, R., Delgado–Jarana, J., Rincón, A.M., Codón, A.C., Benítez, T., 2004. Increased antifungal and chitinase specific activities of *Trichoderma harzianum* CECT 2413 by addition of a cellulose binding domain. Appl. Microbiol. Biotechnol. 64, 675–685.

Liza, B., Bora, B.C., 2009. Comparative efficacy of *Thicoderma harzianum* and *Pseudomonas fluorescens* against *Meliodogyne incognita* and *Ralstonia solanacearum* complex on Brinjal. Indian J. Nematol. 39 (1), 29–34.

Lopes, F.A., Steindorff, A.S., Geraldine, A.M., Brandao, R.S., Monteiro, V.N., Lobo Jr., M., Coelho, A.S., Ulhoa, C.J., Silva, R.N., 2012. Biochemical and metabolic profiles of *Trichoderma* isolates isolated from common bean crops in the Brazilian Cerrado, and potential antagonism against *Sclerotinia sclerotiorum*. Fung. Biol. 116, 815–824.

Lopez–Bucio, J., Pelagio–Flores, R., Herrera–Estrella, A., 2015. *Trichoderma* as biostimulant, exploiting the multilevel properties of a plant beneficial fungus. Sci. Hortic. 196, 109–123.

Lorito, M., Woo, S.L., 2015. Discussion agronomic. In: Lugtenberg, B. (Ed.), Principles of Plant-Microbe Interactions. Springer International Publishing, Berlin, pp. 345–353.

Lorito, M., Woo, S.L., Harman, G.E., Monte, E., 2010. Translational research on *Trichoderma*, from omics to the field. Annu. Rev. Phytopathol. 48, 395–417.

Luo, Y., Zhang, D.D., Dong, X.W., Zhao, P.B., Chen, L.L., Song, X.Y., et al., 2010. Antimicrobial peptaibols induce defense responses and systemic resistance in tobacco against tobacco mosaic virus. FEMS Microbiol. Lett. 313, 120–126.

Maischak, H., Zimmermann, M.R., Felle, H.H., Boland, W., et al., 2010. Alamethicin–induced electrical long distance signaling in plants. Plant Signal. Behav. 5, 988–990.

Maketon, M., Apisitsantikul, J., Siriraweekul, C., 2008. Greenhouse evaluation of Bacillus subtilis AP-01 and Trichoderma harzianum AP-001 in controlling tobacco diseases. Braz. J. Microbiol. 39 (2), 296–300.

Manczinger, L., Molnár, A., Kredics, L., Antal, Z.S., 2002. Production of bacteriolytic enzymes by mycoparasitic *Trichoderma* strains. World J. Microbiol. Biotechnol. 18, 147–150.

Marfori, E.C., Shinichiro, K.S., Ei–ichiro Fukusaki, E., et al., 2002. Trichosetin, a novel tetramic acid antibiotic produced in dual culture of *Trichoderma harzianum* and *Catharanthus roseus* Callus. Z. Naturforsch. 57, 465–470.

Martínez–Medina, A., Del Mar Alguacil, M., Pascual, J.A., Wees, S.C.M.V., 2014. Phytohormone profiles induced by *Trichoderma* isolates correspond with their biocontrol and plant growth–promoting activity on melon plants. J. Chem. Ecol. 40, 804–815.

Mastouri, F., Bjorkman, T., Harman, G.E., 2012. *Trichoderma harzianum* enhances antioxidant defense of tomato seedlings and resistance to water deficit. Mol. Plant Microbe Interact. 25, 1264–1271.

Mathys, J., De Cremer, K., Timmermans, P., Van Kerckhove, S., Lievens, B., Vanhaecke, M., Cammue, B.P., De Coninck, B., 2012. Genome–wide characterization of ISR induced in *Arabidopsis thaliana* by *Trichoderma hamatum* T382 against *Botrytis cinerea* infection. Front. Plant Sci. 3, 108.

Menezes–Blackburn, D., Jorquera, M.A., Gianfreda, L., Greiner, R., de la Luz Mora, M., 2014. A novel phosphorus biofertilization strategy using cattle manure treated with phytase– nanoclay complexes. Biol. Fertil. Soils 50, 583–592.

Mishra, B.K., Mishra, R.K., Tiwari, A.K., Yadav, R.S., Dikshit, A., 2011. Biocontrol efficacy of *Trichoderma viride* isolates against fungal plant pathogens causing disease in *Vigna radiata* L. Arch. Appl. Sci. Res. 3, 361–369.

Mishra, S., Singh, B.R., Naqvi, A.H., Singh, H.B., 2016. Potential of biosynthesized silver nanoparticles using *Stenotrophomonas* sp. BHU–S7 (MTCC5978) for management of soil borne and foliar phytopathogens. Sci. Rep. 7, 45154.

Mondejar, R., Ros, M., Pascual, J.A., 2011. Mycoparasitism–related genes expression of *Trichoderma harzianum* isolates to evaluate their efficacy as biological control agent. Biol. Contr. 56, 59–66.

Monte, E., 2001. Understanding *Trichoderma*, between biotechnology and microbial ecology. Int. Microbiol. 4, 1–41.

Montero, M., Sanz, L., Rey, M., Llobell, A., Monte, E., 2007. Cloning and characterization of bgn16·3., coding for a β–1., 6–glucanase expressed during *Trichoderma harzianum* mycoparasitism. J. Appl. Microbiol. 103, 1291–1300.

Morones, J., Elechiguerra, J.L., Camacho, A., Holt, K., Kouri, J.B., Ramirez, J.T., Yacaman, M.J., 2005. The bactericidal effect of silver nanoparticles. Nanotechnology 16, 2346–2353.

Mukherjee, P.K., Wiest, A., Ruiz, N., Keightley, A., Moran–Diez, M.E., McCluskey, K., Pouchus, Y.F., Kenerley, C.M., 2011. Two classes of new peptaibols are synthesized by a single non– ribosomal peptide synthetase of *Trichoderma virens*. J. Biol. Chem. 286, 4544–4554.

Mukherjee, P.K., Buensanteai, N., Moran–Diez, M.E., Druzhinina, I.S., Kenerley, C.M., 2012a. Functional analysis of non–ribosomal peptide synthetases (NRPSs) in *Trichoderma virens* reveals a polyketide synthase (PKS)/NRPS hybrid enzyme involved in the induced systemic resistance response in maize. Microbiol. 158, 155–165.

Mukherjee, P.K., Horwitz, B.A., Kenerley, C.M., 2012b. Secondary metabolism in *Trichoderma* – a genomic perspective. Microbiol. 158, 35–45.

Mukherjee, P.K., Horwitz, B.A., Herrera–Estrella, A., Schmoll, M., Kenerley, C.M., 2013. *Trichoderma* research in the genome era. Annu. Rev. Phytopathol. 51, 105–129.

Murali, M., Jogaiah, S., Amruthesh, K.N., Ito, S.I., Shekar Shetty, H., 2013. Rhizosphere fungus *Penicillium chrysogenum* promotes growth and induces defense– related genes and downy mildew disease resistance in pearl millet. Plant Biol. 15, 111–118.

Naher, L., Yusuf, U., Ismail, A., Hossain, K., 2014. *Trichoderma* Spp, A biocontrol agent for sustainable management of plant diseases. Pakistan J. Bot. 46 (4), 1489–1493.

Narasimhamurthy, K., Srinivas, C., 2013a. Efficacy of *Trichoderma asperellum* against *Ralstonia solanacearum* under greenhouse conditions. Ann. Plant Sci. 2, 342–350.

Narasimhamurthy, K., Uzma, F., Srinivas, C., 2013b. Induction of systemic resistance by *Trichoderma asperellum* against bacterial wilt of tomato caused by Ralstonia solanacearum. Int. J. Adv. Res. 1, 181–194.

Narasimhamurthy, K., Soumya, K., Chandranayak, S., Niranjana, S.R., Srinivas, C., 2018. Evaluation of biological efficacy of *Trichoderma asperellum* against tomato bacterial wilt caused by *Ralstonia solanacearum*. Egypt. J. Biol. Pest. Cont. 28, 63.

Narayanan, K.B., Sakthivel, N., 2010. Biological synthesis of metal nanoparticles by microbes. Adv. Coll. Interface Sci. 156, 1–13.

Nascimento, F.X., Rossi, M.J., Soares, C.R., McConkey, B.J., Glick, B.R., 2014. New insights into 1– aminocyclopropane–1–carboxylate (ACC) deaminase phylogeny, evolution and ecological significance. PloS One 9, e99168.

Nawrocka, J., Małolepsza, U., 2013. Diversity in plant systemic resistance induced by *Trichoderma*. Biol. Contr. 67, 149–156.

Nelson, E.B., 1990. Exudate molecules initiating fungal responses to seed seeds and roots. Plant Soil 129, 61–73.

Norman, S., Hongda, C., 2013. IB in depth special section on nanobiotechnology, part 2. Ind. Biotechnol. 9, 17–18.

Oliveira, A.G., Chagas Jr., A.F., Santos, G.R., Miller, L.O., Chagas, L.F.B., 2012. Potential phosphate solubilization and AIA production of *Trichoderma* spp. Green J. Agroecol. Sust. Dev. 7, 149–155.

Pallavi, C.M., Srivastava, R., Arora, S., Sharma, A.K., 2016. Impact assessment of silver nanoparticles on plant growth and soil bacterial diversity. 3 Biotech 6, 254.

Perazzoli, M., Moretto, M., Fontana, P., Ferrarini, A., Velasco, R., Moser, C., Delledonne, M., Pertot, I., 2012. Downy mildew resistance induced by *Trichoderma harzianum* T39 in susceptible grape–vines partially mimics transcriptional changes of resistant genotypes. BMC Genom. 13, 660.

Pieterse, Corne, M.J., Zamioudis, Christos, Berendsen, Roeland, L., Weller, David, M., Van Wees, Saskia, C.M., Bakker, P.A.H.M., 2014. Induced systemic resistance by beneficial microbes. Annu. Rev. Phytopathol. 52, 347–375.

Prabahu, N., Ayisha Siddiqua, S., Yamuna Gowri, K., Divya, T.R., 2010. Antifungal activity of silver fungal nanoparticles – a novel therapeutic approach. Arch. Pharm. Res. (Seoul) 2, 355–359.

Raaijmakers, J.M., Vlami, M., de Souza, J.T., 2002. Antibiotic production by bacterial biocontrol agent. Antonie Leeuwenhoek 81, 537–547.

Raghavendra, V.B., Siddalingaiah, L., Sugunachar, N.K., Nayak, C., Ramachandrappa, N.S., 2013. Induction of systemic resistance by biocontrol agents against bacterial blight of cotton caused by *Xanthomonas campestris* pv. Malvacearum. Int. J. Phytopathol. 2 (1), 59–69.

Raghu, S., Ravikumar, M.R., Santosh Reddy, M., Basamma, B.K., Benagi, V.I., 2013. In vitro evaluation of antagonist micro-organisms against *Ralstonia solanacearum*. Ann. Pl. Prot. Sci. 21 (1), 176–223.

Rajput, A.Q., Khanzada, M.A., Shahzad, S., 2014. Effect of different substrates and carbon and nitrogen sources on growth and shelf life of *Trichoderma pseudokoningii*. Int. J. Agric. Biol. 16, 893–898.

Ramamoorthy, V., Raguchander, T., Samiyappan, R., 2002. Enhancing resistance of tomato and hot pepper to Pythium diseases by seed treatment with fluorescent Pseudomonads. Eur. J. Plant Pathol. 108, 429.

Rebuffat, S., El Hajji, M., Hennig, P., Davoust, D., Bodo, B., 1989. Isolation, sequence and conformation of seven trichorzianines B from *Trichoderma harzianum*. Int. J. Pept. Protein Res. 34, 200–210.

Reino, J.L., Guerrero, R.F., Hernandez–Galan, R., Collado, I.G., 2008. Secondary metabolites from species of the biocontrol agent *Trichoderma*. Phytochemistry Rev. 7, 89–123.

Roop, S., Jagtap, G.P., 2017. Effect of selected plant extracts on in vitro growth of Ralstonia solanacearum (Smith) the causal agent of bacterial wilt of Ginger. TECHNOFAME-. A J. Multidiscipl. Adv. Res. 6, 32–39.

Rouphael, Y., Cardarelli, M., Bonini, P., Colla, G., 2017. Synergistic action of a microbial–based biostimulant and a plant derived–protein hydrolysate enhances lettuce tolerance to alkalinity and salinity. Front. Plant Sci. 8, 131.

Roy, S., Mukherjee, T., Chakraborty, S., Das, T.K., 2013. Biosynthesis, characterisation and antifungal activity of silver nanoparticles synthesized by the fungus *Aspergillus* Foetidus Mtcc8876. Digest J. Nanomat. Biost. 8, 197–205.

Ruano-Rosa, D., Cazorla, F.M., Bonilla, N., Martín-Pérez, R., Vicente, A., López-Herrera, C.J., 2014. Biological control of avocado white root rot with combined applications of Trichoderma spp. and Rhizobacteria. Eur. J. Plant Pathol. 138, 751–762.

Saba, H.D.V., Manisha, M., Prashant, K.S., Farham, H., Tauseff, A., 2012. *Trichoderma* – a promising plant growth stimulator and Biocontrol agent. Mycosphere 3 (4), 524–531.

Sadykova, V.S., Kurakov, A.V., Kuvarina, A.E., Rogozhin, E.A., 2015. Antimicrobial activity of fungi strains of *Trichoderma* from middle siberia. Appl. Biochem. Microbiol. 51, 355–361.

Saksirirat, W., Chareerak, P., Bunyatrachata, W., 2009. Induced systemic resistance of biocontrol fungus, *Trichoderma* spp. against bacterial and gray leaf spot in tomatoes. Asian J. Food Agro–Industry. 99–104.

Salas–Marina, M.A., Isordia–Jasso, M., Islas–Osuna, M.A., Delgado–Sánchez, P., Jiménez–Bremont, J.F., et al., 2015. The Epl1 and Sm1 proteins from *Trichoderma atroviride* and *Trichoderma virens* differentially modulate systemic disease resistance against different life style pathogens in *Solanum lycopersicum*. Front. Plant Sci. 23, 77.

Salazar–Badillo, F.B., Sanchez–Rangel, D., Becerra–Flora, A., Lopez–Gomez, M., Nieto–Jacobo, F., et al., 2015. *Arabidopsis thaliana* polyamine content is modified by the interaction with different *Trichoderma* species. Plant Physiol. Biochem. 95, 49–56.

Saloheimo, M., Paloheimo, M., Hakola, S., Pere, J., Swanson, B., Nyyssönen, E., Bhatia, A., Ward, M., Swollenin, M.P., 2002. A Trichoderma reesei protein with sequence similarity to the plant expansins, exhibits disruption activity on cellulosic materials. Eur. J. Biochem. 269, 4202–4211.

Samules, G.J., Hebbar, P.K., 2015. *Trichoderma*, Identification and Agricultural Applications. The American Phytopathological Society, USA.

Sandhya, C., Adapa, L.K.K., Nampoothri, M., Binod, P., Szakacs, G., Pandey, A., 2004. Extracellular chitinase production by *Trichoderma harzianum* in submerged fermentation. J. Basic Microbiol. 44, 49–58.

Sandipan, P.B., Chaunhary, R.F., Shanadre, C.M., Rathod, N.K., 2015. Appraisal of diverse bioagents against soft rot bacteria of potato (*Solanum tuberosum* L.) coused by Erwinia carotovora subsp. carotovora under in-vitro test. Eur. J. Pharma. Med. Res. 2 (4), 495–500.

Santoyo, G., Orozco–Mosqueda, M.D., Govindappa, M., 2012. Mechanisms of biocontrol and plant growth–promoting activity in soil bacterial species of *Bacillus* and *Pseudomonas*, a review. Biocontrol Sci. Technol. 22, 855–872.

Satapute, P., Paidi, M.K., Kurjogi, M., Jogaiah, S., 2019a. Physiological adaptation and spectral annotation of Arsenic and Cadmium heavy metal–resistant and susceptible strain *Pseudomonas taiwanensis*. Environ. Pollut. 251, 555–563.

Satapute, P., Milan, V.K., Shivakanthkumar, S.A., Jogaiah, S., 2019b. Influence of triazole pesticides on tillage soil microbial populations and metabolic changes, 651, 2334–2344.

Saxena, A., Raghuwanshi, R., Singh, H.B., 2015. *Trichoderma* species mediated differential tolerance against biotic stress of phytopathogens in *Cicer arietinum* L. J. Basic Microbiol. 55, 195–206.

Scharf, D.H., Brakhage, A.A., Mukherjee, P.K., 2016. Gliotoxin e bane or boon? Environ. Microbiol. 18, 1096–1109.

Segarra, G., Van der Ent, S., Trillas, I., Pieterse, C.M.J., 2009. MYB72, a node of convergence in induced systemic resistance triggered by a fungal and bacterial beneficial microbe. Plant Biol. 11, 90–96.

Shahid, M., Srivastava, M., Singh, A., Kumar, V., Rastogi, S., Pathak, N., Srivastava, A., 2014. Comparative study of biological agents, *Trichoderma harzianum* (Th Azad) and *Trichoderma viride* (01PP) for controlling wilt disease in pigeon pea. J. Microb. Biochem. Technol. 6, 110–115.

Sharma, R.A., 2012. Brief review on mechanism of *Trichoderma* fungus use as biological control agents. Int. J. Innovat. Bio Sci. 2, 200–210.

Sharma, D.K., 2018. Bio-efficacy of fungal and bacterial antagonists against pv. *Xanthomonas axonopodis* vesicatoria Capsicum (Doidge) Dye in chilli (spp.) grown in Rajasthan. Asian J. Pharm. Pharmacol. 4 (2), 207–213.

Sharma, P., Patel, A.N., Saini, M.K., Deep, S., 2012. Field demonstration of *Trichoderma harzianum* as a plant growth promoter in wheat (*Triticum aestivum* L). J. Agric. Sci. 4, 65–73.

Sharon, M., Choudhary, A.K., Kumar, R., 2010. Nanotechnol. Agric. Dis. & Food Saf. 2, 83–92.

Shoresh, M., Yedidia, I., Chet, I., 2005. Involvement of jasmonic acid/ethylene signaling pathway in the systemic resistance induced in cucumber by *Trichoderma asperellum* T203. Phytopathology 95, 76–84.

Shoresh, M., Harman, G.E., Mastouri, F., 2010. Induced systemic resistance and plant responses to fungal biocontrol agents. Annu. Rev. Phytopathol. 48, 21–43.

Shukla, N., Awasthi, R.P., Rawat, L., et al., 2012. Biochemical and physiological responses of rice (*Oryza sativa* L.) as influenced by *Trichoderma harzianum* under drought stress. Plant Physiol. Biochem. 54, 78–88.

Singh, H.B., 2014. Management of plant pathogens with microorganisms. Proc. Indian Nat. Sci. Acad. 80, 443–454.

Singh, H.B., Singhm, A., Sarmam, B.K., 2014. *Trichoderma viride* 2% WP formulation suppresses tomato wilt caused by Fusarium oxysporum f. sp. lycopersici and chilli damping–off caused by Pythium aphanidermatum effectively under different agroclimatic conditions. Int. J. Agric. Environ. Biotechnol. 7, 313–320.

Singh, H.B., Sarma, B.K., Keswani, C., 2016. Agriculturally Important Microorganisms, Commercialization and Regulatory Requirements in Asia, first ed. Springer Nature, Singapore. https://doi.org/10.1007/978–981–10–2576–1.

Srivastava, M., Shahid, M., Pandey, S., Singh, A., Kumar, V., Gupta, S., Maurya, M., 2014. *Trichoderma* genome to genomics, a review. J. Data Min. Genom. Proteonomics 5, 1000172.

Stewart, A., Hill, R., 2014. Applications of *Trichoderma* in plant growth promotion. In: Biotechnology and Biology of *Trichoderma*. Elsevier B.V, pp. 415–428.

Strakowska, J., Błaszczyk, L., Chełkowski, J., 2014. The significance of cellulolytic enzymes produced by *Trichoderma* in opportunistic lifestyle of this Fungus. J. Basic Microbiol. 54, 1–12.

Suarez-Estrella, F., Arcos-Nievas, M.A., Lopez, M.J., Vargas-Garcia, M.C., Moreno, J., 2013. Biological control of plant pathogens by microorganisms isolated from agro-industrial composts. Biol. Control 67, 509–515.

Sudisha, J., Amruthesh, K.N., Deepak, S.A., Shetty, N.P., Sarosh, B.R., Shekar Shetty, H., 2005. Comparative efficacy of strobilurin fungicides against downy mildew disease of pearl millet. Pestic. Biochem. Physiol. 81, 188–197.

Sudisha, J., Mostafa, A., Lam Son, P.T., Ito, S., 2013. Characterization of rhizosphere fungi that mediate resistance in tomato against bacterial wilt disease. J. Exp. Bot. 64, 3829–3842.

Sultana, J.N., Pervez, Z., Rahman, H., Islam, M.S., 2012. In vitro evaluation of different strains of *Trichoderma harzianum* and *Chaetomium globosum* as biological control agents on seedling mortality of chilli. Bangladesh Res. Pub. J. 6, 305–310.

Sundaramoorthy, S., Balabaskar, P., 2013. Biocontrol efficacy of *Trichoderma* spp. against wilt of tomato caused by *Fusarium oxysporum* f. sp. lycopersici. J. Appl. Biol. Biotechnol. 1, 36–40.

Tijerino, A., Cardoza, R.E., Moraga, J., Malmierca, M.G., Vicente, F., et al., 2011. Overexpression of the trichodiene synthase gene tri5 increases trichodermin production and antimicrobial activity in *Trichoderma brevicompactum*. Fungal Gen. Biol. 48, 285–296.

Tripathi, P., Singh, P.C., Mishra, A., Chauhan, P.S., Dwivedi, S., Bais, R.T., Tripathi, R.D., 2013. *Trichoderma*, a potential bioremediator for environmental clean–up. Clean Technol. Environ. Policy 15, 541–550.

Tucci, M., Ruocco, M., De Masi, L., De Palma, M., Lorito, M., 2011. The beneficial effect of Trichoderma spp. on tomato is modulated by the plant genotype. Mol. Plant Pathol. 12, 341–354.

Utkhede, R., Koch, C., 2004. Biological treatments to control bacterial canker of greenhouse tomatoes. Biocontrol 49, 305–313.

Vahabi, K., Mansoori, G.A., Karimi, S., 2011. Biosynthesis of silver nanoparticles by fungus *Trichoderma reesei*, A route for large scale production of AgNPs. Insciences J 1, 65–79.

Verma, M., Brar, S.K., Tyagi, R.D., Surampalli, R.Y., Valero, J.R., 2007. Antagonistic fungi., *Trichoderma* spp. panoply of biological control. Biochem. Eng. J. 37, 1–20.

Verma, V., Ravindran, P., Kumar, P.P., 2016. Plant hormone–mediated regulation of stress responses. BMC Plant Biol. 16, 86.

Vinale, F., Sivasithamparam, K., Ghisalberti, E., Marra, R., Woo, S., Lorito, M., 2008a. *Trichoderma*–plant–pathogen interactions. Soil Biol. Biochem. 40, 1–10.

Vinale, F., Sivasithamparam, K., Ghisalberti, E.L., Marra, R., Barbetti, M.J., Li, H., et al., 2008b. A novel role for Trichoderma secondary metabolites in the interactions with plants. Physiol. Mol. Plant Pathol. 72, 80–86.

Vinale, F., Ghisalberti, E.L., Sivasithamparam, K., Marra, R., Ritieni, A., Ferracane, R., Woo, S., Lorito, M., 2009. Factors affecting the production of *Trichoderma harzianum* secondary metabolites during the interaction with different plant pathogens. Lett. Appl. Microbiol. 48, 705–711.

Vinale, F., Sivasithamparam, K., Ghisalberti, E.L., Ruocco, M., Woo, S., Lorito, M., 2012. *Trichoderma* secondary metabolites that affect plant metabolism. Nat. Prod. Commun. 7, 1545–1550.

Vinale, F., Nigro, M., Sivasithamparam, K., Flematti, G., Ghisalberti, E.L., Ruocco, M., Varlese, R., Marra, R., Lanzuise, S., Eid, A., Woo, S.L., Lorito, M., 2013. Harzianic acid, a novel siderophore from *Trichoderma harzianum*. FEMS Microbiol. Lett. 347, 123–129.

Vinale, F., Sivasithamparam, K., Ghisalberti, E.L., Woo, S.L., Nigro, M., Marra, R., Lombardi, N., et al., 2014. *Trichoderma* secondary metabolites active on plants and fungal pathogens. Open Mycol. J. 8, 127–139.

Viterbo, A., Chet, I., 2006. TasHyd1, a new hydrophobin gene from the biocontrol agent Trichoderma asperellum, is involved in plant root colonization. Mol. Plant Pathol. 7, 249–258.

Viterbo, A., Ramot, O., Chemin, L., Chet, I., 2002. Significance of lytic enzymes from *Trichoderma* sp. in the biocontrol of fungal plant pathogens. Ant. Van. Leeuw. 81, 549–556.

Viterbo, A., Wiest, A., Brotman, Y., Chet, I., Kenerley, C.M., 2007. The 18 merpeptaibols from *Trichoderma virens* elicit plant defence responses. Mol. Plant Pathol. 8, 737–746.

Viterbo, A., Landau, U., Kim, S., Chernin, L., Chet, I., 2010. Characterization of ACC deaminase from the biocontrol and plant growth–promoting agent *Trichoderma asperellum* T203. Microbiol. Lett. 305, 42–48.

Vos, C.M., DeCremer, K., Cammue, B.P.A., DeConinck, B., 2015. The tool box of *Trichoderma* sp. in the biocontrol of *Botrytis cinerea* disease. Mol. Plant Pathol. 16, 400–412.

Wang, Y., Lim, L., Diguistini, S., Robertson, G., Bohlmann, J., Breuil, C., 2013. A specialized ABC efflux transporter GcABC-G1 confers monoterpene resistance to *Grosmannia clavigera*, a bark beetle-associated fungal pathogen of pine trees. New Phytol. 197, 886–898. https://doi.org/10.1111/nph.12063.

Wilhite, S.E., Lumsden, R.D., Straney, D.C., 2001. Peptide synthetase gene in *Trichoderma virens*. Appl. Environ. Microbiol. 67, 5055–5062.

Woo, S.L., Scala, F., Ruocco, M., Lorito, M., 2006. The molecular biology of the interactions between *Trichoderma* sp., phytopathogenic fungi and plants. Phytopathology 96, 181–185.

Woo, S.L., Ruocco, M., Vinale, F., Nigro, M., Marra, R., Lombardi, N., Pascale, A., Lanzuise, S., Manganiello, G., Lorito, M., 2014. *Trichoderma*–based products and their widespread use in agriculture. Open Mycol. J. 8, 71–126.

Xiao-Yan, S., Qing-Tao, S., Shu-Tao, X., Xiu-Lan, C., Cai-Yun, S., Yu-Zhong, Z., 2006. Broad-spectrum antimicrobial activity and high stability of Trichokonins from *Trichoderma koningii* SMF2 against plant pathogens. FEMS Microbiol. Lett. 260, 119–125.

Yadav, R.N., Mishra, D., Zaidi, N.W., Singh, U.S., Singh, H.B., 2018. Bio–control efficacy of *Trichoderma* spp. against two major pathogens of rice (*Oryzae sativa* L.). Int. J. Agric. Environ. Biotechnol. 11, 543–548.

Yedida, I., Shoresh, M., Kerem, Z., Benhamou, N., Kapulnik, Y., Chet, I., 2003. Concomitant Induction of systemic resistances to *Pseudomonas syringae* pv. lachrymans in cucumber by *Trichoderma asperrelum* (T-203) and accumulation of phytoalexins. Appl. Env. Microbiol. 69, 7343–7353.

Yedidia, I., Srivastva, A.K., Kapulnik, Y., Chet, I., 2001. Effect of *Trichoderma harzianum* on microelement concentrations and increased growth of cucumber plants. Plant Soil 235, 235–242.

Yendyo, S., Ramesh, G.C., Pandey, B.R., 2018. Evaluation of Trichoderma spp., *Pseudomonas fluorescens* and *Bacillus subtilis* for biological control of Ralstonia wilt of tomato. F1000Research 6, 2028.

Yoshioka, Y., Ichikawa, H., Naznin, H.A., Kogure, A., Hyakumachi, M., 2012. Systemic resistance induced in *Arabidopsis thaliana* by *Trichoderma asperellum* SKT–1, a microbial pesticide of seed–borne diseases of rice. Pest Manag. Sci. 68, 60–66.

Yuan, S., Li, M., Fang, Z., Liu, Y., Shi, W., Pan, B., Wu, K., Shi, J., Shen, B., Shen, Q., 2016. Biological control of tobacco bacterial wilt using *Trichoderma harzianum* amended bioorganic fertilizer and the arbuscular mycorrhizal fungi *Glomus mosseae*. Biol. Control 92, 164–171.

Yue, J., Hu, X., Huang, J., 2014. Origin of plant auxin biosynthesis. Trends Plant Sci. 19, 764–770.

Zakharova, O.V., Gusev, A.A., Zherebin, P.M., Skripnikova, E.V., Skripnikova, M.K., Ryzhikh, V.E., Krutyakov, Y.A., 2017. Sodium tallow amphopolycarboxyglycinate-stabilized silver nanoparticles suppress early and late blight of *Solanum lycopersicum* and stimulate the growth of tomato plants. BioNanoScience 7 (4), 692–702.

Zeilinger, S., Gruber, S., Bansal, R., Mukherjee, P.K., 2016. Secondary metabolism in *Trichoderma* – chemistry meets genomics. Fungal Biol. Rev. 30, 74–90.

Zhang, S.W., Gan, Y.T., Xu, B.L., 2014. Efficacy of *Trichoderma longibrachiatum* in the control of *Heterodera avenae*. Biol. Contr. 59, 319–331.

Zuverza–Mena, N., Armendariz, R., Peralta–Videa, J.R., Gardea–Torresdey, J.L., 2016. Effects of silver nanoparticles on radish sprouts, root growth reduction and modifications in the nutritional value. Front. Plant Sci. 7, 90.

Chapter 26

Advance molecular tools to detect plant pathogens

R. Kannan[a], A. Solaimalai[b], M. Jayakumar[c] and U. Surendran[d]

[a]*Department of Plant Pathology, Faculty of Agriculture, Annamalai University, Chidambaram, Tamil Nadu, India;* [b]*Department of Agronomy, ARS, Tamil Nadu Agricultural University, Kovilpatti, Tamil Nadu, India;* [c]*Department of Agronomy, Regional Coffee Research Station, Dindigul, Tamil Nadu, India;* [d]*Water Management (Agriculture) Division, Centre for Water Resources Development and Management (CWRDM), Kunnamangalam, Kerala, India*

26.1 Introduction

Plant diseases cause major production and economic losses in agriculture and forestry. For example, soybean rust (a fungal disease in soybeans) has caused a significant economic loss and just by removing 20% of the infection, the farmers may benefit with an approximately 11 million-dollar profit (Roberts et al., 2006). It is estimated that the crop losses due to plant pathogens in United Stated result in about 33 *billion* dollars every year. Of this, about 65% (21 billion dollars) could be attributed to non-native plant pathogens (Pimentel et al., 2005). Some of the diseases caused by introduced pathogenic species are chestnut blight fungus, Dutch elm disease, and huanglongbing citrus disease (Pimentel et al., 2005; Li et al., 2006). The bacterial, fungal, and viral infections, along with infestations by insects result in plant diseases and damage. There are about 50,000 parasitic and non-parasitic plant diseases in United States (Pimentel et al., 2005). Upon infection, a plant develops symptoms that appear on different parts of the plants causing a significant agronomic impact (Lópezet al., 2003). Many such microbial diseases with time spread over a larger area in groves and plantations through accidental introduction of vectors or through infected plant materials.

Presently, the plant disease detection techniques available are enzyme-linked immunosorbent assay (ELISA), based on proteins produced by the pathogen, and polymerase chain reaction (PCR), based on specific deoxyribose nucleic acid (DNA) sequences of the pathogen (Prithiviraj et al., 2004; Das, 2004; Li et al., 2006; Saponari et al., 2008; Ruiz-Ruiz et al., 2009; Yvon et al., 2009). In spite of availability of these techniques, there is a demand for a fast, sensitive, and selective method for the rapid detection of plant diseases. Disease detection techniques can be broadly classified into direct and indirect methods. Summarizes some of these methods of disease detection. An advanced plant disease detection technique can provide rapid, accurate, and reliable detection of plant diseases in early stages for economic, production, and agricultural benefits. In the present paper, advanced techniques of ground-based disease detection that could be possibly integrated with an automated agricultural vehicle are reviewed. In ground-based disease detection studies, both field-based and laboratory-based experiments are discussed in this paper. The field-based studies refer to studies that involve spectral data collection under field conditions, whereas laboratory-based studies refer to data collection under laboratory conditions. The laboratory-based experiments provide strong background knowledge (such as the experimental protocol and statistical algorithm for classification) for the field-based applications.

26.2 Molecular techniques of plant disease detection

In recent years, molecular techniques of plant disease detection have been well established. The sensitivity of the molecular techniques refers to the minimum amount of microorganism that can be detected in the sample. Lópezet al. (2003) reported that the sensitivity of the molecular techniques for detecting bacteria ranged from 10 to 106 colony forming units/mL. The commonly used molecular techniques for disease detection are ELISA and PCR (PCR and real-time PCR). Other molecular techniques include immune fluorescence (IF), flow cytometry, fluorescence in situ hybridization (FISH), and DNA

Biopesticides. https://doi.org/10.1016/B978-0-12-823355-9.00008-0

microarrays. In the ELISA-based disease detection, the microbial protein (antigen) associated with a plant disease is injected into an animal that produces antibodies against the antigen. These antibodies are extracted from the animal's body and used for antigen detection with a fluorescence dye and enzymes. In presence of the disease causing microorganism (antigen), the sample would fluoresce, thus confirming the presence of a particular plant disease. In PCR-based disease detection, the genetic material (DNA) of the disease-causing microorganism is extracted, purified, and amplified before performing the gel electrophoresis. The presence of a specific band in gel electrophoresis confirms the presence of the plant-disease causing organism.

Different types of immunological and PCR techniques are described by López et al. (2003). There are number of studies on disease detection using the molecular techniques. Efforts are ongoing to improve the efficiency of these techniques. Table 26.1 summarizes few studies on plant disease detection using molecular techniques. López et al. (2003) reviewed the various molecular techniques used for detecting pathogenic viruses and bacteria in plants. Their review paper elaborates the molecular methods of plant disease detection, including different types of PCR and ELISAbased techniques. Schaad and Frederick (2002), and Henson and French (1993) described the applications of PCR technique for the diagnosis of plant diseases. Alvarez (2004) reported that there are about 97 commercially available immunodiagnostic test kits for the detection of bacterial pathogens in plants.

Some of the limitations of the molecular techniques are that they are time-consuming and labor-intensive, and require an elaborate procedure, especially during sample preparation (collection and extraction) to obtain reliable and accurate results on plant disease detection. In addition, these techniques require consumable reagents that must be tailored to detect each specific pathogen (e.g. sequence-specific primers for PCR). The molecular techniques could be used as robust tool to ensure the presence of plant diseases, but cannot be used as a preliminary screening tool for processing large number of plant samples due to the time involved in the process.

26.3 Spectroscopic and imaging techniques

Recent developments in agricultural technology have led to a demand for a new era of automated non-destructive methods of plant disease detection. It is desirable that the plant disease detection tool should be rapid, specific to a particular disease, and sensitive for detection at the early onset of the symptoms (López et al., 2003). The spectroscopic and imaging techniques are unique disease monitoring methods that have been used to detect diseases and stress due to various factors, in plants and trees. Current research activities are toward the development of such technologies to create a practical tool for a large-scale real-time disease monitoring under field conditions. Various spectroscopic and imaging techniques have been studied for the detection of symptomatic and asymptomatic plant diseases. Some the methods are: fluorescence imaging (Bravo et al., 2004; Moshou et al., 2005; Chaerle et al., 2007), multispectral or hyperspectral imaging (Moshou et al., 2004; Shafri and Hamdan, 2009; Qin et al., 2009), infrared spectroscopy (Spinelli et al., 2006; Purcell et al., 2009), fluorescence spectroscopy (Marcassa et al., 2006; Belasque et al., 2008; Lins et al., 2009), visible/multiband spectroscopy (Yang et al., 2007; Delalieux et al., 2007; Chen et al., 2008), and nuclear magnetic resonance (NMR) spectroscopy (Choi et al., 2004).

TABLE 26.1 Comparison of the current methods for detecting plant diseases resulting from bacterial pathogens.

Techniques	Limit of detection (CFU/mL)	Advantages	Limitations
PCR	10^3–10^4	Mature and common technology, portable, easy to operate.	Effectiveness is subjected to DNA extraction, inhibitors, polymerase activity, concentration of PCR buffer and deoxynucleoside triphosphate.
FISH	10^3	High sensitivity	Autofluorescence, photobleaching.
ELISA	10^5–10^6	Low cost, visual color change can be used for detection.	Low sensitivity for bacteria.
IF	10^3	High sensitivity, target distribution can be visualized.	Photobleaching.
FCM$^{S''}$	10^4	Simultaneous measurement of	High cost, overwhelming
		Several parameters, rapid detection.	Unnecessary information.

CFU, colony forming unit; *ELISA*, enzyme-linked immunosorbent assay; *FCM*, flow cytometry; *FISH*, fluorescence in-situ hybridization; *IF*, immunofluorescence; *PCR*, polymerase chain reaction.

Hahn (2009) reviewed multiple methods (sensors and algorithms) for pathogen detection, with special emphasis on postharvest diseases. The spectroscopic and imaging techniques could be integrated with an autonomous agricultural vehicle that can provide information on disease detection at early stages to control the spread of plant diseases. This technology can also be applied to identify stress levels and nutrient deficiencies in plants. In regard to plant disease detection, significant research is ongoing on the prospective of this technology from last few decades. Spectroscopic technology has been successfully applied for plant stress detection such as water-stress detection and nutrient-stress detection. In addition, there have also been significant applications for monitoring the quality of postharvest fruits and vegetables. Some of the commonly used spectroscopic and imaging techniques are described in the following sections.

26.4 Fluorescence spectroscopy

Fluorescence spectroscopy refers to a type of spectroscopic method, where the fluorescence from the object of interest is measured after excitation with a beam of light (usually ultraviolet spectra). For the last twenty years, the laser-induced fluorescence has been used for vegetative studies, such as to monitor stress levels and physiological states in plants (Belasque et al., 2008). Two types of fluorescence: (i) blue-green fluorescence in about 400–600 nm range, and (ii) chlorophyll fluorescence in about 650–800 nm range, are produced by green leaves. The fluorescence spectroscopy can be utilized to monitor nutrient deficiencies, environmental conditions based stress levels, and diseases in plants (Cerovic et al., 1999; Belasque et al., 2008). Belasque et al. (2008) employed fluorescence spectroscopy to detect stress caused by citrus canker (bacterial disease caused by *Xanthomonascitri–X. axonopodis* pv. citri) and mechanical injury.

A portable fluorescence spectroscopy system was taken to the greenhouse and the measurement probe was placed 2 mm above the leaf (attached to greenhouse plants) for collecting data from different samples during the period of study (60 days). The spectral data were further processed and analyzed in the laboratory. A 532 nm 10 mW excitation laser was used for excitation and ratios between fluorescence at different wavelengths were employed to monitor the stress caused by bacterial infection. The samples of leaves collected from the field (detached leaves) as well as leaves from greenhouse plants (attached leaves) were analyzed using the system. The three ratios used were: (i) ratio between fluorescence intensity at 452 and 685 nm, (ii) ratio between fluorescence intensity at 452 nm and 735 nm, and (iii) ratio between fluorescence intensity at 685 nm and 735 nm. Fluorescence of citrus leaves was monitored for 60 days under four different conditions: leaves with no stress, leaves with mechanical stress, leaves with disease, and leaves with disease and mechanical stress.

The studies reported the potential of fluorescence spectroscopy for disease detection and discrimination between the mechanical and diseased stress. A similar approach was taken to detect water stress and differentiate citrus canker leaves from variegated chlorosis leaves (Marcassa et al., 2006). The above studies could classify healthy from citrus canker-affected leaves, but were unable to identify water stress and distinguish between variegated chlorosis and citrus canker-infected leaves. The authors did not yet present any statistical analysis to evaluate the ability of the technique to discriminate or classify different plant conditions. Lins et al. (2009) conducted field experiments to discriminate citrus canker-stressed leaves from chlorosis-infected (caused by *Xylella fastidiosa* bacteria) and healthy leaves. Methods such as principal component analysis (PCA), discriminant analysis, and neural network-based classification algorithms can be applied to analyze the results obtained from fluorescence spectroscopy. Methods such as PCA, parallel factor analysis, cluster analysis, partial least square (PLS) regression, and Fischer's linear discriminant analysis (LDA) can be applied for classifying fluorescent spectrometric data having two or more classes (Guimet, 2005).

26.5 Visible and infrared spectroscopy

Similar to fluorescence spectroscopy, visible and infrared spectroscopy have been used as a rapid, non-destructive, and cost-effective method for the detection of plant diseases. It is a fast developing technology used for varied applications (Ramon et al., 2002; Delwiche and Graybosch, 2002; Pontius et al., 2005; Gomez et al., 2006; Zhang et al., 2008a,b; Guo et al., 2009; Sundaram et al., 2009). Studies have also been conducted on the detection of stress, injury, and diseases in plants using this technology (Polischuk et al., 1997; Spinelli et al., 2006; Naidu et al., 2009). The visible and infrared regions of the electromagnetic spectra are known to provide the maximum information on the physiological stress levels in the plants (Muhammed, 2002, 2005; Xu et al., 2007) and thus, some of these wavebands specific to a disease can be used to detect plant diseases (West et al., 2003), even before the symptoms are visible. In general, visible spectroscopy is used for disease detection in plants in combination with infrared spectroscopy (Malthus and Madeira, 1993; Bravo et al., 2003; Huang et al., 2004; Larsolle and Muhammed, 2007).

Spinelli et al. (2006) assessed the near infrared (NIR)-based technique for detecting fire blight disease in the asymptomatic pear plants under greenhouse conditions. The NIR technique did not exhibit potential for classifying infected plants from that of healthy ones, while electronic nose system showed a better potential to classify diseased plants. The authors reported that the possible reason for the inability of the NIR based technique to distinguish diseased from healthy plants could be due to a very small leaf scan area (2 mm^2 in this study). Purcell et al. (2009) investigated the application of NIR spectroscopy for the determination and rating of sugarcane resistance against Australian sugarcane disease, Fiji leaf gall. The leaf samples from the cane stalks were analyzed with a Fourier transform (FT)−NIR instrument in 2−4 days after the sugarcane stalks were removed. The signal in the spectral range of 11,000−4000 cm^{-1} was procured.

Principal component analysis and PLS-based statistical methods were used to analyze the data. The second derivative of the signal in the spectral range was also determined to verify if the signal is better represented the disease for analysis. The authors reported that the PLS-based method was effective in predicting the disease rating in sugarcane. The data analysis procedure used for the classification of postharvest food products can be applicable for plant disease detection. Thus, the knowledge from one application (on statistical algorithm, data processing, experimental protocol, etc.) can be transferred to other possible applications. For example, Sirisomboon et al. (2009) used visible spectra along with NIR spectra (600−1100 nm) to identify defective pods during soybean processing.

In addition to PCA, soft independent modeling of class analogy (SIMCA) and PLS-discriminant analysis were used to classify the groups. The authors reported that the SIMCA-based model was able to discriminate different groups of soybean better than the PLS-based model. Naidu et al. (2009) used leaf spectral reflectance to identify viral infection (under field conditions) in grapevines (*Vitisvinifera* L.) that cause grapevine leafroll disease. A portable spectrometer was used to collect reflectance data from each leaf of the plant using a plant-probe attachment device having a leaf clip. In addition to the green, near infrared, and mid infrared region of the spectra, vegetative indices were used to assess the applicability of spectral reflectance in identifying the disease.

Discriminant analysis was performed to classify the infected leaves with and without symptoms with that of non-infected leaves. The different categories of leaves could be clearly differentiated with improved accuracies when both the vegetative indices and individual reflectance bands were used. A maximum of 75% accuracy was achieved in the study. Huang and Apan (2006) collected hyperspectral data using portable spectrometer under field conditions to detect Sclerotinia rot disease in celery. PLS regression analysis was performed to analyze the spectral reflectance data. The first and second derivatives were estimated to test their effectiveness in reducing the root mean square error during the validation of the developed model.

The raw data-based model produced lower root mean square errors than the first and second derivatives. The authors also stated that the reflectance in the visible and infrared range from 400 to 1300 nm were sufficient in acquiring similar results as that of entire spectra (400−2500 nm). The cross-validation results using raw, first derivative, and second derivative data provided a prediction error of 11%−13%. Chen et al. (2008) investigated the application of hyper spectral reflectance to identify cotton canopy infected with Verticillium wilt. The data were collected using a portable spectroradiometer under field conditions and it was analyzed in the laboratory.

The authors reported that among the visible and infrared spectra, the first derivative of the infrared spectra in the wavelength range between 731 and 1317 nm were most effective in predicting the Verticillium wilt in cotton canopy accurately based on the developed models. Other sensitive regions for prediction of infection severity levels were found to be from 780 to 1300 nm and the first derivative of the spectra from 680 to 760 nm. Yang et al. (2007) studied brown planthoppers and leaf-folder infestations in rice plants. The infested conditions of the plants were ranked and efforts were made to identify the extent of infestations using spectroscopic reflectance (350−2400 nm) data collected under field conditions. The results indicated that the spectral range from 426 to 1450 nm showed the maximum correlation intensity.

The changes in spectral properties were low in visible and ultraviolet (UV) range, whereas the infrared region (740−2400 nm) yielded the maximum change in spectral signature. Delalieux et al. (2007) used hyper spectral reflectance data (350−2500 nm) to detect apple scab caused by *Venturia inaequalis*. The study involved the identification of infected trees and selection of wavelengths best suited for classifying the infected leaves from those of the healthy leaves. The spectral data were analyzed using methods as LDA, logistic regression analysis (for each wavelength), partial least squares logistic discriminant analysis (PLS-LDA), and tree-based modeling for classifying the infected and healthy leaves. The paper reported that the spectral features from 1350 to 1750 nm and 2200−2500 nm were effective for the classification of the infected leaves from healthy leaves at early stages, whereas 580−660 nm and 688−715 nm were effective in identifying infected leaves at their developed stages of infection. Among the statistical methods, logistic regression analysis, PLS-LDA, and tree-based modeling were preferred for classification.

The authors recommended PLS-LDA and tree-based modeling methods as they are simpler, and less computationally and time-intensive. Kobayashi et al. (2001) utilized multispectral radiometer and airborne multispectral scanner for the identification of panicle blast in rice. The spectral range for airborne multispectral scan was selected based on ground experiments. The four spectral bands of 400–460 nm, 490–530 nm or 530–570 nm, 650–700 nm, and 950–1100 nm were utilized for scanning (instantaneous field of view = 2.5 mrad, ground resolution = 0.94 m at 300 m height). The magnitude of reflectance ratios (R470/R570, R520/R675, and R570/R675) decreased with an increase in frequency of panicle blast occurrence.

Wang et al. (2002) used PLS and artificial neural network (ANN) models on the visible-IR reflectance data for classifying damaged soybean seeds. The authors reported that the ANN yielded higher overall as well as individual-class classification accuracies than PLS models. Various studies have used different methods/models for the classification of diseases/conditions of plants based on spectral data. For an instance, Roggo et al. (2003) utilized eight classification models (linear discriminant analysis, k-nearest neighbors, soft independent modeling of class analogy, discriminant partial least squares (DPLS), procrustes discriminant analysis (PDA), classification and regression tree, probabilistic neural network, and learning vector quantization-based neural network) for qualitative determination of sugarbeet. They found that SIMCA, DPLS, and PDA yielded the highest classification accuracies that those of other models.

Wu et al. (2008) used PCA-based back-propagation neural network (BPNN) model and PLS wavelength-based BPNN for detection of *Botrytis cinerea* affected eggplant leaves prior to the visibility of symptoms under laboratory conditions. The BPNN model yielded a maximum of 85% classification accuracy for predicting fungal infections. In addition to the statistical models for classification, many spectroscopy-based studies use different vegetative indices for evaluating the change in spectral reflectance at different plant conditions (diseased or healthy plant).

26.6 Fluorescence imaging

The change in blue-green fluorescence and chlorophyll fluorescence of the plants upon ultraviolet excitation could provide the status of physiological condition of the plant (Belasque et al., 2008). Fluorescence imaging is an advancement of fluorescence spectroscopy, where fluorescence images (rather than single spectra) are obtained using a camera. A xenon or halogen lamp is used as a UV light source for fluorescence excitation, and the fluorescence at specific wavelengths are recorded using the charge coupled device (CCD)-based camera system (Bravo et al., 2004; Lenk and Buschmann, 2006; Chaerle et al., 2007; Lenk et al., 2007). The regions of electromagnetic spectra that are commonly used for fluorescence imaging are blue (440 nm), green (520–550 nm), red (690 nm), far red (740 nm), and near infrared (800 nm) (Lenk and Buschmann, 2006; Chaerle et al., 2007).

Lenk et al. (2007) described the multispectral fluorescence and its possible application in monitoring fruit quality, photosynthetic activities, tissue structures, and disease symptoms in plants; and the basic instrumentation required. The chlorophyll fluorescence imaging can be an effective tool in monitoring leaf diseases (Chaerle et al., 2004; Scharte et al., 2005; Lenk et al., 2007). Chaerle et al. (2007) used blue-green fluorescence to evaluate the effectiveness of this technique in observing the development of tobacco mosaic virus (TMV) infection in tobacco plants. A temporal effect of TMV infection on the fluorescence (blue-green and chlorophyll fluorescence) of infected plants was observed. The reflectance image at 550 and 800 nm were acquired and considered as the reference images. The authors reported an increase in blue, green, and chlorophyll fluorescence after about 40–55 h upon inoculation of TMV.

The fluorescence imaging demonstrated a visible difference between the infected and non-infected leaves in short period of time (50 h) in comparison to the reference images (14 days for visible symptoms of infection). Bravo et al. (2004) used fluorescence imaging for detecting yellow rust in winter wheat. They acquired two fluorescence images: a background image without the xenon lamp source and a fluorescence image with the xenon lamp source during the experiments. The fluorescence image utilized for the analysis was obtained by subtracting fluorescence image from the background image. The fluorescence was measured at 450, 550, 690, and 740 nm. The authors stated that the difference between the fluorescence at 550 and 690 nm were higher in the diseased portion of the leaves, while it was very low for healthy regions of the leaves.

Moshou et al. (2005) investigated the applicability of hyper spectral reflectance imaging in combination with multispectral fluorescence imaging through sensor fusion to detect yellow rust (*Puccinia striiformis*) disease of winter wheat. The hyper spectral imaging was performed under ambient condition in winter wheat plots, whereas fluorescence images were procured upon UV excitation. The authors reported that when the sensor information from the fluorescence and multispectral imaging were combined, QDA-based classification accuracy of the healthy plants improved from 71% to 97%. The classification accuracy of the diseased plants and healthy plants further improved to 98.7% and 99.4%, respectively, when the self-organizing map (SOM) based neural network was used for the plant classification. The above studies indicate the possibility for using imaging techniques for disease identification.

26.7 Hyper spectral imaging

In recent years, hyper spectral imaging is gaining considerable interest for its application in precision agriculture (Okamoto et al., 2009). In the hyper spectral imaging, the spectral reflectance of each pixel is acquired for a range of wavelengths in the electromagnetic spectra. The wavelengths may include the visible and infrared regions of the electromagnetic spectra. The hyper spectral imaging is similar to multispectral imaging, the difference being a broader range of wavelengths (more number of spectral bands) being scanned for each pixel in the hyper spectral imaging. The resulting information is a set of pixel values (intensity of the reflectance) at each wavelength of the spectra in the form of an image.

Hyper spectral imaging is often used for monitoring the quality of food products (Kim et al., 2001, 2002; Mehl et al., 2004; Yao et al., 2005; Lee et al., 2005; Tallada et al., 2006; Gowen et al., 2007; Mahesh et al., 2008; Sighicelli et al., 2009). Aleixos et al. (2002) used multispectral imaging of citrus fruits to assess the quality of the fruits for developing a machine vision system. Gowen et al. (2007) revealed that the application of hyper spectral imaging for food quality control and food safety applications. The authors discussed the components of hyper spectral imaging system, different image processing techniques, and various applications in food quality and safety.

Some of the major challenges in hyper spectral imaging-based plant disease detection are the selection of disease-specific spectral band and selection of statistical classification algorithm for a particular application, which depends on the data acquisition setup under field conditions. For an example, Lu (2003), Xing and Baerdemaeker (2005), Xing et al. (2005), Nicolai et al. (2006) and ElMasry et al. (2008) reported that used hyper spectral imaging for the detection of bruises in apples and acquired different results. Lu (2003) reported that 1000 nm−1340 nm were best for bruise detection, whereas Xing et al. (2005) and ElMasry et al. (2008) reported bands within range 558−960 nm were suitable for the identification of bruises in apple. Blasco et al. (2007) indicated that applied multi-spectral computer vision using non-visible (ultraviolet, IR, and fluorescence) and visible multiple spectra for citrus sorting. The anthracnose was classified better with NIR images (86%), whereas green mold was more accurately classified with fluorescence imaging (94%).

The stem-end injury was classified up to 100% using the ultraviolet spectra in this study. This study showed the utilization of hyper spectral bands for detecting different aspects of a single problem. Similarly, the hyper spectral imaging could be used for detecting different features within a plant to identify diseases. Each spectral region provides unique information about the plant. For instance, the reflectance at visible wavelength provides the information on the leaf pigmentations while, reflectance at infrared wavelength provides the physiological condition of the plant (Huang et al., 2007). Much attention has been drawn toward utilizing this technology for the plant disease detection for precision agriculture-based applications.

Bravo et al. (2003) investigated that the application of visible-NIR hyper spectral imaging for the early detection of yellow rust disease (*Puccinia striiformis*) in winter wheat. A discrimination model was developed using quadratic discriminant analysis for the classification of diseases from the healthy plants. The classification model yielded about 92%−98% classification accuracy while classifying diseased plants. Similarly, Moshou et al. (2004) utilized a spectrograph to acquire spectral images from 460 to 900 nm to detect yellow rust in wheat. Shafri and Hamdan (2009) reported that used air-borne hyper spectral imaging for the detection of ganoderma basal stem rot disease in oil palm plantations. The authors used various vegetative indices and red edge techniques to classify the diseased from healthy plantations. The classification reported accuracies using different methods ranged from 73 to 84%. The results indicated that an aerial hyper spectral imaging could be used for disease detection and management of plantations in large scale.

Qin et al. (2009) obtained that hyper spectral image in the wavelength range 450−930 nm to detect citrus canker and other damages to Ruby red grapefruit. A spectral information divergence (SID)-based classification method yielded about 96% classification accuracy for discriminating the diseased, damaged, and healthy fruits. Similarly, Lee et al. (2008) investigated the applicability of aerial hyper spectral imaging for the detection of greening in citrus plantation.

26.8 Other imaging techniques

The other imaging techniques that can used for detecting plant diseases are infrared thermography, terahetz spectroscopy, NMR spectroscopy, and X-ray imaging. As these techniques are not cost effective, the present paper provides an overview of these methods and does not discuss them in detail. Infrared thermography refers to an imaging technique that utilizes the thermal energy of the infrared band and transforms the procured information into a visible image. Infrared thermography similar to other imaging techniques can be used for non-destructive monitoring of physiological status of plants (Chaerle et al., 1999, 2001). Lenthe et al. (2007) used infrared thermography to examine the possible relationship between the leaf microclimate and fungal diseases in wheat fields. Although microclimate can be determined from infrared thermography, direct diseased leaf area identification could not be established. Chaerle et al. (2003) and Chaerle and Van Der Straeten (2000) reviewed the application of imaging techniques in agronomy and stress detection. They reported that the stomatal changes in the leaves of the plants upon pathogen infection could be monitored by thermography (e.g. hydrogen peroxide produced by *Pseudomonas syringae* induces stomata in the leaves to close).

Similarly, local temperature changes due to plant defense mechanisms against diseases can also help in monitoring plant diseases. Tobacco leaves produces salicylic acid (promoting thermal and stomatal change) that can be monitored using thermography. Other studies include presymptomatic detection of cucumber downy mildew-*Pseudoperonospora cubensis* (Lindenthal et al., 2005; Oerke et al., 2005, 2006), fungal infection by *Cercospora beticola* in sugarbeet (Chaerle et al., 2004), and *Brassica napus* infection with Phoma lingam (Lamkadmi et al., 1996) among others. In recent years, terahertz (THz) frequencies (0.1−10 THz) are being utilized for measuring the water content in leaves.

The water stress can be observed by utilizing this frequency, as terahertz frequency is absorbed greatly by water molecules (Hadjiloucas et al., 2009). Nuclear magnetic resonance and X-ray-based imaging techniques can also be used for detecting infections, different types of stress, and other health conditions in trees/fruits (Goodman et al., 1992; Williamson et al., 1992; Karunakaran et al., 2004; Pearson and Wicklow, 2006). Goodman et al. (1992) used NMR microscopic imaging for identification of the fungal pathogen *Botrytis cinerea* in red raspberry. Narvankar et al. (2009) applied X-ray imaging to identify fungal infections in wheat. The authors employed statistical discriminant models and ANN to classify the images of the wheat kernels. Statistical classifiers, especially the Mahalanobis discriminant classifier performed better (92%−99% accuracy) than the ANN based classification.

26.9 Profiling of plant volatile organic compounds

The volatile organic compounds (VOC) released by plants and trees contribute about two-thirds of the total VOC emissions present in the atmosphere (Guenther, 1997). There are number of factors that affect the volatile metabolic profile of a plant or tree. The VOCs released by the plants depend on various physico-chemical factors such as humidity, temperature, light, soil condition, and fertilization, as well as biological factors such as growth and developmental stage of the plant, insects, and presence of other herbs (Vallat et al., 2005; Vuorinen et al., 2007).

These plant volatiles in turn influence their relationship between the plants and other organisms including pathogens (Vuorinen et al., 2007). For example, acetaldehyde is released by the leaves of young poplar trees are controlled by the transfer of ethanol to leaves through transpiration. Dudareva et al. (2006) reviewed a range of plant volatiles released by the plants due to biotic and abiotic interactions. Some of the commonly found secondary plant volatiles are terpeniods, volatile fatty acids (such as trans-2-hexenal, cis-3-hexenol and methyl jasmonate), phenylpropanoids and benzenoids, and amino acid volatiles (such as aldehydes, alcohols, esters, acids, and nitrogen- and sulfur-containing volatiles derived from amino acids).

Cevallos-Cevallos et al. (2009) reported that the compounds (extracted from leaves) such as hesperidin, naringenin, and quercetin present in the leaves could be used as a biomarker to identify huanglongbing diseases in citrus trees. The volatiles of these compounds could be tested in the atmosphere near the citrus plantations to detect the presence of huanglongbing disease. Tholl et al. (2006) reviewed practical methods to study the plant volatiles suitable for various applications. The authors described the methods for VOC sampling and analysis for in-situ experiments as well as some for field experiments. The focus of the present paper is toward the application of plant VOC profile monitoring for detecting diseases in plants. The VOCs released by the plants change when the plant is infected with a disease due to change in its physiology. These emissions are expected to vary from the VOCs released under normal plant health conditions.

26.10 Electronic nose system

An electronic nose system consists of a series of gas sensors that are sensitive to a range of organic compounds. As each sensor has specific sensitivities, the sensitivities of a series of sensors could be used to discriminate different compounds present in the atmosphere. Electronic nose systems have been used for multiple applications. They have been used to determine food quality (Evans et al., 2000; Di Natale et al., 2001; Zhang et al., 2008a,b), identify diseases in humans (Gardner et al., 2000; Lin et al., 2001; Dragonieri et al., 2007), and detect microorganisms in food products (Falasconi et al., 2005; Rajamaki et al., 2006; Balasubramanian et al., 2008; Concina et al., 2009) among others. The application of electronic nose systems for identifying plant diseases is relatively new domain for its application. Li et al. (2009) used a Cyranose @ 320 (an array of 32 conducting polymer-based sensors) to detect postharvest fungal disease in blueberries in a controlled environment.

Blueberries were disinfected with ethanol to eliminate any naturally present fungal spores and bacteria. Once the blueberries were rinsed with distilled water to remove residual ethanol, they were inoculated with spore suspensions of three fungal species: *Botrytis cinerea*, *Colletotrichum gloeosporioides*, and *Alternaria* spp. that cause graymold, anthracnose, and Alternaria fruit rot in postharvest blueberries, respectively. The berries were placed in a 500 mL bottle and headspace gases were tested using Cyranose @ 320. GC–MS (Gas chromatography–mass spectroscopy) analysis was also performed to identify specific compounds that could be related to fungal diseases. Principal component analysis plots indicated a clear delineation between the control (fresh berries) and berries with fungal infections. The berries with *C. gloeosporioides* could be distinctively differentiated from the other groups, though there was some overlap in the VOC profiles of the berries infected with *B. cinerea* and *Alternaria* spp.

Laothawornkitkul et al. (2008) evaluated the potential of plant volatile signature for pest and disease monitoring in cucumber, pepper, and tomato plants. Similar to Li et al. (2009), an electronic nose system and GC–MS were used to identify and distinguish the volatiles released by the plants under different conditions. The authors used Bloodhound @ ST214 electronic nose (an array of 14 conducting polymer-based sensors with one sensor as a reference) for determining sensors' responses to VOCs released by the plant leaves in experiments conducted in a greenhouse. Markom et al. (2009) used an electronic nose system to detect basal stem rot disease in oil palm plantation during field experiments. The authors reported that two principal components based on PCA were able to account for 99.32% variability in the data, and MLP-based artificial neural network could classify infected from healthy trees with high accuracy.

26.11 GC–MS

The GC–MS is commonly used technique for a qualitative as well as quantitative analysis of volatile metabolites released by plants/trees in different environmental and physiological conditions. The GC–MS studies have been performed to evaluate the change in volatiles caused by bacterial or fungal infection in various food products. Prithiviraj et al. (2004) assessed the variability in the volatiles released from onion bulbs infected with bacterial (*Erwiniacarotovora* causing soft rot) and fungal species (*Fusarium oxysporum* and *Botrytis allii* causing basal and neck rots) using HAPSITE, commercial portable GC–MS instrument. The study indicated that 25 volatile compounds (among the 59 consistently detected compounds) released from onion can be used to identify the disease based on VOC profiling. Although no statistical analyses were performed to determine the discriminatory ability of an algorithm in classifying the VOC profiles for disease detection, model development and software development were recommended for the purpose.

Similar studies on potato tubers inoculated with *Erwinia carotovora* subsp. *carotovora*, *E. carotovora* subsp. *atroseptica*, *Pythium ultimum*, *Phytophthora infestans*, or *Fusarium sambucinum* using solid phase microextraction (SPME) fiber along with GC-flame ionization detector (FID) indicated the potential of the VOC profiling for disease detection (Kushalappa et al., 2002). The amount of volatiles increased with an increase in disease severity. A BPNN model was applied to classify the volatile metabolite profiles with respect to the diseases. The gas retention time of the volatile compounds (GC feature) was used as the input data and two hidden layers were used for cross-validation. The cross-validation probabilities (using BPNN) were >67% (67%–75%) for all groups except potato tubers infected with *Phytophthora infestans*. Unlike other studies, this study did not determine the specific compounds that resulted in VOC peaks in the FID. Lui et al. (2005) inoculated potato tubers with *Phytophthora infestans*, *Pythium ultimum*, or *Botrytis cinerea* and analyzed the VOC profile using GC–MS. The compounds in the headspace of the potatoes were identified and their abundance in terms of peak area was determined. Stepwise discriminant analysis was performed using the 32 compounds that were consistently present in the headspace and mass ions from the MS data as the input. The developed discriminant analysis models categorized the diseases with a classification accuracy of 13%–100%. Comparing studies conducted by Lui et al. (2005) and Kushalappa et al. (2002), BPNN-based classification provided higher classification accuracy than discriminant analysis based models.

Vuorinen et al. (2007) utilized VOC emission pattern of silver birch to determine whether the plants were damaged by larvae (herbivore arthropod *Epirrita autumnata*), infected with pathogenic leaf spot (*Marssonina betulae*) or if they were healthy. The VOCs were collected from pathogen-inoculated leaves, herbivore damaged leaves, and undamaged leaves from the top of the branches as well as undamaged detached twigs. Different plant conditions produced specific patterns of VOCs. The herbivore damaged leaves released VOCs such as methyl salicylate, linalool etc., especially after 72 h of feeding. Staudt and Lhoutellier (2007) evaluated the VOC release profile of the damaged and undamaged leaves of holm oak tree infested by gypsy moth larvae. The researchers stated that the leaves released linalool, homoterpene (E)-4, 8-dimethyl1,3,7-nonatriene, germacrene D, caryophyllene, and several other sesquiterpenes upon days of caterpillar growth on the leaves. These gases were not present in the VOCs released by the control plants. Moalemiyan et al. (2006) employed VOC profiling to detect fungal diseases (Lasiodiplodia theobromae causing stemend rot and

Colletotrichum gloeosporioides causing anthracnose) in mangoes. Discriminant analysis models were used to classify the groups using the gaseous metabolite profile and mass ions. Though the methods showed potential for detecting fungal diseases, the classification accuracy of the models needs to be further increased to make it a feasible method for postharvest disease detection.

26.12 Fluorescence in-situ hybridization

Another type of molecular detection technique is fluorescence in-situ hybridization (FISH), which is applied for bacterial detection in combination with microscopy and hybridization of DNA probes and target gene from plant samples. Due to the presence of pathogen-specific ribosomal RNA (rRNA) sequences in plants, recognizing this specific information by FISH can help detect the pathogen infections in plants. In addition to bacterial pathogens, FISH could also be used to detect fungi and viruses and other endosymbiotic bacteria that infect the plant. The high affinity and specificity of DNA probes provide high single-cell sensitivity in FISH, because the probe will bind to each of the ribosomes in the sample. However, the practical limit of detection lies in the range of around 10^3 CFU/mL. In addition to the detection of culturable microorganisms that cause the plant diseases, FISH could also be used to detect yet-to-be cultured (so called unculturable) organisms in order to investigate complex microbial communities.

26.13 Hyper spectral techniques

Hyper spectral imaging can be used to obtain useful information about the plant health over a wide range of spectrum between 350 and 2500 nm. Hyper spectral imaging is increasingly being used for plant phenotyping and crop disease identification in large scale agriculture. The technique is highly robust and it provides a rapid analysis of the imaging data. Furthermore, hyper spectral imaging cameras facilitate the data collection in three dimension, with X- and Y− axes for spatial and Z-for spectral, which contributes to more detailed and accurate information about plant health across a large geographic area. Hyper spectral techniques have been widely used for plant disease detection by measuring the changes in reflectance resulting from the biophysical and biochemical characteristic changes upon infection. *Magnaporthe grisea* infection of rice, *Phytophthora infestans* infection of tomato and *Venturia inaequalis* infection of apple trees have been identified and reported using hyperspectral imaging techniques.

26.14 Biosensor platforms based on nonmaterials

For biosensing application, the limit of detection and the overall performance of a biosensor can be greatly improved by using nonmaterials for their construction. The popularity of nonmaterial's for sensor development could be attributed to the friendly platform it provides for the assembly of bio-recognition element, the high surface area, high electronic conductivity and plasmonic properties of nonmaterial's that enhance the limit of detection. Various types of nanostructures have been evaluated as platforms for the immobilization of a bio-recognition element to construct a biosensor. The immobilization of the biorecognition element, such as DNA, antibody and enzyme, can be achieved using various approaches including biomolecule adsorption, covalent attachment, encapsulation or a sophisticated combination of these methods. The nanomaterials used for biosensor construction include metal and metal oxide nanoparticles, quantum dots, carbon nanomaterials such as carbon nanotubes and graphene as well as polymeric nanomaterials. Nanoparticles have been utilized with other biological materials such as antibody for detecting *Xanthomonas axonopodis* that causes bacterial spot disease. Fluorescent silica nanoparticles (FSNPs) combined with antibody as a biomarker has been studied as the probe, which successfully detected plant pathogens such as *Xanthomonas axonopodis* pv. *Vesicatoria* that cause bacterial spot disease in Solanaceae plant.

26.15 Affinity biosensors

Compared to the non-specific nanoparticle-based biosensors, inclusion of a bio-recognition element can greatly increase the specificity of the sensor. Consequently, other types of biosensors have been developed and among them affinity biosensors are popular. In affinity biosensors, the sensing is achieved based on the reaction of the bio-recognition element and the target analyte. Affinity biosensors can be developed using antibody and DNA as recognition elements.

26.16 Antibody-based biosensors

Antibodies are versatile and are suitable for diverse immune sensing applications. Antibody-based biosensor allows rapid and sensitive detection of a range of pathogens especially for food borne diseases and this technique has already been developed for food safety monitoring. The antibody-based biosensors provide several advantages such as fast detection, improved sensitivity, real-time analysis and potential for quantification. Antibody-based biosensors hold great value for agricultural plant pathogen detection. The biosensors enable the pathogen detection in air, water and seeds with different platforms for greenhouses, on-field and postharvest storages of processors and distributors of crops and fruits.

26.17 DNA/RNA-based affinity biosensor

A recently developed new type of affinity biosensor uses nucleic acid fragments as elements for pathogen detection. The detection of specific DNA sequence is of significance in a variety of applications such as clinical human disease detection, environmental, horticulture and food analysis. Due to the possibility of detection at a molecular level, the DNA-based biosensor enables early detection of diseases before any visual symptoms appear/The application of specific DNA sequences has been widely used for detection of bacteria, fungi and genetically modified organisms. Based on the specific nucleic acid hybridization of the immobilized DNA probe on the sensor and the analyte DNA sequence, DNA-based biosensor allows rapid, simple and economical testing of genetic and infectious diseases. The most commonly adopted DNA probe is single stranded DNA (ssDNA) on electrodes with electroactive indicators to measure hybridization between probe DNA and the complementary DNA analyte. There are four major types of DNA-based biosensors depending on their mode of transduction: optical, piezoelectric, strip type and electrochemical DNA biosensors. Optical DNA biosensors transduce the emission signal of a fluorescent label. The detection of DNA analyte is realized through a variation in physiochemical properties such as mass, temperature, optical property and electrical property as a result of double-stranded DNA (dsDNA) hybridization occurs during the analyte recognition.

26.18 Enzymatic electrochemical biosensors

The use of enzyme as bio-recognition element can provide highly selective detection of the target analyte due to the high specificity of enzymes toward the analyte. An enzyme specific for the analyte of interest is immobilized on the nanomaterial modified-electrode. The amperometric detection is based on the bio-electro catalytic reaction between the target analyte and electrode, which results in an electrical signal (current) that can be used for quantitative detection of the analyte. The amperometric signal can be obtained through either direct or mediated electron transfer based electrochemical reactions. Unlike other types of biosensors, which are not widely commercialized, the enzymatic electrochemical biosensors have been successfully commercialized, thanks to the invention of glucose biosensors, which are widely used in personal diabetes monitors; a similar biosensing methodology can be adopted for plant pathogen detection, food quality detection and environmental monitoring (Table 26.2). For plant pathogen detection, enzymatic biosensors could be used if the target VOC could be collected in the form of a liquid sample.

26.19 Bacteriophage based biosensors

Bacteriophage is a virus, composed of protein capsid that encapsulates a DNA or RNA genome. It infects the bacteria and replicates within the bacteria and finally lyses the bacterial host to propagate. Being able to lyse the bacteria, bacteriophage

TABLE 26.2 VOCs emitted from whole, intact tomato plants or detached leaves, and biotic stress causing agents responsible for increases in VOC emissions.

Volatiles	Biotic stress causing agents that increase VOC emissions
Cis-3-hexen-1-ol	*Botrytis cinerea, Spodoptera littoralis, Lirimyza huidobrensis, Spodoptera exigua, Manduca sexta, Macrosiphum eurhorbiae, Helicoverpa armigera*
Trans-2-hexanal	*Botrytis cinerea, Spodoptera littoralis, Lirimyza huidobrensis, Spodoptera exigua, Manduca sexta, Helicoverpa armigera*
Methyl salicylate	*Botrytis cinera, Spodoptera littoralis, Tetranychus urticae, Manduca sexta, Macrosiphum euphoria, Tobacco mosaic virus.*

has been widely studied and used in phage therapy to cure bacterial infections. Phage therapy has been used for not only human diseases, but also plant disease control. In addition to phage therapy, bacteriophage is also emerging as a promising alternative for pathogen detection due to its high sensitivity, selectivity, low cost and higher thermo stability. Upon the interaction between the bacteriophage and the target analyte, the impedance of charge transfer reactions at the interface changes which are used as a signal for detection.

26.20 Affinity-based biosensors

Receptor proteins can provide some of the same properties as enzymes, and well-selected antibodies provide highly selective and high affinity recognition for almost any soluble analyte. Receptors are not always suitable, often due to lability or requirement for a membrane. Antibodies on the other hand are robust and versatile and have been adapted into diverse immune sensor systems. Affinity recognition need not be based on biomolecules as the recognition element and some robust sensor systems make use of molecular imprinted materials and volume-sensitive hologram-based biosensor surfaces. The most recent addition to the family of affinity-based sensor technologies is the development of nucleic acid fragments as recognition species, especially the use of selected or designed aptamers (e.g., short 20–40 nucleotide, single stranded DNA) which bind the analyte.

26.21 Genetically-encoded biosensors

The optical biosensor field has been led and enlightened, in particular, by calcium sensitive reporters such as aequorin and its successors the cameleons. The cameleons utilize the large conformational rearrangement induced by calcium binding to calmodulin to change the proximity of paired FPs fused at either end, frequently CFP and YFP or their variants, and the sensor is interrogated using FRET microscopy. Second generation versions were improved for selectivity by using engineered versions of calmodulin and a calmodulin-binding peptide in the bridge between FPs so that on binding calcium the sensor folds up on itself to promote FRET.

26.22 Spectroscopic and imaging techniques

Recent developments in agricultural technology have led to a demand for a new era of automated non-destructive methods of plant disease detection. It is desirable that the plant disease detection tool should be rapid, specific to a particular disease, and sensitive for detection at the early onset of the symptoms (Lopez et al., 2003). The spectroscopic and imaging techniques are unique disease monitoring methods that have been used to detect diseases and stress due to various factors, in plants and trees. Current research activities are toward the development of such technologies to create a practical tool for a large-scale real-time disease monitoring under field conditions.

26.22.1 Fluorescence spectroscopy

Fluorescence spectroscopy refers to a type of spectroscopic method, where the fluorescence from the object of interest is measured after excitation with a beam of light (usually ultraviolet spectra).For the last twenty years, the laser-induced fluorescence has been used for vegetative studies, such as to monitor stress levels and physiological states in plants (Belasque et al., 2008). Two types of fluorescence: (i) blue-green fluorescence in about 400–600 nm range and (ii) chlorophyll fluorescence in about 650–800 nm range, are produced by green leaves. The fluorescence spectroscopy can be utilized to monitor nutrient deficiencies, environmental conditions based stress levels and diseases in plants (Cerovic et al., 1999; Belasque et al., 2008).

26.22.2 Visible and infrared spectroscopy

Similar to fluorescence spectroscopy, visible and infrared Spectroscopy have been used as a rapid, non-destructive, and cost-effective method for the detection of plant diseases. It is a fast-developing technology used for varied applications (Ramon et al., 2002; Delwiche and Graybosch, 2002; Pontius et al., 2005; Gomez et al., 2006; Zhang et al., 2008a,b; Guo et al., 2009; Sundaram et al., 2009). Studies have also been conducted on the detection of stress, injury, and diseases in plants using this technology (Polischuk et al., 1997; Spinelli et al., 2006; Naidu et al., 2009). The visible and infrared regions of the electromagnetic spectra are known to provide the maximum information on the physiological stress levels in the plants (Muhammed, 2002, 2005; Xu et al., 2007) and thus, some of these wavebands specific to a disease can be used to detect plant diseases (West et al., 2003), even before the symptoms are visible. In general, visible spectroscopy is used for disease detection in plants in combination with infrared spectroscopy (Malthus and Madeira, 1993; Bravo et al., 2003; Huang et al., 2004; Larsolle and Muhammed, 2007).

26.23 Conclusion

Developing detection methods is both an art and an ever-ending story, and the concept of accurate detection of plant pathogenic bacteria and viruses, is moving from conventional methods to molecular techniques, included in integrated approaches. PCR and especially real-time PCR are the methods of choice for rapid and accurate diagnosis of plant pathogenic bacteria but conventional serological methods, such as immune fluorescence, are still widely used and ELISA is the most frequently applied method for virus detection. Consequently, trying to answer the question: "Are nucleic-acid tools solving the challenges posed by specific and sensitive detection of plant pathogenic bacteria and viruses?" we could answer yes, when a rapid analysis of a reduced number of samples is performed and the protocols have been suitably optimized and also when the presence of false positives is not crucial because the main goal is the quality of the negatives. However, for quarantine pathogens or in critical cases of export-import, experience advises the use of more than one technique.

References

Aleixos, N., Blasco, J., Navarron, F., Molto, E., 2002. Multispectral inspection of citrus in real-time using machine vision and digital signal processors. Comput. Electron. Agric. 33 (2), 121–137.

Alvarez, A.M., 2004. Integrated approaches for detection of plant pathogenic bacteria and diagnosis of bacterial diseases. Annu. Rev. Plant Physiol. 42, 339–366.

Balasubramanian, S., Panigrahi, S., Logue, C.M., Doetkott, C., Marchello, M., Sherwood, J.S., 2008. Independent component analysis-processed electronic nose data for predicting *Salmonella typhimurium* populations in contaminated beef. Food Contr. 19 (3), 236–246.

Belasque, L., Gasparoto, M.C.G., Marcassa, L.G., 2008. Detection of mechanical and disease stresses in citrus plants by fluorescence spectroscopy. Appl. Opt. 47 (11), 1922–1926.

Blasco, J., Alexios, N., Gomez, J., Molto, E., 2007. Citrus sorting by identification of the most common defects using multispectral computer vision. J. Food Eng. 83 (3), 384–393.

Bravo, C., Moshou, D., West, J., McCartney, A., Ramon, H., 2003. Early disease detection in wheat fields using spectral reflectance. Biosyst. Eng. 84 (2), 137–145.

Bravo, C., Moshou, D., Oberti, R., West, J., McCartney, A., Bodria, L., Ramon, H., December 2004. Foliar disease detection in the field using optical sensor fusion. Manuscript FP 04 008 Agric. Eng. Int. CIGR J. Sci. Res. & Develop. Vol. VI.

Cerovic, Z.G., Samson, G., Morales, F., Tremblay, N., Moya, I., 1999. Ultraviolet induced fluorescence for plant monitoring: present state and prospects. Agronomie 19, 543–578.

Cevallos-Cevallos, J.M., Rouseff, R., Reyes-De-Corcuera, J.I., 2009. Untargeted metabolite analysis of healthy and Huanglongbing-infected orange leaves by CE-DAD. Electrophoresis 30, 1–8.

Chaerle, L., Caeneghem, W.V., Messens, E., Lamber, H., Van Montagu, M., Van Der Straeten, D., 1999. Presymptomatic visualization of plant-virus interactions by thermography. Nat. Biotechnol. 17, 813–816.

Chaerle, L., Van Der Straeten, D., 2000. Imaging techniques and the early detection of plant stress. Trends Plant Sci. 5 (11), 495–501.

Chaerle, L., De Boever, F., Van Montagu, M., Van Der Straeten, D., 2001. Thermographic visualization of cell death in tobacco and *Arabidopsis*. Plant Cell Environ. 24 (1), 15–25.

Chaerle, L., Hulsen, K., Hermans, C., Strasser, R.J., Valcke, R., Hofte, M., Van Der Straeten, D., 2003. Robotized time-lapse imaging to assess in-plant uptake of phenylurea herbicides and their microbial degradation. Phys. Plantarium 118, 613–619.

Chaerle, L., Hagenbeek, D., De Bruyne, E., Valcke, R., Van Der Straeten, D., 2004. Thermal and chlorophyll-fluorescence imaging distinguish plant-pathogen interactions at an early stage. Plant Cell Physiol. 45, 887–896.

Chaerle, L., Lenk, S., Hagenbeek, D., Buschmann, C., Van Der Straeten, D., 2007. Multicolor fluorescence imaging for early detection of the hypersensitive reaction to tobacco mosaic virus. J. Plant Physiol. 164 (3), 253–262.

Chen, B., Wang, K., Li, S., Wang, J., Bai, J., Xiao, C., Lai, J., 2008. Spectrum characteristics of cotton canopy infected with verticillium wilt and inversion of severity level. In: IFIP International Federation for Information Processing. Computer and Computing Technologies in Agriculture, Volume 2, vol. 259. Daoliang Li, Springer, Boston, pp. 1169–1180.

Choi, Y.H., Tapias, E.C., Kim, H.K., Lefeber, A.W.M., Erkelens, C., Verhoeven, J.T.J., Brzin, J., Zel, J., Verpoorte, R., 2004. Metabolic discrimination of *Catharanthus roseus* leaves infected by phytoplasma using 1H-NMR spectroscopy and multivariate data analysis. Plant Physiol. 135, 2398–2410.

Concina, I., Falasconi, M., Gobbi, E., Bianchi, F., Musci, M., Mattarozzi, M., Pardo, M., Mangia, A., Careri, M., Sberveglieri, G., 2009. Early detection of microbial contamination in processed tomatoes by electronic nose. Food Contr. 20 (10), 873–880.

Das, A.K., 2004. Rapid detection of *Candidatus* Liberi bacterasiaticus, the bacterium associated with citrus Huanglongbing (Greening) disease using PCR. Curr. Sci. 87 (9), 1183–1185.

Delalieux, S., van Aardt, J., Keulemans, W., Schrevens, E., Coppin, P., 2007. Detection of biotic stress (*Venturia inaequalis*) in apple trees using hyperspectral data: non-parametric statistical approaches and physiological implications. Eur. J. Agron. 27 (1), 130–143.

Delwiche, S.R., Graybosch, R.A., 2002. Identification of waxy wheat by near-infrared reflectance spectroscopy. J. Cereal. Sci. 35 (1), 29–38.

Di Natale, C., Macagnano, A., Martinelli, E., Paolesse, R., Proietti, E., D'Amico, A., 2001. The evaluation of quality of post-harvest oranges and apples by means of an electronic nose. Sensor. Actuator. B Chem. 78 (1–3), 26–31.

Dragonieri, S., Schot, R., Mertens, B.J.A., Le Cessie, S., Gauw, S.A., Spanevello, A., Resta, O., Willard, N.P., Vink, T.J., Rabe, K.F., Bel, E.H., Sterk, P.J., 2007. An electronic nose in the discrimination of patients with asthma and controls. J. Allergy Clin. Immunol. 120 (4), 856–862.

Dudareva, N., Negre, F., Nagegowda, D.A., Orlova, I., 2006. Plant volatiles: recent advances and future perspectives. Crit. Rev. Plant Sci. 25, 417–440.

ElMasry, G., Wang, N., Vigneault, C., Qiao, J., ElSayed, A., 2008. Early detection of apple bruises on different background colors using hyper spectral imaging. LWT Food Sci. & Technol. 41 (2), 337–345.

Evans, P., Persaud, K.C., McNeish, A.S., Sneath, R.W., Hobson, N., Magan, N., 2000. Evaluation of a radial basis function neural network for the determination of wheat quality from electronic nose data. Sensor. Actuator. B Chem. 69 (3), 348–358.

Falasconi, M., Gobbi, E., Pardo, M., Della Torre, M., Bresciani, A., Sberveglieri, G., 2005. Detection of toxigenic strains of *Fusarium verticillioides* in corn by electronic olfactory system. Sensor. Actuator. B Chem. 108 (1–2), 250–257.

Gardner, J.W., Shin, H.W., Hines, E.L., 2000. An electronic nose system to diagnose illness. Sensor. Actuator. B Chem. 70 (1–3), 19–24.

Gomez, A.H., He, Y., Garcia Pereira, A., 2006. Non-destructive measurement of acidity, soluble solids and firmness of Satsuma Mandarin using Vis/NIR spectroscopy techniques. J. Food Eng. 77 (3), 313–319.

Goodman, B.A., Williamson, B., Chudek, A., 1992. Non-invasive observation of the development of fungal infection in fruit. Protoplasma 166, 107–109.

Gowen, A.A., O'Donnell, C.P., Cullen, P.J., Downey, G., Frias, J.M., 2007. Hyper spectral imaging—an emerging process analytical tool for food quality and safety control. Trends Food Sci. Technol. 18 (12), 590–598.

Guenther, A., 1997. Seasonal and spatial variations in natural volatile organic compound emissions. Ecol. Appl. 7 (1), 34–45.

Guimet, F., 2005. Olive Oil Characterization Using Excitation-Emission Fluorescence Spectroscopy and Three-Way Methods of Analysis. Ph.D. thesis. Rovirai Virgili University, Spain.

Guo, T.T., Guo, L., Wang, X.H., Li, M., 2009. Application of NIR spectroscopy in classification of plant species. In: I. nternational Workshop on Education Technology and Computer Science, Wuhan, Hubei, China, vol. 3, pp. 879–883.

Hadjiloucas, S., Walker, G.C., Bowen, J.W., Zafiropoulos, A., 2009. Propagation of errors from a null balance terahertz reflect meter to a sample's relative water content. J. Phys.: Conf. Series Sensor & Appl. XV 178, 1–5, 012012.

Hahn, F., 2009. Actual pathogen detection: sensors and algorithms—a review. Algorithms 2, 301–338.

Henson, J.M., French, R., 1993. The polymerase chain reaction and plant disease diagnosis. Annu. Rev. Plant Physiol. 31, 81–109.

Huang, J.F., Apan, A., 2006. Detection of Sclerotinia rot disease on celery using hyper spectral data and partial least squares regression. Spatial Sci. 51 (2), 129–142.

Huang, M.Y., Huang, W.H., Liu, L.Y., Huang, Y.D., Wang, J.H., Zhao, C.H., Wan, A.M., 2004. Spectral reflectance feature of winter wheat single leaf infested with stripe rust and severity level inversion. Trans. CSAE 20 (1), 176180.

Huang, W., Lamb, D.W., Niu, Z., Zhang, Y., Liu, L., Wang, J., 2007. Identification of yellow rust in wheat using in-situ spectral reflectance measurements and airborne hyper spectral imaging. Precis. Agric. 8, 187–197.

Karunakaran, C., Jayas, D.S., White, N.D.G., 2004. Identification of wheat kernels damaged by the red flour beetle using X-ray images. Biosyst. Eng. 87 (3), 267–274.

Kim, M.S., Chen, Y.R., Mehl, P.M., 2001. Hyper spectral reflectance and fluorescence imaging system for food quality and safety. Trans. ASAE 44 (3), 721–729.

Kim, M.S., Lefcourt, A.M., Chao, K., Chen, Y.R., Kim, I., Chan, D.E., 2002. Multispectral detection of fecal contamination on apples based on hyper spectral imagery: Part I. Application of visible and near-infrared reflectance imaging. Trans. ASAE 45 (6), 2027–2037.

Kobayashi, T., Kanda, E., Kitada, K., Ishiguro, K., Torigoe, Y., 2001. Detection of rice panicle blast with multispectral radiometer and the potential of using airborne multispectral scanners. Phytopathology 91 (3), 316–323.

Kushalappa, A.C., Lui, L.H., Chen, C.R., Lee, B., 2002. Volatile fingerprinting (SPMEGCFID) to detect and discriminate diseases of potato tubers. Plant Dis. 86, 131–137.

Lamkadmi, Z., Esnault, M.A., Le Normand, M., 1996. Characterization of a 23 kDa polypeptide induced by Phoma lingam in *Brassica napus* leaves. Plant Physiol. Biochem. 34 (4), 589–598.

Laothawornkitkul, J., Moore, J.P., Taylor, J.E., Possell, M., Gibson, T.D., Hewitt, C.N., Paul, N.D., 2008. Discrimination of plant volatile signatures by an electronic nose: a potential technology for plant pest and disease monitoring. Environ. Sci. Technol. 42, 8433–8439.

Larsolle, A., Muhammed, H.H., 2007. Measuring crop status using multivariate analysis of hyper spectral field reflectance with application to disease severity and plant density. Precis. Agric. 8 (1–2), 37–47.

Lee, K.J., Kang, S., Kim, M.S., Noh, S.H., July, 2005. Hyper spectral imaging for detecting defect on apples. ASABE Paper No. 053075. In: 2005 ASAE Annual International Meeting, Tampa, FL, pp. 17–20.

Lee, W.S., Ehsani, R., Albrigo, L.G., 2008. Citrus greening disease (Huanglongbing) detection using aerial hyper spectral imaging. In: The Proceedings of the 9th International Conference on Precision Agriculture, July 20–23, 2008, Denver, CO.

Lenk, S., Buschmann, C., 2006. Distribution of UV-shielding of the epidermis of sun and shade leaves of the beech (*Fagus sylvatica* L.) as monitored by multi-colour fluorescence imaging. J. Plant Physiol. 163 (12), 1273–1283.

Lenk, S., Chaerle, L., Pfundel, E.E., Langsdorf, G., Hagenbeek, D., Lichtenthaler, H.K., Van Der Straeten, D., Buschmann, C., 2007. Multispectral fluorescence and reflectance imaging at the leaf level and its possible applications. J. Exp. Bot. 58 (4), 807–814.

Lenthe, J.H., Oerke, E.C., Dehne, H.W., 2007. Digital infrared thermography for monitoring canopy health of wheat. Precis. Agric. 8 (1–2), 15–26.

Li, W., Hartung, J.S., Levy, L., 2006. Quantitative real-time PCR for detection and identification of *Candidatus* Liberi bacter species associated with citrus Huanglongbing. J. Microbiol. Methods 66 (1), 104–115.

Li, C., Krewer, G., Kays, S.J., June . Blueberry postharvest disease detection using an electronic nose. ASABE Paper No. 096783. In: ASABE Annual International Meeting, Reno, NV.

Lin, Y.J., Guo, H.R., Chang, Y.H., Kao, M.T., Wang, H.H., Hong, R.I., 2001. Application of the electronic nose for uremia diagnosis. Sensor. Actuator. B Chem. 76 (1–3), 177–180.

Lindenthal, M., Steiner, U., Dehne, H.W., Oerke, E.C., 2005. Effect of downy mildew development on transpiration of cucumber leaves visualized by digital infrared thermography. Phytopathology 95 (3), 233–240.

Lins, E.C., Belasque Junior, J., Marcassa, L.G., 2009. Detection of citrus canker in citrus plants using laser induced fluorescence spectroscopy. Precis. Agric. 10, 319–330.

Lopez, M.M., Bertolini, E., Olmos, A., Caruso, P., Gorris, M.T., Llop, P., Penyalver, R., Cambra, M., 2003. Innovative tools for detection of plant pathogenic viruses and bacteria. Int. Microbiol. 6, 233–243.

Lu, R., 2003. Detection of bruises on apples using near-infrared hyper spectral imaging. Trans. ASAE 46 (2), 523–530.

Lui, L., Vikram, A., Hamzehzarghani, H., Kushalappa, A.C., 2005. Discrimination of three fungal diseases of potato tubers based on volatile metabolic profiles developed using GC/MS. Potato Res. 48, 85–96.

Mahesh, S., Manickavasagan, A., Jayas, D.S., Paliwal, J., White, N.D.G., 2008. Feasibility of near-infrared hyper spectral imaging to differentiate Canadian wheat classes. Biosyst. Eng. 101 (1), 50–57.

Malthus, T.J., Madeira, A.C., 1993. High resolution spectro radiometry: spectral reflectance of field bean leaves infected by *Botrytis fabae*. Rem. Sens. Environ. 45, 107–116.

Marcassa, L.G., Gasparoto, M.C.G., Belasque Junior, J., Lins, E.C., Dias Nunes, F., Bagnato, V.S., 2006. Fluorescence spectroscopy applied to orange trees. Laser Phys. 16 (5), 884–888.

Markom, M.A., MdShakaff, A.Y., Adom, A.H., Ahmad, M.N., WahyuHidayat, Abdullah, A.H., AhmadFikri, N., 2009. Intelligent electronic nose system for basal stem rot disease detection. Comput. Electron. Agric. 66 (2), 140–146.

Mehl, P.M., Chen, Y.R., Kim, M.S., Chan, D.E., 2004. Development of hyper spectral imaging technique for the detection of apple surface defects and contaminations. J. Food Eng. 61 (1), 67–81.

Moshou, D., Bravo, C., West, J., Wahlen, S., McCartney, A., Ramon, H., 2004. Automatic detection of 'yellow rust' in wheat using reflectance measurements and neural networks. Comput. Electron. Agric. 44 (3), 173–188.

Moalemiyan, M., Vikram, A., Yaylayan, V., 2006. Volatile metabolite profiling to detect and discriminate stem-end rot and anthracnose diseases of mango fruits. Pl. Pathol. 55, 792–802.

Moshou, D., Bravo, C., Oberti, R., West, J., Bodria, L., McCartney, A., Ramon, H., 2005. Plant disease detection based on data fusion of hyper-spectral and multi-spectral fluorescence imaging using Kohonen maps. R. Time Imag. 11 (2), 7583.

Muhammed, H.H., 2002. Using hyper spectral reflectance data for discrimination between healthy and diseased plants, and determination of damage-level in diseased plants. In: IEEE: Proceedings of the 31st Applied Imagery Pattern Recognition Workshop, pp. 49–54.

Muhammed, H.H., 2005. Hyper spectral crop reflectance data for characterizing and estimating fungal disease severity in wheat. Biosyst. Eng. 91 (1), 9–20.

Naidu, R.A., Perry, E.M., Pierce, F.J., Mekuria, T., 2009. The potential of spectral reflectance technique for the detection of Grapevine leaf roll-associated virus-3in two red- berried wine grape cultivars. Comput. Electron. Agric. 66, 38–45.

Narvankar, D.S., Singh, C.B., Jayas, D.S., White, N.D.G., 2009. Assessment of soft X-ray imaging for detection of fungal infection in wheat. Biosyst. Eng. 103 (1), 49–56.

Nicolai, B.M., Lotze, E., Peirs, A., Scheerlinck, N., Theron, K.I., 2006. Non-destructive measurement of bitter pit in apple fruit using NIR hyper spectral imaging. Postharvest Biol. Technol. 40 (1), 1–6.

Oerke, E.C., Lindenthal, M., Frohling, P., Steiner, U., 2005. Digital infrared thermography for the assessment of leaf pathogens. In: Stafford, J.V. (Ed.), Precision Agriculture '05. Wageningen University Press, Wageningen, The Netherlands, pp. 91–98.

Oerke, E.C., Steiner, U., Dehne, H.W., Lindenthal, M., 2006. Thermal imaging of cucumber leaves affected by downy mildew and environmental conditions. J. Exp. Bot. 57 (9), 2121–2132.

Okamoto, H., Suzuki, Y., Kataoka, T., Sakai, K., 2009. Unified hyperspectral imaging methodology for agricultural sensing using software framework. Acta Hortic. 824, 49–56.

Pearson, T.C., Wicklow, D.T., 2006. Detection of kernels infected by fungi. Trans. ASABE 49 (4), 1235–1245.

Pimentel, D., Zuniga, R., Morrison, D., 2005. Update on the environmental and economic costs associated with alien-invasive species in the United States. Ecol. Econ. 52 (3), 273–288.

Polischuk, V.P., Shadchina, T.M., Kompanetz, T.I., Budzanivskaya, I.G., Sozinov, A., 1997. Changes in reflectance spectrum characteristic of *Nicotiana debneyi* plant under the influence of viral infection. Arch. Phytopathol. Plant Protect. 31 (1), 115–119.

Pontius, J., Hallett, R., Martin, M., 2005. Assessing hemlock decline using visible and near-infrared spectroscopy: indices comparison and algorithm development. Appl. Spectrosc. 59 (6), 836–843.

Prithiviraj, B., Vikram, A., Kushalappa, A.C., Yaylayam, V., 2004. Volatile metabolite profiling for the discrimination of onion bulbs infected by *Erwiniacarotovora* ssp. *carotovora*, *Fusarium oxysporum* and *Botrytis allii*. Eur. J. Plant Pathol. 110, 371–377.

Purcell, D.E., O'Shea, M.G., Johnson, R.A., Kokot, S., 2009. Near-infrared spectroscopy for the prediction of disease rating for Fiji leaf gall in sugarcane clones. Appl. Spectrosc. 63 (4), 450–457.

Qin, J., Burks, T.F., Ritenour, M.A., Bonn, W.G., 2009. Detection of citrus canker using hyperspectral reflectance imaging with spectral information divergence. J. Food Eng. 93 (2), 183–191.

Rajamaki, T., Alakomi, H.L., Ritvanen, T., Skytta, E., Smolander, M., Ahvenainen, R., 2006. Application of an electronic nose for quality assessment of modified atmosphere packaged poultry meat. Food Contr. 17 (1), 5–13.

Ramon, H., Anthonis, J., Vrindts, E., Delen, R., Reumers, J., Moshou, D., Deprez, K., De Baerdemaeker, J., Feyaerts, F., Van Gool, L., De Winne, R., Van den Bulcke, R., 2002. Development of a weed activated spraying machine for targeted application of herbicides. Aspect Appl. Biol. 66, 147–164.

Roberts, M.J., Schimmelpfennig, D., Ashley, E., Livingston, M., Ash, M., Vasavada, U., 2006. The Value of Plant Disease Early-Warning Systems. Economic Research Service, No. 18, United States Department of Agriculture.

Roggo, Y., Duponchel, L., Huvenne, J.P., 2003. Comparison of supervised pattern recognition methods with McNemar's statistical test: application to qualitative analysis of sugar beet by near-infrared spectroscopy. Anal. Chim. Acta 477 (2), 187–200.

Ruiz-Ruiz, S., Ambros, S., Carmen Vives, M., Navarro, L., Moreno, P., Guerri, J., 2009. Detection and quantification of Citrus leaf blotch virus by TaqMan real-time RTPCR. J. Virol Methods 160 (1–2), 57–62.

Saponari, M., Manjunath, K., Yokomi, R.K., 2008. Quantitative detection of *Citrus tristeza* virus in citrus and aphids by real-time reverse transcription-PCR (TaqMan). J. Virol Methods 147 (1), 43–53.

Schaad, N.W., Frederick, R.D., 2002. Real-time PCR and its application for rapid plant disease diagnostics. J. Indian Dent. Assoc. 24 (3), 250–258.

Scharte, J., Schon, H., Weis, E., 2005. Photosynthesis and carbohydrate metabolism in tobacco leaves during an incompatible interaction with *Phytophthora nicotianae*. Plant Cell Environ. 28, 1421–1435.

Shafri, H.Z.M., Hamdan, N., 2009. Hyper spectral imagery for mapping disease infection in oil palm plantation using vegetation indices and red edge techniques. Am. J. Appl. Sci. 6 (6), 1031–1035.

Sighicelli, M., Colao, F., Lai, A., Patsaeva, S., 2009. Monitoring post-harvest orange fruit disease by fluorescence and reflectance hyper spectral imaging. ISHS Acta Hortic. 817, 277–284.

Sirisomboon, P., Hashimoto, Y., Tanaka, M., 2009. Study on non-destructive evaluation methods for defect pods for green soybean processing by near-infrared spectroscopy. J. Food Eng. 93, 502–512.

Spinelli, F., Noferini, M., Costa, G., 2006. Near infrared spectroscopy (NIRs): perspective of fire blight detection in asymptomatic plant material. Proceeding of 10th International Workshop on Fire Blight Acta Hortic. 704, 87–90.

Staudt, M., Lhoutellier, L., 2007. Volatile organic compound emission from holm oak infested by gypsy moth larvae: evidence for distinct responses in damaged and undamaged leaves. Tree Physiol. 27, 1433–1440.

Sundaram, J., Kandala, C.V., Butts, C.L., 2009. Application of near infrared (NIR) spectroscopy to peanut grading and quality analysis: overview. Sens. & Instrument. Food Qual. & Saf. 3 (3), 156–164.

Tallada, J.G., Nagata, M., Kobayashi, T., July 2006. Detection of bruises in strawberries by hyperspectral imaging. ASABE Paper No. 063014. In: 2006 ASABE, Annual International Meeting, Portland, Oregon, pp. 9–12.

Tholl, D., Boland, W., Hansel, A., Loreto, F., Rose, U.S.R., Schnitzler, J.P., 2006. Practical approaches to plant volatile analysis. Plant J. 45, 540–560.

Vallat, A., Gu, H., Dorn, S., 2005. How rainfall, relative humidity and temperature influence volatile emissions from apple trees in situ. Phytochemistry 66, 1540–1550.

Vuorinen, T., Nerg, A.M., Syrjala, L., Peltonen, P., Holopainen, J.K., 2007. Epirritaautumnata induced VOC emission of silver birch differ from emission induced by leaf fungal pathogen. Arthropod-Plant Interac. 1, 159–165.

Wang, D., Ram, M.S., Dowell, F.E., 2002. Classification of damaged soybean seeds using near-infrared spectroscopy. Trans. ASAE 45 (6), 1943–1948.

West, J.S., Bravo, C., Oberti, R., Lemaire, D., Moshou, D., McCartney, H.A., 2003. The potential of optical canopy measurement for targeted control of field crop disease. Annu. Rev. Phytopathol. 41, 593–614.

Williamson, B., Goodman, B.A., Chudek, J.A., 1992. Nuclear magnetic resonance (NMR) micro-imaging of ripening red raspberry fruits. New Phytol. 120, 21–28.

Wu, D., Feng, L., Zhang, C., He, Y., 2008. Early detection of *Botrytis cinerea* on eggplant leaves based on visible and near-infrared spectroscopy. Trans. ASABE 51 (3), 1133–1139.

Xing, J., Baerdemaeker, J.D., 2005. Bruise detection on 'Jonagold' apples using hyper spectral imaging. Postharvest Biol. Technol. 37 (2), 152–162.

Xing, J., Bravo, C., Jancsok, P.T., Ramon, H., Baerdemaeker, J.D., 2005. Detecting bruises on 'golden delicious' apples using hyper spectral imaging with multiple wavebands. Biosyst. Eng. 90 (1), 27–36.

Xu, H.R., Ying, Y.B., Fu, X.P., Zhu, S.P., 2007. Near-infrared spectroscopy in detecting leaf minor damage on tomato leaf. Biosyst. Eng. 96 (4), 447–454.

Yang, C.M., Cheng, C.H., Chen, R.K., 2007. Changes in spectral characteristics of rice canopy infested with brown planthopper and leaf folder. Crop Sci. 47, 329–335.

Yao, H., Hruska, Z., DiCrispino, K., Brabham, K., Lewis, D., Beach, J., Brown, R.L., Cleveland, T.E., July 2005. Differentiation of fungi using hyper spectral imagery for food inspection. ASAE Paper No. 053127. In: 2005 ASAE Annual International Meeting, Tampa, FL, pp. 17–20.

Yvon, M., Thebaud, G., Alary, R., Labonne, G., 2009. Specific detection and quantification of the phytopathogenic agent '*Candidatus Phytoplasma prunorum*'. Mol. Cell. Probes 23 (5), 227–234.

Zhang, C., Shen, Y., Chen, J., Xiao, P., Bao, J., 2008a. Non-destructive prediction of total phenolics, flavonoid contents and antioxidant capacity of rice grain using near-infrared spectroscopy. J. Agric. Food Chem. 56 (18), 8268–8272.

Zhang, H., Chang, M., Wang, J., Ye, S., 2008b. Evaluation of peach quality indices using an electronic nose by MLR, QPST and BP network. Sensor. Actuator. B Chem. 134 (1), 332–338.

Index

Note: 'Page numbers followed by "*f*" indicate figures and "*t*" indicate tables.'

A

Abiotic stresses management in crop plants, 360–364
  drought stress, 363–364
  salinity stress, 364
  stress of
    extreme temperatures, 364
    heavy metals in soils, 364
Abscisic acid (ABA), 286, 389
Absidia sp., 192
Acacia melanoxylon, 340
Acetic acid
  eco-toxicology, 341
  as potential product, 341
    chemical composition of wood vinegar, 341
    eco-toxicology of pyrolytic products, 341
Acetohydroxyacid synthase (AHAS), 241
Acetyl-coenzyme A carboxylase (ACCase), 241
Achaea janata, 196–197
Achillea
  A. fragrantissima, 145–146
  A. santolina, 145
Acids compound, 341
Acinetobacter, 250
Acorus calamus, 203
Acrothecium carotae, 193
Acrylamide, 204
Actinobacteria, 24, 253
Actinomycetaceae, 41
Actinomycetes spp., 9, 262
Advance molecular tools to detect plant pathogens
  affinity biosensors, 409, 411
  antibody-based biosensors, 410
  bacteriophage based biosensors, 410–411
  biosensor platforms based on nonmaterials, 409
  DNA/RNA-based affinity biosensor, 410
  electronic nose system, 407–408
  enzymatic electrochemical biosensors, 410
  FISH, 409
  fluorescence imaging, 405
  fluorescence spectroscopy, 403
  GC–MS, 408–409
  genetically-encoded biosensors, 411
  hyper spectral
    imaging, 406
    techniques, 409
  molecular techniques of plant disease detection, 401–402
  other imaging techniques, 406–407
  profiling of plant volatile organic compounds, 407
  spectroscopic and imaging techniques, 402–403, 411–412
  visible and infrared spectroscopy, 403–405
Aedes
  A. aegypti, 197
  A. fluviatilis, 25
Aerva lanata, 167–168
Aeschynomene virginica, 10
Affinity biosensors, 409, 411
Agaricus bisporus, 355
Agelastica alni, 192
Ageratum conyzoides, 146
Aggregation pheromones, 12, 184
Agri-food production, 159–160
Agricultural applications, 35
Agricultural intensification, 7
Agricultural pests, 7
  management, 184–185
Agricultural resources, 7
Agricultural systems, 94
Agriculture, 1, 7–8, 19, 32, 183, 203, 237–238, 286, 383
Agro-ecosystem, 1
Agrobacterium, 48–51, 383
  A. radiobacter, 23, 225, 262
  A. radiobactor, 302–304
  A. tumefaciens, 11, 23, 192, 302–304
Agromaterials Management Division (AMD), 57–58
Agrotis ipsilon, 197
Air temperature, 34
Akanthomyces lecanii RCEF 1005, 312–313
Alarm pheromones, 12, 184
Albizia procera, 137
Alcaligenes sp, 23, 250
Alcohol, 348–349
  disadvantages, 348–349
Aldehydes compound, 341
Algae, 325
Algal biopesticides, 309
Algicides, 201
Alhagi maurorum, 196
Aliphatic alcohol, 348
Alisma plantago-aquatica, 140
Alkaline phosphatase (ALP), 120
Alkaloids, test for, 161
Allelopathic mechanisms, 135
Allium sativum, 23, 25, 140–141
Allorhizobium, 48–51
Alternaria alternata (AA), 184, 190, 196, 251, 253
Alternaria spp., 195, 408
  A. solani, 219
Alternative plant protection techniques, 337
Amaranthus
  A. deflexus, 137
  A. retroflexus, 136–137
  A. spinosum, 137, 139
  A. viridis, 137
1-aminocyclopropane-1-carboxylate deaminase (ACCD), 364
1-aminocyclopropane-1-carboxylic acid (ACC), 286, 364
Aminopeptidase-N (APN), 120
Amorphafructicosa, 109
Ampelomyces sp., 262, 383
  A. quisqualis, 185
Amylase, protease, proteinase, and lipase assay, 164–165
Anabaena laxa, 23
Anagallis arvensis L, 136
Anagrapha falcifera, 289
Analysis of variance (ANOVA), 165
Androctonus australis, 47, 111
Aneurinibacillus sp., 185
Animal manure, 339
Anopheles
  A. albimanus, 197
  A. stephensi, 146, 160
  A. sundaicus, 160
Antibiosis, 320, 359, 385–386
  by antimicrobial metabolites, 267–268
  important criterion for development of microbial biopesticides, 288
Antibiotics, 93–94
  production, 385–386
Antibody-based biosensors, 410
Anticarsia gemmatalis, 108
Anticarsia gemmatalis multiple nucleopolyhedrovirus (AgMNPV), 230, 289
Anticarsia gemmatalis multiple nucleopolyhedrovirus isolate 2D (AgMNPV-2D), 230
Antimicrobial(s), 93–94
  activity of *B. thuringiensis* based biopesticides, 192

Antimicrobial(s) (*Continued*)
antibiosis by antimicrobial metabolites, 267–268
Antiparallel β-sheets, 2
Aphanomyces euteiches f. sp. *pisi*, 324
Aphid infestation in crop plants, 239–240
Aphids (*Myzus persicae*), 142
Aphis gossypii, 26
Apiaceae, 25
Apicomplexa, 197
Application technology of biopesticides
biopesticides and adjuvants, 32–33
coverage, 31–32
biological target, 32
biopesticide, 32
crop, 32
theoretical density of droplets, 32t
final considerations, 36
influence of climatic factors on, 33–35
relative humidity and temperature, 33–34, 34f
timing of biopesticides application, 35
wind speed and direction, 34–35
mixture of biopesticides and pesticides, 33
Approximate digestibility (AD), 164
Aproaerema modicella. *See* Leaf miner (*Aproaerema modicella*)
Aqueous suspensions, 82
Arabidopsis thaliana, 386, 388
Aranta arbustorum. *See* Snails (*Aranta arbustorum*)
Araujo moratorium, 11
Arbuscular mycorrhiza (AM), 253
Arbuscular mycorrhizal fungi, 94
Arion lusitanicus. *See* Slugs (*Arion lusitanicus*)
Aristolochia elegans, 229
Arnaranthus powellii, 138
Artemisia annua, 323–324
Arthrobacter, 48–51, 250
Arthropods, 45
insects, 7
Artificial neural network models (ANN models), 405
Ascaris suum, 120
Aschersonia badia, 312–313
Ascochytain pea, 193
Ascogregarina culicis, 197
Ascomycota, 43, 228, 354
Ascomycotina, 195
Ascorbate (AsA), 270
Ascorbate peroxidases, 363
Ascosphaera apis, 312–313
Ascoviridae, 196
Aspergillus, 51, 190, 192, 196, 214
A. flavus, 25, 196, 306–307
A. niger, 23, 262, 286
A. solani, 196
A. versicolor, 56
A. viridae, 219
Asteraceae, 25
Atteva fabriciella, 196
Audouin. *See Plexippus paykulli* (Audouin)
Aureobasidium pullulans, 196
Australian Pesticides and Veterinary Medicines Authority (APVMA), 292
Autographa californica, 289
Autographa californica multiple nucleocapsid nucleopolyhedrovirus (AcMNPV), 47, 230
Avenafatua, 138
Ayurveda, 247
Azadirachta indica, 23, 109, 113, 140–141
Azadirachtin, 26
Azocasein, 164
Azospirillum, 48–51, 93–94, 250
Azotobacter, 48–51, 93–94, 278
A. chroococcum, 279

B

Bacillaceae, 41
members of *Bacilliaceae* as biopesticides, 185–193
antimicrobial activity of *B. thuringiensis* based biopesticides, 192
Bacillus firmus, 191
Bacillus subtilis, 191
Bacillus thuringiens as nano pesticides, 192–193
Bt, 191–192
Lysinibacills sphaericus (*Bacillus sphaericus*), 190
Paenibacillus popilliae (*Bacillus papillae*) and *B. lentimorbus*, 190
mode of action of entomopathogenic microbes, 41f
Bacillus, 1–2, 23, 48–51, 185, 214, 250, 253, 262, 278, 301–302, 325–328, 383
B. anthracis, 290–291
B. bassiana, 280
B. cereus, 118–119, 323–324
B. endophyticus, 140
B. firmus, 191, 288
B. flexus, 140
B. lentimorbus, 190, 323–324
B. licheniformis, 46, 278–279, 302–304
B. licheniformis MML2501, 323
B. megaterium, 122, 279, 364
B. mycoides, 215–216, 279
B. papillae, 190
B. polymyxa, 226–227
B. popilliae, 41, 54, 225, 302–304
B. pumilus, 185, 278–279
IN937b, 323
B. sotto, 121
B. subtilis, 54–55, 122, 185, 191, 225, 228, 278–279, 302–304, 311, 322
BHHU100, 215–216, 324
BS14, 46
QST 713, 191
TM4, 191
B. tequilensis, 286
MML2551, 323
carriers used, 214
commercial names of products, 214
dry formulations, 214
liquid formulations, 214
Bacillus amyloliquefaciens (BA), 251, 253, 278–279
FZB42, 227
IN937a, 323
MBAA3, 323–324
Bacillus sphaericus (Bs), 41, 94, 190, 225–226
Bacillus thuringiensis (Bt), 1, 9, 11, 21–22, 25, 33, 41, 94, 110, 142, 191–192, 203, 210–211, 214, 225–226, 237–239, 288, 290–291, 293, 302–304, 312–313
B. bassiana, 113
B. thuringiensis israelensis, 23–24, 225
B. thuringiensis sphaericus, 23–24
B. thuringiensis tenebriomsis, 23–24, 225
B. velezensis 9D-6, 291
based biopesticides for integrated crop management, 1
early beginning of, 1–2
past last twenty years of, 2–3
present and future of, 3–5, 4f
compatibility with natural enemies and Bt plants, 122–124
final considerations, 124
cotton adoption in India, 239, 239f–240f
DAR 81934, 293
development of Bt formulations, 121–122
genes for insect resistance, 238–239
isolation and epizootic potential of, 118–119
LLP29, 291
mode of action of cry toxins, 120–121
as nano pesticides, 192–193
nomenclature and characterization of cry toxins, 119
strains, 2
toxins, 2, 210–211
transgenic crops, 2
Bacillus thuringiensis subsp. *aizawai* (Bta), 23–24, 225, 280
Bacillus thuringiensis subsp. *kurstaki* (Btk), 23–24, 185, 225, 280, 288
Back-propagation neural network model (BPNN model), 405
Bacteremia, 2
Bacteria, 41–43, 325
Bacillaceae, 41
as biopesticides, 185, 288
entomopathogenic, 225–228
Morganellaceae and enterobacteriaceae, 42
Paenibacillaceae, 42
Pseudomonadaceae, 43
Pseudonocardiaceae, 42
Streptomycetaceae, 42
Yersiniaceae, 43
Bacterial antagonists, 323–324
Bacterial bioherbicide, 9–10
Bacterial bioinsecticides, 9
Bacterial biopesticides, 45, 302–304
mode of parasitism of Bt toxin, 304f
used in plant disease management, 212

Bacterial pesticides, 302–304
Bacterial phytopathogens, 383
Bacteriophage based biosensors, 410–411
Bactrocera dorsalis, 204
Baculoviridae, 45, 196
Baculovirus ChinNPV (Chrysogen®), 35
Baculoviruses (BVs), 10–11, 23, 94, 111, 229, 304–305
  baculovirus-induced *Wipfelkrankheit* disease, 45
Baiting of EPN, 74
Basidiobolus meristosporus, 312–313
Bassiana RNA Virus 1 (BbRV1), 289
Bassianolide, 43–44
Bean golden mosaic virus (BGMV), 241
Beauveria bassiana, 10, 23, 25, 43–44, 56, 147, 185, 194, 228–229, 231, 278, 289–291, 306–307, 322–323
  CPD9, 195
  JEF-007, 293
Beauveria brongniartii, 25, 43–44, 228–229, 289
  RCEF 3172, 312–313
Beauvericin, 43–44
Bee vectoring, 368
Beta proteobacteria, 302–304
β-sheets, 2
Beuveria bassiana, 306–307
*Bgn*13.1 gene, 267
Bio-active compounds, 248
Bioagents, 354
Biochemical biopesticides, 109
Biochemical pesticides, 12–13, 22, 37, 107, 133, 210, 301–302
  pheromones, 12
  plant essential oils, 12–13
Biochemical preparations, 22
Biocides, 100, 337
  acetic acid
    eco-toxicology, 341
    as potential product, 341
  bio-pesticides, 338
  catechol eco-toxicology, 348
  future prospects, 349–350
    prevailing bio-cide models, 345t–347t
  other alcohol, 348–349
  phenol eco-toxicology, 348
  products of pyrolysis, 339–341
  pyrolysis-efficient technology, 338
  pyrolytic feedstock, 339
  quinone eco-toxicology, 348
Biocontrol agents (BCA), 9
  Trichoderma spp. as biocontrol agents for diseases management in crops, 358–360
    antibiosis and lysis, 359
    competition for food, space and nutrients, 358–359
    hyperparasitism/mycoparasitism and predation, 359–360
    induction and exploitation of induced systemic resistance against different stresses, 360
    mechanisms of biological control, 359t
    mode of action, 358–360
Biocontrol mechanisms, 384
  by *Trichoderma* in control of wilt pathogens, 264–270
    competition, 264
Biofertilizer, 364–366
  increase in minerals and yield in crops, 366
  phosphate solubilization, 364
Biofuel, 368
Biofungicides, 27, 325–328, 338
  plant sources and their anti-microbial properties, 339t
Bioherbicides, 14, 27, 94, 135–140, 325–328
  membrane process, 140
  in nanoformulations, 144
  plant extracts and essential oils with herbicidal activity, 135–139
  produced by microorganisms, 139–140
Bioinoculants
  in strawberry, 277–279
  in tomato, 279
Bioinsecticides, 13, 27, 140, 325–328
Biological control, 228–229
  MC, 320–321
    mediated plant defense mechanism in, 322
Biological control agents (BCAs), 262, 286, 302, 319–320, 383
Biological Farmers of Australia (BFA), 202
Biological pest control
  agent, 19
  techniques, 21
Biological Products Industry Alliance (BPIA), 27
Biomass, 339
  pyrolysis, 337–338
Bionematicides, 94
Bionematocides, 27
Biopesticide Industry Alliance (BIP), 57
Biopesticide sand Pollution Prevention Division (BPPD), 20, 57
Biopesticide(s), 31–33, 94–95, 170, 183–184, 202, 207–209, 219–220, 237–238, 262, 285–286, 290–291, 293–294, 301–302, 325, 330–332, 338
  and adjuvants, 32–33
  advantages of application biopesticides in pest management, 107–109, 219–220
    biochemical biopesticides, 109
    microbial biopesticides, 108
    PIP, 108–109
  agriculture and pests, 7–8
    from pesticides to biopesticides, 8
    synthetic pesticides and challenges, 7–8
  application, 2, 8
  area under cultivation with different categories of pesticides, 238f
  and association with growth promoter, 217–218
  bacteria as, 185
  biochemical pesticides, 12–13
  classification, 209–211, 210f
  Coniothyrium minitans as, 195
  constraints for applications of, 110
  consumption of different categories of pesticides, 238f
  current status of biopesticides in India, 309–312
    evolution of microbial biopesticides for management of insect pest in India, 311–312
    registration norms and regulation of microbial biopesticides, 311
  factors for increasing trends toward, 109
  formulations, 19–20, 23–26, 330
    challenges and future perspectives, 27–28
    diversity of, 21–23
    market, 26–27
    regulatory framework, 20–21
    several examples of microbial biopesticides commercially available, 24t
  Gliocladium catenulatum as, 195–196
  in grapes, 279–280
  against harmful arthropodes, 140–144
  inconsistencies in, 8–11
    microbial biopesticides, 9–11
    nematodes biopesticides, 11
  in India, regulatory framework and challenges for, 330–331
  limitations, 220
  for management of arthropod pests and weeds
    bioherbicides, 135–140
    biopesticides against harmful arthropodes, 140–144
    nanoscale biopesticide formulations against arthropod pests and weeds, 144–148
  market, 5, 26–27, 56
    trends of, 109
  market and challenges, 13–14
    bioherbicides, 14
    bioinsecticides, 13
  marketing, factors affecting, 220, 221f
  medicinal plants associated microflora
    medicinal plant diversity in India, 247–248
    niche of microflora, 248
    plant-microbe association, 248–253
    relative factors between microflora and plants, 254
  members of *Bacilliaceae* as, 185–193
  members of *Pseudomonadaceae* and *Enterobacteriaceae* as, 193–195
  nanoformulations of against anthropodes, 145–148
    green-synthesized metal nanoparticles, 146–148
    nanoemulsions of botanical insecticides, 145–146
  need for development biopesticides containing MC of, 325–328
    microbial pesticides, 325–328
  plant-incorporated protectants, 11–12
  producers, 27

Biopesticide(s) (*Continued*)
  research, 96
  role of genetic engineering in context of, 110–113
    Bacillus thuringiensis, 110
    baculoviruses, 111
    botanical biopesticides, 113
    entomopathogenic fungi, 113
    entomopathogenic nematodes, 111
    PIP, 112
    RNAi based biopesticides, 111–112
  in strawberry, 279–280
  in vegetables, 280
Bioremediation, 93, 366
Biosensor platforms based on nonmaterials, 409
Bioshield™, 27
Biotechnology, 93–94
Biotic stresses in crop plants, management of, 360–364
Bipolaris
  B. maydis, 191
  B. spicifera, 190
Black Zira essential oil, 165
Blue cluster, 99
Blue-green fluorescence, 403, 405
Bombyx mori, 121
Bombyx mori cytoplasmic polyhedrosis virus (BmCPV), 45
Botanical(s), 9
  biopesticides, 113
    usage plant pathogen, 219
  insecticides, 146
  pesticides, 25
Bothriochloa
  B. barbinodis, 138
  B. laguroides, 138
Botrytis sp., 195–196
  B. allii, 408
  B. cinerea, 184, 194–195, 215–216, 227, 360, 405, 407–408
  B. fabae, 196
  B. squamosa, 196
Bradyrhizobium, 48–51, 251
Brassica
  B. napus, 407
  B. nigra, 25
  B. oleracea, 163, 177
Brevibacillus spp., 185
  B. laterosporus, 42, 225–226
Budded virus (BVs), 196, 230
Burkholderia, 48–51, 250, 288, 325–328
  B. cepacia, 23
  B. pyrrocinia JK-SH007, 287
  B. rinojensis, 280
Burkholderiacea, 41
Butomus umbellatus. *See* Flowering-rush (*Butomus umbellatus*)

C

Caenorhabditis elegans, 86
Calceolaria sp, 26
Calcium magnesium acetate, 349
Candida spp., 262, 383
  C. albicans, 43, 227
  C. oleophila, 23, 196
Capsicum sp, 25, 140–141
Carbon dioxide emissions, 93
Carbonaceous rot, 196
Carboxy methyl cellulose (CMC), 367
Carcinogenicity, 7–8
Catalases (CAT), 363
Catechol eco-toxicology, 348
Catharanthus roseus, 386
Cattail (*Typha latifolia*), 203
Caulobacter, 48–51
Cell cytoplasm (CPVs), 229–230
Cell wall degrading enzymes (CWDEs), 267, 385–387
Central Insecticides Board (CIB), 57–58, 292
Central Insecticides Board and Registration Committee (CIBRC), 55, 309–311, 328–330, 331f
Central Institute of Cotton Research, 163
Cercospora beticola, 407
Chaetomium spp., 211, 216
  C. cupreum, 211
  C. globosum, 211
Charge coupled device-based camera system (CCD-based camera system), 405
Chemical herbicides, 133
Chemical pesticides, 12, 19, 21, 33, 37
  on environment, 94
Chenopodium
  C. album, 196
  C. ambrosioides, 12–13, 26
ChiA, 290–291
Chickpea (*Cicer arietinum* L.), 191
Chinaberry (*Melia azedarach*), 140–141, 185, 196
Chit33, 267
Chit42, 267
Chitinases enzymes, 45–46
Chitosan (CS), 140–141, 184–185, 204
  encapsulations, 160
Chlamydospores, 355
Chlorella, 309
  C. vulgaris, 23
Chloridea virescens, 120
Chlorophyll fluorescence, 403, 405
Choristoneura fumiferana defective MNPV (CfDefNPV), 230
Chromobacterium, 48–51, 288
  C. subtsugae, 227, 280, 302–304
Chromolaena odorata, 140–141
Chrysanthemum, 185
  C. cinerariaefolium, 25, 109, 113
  C. cinerariifolium, 12–13
  C. cineum, 25
  C. coccineum, 12–13
Chrysodeixis, 33
Chytridiomycota, 228
Cicer arietinum L. *See* Chickpea (*Cicer arietinum* L.)
Ciliophora, 197
Cinnamomum
  C. verum, 25
  C. zeylanicum, 23, 145
Citrobacter freundii, 279
Citrus canker, 403
Citrus reticulata, 203
Citrus-based Ag nanoparticles, 203
Cladosporium, 195
Clavicipitaceae, 43–44
Clay and powder, 85
Clethrionomys rufocanus. *See* Vole (*Clethrionomys rufocanus*)
Cnaphalocrosis medinalis, 94, 196–197
Cochliobolus australiensis, 139
Coleoptera, 12, 81, 120, 191–192, 227, 229
Coleopteran pests, 2
Colias lesbian, 56
Collapse, 196
Colletotrichum, 10
  C. coccodes, 46
  C. gloeosporioides, 139, 192, 226–227, 323, 408
    C. gloeosporioides f. sp. *aeschynomene*, 10
    C. gloeosporioides f. sp. *malvae*, 10
  C. lindemuthianum, 360
  C. orbiculare, 10
  C. truncatum, 10, 322–323
Colony forming units (CFUs), 118, 366
Comamonas testosterone, 279
Commercialization
  advantages of, 231–232
  of microbial based products in India, 329–332
  of microbial pesticides, 292
  of *Trichoderma* spp., 368–370
    isolation, 369
    sensitivity against agro-chemicals, 369
Competition, 320
  with pathogens for space and nutrients, 384–385
Compound Annual Growth rate (CAGR), 37, 109
Concanavalin A (ConA), 240
Condor biopesticide, 3
Conidia, 355
Conidiophores, 354
Coniothyrium spp., 262, 383
  C. minitans, 27, 195
Conjugal plasmid exchange system, 3
Consumption index (CI), 175
Conventional pesticides, 26
Convolvulus arvensis, 139
Copper chitosan nanoparticle, 166–167
Cordyceps
  C. bassiana, 312–313
  C. fumosorosea ARSEF 2679, 312–313
  C. militaris, 291, 312–313
Cordycipitaceae, 43–44
Coriandrum sativum, 25
Cotton plants, 238–239
Cowpea trypsin proteinase (CpTI), 238–239
Crab shells, collection and processing of, 162
CRISPR/Cas9, 241, 293
  nickase, 46

Critical differences (CD), 165
Crops
  biocontrol agents for diseases management in, 358–360
  growth enhancer *Trichoderma* spp. as, 364–366
    biofertilizer, 364–366
    bioremediation, 366
    decomposer of organic matter, 366
    effect of *Trichoderma* harzianum, 365f–366f
  increase in minerals and yield in, 366
  pests, 7
  plants
    abiotic stresses management in, 360–364
    GE approaches for combating aphid infestation in, 239–240
  production, 7, 107
  protection, 7–8
Cry1A gene, 290–291
Cry1Ac7 gene, 290–291
Cryphonectria parasitica, 191
Cryptococuss sp., 383
Crystal (cry), 302–304
  gene, 107
  proteins, 1–2, 142, 226, 238–239, 288
  toxins, 2, 117, 191–192
Culex
  C. quinquefasciatus, 25
  C. tritaeniorhynchus, 197
Cultivated crop fields, mitigating nitrous oxide emissions from, 368
Cuminum cyminum, 25
Curcuma sp, 25
Curvularia sp., 192
  C. lunata, 190, 196
Cyanobacteria, 309
Cyclic AMP accumulation (cAMP accumulation), 120
Cyclodepsipeptides, 43–44
Cydia pomonella granulovirus (CpGV), 108, 196
  matrix metalloprotease, 47
Cylindrocarpon destructans, 194–195
Cymbopogon nardus, 138
Cyperus iria, 136
Cypovirus (CPV), 45
Cytisus scoparius (L.), 138
Cytolytic toxins (Cyt toxins), 2

D

Daphnia magna, 141
Decapoda, 45
Decomposer of organic matter, 366
Delonix regia, 137
Dendrobium nobile, 355–356
Densovirinae, 196
Deoxyribose nucleic acid (DNA), 401
  DNA-based affinity biosensor, 410
  microarrays, 401–402
Department of Biotechnology (DBT), 330–331
Derris elliptica, 109
Deuteromycota, 228
Deuteromycotina, 195
2,4-diacetylphloroglucinol (2,4-DAPG), 287
Diaporthe phaseolorum, 139
Dichlorodiphenyltrichloroethane (DDT), 117, 387
Digitaria sanguinalis, 138
Diguetia canities, 230
Diptera, 45, 120, 191–192, 227, 229
Directorate of Plant Protection, Quarantine and Storage (DPPQS), 328–329
Dirhamnolipid, 139
Discriminant analysis, fluorescence spectroscopy, 403–404
Discriminant partial least squares (DPLS), 405
Diseases
  management, 360–364
    in crops, 358–360
Dispersal pheromones, 12
Displacement of atmospheric air, 34–35
Disruption of insect peripheral nervous system, 24
Ditylenchus sp., 191
Diversity of biopesticides, 21–23
Division of labor (DoL), 325
DNA methyltransferases (DNMTases), 291
δ-endotoxins, 41
Double-stranded DNA (dsDNA), 410
Drip irrigation water, 367
Drosophila melanogaster, 26
Duckweed (*Lemna minor*), 203
Dynamic soil environment, 87

E

Echinochloa crus-galli, 135–136
Eco-toxicology of pyrolytic products, 341
Efficiency of conversion of digested food (ECD), 175
Efficiency of conversion of ingested food (ECI), 175
Eldana saccharina, 290–291
Electronic nose system, 407–408
Eligma narcissus, 196
Encapsulation, 122
Encapsulation chitosan nanoparticles treatment, 160
Endocrine system, 175
Endophytes, microbes, 354
Endophytic fungi as biocontrol agents, 196
Endophytic microbiome association with medicinal plants, 251–253
  effects of endophytic bacteria, 251f
  microbial association with medicinal plants, 252t–253t
  scanning electron micrographs, 254f
Endophytic *Trichoderma* spp., 355–356
Energy dispersive X-ray spectroscopy (EDX), 160, 168, 168f
Enhanced locomotory activity (ELA), 45
Enterobacter sp, 23, 250
  E. cloacae, 279
Enterobacteriaceae, 41–42, 73, 76–79, 193–195
  fungi as biopesticides, 194
  Trichoderma spp. as biopesticide, 194–195
Entomopathogenic bacteria, 41, 225–228
Entomopathogenic fungi, 10, 25, 43, 113, 194, 225–228, 291, 306–307
Entomopathogenic microbes, 9
Entomopathogenic nematodes (EPN), 11, 73, 111, 230–231, 311
  application of EPN genomics to enhance field efficacy, 86–87
  baiting, isolation, multiplication, 74, 74f, 81f
  identification of, 75
    morphometric parameters, 76t
    universal primers, 76t
  liaison between EPNs and mutualistic bacteria and identification, 76–80
    associated species of *Heterorhabditis* nematode and *Photorhabdus* bacteria and distribution, 77t
    associated species of *Steinernema* nematode and *Xenorhabdus* bacteria and distribution, 78t–79t
    characterization parameters of symbiotic bacteria, 79t
  life cycle, pathogenicity and host range of, 80–81
  mass production, formulation development and application, 81–85
    commercial EPN products available in different countries, 83t–84t
    genomic feature of EPN, 85t
    types of formulations, 82–85
Entomopathogenic protozoans, 225–228, 308
Entomopathogenic roles of hypocrealean entomopathogenic fungi, 280–281
Entomopathogenic viruses, 44
Entomopathogens, 85, 118, 225
Entomophthora muscae, 312–313
Entomopoxvirinae, 196
Environment Protection Act (EPA), 57
Environmental Policy Secretary (EPS), 58
Enzymatic electrochemical biosensors, 410
Enzyme-linked immunosorbent assay (ELISA), 401
Ephestia kuehniella, 121
Epicoccum nigrum, 196, 355–356
Epipolythio dioxopi perazines (ETPs), 385–386
Episomus lacerta, 204
Epithelial midgut cells, 2
Epizootic diseases, 118
Erwinia, 48–51, 250
  E. amylovora, 23
  E. carotovora, 226–227, 408
    E. carotovora subsp. *atroseptica*, 408
    E. carotovora subsp. *carotovora*, 408
Escherichia coli, 122, 348
Essential oils (EOs), 135–136, 219, 220t
  characterization of essential oil loaded chitosan nanoparticles, 166–167
  UV-VIS spectral analysis of essential oil loaded chitosan nanoparticles, 166–167
Ethylene (ET), 389

Eucalyptus citriodora, 138
Euphorbia sp, 23
European and Mediterranean Plant Protection Organization (EPPO), 20, 57
European Crop Protection Association (ECPA), 202
European Food Safety Authority (EFSA), 201
European Union (EU), 56, 292
Evaporation process, 34
Extracellular PGPR (ePGPR), 48–51
Extraction
  of chitosan from crab shell, 162
    structure of chitosan, 162, 162f
  of *S. leucantha* essential oil, 160
Exudation process, 47–48

F

Farinocystis tribolii, 197
Fatty acids and retinol binding (FAR), 86
Federal Insecticide, Fungicide, and Rodenticide Act (FIFRA), 20, 292
Fermentation process, 122
Filamentous fungal genus, 354
Fischerella ambigua, 23
Fischer's linear discriminant analysis (LDA), 403
Flavobacterium, 48–51, 250
Flavonoids test for, 161
Flow cytometry technique, 81, 401–402
Flowering-rush (*Butomus umbellatus*), 203
Fluorescence imaging, 405
Fluorescence in situ hybridization (FISH), 401–402, 409
Fluorescence spectroscopy, 403, 411
Fluorescent silica nanoparticles (FSNPs), 409
Foeniculum vulgare, 25
Foliar application, 31
Food
  competition for, 358–359
  products, 406
  systems, 94
  utilization measures, 175
Food and Agriculture Organization (FAO), 20, 207
Food Quality Protection Act (FQPA), 20
Formulations
  development and application, 81–85
  types of, 82–85
    application, 85
    aqueous suspensions, 82
    clay and powder, 85
    gels, 82–85
    sponges, 85
Fourier transform infrared spectroscopy (FTIR), 160
  analysis of essential oil chitosan nanoparticles, 168–169, 169f
Frankia, 48–51
Frankliniella occidentalis, 280
Fulvic acid, 338
Fungal antagonists, 322–324
Fungal bioherbicides, 10
Fungal bioinsecticides, 10
Fungal biopesticides, 216, 306–307
  mode of parasitism of entomopathogenic fungi, 306f
  used in plant disease management, 212
Fungal pathogens, 10
Fungi, 43–44, 354
  as biopesticides, 194, 289
  Clavicipitaceae, 44
  Cordycipitaceae, 43–44
  entomopathogenic, 228–229
  Ophiocordycipitaceae, 44
Fungicides, 1, 19, 201
Fusarium, 51, 192, 195, 214, 262, 354, 383
  F. avenaceum, 194–195
  F. culmorum, 194–195
  F. equiseti, 196
  F. flocciferum, 355–356
  F. fujikuroi, 33, 140
  F. oxysporum, 23, 140, 184, 219, 227, 261–262, 264, 267–268, 408
    F. oxysporoum f. sp. *seseame*, 196
    F. oxysporum f. sp. *cicero*, 191
    F. oxysporum f. sp. *cucumerinum*, 267
    F. oxysporum f. sp. *lycopersici*, 269
    F. oxysporum f. sp. *melonis*, 46, 267
    F. oxysporum f. sp. *radicis-lycopersici*, 185
  F. ploriferatum, 140, 185
  F. sambucinum, 408
  F. solani, 213–214
Fusarium oxysporum f. sp. *neveum*. *See* Wilt (*Fusarium oxysporum* f. sp. *neveum*)

G

Gacposttranscriptional system, 46
Gaeumannomyces graminis, 193
Galanthus nivalis agglutinin (GNA), 240
Galleria mellonella. *See* Greater Wax Moth (*Galleria mellonella*)
γ-aminobutyric acid receptor (GABA receptor), 42
Gas chromatography–mass spectroscopy analysis (GC–MS analysis), 408–409
  analysis of essential oil of *S. leucantha*, 161–162, 165–166
    chemical profile of *Salvia leucantha* essential oil, 167t
    GC–MS chromatogram of the essential oil of *Salvia leucantha*, 166f
  GC–MS specification, 161
Gelatin, 204
Gelleria mellonella, 193
Gels, 82–85
Gene-gene interactions, 110
Genetic
  advantages of genetic manipulation, 231–232
  improvements of microbial pesticides, 290–292
  material, 197, 401–402
  recombination, 215
  tools, 73
Genetic engineering (GE), 110, 237–238
  approaches, 237–238
    for combating aphid infestation in crop plants, 239–240
  intervention
    applications of genome editing to control pests, 241
    applications of RNAi to control pests, 240–241
    biopesticides, 237–238
    Bt cotton adoption in India, 239
    engineering of Bt genes for insect resistance, 238–239
    GE approaches for combating aphid infestation in crop plants, 239–240
  technologies, 47
Genetically engineered crops (GE crops), 142
Genetically modified (GM), 108–109, 197–198, 325
  crops, 197–198, 325
  microbes, 45–47
Genetically modified organisms (GMO), 108–109, 202
Genetically-encoded biosensors, 411
Genome
  editing to control pests, 241
    list of genetically engineered plants as biopesticides, 241t–242t
  of microbial biopesticides, 227
    advantages of genetic manipulation and commercialization, 231–232
    ENP, 230–231
    entomopathogenic bacteria, 225–228
    entomopathogenic fungi, 228–229
    entomopathogenic protozoans, 231
    viral biopesticides, 229–230
  shuffling, 46
Gibbago trianthemae, 140
Gibberellic acid (GA), 286
Gliocladium sp., 383
  G. catenulatum, 195–196, 306–307
    Beauveria bassiana, 195
    endophytic fungi as biocontrol agents, 196
    Lecanicillium (*Verticillium*) *lecanii*, 195–196
    Purpureocillium lilacinum, 195
Gliotoxin, 386
Gliricidia sepium, 140–141
Global market
  for biopesticides, 27
  reports on use of microbial pesticides, 54–56
    Asia Pacific, 55–56
    Europe, 54–55
    Latin America, 56
    Middle East and Africa, 56
    North America, 54
Global pesticide market, 13, 37
Globodera spp., 195
Glomus
  G. aggregatum, 278
  G. etunicatum, 278
  G. intraradices, 278

G. mosseae, 278
Glucose transporter (Gtt1), 385
Glutathione pathway (GSH pathway), 270
Glyceraldehyde-3-phosphate dehydrogenase (GAPDH), 229
Glycoside hydrolase 18 gene (GH18 gene), 291
Glycosides test for, 161
Glyoviridin, 386
Glyphosate, 112
Gossypium hirsutum, 163, 177
Gram pod borer (*Helicoverpa armigera*), 184
Gram-positive bacterium, 41
Granuloviruses (GV), 10, 45, 196–197, 229–230, 289, 304–305
Grapes, biopesticides in, 279–280
Grapevines (*Vitis vinifera L.*), 404
Gravimetric technique, 164
Greater Wax Moth (*Galleria mellonella*), 3, 195, 289–290
Green cluster, 99
Green revolution, 237–238
Green-synthesized metal nanoparticles, 146–148
Gregarina garnhani, 197
Gremmeniella abietina, 355–356
Gut digestive enzymes of *H. armigera, S. litura* and *P. xylostella* larvae, 177, 177t, 178f
Gut paralysis, 12

H

Haemaphysalis longicornis, 143
Haematococcus pluviallis, 23
Halobacillus sp., 185
Harrisina metallica, 280
Harvest treatment, 23
Harzianic acid, 389
Harzianum harzianum, 185
Heartwood (HW), 340
Heavy metals in soils, stress of, 364
Helianthus annus, 229
Helicoverpa, 111
 H. armigera, 31, 140–141, 159–160, 163, 197, 237–238, 288, 312–313
 H. musciformis, 170–171
 H. zea, 55, 197
Helicoverpa armigera. *See* Gram pod borer (*Helicoverpa armigera*)
Heliothis, 111
 H. punctigera, 80
 H. zea nudivirus, 312–313
Helix amphipathic bundle, 2
Helminthosporium sp., 192
Hemiptera, 81, 120, 227
Herbal biopesticides, 22
Herbal pesticides, 37
Herbicides, 1, 19, 100, 201
 glyphosate, 33
Herbivore-induced plant volatiles (HIPV), 143
Heterocyclic compound, 341
Heterodera spp., 191, 195
Heterohabditis
 H. bacteriophora, 23
 H. downesi, 23
Heterorhabditidae, 11, 73
Heterorhabditids, 11
Heterorhabditis, 73, 75, 80, 111, 307–308
 H. bacteriophora, 11, 86, 111, 289–290, 307–308, 312–313
 H. indica, 81
 H. megidis, 307–308
High molecular weight lignin compounds (HMWL compounds), 341
Hirsutella spp., 228–229
 H. minnesotensis, 312–313
 H. rhossiliensis OWVT, 312–313
 H. thompsonii, 291, 312–313
 H. vermicola AS3. 7877, 312–313
Holotrichia consanguinea, 184
Hormonal system disruption, 7–8
Host cell detection, 111
Host marking pheromones, 184
Host range
 of EPN, 80–81
 expansion, 14
Host-parasite relationship, 81
Host-pathogen interactions, 110
Humic acid, 278
Hyadaphis foeniculi, 280
Hyalomma lusitanicum, 140–141
Hyblea purea, 196
Hydro distillation method, 160
Hydrogen cyanide (HCN), 287, 320–321
Hydroponic systems, 368
Hymenoptera, 45, 120, 229
Hyper spectral imaging, 406
Hyper spectral techniques, 409
Hypocrea, 355
 H. lixii, 355–356
 H. orientalis, 355
Hypocreaceae, 43, 354
Hypocrealean entomopathogenic fungi, non-entomopathogenic roles of, 280–281
Hypocreales, 354
Hyssopus officinalis, 140–141

I

Immune fluorescence (IF), 401–402
Impatiens glandulifera, 11
India
 Bt cotton adoption in, 239
 medicinal plant diversity in, 247–248
Indian biopesticide sector
 consumption of synthetic pesticides and biopesticides in India, 329f
 current status of, 328–329
 current structure of biocontrol laboratories and units working in India, 328f
 share of different categories of biopesticides, 329f
Indian Council of Agricultural Research (ICAR), 57–58, 328
Indian Himalayan Region (IHR), 247
Indole acetic acid (IAA), 227, 286
Induced systemic resistance (ISR), 218, 263, 268–270, 321, 389–391
 against different stresses, 360, 361t
 effect of different *Trichoderma* sp. against bacterial phytopathogens, 390t–391t
 effect of *T. asperellum* seed treatment, 389f
Industrial bioreactors, 368
Infective juveniles (IJ), 11, 73
Infrared spectroscopy, 403–405, 411–412
Insect
 baiting technique, 74
 nervous system, 12–13
 pests, 7, 10, 112
  management, 73, 110
 sex pheromones, 12
 viruses as biopesticides, 196–197
Insect growth regulators (IGRs), 286
Insecticidal crystalline proteins (ICPs), 94–95
Insecticidal proteins, 1
Insecticide Act, 292
Insecticides, 1, 19
Institute for the Control of Agrochemicals (ICAMA), 57–58
Intact-cell MALDI-TOF MS (ICMSs), 356–357
Integral pest control management, 94
Integrated Crop Management (ICM), 1
Integrated insect pest management nanotechnology, 160
Integrated Pest and Disease Management, 19
Integrated Pest Management (IPM), 1, 8, 19–20, 57–58, 95, 117, 143, 202, 209, 279, 301–302, 337
Intensify sustainability, 94
Internal transcribed spacers (ITSs), 356
International Biocontrol Manufacturer's Association (IBMA), 9
International Organization for Biological Control (IOBC), 9, 20, 57
International Patent Classification (IPC), 100
International power units (IUs), 121
International Year of Plant Health (2020), 353
Intracellular PGPR (iPGPR), 48–51
Ipomoea grandifolia, 139
Iridoviridae, 44
Iron (Fe), 287
Isaria, 25, 228–229
 I. farinosa, 312–313
 I. fumosorosea, 44, 291
 I. javanica, 312–313
Isolation
 of chitosan from crab shell, 162
 and epizootic potential of *Bacillus thuringiensis*, 118–119
 of EPN, 74
Isoptera, 120
Ixodes scapularis, 289–290

J

JASCO spectrophotometer, 163
Jasmonic acid (JA), 263, 286, 389
Journal of Citation Reports (JCR), 96–97

K

Kenya Plant Health Inspectorate Service Act, 58
Knowledge Based Bio-Economy (KBBE), 337

L

Lactases, 387
Lactuca sativa, 137
Lagenidium giganteum, 306–307
Lamiaceae, 25, 134, 247
Lamiacease, 25
Larvicidal toxicity against *H. armigera*, *S. litura* and *P. xylostella*, 170–173, 170t–173t
Lasiodiplodia pseudotheobromae, 139
Lauraceae, 25
Lavandula officinalis, 185
Lead acetate, 161
Leaf miner (*Aproaerema modicella*), 184
Lecamicillium lecanii, 44
Lecanicillium, 25
Lecanicillium spp., 228–229, 289
  L. fungicola, 312–313
  L. lecanii, 194–196
  L. longisporum, 54–55
  L. muscarium, 289
  L. psalliotae, 289
  L. saksenae, 312–313
Lemna minor. *See* Duckweed (*Lemna minor*)
Lens esculenta, 355–356
Lepidium sativum, 137
Lepidoptera, 12, 45, 81, 120, 191–192, 227, 229
Lepidopteran insects, 1–2
Lepidopteran pests, 2
  control, 142
Leptinotarsa decemlineata, 25
Leptomonas
  L. seymouri, 291–292
  L. wallacei, 290
Lignin, 337–338
  lignin-derived components, 340
Lignocellulosic biomass, 338
Lippia alba, 145
Liquid formulations, 213–214, 370
  Bacillus sp., 214
Locusta migratoria, 290
Lolium rigidum, 137
Low molecular weight lignin compounds (LMWL compounds), 341
Ludwigia hyssopifolia, 136–137
Lure-kill systems, 12
Lygus hesperus, 279
Lymantria dispar, 230
Lysinibacillus sphaericus, 142–143, 190, 288
Lysis, 359
Lytic enzymes, 321, 387

M

Macrophomina, 358
  M. phaseolina, 196, 280–281
Maerua edulis, 141
Magnaporthe oryzae, 361
Makes caterpillars floppy (Mcf), 42, 193
"MALDI-TOF MS" technique, 356–357
Malva pusilla, 10
Mandula sexta, 193
Mangifera indica, 137
Matricaria chamomilla L, 146
Mattesia trogodermae, 197
Maximum residue levels (MRL), 8
Maydis leaf blight (MLB), 191
Mayer's Reagent, 161
Medicago polymorpha L, 137–138
Medicinal plants
  associated microflora
    medicinal plant diversity in India, 247–248
    niche of microflora, 248
    plant-microbe association, 248–253
    relative factors between microflora and plants, 254
  diversity in India, 247–248
  endophytic microbiome association with, 251–253
Medinilla magnifica, 136
Melanin, 2
Melia azedarach. *See* Chinaberry (*Melia azedarach*)
Melissa officinalis, 137
Meloidogyne spp, 44, 191, 195
  M. incognita, 191
  M. javanica, 323
Mesorhizobium sp., 48–51, 322
Metacordyceps brittlebankisoides, 312–313
Metallic salts, 203
Metarhizium acridium PacC (MaPacC), 291
Metarhizium sp., 113, 291, 312–313
  M. acridium, 44, 291
  M. anisopilae, 25, 44, 46, 56, 185, 194, 228–229, 231, 289, 291, 306–307, 312–313, 322–323
    ARSEF 8319, 280
    CPD 5, 195
  M. robertsii, 46, 291
Methyl viologen (MV), 363
Micro bialinoculants, 93–94
Micro-biopesticides, 203
Microalga biomass, 25
Microalgae, 309
Microbe associated molecular patterns (MAMPs), 384
Microbes, 37, 135
  benefits of microbes in rhizosphere, 354
  phyllospheric association of, 251
  rhizospheric association of, 248–250
  supporting plants, 47–51
    PGPF, 51
    PGPR, 48–51
  supporting soil health, 47–51
Microbial associated molecular patterns (MAMPs), 219
Microbial based products in India
  future prospects, 331–332
  hurdles in commercialization of, 329–332
  regulatory framework and challenges for biopesticides in India, 330–331
Microbial bio-pesticide, 9–11, 21, 108, 185, 186t–190t, 211, 211t, 237–238, 285–286. *See also* Microbial pesticides; Nano bio pesticide
  advantages of, 198
  bacteria as biopesticides, 185
  bacterial bioherbicide, 9–10
  bacterial bioinsecticides, 9
  bacterial species used as biopesticides, 212t
  biochemical pesticides, 184
    insect pheromones, 184
  Coniothyrium minitans, 195
  development of, 286–288
    antibiosis, an important criterion for development of, 288
    plant growth regulators play crucial role in, 286
    siderophores causes iron limiting conditions for many pathogenic pests, 287
  disadvantages of, 198
  examples of pheromones used in agricultural pest management, 184–185
  fungal bioherbicides, 10
  fungal bioinsecticides, 10
  Gliocladium catenulatum, 195–196
  for management of insect pest in India, 311–312, 312t
  members of
    Bacilliaceae as biopesticides, 185–193
    Pseudomonadaceae and *Enterobacteriaceae* as biopesticides, 193
  plant incorporated protectants, 184, 197–198
  registration norms and regulation of, 311
  for sustainable agricultural practices, 301–309
    algal biopesticides, 309
    bacterial biopesticides, 302–304
    current advancement in, 312–313
    current status of biopesticides in India, 309–312
    fungal biopesticides/mycopesticides, 306–307
    microbial products in biopesticides, 309
    nematode biopesticides, 307–308
    protozoan biopesticides, 308–309
    viral biopesticides, 304–305
  types of microbial biopesticides that have been approved by EPA, 108t
  viral bioherbicides, 11
  viral bioinsecticides, 10–11
  yeast as biocontrol agents, 196–197
Microbial consortium (MC), 215, 319, 321
  effect of application, 322f
  biological control, 320–321
  characteristics of, 322
  current status of Indian biopesticide sector, 328–329
  different types of, 322–325
    algae and bacteria, 325

bacterial and bacterial, 323
fungal and bacterial, 323–324
fungal and fungal, 322–323
hurdles in commercialization of microbial based products in India, 329–332
microbial consortium mediated plant defense mechanism in biological control, 322
need for development of biopesticides containing, 325–328
Microbial Pest control Agents (MPCA), 22
Microbial pest management organisms, 228
Microbial pesticides, 37–51, 107, 133, 210, 225, 286, 287f, 288–290, 325–328. *See also* Microbial biopesticides
bacteria as biopesticides, 288
development of microbial biopesticide, 286–288
fungi as biopesticides, 289
future prospects, 293–294
general mode of actions of microbial pesticides against plant pathogens, 216
genetic improvements of, 290–292
nematodes as biopesticides, 289–290
in post-genomic era, 293
protozoan as biopesticides, 290
registration and regulation of, 57–58
regulation and commercialization of, 292
trends and market demand for, 53–56
types of, 37–51
bacteria, 41–43
fungi, 43–44
genetically modified microbes, 45–47
microbes supporting plants and soil health, 47–51
microsporidia, 44
and target organisms, 38t–40t
virus, 44–45
viruses as biopesticides, 289
Microbial plant protection, 21
Microbial products in biopesticides, 309, 310t–311t
Microbial secretions, 110
Microbial toxin, 225
Microbiological pesticides, 231–232
Microbiological preparations, 22
Micrococcus, 48–51
Microemulsions (ME), 144
Microflora, niche of, 248
Microorganisms, 21
Microspora, 197
Microsporidia, 44, 290
Midgut tissues, 164
Moelleriella libera, 312–313
RCEF 2490, 312–313
Mokusaku-eki compound, 338
Molecular genetic protein engineering product, 3
Molecular techniques of plant disease detection, 401–402
comparison of current methods for detecting plant diseases resulting, 402t
Monilina laxa, 184
Morganellaceae, 42
Moringa oleifera, 137, 140–141
Mosquito-borne diseases, 134
Mosquitocidal toxins (Mtx), 225–226
Motility, 46
Multiplication of EPN, 74
Municipal solid waste (MSW), 339
Murraya koenigii, 25
Musca domestica, 56, 204
Muscodor albus, 25
Mutation, 215
Mycelium, 354
Mycobacterium marinum, 43
Mycoparasitism, 216, 320
Myrtaceae, 25
Myzus persicae, 26, 46, 280, 289
Myzus persicae. *See* Aphids (*Myzus persicae*)

N

N-acetylglucosaminidase, 267
Nano bio pesticide, 201, 203, 216. *See also* Microbial bio-pesticide
categories of nano-pesticides and nano-biopesticides, 217f
chemical pesticides, 201–202
diagrammatic representation of silver nanoparticle synthesis, 204f
efficiency and effectiveness, 203
examples of biopesticides recommendations in India, 218t
list of nanobiopesticides formulations, 219t
most commonly used biopesticides active ingredient, 217t
plant protection products, 202
steps in production and evaluation, 217f
Nano pesticides, *Bacillus thuringiens* is used as, 192–193
Nanoemulsions (NEs), 144
of botanical insecticides, 145–146
Nanoformulations, 134
Nanoparticles, 160, 192, 203, 391–392
factories, 203
of metal oxides, 134
of metalloids, 134
of metals, 134
Nanoparticulate systems, 3
Nanopesticides, 95
Nanoscale biopesticide formulations against arthropod pests and weeds, 144–148
bioherbicides in nanoformulations, 144
biopesticide nanoformulations of against anthropodes, 145–148
Nanotechnology, 134
National Agency for Food and Drug Administration and Control (NAFDAC), 58
National Agricultural Research System (NARS), 57–58, 330–331
National Agricultural Technology Project (NATP), 57–58
National Agriculture Department (NAD), 58
National Board of Accreditation (NBA), 330–331
National Bureau of Agricultural Insect Resources (NBAIRs), 328
National Bureau of Agriculturally Important Microorganisms (NBAIMs), 312
National Center for IPM (NCIPM), 330–331
National Farmer Policy, 57–58
National Health Surveillance Agency, 58
National Registration Scheme (NRS), 57–58
Near infrared-based technique (NIR-based technique), 404
Neem (*Azadirachta indica*), 25, 185
Neisseriaceae, 41
Nematodes
biopesticides, 11, 289–290, 307–308, 307f
entomopathogenic, 230–231
nematode-insect relationship, 80
Neosteinernema, 75, 80
Nicotiana tabacum, 140–141, 195
Nicotine, 26
Nigrospora sp., 192
Nitrogen use efficiency (NUE), 360
Nitrosoguanidine mutagenesis, 46
N–methyl-N–nitrosoguanidine (NTG), 215
Nomuraea rileyi, 146, 194, 228, 306–307
Non-inclusion viruses (NIV), 10
Non-ribosomal peptides (NRPs), 267–268, 385
Non-toxic mechanisms, 12, 22
Nonmaterials, biosensor platforms based on, 409
Nonribosomal peptide synthetases (NRPSs), 387
North American Free Trade Agreement (NAFTA), 56
Nosema spp., 197, 290
N. acridophagous, 197
N. adaliae, 290
N. algerae, 197
N. bombycisisis, 308
N. cuneatum, 197
N. fumifueranae, 197
N. heliothidis, 197
N. locustae, 197, 231, 290, 308–309
N. pyrausta, 197, 231, 308–309
Nosematidae, 44
Nostoc sp, 23
Nuclear polyhedrosis viruses (NPV), 111, 291, 305
Nuclearmagnetic resonance spectroscopy (NMR spectroscopy), 401–402
Nuclearpolyhedro viruses (NPV), 10, 23, 45, 55, 196–197, 229–230, 289, 304–305
Nursery soil treatment, 367
Nutrient broth (NB), 213–214
Nutrients, competition for, 358–359

O

Obuda pepper virus, 11
Occlusion body, 291
Occlusion-derived viruses (ODVs), 196, 230, 305
Ocimum
O. basilicum, 137, 145
O. gratissimum, 140–141

Office of Chemical Safety and Pollution Prevention (OCSPP), 57
Office of Gene Technology Regulator (OGTR), 292
Office of Pesticide Programs (OPP), 57
Oil encapsulations, 160
Oligomerization, 2
Oomycota, 228
Ophiocordyceps sinensis, 291, 312–313
Ophiocordycipitaceae, 43–44
Organization for Economic and Co-operative Development (OECD), 20, 57, 185, 225
Orography, 35
Orthoptera, 120, 229
Oryza cystatin (OC), 240
Oryza sativa, 139
Ostrinia nubilalis, 121, 197, 289
Overlapping open reading frames (ORFs), 230
Oxidative pentose phosphate pathway (OPPP), 270

P

P-nitrophenyl palmitate (pNPP), 164–165
Pachybasium, 355
Paecilomyces, 228–229
  P. farinosus, 194, 228, 306–307
  P. fumosoroseus, 194, 197
  P. lilacinus, 185
    KIA, 323
Paenibacillaceae, 41–42
Paenibacillus spp, 142–143, 185, 192, 278
  P. polymyxa, 226–228
    HY96–2, 227
  P. popilliae, 42, 190
  P. yonginensis DCY84T, 227
Panax notoginseng, 355–356
Papaya ringspot, 241
Parallel factor analysis, fluorescence spectroscopy, 403
Paranosema locustae, 142
Partial least square regression (PLS regression), 403, 405
Partial least squares logistic discriminant analysis (PLS-LDA), 404–405
Parvoviridae, 44
Paspalum notatum, 138
Paspalumurvillei, 138
Pasteuria, 312–313
Pasteuriaceae, 41
Patent Cooperation Treaty (PCT), 100, 292
Pathogenesis related proteins genes (PRs genes), 389
Pathogenicity, 80–81
  of EPN, 80–81
Pathogens, 7–8
  of lepidopteran insects, 42
Pathogens and beneficial microorganisms (PGPMs), 354
Paucimonas lemoignei, 192
Pelargonium graveolens, 196
Penicillium, 51, 192, 195–196, 383
  P. chrysogenum, 196
  P. digitatum, 219
Pennisetum purpureum, 139
6-pentyl-a-pyrone (6PP), 385–386
Pepino mosaic virus, 11
Peptaiboles, 385–386
Peptaibols, 360
Pericallia ricini, 196–197
Periplaneta americana, 146–147
Peroxidase, 387
Pest Control Product Board (PCPB), 58
Pesticide Registration Improvement Act (PRIA), 20
Pesticide Registration Improvement Extension Act (2012), 20
Pesticide(s), 100, 183, 201, 207, 337
  application technology, 31
  to biopesticides, 8
  market, 109
Pests, 7–8, 107
  control, 117
  control
    applications of genome editing to, 241
    applications of RNAi to, 240–241
  management, 11, 41, 110
Pezizomycotina, 354
Phakopsora pachyrhizi, 191
Phenol
  compound, 341
  eco-toxicology, 348
Phenolics test for, 161
Pheromones, 23, 184
  used in agricultural pest management, 184–185
    chitosan, 184–185
    plant extract biopesticides, 185
Phialides, 354
Phlebiopsis sp., 383
Phoma, 51
  P. herbarum, 355–356
Phomopsis longicolla, 196
Photorhabdus insect-related proteins (Pir proteins), 42, 193
Photorhabdus spp, 11, 73, 76–80, 226, 230–231
  P. asymbiotica, 79–80
  P. luminescens, 42, 86–87, 312–313
  P. luminescens, 79–80
  P. temperata, 42, 79–80
Photorhabdus virulence cassettes (PVCs), 42, 193
Photosystem (PS), 137
Phthorimaea operculella, 196–197
Phyllospheric association of microbes, 251
Phyllospheric microbial communities, 251
Phyllospheric microbiome, 251
Physio-metabolic process, 171–173
Phythium sp., 383
Phytochemical analysis, 161
Phytochemical screening for essential oil of *Salvia leucantha*, 165, 165t
Phytohormones, 93–94
Phytopathogens, 321
Phytophagous insects, 175
Phytophthora spp., 362
  P. infestans, 46, 193, 196, 408
Pichia
  P. anomala, 196
  P. guilermondii, 215–216
Picornaviridae, 44
Pieris brassicae, 195
Pinaceae, 134
Pinellia ternate agglutinin (PTA), 240
Pinus halepensis, 355–356
Piper nigrum, 25
Piriformospora, 51
  P. indica, 23, 229
Pituranthus tortuosus L, 137–138
Plant diseases, 201, 401
  advantages and limitations of biopesticides, 219–220
  biopesticides, 209
    and association with growth promoter, 217–218
  botanical biopesticides usage plant pathogen, 219
  classification of biopesticides, 209–211
  control methods, 383
  current global scenario, 208–209
    annual pesticide consumption, 209t
  detection techniques, 401
  development of compatible consortia for improvement of formulation efficiency, 215–216
    combining different mode of action, 216
    combining various microbes, 215–216
    development of strain mixtures, 216
  EOs, 219
  factors affecting biopesticides marketing, 220
  formulations for *Pseudomonas fluorescens*, 213
  general mode of actions of microbial pesticides against plant pathogens, 216
  history of synthetic pesticides used in, 208
  improvement of formulation efficacy, 214
  inducer of systemic resistance in plant against plant pathogen, 218–219
  insight into popular fungal and bacterial biopesticides used in, 212
    Trichoderma spp., 212
  liquid formulation, 213–214
  mass production of *Trichoderma* for commercial purpose, 213
  methods, 213
    powder formulations, 213
  microbial biopesticides, 211
  molecular approach for improvement of formulation efficacy, 214–215
    genetic recombination, 215
    mutation, 215
    protoplast fusion, 214–215
  molecular techniques of plant disease detection, 401–402
  nanobiopesticides, 216
  processes responsible for fate of applied pesticides in environment, 208f
Plant growth
  delivery system for, 366–368

promotion, 321, 387–389
Plant growth promoters (PGPs), 286
Plant growth promoting bacteria (PGPB), 215
Plant growth promoting fungi (PGPF), 387
Plant growth promoting rhizobacteria (PGPR), 48–51, 217–218, 250, 321
biocontrol and plant growth promoting activities, 48t–50t
Plant growth regulators (PGRs), 286
in development of biopesticides, 286
Plant growth-promoting fungi (PGPF), 51, 217–218
biocontrol and Plant growth promoting activities of PGPF, 51t–53t
Plant growth-promoting microbes (PGPMs), 217–218
Plant parasitic nematodes (PPN), 73
Plant Protection Organization, 58
Plant protection products (PPP), 22
Plant viral coat proteins plant-incorporated-protectants (PVCP-PIPs), 210–211
Plant-incorporated protectants (PIP), 9, 11–12, 22, 37, 94, 107–109, 112, 133, 202, 209–211, 285–286, 301–302
Planting materials, treatment of, 367
Plants, 211
biopesticides, 185
extracts, 23
inducer of systemic resistance in plant against plant pathogen, 218–219
material, 160
microbiome, 248
pathogens, 7, 108, 354–371, 383, 401, 410
botanical biopesticides usage, 219
inducer of systemic resistance in, 218–219
pesticides, 201
plant-microbe association, 248–253
endophytic microbiome association with medicinal plants, 251–253
phyllospheric association of microbes, 251
rhizospheric association of microbes, 248–250
probiotics, 288
protection, 349–350
reproduction, 100
rhizosphere, 354
treatment, 367
VOC, 407
Plasmodiophora brassica, 195
Plasmopara viticola, 360
Plectosporium alismatis, 140
Pleurotus ostreatus, 355
Plexippus paykulli (Audouin), 142
Plicaria, 195
Plumpox virus, 241
Plutella xylostella, 159–160, 177, 193, 195
Poaceae, 134
Pochonia chlamydosporia, 322–323
Podisus nigrispinus, 123
Podosphaera leucotricha, 203
Poecilia reticulata, 160
Polar hydrophilic compounds, 340
Poly-N-isopropyl, 204
Polycyclic aromatic hydrocarbons (PAHs), 337–338
Polydnaviridae, 196
Polyethylene glycol (PEG), 146
Polygonum hydropiper. *See* Water-pepper (*Polygonum hydropiper*)
Polymerase chain reaction (PCR), 119, 401
Polymers, 204
Popillia japonica, 190
Portable fluorescence spectroscopy system, 403
Portulaca oleracea, 136, 139
Post-harvest treatment, 23
Potato dextrose agar (PDA), 354
Poxviridae, 44
Pratylenchus sp., 195
*prb*1 gene, 267
Principal component analysis (PCA), 403–404
Procrustes discriminant analysis (PDA), 405
Protein kinase A (PKA), 120
Proteinase inhibitors (PIs), 240
Proteomics, 110
Protonmotive force (PMF), 348
Protoplast fusion, 214–215
Protozoan
as biopesticides, 290, 308–309, 308f
pathogens, 231
Pseudomonadaceae, 41, 43, 193
Pseudomonas, 9, 48–51, 193, 227, 250, 253, 262, 278, 301–302, 323, 325–328, 383
P. aeruginosa, 41, 225, 288, 302–304, 322
MBAA1, 323–324
MML2212, 323
PJHU15, 324
P. aurantiaca strain B-162, 46
P. cepacia, 225, 302–304
P. chlororaphis, 225, 302–304
P. entomophila, 43, 226, 312–313
P. putida, 279
P. savastanoi, 192
P. solanacearum, 225, 302–304
P. stutzeri, 287
P. syringae, 192, 225, 302–304
Pseudomonas fluorescens (PF), 3, 23, 43, 122, 185, 213–214, 216, 225, 251, 253, 288, 290–291, 302–304, 324
BRG100 strain, 9–10
D7 strain, 9–10
EBC5, 323
EBC6, 323
formulations for, 213
inorganic carriers, 213
organic carriers, 213
Pseudonocardiaceae, 41–42
Pseudoperonospora cubensis, 227, 407
Pterodon emarginatus, 146
Puccinia striiformis, 405–406
Pupal mortality, 163
Pupicidal toxicity against *H. armigera*, *S. litura* and *P. xylostella*, 170–173
Purpureocillium lilacinum, 44, 195, 322–323
Pyrethrin effect, 25
Pyrethroids, 26
Pyrethrum, 25
Pyroligneous acid (PA), 337–338
Pyrolysis, 337–338
efficient technology, 338
products of, 339–341
Pyrolytic feedstock, 339, 340f
Pythium spp., 195, 354, 358
P. aphanidermatum, 184, 323
P. oligandrum, 23
P. ultimum, 215, 408

Q

Quantitative food utilization efficiency measures, 164
Quantitative reverse transcription and polymerase chain reaction (qRT-PCR), 267
Quinone eco-toxicology, 348

R

Radicinin, 139
Ragmus morosus, 204
Ralstonia solanacearum, 261–262, 269, 323, 389
Random amplified polymorphic DNA (RAPD), 356
Ranunculaceae, 247
Raw data-based model, 404
Reactive nitrogen species (RNS), 270
Reactive oxygen species (ROS), 135, 219, 270, 291, 387
Rearing of *P. xylostella*, 163
Recombinant DNA technology (rDT), 3, 290–291
Registration Committee (RC), 57–58
Regulation of microbial pesticides, 292
Reichardia tingitana, 136
Relative growth rate (RGR), 175
Relative humidity (RH), 33–34, 163
lifetime and fall distance of water droplets, 35t
and temperature, 33–34
Reoviridae, 45, 196, 229–230
Resistance process, 20
Rhabdoviridae, 44
Rhipicephalus, 289
Rhizo-microbiome, 47–48, 51
Rhizobium, 48–51, 250–251, 323
Rhizoctonia, 51, 192, 195, 214, 354
R. fragariae, 196
R. solani, 45–46, 196, 219, 226–227, 264
Rhizopus sp., 196, 219
Rhizosphere
colonization, 46
competent, 386
Rhizospheric association of microbes, 248–250, 249t–250t

Rhopalosiphum
R. erysimi, 142
R. padi, 140–141
R. rufiabdominale, 280
Rhyzoperta dominica, 204
Ribosomal RNA (rRNA), 409
Ricinius communis, 163, 177
RNA interference (RNAi), 117, 197, 240–241
RNAi based biopesticides, 111–112
RNA-based affinity biosensor, 410
Rodopholus similis, 191
Rosaceae*Saccharomyces*, 196, 247
S. cerevisiae, 196, 227, 278
Rossi. *See Theyne imperialis* (Rossi)

S

Saccharopolyspora spinosa, 24, 42, 227, 302–304
Salicylic acid (SA), 263, 286, 389, 407
Salinity stress, 364
Salt stress, 364
Salvia leucantha essential oil (SlEO), 160
Salvia leucantha essential oil encapsulated in chitosan nanoparticles
amylase, protease, proteinase, and lipase assay, 164–165
characterization of essential oil loaded chitosan nanomaterials, 163
characterization of essential oil loaded chitosan nanoparticles, 166–167
chitosan nanoparticles preparation with essential oil, 162–163
collection and processing of crab shells, 162
energy-dispersive X-ray spectroscopy analysis, 168
food utilization measures, 175
FTIR analysis of essential oil chitosan nanoparticles, 168–169
GCMS analysis of essential oil of *S. leucantha*, 161–162
gut digestive enzymes of *H. armigera, S. litura* and *P. xylostella* larvae, 177
H. armigera and *S. litura* rearing, 163
impact on longevity and fecundity of *H. armigera, S. litura* and *P. xylostella*, 164
isolation and extraction of chitosan from crab shell, 162
larvicidal and pupicidal toxicity against *H. armigera, S. litura* and *P. xylostella*, 170–173
materials and methods, 160
extraction of *S. leucantha* essential oil, 160
plant material, 160
qualitative analysis, 161
phytochemical analysis, 161
quantitative food utilization efficiency measures, 164
rearing of *P. xylostella*, 163
results and discussion, 165
phytochemical screening for essential oil of *Salvia leucantha*, 165
impact of *Salvia leucantha* essential oil on insect longevity and fecundity, 173–175
SEM analysis, 167–168
statistical analysis, 165
test for
alkaloids, 161
flavonoids, 161
glycosides, 161
phenolics, 161
saponins, 161
tannins, 161
terpenoids, 161
toxicity against the *H. armigera, S. litura* and *P. xylostella*, 163
zeta potential measurements, 169–170
Salvia officinalis, 137
Santolina chamaecyparissus, 140–141
Saponins test for, 161
Saprolegniaparasitica, 185
Sapwood (SW), 340
Sarcomastigophora, 197
Sargassum polycystum, 147
Scanning electron microscopes (SEM), 75, 160, 163, 167–168, 168f
Scerotinia sp., 190
Schistocera gregaria, 197
Schoenocaulon officinale, 185
Sclerotinia, 10, 27, 354
S. sclerotiorum, 196, 215–216, 219, 322
Sclerotium, 354, 358
S. cepivorum, 194–195
S. rolfsii, 192, 322–323
Scylla serrata, 162
SDGs. *See* UN sustainable development goals (SDGs)
Secondary metabolites (SMs), 385
Seed
bacterization, 140
biopriming, 367
treatment, 31, 367
Seedling
dipping application, 31
root dip, 367
Self-organizing map (SOM), 405
Semiochemicals, 9
Sennaoccidentalis, 139
Sequential binding model, 120
Serotyping, 119
Serratia, 43, 48–51, 226, 250
S. entomophila, 27, 142–143
S. marcescens, 219, 225, 290–291, 302–304, 312–313
S. marcescens, 43, 45–46
S. plymuthica C48, 321
Servicio Nacional de Sanidad y Calidad Agroalimentaria (SENASA), 58
Sesquiterpene lactones (STL), 136
Setaria viridis, 138
Sex pheromones, 12, 184
Shenginmycin, 193
Siderophores, 287
Silver nanoparticles (AgNPs), 160, 170–171, 203, 392
Sinapis alba, 138
Sinapis arvensis, 138
Single stranded DNA (ssDNA), 410
Sitophilus oryzae, 204
Small secreted proteins (SSPs), 229
Small to medium enterprises (SMEs), 348–349
Sodium carbonate (Na_2CO_3), 164–165
Sodium Tripolyphosphate (TPP), 162–163
Soft independent modeling of class analogy (SIMCA), 404
Soil
borne diseases, 354–371
borne plant diseases, 261
entomopathogenic nematodes, 42
fertility, 47–48
microorganisms, 319
treatment, 367
Soil Association in United Kingdom, 202
Soilless systems, 368
Solanum trilobatum, 148
Solanum vivarium, 11
Solanumnigrum, 11
Solidago virgaurea, 137
Sonchus oleraceus, 138
Sordariomycetes, 354
Space, competition for, 358–359
Spectral information divergence (SID), 406
Spectroscopic and imaging techniques, 402–403, 411–412
fluorescence spectroscopy, 411
visible and infrared spectroscopy, 411–412
Spider venoms, 142
Spilanthes calva, 229
Spinosad, 42
Spinosins, 24
Spirulina platensis, Lyngbya sp, 23
Spo0A protein, 228
Spo0B protein, 228
Spodoptera
S. frugiperda, 26, 143, 290–291
S. frugiperda, 123
S. littoralis, 140–141, 286
S. litura, 159–160, 197, 237–238
rearing, 163
Spodoptera littoris nucleo polyhedro viruses (SpliNPV), 196–197
Spodoptera litura. See Tobacco caterpillar (*Spodoptera litura*)
Sponges, 85
Sporothrix
S. globosa, 312–313
S. insectorum, 312–313
S. pallida, 312–313
S. schenckii, 312–313
Spray drift, 35
Spraying process, 34
Stabilized organic materials (SOMs), 339
Stachybotryaceae, 43
Stachybotrys charatum, 192
Staphylococcus aureus, 227, 348
Starch, 204
State Agricultural Universities (SAUs), 328
Steinemernatidae, 73

Steinernema, 73, 75, 79–80, 111, 230–231, 307–308, 312–313
S. borjomiensis n. sp., 312–313
S. carpocapsae, 81, 86, 289–290, 307–308
S. feltiae, 23, 86, 312–313
S. glaseri, 11, 81, 86, 289–290, 307–308, 312–313
S. kraussei, 23
S. monticolum, 86
S. monticolum, 312–313
S. riobrave, 11, 307–308
S. riobravis, 289–290
S. scapterisci, 11, 81, 86, 307–308, 312–313
Steinernematidae, 11
Stellaria media, 136
Stem-end injury, 406
Stemphylium botryosum, 287
Sterilization, 122
Stratified sampling method, 74
Strawberry
bioinoculants in, 277–279
biopesticides in, 279–280
Streptomyces, 23, 42, 278, 383
S. fradiae MML1042, 323
S. griseoviridis, 23
S. griseoviridis, 302–304
S. lydicus, 23
S. melanosporofaciens EF76, 46
S. scabies, 46
Streptomycetaceae, 42
Superoxide dismutases (SOD), 363
SOD1 gene, 291
Sustainable agricultural practices, microbial biopesticides for, 301–302
Sustainable agricultural solutions
biopesticides: benefits and challenges, 96f
discussion and analysis of results, 96–99
method, 95–96
scientific trajectory, 96–99
technological trajectory, 100–101
Sustainable agriculture, 93, 268–269
Sustainable environment in *Trichoderma* spp, 368
industrial bioreactors, 368
mitigating nitrous oxide emissions from cultivated crop fields, 368
wood preservation, 368
Sustainable food production, strategies and implications for, 281
Sustainable pest management, 140
Symbionts, microbes, 354
Synthetic pesticides, 19
and challenges, 7–8
Systemic acquired resistance (SAR), 218, 263, 357, 357f, 389
Syzygium aromaticum, 25

T

Tagetes minuta, 146
Tanacetum cinerariifolium, 146
Tannins test for, 161
Technological trajectory, 100–101
patent applications by
applicant, 101f
country or territory, 101f
IPC code, 102t
publication date, 102f
tef1-gene, 356
Tegenaria agrestis, 230
Teleomorphs, 355
Tenebrio molitor, 197
Tephrosia vogeli, 109, 140–141
Teratogenicity, 7–8
Termites, 204
Terpenoids, 12–13
Tetraclinis articulata, 138
Tetranychus urticae, 140–141
The elongation factor 1-α (TEF-α), 229
Theobroma cacao, 355–356
Theyne imperialis (Rossi), 142
Thrips palmi, 195
Thymus vulgaris, 137
Tissue colonization, 2
Tithonia diversifolia, 140–141
Tithonia urticae, 141
Tobacco caterpillar (*Spodoptera litura*), 184
Toddalia asiatica, 203
Tolypocladium spp., 194
T. cylindrosporum, 312–313
T. inflatum, 312–313
T. paradoxum, 312–313
Tomato, bioinoculants in, 279
Toxin protein 40 (Txp40), 42
Toxins complex (Tc), 42, 230–231
Transcinnamic acid (TCA), 193
Transgene products, 9
Transgenic plants, 11, 368
Transposition, 3
Tree-based modeling methods, 404–405
Trianthemaportulacastrum L, 140
Triatoma infestans, 56
Tribolium castaneum, 197
Tricalcium phosphate (TCP), 364
TrichoBLAST tool, 356
Trichoderma sp, 23, 51, 212, 216, 219, 253, 262, 278, 289, 301–302, 312–313, 323–328, 355–356, 383–385
agricultural applications, 353, 357–368
achieving UN SDGs, 353
advantages, challenges, constaints in sustainability of *Trichoderma*, 370–371
benefits of microbes in rhizosphere, 354
biocontrol agents for management of diseases in crops, 358–360
commercialization of *Trichoderma* spp., 368–370
fungus of unique characteristics, 354–357
management of biotic and abiotic stresses in crop plants, 360–364, 362t–363t
pesticides consumption in management of pests, 353
soil borne diseases and plant pathogens, 354–371
sustainable environment, 368
antibiosis by antimicrobial metabolites, 267–268
biocontrol mechanism by, 264–270
biocontrol mechanisms of, 263f
as biopesticide, 194–195
commercialization of, 368–370
in control of wilt disease, 263–264
as crops growth enhancer, 364–366
CWDEs, 267
delivery systems for stresses management and plant growth, 366–368
bee vectoring, 368
through drip irrigation water, 367
nursery soil treatment, 367
plant treatment, 367
seed biopriming, 367
seed treatment, 367
seedlings root dip, 367
soil treatment, 367
soilless and hydroponic systems, 368
transgenic plants, 368
treatment of planting materials, 367
description, 354–355
chlamydospores, 355
conidia, 355
conidiophores, 354
growth, 354
mycelium, 354
phialides, 354
teleomorphs, 355
ecology and biodiversity, 355–356
distribution, 355
habitat, 355
Trichoderma as endophytes, 355–356
as endophytes, 355–356
formulations of, 369–370
carriers based formulations of, 370
liquid formulations, 370
fungus of unique characteristics, 354–357
identification of *Trichoderma* spp. and strains, 356–357
molecular, 356–357
morphological, 356
induced systemic resistance, 268–270
mechanisms, 384–391, 384f
antibiosis, 385–386
competition with pathogens for space and nutrients, 384–385
CWDEs, 386–387
ISR, 389–391
plant growth promotion, 387–389
mycoparasitism, 264–267
standardization of mass production of, 369
liquid state fermentation, 369
solid state fermentation, 369
substrate/carrier materials for solid formulationns, 369
in sustainable environment, 368
T. aggressivum, 355
T. amazonicum, 355–356
T. asperella, 322–323
T. asperelloides, 268, 355–356, 388
T. asperellum, 194–195, 267–270, 322–323, 355–356, 360, 385–386, 388–389, 392
T. atroviride, 360, 385–386

Trichoderma sp (*Continued*)
T. atrovirie, 392
T. brevicompactum, 268
T. chlorosporum, 355–356
T. citrinoviride, 194–195, 323–324, 386
T. eijii, 355
T. flagellatum, 355–356
T. fulvidum, 355
T. gamsii, 355–356, 385
T. hamatum, 262, 269, 355–356, 385–386
T. harzianum, 25, 46–47, 56, 142–143, 262, 264, 267–269, 278–279, 322–324, 355–356, 360, 385–386, 388, 392
AK 20G, 364
CECT 2413, 385, 387
NF-9, 386
TNHU27, 215–216, 324
T. koningii, 262, 268–269, 386
T. koningiopsis, 140, 268
T. lignorum, 264
T. longibrachiatum, 262, 267–268, 355, 386–388, 392
T. martiale, 355–356
T. peudokoningii, 269
T. pleurotophilum, 355
T. polysporum, 262, 385
T. pseudoharzianum, 268
T. pseudokoningii, 262, 268–269, 392
T. pseudolacteum, 355
T. reesei, 385
T. spirale, 355
T. stromaticum, 355
T. virens, 262, 268, 385, 392
T. viridae, 185, 262, 264, 268–269, 311, 322–323, 355, 360, 386–388
trichogenic-nanoparticles and its application in crop protection, 391–392
Trichogenic-nanoparticles and its application in crop protection, 391–392
TrichOKEY BarCode, 356
*Trichophyton*sp., 190
Trichorderma spp, 54–55
Trigonella foenumgraecum, 25
Triticum aestivum, 137–138
Trogoderma granarium, 197
Tropical Pesticide Research Institute (TPRI), 58
Trypsin assays, 164
Tryptophol compounds, 388
Tulsi (*Ocimum basilicum*), 185
Txp40. *See* Toxin protein 40 (Txp40)
Tylenchulus semipenetrans, 191
Typha latifolia. *See* Cattail (*Typha latifolia*)

U

Ulex europaeus L, 138
Ultrasonication process, 145
Ultraviolet (UV), 309
light, 26
radiation, 122
range, 404
UV–visible spectroscopy, 160, 166–167
UN sustainable development goals (SDGs), 353
US Environment Protection Agency (USEPA), 1, 8–9, 19–20, 22, 109, 183–184, 193, 202, 209–211, 285–286
US Natural Insurance Organization (USEPA), 325

V

Vairimorpha necatrix, 44, 197, 231
Vegetables, biopesticides in, 280
Venturia inaequalis, 203, 404
Vernonia amygdalina, 140–141
Verticillium, 25
V. chlamydosporium, 23
V. lecanii, 23, 185, 194, 228–229, 306–307
as biopesticide, 195–196
Verticilliun lecanii, 27
Viral bioherbicides, 11
Viral bioinsecticides, 10–11
Viral biopesticides, 229–230, 304–305
mode of parasitism of baculoviruses againstlepidopteran insects, 305f
Virus, 44–45
Baculoviridae, 45
Reoviridae, 45
Virus induced gene silencing (VIGS), 197
Visible spectroscopy, 403–405, 411–412
Vitex negundo, 203
Vitis vinifera L. See Grapevines (*Vitis vinifera L.*)
Volatile fatty acids, 407
Volatile organic compounds (VOC), 138, 159–160, 407

W

Wagner's Reagent, 161
Water, 32
compound, 341
stress, 403, 407
Water-pepper (*Polygonum hydropiper*), 203
Weed, 7–8
management, 135
Wettable powder (WP), 121, 212
Wheat germ agglutinin (WGA), 240
Wilt (*Fusarium oxysporum* f. sp. *neveum*), 196
Wilt diseases, 261
Trichoderma in control of, 263–264, 265t–266t
Wilt pathogens, biocontrol mechanism by *Trichoderma* in control of, 264–270
Wind
direction, 34–35
speed, 34–35
Withania somnifera, 251
Wood
preservation, 368
vinegar, 337–338
chemical composition of, 341
World Health Organization (WHO), 20, 248
World Intellectual Property Organization (WIPO), 96

X

Xanthium strumarium, 138
Xanthomonas
X. axonopodis, 409
X. axonopodis pv. *Vesicatoria*, 409
X. oryzae, 386
Xenorhabdus nematophila GroEL (XnGroEL), 42
Xenorhabdus nematophilus, 80
Xenorhabdus protein toxins (Xpt), 42
Xenorhabdusalpha-xenorhabdolysin A, B (XaxAB), 42
Xenorhabus sp., 11, 73, 76–80, 193, 226, 230–231
Xenorhabdus cabanillasii JM26, 193
Xenorhabdus nematophila, 42
Xenorhabdus stockiae PB09, 193
Xhantomonas campestris, 389, 392
Xiphinema index, 191

Y

Yeast as biocontrol agents, 196–197
insect viruses as biopesticides, 196–197
protozoans as biopesticides, 197
Yersinia entomophaga, 226
Yersiniaceae, 41, 43
Yield reducers, 7

Z

Zea mays, 140
Zeta potential (ZP), 169–170, 169f
Zingiber officinale, 25
Zucchini Yellow Mosaik Virus, 23
Zygomycota, 228
Zygosaccharomyces, 196

Printed in the United States
by Baker & Taylor Publisher Services